Lecture Notes in Computer Science 9010

Commenced Publication in 1973
Founding and Former Series Editors:
Gerhard Goos, Juris Hartmanis, and Jan van Leeuwen

More information about this series at http://www.springer.com/series/7412

C.V. Jawahar · Shiguang Shan (Eds.)

Computer Vision – ACCV 2014 Workshops

Singapore, Singapore, November 1–2, 2014
Revised Selected Papers, Part III

 Springer

Editors
C.V. Jawahar
Center for Visual Information Technology
International Institute of Information
    Technology
Hyderabad
India

Shiguang Shan
Institute of Computing Technology
Chinese Academy of Sciences
Beijing
China

Videos to this book can be accessed at
http://www.springerimages.com/videos/978-3-319-16633-9

ISSN 0302-9743   ISSN 1611-3349 (electronic)
Lecture Notes in Computer Science
ISBN 978-3-319-16633-9   ISBN 978-3-319-16634-6 (eBook)
DOI 10.1007/978-3-319-16634-6

Library of Congress Control Number: 2015934895

LNCS Sublibrary: SL6 – Image Processing, Computer Vision, Pattern Recognition, and Graphics

Springer Cham Heidelberg New York Dordrecht London

Printed on acid-free paper

Springer International Publishing AG Switzerland is part of Springer Science+Business Media
(www.springer.com)

Preface

The three-volume set of LNCS contains the carefully reviewed and selected papers presented at the 15 workshops that were held in conjunction with the 12th Asian Conference on Computer Vision, ACCV 2014, in Singapore, during November 1–2, 2014. These workshops were carefully selected from a large number of proposals received from almost all the continents.

This series contains 153 papers selected from 307 papers submitted to all the 15 workshops as listed below. A list of organizers for each of these workshops is provided separately.

1. Human Gait and Action Analysis in the Wild: Challenges and Applications
2. The Second International Workshop on Big Data in 3D Computer Vision
3. Deep Learning on Visual Data
4. Workshop on Scene Understanding for Autonomous Systems
5. RoLoD: Robust Local Descriptors for Computer Vision
6. Emerging Topics In Image Restoration and Enhancement
7. The First International Workshop on Robust Reading
8. The Second Workshop on User-Centred Computer Vision
9. International Workshop on Video Segmentation in Computer Vision
10. My Car Has Eyes - Intelligent Vehicles with Vision Technology
11. Feature and Similarity Learning for Computer Vision
12. The Third ACCV Workshop on e-Heritage
13. The Third International Workshop on Intelligent Mobile and Egocentric Vision
14. Computer Vision for Affective Computing
15. Workshop on Human Identification for Surveillance

Workshops in conjunction with ACCV have been emerging as a forum to present focused and current research in specific areas of interest within the broad scope of ACCV. This year, the workshops covered diverse research topics including both conventional ones (such as robust local descriptor) and newly emerging ones (such as deep feature learning). Besides direct submissions to the workshops, submissions rejected by the main conference were provided the opportunity to co-submit to the workshops, following the policy of previous ACCVs.

We would like to thank many people for their efforts in making this publication possible. General Chairs, Publication Chairs, and Local Organizing Chairs helped a lot in smoothly organizing the workshops and coming out with this proceedings. Reviewers of the individual workshops did an excellent job of selecting quality papers for the final presentation. They deserve credit for the excellent quality of the papers in this proceedings.

It is our pleasure to place these volumes in front of you..

November 2014

C.V. Jawahar
Shiguang Shan

Organization

ACCV 2014 Workshop Organizers

1. Human Gait and Action Analysis in the Wild: Challenges and Applications

Mark Nixon   University of Southampton, UK
Liang Wang   Chinese Academy of Sciences, China
Jian Zhang   University of Technology, Sydney, Australia
Qiang Wu   University of Technology, Sydney, Australia
Zhaoxiang Zhang   Beihang University, China
Yasushi Makihara   Osaka University, Japan

2. Second International Workshop on Big Data in 3D Computer Vision

Jian Zhang   University of Technology, Sydney, Australia;
                      National ICT Australia, Australia
Mohammed Bennamoun   University of Western Australia, Australia
Fatih Porikli   NICTA, Australia
Ping Tan   National University of Singapore, Singapore
Hongdong Li   Australian National University, Australia
Lixin Fan   Nokia Research Centre, Finland
Qiang Wu   University of Technology, Sydney, Australia

3. Deep Learning on Visual Data

Wanli Ouyang   The Chinese University of Hong Kong, China
Xiaogang Wang   The Chinese University of Hong Kong, China
Kai Yu   Baidu, China
Quoc Le   Google, USA
Shuicheng Yan   National University of Singapore, Singapore

4. Workshop on Scene Understanding for Autonomous Systems (SUAS)

Sebastian Ramos   CVC, Universitat Autònoma de Barcelona, Spain
Raquel Urtasun   University of Toronto, Canada
Antonio Torralba   Massachusetts Institute of Technology, USA

Nick Barnes   NICTA and Australian National University,
                         Australia
Markus Enzweiler   Daimler AG, Germany
David Vazquez   CVC, Universitat Autònoma de Barcelona, Spain
Antonio M. Lopez   CVC, Universitat Autònoma de Barcelona, Spain

5. RoLoD: Robust Local Descriptors for Computer Vision

Jie Chen   CMV, University of Oulu, Finland
Zhen Lei   NLPR, Chinese Academy of Sciences, China
Li Liu   VIP, University of Waterloo, Canada
Guoying Zhao   CMV, University of Oulu, Finland
Matti Pietikäinen   CMV, University of Oulu, Finland

6. Emerging Topics In Image Restoration and Enhancement

Zhe Hu   University of California Merced, USA
Oliver Cossairt   Northwestern University, USA
Yu-Wing Tai   KAIST, Korea
Sunghyun Cho   Samsung Electronics, Korea
Chih-Yuan Yang   University of California Merced, USA
Robby Tan   SIM University, Singapore

7. First International Workshop on Robust Reading

Masakazu Iwamura   Osaka Prefecture University, Japan
Dimosthenis Karatzas   CVC, Universitat Autònoma de Barcelona, Spain
Faisal Shafait   University of Western Australia, Australia
Pramod Sankar Kompalli   Xerox Research India, India

8. Second Workshop on User-Centred Computer Vision (UCCV 2014)

Gregor Miller   University of British Columbia, Canada
Darren Cosker   University of Bath, UK
Kenji Mase   Nagoya University, Japan

9. International Workshop on Video Segmentation in Computer Vision

Michael Ying Yang   Leibniz University Hannover, Germany
Jason Corso   University of Michigan, Ann Arbor, USA

10. My Car Has Eyes - Intelligent Vehicles with Vision Technology

Xue Mei	Toyota Research Institute North America, Ann Arbor, USA
Andreas Geiger	Max Planck Institute for Intelligent Systems, Germany
Michael James	Toyota Research Institute North America, Ann Arbor, USA
Yi-Ping Hung	National Taiwan University, Taiwan
Fatih Porikli	Australian National University, Australia
Danil Prokhorov	Toyota Research Institute North America, Ann Arbor, USA

11. Feature and Similarity Learning for Computer Vision

Jiwen Lu	Advanced Digital Sciences Center, Singapore
Shenghua Gao	ShanghaiTech University, China
Gang Wang	Nanyang Technological University, Singapore
Weihong Deng	Beijing University of Posts and Telecommunications, China

12. Third ACCV Workshop on e-Heritage

Takeshi Oishi	University of Tokyo, Japan
Ioannis Pitas	Aristotle University of Thessaloniki, Greece
Bo Zheng	University of Tokyo, Japan
Manjunath Joshi	DA-IICT, Gandhinagar, India
Anupama Mallik	Indian Institute of Technology, Delhi, India

13. Third International Workshop on Intelligent Mobile and Egocentric Vision (IMEV2014)

Chu-Song Chen	Academia Sinica, Taiwan, China
Mohan Kankanhalli	National University of Singapore, Singapore
Shang-Hong Lai	National Tsing Hua University, Taiwan, China
Joo Hwee Lim	Institute for Infocomm Research, Singapore
Vijay Chandrasekhar	Institute for Infocomm Research, Singapore
Liyuan Li	Institute for Infocomm Research, Singapore
Yu-Chiang Frank Wang	Academia Sinica, Taiwan, China
Shuicheng Yan	National University of Singapore, Singapore

14. Computer Vision for Affective Computing (CV4AC)

Abhinav Dhall	University of Canberra/Australian National University, Australia

Roland Goecke   University of Canberra/Australian National
                 University, Australia
Nicu Sebe   University of Trento, Italy

15. Workshop on Human Identification for Surveillance (HIS)

Tao Xiang   Queen Mary University of London, UK
Nalini K. Ratha   IBM Research, USA
Venu Govindaraju   University at Buffalo, USA
Meina Kan   Chinese Academy of Sciences, China
Wei-Shi Zheng   Sun Yat-sen University, China
Marco Cristani   University of Verona, Italy

Contents – Part III

**Third International Workshop on Intelligent Mobile and Egocentric
Vision (IMEV2014)**

Workshop on Human Identification for Surveillance (HIS)

Feature and Similarity Learning for Computer Vision (FSLCV)

Discovering Multi-relational Latent Attributes by Visual Similarity Networks

Fatemeh Shokrollahi Yancheshmeh[✉],
Joni-Kristian Kämäräinen, and Ke Chen

Department of Signal Processing,
Tampere University of Technology, 33101 Tampere, Finland
fatemeh.shokrollahiyancheshmeh@tut.fi
http://vision.cs.tut.fi

Abstract. The key problems in visual object classification are: learning discriminative feature to distinguish between two or more visually similar categories (e.g. dogs and cats), modeling the variation of visual appearance within instances of the same class (e.g. Dalmatian and Chihuahua in the same category of dogs), and tolerate imaging distortion (3D pose). These account to within and between class variance in machine learning terminology, but in recent works these additional pieces of information, *latent dependency*, have been shown to be beneficial for the learning process. Latent attribute space was recently proposed and verified to capture the latent dependent correlation between classes. Attributes can be annotated manually, but more attempting is to extract them in an unsupervised manner. Clustering is one of the popular unsupervised approaches, and the recent literature introduces similarity measures that help to discover visual attributes by clustering. However, the latent attribute structure in real life is multi-relational, e.g. two different sport cars in different poses vs. a sport car and a family car in the same pose - what attribute can dominate similarity? Instead of clustering, a network (graph) containing multiple connections is a natural way to represent such multi-relational attributes between images. In the light of this, we introduce an unsupervised framework for network construction based on pairwise visual similarities and experimentally demonstrate that the constructed network can be used to automatically discover multiple discrete (e.g. sub-classes) and continuous (pose change) latent attributes. Illustrative examples with publicly benchmarking datasets can verify the effectiveness of capturing multi- relation between images in the unsupervised style by our proposed network.

1 Introduction

Active research on visual object class detection and classification during the last ten years has produced novel approaches and many effective methods. At the same time, the main benchmark has switched from the Caltech-4 dataset of 4 categories and 3k images to the ImageNet [3] LSVRC challenge of 200 classes and 450k images (in ILSVRC 2014). The only change is not the increased number of

© Springer International Publishing Switzerland 2015
C.V. Jawahar and S. Shan (Eds.): ACCV 2014 Workshops, Part III, LNCS 9010, pp. 3–14, 2015.
DOI: 10.1007/978-3-319-16634-6_1

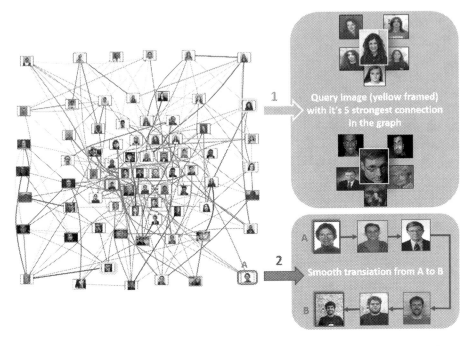

Fig. 1. A visual similarity network of ImageNet faces constructed using our procedure. The network can be used to find similar examples (strong links) or gradual change path from one example to another via the shortest path.

images and classes but also a more realistic problem setting. That is, instead of a single well-captured object in a fixed pose, ILSVRC contains multiple objects with severe 3D pose changes, occlusions and background clutter. These result great problems to the standard monolithic 2D methods and therefore novel learning paradigms, such as attribute learning [7,22], have recently gained momentum. Visual class detection can benefit from discovered latent visual attributes by learning multiple "fine-grained" classifiers [4].

Many attribute learning works utilise manually annotated attributes [2,7], which are not suitable for large-scale problems due to the expensive manpower involved in the annotation. In view of this, unsupervised approaches that can automatically discover latent attributes [1,4,9,29] are considered and adopted. The popular unsupervised tool is clustering, but it omits the fact that often latent attributes are multi-relational and thus breaking them to discrete "modes" is not sensible – what is anyway the more dominant attribute, car pose or model?

In this work, we adopt a network structure that can unsupervisely learn and represent multi-relational attributes simultaneously. For network construction, we propose a pairwise similarity measure and in the experimental part demonstrate that the network can be used to automatically discover multiple discrete (e.g. sub-classes) and continuous (pose change) latent attributes (see Fig. 1).

Our network establishes a structure which can be used in visual object categorisation to learn and represent multiple complex attribute-interpreted interconnections and benefits from more and more data. Our main contributions are as follows:

– A novel similarity measure, which combines descriptor based local appearance similarity and part-based constellation similarity into a unique similarity score, is proposed for constructing a visual similarity network based on pairwise matches of images.
– Experimental results where multiple multi-relational latent attributes are discovered using a network (sub-categories, gradual change between examples in a same category and continuous attributes such as 3D pose change).

All source codes and data will be made publicly available[1].

2   Related Work

Attribute learning – Explicit learning of visual attributes was first proposed by Ferrari and Zisserman [7]. Their method learned visual models of various attributes via weakly supervised setting where the training set was produced by pre-defined Web searches. Recently, unsupervised attribute discovery has gained more attention owing to its superiority to saving the involvement of manpower. Methods for completely unsupervised visual object classification (no labels or bounding boxes) have been proposed [9,10,26], but due to their large accuracy gap to the state-of-the-art supervised methods [5,11,24] they have not received enough attention. Attributes may still be beneficial in certain cases, such as zero-shot learning [13], with only a small number of training images [2], fine-grained classification [8] or utilising scene attributes to improve detection [15,18].

Visual networks – Methods employing a network (graph) structure to represent visual relationships have recently been proposed [4,9,18,21,28,29]. Most of these algorithms aim at finding classes or a specific object automatically [9,21,28]. The core element of the methods is to introduce a proper similarity measure and tailoring it for a problem-specific goal. In recent works, Aghazadeh et al. [1] and Dong et al. [4] establish their similarity measures using classification scores of the exemplar SVM [16] which forms own classifier for each sample. Another similarity measure was proposed [29] using feature's tree distances in unsupervised random clustering forests. Learning similarity measures can be time consuming since they may change if new images are added and therefore our measure will be based on pairwise structural similarity combining local part appearance and part configuration.

3   Visual Similarity Networks

Given a set of images containing objects from visual classes and with appearance and pose variation and imaging distortions, we aim to create an image

---

[1] The codes can be downloaded from: https://bitbucket.org/kamarainen/imgalign/code.

network where link strengths represent the visual similarity between two network nodes (images). The network, or termed graph, consists of nodes (images), edges between the nodes and weights representing the visual similarity (distance) between two nodes. The graph is directed if the used similarity measure does not commute, or undirected if it does.

In our proposed network-based framework, the assumption is that such network can be constructed from pairwise similarities by forming a similarity matrix that represents a full-connected network. If the matrix is symmetric, the network is undirected. The core elements in this procedure are *(i)* a pairwise image visual similarity measure (Sect. 3.1) and *(ii)* a network construction strategy (Sect. 3.2).

3.1   A Pairwise Visual Similarity Measure

Explicit visual measures, such as the simple pixel-wise difference, have the problem that they cannot tolerate well standard imaging distortion and typically behave well only close to a perfect match. Therefore, the recent trend in measuring visual similarity is to use "learning metrics" that establish a computational measure via ad hoc learning [1,4,29]. With learning metrics, the problem is that they depend on used training images and a selected objective function, and it is unclear how they generalise beyond images in the training set.

To measure pairwise similarity on object class level, we adopt structural visual similarity based on the part-based models of visual classes that has been particularly successful in object class detection [5,6,12]. In the simplest form, we describe $j = 1, 2, \ldots, M$ parts of an image I_i by feature descriptors $F_{i,j}$ (e.g. SIFT). Every descriptor is associated with a spatial location $\boldsymbol{x}_j = (x, y)_j$. The part-based visual similarity of two images I_a and I_b can be defined as

$$s(I_a, I_b) = s\left(\{< F_{a,j}, \boldsymbol{x}_j >_{j=1\ldots M_a}\}, \{< F_{b,j}, \boldsymbol{x}_j >_{j=1\ldots M_b}\}\right) \tag{1}$$

The problem is that the two images are related to each other by unknown geometric transformation $\mathbf{T} : \{< F, \boldsymbol{x} >\} \mapsto \{< F', \boldsymbol{x}' >\}$ that aligns the object parts (and affects also to the part descriptors if these are not invariant to the selected transformation type). The similarity measure of two unaligned images must therefore include the transformation term

$$s(I_a, I_b) = \max_{\mathbf{T}} s(I_a, \mathbf{T}(I_b)) \; . \tag{2}$$

While the parts may have false matches due to background clutter, self-similarity or descriptor mismatch, or no matches due to occlusion, practical implementation of the similarity function becomes very complex.

In order to match two images I_a and I_b well, there should be good matches between the descriptors $F_{a,j} \sim F_{b,j'}$ and the matching descriptors ($j \leftrightarrow j'$) should locate spatially close $\boldsymbol{x}_{a,j} \sim \boldsymbol{x}_{b,j'}$ under the transformation $\mathbf{T}$ (e.g., 2D scaling, translation and rotation). To avoid the complex approximation, we construct the similarity matching step-wise: first we construct the part appearance similarity matrix $\mathbf{D}$ and then using the sparse binary $\mathbf{D}$ we sample transformations $\mathbf{T}$ such that the geometric matching of feature locations is maximised. Our visual measure combines part-based appearance similarity and parts' constellation based structural similarity into a single novel measure.

Part-based appearance similarity – As a standard procedure, we compute dense SIFT descriptors using the VLFeat library [27] for every image scaled into the same image resolution (320×200, 640×400). Scaling makes the method resolution independent such that proportionally same size objects are matched against each other. We can achieve scaling invariance by using multi-resolution pyramid grids scaling invariance, but we did not find it necessary with the standard benchmark datasets (e.g. Caltech-101/256, Pascal VOC, ImageNet). For pairwise similarity of images a and b, the full descriptor distance matrix is computed between all features $\{F_a\}_i$ and $\{F_b\}_j$ forming the descriptor distance matrix $\mathbf{D}_{M_a \times M_b}$. We convert the distance matrix into a sparse binary form by assigning 1 to the five best matches and 0 for the rest. The five best is justified by the class level matching which is much weaker than between two views of a same object. 2–5 best matches were found to improve the matching considerably while beyond 5 improvement saturated quickly.

Structural similarity – The part-based similarity (e.g. summing N best descriptor distances) would somehow resemble the *visual bag of words* approach which has been used in graph construction [28]. However, the problem with that measure is that it does not constrain the accepted matches to be spatially consistent. Instead, we propose a similarity scoring procedure similar to used in specific image matching [14,20,25], but in our case for multiple candidates and not restricted by some fixed number of inliers. Due to enormous number of potential matches for exhaustive search, we repeat a random sampling procedure that selects two (minimum for similarity transform) features from a, their best candidates within the five best in $D_{M_a \times M_b}$, transforms all features in a to b and counts how many matches were found within the descriptor ellipses [17]. This procedure is repeated R times (100 found sufficient in our experiments for images of size 320×200) and the highest number used as the similarity measure between the images from a to b. It is noted that we do not restrict the transformation although it would improve the results with the standard datasets where objects are typically captured in a few standard poses (e.g. pictures of horizontal and vertical guitars in ImageNet). Moreover, the similarity matrix $S_{N \times N}$ is non-symmetric since matching from a to b can be different from b to a.

Similarity matrix – The output of the structural similarity is the number of parts that match both by their appearance and their configuration between the two images a and b. To establish a similarity matrix, the procedure is executed between all images making our method's computational complexity $O(N^2)$ for N images. At this stage, we make one more trick that prevents classes with plenty of salient parts to dominate by their high similarity scores and transform the actual scores to match ranks, i.e. the highest rank N is assigned to the best matching image and 1 to the worst matching:

$$S'\left(i = 1 \dots N, j = 1 \dots N\right) = row\_wise\_rank(S(i, j), S_{N \times N}) \ . \tag{3}$$

3.2   Network Construction

The network construction is straightforward if we change the structural visual similarity scores given as rank numbers to distances by

$$\hat{S}\left(i = 1 \dots N, j = 1 \dots N\right) = \frac{N}{S'(i, j)} \ . \tag{4}$$

Moreover, to make undirected graph algorithms available we convert the distance matrix symmetric by

$$\hat{S}\left(i, j\right) = min\left(\hat{S}(i, j), \hat{S}(j, i)\right) \ . \tag{5}$$

The visual similarity network of images is constructed as a graph $G = (V, E, W)$, where $V = \{V_1, ..., V_N\}$ denotes the set of vertices (nodes), $E_{N \times N}$ denotes the set of edges or links, and $W_{N \times N}$ is the set of edge weights. We assign each image $I_q \in \{I_1, ..., I_N\}$ to the corresponding vertice V_q and set the computed $\hat{S}_{N \times N}$ as the edge weights such that $W(i, j) = S(i, j)$. Therefore the edge weights would reflect the visual similarity such that low weights are assigned between similar images. It should be noted that we set the diagonal of W to 0 to remove self-references. The full network consists of $N \times (N - 1)$ links. An example of the constructed network is illustrated in Fig. 1.

4   Examples and Experiments

In the experiments, we used images from various synsets of the ImageNet Large Scale Visual Recognition Challenge 2014 (ILSVRC[2]), EPFL GIMS08 [19], and 10 categories of 3D object [23]. The only input for the network construction was images and therefore only their visual content affects the results.

4.1   Discovering Classes and Sub-classes

At first, it is attempting to use the network structure to automatically discover the most apparent discrete attributes such as visual classes (synsets) [9,28] or their sub-classes [4]. Whether our network structure can represent inter-class relationships we selected three distinct ImageNet classes: cars, motorbikes and airplanes. Using the Prim's algorithm, we constructed a minimum spanning tree (MST) shown in Fig. 2. From the tree and its closeups, it is evident that the classes have distinct branches that also represent sub-class information (scooters form their own branch in the motorbikes).

---

[2] http://image-net.org/challenges/LSVRC/2014/.

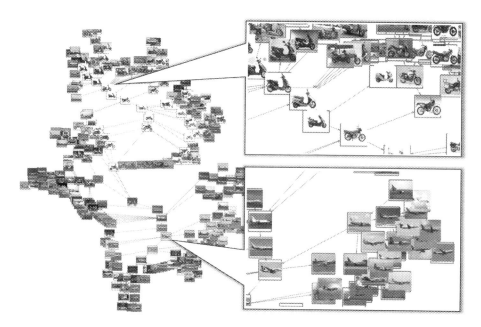

Fig. 2. A minimum spanning tree of a network constructed using our pairwise similarity measure for ImageNet cars, motorbikes and airplanes with closeups demonstrating the class branches.

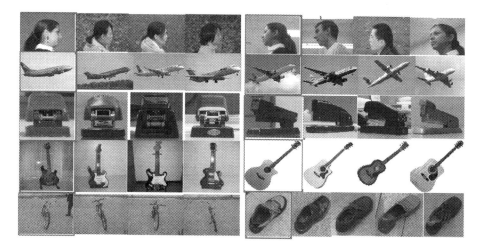

Fig. 3. Single network node images (denoted by the red rectangular) and for each 3–4 neighbour images with the strongest connection links (similarity scores). The results are more dominated by a specific pose rather a specific object.

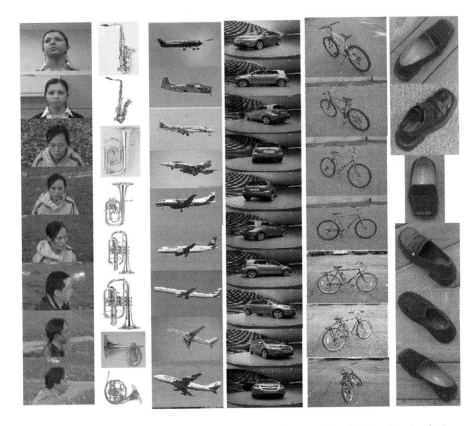

Fig. 4. "Source" images selected from 3D object dataset (head, bicycle, shoe), ImageNet (sax, airplane), EPFL GIMS08 (car) on the top and "sink" images at the bottom. The other images represent nodes, "smooth transition", between the source and sink within the minimum spanning tree. Note that, the path encodes gradual change in pose (face, car, bike and shoe) or appearance mixed with pose (music instrument, airplane). The shortest path depends on all images and there can be multiple almost equally good paths.

4.2   From Classes to Objects in Similar Pose

The interpretation that our network can be used to unsupervised discover classes and sub-classes as shown in Fig. 2 is not accurate. It can be used for dedicatedly selected examples, but if we add more ImageNet images that span almost full 3D poses, pairwise similarities do not anymore code class level similarity but combined pose and class similarity, and which dominates depends of a specific image pair. This finding is demonstrated in Fig. 3 where we generated the graphs for a large number of images from multiple ImageNet synsets (guitar, airplane), 3D object dataset (head, bicycle, shoe, stapler). In these examples, the pose dominates the pairwise similarity and not the specific object example. The finding conflicts with the works that try to discover classes and sub-classes in a graph

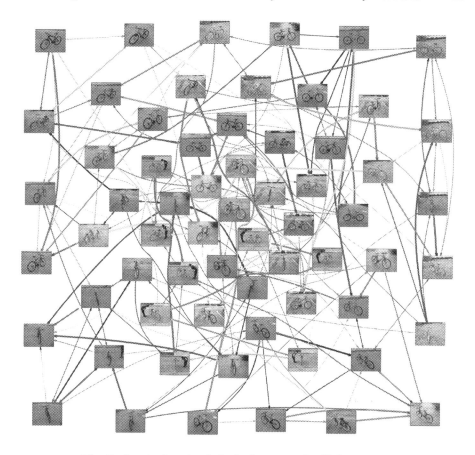

Fig. 5. A pairwise visual similarity network of bike images.

structure [4, 9, 28], but supports works learning multiple classifiers for different poses [1]. Overall, our examples illustrate the fact that pairwise visual similarity graph represents multiple multi-relational latent attributes at the same time.

4.3 Traveling Graph in Latent Continuous Attribute Space

The previous examples illustrated how a single graph can represent multiple "competing" attributes simultaneously and therefore methods based on strict boundaries, such as clustering, are deemed to fail. In a single network, there can be several almost equally good paths between two nodes and the transition "smoothness" depends on how many images there are in a network. A single shortest path can be selected by, for example, MST algorithm. In Fig. 4 we demonstrate continuous attributes by selecting "source" and "sink" images, and traveling the graph from the source to the sink via the MST's path. The images in nodes within the path represent the active attributes.

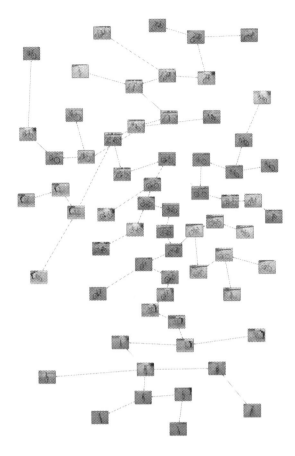

Fig. 6. A minimum spanning tree of the network in Fig. 5.

5   Conclusions

In this work, we proposed to use pairwise "structural visual similarity" between images to construct a network of images. The structural visual similarity is based on a simple principle that local regions of the two images match and are in similar spatial configuration (constellation). By using the similarity measure in network construction we noticed that the network can represent multi-relational attributes of discrete type (e.g. class/sub-class) and continuous type (e.g. 3D pose) (Fig. 5). However, with a large number of images, such as in the largest ImageNet synsets, certain attributes, such as the pose, may dominate other attributes and therefore any unsupervised graph-based attribute search using tight boundaries, such as clustering or minimum spanning tree (Fig. 6), is deemed to fail. Therefore, we believe that the network structure representing images as nodes and their similarities as connecting edges is a natural presentation of visual data. In this sense, even the attribute classifiers may share some images. Our future work will exploit the network structure further and the immediate interests are two-fold: (1) how to boost our network algorithm of the computational complexity $O(N^2)$ (similarity matrix construction) more efficient to cope with millions of images; and (2) how to unsupervised discover all active multi-relational attributes.

References

1. Aghazadeh, O., Azizpour, H., Sullivan, J., Carlsson, S.: Mixture component identification and learning for visual recognition. In: Fitzgibbon, A., Lazebnik, S., Perona, P., Sato, Y., Schmid, C. (eds.) ECCV 2012, Part VI. LNCS, vol. 7577, pp. 115–128. Springer, Heidelberg (2012)
2. Akata, Z., Perronnin, F., Harchaoui, Z., Schmid, C.: Label-embedding for attribute-based classification. In: CVPR (2013)
3. Deng, J., Dong, W., Socher, R., Li, L.-J., Li, K., Fei-Fei, L.: ImageNet: a large-scale hierarchical image database. In: CVPR (2009)
4. Dong, J., Xia, W., Chen, Q., Feng, J., Huang, Z., Yan, S.: Subcategory-aware object classification. In: CVPR (2013)
5. Felzenszwalb, P.F., Girshick, R.B., McAllester, D., Ramanan, D.: Object detection with discriminatively trained part-based models. IEEE Trans. Pattern Anal. Mach. Intell. **32**(9), 1627–1645 (2010)
6. Felzenszwalb, P.F., Huttenlocher, D.P.: Pictorial structures for object recognition. Int. J. Comput. Vis. **61**(1), 55–79 (2005)
7. Ferrari, V., Zisserman, A.: Learning visual attributes. In: Advances in Neural Information Processing Systems (NIPS) (2007)
8. Gavves, E., Fernando, B., Snoek, C.G.M., Smeulders, A.W.M., Tuytelaars, T.: Fine-grained categorization by alignments. In: ICCV (2013)
9. Kim, G., Faloutsos, C., Hebert, M.: Unsupervised modeling of object categories using link analysis techniques. In: CVPR (2008)
10. Kinnunen, T., Kamarainen, J.-K., Lensu, L., Kälviäinen, H.: Unsupervised object discovery via self-organisation. Pattern Recogn. Lett. **33**(16), 2102–2112 (2012)
11. Krizhevsky, A., Sutskever, I., Hinton, G.: ImageNet classification with deep convolutional neural networks. In: NIPS (2012)
12. Kumar, M., Zisserman, A., Torr, P.: Efficient discriminative learning of parts-based models. In: ICCV (2009)
13. Lampert, C.H., Nickisch, H., Harmeling, S.: Attribute-based classification for zero-shot visual object categorization. IEEE Trans. Pattern Anal. Mach. Intell. **36**(3), 453–465 (2014)
14. Lankinen, J., Kamarainen, J.-K.: Local feature based unsupervised alignment of object class images. In: BMVC (2011)
15. Malisiewicz, T., Efors, A.: Beyond categories: the visual memex model for reasoning about object relationships. In: NIPS (2009)
16. Malisiewicz, T., Gupta, A., Efors, A.: Ensemble of exemplar-SVMs for object detection and beyond. In: ICCV (2011)
17. Mikolajczyk, K., Schmid, C.: A performance evaluation of local descriptors. IEEE PAMI **27**(10), 1615–1630 (2005)
18. Myeong, H., Chang, J.Y., Lee, K.M.: Learning object relationships via graph-based context model. In: CVPR (2012)
19. Ozuysal, M., Lepetit, V., Fua, P.: Pose estimation for category specific multiview object localization. In: CVPR (2009)

20. Philbin, J., Chum, O., Isard, M., Sivic, J., Zisserman, A.: Object retrieval with large vocabularies and fast spatial matching. In: CVPR (2007)
21. Philbin, J., Sivic, J., Zisserman, A.: Geometric latent dirichlet allocation on a matching graph for large-scale image datasets. Int. J. Comput. Vis. **95**(2), 138–153 (2011)
22. Russakovsky, O., Fei-Fei, L.: Attribute learning in large-scale datasets. In: Kutulakos, K.N. (ed.) ECCV 2010 Workshops, Part I. LNCS, vol. 6553, pp. 1–14. Springer, Heidelberg (2012)
23. Savarese, S., Li, F.-F.: 3d generic object categorization, localization and pose estimation. In: ICCV, pp. 1–8 (2007)
24. Simonyan, K., Vedaldi, A., Zisserman, A.: Deep fisher networks for large-scale image classification. In: NIPS (2013)
25. Tuytelaars, T., Gool, L.V.: Wide baseline stereo matching based on local, affinely invariant regions. In: BMVC (2000)
26. Tuytelaars, T., Lampert, C., Blaschko, M., Buntine, W.: Unsupervised object discovery: a comparison. Int. J. Comput. Vis. **88**(2), 284–302 (2010)
27. Vedaldi, A., Fulkerson, B.: VLFeat: an open and portable library of computer vision algorithms (2008). http://www.vlfeat.org/
28. Xia, S., Hancock, E.R.: Incrementally discovering object classes using similarity propagation and graph clustering. In: Zha, H., Taniguchi, R., Maybank, S. (eds.) ACCV 2009, Part III. LNCS, vol. 5996, pp. 373–383. Springer, Heidelberg (2010)
29. Zhu, X., Loy, C., Gong, S.: Constructing robust affinity graph for spectral clustering. In: CVPR (2014)

Blur-Robust Face Recognition
via Transformation Learning

Jun Li, Chi Zhang, Jiani Hu, and Weihong Deng[✉]

Beijing University of Posts and Telecommunications, Beijing, China
whdeng@bupt.edu.cn

Abstract. This paper introduces a new method for recognizing faces degraded by blur using transformation learning on the image feature. The basic idea is to transform both the sharp images and blurred images to a same feature subspace by the method of multidimensional scaling. Different from the method of finding blur-invariant descriptors, our method learns the transformation which both preserves the manifold structure of the original shape images and, at the same time, enhances the class separability, resulting in a wide applications to various descriptors. Furthermore, we combine our method with subspace-based point spread function (PSF) estimation method to handle cases of unknown blur degree, by applying the feature transformation corresponding to the best matched PSF, where the transformation for each PSF is learned in the training stage. Experimental results on the FERET database show the proposed method achieve comparable performance against the state-of-the-art blur-invariant face recognition methods, such as LPQ and FADEIN.

1 Introduction

Face recognition plays an important role in the field of computer vision. Previous works on this topic [1–4] mainly focuses on recognizing faces under controlled imaging conditions. However, in practice, the performance of recognition algorithms tends to suffer from image degradation [5]. One example of degradation is blur as a consequence of out-of-focus lens, atmospheric turbulence, and relative motion between the camera and the objects. The process of such image blur could be modeled as:

$$g(n_1, n_2) = (I * H)(n_1, n_2) + n(n_1, n_2) \tag{1}$$

where (n_1, n_2) is the pixel location at which a convolution $*$ is performed between the original sharp image I and a point spread function (**PSF**) H, which has the same size with image I. n denotes the additive image noise coming from quantization, or other camera-induced errors.

The performances of face recognition could suffer from blur for the fact that blur leads to two main problems [6]:

(i) the facial appearance of an individual changes drastically due to blur.
(ii) different individuals tend to appear more similar when blurred.

© Springer International Publishing Switzerland 2015
C.V. Jawahar and S. Shan (Eds.): ACCV 2014 Workshops, Part III, LNCS 9010, pp. 15–29, 2015.
DOI: 10.1007/978-3-319-16634-6_2

A few existing approaches have been proposed to handle these problems. All of them are based on the formulation of the process of blur,and could be roughly classified into three categories: (i)converting sharp images to match the blurred ones via blurring methods, (ii)converting blurred images to match sharp images via deblurring methods, (iii)finding blur-invariant descriptors.

One typical method of class (i) could be found in [7], where Stainvas & Intrator artificially blurred the sharp gallery images to match a blurred query image through a hybrid network architecture. Although problem (i) can be thoroughly handled, problem (ii) still remains. Besides, the gallery images may be blurred themselves.

The deblurring approaches are widely used to estimate I from the observed blurred image g, which means solving the inverse problem of formulation (1). But this is a difficult task considering unknown types of blur kernel. For instance, the PSF H of ideal motion and out-of-focus blur tends to be rectangular according to [8,9], while atmospheric and optical blurs are more likely to be Gaussian blur [9,10]. These approaches could also been treated as a way of image restoration. When the model for blur is known, known as non-blind deconvolution [11], Levin et al. [12] learn priors on clean image statistics, Fergus [13] use coded-computational photography to remove motion blur. Blind image deconvolution assumes nothing about the blur kernels is known [14]. Most of the existing works perform blind image deconvolution based on a single image [15]. A PSF is inferred through total variation regularization [16], the variation of Gaussian scale in the edges [17–19], or variation in the wavelet domain [20,21]. Recently, M. Nishiyama [6] has revealed that deblurring from a single image is an ill-posed problem and these deblurring methods are insufficient for accurate face recognition. Other methods deduce a PSF using multiple images [22,23], and M. Nishiyama et al. [6,24] propose to build multiple PSFs and use the best match as the final PSF. Even if the PSF could be correctly estimated, deblurring could not be very robust due to unknown noise n.

As regards blur-invariant descriptors, researchers have tried to find blur insensitive descriptors. Useful descriptors include local binary pattern 'LBP' [25, 26], the subspace [27], manifold [28], sparse representation [29] and the local phase quantization 'LPQ' [30]. The 'LPQ' descriptor performs best when the blur is centrally symmetric [31], which is invariant with respect to blur effects. Recently, Gopalan et al. [32] proposed a new blur-robust descriptor using subspace techniques. These blur-invariant-descriptor relevant methods could also be viewed as direct methods.

Our Contributions: All of the approaches mentioned above could be viewed as preprocessing for feature extraction. We offer a new perspective that handling the effects of blur and extracting features at the same time via Transformation Learning. Based on the intuition that face recognition algorithms perform best when the gallery images and their corresponding blurred ones are in the same kind of feature spaces, we transform the features of sharp gallery images and blurred query images to a same space. We adopted Multidimensional Scaling (MDS) [33] to learn the desired transformation from training images using iterative majorization algorithm [34]. Extensive experiments on FERET and CMU-PIE datasets show that our MDS-based feature extraction could efficiently weaken the degradation caused by blur.

Outline of the Paper: The rest of the paper is organized as follows. In Sect. 2, we present our method for blur-robust face recognition. In Sect. 3, we describe the detailed experiments and show that our method not only outperforms others, but also demonstrates excellent capability of generalization. Finally, a brief summary and discussion is provided in Sect. 4.

2 Our Approaches

We try to find a subspace into which the sharp original images and their blurred counterparts are mapped in the same way. What's more, the performance of face recognition algorithms has been taken into consideration by combining classifiers.

We introduce Multidimensional Scaling (MDS) method to conduct transformation learning to solve the blur problem. It should be noticed that MDS assumes already knowing the blur degree, as the transformation is performed between the sharp images and those with certain blur degrees.

For an image with unknown blur degree, we first infer the PSF with the method in [6]. Then, we select the corresponding transformation matrix from the previously learned MDS matrices. Finally, classification methods are applied in the transformed space.

An overview of our method could be found in Fig. 1.

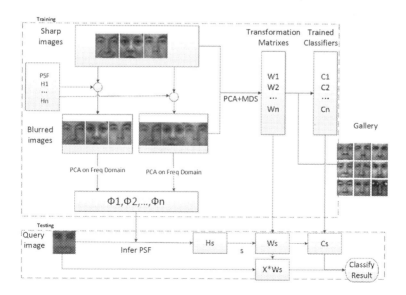

Fig. 1. Overview of our method

Recently, Biswas et al. [35] also adopted MDS to solve various pose and resolution problem in face recognition, but we handle clearly different tasks in face recognition. Moreover, our work differs from [35] in the following two respects:

(i) In [35], an uniform W is trained for all kinds of variations in pose and resolution, while we differentiate different blur levels, and train a series of transformations $W_1 \ldots W_n$. This enables us to acquire the most accurate transformation for the blurred query image from the series of matrices (see Fig. 1). For the blur problem, all kinds of blur have certain blur kernels [2], which means detailed PSF quantization is feasible and effective for solving the problem. (ii) Traditional multidimensional scaling is adopted in [35] to transform data from a high dimension to a low dimension. In contrast, we transform the data to a higher dimension, which is indeed a linear kernel trick.

2.1   PSF Inference

We adopt the methods proposed in [6] to infer the PSF. The basic idea is to find a subspace which is much more sensitive to the variance caused by blur than differences between individuals. The algorithm is shown as follows, which finally gives the subscript s of the inferred PSF:

Algorithm 1. PSF Inference

1: Extract feature image $x(\epsilon', \eta')$ from a blurred image $g(u, v)$ as:

$$x(\epsilon', \eta') = [log(|g(\epsilon, \eta)|)] \downarrow \qquad (2)$$

where $g(\epsilon, \eta)$ is the Fourier transform of $g(u, v)$, $||$ takes the amplitude, and $[] \downarrow$ stands for down-sampling. It should be mentioned that x is normalized so that $||x||_2 = 1$.

2: Running PCA in the training set. Training images are blurred with each PSF H_i, We extract the feature images for the whole blurred training set, forming correlation matrix $A_i = \frac{1}{M} \Sigma_{k=1}^M x'_{ik}(x'_{ik})^T$, where x'_{ik} represents the feature image extracted from image g_{ik} which is blurred with H_i. A subspace $\theta_i = \{b_{ij}\}_{j=1}^D$ is got with the first D eigenvectors by decreasing eigenvalue.

3: Infer the PSF. For a new query image with unknown blur, we calculate the closest subspace with cosine distance to determine the PSF H_s:

$$s = arg \max_i \Sigma_{j=1}^D (b_{ij}^T x)^2 \qquad (3)$$

2.2   Feature Selection

Feature selection is an issue worth considering for blurred face recognition. Previous researchers attempted to find a blur invariant descriptor, however, this is an ill-posed problem and we could only find approximately invariant ones. For instance, Local Phase Quantization(LPQ) is such a descriptor, and only when the facial image is boundless could it be invariant to blur. More details about how LPQ is applied in face recognition can be found in [30]. In fact, blur is closely related to scaling, making global descriptors inappropriate for the problem. Instead, local descriptors like LPQ, LBP and EF describe a pixel in an image with its neighborhood content, and are therefore capable of handling the invariants of blur to some extent. In our experiments, we first choose uniformed LBP as our feature. For a new coming face image, we first resize it to 64×64. Then, we compute the uniformed LBP(ranging from 0 to 58) for the pixels of the whole image. Later, each image is divided into 8×8 cells, and we would obtain the histograms for each cell. Combining these histograms together will result in a feature with dimension $59 \times 8 \times 8$. Besides, we also extract LPQ features similarly, while the cell size is set to be 6×6 and cell is slid with a gap of 4 pixels.

2.3   Transformation Learning

High-dimensional transformation learning is always terribly high in time and space complexity. For instance, the dimension of using uniformed LBP is 3776, and the raw pixel dimension is 4096. Therefore, before we perform transformation learning, PCA (Principal Component Analysis), which is a standard technique to create low-dimensional representation of high-dimensional data, is always required. It helps to reduce the complexity significantly while preserving the performance of Transformation Learning. Besides, PCA is also effective for reducing noises. When the training set contains enough images for each individual, LDA (Linear Discriminant Analysis) is performed to make the distances between facial images of different classes as large as possible while preserving the distances between same-class images. Based on this, LDA further reduces the dimensions of the data. However, later we would demonstrate that PCA and LDA only help promote the recognition performance – it is the multidimensional scaling that really counts.

Objective Function. We denote f as the transformation from R^l to R^t, here l represents the dimension of the input space, while t is the dimension of the transformed space. f could be written as:

$$f(x) = W^T x \tag{4}$$

where $x \in R^l$ is the feature vector we got from LBP, PCA and LDA. W is a $l \times t$ matrix, denoting the weights to be learned. Traditional metric multidimensional scaling tends to set $t = 2$ or 3 for visualization; while on the contrary, we find that mapping data to a higher dimension may lead to better separability, just as Support Vector Machine reveals. Suppose we have N original training images, all of which are processed with feature extraction (always followed with PCA for high dimensional feature), and we denote them as $\{x_1^s, x_2^s, \ldots x_N^s\}$, and their corresponding blurred images are $\{x_1^b, x_2^b, \ldots x_N^b\}$. We apply the objective function of [34] to solve the problem:

$$\min J(W) = \lambda J_{SP}(W) + (1 - \lambda) J_{CS}(W) \tag{5}$$

where $J_{SP}(W)$ is a structure-preserving term. Our objective is to make sure the distance of transformed feature space between the original sharp image $f(x_i^s)$ and the blurred image $f(x_j^b)$ approximates the distance of input space between original sharp images d_{ij}^s as well as possible. We define $J_{SP}(W)$ as:

$$J_{SP}(W) = \sum_{i=1}^{N} \sum_{j=1}^{N} (q_{ij}(W) - d_{ij}^s)^2 \tag{6}$$

where

$$q_{ij}(W) = |f(x_i^s) - f(x_j^b)| = |W^T(x_i^s - x_j^b)| \tag{7}$$

stands for the distance between the sharp and the blurred images in the transformed space. d_{ij}^s is the distance between two sharp images in the input space, usually defined as $|x_i^s - x_j^s|$. This formulation could effectively handle the two main problems we have presented.

$J_{CS}(W)$ is the class-separation term. Similar to Koontz and Fukunaga, we propose to define the separability term as:

$$J_{CS}(W) = \sum_{i} \sum_{j} \delta_{ij} q_{ij}^2(W) \tag{8}$$

However, our term differs from Koontz and Fukunaga in that both i and j are taken from 1 to N, and we define δ_{ij} as:

$$\delta_{ij} = \begin{cases} 0 & \text{if } w_i \neq w_j \\ 1 & \text{if } w_i = w_j(i \neq j) \\ 2 & \text{if } i = j \end{cases} \tag{9}$$

where w_i stands for the class label of image i. λ is the balance parameter between the structure-preserving term and class-separation term, whose value is taken in the range $[0, 1]$. Combining the Eqs. (5)–(9), we could transform the objective function as:

$$J(W) = \sum_{i=1}^{N} \sum_{j=1}^{N} \alpha_{ij}(q_{ij}(W) - \beta_{ij} d_{ij}^s)^2 \tag{10}$$

where

$$\alpha_{ij} = (1 - \lambda)\delta_{ij} + \lambda \tag{11}$$

and

$$\beta_{ij} = \lambda/\alpha_{ij}. \tag{12}$$

We then adopt the iterative majorization algorithm [33, 34] to solve the minimization problem (10).

2.4   Iterative Majorization

Here, we quickly review the iterative majorization algorithm, which is also called stress majorization. If we define:

$$J_m^2(W, V) = Tr\{W^T A W\} + \sum_{i=1}^{N} \sum_{j=1}^{N} \alpha_{ij} d_{ij}^2$$
$$-2Tr\{V^T D(V) W\}$$
(13)

where

$$A = \sum_{i=1}^{N} \sum_{j=1}^{N} \alpha_{ij} (x_i - x_j)(x_i - x_j)^T$$
(14)

and

$$D(V) = \sum_{i=1}^{N} \sum_{j=1}^{N} C_{ij}(V)(x_i - x_j)(x_i - x_j)^T$$
(15)

with

$$C_ij(V) = \begin{cases} \lambda d_{ij}^s / q_{ij}(V) & \text{if } q_{ij}(V) > 0 \\ 0 & \text{if } q_{ij}(V) = 0 \end{cases}$$
(16)

The solution for W that minimizes $J_m^2(W, V)$ satisfies

$$AW = D(V)V$$
(17)

based on the theory that $J(W) \leq J_m^2(W, V) \leq J(V)$. The detailed minimization procedure is:

Algorithm 2. Iterative Majorization

1: set $t = 0$, initial convergence precision ξ with a predefined value, initial W with random value in the range $[-1, 1]$.
2: set $V = W^t$
3: update W^t to W^{t+1} as

$$W^{t+1} = A^{-1} D(V) V$$
(18)

where A^{-1} stands for the Moore-Penrose pseudoinverse of A.
4: check for convergence. If $|W^{t+1} - W^t| < \xi$, stop the iteration and output W; If otherwise, set $t = t + 1$ and go to step 2.

2.5   Recognition Across Blur

The transformation above gives a common feature space for sharp and blurred images, in which distance metrics such as Euclidean distance and cosine distance can be utilized for recognition. Here we choose cosine distance for verification, considering its best performance in verification, or choose Euclidean distance for face recognition combined with an NN classifier.

The gallery set and the query set consist of sharp and blurred images respectively. A query blurred image could be compared to all of the sharp gallery images using cosine distance metrics for verification, or be put into the trained classifier C_s with Euclidean distance for recognition.

3   Experiments and Results

The following experiments are conducted to demonstrate the effectiveness of our method. We assume that the blur degrees of the test images are the same and already known, thus we can construct training images with the same blur effect. We shall discuss how to deal with unknown degree of blur in the Conclusion part.

3.1   Databases

We perform different experiments based on the following two databases. It should be noted that blur does not matter much in face detection, as the current face detector could detect faces as small as 15×15 pixels.

CMU-PIE Database. Reference [36] contains 68 subjects with 41368 images on the whole. For each subject, the illumination subset (C27), which contains 21 distinct sources of lights, was used in our experiment. For both datasets, all the images were first normalized by a similarity transformation that sets the inter-eye line horizontal and the two eyes 70 pixels apart, and then cropped to the size of 128×128 with the centers of the eyes located at $(29, 34)$ and $(99, 34)$ to extract the pure face region. No additional preprocessing procedure is required in our experiments.

FERET Database. Reference [37] is a standard database used for algorithm development and testing, which is divided into development portion and sequestered portion. The development portion provides a common database for designing algorithms, and the sequestered portion is for testing and evaluating face recognition algorithms. In our experiments, we select three subsets, 'bk','bj', and 'ba', from the development portion, and two subsets, 'fa' and 'fb', from the sequestered portion. The faces of all subsets are detected with a V-J face detector and the facial area is cropped to the size of 128×128.

3.2   Transformation Learning: Intuition

To give an intuition as to how Transformation Learningworks, we select two individuals from database CMU-PIE, and each individual consists of 21 images. First of all, we resize the images to 64×64 bilinearly. Then all the images are artificially blurred with Gaussian blur PSF: $H(u, v) = \frac{1}{Z} exp(\frac{-(u^2+v^2)}{2\sigma^2})$, where σ denotes the standard deviation, and Z is a normalization term. In our experiment, we set

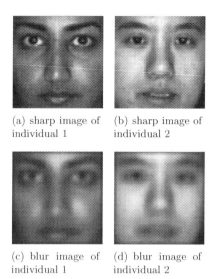

(a) sharp image of
individual 1

(b) sharp image of
individual 2

(c) blur image of
individual 1

(d) blur image of
individual 2

Fig. 2. Example images from 2 different individuals. Each one is artificially blurred with a Gaussian blur filter whose size is 5×5 and the standard deviation is 3.

$\sigma = 3$ and the filter size is fixed to be $HSIZE = [5\,5]$. Typical images of the two individuals and their corresponding blurred ones are shown in Fig. 2.

To exclude the effects of other factors, we simply choose raw pixel data as the feature vector. We use data of another 34 individuals to learn a PCA matrix and a transformation matrix. Then dimension reductions are conducted on all the original sharp images and their corresponding blurred ones, to 100 dimension after PCA and then to 900 dimension after MDS. The reason why PCA is used and why the dimension is 100 are that we want to perform MDS in a rapid way and to give a convenient visualization at the same time. The first 3 dimensions (the biggest 3 principals) are chosen for visualization. The result is shown in Fig. 3.

It's obvious that the non-separable data in the original data space becomes linear-separable in the transformed space through MDS, which means a simple linear classifier would be sufficient to work well.

3.3 Face Identification with and Without MDS

Next, we conduct a face identification experiment on the CMU-PIE database. This time, all the cropped images are used. We randomly select 34 individuals for training, the remainings are used as gallery set. The training set images are blurred using Gaussian PSF, where σ is set to be 3, and $HSIZE$ is set to be [33]. After resizing the images to 64×64, we extract the LBP features of the images and then conduct PCA, LDA to reduce the dimension to 100. We use the training set (34×21 images) to train PCA and MDS transformation matrix. Note that we set the initial MDS transformation matrixes W as equal values to give a fair comparison. Finally, given the test images and gallery ones, we adopt Cosine distance rather

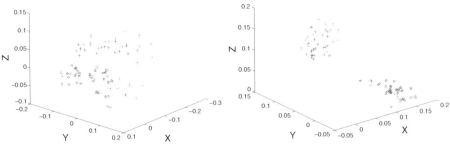

(a) images visualization without MDS

(b) image visualization with MDS

Fig. 3. Visualization of the effectiveness of our method. Different shapes denote different individuals with 'cross' represents individual 1, and 'star' represents individual 2. And we mark the sharp images in blue color and the blurred ones in red (Color figure online).

than Euclidean distance as the distance metric to get the identification results. See Eq. 19.

$$\tilde{d}(i,j) = \frac{\tilde{x}_i \tilde{x}_j}{|x_i||x_j|} \tag{19}$$

The nature of the results would be affected by the threshold value. Take false detection rate as x axis, and true positive rate as y axis, and if we change the threshold value continuously, an ROC curve would be acquired. In our experiment, we change threshold value of identification for 505521 times, meaning that 505521 points (fp_i, tp_i) would be acquired. Here fp_i stands for false positive rate, while tp_i denotes true positive rate. The result is shown in Fig. 4.

The result indicates that PCA and LDA help to improve the performance of identification, yet only marginally. The next experiment may reveal that PCA could even degrade the recognition rate sometimes. We calculate the Area Under Curve (AUC) of the ROC curve with the Trapezoidal Rule:

$$AUC = \sum_{i=1}^{n-1} (fp_{i+1} - fp_i)\frac{tp_{i+1} + tp_i}{2} \tag{20}$$

Here $n = 505521$, which denotes the number of sample points.

As shown in Table 1, MDS significantly improves the performance of face identification, as the AUC value is much higher when MDS is employed.

3.4　Comparison with State-of-Art Methods

The state-of-art method is FADEIN+LPQ [6]. To make a fair comparison, we follow the same approaches. FERET database is adopted where all images could be

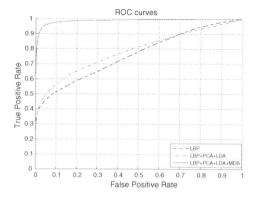

Fig. 4. Comparision of ROC curves.

Table 1. Comparison of methods with and without MDS. We calculate the accuracy with false positive rate of 0.01. And the AUC values are based on the ROC curve shown in Fig. 4

Method	Accuracy (false positive rate=0.01)	AUC
LBP	37.66 %	0.762
LBP+PCA+LDA	37.64 %	0.787
LBP+PCA+LDA+MDS	**87.36 %**	**0.989**

considered sharp. We choose 10001 images of 1001 individuals from subset 'fa' as the gallery set. Meanwhile, their corresponding images in 'fb' subset are selected to build up the target set, which have been filtered with Gaussian blur and added with 30 dB Gaussian white noise. It should be mentioned that Gopalan et al. [32] improperly set the filter sizes to be the same ($hsize = 5$) for different standard deviations, while in fact, the filter size should increase with the standard deviation. Specifically, we set $hsize = 1.5 * \delta$, then around 87 % energy would be preserved for all filters. Our objectives imply that only one image for each individual is not enough for learning the transformation matrix, so we combine 'bk','bj','ba' subsets together for training, with 3 images for each individual. We artificially blur these images as we did to those in the 'fb' subset. For the feature, we choose uniformed LPQ, where the cell size is set to be 6×6 and the sliding step to be 4. PCA is performed on the extracted features to reduce the dimension to 400, and a transformation to a dimension of 900 is learned through our MDS method, where we set $\lambda = 0.2$. Here we simply denote 'LPQ+PCA+MDS' as 'LPQ+MDS'. Finally, NN classifier is used for recognition, and the result is shown in Fig. 5.

It can be seen clearly that our method could compete with the FADEIN+LPQ method. Both of them present considerable improvement over the pure LPQ method, which means that both of our MDS method (without deblurring) and FADEIN (with deblurring) are effective for improving the recognition accuracy.

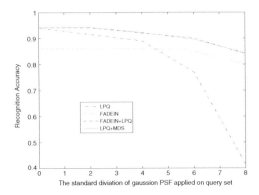

Fig. 5. Comparision with state-of-art methods.

3.5   Performance on Mixed Blur Settings

The experiments above assume that all the training images and query images are blurred by PSFs with a certain standard deviation. The question is, what if there is no information available about the PSF? We solve this problem by adopting the PSF inference method by M. Nishiyama and A. Hadid [6]. While our experiment considers only Gaussian blur, the method could handle all kinds of blur indeed. First of all, we test the PSF inference accuracy on FERET database. The chosen datasets are the same as in Sect. 2.4, and we still set $hsize = 1.5 * \sigma$. We blur the whole query set with different Gaussian PSFs each time. The result is shown in Table 2.

Table 2. PSF inference accuracy for different standard deviation.

Standard deviationσ	Inference accuracy
0	99.80 %
2	98.49 %
4	97.68 %
6	98.39 %
8	98.19 %

The average inference accuracy is 98.43 %, proving the effectiveness of the PSF inference method.

Then, we blur the images with random standard deviation chosen from $\{0, 2, 4, 6, 8\}$, for each query image, we infer the PSF first, and then choose the corresponding transformation matrix, using transformed feature to recognize with the corresponding classifier. The result is shown in Table 3.

Table 3. PSF inference accuracy for different standard deviation.

Method	Recognition rate
LPQ	83.56 %
FADEIN	82.45 %
FADEIN+LPQ	88.12 %
LPQ+PCA+MDS	**88.56 %**

It can be noticed that without deblurring, our method performs slightly better than FADEIN combined with LPQ, thus demonstrating the effectiveness of our method.

3.6 Time Costs

It is the training procedure that takes much time. When we set the convergence precision to be 0.1, it would take around 30 iteration steps to complete the transformation from 400 to 900 dimensions. In our Matlab implemention on 3.2 GHZ core i5 CPU, about 3 h was consumed to compute the matrix W. After obtaining W, however, it takes very short time for testing, thus making it suitable for real-time face recognition applications.

4   Conclusion

In this paper, we solve the blur problem of face recognition based on a new point of view, which is transforming both the sharp images and blurred ones to a common subspace. Our approach could be regarded as a procedure coming after feature extraction, thus compatible with various existing methods. One problem with our method is that we assume the blur degree to be already known, which could be solved by training a set of W with different blurs. Concretely, when a new test image is coming, we could first infer the blur degree with PSF inference methods, and then select the best matched W. Further research may focus on improving the transformation learning matrix through finding a subspace for various degrees of blur.

Acknowledgement. This work was partially sponsored by National Natural Science Foundation of China (NSFC) under Grant No. 61375031, No. 61471048, and No. 61273217. This work was also supported by the Fundamental Research Funds for the Central Universities, Beijing Higher Education Young Elite Teacher Project, and the Program for New Century Excellent Talents in University.

References

1. Deng, W., Hu, J., Guo, J.: Extended src: Undersampled face recognition via intraclass variant dictionary. IEEE Trans. Pattern Anal. Mach. Intell. **34**, 1864–1870 (2012)

2. Deng, W., Hu, J., Guo, J., Cai, W., Feng, D.: Robust, accurate and efficient face recognition from a single training image: a uniform pursuit approach. Pattern Recogn. **43**, 1748–1762 (2010)
3. Deng, W., Hu, J., Zhou, X., Guo, J.: Equidistant prototypes embedding for single sample based face recognition with generic learning and incremental learning. Pattern Recogn. **47**, 3738–3749 (2014)
4. Deng, W., Hu, J., Lu, J., Guo, J.: Transform-invariant pca: a unified approach to fully automatic face alignment, representation, and recognition. IEEE Trans. Pattern Anal. Mach. Intell. **36**, 1275–1284 (2014)
5. Deng, W., Hu, J., Guo, J., Cai, W., Feng, D.: Emulating biological strategies for uncontrolled face recognition. Pattern Recogn. **43**, 2210–2223 (2010)
6. Nishiyama, M., Hadid, A., Takeshima, H., Shotton, J., Kozakaya, T., Yamaguchi, O.: Facial deblur inference using subspace analysis for recognition of blurred faces. IEEE Trans. Pattern Anal. Mach. Intell. **33**, 838–845 (2011)
7. Stainvas, I., Intrator, N., Moshaiov, A.: Blurre face recognition via a hybrid network architecture. Pattern Recogn. **2**, 805–808 (2000)
8. Cannon, M.: Blind deconvolution of spatially invariant image blurs with phase. IEEE Trans. Acoust. Speech Signal Process. **24**, 58–63 (1976)
9. Katsaggelos, A.K.: Digital image restoration. Springer Publishing Company (2012, Incorporated)
10. Kimia, B.B., Zucker, S.W.: Analytic inverse of discrete gaussian blur. Opt. Eng. **32**, 166–176 (1993)
11. Yuan, L., Sun, J., Quan, L., Shum, H.Y.: Progressive inter-scale and intra-scale non-blind image deconvolution. In: ACM Transactions on Graphics (TOG), vol. 27, p. 74. ACM (2008)
12. Levin, A., Lischinski, D., Weiss, Y.: A closed-form solution to natural image matting. IEEE Trans. Pattern Anal. Mach. Intell. **30**, 228–242 (2008)
13. Fergus, R., Singh, B., Hertzmann, A., Roweis, S.T., Freeman, W.T.: Removing camera shake from a single photograph. In: ACM Transactions on Graphics (TOG), vol. 25, pp. 787–794. ACM (2006)
14. Levin, A., Weiss, Y., Durand, F., Freeman, W.T.: Efficient marginal likelihood optimization in blind deconvolution. In: 2011 IEEE Conference on Computer Vision and Pattern Recognition (CVPR), pp. 2657–2664 (2011)
15. Kundur, D., Hatzinakos, D.: Blind image deconvolution. IEEE Signal Process. Mag. **13**, 43–64 (1996)
16. Chan, T.F., Wong, C.K.: Total variation blind deconvolution. IEEE Trans. Image Process. **7**, 370–375 (1998)
17. Hu, H., de Hann, G.: Low cost robust blur estimator. In: 2006 IEEE International Conference on Image Processing, pp. 617–620. IEEE (2006)
18. Elder, J.H., Zucker, S.W.: Local scale control for edge detection and blur estimation. IEEE Trans. Pattern Anal. Mach. Intell. **20**, 699–716 (1998)
19. Marziliano, P., Dufaux, F., Winkler, S., Ebrahimi, T.: A no-reference perceptual blur metric. In: 2002 International Conference on Image Processing, Proceedings, vol. 3, pp. III–57. IEEE (2002)
20. Rooms, F., Pizurica, A., Philips, W.: Estimating image blur in the wavelet domain. In: IEEE International Conference on Acoustics Speech and Signal Processing 1999, vol. 4, pp. 4190–4190. IEEE (2002)
21. Tong, H., Li, M., Zhang, H., Zhang, C.: Blur detection for digital images using wavelet transform. In: 2004 IEEE International Conference on Multimedia and Expo, ICME 2004, vol. 1, pp. 17–20. IEEE (2004)

22. Yuan, L., Sun, J., Quan, L., Shum, H.Y.: Image deblurring with blurred/noisy image pairs. In: ACM Transactions on Graphics (TOG), vol. 26, p. 1. ACM (2007)
23. Ancuti, C., Ancuti, C.O., Bekaert, P.: Deblurring by matching. Computer Graphics Forum **28**, 619–628 (2009). Wiley Online Library
24. Nishiyama, M., Takeshima, H., Shotton, J., Kozakaya, T., Yamaguchi, O.: Facial deblur inference to improve recognition of blurred faces. In: IEEE Conference on Computer Vision and Pattern Recognition, CVPR 2009, pp. 1115–1122. IEEE (2009)
25. Ojala, T., Pietikainen, M., Maenpaa, T.: Multiresolution gray-scale and rotation invariant texture classification with local binary patterns. IEEE Trans. Pattern Anal. Mach. Intell. **24**, 971–987 (2002)
26. Ahonen, T., Hadid, A., Pietikainen, M.: Face description with local binary patterns: Application to face recognition. IEEE Trans. Pattern Anal. Mach. Intell. **28**, 2037–2041 (2006)
27. Deng, W., Liu, Y., Hu, J., Guo, J.: The small sample size problem of ica: a comparative study and analysis. Pattern Recogn. **45**, 4438–4450 (2012)
28. Deng, W., Hu, J., Guo, J., Zhang, H., Zhang, C.: Comments on "globally maximizing, locally minimizing: unsupervised discriminant projection with applications to face and palm biometrics". IEEE Trans. Pattern Anal. Mach. Intell. **30**, 1503–1504 (2008)
29. Deng, W., Hu, J., Guo, J.: In defense of sparsity based face recognition. In: 2013 IEEE Conference on Computer Vision and Pattern Recognition (CVPR), pp. 399–406. IEEE (2013)
30. Ahonen, T., Rahtu, E., Ojansivu, V., Heikkila, J.: Recognition of blurred faces using local phase quantization. In: 19th International Conference on Pattern Recognition, ICPR 2008, pp. 1–4. IEEE (2008)
31. Ojansivu, V., Heikkilä, J.: Blur insensitive texture classification using local phase quantization. In: Elmoataz, A., Lezoray, O., Nouboud, F., Mammass, D. (eds.) ICISP 2008 2008. LNCS, vol. 5099, pp. 236–243. Springer, Heidelberg (2008)
32. Gopalan, R., Taheri, S., Turaga, P., Chellappa, R.: A blur-robust descriptor with applications to face recognition. IEEE Trans. Pattern Anal. Mach. Intell. **34**, 1220–1226 (2012)
33. Cox, T.F., Cox, M.A.: Multidimensional Scaling. CRC Press, Boca Raton (2000)
34. Webb, A.R.: Multidimensional scaling by iterative majorization using radial basis functions. Pattern Recogn. **28**, 753–759 (1995)
35. Biswas, S., Aggarwal, G., Flynn, P.J., Bowyer, K.W.: Pose-robust recognition of low-resolution face images. IEEE Trans. Pattern Anal. Mach. Intell. **35**, 3037–3049 (2013)
36. Sim, T., Baker, S., Bsat, M.: The cmu pose, illumination, and expression (pie) database. In: Fifth IEEE International Conference on Automatic Face and Gesture Recognition 2002, Proceedings, pp. 46–51. IEEE (2002)
37. Phillips, P.J., Moon, H., Rizvi, S.A., Rauss, P.J.: The feret evaluation methodology for face-recognition algorithms. IEEE Trans. Pattern Anal. Mach. Intell. **22**, 1090–1104 (2000)

Spectral Shape Decomposition by Using a Constrained NMF Algorithm

Foteini Fotopoulou and Emmanouil Z. Psarakis[(✉)]

Department of Computer Engineering and Informatics,
University of Patras, Rion, Patras, Greece
psarakis@ceid.upatras.gr

Abstract. In this paper, the shape decomposition problem is addressed as a solution of an appropriately constrained Nonnegative Matrix Factorization Problem (NMF). Inspired from an idealization of the visibility matrix having a block diagonal form, special requirements while formulating the NMF problem are taken into account. Starting from a contaminated observation matrix, the objective is to reveal its low rank almost block diagonal form. Although the proposed technique is applied to shapes on the MPEG7 database, it can be extended to 3D objects. The preliminary results we have obtained are very promising.

1 Introduction

Shape decomposition constitutes a vital procedure in the field of computer vision, that is able to distinguish the different components of the original object and split it into meaningful ones. Meaningful components are defined as parts that can be perceptually distinguished from the remaining object. In this paper the shape decomposition problem is addressed and a novel decomposition technique is proposed, which solves the above mentioned problem as a special case of the well known NMF situation, using spectral analysis as a head-start. From algebraic perspective, the formulation of NMF can be regarded as decomposing the original matrix into two factor matrices, incorporating the nonnegativity requirement. Far beyond this mathematical exploration, the notion underlying the NMF is closely related to the human perception mode, as perception of the whole is achieved by perception of its parts [1]. For an extended review in the NMF we urge the interested reader to look at [2].

Let us consider a *plane curve*, that describes a shape boundary, to be defined from the path traced by the following N position vectors:

$$\mathbf{r}(i) = (x(i),\ y(i)),\ i = 1, 2, \cdots, N. \tag{1}$$

Then, we can construct the following Visibility Graph $\mathcal{G}_V = (\mathcal{V}, \mathcal{E}, \mathcal{W})$, where $\mathcal{V}, \mathcal{E}, \mathcal{W}$ are the nodes and edges sets and a binary weighted matrix respectively. More precisely, in this graph model of the *plane curve*, nodes' set $\mathcal{V}$ is defined as follows:

$$\mathcal{V} = \{\mathbf{r}(i),\ i = 1, 2, \cdots, N\}, \tag{2}$$

© Springer International Publishing Switzerland 2015
C.V. Jawahar and S. Shan (Eds.): ACCV 2014 Workshops, Part III, LNCS 9010, pp. 30–43, 2015.
DOI: 10.1007/978-3-319-16634-6_3

and the w_{ij} element of the $N \times N$ matrix $\mathcal{W}$ can be defined as follows:

$$w_{ij} = \begin{cases} 1, & \text{if nodes } i, j \text{ are visible} \\ 0, & \text{otherwise} \end{cases} \tag{3}$$

where nodes i, j are considered as visible if the following *Visibility Rule* holds:

– $\mathcal{VR}$: The connecting edge ϵ_{ij} is totally located inside the *plane curve*.

The $\mathcal{G}_V$ of the camel (Fig. 1(a)), obtained by the application of the above mentioned *Visibility Rule*, is depicted in Fig. 1(b). As it is obvious the structure of this matrix does not facilitate shape partitioning. An ideal matrix for shape decomposition would have the form of an almost block-diagonal similarity matrix [3], where its non-overlapping blocks could represent the shape's parts, in a sequential manner as it is shown in Fig. 1(c). The potential gaps between the blocks denote parts of the shape that do not constitute a group. Such parts are for example those between the camel's hunches, legs etc. (see Fig. 1(d)). The basic idea behind this idealization is that each shape component can be represented by a block in the respective block-diagonal matrix. Thus, the shape decomposition problem can be restated as follows: Given the visibility matrix $\mathcal{W}$ construct a block diagonal matrix that best approximates the desired form, which is the objective of our paper.

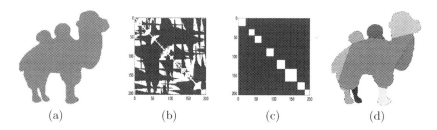

Fig. 1. The camel-shape (a). The corresponding $\mathcal{G}_V$ (b). An ideal block diagonal matrix for the camel shape decomposition (c), and the resulting shape decomposition (d).

In an attempt to achieve a rough approximation of the desired block diagonal form, a restriction is imposed to the $\mathcal{W}$ which allows visibility only to a neighborhood of size n on both sides of its main diagonal. In order to calculate a proper radius- n it can be adopted any hierarchical method or the method of [3], which we will adopt in this paper.

Constrained $\mathcal{G}_V$ resulting from the original $\mathcal{G}_V$ of the camel shape depicted in Fig. 2(a), are shown in Fig. 2(b)–(d) for three (3) different values of radius n. We are expecting that different values of neighborhood radius, result in different decompositions of the candidate shape, which becomes obvious in Fig. 3. It is apparent that none of them is perceptually meaningful, as in none of them the

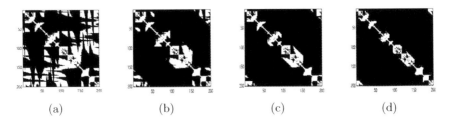

(a)   (b)   (c)   (d)

Fig. 2. Initial $\mathcal{G}_V$ (a) and n-conditioned $\mathcal{G}_V$ for different values of n: 40 (b), 30 (c) and 10 (d).

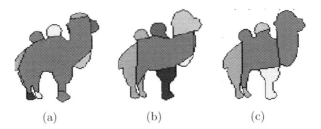

(a)   (b)   (c)

Fig. 3. Different decompositions of the camel shape corresponding to different values of neighborhood radius n (10, 30 and 40 respectively).

"optimal" n value is used. For further details, please refer to [3]. From now and on we will refer to the redefined $\mathcal{G}_V$ as $\hat{X}$.

The remainder of the paper is organized as follows: In Sect. 2 related work is briefly reviewed. In Sect. 3 the proposed shape decomposition problem is formulated, while in Sect. 4 the proposed shape decomposition method is developed. In Sect. 5 the results of the experiments we have conducted in the MPEG7 shape database are presented. In addition a comparison to the matrix completion problem is provided. Finally, the paper concludes at Sect. 6.

2 Related Work

The task of determining shape's parts is a difficult one due to the involvement of the human perception. However, there exist some generic perception rules examined in psychologist science, with the short cut rule [4], the minima rule [5] and the convexity rule [6,7] to be the most popular among them. Methods for shape decomposition such as [6,8–11] are based on the above mentioned rules. Besides these three popular perception rules, the authors of [12] propose a new one called part-similarity rule. This rule is based on the observation that similar parts of objects have to be decomposed with the same way, although they may look different due to deformations. A method based on differential geometry of smoothed local symmetries taylored for decomposing a shape into its meaningful parts was proposed in [13]. Method [14] suggests an hierarchical segmentation by exploiting the multiscale properties of diffusion distance. In [15] the use of a weak-convexity rule, based on "lines-of-sight" is suggested and the shape decomposition problem is solved by using a spectral clustering algorithm. Finally, the method proposed in [3] originates from a visibility graph, which captures the local neighborhoods and uses an iterative procedure to transform it into a block diagonal matrix.

As already stated in the Introduction, the shape decomposition problem can be formulated as a special case of the NMF problem. The idea of NMF was initiated by Paatero and Taper [16,17] together with Lee and Seung [18], who demonstrated the NMF potential use in parts based representation. As in this paper we will focus on the constrained NMF situation, and especially in cases where the orthogonality constraint is imposed we proceed with a brief discussion among this category.

The orthogonality principle was first employed in [19] and then in [20] the concept of orthogonal NMF was explicitly expressed. Moreover in [21] the equivalence of the orthogonal NMF to k-means clustering, its formulation as a constrained optimization problem and its solution by using a Lagrange multiplier approach were presented. However the resulting multiplicative update rule, suffers from the zero locking problem. One solution than ensures robust convergence of the algorithm thus solving the above mentioned problem was proposed in [22]. Finally, in [23] an orthogonal NMF algorithm with improved clustering capabilities, based on the original Lee and Seung algorithm and [22], was presented.

Finally, several schemes for the initialization of the factor matrices which affect both the convergence rate as well as the quality of the final solution, have been proposed. In particular, alternatives to the random seed initialization scheme [18] based on k-means [20] and svd [24], have been reported in the literature.

3   Problem Formulation

The original symmetric nonnegative matrix factorization problem with orthogonality constraints is already known and can be stated as follows:

Given a symmetric nonnegative data matrix $X \in \Re^{N \times N}$, the goal is to decompose it into a product of two nonnegative matrices, so that $X = VSV^T$, where V additionally satisfies the orthogonality constraint and S is added to absorb the different scales of the matrices. More formally, the orthogonal NMF problem can be stated by the following optimization problem:

$$\min_{V} ||X - VSV^T||_F^2, \text{ s.to } V \geq 0 \text{ and } V^TV = I, \qquad (4)$$

where $||A||_F^2$ denotes the squared Frobenius norm of matrix A. The aim of this paper is to appropriately reformulate the shape decomposition problem into a constrained NMF problem and solve it. Due to the specific binary form of the data matrix $\hat{X}$ the shape decomposition problem is differentiated from the above mentioned NMF problem mainly due to the special form of its low rank component. In particular we would like to decompose the binary n-constrained visibility matrix $\hat{X}$ into a low rank component, represented by the binary matrix X, and a sparse component E, that is: $\hat{X} = X + E$, where matrix's E elements take values from the set $\{1, 0, -1\}$.

At a first glance, the above mentioned problem resembles to the well known matrix completion problem [25]. However, there exist some significant differences. At first, our input matrix is binary and the desired output low rank matrix must be strictly binary too. Continuing, it is important to highlight that although a matrix completion algorithm aims at a low rank matrix of general form, we are seeking for a block diagonal one. Indeed, as we are going to see in Sect. 5, the matrix completion problem is not well fitted to shape decomposition problem. Consequently, our aim is to recover a special form of X, from noisy binary observations contained in the matrix $\hat{X}$. By taking into account all the above mentioned, the shape decomposition problem can be stated as follows:

$$\min_{V,k} ||\hat{X} - V_k V_k^T||_F^2, \tag{5}$$

subject to the following $N + k$ constraints:

$$\mathcal{C}_1 : \mathbf{c}_j^T \mathbf{i}_{N+1} = 2, \ j = 1, 2, \ldots, k \ \text{with} \ c_{ji} = |v_{ji} - v_{j(i-1)}|, \ v_{j0} = v_{j(N+1)} = 0$$
$$\mathcal{C}_2 : V_k \mathbf{i}_k = \mathbf{i}_N^\epsilon$$

where c_{ji} denotes the absolute forward difference of the elements of the zero padded eigenvector $\mathbf{v}_j$, $\mathbf{i}_M$ the all one's vector of length M and $\mathbf{i}_M^\epsilon$ an ϵ-perturbed version of this vector with the value of ϵ expressing the percentage of its zero elements. This special form of $\mathbf{i}_M$ vector allows the decomposition matrix to be almost block diagonal if it is required. This fact is in fully accordance with the ideal matrix description mentioned in the Introduction. Note also that because $v_{ji} \in \{0, 1\}$, each one of the constraints $\mathcal{C}_1$ imposes the desired form of the eigenvectors, while the constraints $\mathcal{C}_2$ express the orthogonality of the columns of the binary matrix V. Note finally, that the objective function of the constrained optimization problem does not contain the matrix S anymore.

Although, the above stated problem is NP hard, we are going to overcome this difficulty by starting with the eigenanalysis of matrix $\hat{X}$ and properly imposing in the produced eigenvectors the already stated requirements.

4   The Proposed Spectral Decomposition Method

Let us consider that the $\hat{X}$ matrix is given. Then, the proposed method aims at appropriately transforming the original visibility matrix into an almost block diagonal one, which can be easily used for the visually meaningful decomposition of the candidate shape.

The first step of the proposed decomposition method consists of the eigenanalysis of the given matrix $\hat{X}$. Specifically, since this matrix is binary and symmetric, from the finite dimensional spectral theorem we know that its eigenvalues are real and that it can be diagonalized by an orthonormal matrix U, that is $U^T U = I$, as follows:

$$\hat{X} = U \Lambda U^T \tag{6}$$

where matrix Λ contains the real eigenvalues of the matrix.

Let:

$$\Lambda = \Lambda_+ + \Lambda_- \qquad (7)$$

be two diagonal matrices containing the non negative and negative eigenvalues of matrix $\hat{X}$, respectively. Then, the original matrix can be written as follows:

$$\hat{X} = \hat{X}_+ + \hat{X}_-, \qquad (8)$$

where $\hat{X}_\pm = U_\pm \Lambda_\pm U_\pm^T$ and U_+, U_- matrices that contain the eigenvectors which correspond to the non negative and the negative eigenvalues respectively. The number of columns of these matrices are denoted by N_+ and N_- respectively with their sum to be equal to N. Based on the orthogonality of the above defined matrices $\hat{X}_\pm$, the following relation holds:

$$||\hat{X}||_F^2 = ||\hat{X}_+||_F^2 + ||\hat{X}_-||_F^2. \qquad (9)$$

Note that matrix $\hat{X}_+$ constitutes the optimal non-negative definite symmetric approximation of the original matrix $\hat{X}$. In addition, its Singular Value Decomposition coincides with the following decomposition of the matrix $\hat{X}_+$:

$$\hat{X}_+ = U_+ \Lambda_+ U_+^T. \qquad (10)$$

Although, the above mentioned matrix is not a non-negative matrix as the desired one, it constitutes a better approximation to our ultimate goal which is a block diagonal matrix. Therefore, we are going to use this matrix in the next step of the proposed method.

In the second step of the proposed algorithm the eigenvectors are sorted in descending order according to the absolute value of their projection onto the vector $\mathbf{i}_N$. Specifically, let:

$$\mathbf{p} = U_+^T \mathbf{i}_N \qquad (11)$$

be the projection of the vector $\mathbf{i}_N$ onto the matrix U_+.

By taking into account the definition of the inner product, the unit norm of each eigenvector and the specific form of vector $\mathbf{i}_N$, each element of the above defined vector $\mathbf{p}$ can be expressed as follows:

$$p_j =< \mathbf{u}_j, \ \mathbf{i}_N >= \sqrt{N} \cos(\theta_j), \ j = 1, 2, \ldots, N_+, \qquad (12)$$

with $< \mathbf{a}, \mathbf{b} >$ and θ_j denoting the inner product of vectors $\mathbf{a}$, $\mathbf{b}$ and the existing angle between them respectively.

By defining the following matrices:

$$S = diag\{sign(\cos(\theta_j)), \ j = 1, 2, \ldots, N_+\}$$
$$\hat{U}_+ = U_+ S \qquad (13)$$

Equation (11) can be equivalently rewritten as follows:

$$|\mathbf{p}| = \hat{U}_+^T \mathbf{i}_N. \qquad (14)$$

Note also that by taking into account the fact that the Singular Value Decomposition is only unique up to a reflection of each eigenvector, the decomposition of Eq. (10) can be equivalently rewritten as follows:

$$\hat{X}_+ = \hat{U}_+ \Lambda_+ \hat{U}_+^T. \tag{15}$$

Let us now concentrate ourselves on the vector $|\mathbf{p}|$ defined in Eq. (14). As it is clear, each element of this vector constitutes the projection of the redefined eigenvector $\hat{\mathbf{u}}_j$ (or equivaently the j-th column of the matrix $\hat{U}_+$ defined in Eq. (13)) to the all one's vector $\mathbf{i}_N$. Thus, it makes sense to assume that the value of each element specifies the contribution of the corresponding eigenvector in the reconstruction of this vector. Consequently, let us sort them in descending order into the vector $|\mathbf{p}_S|$, and rearrange accordingly the columns of matrix $\hat{U}_+$ to obtain its desired sorted counterpart $\hat{U}_{S+}$.

In the next step, we are going to replace all these N_+ eigenvectors by their binary equivalents. In order to achieve our goal, for each one of the N_+ eigenvectors, a hard thresholding procedure is applied. To this end, let:

$$\mathcal{I}_j = \{\min(\hat{\mathbf{u}}_{Sj}) : \frac{\max(\hat{\mathbf{u}}_{Sj}) - \min(\hat{\mathbf{u}}_{Sj})}{L-1} : \max(\hat{\mathbf{u}}_{Sj})\}, \tag{16}$$

be a sequence of length L, resulting from the uniform sampling of the range of the j-th column of matrix $\hat{U}_{S+}$, that is the range of the eigenvector $\hat{\mathbf{u}}_{Sj}$ and let us define the following sequence (of the same length) of binary vectors:

$$\mathcal{U}_j = \{\mathbf{v}_{ji} = sign\left((\hat{\mathbf{u}}_{Sj} - T_i) > 0\right), \ T_i \in \mathcal{I}_j\}. \tag{17}$$

Note that the above defined set contains all the binary versions of eigenvector $\hat{\mathbf{u}}_{Sj}$ after their hard threshoding by T_i.

Let us now denote by $\mathcal{U}_{Aj}$ the subset of the binary vectors that belong into the set $\mathcal{U}_j$ defined in Eq. (17) and strictly satisfy the $\mathcal{C}_1$ constraints of the optimization problem (5). It is clear that this subset contains all the admissible binary representations of the eigenvector $\hat{\mathbf{u}}_{Sj}$. In order to isolate the most characteristic ones, let us compute the l_0 norm of each admissible vector, i.e.:

$$l_i = ||\mathbf{v}_{ji}||_0, \ i = 1, 2, \ldots, |\mathcal{U}_{Aj}| \tag{18}$$

where $|\mathcal{U}_{Aj}|$ denotes the cardinality of set $\mathcal{U}_{Aj}$. Then, find out the most frequent element of the above defined sequence $\mathbf{v}_j$ and if its l_0 norm is greater than a predefined minimum admissible value m_{ℓ_0}, consider it as the most representative binary vector of the specific eigenvector. We must stress at this point that the latter restriction puts a down limit to the size of the smallest permitted shape component. Repeat the above described procedure for each one of the N_+ sorted eigenvectors contained in the matrix $\hat{U}_{S+}$. It is clear that after this we end up with a subset of M out of the $N+$ binary eigenvectors which strictly satisfy the constraints $\mathcal{C}_1$ and their l_0 norms are greater than the minimum admissible value m_{ℓ_0}. Finally, sort these binary eigenvectors in descending order according to their l_0 norm.

Having defined the binarized form of the M eigenvectors, in the last step of the proposed algorithm we impose all the necessary constraints onto the binary eigenvectors in order to satisfy the constraint $\mathcal{C}_2$ of the constrained optimization problem (5). Specifically, each one of the M selected binary eigenvectors is sequentially examined for the satisfaction of the above mentioned constraint. In particular, for each pair of successive binary vectors their intersection is identified and if it is not empty is it substracted from the eigenvector with the smaller l_0 norm. Keep the resulting vector if its l_0 norm is greater than m_{l_0}. Note that by repeating the above described procedure over all the vectors, we end up with $k \leq M$ mutually exclusive binary vectors which sum up to a $\mathbf{i}_N^\epsilon$ vector and this concludes the proposed technique.

An outline of the proposed algorithm follows.

Algorithm 1. NMF Spectral Shape Decomposition

---

1. Input $\hat{X}$ and m_{l_0}
2. Make the eigenanalysis of matrix $\hat{X}$: $\hat{X} = U\Lambda U^T$
3. Form its non-negative definite part: $\hat{X}_+ = U_+\Lambda_+ U_+^T$
4. Compute the projection vector $\mathbf{p}$ of Eq. (11)
5. Use Eq. (12) to compute matrices S and $\hat{U}_+$ of (13)
6. Sort vector $|\mathbf{p}|$ and form the sorted matrix $\hat{U}_{S+}$
7. Use Eqs. (16–18) to isolate the binary equivalents of the eigenvectors $\hat{u}_{Sj}$, $j = 1, 2, \ldots N+$ which strictly satisfy the constraints $\mathcal{C}_1$ of the optimization problem (5), and their l_0 norm is greater than m_{l_0}. Sort them according to their l_0 norm.
8. Impose constraints to satisfy the constraints $\mathcal{C}_2$ of the optimization problem (5).

Having completed the presentation of the proposed technique, in the next section we are going to apply it in a number of 2D shapes and compare its performance against other well-known techniques.

5 Experimental Results

5.1 Shape Decomposition

In this section we present comparative results obtained by applying the proposed method to several shapes of the MPEG7 shape database part B [26]. All shape contours we used in our experiments were sampled uniformly at 200 points and the size of the smallest permitted shape component, controlled by the value of the parameter m_{l_0}, was set to 5. A sample of 2D decomposed shapes, using the introduced method, is shown in Fig. 4.

In our opinion the proposed method seems to be insensitive to the number and the complexity of the shape components (see for example the decomposition of the mouse or that of the butterfly shown in Fig. 4). It is evident, that for most shapes the decomposition is meaningful, while in some situations, parts

Fig. 4. A sample of decomposed shapes of the MPEG7 database using the proposed method.

that obviously could not be separated by a human being, are splitted by the algorithm. Specifically, the dog shapes are decomposed into their main parts (i.e. the head, the tail and the legs) in most cases. Moreover, in the mouse the ears, the tails, the hands and the legs are successfully separated. Finally, in the butterfly shapes, the antennas are defined as separate parts, except from the third one. In addition, the wings are well defined, too. An exception is made to the first and third butterfly figures, where the wings are distorted. It is worth mentioning, that the spectral decomposition proposed in this paper, provides satisfactory results achieving at a great degree to recognize most meaningful articulated shape parts, even if they are depicted in different poses, which are acceptable compared to other methods.

To further demonstrate the effectiveness of the proposed decomposition method, we will proceed in showing some comparative results. In particular, we will compare our method to [9–11, 27]. The above mentioned results are shown in Fig. 5. Since human perception is essential for the evaluation of the produced results, human decompositions are shown in the first column of this figure. For each of the categories (see Fig. 5) for which the experiment conducted, humans were asked to decompose the shapes manually into meaningful parts. The results of this experiment were borrowed from [11]. In addition, we should mention that in the fifth column of the Fig. 5, where the results of the [11] are depicted, the decomposition does not include the straightening process. As it is apparent, our proposed method succeeds in most cases to successfully approximate the results of the first column. In addition, although in some cases the introduced method identifies some extra components, it succeeds in capturing other ones, that none

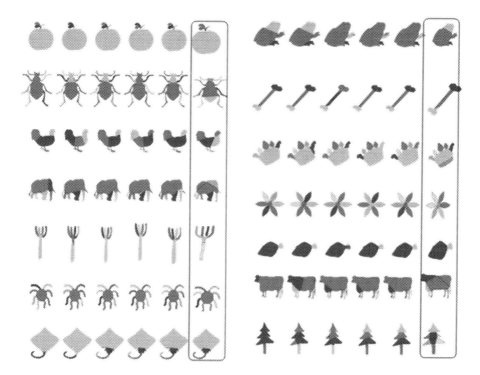

Fig. 5. Examples of the decomposition for 14 categories of the MPEG7 shape database. Human decompositions (1st column) and the results obtained from the application of, [27] (2nd column), [9] (3rd column), [10] (4th column), [11] (5th column), the proposed decomposition technique (last column).

of the compared methods can do. For example, in the elephant shape we have to mention that although its spine is found as a separate shape part, the head is also found as a component, which is in accordance with the perceptual decomposition shown in the first column and in most of the other methods is missed. The same observation holds for the cow's and frog's head. Finally, regarding the beetle shape, our method is the only one among the compared ones that decomposes its legs as separate parts, which confirms the effectiveness of our method to bendable shape parts. Concluding, although the performance of the proposed technique can be characterized as quite good, it could be further improved if a more sophisticated scheme, than that we adopted from [3] is used for structuring the constrained visibility matrix $\hat{X}$. This point is currently under investigation.

5.2   Relevance to Other Closely Related Techniques

Visibility Shape Decomposition: As already mentioned, the idea of the ideal form of an ideal visibility matrix and the use of a constrained visibility matrix as a head-start were borrowed from the VSD method proposed in [3]. Although

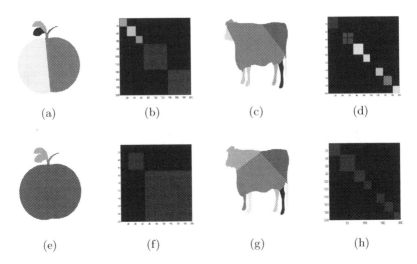

(a)　　　　　(b)　　　　　(c)　　　　　(d)

(e)　　　　　(f)　　　　　(g)　　　　　(h)

Fig. 6. Decompositions and block diagonal matrices of the apple and cow shapes resulting from the application of VSD (a–d) and the proposed technique (e–h)

in [3] the decomposition problem is solved in a totally different way, we consider it essential to provide a short comparison between the decomposition results provided by these two methods. Although the results obtained by the VSD are very promising, there exist many occasions where the resulting decomposition leads to over segmented shapes. This problem is due to the iterative procedure which sometimes fails to capture whole segments as a compact block in the final almost-block diagonal matrix. Therefore, we can observe decompositions where a perceptually expected group is splitted into two smaller ones, with no physical meaning. In Fig. 6 two indicative examples of the above stated problem, accompanied by their corresponding block diagonal matrices are shown. As we can observe, the cow's head is over segmented and the same holds for the round part of the apple. On the other side regarding the performance of the proposed method, all the comments we made in the previous subsection are still valid.

Matrix Completion Problem: Given a corrupted data matrix D, the process of matrix completion is to decompose the matrix into a sum of a low-rank matrix X and a sparse matrix E. In [25] is shown that, under broad conditions, the optimal solution to the completion problem is given from the solution of the following convex optimization problem:

$$\min_{X,E} ||X||_\star + \gamma||E||_1, \ s.t. \ D = X + E \tag{19}$$

where $\gamma > 0$ is a parameter that is used to control the sparseness of the matrix E.

There exist many well-known techniques that can be used for the solution of the constrained optimization problem defined in (12), such as Augmented Lagrange Multiplier (ALM) Method [28], the Accelerated Proximal Gradient

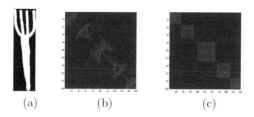

(a)   (b)   (c)

Fig. 7. The fork shape (a), the rank 193 visibility matrix (b) and the rank 5 block diagonal matrix resulting from the application of the proposed technique (c)

(APG) [29], the Dual Method [29], the Singular Value Thresholding [30] to name a few. In all the experiments we have conducted we used the IALM and the SVT algorithms,[1] and the results we have obtained confirm that although the Robust PCA [31] methods can be successfully used for solving the matrix completion problem, for the shape decomposition one they seem to be inappropriate.

As it is apparent from (19), the greater the value of γ is, the sparser the matrix becomes. From the Fig. 8 we can observe that as the value of the parameter γ increases, the matrix's rank is increased and the resulting matrix escapes the block diagonal form. However, for small values of γ some groups seem to reveal, but the desired block diagonal form is not achieved. Specifically, as we can see from Fig. 8(a), although the rank of the produced matrix is exactly the same as the rank of the block diagonal matrix resulting from the application of our method (see Fig. 7(c)), the IALM algorithm produces a matrix that is not in the desired form. Moreover, by increasing the parameter γ we can see in Fig. 8(b)–(d) that the constructed matrix seems to somehow approximate the block diagonal form, however by escaping the low rank constraint. These observations confirm the fact that the completion algorithms in their original form seem to be unsuitable for the shape decomposition problem, as they result in a

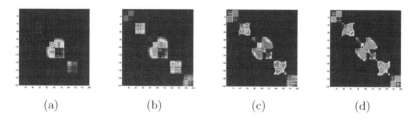

(a)   (b)   (c)   (d)

Fig. 8. IALM decompositions results and the rank of the obtained low rank matrix X for the fork shape for different values of the parameter γ. $\gamma = 0.05$, rank $= 5$ (a), $\gamma = 0.06$, rank $= 13$ (b), $\gamma = 0.08$, rank $= 25$ (c), $\gamma = 0.1$, rank $= 46$ (d)

---

[1] In our experiments for the implementation of the matrix completion techniques we have used the matlab codes from http://perception.csl.illinois.edu/matrix-rank/sample_code.html.

low rank matrix which has not necessarily the special desired form. The same remarks hold for the SVT completion algorithm.

6   Conclusions

In this work, a novel perspective on the shape decomposition issue was proposed. Originating from an eigenanalysis of a constrained visibility matrix, the shape decomposition problem was formulated as a constrained orthogonal NMF one. From the results we obtained it seems that the introduced technique results in perceptually more meaningful decomposition than more of the existing methods. However, the structuring of the constrained visibility matrix, the relationship of the proposed technique to spectral graph theory as well as its extension in 3D shapes are some topics which are currently under investigation.

References

1. Ullman, S.: High-Level Vision: Object Recognition and Visual Cognition. MIT press, Cambridge (2000)
2. Wang, Y.X., Zhang, Y.J.: Nonnegative matrix factorization: a comprehensive review. IEEE Trans. Knowl. Data Eng. **25**, 1336–1353 (2013)
3. Fotopoulou, F., Psarakis, E.Z.: A visibility graph based shape decomposition technique. In: VISAPP (1), pp. 515–522 (2014)
4. Singh, M., Seyranian, G.D., Hoffman, D.D.: Parsing silhouettes: the short-cut rule. Percept. Psychophys. **61**, 636–660 (1999)
5. Hoffman, D.D., Richards, W.A.: Parts of recognition. Cognition **18**, 65–96 (1984)
6. Latecki, L.J., Lakämper, R.: Convexity rule for shape decomposition based on discrete contour evolution. Comput. Vis. Image Underst. **73**, 441–454 (1999)
7. Walker, L.L., Malik, J.: Can convexity explain how humans segment objects into parts? J. Vis. **3**, 503–503 (2003)
8. Lien, J.M., Amato, N.M.: Approximate convex decomposition of polygons. In: Proceedings of the Twentieth Annual Symposium on Computational Geometry, pp. 17–26. ACM (2004)
9. Liu, H., Liu, W., Latecki, L.J.: Convex shape decomposition. In: 2010 IEEE Conference on Computer Vision and Pattern Recognition (CVPR), pp. 97–104. IEEE (2010)
10. Ren, Z., Yuan, J., Li, C., Liu, W.: Minimum near-convex decomposition for robust shape representation. In: 2011 IEEE International Conference on Computer Vision (ICCV), pp. 303–310. IEEE (2011)
11. Jiang, T., Dong, Z., Ma, C., Wang, Y.: Toward perception-based shape decomposition. In: Lee, K.M., Matsushita, Y., Rehg, J.M., Hu, Z. (eds.) ACCV 2012, Part II. LNCS, vol. 7725, pp. 188–201. Springer, Heidelberg (2013)
12. Ma, C., Dong, Z., Jiang, T., Wang, Y., Gao, W.: A method of perceptual-based shape decomposition. In: IEEE International Conference on Computer Vision, ICCV 2013, Sydney, Australia, 1–8 December 2013, pp. 873–880 (2013)
13. Mi, X., DeCarlo, D.: Separating parts from 2d shapes using relatability. In: IEEE 11th International Conference on Computer Vision, 2007, ICCV 2007, pp. 1–8. IEEE (2007)

14. De Goes, F., Goldenstein, S., Velho, L.: A hierarchical segmentation of articulated bodies. Comput. Graph. Forum **27**, 1349–1356 (2008). Wiley Online Library
15. Asafi, S., Goren, A., Cohen-Or, D.: Weak convex decomposition by lines-of-sight. Comput. Graph. Forum **32**, 23–31 (2013). Wiley Online Library
16. Paatero, P., Tapper, U.: Positive matrix factorization: a non-negative factor model with optimal utilization of error estimates of data values. Environmetrics **5**, 111–126 (1994)
17. Paatero, P.: Least squares formulation of robust non-negative factor analysis. Chemom. Intell. Lab. Syst. **37**, 23–35 (1997)
18. Lee, D.D., Seung, H.S.: Algorithms for non-negative matrix factorization. In: Leen, T.K., Dietterich, T.G., Tresp, V. (eds.) Advances in Neural Information Processing Systems, pp. 556–562. MIT Press, Cambridge (2000)
19. Li, S.Z., Hou, X., Zhang, H., Cheng, Q.: Learning spatially localized, parts-based representation. In: Proceedings of the 2001 IEEE Computer Society Conference on Computer Vision and Pattern Recognition, 2001, CVPR 2001, vol. 1, p. I-207. IEEE (2001)
20. Ding, C., Li, T., Peng, W., Park, H.: Orthogonal nonnegative matrix tri-factorizations for clustering. In: Proceedings of the 12th ACM SIGKDD International Conference on Knowledge Discovery and Data Mining, pp. 126–135. ACM (2006)
21. Ding, C.H., He, X., Simon, H.D.: On the equivalence of nonnegative matrix factorization and spectral clustering. In: SDM, vol. 5, pp. 606–610. SIAM (2005)
22. Lin, C.J.: On the convergence of multiplicative update algorithms for nonnegative matrix factorization. IEEE Trans. Neural Netw. **18**, 1589–1596 (2007)
23. Mirzal, A.: A convergent algorithm for orthogonal nonnegative matrix factorization. J. Comput. Appl. Math. **260**, 149–166 (2014)
24. Boutsidis, C., Gallopoulos, E.: Svd based initialization: a head start for nonnegative matrix factorization. Pattern Recogn. **41**, 1350–1362 (2008)
25. Candès, E.J., Li, X., Ma, Y., Wright, J.: Robust principal component analysis? J. ACM (JACM) **58**, 11 (2011)
26. Latecki, L.J., Lakamper, R., Eckhardt, T.: Shape descriptors for non-rigid shapes with a single closed contour. In: IEEE Conference on Computer Vision and Pattern Recognition, 2000, Proceedings, vol. 1, pp. 424–429. IEEE (2000)
27. Gopalan, R., Turaga, P., Chellappa, R.: Articulation-invariant representation of non-planar shapes. In: Daniilidis, K., Maragos, P., Paragios, N. (eds.) ECCV 2010, Part III. LNCS, vol. 6313, pp. 286–299. Springer, Heidelberg (2010)
28. Lin, Z., Chen, M., Ma, Y.: The augmented lagrange multiplier method for exact recovery of corrupted low-rank matrices (2010). arXiv preprint arXiv:1009.5055
29. Lin, Z., Ganesh, A., Wright, J., Wu, L., Chen, M., Ma, Y.: Fast convex optimization algorithms for exact recovery of a corrupted low-rank matrix. Comput. Adv. MultiSens. Adapt. Process. (CAMSAP) **61** (2009)
30. Cai, J.F., Candès, E.J., Shen, Z.: A singular value thresholding algorithm for matrix completion. SIAM J. Optim. **20**, 1956–1982 (2010)
31. Wright, J., Ganesh, A., Rao, S., Peng, Y., Ma, Y.: Robust principal component analysis: exact recovery of corrupted low-rank matrices by convex optimization. In: Proceedings of Neural Information Processing Systems, vol. 3 (2009)

A Simple Stochastic Algorithm for Structural Features Learning

Jan Mačák[✉] and Ondřej Drbohlav

Faculty of Electrical Engineering, Department of Cybernetics, Center for Machine Perception, Czech Technical University in Prague, Prague, Czech Republic
{macakj1,drbohlav}@cmp.felk.cvut.cz

Abstract. A conceptually very simple unsupervised algorithm for learning structure in the form of a hierarchical probabilistic model is described in this paper. The proposed probabilistic model can easily work with any type of image primitives such as edge segments, non-max-suppressed filter set responses, texels, distinct image regions, SIFT features, etc., and is even capable of modelling non-rigid and/or visually variable objects. The model has recursive form and consists of sets of simple and gradually growing sub-models that are shared and learned individually in layers. The proposed probabilistic framework enables to exactly compute the probability of presence of a certain model, regardless on which layer it actually is. All these learned models constitute a rich set of independent structure elements of variable complexity that can be used as features in various recognition tasks.

1 Introduction

Unsupervised learning of object appearance has been a challenging task since the very beginning of computer vision. There are many approaches to this problem, but considering a huge number of visual categories, low computational requirements, easy extension-ability requirements, the most promising object representations seem to be hierarchic/compositional ones [1,2]. Focusing on these, there are various hierarchically organized models proposed, taking inspiration from different fields of science. There are nature inspired designs [3,4], there are models using grammars [5–7], there are very successful approaches using neural networks [8,9]. Learning strategies of such structures are similarly diverse, ranging from semi-automatic methods when the structure is given by human and only its parameters are learned [5] over sophisticated supervised/unsupervised methods of deep learning [8–10].

The core question this work shall answer is that of whether it is possible to learn both structure and parameters of a generative compositional probabilistic model using a very simple algorithm based on the *Expectation-Maximization* principle, that means by random initialization and iterative updating. Noticeable difference from the deep learning is that unlike learning deep belief [10] or convolutional neural network [8] this method gives an explicit structure model similar to image grammars and requires less hyper-parameter/design choices – there is

© Springer International Publishing Switzerland 2015
C.V. Jawahar and S. Shan (Eds.): ACCV 2014 Workshops, Part III, LNCS 9010, pp. 44–55, 2015.
DOI: 10.1007/978-3-319-16634-6_4

actually just one hyper-parameter that needs to be set and that is the maximal allowed portion of non-modelled data. Furthermore, its individual learned compositions can be used as features in more sophisticated classification framework such as SVM or AdaBoost. Similar approach [11] has been recently shown to be very efficient in domain transfer task[1]. This indicates that structural approach has very good generalization capabilities.

2   Concepts

Given a dataset which consists of a set of detected features (denoted $\mathcal{D}$ further on) per image, the task is to construct a set of generative probabilistic models that would be able to generate the dataset. This construction is designed to work in an unsupervised manner. Following the idea of Occam's razor or its modern version represented by the *MDL* approach [12], the set of models is to be as simple as possible. Significant reduction of the model complexity can be achieved by sharing of model parts. This has also other practical side-effect benefits such as computation cost reduction, smaller memory consumption, etc. A natural form of model that allows for immediate and efficient sub-parts sharing is a hierarchical model where the root node (in this paper the root is always at the top) generates a number of children nodes, these children again generate sets of its children and this scheme is repeated until the last (lowest) layer is reached, the lower a node is the smaller its working radius is. The advantage of this form of the model organization also is the fact that it can very naturally model non-rigid data. For example, if the modelled object is a human body, the model can consist of rigid sub-models of individual limbs (more precisely their rigid pieces) on a certain layer and then on higher layers define their spatial relations including the rotations.

In the case that the learning shall proceed without supervision, it is reasonable to start from the most local properties and have the complexity grown while proceeding to higher layers. At each layer a set of *compositions* capable of generating the given data is acquired using random sampling and *Expectation-Maximization* parameter learning. The word *composition* is used to represent a single shareable model (actually on an arbitrary layer) which defines spatial relations between composition's root node and its children nodes.

2.1   Probabilistic framework

The form of the probabilistic model, which would be a member of the set of models representing the data, is shown in the Fig. 1. In the illustration, there are actually two *compositions* explicitly shown, however, the nodes b_{12}, b_{13} are similar *compositions*. The model itself is a directed acyclic graph (*DAG*), the *compositions* on lower layers might have multiple parents though – in a sense that a *composition* can be shared by more than one higher layer models. Because the

---

[1] Training and classification of same object classes in different datasets.

model structure is recursive, it is sufficient to describe only a single *composition* in detail. The following description refers to the model comp.I. in the Fig. 1. It can be interpreted this way[2]: the node b with probability m_i generates *composition* at a random position c_i. There can be an arbitrary number of such structural components, but the reasonable number is ca 2–8 [13].

Mathematically, the node m_i has two discrete states

$$P(m_i|b) = \begin{cases} p_i & \text{if } m_i = 1, \\ 1 - p_i & \text{if } m_i = 0 \end{cases} \tag{1}$$

where state $m_1 = 1$ means that i-th component is generated, the state $m_1 = 0$ means that nothing is generated in that branch. The probability model for the position of the generated underlying child *composition* is normal distribution

$$P(c_i|m_i = 1) = \mathcal{N}(\mu_i, \Sigma_i), \tag{2}$$

in the case that nothing is generated, the model branch is terminated with constant probability

$$P(\{\}|m_i = 0) = 1. \tag{3}$$

Except these so called *structural* components, each node is also equipped with ability of generating random patterns. This mechanism is incorporated in the right-most branch with the node e and double-bordered node c_e. The node e is again discreet-state node and its state can be any natural number k meaning that the model generates k independent random patterns – either arbitrary *composition* from lower layer or so called *non-structural random pattern* which basically means that it is a bunch of data that can not be explicitly modelled using the structural model. The probability model of k is a standard *Poisson distribution*

$$P(e|b) = \frac{\lambda^k}{k!} e^{-\lambda}. \tag{4}$$

The reason for the double border of the node c_e is that it denotes that there can actually be a number of such nodes, depending on the state k. However, each such node is of the same form, there is a distribution for its position

$$P(c_e|e) = \mathcal{N}(\mu_e, \Sigma_e) \tag{5}$$

and also the distribution over *compositions* that can be generated plus the *non-structural random pattern* ϵ is

$$P(c^t|c_e) = p^t, \quad \sum_{i=1}^{|c|} p^i + p^\epsilon = 1 \tag{6}$$

Due to the fact that this internal state c^t is always marginalized, it is not drawn in the picture explicitly.

---

[2] For the sake of clarity, the unnecessary indices are ommited.

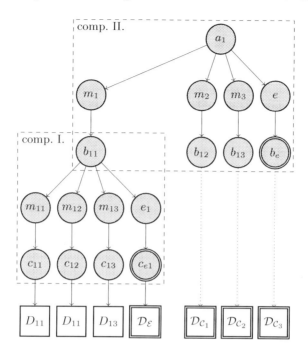

Fig. 1. The structure of the probabilistic model of a composition. The rectangular nodes denoted with letter D are individual input data, the round nodes (a, b, c, m, e) are internal nodes. Nodes b (c) model spatial relations of children and parent, m model the decision if the child is present. Nodes e are discrete-state and model the number of random structure fragments, nodes b_e and c_e model the location of each such fragment. Obviously, being dependent on the actual state of parent e node (non-negative number k), there are k such nodes, which is graphically denoted by double border. Similarly, the double border of the D nodes indicates that the nodes are actually disjoint sets of input data and for whole data holds $\mathcal{D} = D_{11} \cup D_{12} \cup D_{13} \cup \mathcal{D}_{\mathcal{E}} \cup \mathcal{D}_{\mathcal{C}_1} \cup \mathcal{D}_{\mathcal{C}_2} \cup \mathcal{D}_{\mathcal{C}_3}$. Dashed rectangles delimit exemplar individual compositions.

The marginalized probability for *non-structural random pattern* data generated by node b can be written as

$$P(\mathcal{D}_{\mathcal{E}}|b) = \sum_{k=1}^{\infty} \frac{\lambda^k}{k!} e^{-\lambda} \prod_{i=1}^{k} \int_{c_{e_k}} P(c_{e_k}|e) \sum_{t \in T} P(\mathcal{D}_{\mathcal{E}_i}|c^t) P(c^t|c_{e_k}) dc_{e_k}. \quad (7)$$

The marginalized probability of all data $\mathcal{D}_b$ generated by node b then is

$$P(\mathcal{D}_b|b) = \left(\prod_{i=1}^{3} \int_{c_i} P(D_i|c_i) P(c_i|m_i) P(m_i = 1|b) dc_i + P(m_i = 0|b) \right) P(\mathcal{D}_{\mathcal{E}}|b). \quad (8)$$

When evaluating higher layer *composition*, the formula is recurrent and structurally just the same, so it can be computed easily using the *Message Passing* algorithm [14].

2.2   Inference

As the probabilistic model models the simplest and smallest parts of the data at the lowest layer, it is reasonable to start inference from bottom and proceed up – if no bottom *compositions* are found then there is no reason for continuing in searching for more complex objects, since these are build of those simple ones. Due to the recursive character of the probabilistic model, the mechanism is just the same on each layer and therefore it is sufficient to describe only the transition from the layer $n-1$ to the layer n.

Suppose that the given data is organized into non-overlapping groups $\mathcal{A} = \{A_1, \ldots, A_n\}$ and each group contains some instances:

$$\mathcal{D}' = \mathcal{D}_{A_1} \cup \mathcal{D}_{A_2} \cup \cdots \cup \mathcal{D}_{A_n} =$$
$$= \left\{ {}_1\mathrm{d}^1_{1,1}\, \mathrm{d}^1_{2,1}\, \mathrm{d}^2_1, \ldots, {}_1\mathrm{d}^\epsilon \right\} \cup \left\{ {}_2\mathrm{d}^1_{1,2}\, \mathrm{d}^1_{2,2}\, \mathrm{d}^\epsilon \right\} \cup \ldots \left\{ {}_3\mathrm{d}^2_{1,3}\, \mathrm{d}^2_{2,3}\, \mathrm{d}^2_3, \ldots, {}_3\mathrm{d}^\epsilon \right\} \quad (9)$$

– the set of instances from the layer $n-1$, each of the known probability $P(\mathcal{D}_i|_i\mathrm{d}^c_k)$, where i is the data group index, c indicates which *composition* the instance is of and index k is the number of variant (there can be more than one instance of a *composition*) – and the set of *compositions* $\mathcal{C} = \{c^1, c^2, \ldots, c^n, c^\epsilon\}$, the task of inference algorithm is to find a set of instances $\mathcal{I} = \{c^1_1, c^1_2, c^2_1, \ldots, c^\epsilon\}$ of *compositions* that model the data with reasonably high probability $P(\mathcal{D}'|c^c_k)$. Such scenario is advantageous from at least two viewpoints: (i) if the *composition* instance c^c_k use any instance from each group of data, it is assured that the instance c^c_k models all assigned data, (ii) as the data groups are mutually share-free, by choosing precisely one instance from each group, the algorithm can not produce cyclic structure.

Due to the limited maximal complexity of compositions at each layer and the limited receptive field, it is feasible to find the globally best instances by brute-force enumeration of all consistent proposals in each grouping. After inferring instances of the layer n, the grouping of the layer $n-1$ becomes obsolete.

This grouping approach does have a disadvantage, though. It is apparent that the grouping can not be optimal with respect to all candidate *compositions* and consequently the approach produces sub-optimal instance when the grouping is not in favour[3] of that particular *composition*. However, this problem can be overcome by involving a second mechanism – *top-down* optimization of instances of low probabilities caused by missing or dislocated components similar to [15,16]. This *top-down* mechanism can either search for more suitable already existing instances or can come up with completely new instances. It is also capable of changing the layer n grouping when it is beneficial.

2.3   Learning

Besides the probability model already introduced, the core of this work is the learning method. First, it is necessary to return to the *grouping* mentioned in previous section, these groups are referred as *area* from now on. To give more

---

[3] The grouping is actually generated randomly.

1) get layer 1 instances from an image **for** $i \leftarrow 1$ **to** N **do**
  2) find random grouping of instances of previous layer;
  3) infer instances of current layer (*bottom-up* process);
  4) merge the partitioning in previous layer;
  5) improve the instances - find missing parts (*top-down* process);
end

Algorithm 1. The sketch of the inference algorithm.

precise description, an *area* is an artificial container which temporarily owns a subset of instances of *compositions* from the lower layer (see Eq. 9). As a consequence, an *area* always represents a well-defined subset of data, see the Fig. 2 for an illustrative example. All inferred instances (the set $\mathcal{I}$) in the *area* has to have assigned all the content of the *area* – either as a part of the structural model or as a part of *non-structural random pattern*. This makes the inferred instances comparable and allows for computing the posterior probabilities of individual instances given the *area* content using the Bayes formula

$$P(c_k^c|\mathcal{D}') = \frac{P(\mathcal{D}'|c_k^c)P(c_c)}{\sum_{c,k} P(\mathcal{D}'|c_k^c)P(c_c)}. \tag{10}$$

The learning itself is iterative and uses these probabilities $P(c_k^c|\mathcal{D}')$ as weights for computing updated values of model parameters. First, the data is randomly partitioned into *areas* of pre-set size. Then a *composition* is randomly created by sampling from one of the *areas*. The inference algorithm over the whole data (all areas) is run as to get all instances of the *composition*. These instances are then used for update of the *composition* parameters in a *Maximum-likelihood* (further referred as *ML*) manner. The model branches are conditionally independent and therefore can be optimized separately. It can be shown that for example a *ML* estimate of any of *composition*'s component position is

$$\mu_{\text{comp}} = \frac{\sum_{\mathcal{A},k} \mu_k P(b_k^c|\mathcal{D}_\mathcal{A})}{\sum_{\mathcal{A},k} P(b_k^c|\mathcal{D}_\mathcal{A})}, \tag{11}$$

where $\mathcal{A}$ is the set of all areas and k is the index of the set of instances of the *composition* c, and analogically for the other parameters. Only the parameter λ is learned differently. The *ML* estimate for this parameter of the *Poisson distribution* is the mean of the modelled values. In this case, it is the mean value of the number of lower layer *areas* contained in the current *areas*. This value is used directly as λ_ϵ, for each *composition* is the parameter lambda different and it is set as

$$\lambda^c = \max(\lambda_{\min}, \lambda_\epsilon - N), \tag{12}$$

where N is the number of components in the *composition*. This reflects the intent that each *composition* generates on average a similar amount of data.

Besides the *ML* updating the parameters of the already assigned *composition* components, also the positions and types of neighbouring non-explained

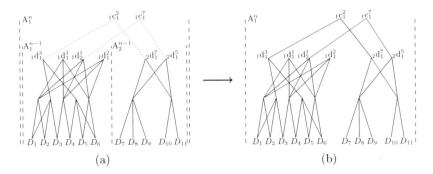

Fig. 2. An illustration of groups and their life cycle. The (a) shows the situation before inference of new instances on the layer n, the (b) shows the same situation after inference. Noticeable changes are: (i) two new instances, $_1c_1^2, _1c_1^7$, were created, each of them is built of different lower layer instances $(_1d_1^3, _2d_1^7)$ vs. $(_1d_2^1, _2d_1^5)$, (ii) partitioning on the layer $n-1$ became ineffective and (iii) both new layer n instances model all underlying data.

instances are tracked. When a significant cluster is discovered, it can be added to *composition* as a new component. Analogously, if any of the components prove to be useless, it can be removed. These mechanisms enable *composition* to travel within the configuration space towards at least locally most stable and frequent form.

In the second and further iterations, the *compositions* are updated and one new random *composition* is always added, unless the fraction of non-modelled data drops below a given treshold. After this event has happened, the algorithm keeps running for a predefined number of iteratiors and in each iteration one new spare *composition* is sampled and if turns out to be more useful, it replaces the least useful *composition* from the final set. The estimated prior probability of a *composition* is taken as the usefulness measure. This can be viewed as a simple restarting scheme in order not to end up in the first local extrema. When the learning of one layer finishes, the final set of *compositions* $\mathcal{C}_{\text{final}}$ is selected as the N highest probability models following the condition

$$\sum_{c \in \mathcal{C}_{\text{final}}} p^c \geq T \tag{13}$$

and the learning proceeds to higher layer and ends when no new layer can be built. This happens when there is a single area in each piece of dataset (i.e. in an image), because nothing can be learned from such data.

When sampling an *area* to be taken as an initial *composition*, the candidates are weighted according to the probability $P(b^e | \mathcal{D}_\mathcal{A})$ – that is by the probability that the *area* is not well modelled by any *composition*. By this, the algorithm softly focuses on yet not-modelled data.

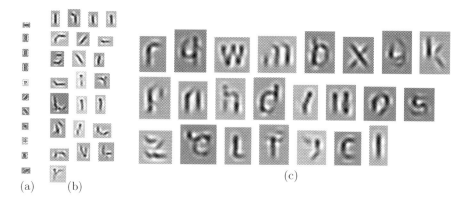

(a)   (b)   (c)

Fig. 3. The complete library of *compositions* that has been learned on the alphabet letters data. The library consists of three compositional layers in this case. The (a) shows the first compositional layer, the (b) middle layer and the (c) shows the top layer. All layers are plotted in the same scale, the apparent growing fuzziness is due to the increasing uncertainty of positions of components.

3 Experiments

3.1 Learning the Alphabet

The functionality of the learning algorithm was studied on a simple yet interesting dataset of rendered letters. This dataset consisted of 26 black and white images of small alphabet letters, where each letter was present exactly once. The noticeable fact is, that some letters differ from each other by just a tiny part and while a human eye is very sensitive to these small differences, the proposed method does not have any information on the meaning or importance of individual parts of the content of images and works completely unsupervised. This observation indicates that alphabet letters exhibit a high degree of sharing of shape segments - there are actually only straight vertical, horizontal or slanted lines and arcs, so the learned library of *composition* shall be rather small. But it can not be too small, because if there is a strong stress on obtaining an as compact representation as possible, there is also an imminent danger of losing discriminative accuracy. Naturally, the smaller the library is the less discriminative the learned model is – meaning that the model is definitely not capable of telling some letters apart. If this is to happen on the alphabet dataset, the most expected candidates are the letters 'l', 'i' and possibly 'j'. Practical challenging aspect was the fact that statistical approaches like random partitioning might not work with so small number of images.

The data entering into the learning are not preprocessed in any way, except for finding the edges and their orientations and random selection of edgels reasonably distant from each other. The edgels orientations are discretized into 8 categories representing orientations of 0, 45, 90, 135, 180, 225, 270 and 315 degrees.

The output of the learning is shown in the Fig. 3, learned *compositions* are shown in groups per layers. As can be seen, all letters do have a model and the result also follows the expectation that there might be some letters sharing one model. It happened for the letters 'i', 'j' and 'l' which are all represented as a vertical line segment. The Fig. 4 shows an example of how the data is decomposed and hierarchically organized.

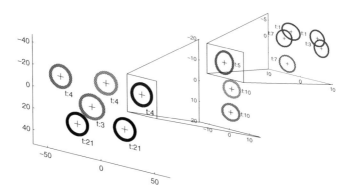

Fig. 4. A detailed view of one branch of the learned structure for the letter 'w', the ellipses show the covariance matrix encoding the position uncertainty of individual components, the gray-level of the ellipses depicts the probability that the corresponding component is present/generated.

3.2   Learning Cat Faces

Second experiment had a different scenario, it was done on a single category set of real images of cat faces from the LHI-Animal-Faces dataset [17]. The subset of cat faces consisted of 89 images roughly on a same scale. A small selection of images from this set is shown in the Fig. 5(a). The images were preprocessed before entering the learning algorithm in order to extract oriented edge segments. This was done by convolving the image with a four-item set of *Gabor filters* and taking edges above a threshold of $(T = 0.25 \cdot \max_{x,y} I(x, y)$ – individual for each image). Also *nonmax-suppression* was applied to sparsify the data. The final edge maps for a few images is shown in the Fig. 5(b). A dense edge segment set was acquired of this data and was used for learning. In fact, any other suitable edge/texture kernel or image descriptor could be used instead of the edge segments.

The learning algorithm succeeded and found a three-layered set of *compositions* capable of modelling a cat face which can be seen in the Fig. 6 showing a few selected learned *compositions*. Looking at the over-all statistics of the second layer, there are 52 models of which about a 1/3 are models for ears, there is also a model for pair of eyes and some face fractions. On the top layer, there are 26 models of which about eight represent full faces, there are some non-complete faces and there are also a few models covering the 'random' fur texture patterns.

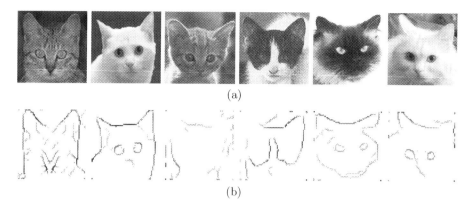

Fig. 5. An illustrative selection of the input data for the second experiment. In the (a) there are original pictures shown, what the learning algorithm really sees is shown in the (b).

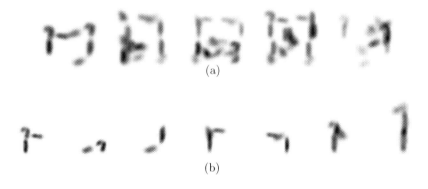

Fig. 6. Some of the learned cat faces are shown in the (a). In the (b), there is a selection of lower layer *compositions* plotted. The first three models from the left are the building blocks of the left-most head in the (a).

This results indicate that the proposed probabilistic model would benefit from higher expresivity in a sense of allowing multiple mutually exclusive components in the individual *branches* – an analogy to the *OR* nodes in [5] – that would merge some *compositions* together yielding a more compact model, but at the risk of losing discriminability. This can be achieved by rather simple modification of the m node of the model, see Fig. 1, more specifically, by changing it from two-state to multiple-state with appropriate probabilities.

4   Summary

In this paper, a very simple stochastic algorithm for unsupervised joint learning of structure and parameters is described. The probabilistic model is generative and its structure is compositional. The learning of the model is done gradually per individual layers of compositions exploiting the *maximum-likelihood* principle and *expectation-maximization*. The produced learned compositions can be seen as structure elements and can be used as features in various computer vision tasks.

The functionality is demonstrated on two experiments. The first was learning a compositional representation of alphabet letters – a compact representative of a dataset which exhibit both compositionality and presence of multiple categories. The algorithm succeeded in both aspects. The second experiment was on a single category dataset of cat faces with significant within-category visual diversity.

Acknowledgement. The authors were supported by the Czech Science Foundation under the Project P103/12/1578.

References

1. Tsotsos, J.K.: Analyzing vision at the complexity level. Behav. Brain Sci. **13**, 423–445 (1990)
2. Bienenstock, E., Geman, S., Potter, D.: Compositionality, MDL priors, and object recognition. In: Mozer, M., Jordan, M.I., Petsche, T. (eds.) NIPS, pp. 838–844. MIT Press, Cambridge (1996)
3. Fukushima, K.: Neocognitron: a hierarchical neural network capable of visual pattern recognition. Neural Netw. **1**, 119–130 (1988)
4. Serre, T., Wolf, L., Bileschi, S., Riesenhuber, M., Poggio, T.: Robust object recognition with cortex-like mechanisms. IEEE Trans. Pattern Anal. Mach. Intell. **29**, 411–426 (2007)
5. Zhu, S.C., Mumford, D.: A stochastic grammar of images. Found. Trends. Comput. Graph. Vis. **2**, 259–362 (2006)
6. Zhu, L., Chen, Y., Yuille, A.L.: Learning a hierarchical deformable template for rapid deformable object parsing. IEEE Trans. Pattern Anal. Mach. Intell. **32**, 1029–1043 (2010)
7. Si, Z., Zhu, S.C.: Learning and-or templates for object recognition and detection. IEEE Trans. Pattern Anal. Mach. Intell. **35**, 2189–2205 (2013)
8. Lee, H., Grosse, R., Ranganath, R., Ng, A.Y.: Unsupervised learning of hierarchical representations with convolutional deep belief networks. Commun. ACM **54**, 95–103 (2011)
9. Krizhevsky, A., Sutskever, I., Hinton, G.E.: Imagenet classification with deep convolutional neural networks. In: Pereira, F., Burges, C., Bottou, L., Weinberger, K. (eds.) Advances in Neural Information Processing Systems 25, pp. 1097–1105. Curran Associates, Inc., Red Hook (2012)
10. Hinton, G.E., Osindero, S.: A fast learning algorithm for deep belief nets. Neural Comput. **18**, 1527–1554 (2006)
11. Dai, J., Hong, Y., Hu, W., Zhu, S.C., Wu, Y.N.: Unsupervised learning of dictionaries of hierarchical compositional models. In: Proceedings of the IEEE Conference on Computer Vision and Pattern Recognition (CVPR) (2014)
12. Grünwald, P.D.: The Minimum Description Length Principle. MIT Press, Cambridge (2007)

13. Fidler, S., Berginc, G., Leonardis, A.: Hierarchical statistical learning of generic parts of object structure. In: Proceedings of the IEEE conference of CVPR, pp. 182–189 (2006)
14. Pearl, J.: Probabilistic Reasoning in Intelligent Systems: Networks of Plausible Inference. Morgan Kaufmann Publishers Inc., San Francisco (1988)
15. Fidler, S., Leonardis, A.: Towards scalable representations of object categories: learning a hierarchy of parts. In: Proceedings of the IEEE conference of CVPR (2007)
16. Zhu, L(.L)., Lin, C., Huang, H., Chen, Y., Yuille, A.L.: Unsupervised structure learning: hierarchical recursive composition, suspicious coincidence and competitive exclusion. In: Forsyth, D., Torr, P., Zisserman, A. (eds.) ECCV 2008, Part II. LNCS, vol. 5303, pp. 759–773. Springer, Heidelberg (2008)
17. Si, Z., Zhu, S.C.: Learning hybrid image templates (HIT) by information projection. IEEE Trans. Pattern Anal. Mach. Intell. 34, 1354–1367 (2012)

Inter-Concept Distance Measurement with Adaptively Weighted Multiple Visual Features

Kazuaki Nakamura$^{(\boxtimes)}$ and Noboru Babaguchi

Graduate School of Engineering, Osaka University,
2-1 Yamadaoka, Suita, Osaka 565-0871, Japan
k-nakamura@comm.eng.osaka-u.ac.jp

Abstract. Most of the existing methods for measuring the inter-concept distance (ICD) between two concepts from their image instances use only a single kind of visual feature extracted from each instance. However, a single kind of feature is not enough for appropriately measuring ICDs due to a wide variety of perspectives for similarity evaluation (e.g., color, shape, size, hardness, heaviness, and functions); the relationships between different concept pairs are more appropriately modeled from different perspectives provided by multiple kinds of features. In this paper, we propose extracting two or more kinds of visual features from each image instance and measuring ICDs using these multiple features. Moreover, we present a method for adaptively weighting the visual features on the basis of their appropriateness for each concept pair. Experiments demonstrated that the proposed method outperformed a method using only a single kind of visual feature and one combining multiple kinds of features with a fixed weight.

1 Introduction

Inter-concept distance measurement (ICDM), which is a problem of computing the distance between two concepts, plays an important role in various vision applications including image retrieval, automatic image annotation, indexing, and clustering. For instance, for image retrieval, query suggestion and expansion based on ICDM can help to bridge the *intention gap* [1] between user's search intent and input queries. Also, for automatic image annotation, consistency of annotation tags associated to an image can be improved by ICDM.

In the past decade, several methods for measuring inter-concept distances (ICD) have been proposed [2–9]. These methods can be roughly divided into three categories: ontology-based, text-based, and image instance-based. The last one, on which we focus in this paper, has been most actively studied in recent years. The general procedure of image instance-based ICDM for two concepts u and v is as follows: First, a set of images annotated with u and another annotated with v are gathered. Then visual models of u and v are constructed by using the respective image set. Finally, the dissimilarity score between the two visual models is calculated and used as the ICD between u and v.

© Springer International Publishing Switzerland 2015
C.V. Jawahar and S. Shan (Eds.): ACCV 2014 Workshops, Part III, LNCS 9010, pp. 56–70, 2015.
DOI: 10.1007/978-3-319-16634-6_5

In the above procedure, a visual model of each concept is generally constructed using visual features extracted from each image instance. Local descriptors including scale invariant feature transform (SIFT) [10] and speeded up robust features (SURF) [11] are often used for this purpose, but any kind of visual feature such as color, shape, edge, and texture can also be used. Nevertheless, most image instance-based methods use only a single kind of visual feature for all concept pairs [7,8]. Actually, there are various perspectives for measuring ICDs. For instance, the ICDs between concrete concepts such as object (e.g., animals and vehicles) can be measured from the perspectives of color, shape, size, heaviness, functions, and so on. ICDM methods should take into account these various perspectives as widely as possible. However, a single kind of visual feature can only provide a single perspective.

Humans adaptively select appropriate perspectives for measuring the ICD for each concept pair. For instance, humans would give a low distance to the concept pair (*cat, tiger*) from the perspective of shape, although the sizes of cats and tigers are quite different. This fact indicates that only using multiple kinds of visual features is not enough for measuring good ICDs; it is necessary to adaptively select or weight the features in accordance with their appropriateness for each concept pair. Fan et al. have proposed an ICDM method with multiple kinds of visual features [9], but it equally uses all kinds of visual features with the same level of importance.

In this paper, we propose the combined use of multiple kinds of visual features extracted from each image instance for ICDM. Moreover, we propose a method for adaptively weighting each kind of visual feature based on its appropriateness for each concept pair. This work makes following two novel contributions.

(1) A variety of perspectives are considered due to the use of multiple kinds of visual features.
(2) The most appropriate perspective for measuring the ICD is automatically selected for each concept pair due to the adaptively weighted combination of the multiple visual features.

The remainder of this paper is organized as follows: Sect. 2 briefly reviews existing ICDM methods, especially focusing on image instance-based ones. Section 3 describes the principle of image instance-based ICDM with the simplest case, where only a single kind of visual feature is used. Section 4 describes the proposed method in detail, which uses an adaptively weighted combination of multiple visual features. Section 5 shows some experimental results, and finally Sect. 6 concludes this paper.

2   Related Works

As mentioned in Sect. 1, existing ICDM methods can be roughly divided into three types: ontology-based, text-based, and image instance-based. In every type, each concept is described by a corresponding text word, but their measuring procedures differ completely. We briefly review each type of methods in this section. In the remainder of this paper, we refer to the ICD between two concepts u and v as $ICD(u, v)$.

2.1   Ontology-Based ICDM

The ontology-based ICDM [2,3] uses a human-defined ontology in which semantic hierarchical relationships between various text words are represented as a tree structure and regards the length of the path between two text words u and v in the tree as $ICD(u, v)$. One of the most common ontologies used for this purpose is WordNet[1]. ICDs computed by WordNet-based methods are close to human perception since the WordNet is structured by human experts. However, the ontology-based methods can handle only a limited number of concepts, i.e., those that are included in the ontology. To overcome this drawback, the text-based and the image-instance based ICDM use web-scale data sources.

2.2   Text-Based ICDM

The text-based ICDM [4–6] counts the co-occurrence frequency of two text words u and v in a web-scale corpus and regards the inverse of the co-occurrence frequency as $ICD(u, v)$. Since each concept is described by a text word, we can say the text-based methods are more direct than the image instance-based methods. However, text-based methods cannot handle some inter-concept relationships that are trivial for people to handle [8]; two concepts that have a trivial relationship do not always co-occur as text words in a corpus because such relationships have no need to be verbalized. To overcome this drawback, image instance-based ICDM have been studied more actively in recent years as they are expected to be able to handle various kinds of inter-concept relationships implicitly contained in images.

2.3   Image Instance-Based ICDM

The image instance-based ICDM is generally achieved by the following two steps:

i. For each concept c, construct its visual model $\mathcal{M}(c)$ by using a set of images annotated with c.
ii. For each concept pair (u, v), compute the dissimilarity score between $\mathcal{M}(u)$ and $\mathcal{M}(v)$ and regard it as $ICD(u, v)$.

Step i. is the process called "Visual Concept Learning (VCL)."

There have been a number of previous studies on VCL [9,12–16]. Most of them define a VCL task as a problem of learning a classifier for classifying the presence or absence of each concept c in images. In general, the learning process for each concept c is performed independently of other concepts in a supervised or semi-supervised manner using a number of positive images (those tagged

---

[1] WordNet 3.0, http://wordnet.princeton.edu/.

with c) and negative images (those not tagged with c). Discriminative models such as Support Vector Machines (SVM) are often applied for this purpose. Sjöberg et al. use a linear SVM which is computationally light and fast, aiming at real-time applications [15]. Yang et al. use a non-linear SVM with per-sample Multiple Kernel Learning (MKL), which is an expansion of traditional MKL, for achieving better performance in VCL [13]. Zhu et al. also use SVMs and learn a classifier for each concept in the space spanned by concatenating different kinds of visual features [12]. This allows the classifier to implicitly give an adaptive weight to each kind of visual features. Like Zhu's method, combining multiple kinds of visual features with adaptively determined weights is widely used for VCL.

However, the above framework for VCL is not suitable for ICDM. IDCM is a task of computing the distance between two concepts. Therefore, the appropriateness of each kind of visual features is determined not for each concept c but for each concept pair (u, v). In other words, the appropriateness of a certain feature for the concept u would vary depending on the counterpart concept v. For instance, the distance between *leaf* and *tree* seems to be low because *leaf* is obviously a meronym of *tree*, and it would be appropriately measured with color feature because the color of leaves and trees are both green in most cases. On the other hand, the distance between *leaf* and *flower* also seems to be low because both are organs of plants, but it cannot be appropriately measured with color feature because flowers have a wide variety of colors in contrast to leaves. Rather than the color feature, local texture feature would be more appropriate for the concept pair (*leaf*, *flower*). Nevertheless, a visual model of *leaf* constructed by the above VCL framework always gives the same weights to each kind of visual features, regardless of relations with any other concepts including *tree* and *flower*.

There are a few VCL methods taking into account inter-dependence between concepts [9,14], but these methods generally do not consider the use of adaptively weighted combination of multiple visual features. For instance, Qi et al. have proposed a technique of cross-category transfer learning for VCL, which learns a classifier

$$f_u(\boldsymbol{x}; v) = \frac{1}{|\mathcal{X}^v|} \sum_{\boldsymbol{x}_i^v \in \mathcal{X}^v} z_i^v \boldsymbol{x}^{\mathrm{T}} S \boldsymbol{x}_i^v \tag{1}$$

for a target concept u using an image instance set $\mathcal{X}^v = \{\boldsymbol{x}_i^v | i = 1, 2, \cdots\}$ of a source concept $v \neq u$, where $\boldsymbol{x}_i^v$ is a visual feature extracted from the i-th image instance of the concept v, $z_i^v \in \{1, -1\}$ is the ground truth label of $\boldsymbol{x}_i^v$, and S is a parametric matrix optimized in the learning process [14]. S can be interpreted as the correlation matrix between the target concept u and the source concept v, so it would be related to ICD(u, v) implicitly. However, since the relation between S and ICD(u, v) is non-trivial, it is not straightforward to compute ICD(u, v) based on S. Moreover, their method only uses a single kind of visual feature, i.e., bag of SIFT descriptors. Fan et al. directly compute inter-concept similarity $\gamma(u, v)$ for each concept pair (u, v) and add it to the classifier learning process of SVM for increasing the discrimination power of the classifier [9]. Their inter-concept similarity score $\gamma(u,v)$ can be easily applied to ICDM by computing $\mathrm{ICD}(u,v)$ as $1/\gamma(u,v)$. However, their algorithm for computing $\gamma(u,v)$, in which multiple kinds of visual features are extracted for each image, equally uses all kinds of visual features with the same weight.

One more drawback of the above VCL framework for ICDM is to use discriminative models which often provide complex discriminant hypersurfaces (e.g. non-linear SVM). Actually, it is not a straightforward problem to compute the dissimilarity score between two complex hypersurfaces. Therefore, generative models are more suitable for ICDM. There are a few VCL methods using not discriminative models but generative models. One of them is the method proposed by Zhuang et al. [16], in which semi-supervised Latent Dirichlet Allocation is used instead of SVM. Their method constructs visual models of all concepts as a full probability distribution $p(\boldsymbol{x}, c)$ over concept c and visual feature vector $\boldsymbol{x}$. This construction way for visual models is also adopted in many existing image-instance based ICDM methods [7,8]. More precisely, the existing ICDM methods construct a visual model of each concept c as $p(\boldsymbol{x}|c) = p(\boldsymbol{x}, c)/p(c)$ with the assumption that prior $p(c)$ is constant. Wu et al. model the $p(\boldsymbol{x}|c)$ as

$$p(\boldsymbol{x}|c) = \sum_{h} p(\boldsymbol{x}|h)p(h|c) \tag{2}$$

with probabilistic Latent Sematic Analysis (pLSA), where h is a hidden state [8]. Kawakubo et al. also use pLSA, but they model each concept c with not $p(\boldsymbol{x}|c)$ but $p(h|c)$ [7]. In the method of Kawakubo et al., a set of hidden states is shared by all concepts. In these methods, the dissimilarity score between two visual models $p(\boldsymbol{x}|u)$ and $p(\boldsymbol{x}|v)$ is easily computed by information divergence measure. However, in the existing ICDM methods [7,8], only a single kind of visual feature is used as the vector $\boldsymbol{x}$.

In contrast to the existing methods, we introduce an adaptively weighted combination of multiple visual features into image instance-based ICDM, which is widely considered in many VCL methods as mentioned above.

3 Image Instance-Based ICDM with a Single Kind of Visual Feature

To comprehensively describe the principle of image instance-based ICDM, we start with the simplest case, where only a single kind of visual feature is used.

3.1 Notation

Let $\mathcal{C}$ denote the finite set of concepts on which we focus. For each concept $c \in \mathcal{C}$, let I_i^c $(i = 1, \cdots, n_c)$ denote the i-th image instance of concept c, where n_c is the total number of image instances of concept c. Each image is converted into the same kind of d-dimensional feature vector $\boldsymbol{x} \in \mathbb{R}^d$ by some feature extractor. Let $\boldsymbol{x}_i^c$ denote the actual value of feature vector $\boldsymbol{x}$ extracted from image I_i^c. In addition, let $\mathcal{X}^c$ denote the set of feature vectors for concept c; that is, $\mathcal{X}^c = \{\boldsymbol{x}_i^c | i = 1, \cdots, n_c\}$.

3.2   Visual Concept Learning with a Generative Model

As mentioned in Sect. 2.3, most existing ICDM methods use the following two-step algorithm:

i. For each concept $c \in \mathcal{C}$, construct its visual model as conditional distribution $p(\boldsymbol{x}|c)$ by using $\mathcal{X}^c$.
ii. For each concept pair $(u, v) \in \mathcal{C}^2$, compute the dissimilarity score between $p(\boldsymbol{x}|u)$ and $p(\boldsymbol{x}|v)$ by information divergence measure.

In step i., the probability $p(\boldsymbol{x}|c)$ is generally assumed to be well modeled by a parametric distribution such as a mixture of Gaussians or a categorical distribution. We therefore use $p(\boldsymbol{x}|\boldsymbol{\Theta}^c)$ instead of $p(\boldsymbol{x}|c)$, where $\boldsymbol{\Theta}^c$ is the parameter vector of the parametric distribution.

According to probability theory, modeling each concept c as $p(\boldsymbol{x}|\boldsymbol{\Theta}^c)$ boils down to estimating parameter $\boldsymbol{\Theta}^c$ from the feature vector set $\mathcal{X}^c$. On the basis of MAP estimation, $\boldsymbol{\Theta}^c$ can be estimated as

$$\boldsymbol{\Theta}^c = \operatorname{argmax}_{\boldsymbol{\Theta}} \left[p\left(\boldsymbol{\Theta}\,|\mathcal{X}^c\right) \right] \tag{3}$$

by using parameter variable $\boldsymbol{\Theta}$.

According to Bayes' rule, $p\left(\boldsymbol{\Theta}\,|\mathcal{X}^c\right)$ can be decomposed as

$$p\left(\boldsymbol{\Theta}\,|\mathcal{X}^c\right) = \frac{1}{\lambda} p(\boldsymbol{\Theta}) p\left(\mathcal{X}^c|\,\boldsymbol{\Theta}\right), \tag{4}$$

where λ is the normalization constant for satisfying condition $\int p\left(\boldsymbol{\Theta}\,|\mathcal{X}^c\right) d\boldsymbol{\Theta} = 1$. For convenience of computation, we assume that all image instances are independently observed. This assumption means that each feature vector $\boldsymbol{x}_i^c$ is independent of any other feature vector, so

$$p\left(\mathcal{X}^c|\,\boldsymbol{\Theta}\right) = \prod_{i=1}^{n_c} p(\boldsymbol{x}_i^c|\boldsymbol{\Theta}) . \tag{5}$$

On the basis of Formulas (3), (4), and (5), $\boldsymbol{\Theta}^c$ can be estimated as

$$\boldsymbol{\Theta}^c = \operatorname{argmax}_{\boldsymbol{\Theta}} \left[p(\boldsymbol{\Theta}) \prod_{i=1}^{n_c} p(\boldsymbol{x}_i^c|\boldsymbol{\Theta}) \right] . \tag{6}$$

Note that λ can be ignored in the above maximization since it is constant with respect to $\boldsymbol{\Theta}$. This formula can be rewritten as

$$\boldsymbol{\Theta}^c = \operatorname{argmax}_{\boldsymbol{\Theta}} \left[\log p(\boldsymbol{\Theta}) + \sum_{i=1}^{n_c} \log p(\boldsymbol{x}_i^c|\boldsymbol{\Theta}) \right] \tag{7}$$

because of the monotonicity of the log function.

In order to estimate parameter $\boldsymbol{\Theta}^c$ for each concept c based on Formula (7), we model $p(\boldsymbol{\Theta})$ and $p(\boldsymbol{x}|\boldsymbol{\Theta})$ with some parametric distributions.

First, we model $p(\boldsymbol{x}|\boldsymbol{\Theta})$ with a K-dimensional categorical distribution, quantizing feature vector $\boldsymbol{x}$ into K representative values $\{\boldsymbol{y}_k|k = 1, \cdots, K\}$ by using a vector quantization method. Actually, we can employ more sophisticated approaches for modeling $p(\boldsymbol{x}|\boldsymbol{\Theta})$ like Kawakubo et al. [7] and Wu et al. [8], but it is not our focus, so that we employ a categorical distribution in this study. In a K-dimensional categorical distribution, model parameter $\boldsymbol{\Theta}$ is defined as K-dimensional real vector $(\theta_1 \cdots \theta_K)^T$, each of whose element θ_k is corresponding to the probability of $\boldsymbol{x} = \boldsymbol{y}_k$, that is, $p(\boldsymbol{x} = \boldsymbol{y}_k|\boldsymbol{\Theta}) = \theta_k$. Trivially,

$$\sum_{k=1}^{K} \theta_k = 1 \tag{8}$$

is satisfied. Distribution $p(\boldsymbol{x}|\boldsymbol{\Theta})$ can be formulated as

$$p(\boldsymbol{x}|\boldsymbol{\Theta}) = \prod_{k=1}^{K} (\theta_k)^{\delta(\boldsymbol{x}, \boldsymbol{y}_k)} , \tag{9}$$

where δ is Kronecker's delta. Taking the log of both sides, the above Formula (9) is rewritten as

$$\log p(\boldsymbol{x}|\boldsymbol{\Theta}) = \sum_{k=1}^{K} \delta(\boldsymbol{x}, \boldsymbol{y}_k) \log \theta_k . \tag{10}$$

Next, we model $p(\boldsymbol{\Theta})$ with a K-order Dirichlet distribution, which is the conjugate prior of the K-dimensional categorical distribution. With a parameter $\boldsymbol{\alpha} = (\alpha_1 \cdots \alpha_K)^T$, distribution $p(\boldsymbol{\Theta})$ is formulated as

$$p(\boldsymbol{\Theta}) = \frac{1}{Z} \prod_{k=1}^{K} (\theta_k)^{(\alpha_k - 1)} , \tag{11}$$

where Z is the normalization constant for satisfying condition $\int p(\boldsymbol{\Theta})d\boldsymbol{\Theta} = 1$. Taking the log of both sides, the above Formula (11) is rewritten as

$$\log p(\boldsymbol{\Theta}) = -\log Z + \sum_{k=1}^{K} (\alpha_k - 1) \log \theta_k . \tag{12}$$

On the basis of Formulas (7), (10), and (12), $\boldsymbol{\Theta}^c$ is finally estimated as

$$\boldsymbol{\Theta}^c = \operatorname{argmax}_{\boldsymbol{\Theta}} \left[\sum_{k=1}^{K} \log \theta_k \left\{ \alpha_k - 1 + \sum_{i=1}^{n_c} \delta(\boldsymbol{x}_i^c, \boldsymbol{y}_k) \right\} \right] . \tag{13}$$

Note that $-\log Z$ can be ignored in the above maximization since it is constant with respect to θ_k. Maximization problem (13) subject to constraint (8) can be solved by Lagrange multiplier method, whose solution $\boldsymbol{\Theta}^c = (\theta_1^c \cdots \theta_K^c)^T$ is explicitly described as

$$\theta_k^c = \frac{1}{\xi} \left\{ \alpha_k - 1 + \sum_{i=1}^{n_c} \delta(\boldsymbol{x}_i^c, \boldsymbol{y}_k) \right\} \tag{14}$$

for each $k \in \{1, \cdots, K\}$, where ξ is the normalization constant for satisfying the constraint (8).

3.3   Dissimilarity Score Between Two Visual Models

The dissimilarity score between two distributions $p(\boldsymbol{x}|\boldsymbol{\Theta}^u)$ and $p(\boldsymbol{x}|\boldsymbol{\Theta}^v)$ is generally computed by information divergence measures. A commonly used one is Kullback-Leibler (KL) divergence, which is defined as

$$D_{\mathrm{KL}}[\phi||\psi] = \int p(\boldsymbol{x}|\phi) \log \frac{p(\boldsymbol{x}|\phi)}{p(\boldsymbol{x}|\psi)} d\boldsymbol{x} \qquad (15)$$

for two distributions $p(\boldsymbol{x}|\phi)$ and $p(\boldsymbol{x}|\psi)$. However, KL divergence has an undesirable property as a dissimilarity measure, i.e., asymmetry. Therefore, Jensen-Shannon (JS) divergence [17], which is symmetric, has also been used as a dissimilarity measure [7,8]. The JS divergence between $p(\boldsymbol{x}|\phi)$ and $p(\boldsymbol{x}|\psi)$ is defined as

$$D_{\mathrm{JS}}[\phi||\psi] = \frac{1}{2}\left(D_{\mathrm{KL}}[\phi||\eta] + D_{\mathrm{KL}}[\psi||\eta]\right), \qquad (16)$$

where

$$p(\boldsymbol{x}|\eta) = \frac{1}{2}\left\{p(\boldsymbol{x}|\phi) + p(\boldsymbol{x}|\psi)\right\}. \qquad (17)$$

We also use JS divergence and compute the dissimilarity score between two concepts u and v as

$$\mathrm{Dissim}(u,v) = D_{\mathrm{JS}}[\boldsymbol{\Theta}^u||\boldsymbol{\Theta}^v]. \qquad (18)$$

More specifically, for two model parameters $\boldsymbol{\Theta}^u = (\theta_1^u \cdots \theta_K^u)^T$ and $\boldsymbol{\Theta}^v = (\theta_1^v \cdots \theta_K^v)^T$, we compute the dissimilarity score $\mathrm{Dissim}(u,v)$ as

$$\mathrm{Dissim}(u,v) = \sum_{k=1}^{K}\left\{\frac{\theta_k^u}{2}\log\frac{2\theta_k^u}{\theta_k^u+\theta_k^v} + \frac{\theta_k^v}{2}\log\frac{2\theta_k^v}{\theta_k^u+\theta_k^v}\right\}. \qquad (19)$$

4   Image Instance-Based ICDM with Multiple Kinds of Visual Features

Here we consider the case in which multiple kinds of visual features are used in conjunction with each other. We use the symbol $\mathcal{S}$ to represent a set of visual feature extractors such as SIFT descriptor and color histogram descriptor. The vector $\boldsymbol{x}_i^c$ is rewritten as $\boldsymbol{x}_i^{c,s}$, which represents the feature vector extracted from image instance I_i^c by extractor $s \in \mathcal{S}$, and the feature vector set $\mathcal{X}^c$ is redefined as $\mathcal{X}^{c,s} = \{\boldsymbol{x}_i^{c,s}|i = 1, \cdots, n_c\}$. The scoring function $\mathrm{Dissim}(u,v)$ in Formula (19) is redefined as $\mathrm{Dissim}(u,v;s)$, which represents the dissimilarity score between concepts u and v computed using feature vector sets $\mathcal{X}^{u,s}$ and $\mathcal{X}^{v,s}$.

4.1 Distance as Weighted Sum of Dissimilarity Scores

There are two popular strategies to fuse multiple kinds of visual features: feature-level fusion and decision-level fusion. In the context of ICDM, the feature-level fusion is achieved by defining a new large vector

$$\boldsymbol{X}_i^c = \left(w(s_1)(\boldsymbol{x}_i^{c,s_1})^{\mathrm{T}} \quad w(s_2)(\boldsymbol{x}_i^{c,s_2})^{\mathrm{T}} \quad \cdots \right)^{\mathrm{T}} \tag{20}$$

with a weight set $\mathcal{W} = \{w(s_j)|j = 1, 2, \cdots\}$ for all $c \in \mathcal{C}$ and $i \in \{1, \cdots, n_c\}$, where $s_j \in \mathcal{S}$ for all j, and applying the method of Sect. 3 on a set of the new vectors $\{\boldsymbol{X}_i^c|i = 1, \cdots, n_c\}$. However, this strategy is not suitable for adaptive weight control, because the weight set $\mathcal{W}$ only can affect the scale of each dimension of the new vector $\boldsymbol{X}$; the general shape of the distribution $p(\boldsymbol{X}|c)$ is not affected by $\mathcal{W}$, which means the information divergence between $p(\boldsymbol{X}|u)$ and $p(\boldsymbol{X}|v)$ for any pair $(u, v) \in \mathcal{C}^2$ does not change with $\mathcal{W}$. Therefore, in this paper, we employ the other strategy, the decision-level fusion, which is achieved by computing $\mathrm{Dissim}(u, v; s)$ separately for each $s \in \mathcal{S}$ and fusing the set $\{\mathrm{Dissim}(u, v; s)|s \in \mathcal{S}\}$ into a single distance measure. For this purpose, we use a weighted-sum method, a simple yet commonly used fusion method.

As shown in Fig. 1, our fused distance measure gives $\mathrm{ICD}(u, v)$ as

$$\mathrm{ICD}(u, v) = \sum_{s \in \mathcal{S}} w(s; u, v)\mathrm{Dissim}(u, v; s) \tag{21}$$

by using the weighted sum scheme, where $w(s; u, v)$ denotes the weight for $\mathrm{Dissim}(u, v; s)$. The weight set $\{w(s; u, v)|s \in \mathcal{S}\}$ always satisfies

$$\sum_{s \in \mathcal{S}} w(s; u, v) = 1, \tag{22}$$

but it is determined differently for each concept pair (u, v). If the feature vector set extracted using extractor s is more appropriate for modeling the relationship between u and v, a larger value is assigned to $w(s; u, v)$.

4.2 Weight Determination Based on Dissimilarity Itself

Now the problem is how to determine weight $w(s; u, v)$ for each concept pair $(u, v) \in \mathcal{C}^2$ and each extractor $s \in \mathcal{S}$. In other words, the problem is how to evaluate the appropriateness of each extractor s for modeling the relationship between concepts u and v.

In the context of VCL, "appropriateness" can be simply defined as "degree of correlation." This is because if the feature vectors extracted by extractor s are deeply correlated with the presence or absence of concept c, the extractor s is appropriate for the concept c, and vice versa. Hence the appropriateness of each extractor s for each concept c is implicitly given in the process of supervised learning for the concept c. However, in the context of ICDM, the "appropriateness" is determined not for each concept but for each concept pair, as mentioned

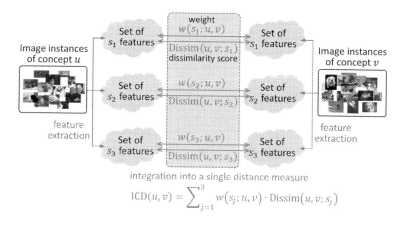

Fig. 1. Overview of method proposed for measuring ICDs

in Sect. 2.3. In this case, "appropriateness" differs from "degree of correlation," and therefore it can hardly be given by supervised learning. We therefore employ a non-learning based approach in this study.

As mentioned in Sect. 1, there are a wide variety of perspectives for evaluating the similarity between two concepts. This means that the similarity score between two concepts would vary depending on the perspective. For instance, the similarity between *car* and *train* would likely be high from the perspective of function, but it would likely be low from the perspective of shape. On the other hand, the similarity between *cat* and *tiger* would likely be high from the perspective of shape, but it would likely be low from the perspectives of size and color. For such concept pairs, how should the semantic distance between them be determined? Generally speaking, humans would give a low distance to such concept pairs. *Car* and *train* would likely be considered similar in spite of their dissimilarity in shape, and *cat* and *tiger* would likely be similar in spite of their dissimilarity in size and color. Analogizing from these instances, we hypothesize that the extractor s having lower $\mathrm{Dissim}(u, v; s)$ is more important for modeling the relationship between u and v. In other words, $\mathrm{Dissim}(u, v; s)$ itself can be used as the indicator for evaluating the appropriateness of extractor s for concept pair (u, v).

In our method, we define appropriateness indicator $a(s; u, v)$ of extractor s for concept pair (u, v) as

$$a(s; u, v) = \frac{1}{\epsilon + \mathrm{Dissim}(u, v; s)} \,, \tag{23}$$

where $\epsilon\ (> 0)$ is a regularization term for avoiding division by zero. The lower the $\mathrm{Dissim}(u, v; s)$, the higher the $a(s; u, v)$ due to their inverse relationship. If $a(s; u, v)$ is high, extractor s is considered to be appropriate for modeling the relationship between u and v. Using $a(s; u, v)$, we determine weight $w(s; u, v)$ as

$$w(s; u, v) = \frac{w'(s; u, v)}{\sum_{s \in \mathcal{S}} w'(s; u, v)} \ , \tag{24}$$

where

$$w'(s; u, v) = (a(s; u, v))^{\beta} \ . \tag{25}$$

Parameter $\beta \, (\geq 0)$ is used to adjust the balance between weights; the smaller the β, the more uniform the weights. If $\beta = 0$, the same weight is given to all descriptors. If $\beta = +\infty$, $w(\hat{s}; u, v) = 1$ is given to descriptor $\hat{s}$ such that $\hat{s} = \mathrm{argmax}_{s \in \mathcal{S}}\{a(s; u, v)\}$, and zero weight is given to all other descriptors.

5 Evaluation

We evaluated the effectiveness of the proposed method experimentally in comparison with two other methods: (A) using only a single kind of visual feature and (B) combining multiple kinds of visual features with the same fixed-weight used for all descriptors. We refer to the comparative method (B) as "Simple avg." method in this section. The performance of each method was evaluated on the basis of correlation coefficient with JCN [2], which we refer to as CCJ in this section. JCN is a representative ontology-based ICDM method using WordNet and can provide ICDs close to human perception; so JCN has been often used as the basis for performance evaluation in previous works [6, 8].

5.1 Experimental Setting

We first gathered 2,000 commonly used English nouns from the *Corpus of Contemporary American English*[2] as a candidate set of concepts. The nouns not included in WordNet were excluded. Next, the 400 nouns that had been most frequently used as annotation tags in Flickr[3] were selected from the candidate set and used to construct concept set $\mathcal{C}$. Then, for each of the 400 concepts, the corresponding image instances were collected from Flickr. The number of collected images ranged from 249 to 1868 per concept, and the total number was 379,294. From each of the collected images, we extracted four kinds of visual feature: HSV color histogram (HSV Histo), GIST [18], GLCM [19], and bag of visual words (BoVW) using the local descriptors proposed by Wu et al. [8]. Note that the comparative method using only BoVW simulates the result of Wu's method. Using these 4 kinds of visual features, we computed the ICDs for all combination of the 400 concepts. The total number of the considered concept pairs was $(400 \times 399)/2 = 79800$.

The proposed method has four parameters: K, $\boldsymbol{\alpha} = (\alpha_1 \ \cdots \ \alpha_K)^T$, β, and ϵ. Empirically, K was set to 100, β was set to 8, and ϵ was set to 0.01 in this experiment. As to $\boldsymbol{\alpha}$, we assign the same value $\bar{\alpha}$ to all α_k and set the $\bar{\alpha}$ to 50.

---

[2] Corpus of Contemporary American English, http://www.wordfrequency.info/.
[3] Flickr, http://www.flickr.com/.

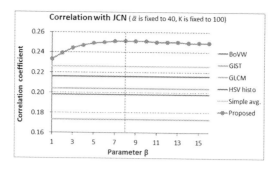

Fig. 2. Experimental results with several settings of β

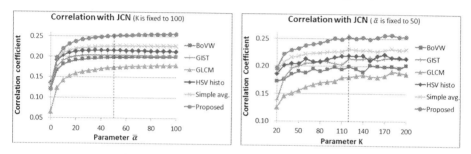

Fig. 3. Results of additional experiments with several settings of K and $\bar{\alpha}$. K is fixed to 100 in the left figure and $\bar{\alpha}$ is fixed to 50 in the right figure.

5.2   Results and Discussion

The CCJ of the proposed method and that of each comparative method are plotted in Fig. 2, in which the horizontal axis represents parameter β. Since β is not related to comparative methods (A) and (B), their results are represented as lines parallel to the horizontal axis. The CCJs of comparative methods (A) ranges from 0.17 to 0.21. These were outperformed by Simple avg. method, whose CCJ is 0.226. This indicates that it is important for ICDM to consider a variety of perspectives which can be provided by the use of multiple kinds of visual features. Moreover, Simple avg. method was outperformed by the proposed method, whose CCJ is 0.251 for $\beta = 8$. This demonstrates that using an adaptively weighted combination of multiple visual features is further effective than the simple combination with the same fixed-weight.

The CCJ of the method using only BoVW, which simulates the result of the method of Wu et al. [8], is 0.198. This is much lower than the result reported in [8]. This is because the method described in Sect. 3 does not adopt a complex way for modeling $p(\boldsymbol{x}|\boldsymbol{\Theta})$ unlike Wu's study. Using more sophisticated approaches for modeling $p(\boldsymbol{x}|\boldsymbol{\Theta})$ for each kind of visual feature would improve the effectiveness of the proposed method.

We examined the effects of parameters K and $\bar{\alpha}$ in additional experiments. Figure 3 (left) shows the CCJ of each method with several settings of $\bar{\alpha}$ and

Table 1. Average rank of 80 concept pairs that are judged as similar by humans

BoVW	GIST	GLCM	HSV histo	Simple avg	Proposed
11183.2	10946.6	12132.8	10730.0	9940.7	**9438.9**

fixed K. All of the shown CCJ is seriously decreased for small $\bar{\alpha}$, especially for $\bar{\alpha} \leq 20$. This indicates the possibility that the size of collected image set was not enough. The method described in Sect. 3 tends to cause the overfitting of Θ^c to image set $\mathcal{X}^c$ if the size of $\mathcal{X}^c$ is too small. Large $\bar{\alpha}$ is helpful for avoiding this overfitting problem. Hence, the CCJ of each method increases with larger $\bar{\alpha}$. However, if the size of $\mathcal{X}^c$ is enough large, the overfitting problem does not occur so seriously. On the other hand, Fig. 3 (right) shows the CCJ of each method with several settings of K and fixed $\bar{\alpha}$. Roughly speaking, the larger the K, the more improved the performance of each method. However, for $K \geq 120$, we can see little change in the CCJ of each method. This indicates that K should be set to more than $1/8$ of the size of a training set since the number of collected image instances per concept in this experiment is 948 on an average. Note that the proposed method outperformed the comparative methods (A) and (B) for any settings of parameters K and $\bar{\alpha}$.

In the above experiments, all methods including the proposed one had not high CCJ: at most 0.26. This is mainly because the image instance sets gathered from Flickr included a non-negligible number of junk images, i.e., ones unrelated to the annotation tags. Removing such junk images from the image instance sets [20] as a pre-process would improve the effectiveness of the proposed method.

5.3   Comparison with Human Perception

Although JCN can provide ICDs close to human perception, there is not complete agreement between them. Therefore we also experimentally compared the result of each method with human perception.

In this experiment, we first picked up 783 concept pairs from all the 79800 concept pairs mentioned in Sect. 5.1. Next, we showed the 783 pairs to 4 people and instructed them to judge whether each pair is semantically similar or not. As a result, 80 pairs out of the 783 pairs were judged as similar by all the 4 people. On the other hand, we ranked the original 79800 concept pairs in order of ICD computed by the proposed method, and created a ranked list. For comparative methods (A) and (B), we created ranked lists in the same way. Then we calculated the average rank of the above 80 concepts in each ranked list. The average rank would be low if the corresponding method is effective. Table 1 shows the result.

The result shown in Table 1 strongly supports the discussion in Sect. 5.2. The average rank of Simple avg. method is lower than that of comparative methods (A), and that of the proposed method is lowest. This also demonstrates that using an adaptively weighted combination of multiple visual features is effective.

6   Conclusion

We proposed measuring image instance-based inter-concept dissimilarity by using multiple kinds of visual features and combining them into a single distance measure using adaptively determined weights. Analogous to how people judge ICDs, it determines the weight for each feature s in accordance with the dissimilarity score calculated using s itself. The experimental results show that the proposed method outperforms a method using only a single kind of feature and one combining multiple kinds of features with a fixed weight. We will further improve its performance by devising and using a method for removing junk images from the image instance sets.

Acknowledgement. This work was supported in part by a Grants-in-Aid for Scientific Research from the Japan Society for the Promotion of Science.

References

1. Zha, Z.J., Yang, L., Mei, T., Wang, M., Wang, Z., Chua, T.S., Hua, X.S.: Visual query suggestion: Towards capturing user intent in internet image search. ACM Trans. Multimedia Comput., Commun., Appl. **6** (2010)
2. Pedersen, T., Patwardhan, S., Michelizzi, J.: Wordnet::similarity - measuring the relatedness of concepts. In: Proceedings of 5th Annual Meeting of the North American Chapter of the Association for Computational Linguistics (NAACL-04), pp. 38–41 (2004)
3. Budanitsky, A., Hirst, G.: Evaluating wordnet-based measures of lexical semantic relatedness. Comput. Linguist. **32**, 13–47 (2006)
4. Cilibrasi, R.L., Vitányi, P.M.: The google similarity distance. IEEE Trans. Knowl. Data Eng. **19**, 370–383 (2007)
5. Liu, D., Hua, X.S., Yang, L., Wang, M., Zhang, H.J.: Tag ranking. In: Proceedings of the 18th International Conference on World Wide Web, pp. 351–360 (2009)
6. Mousselly-Sergieh, H., Döller, M., Egyed-Zsigmond, E., Gianini, G., Kosch, H., Pinon, J.M.: Tag relatedness using laplacian score feature selection and adapted jensen-shannon divergence. In: Gurrin, C., Hopfgartner, F., Hurst, W., Johansen, H., Lee, H., O'Connor, N. (eds.) MultiMedia Modeling. LNCS, vol. 8325, pp. 159–171. Springer, Basel (2014)
7. Kawakubo, H., Akima, Y., Yanai, K.: Automatic construction of a folksonomy-based visual ontology. In: Proceedings of the 6th IEEE International Workshop on Multimedia Information Processing and Retrieval, pp. 330–335 (2010)
8. Wu, L., Hua, X.S., Yu, N., Ma, W.Y., Li, S.: Flickr distance: A relationship measure for visual concepts. IEEE Trans. Pattern Anal. Mach. Intell. **34**, 863–875 (2012)
9. Fan, J., He, X., Zhou, N., Peng, J., Jain, R.: Quantitative characterization of semantic gaps for learning complexity estimation and inference model selection. IEEE Trans. Multimedia **14**, 1414–1428 (2012)
10. Lowe, D.G.: Distinctive image features from scale-invariant keypoints. Intern. J. Comput. Vis. **60**, 91–110 (2004)
11. Bay, H., Tuytelaars, T., Gool, L.V.: Surf: Speeded up robust features. In: Proceedings of the 9th European Conference on Computer Vision, pp. 404–417 (2006)

12. Zhu, S., Wang, G., Ngo, C.W., Jiang, Y.G.: On the sampling of web images for learning visual concept classifiers. In: Proceedings of the 9th ACM International Conference on Image and Video Retrieval, pp. 50–57 (2010)

13. Yang, J., Li, Y., Tian, Y., Duan, L.Y., Gao, W.: Per-sample multiple kernel approach for visual concept learning. EURASIP J. Image Video Process. **2010**, 1–14 (2010)

14. Qi, G.J., Aggarwal, C., Rui, Y., Tian, Q., Chang, S., Huang, T.: Towards cross-category knowledge propagation for learning visual concepts. In: Proceedings of the 2011 IEEE Conference on Computer Vision and Pattern Rrecognition, pp. 897–904 (2011)

15. Sjöberg, M., Koskela, M., Ishikawa, S., Laaksonen, J.: Real-time large-scale visual concept detection with linear classifiers. In: Proceedings of the 21st International Conference on Pattern Recognition, pp. 421–424 (2012)

16. Zhuang, L., Gao, H., Luo, J., Lin, Z.: Regularized semi-supervised latent dirichlet allocation for visual concept learning. Neurocomputing **119**, 26–32 (2013)

17. Fuglede, B., Topsøe, F.: Jensen-shannon divergence and hilbert space embedding. In: Proceedings of the 2004 International Symposium on Information Theory 30 (2004)

18. Oliva, A., Torralba, A.: Building the gist of a scene: The role of global image features in recognition. Prog. Brain Res. **155**, 23–26 (2006)

19. Soh, L.K., Tsatsoulis, C.: Texture analysis of sar sea ice imagery using gray level co-occurrence matrices. IEEE Trans. Geosci. Remote Sens. **37**, 780–795 (1999)

20. Zhu, S., Ngo, C.W., Juang, Y.G.: Sampling and ontologically pooling web images for visual concept learning. IEEE Trans. Multimedia **14**, 1068–1078 (2012)

Extended Supervised Descent Method
for Robust Face Alignment

Liu Liu, Jiani Hu, Shuo Zhang, and Weihong Deng[⊠]

Beijing University of Posts and Telecommunications, Beijing, China
whdeng@bupt.edu.cn

Abstract. Supervised Descent Method (SDM) is a highly efficient and accurate approach for facial landmark locating/face alignment. It learns a sequence of descent directions that minimize the difference between the estimated shape and the ground truth in HOG feature space during training, and utilize them in testing to predict shape increment iteratively. In this paper, we propose to modify SDM in three respects: (1) Multi-scale HOG features are applied orderly as a coarse-to-fine feature detector; (2) Global to local constraints of the facial features are considered orderly in regression cascade; (3) Rigid Regularization is applied to obtain more stable prediction results. Extensive experimental results demonstrate that each of the three modifications could improve the accuracy and robustness of the traditional SDM methods. Furthermore, enhanced by the three-fold improvements, the extended SDM compares favorably with other state-of-the-art methods on several challenging face data sets, including LFPW, HELEN and 300 Faces in-the-wild.

1 Introduction

Facial landmark locating is key to many visual tasks such as face recognition, face tracking, gaze detection, face animation, expression analysis etc. It is also a face alignment procedure, if we regard all landmarks together as a face shape. Previous works on face recognition has proven that the recognition performance highly depend on the preciseness of the image alignment process [1–4]. An excellent facial landmark locating approach should be fully automatic, efficient and robust in unconstrained environment with large variations in facial expression, appearance, poses, illuminations, etc. Also, the number of feature points that are needed to be located varies with the intended application. For example, for face recognition or detection, 10 points including nose tip, four eye corners and two mouth corners are enough. However, for higher level applications such as facial animation and 3D reconstruction, 68 or even more landmarks are preferred. To facilitate the assessment of locating algorithm in complex environment, datasets annotated with various number of landmarks have been released, such as LFPW(29) [5], HELEN(194) [6] and IBUG(68) [7]. The images in these datasets are almost real-world, cluttered images which are mainly collected from the Internet and exhibit large variations in pose, appearance, expression, resolution etc. When dealing with large amounts of

© Springer International Publishing Switzerland 2015
C.V. Jawahar and S. Shan (Eds.): ACCV 2014 Workshops, Part III, LNCS 9010, pp. 71–84, 2015.
DOI: 10.1007/978-3-319-16634-6_6

landmarks, a global shape constraint is essential for the landmark detector, since some points such as those on the chin and contours are difficult to be characterized alone by the local appearance and may need clues from the correlation between landmarks or the facial structure.

The normal process to achieve landmark locating is first learning some discriminative information such as principal components [8–10], static probability model [5,11–13] or regressors [13–19] from the training data, and then when given a test image, making use of the learned information or model as well as the property of the test image (feature or appearance) to predict the shape (increment) or parameter, aiming at aligning the estimated shape to the true shape. Always the estimation procedure could not be accomplished in one step, due to both intrinsic factors such as variability between individuals and extrinsic factors such as pose, partial occlusion, illumination and camera resolution differences and so forth, which means that usually feature point locating is an iterative coarse-to-fine procedure.

In recent years, discriminative shape regression has emerged as the leading approach for accurate and robust face alignment. Among them, SDM [15] proposed by Xiong et al. is a representative for its natural derivation from Newton Descent Method as well as its high efficiency and accuracy. Furthermore, the authors put forward the error function for discriminative methods, connecting it with parameterized appearance models. In this paper, based on SDM and the coarse-to-fine principle in face alignment procedure, we make three modifications to the original algorithm: (1) Multi-scale HOG features are applied orderly as a coarse-to-fine feature detector; (2) Global to local constraints of the facial features are considered orderly in regression cascade; (3) Rigid Regularization is applied to obtain more stable prediction results. We show the improvement in locating accuracy by testing these modifications on several challenging datasets.

2   Related Work

Pioneering work on landmark locating/shape estimation includes ASM and AAM [8]. These methods learn shape and appearance principal components from training data to guide the estimation of parameter such as reconstruction coefficients and geometric transformation to minimize the differences between the query and model face. The model face is reconstructed by projecting the query face to the trained principal component subspace. Due to the their elegant mathematical formulation and efficient computation, a number of such AAM/ASM-based approaches have been proposed. A representative example is STASM [9], it makes extensions to the original ASM including stacking two ASM in series, using two-dimensional profile, loosening up the shape model etc. Although various improvements have been proposed, the drawback of AAM/ASM-based methods remains apparent: the expressive power of the eigen space is still limited even when a much higher-dimensional appearance subspace is employed, therefore these methods show unsatisfactory performance on unseen images or images with large variation from the training data. Moreover, due to their use of gradient descent optimization, the result given by these approaches depends heavily on initialization, which is usually considered undesirable.

Other popular modern approaches prefer to perform alignment by maximizing the posterior likelihood of each facial part or point. They usually construct specific fiducial point detectors for each point or part, which predict their probability distributions given the local features, and combine them with a global shape prior to exert constraint. In [5], a Bayesian framework unifies the probability of a global shape in order to select similar samples from training data for query images. In [19], seven fiducial landmarks are firstly detected by priori probability, then MRF is employed to construct global shape to further estimate fewer fiducial landmarks. Besides, methods based on convolutional network [20] have also been evaluated and proposed in recent years.

In this study, we focus on the regression-based methods [14,15,17,21–24], which are the leading approaches in facial landmark locating in recent years because of their higher accuracy and efficiency over other methods. The good performance of regression-based methods mainly owe to their ability to adaptively enforce shape constraints, the strong learning capacity from large training datasets as well as the precise objective function.

In general, regression-based methods learn a mapping from image features to the face shape or shape increments.

Two representative regression-based approaches are explicit shape regression (ESR) and supervised descent method (SDM). In [22], Cao et al. make use of "shape indexed feature" and learn a weak regressor between the shape increment ΔS and the features in a cascade manner, and the features are regenerated in each iteration. During test, starting with an initial shape, shape increments are predicted step-by-step to refine the shape.

Xiong et al. proposed Supervised Descent Method (SDM) [15] which directly regresses the shape increment by applying linear regression on HOG feature map. Different from boosted regression based methods as ESR, they establish the objective function of aligning image in feature space and infer the approach from Newton Descent Method to give an explanation to linear regressor and feature re-generation, which is quite simple, effective and well-understood.

3   Method

To make this paper self-contained, we first give a brief introduction to the SDM approach [15]. Then we describe the modifications proposed in detail.

3.1   Supervised Descent Method(SDM)

Given an image $\mathbf{d} \in \mathfrak{R}^{m \times 1}$ of m pixels, $\mathbf{d}(\mathbf{x}) \in \mathfrak{R}^{2p \times 1}$ represents the vector of landmarks (face shape) coordinate, where p is the number of landmarks. $\mathbf{h}(\bullet)$ is feature extraction function. The objective function for face alignment is

$$f(x_0 + \Delta x) = \|\mathbf{h}(\mathbf{d}(x_0 + \Delta x)) - \phi_*\|_2^2 \tag{1}$$

where $\phi_* = \mathbf{h}(\mathbf{d}(x_*))$ represents the feature(HOG) of the true shape.

The target is to calculate the shape increment Δx recursively to minimize the difference between the feature of estimated shape and that of the true shape, i.e. align the test image w.r.t. a template ϕ_* in feature space. HOG feature extracted from patches around the landmarks is used to achieve robustness against illumination and appearance variations. After a series of derivation and approximation such as second order Taylor expansion, similar to Newton's method [25], the update for x is derived: $\Delta x_1 = -2H^{-1}J_h^T(\phi_0 - \phi_*)$. Considering that ϕ_* is unknown in testing, ϕ_* is not used in training, and the equation is rewritten as a generic combination of feature vector ϕ_0 plus a bias term b_0.

$$\Delta x_1 = \mathbf{R}_0\phi_0 + b_0 \tag{2}$$

where $\mathbf{R}_0$ is referred as a descent direction.

Due to the large variations of face, alignment can hardly be accomplished in a single step. To cope with the non-quadratic function, a sequence of generic descent directions are learned by applying the update rule in Eq. 3.

$$
\begin{aligned}
x_k &= x_{k-1} + \mathbf{R}_{k-1}\phi_{k-1} + b_{k-1} \\
\Delta x_k &= x_k - x_* \\
\phi_k &= \mathbf{h}(\mathbf{d}(x_k))
\end{aligned}
\tag{3}
$$

In training, for each iteration, Δx_* and ϕ_k are created applying the update rule in Eq. 3, then generic descent directions R_k and bias terms b_k are learned through minimizing the difference between the true shape and the predicted shape in feature space, which can be solved by linear regression, as expressed in Eq. 4. Then the current estimated shape is updated using Eq. 3. After several steps, the estimated shape x_k converges to the true shape x_* for all images in the training set, which means the training process is completed. The main step of training is summarized in Algorithm 1.

$$\arg\min_{R,b} \|\Delta x_*^i - R_k\phi_k - b_k\|_2^2 \tag{4}$$

During testing, given a test image, the learned regressors $\{R_i, b_i\}_{i=0}^N$ are employed to compute the shape increment recursively using Eq. 3, where N indexes the number of training stage. The main steps are summarized in Algorithm 2.

Apart from the above, training data augmentation (i.e. initializing each training image from different estimate) is also mentioned to enlarge the training set and improve the generalization ability.

3.2   Modifications of Original SDM Algorithm

The main step of SDM is to predict the shape increment making use of features extracted in current iteration by linear regression. Therefore, the performance depends on both the discrimination of extracted local features and the effectiveness of linear regression. Based on this, we propose three modifications to the original supervised descent method, which are described in detail below.

Algorithm 1. SDM training

Input:
  a set of K face images $\{\mathbf{d}^i\}_{i=1}^{K}$
  corresponding true shape $\{x_*^i\}$
  initial estimate shape x_0
Output:
  regressor $\{R_i, b_i\}_{i=0}^{N}$
1: **while** $i < N$ **do**
2:    compute regress target shape $\Delta x_* = x_* - x_{i-1}$
3:    extract HOG feature ϕ_i from current shape x_{i-1}
4:    learn R_k and b_k using Eq. 4
5:    update estimated shape x_i using Eq. 2
6:    $i \leftarrow i + 1$
7: **end while**

Algorithm 2. SDM testing

Input:
  test image I
  regressor $\{R_i, b_i\}_{i=0}^{N}$
  initial shape S_0
Output:
  final predicted shape S^N
1: **while** $i < N$ **do**
2:    extract feature ϕ_i in current shape S^i
3:    compute shape increment $\Delta S = R_i \phi_i + b_i$
4:    update shape S^{i+1} using $S^{i+1} = S^i + \Delta S$
5:    $i \leftarrow i + 1$
6: **end while**

Adaptive Feature Block. In the original SDM algorithm [15], the SIFT features of landmarks are extracted in fixed-size blocks to predict the shape increment for each iteration. Intuitively, the feature extraction block size is related with the value of shape increment. If the shape increments of all training samples are scattered widely, a larger feature extraction block is preferred, thus our first modification is to replace the original fixed-size block by an adaptive feature extracting block.

In initial stages, when the estimated shape is far from the ground truth, it is favorable to extract features in bigger patches around predicted landmarks to utilize more useful information, which is beneficial for handling large shape variations and guaranteeing robustness, as the first image in Fig. 1 indicates. The red points stand for the true shape, green ones for currently estimated shape, and green circles are the range of feature extraction block, where we only draw five feature point for clarity. In training, the difference between the predicted shape and the true shape decreases step-by-step. Extracting features in a gradually-shrinking region helps to grasp the discriminative features effectively. Especially in late stages, where extracting feature in small-size blocks tends to reduce the proportion of noise, thus ensuring accuracy, as is shown in the third picture in Fig. 1. Therefore, we recommend the progressively-reducing block size other than fixed size to achieve better feature extraction performance.

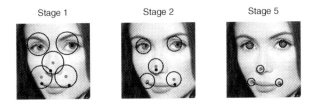

Stage 1   Stage 2   Stage 5

Fig. 1. An illustration of adaptive feature extraction block mechanism. The red points represents the true shape, green for estimated landmarks in corresponding stage, green circles are radius of extraction patch. Along with approximation to true shape, the radius shrinks (Colour figure online).

Adaptive Regression. The second modification is intended to modify the regression mode.

During the training process, features of all fiducial points constitute the shape feature ϕ. In the same way, the displacement of all landmarks consist of the shape increment Δx. R and b are obtained by minimizing Eq. 2 through linear regression in 4 or 5 steps. ϕ and Δx are both an ensemble of all landmarks. We denote this mode as global regression. Global regression enforces a shape constraint to guarantee robustness, avoiding divergency in iteration, which is essential when the prediction is far from the target.

Though variations in face geometry and expression are considered, global regression would sacrifice the accuracy to guarantee robustness at times. Other regression methods involve part regression, which is regressing different regions of face separately, and local regression, to regress each landmark individually. By operating locally at each time, these methods could avoid influence from other part or landmark of the image.

To enhance the capability of handling large variation, we recommend replacing the original global regression in all stages by a more flexible mode of adaptive regression: global regression$\Rightarrow$part regression$\Rightarrow$local regression mechanism. The regression process starts from the mean shape. At initial stages, where the estimated shape is unreliable, global regression is preferred to enforce a shape constraint to guarantee robustness. At middle stages, however, thanks to the restricted model in previous stages, the majority of the landmarks have been at or near their optimal position except for some particular regions. For instance, as shown in Fig. 2(a), in the current estimated shape, the mouth region is the furthest from the true shape than any other regions of the face. At this time, regressing different regions independently may improve the generalization ability of the regression model.

In late stages, on the premise of proper operations in the previous step, only small refinement on a few of the points is required, i.e., the shape constraint is now exhausted. Hence, regressing each landmark increment individually is

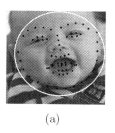 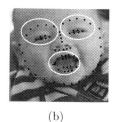 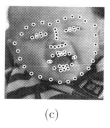

(a)   (b)   (c)

Fig. 2. A graphical representation of adaptive regression mechanism.

beneficial, which means that employing local regression at late stages helps to further improve the locating accuracy.

The choice of landmarks in part regression is alterable, which depends on the distribution and amount of landmarks. A normal way is dividing the face into five parts: two eyes region, nose region, mouth region and contour, which is adopted in our method. Similarly, the configuration of regressions is also alterable. When the target is complicated, we tend to properly increase the proportion of global regression, otherwise, the number of local or part regression could be set higher to achieve better accuracy. In our experiment, we always adopt the setting of $2\ global \Rightarrow 2\ part \Rightarrow 3\ local$ regression. It should be noticed that adaptive regression is compatible with the adaptive feature block scheme above, which jointly improve the locating accuracy.

Rigid Regularization. Minimizing Eq. 4 is the well-known least squares problem, which can be solved by linear regression. A straightforward way to improve performance is to impose regularization, which can be expressed using Eq. 5, where λ controls the regularization strength, the value of which is adaptive. The dimension of HOG feature of a single point is 128; for a face shape consisting of 68 landmarks, the number could be as large as $10K+$. Without regularization, the linear regression model is subject to over-fitting, which can be observed clearly in the experiment. Furthermore, regularization gives rise to more iterations and slows down the convergence rate, thus making the first two modifications possible. The regularization strength should be adaptive in order to work with our coarse-to-fine principle. Concretely, in early stages, strong regularization is preferred to construct a robust model, while later on, a lower regularization strength is favorable to better fit the true shape S_*. In short, with regularization, we achieve a more flexible and robust model.

$$\arg\min_{R,b}(\|\Delta x_*^i - R_k\phi_k - b_k\|_2^2 + \lambda\|R_k\|_2^2) \tag{5}$$

3.3 Summary

To sum up, we propose three modifications to SDM in total: 1. adaptive feature extracting block; 2. adaptive regression mechanism; 3. rigid regularization. With these extensions, SDM better satisfies the coarse-to-fine criterion of landmark locating. Essentially, they are inter-related rather than independent.

The involvement of regularization which slows down convergence, allows an adaptive regression mechanism as well as finer selection of block sizes. Also, the strength of regularization should decrease with the global-to-local regressions going on. It is worth mentioning that the computation complexity is reduced by applying local regression instead of simply global regression. When tackling faces with p landmarks, the dimension of one-level regressor R is $2p \times (128p+1)$ (ignoring the bias term b), while the sum of all part (local) regressor is $\Sigma_{i=1}^{k} 2p_i \times (128p_i + 1)$, $\Sigma_{i=1}^{k} p_i = p$.

4   Experiment and Comparison

In this section, we evaluate the performance of our method from two aspects. First of all, we validate the effectiveness of each modification on SDM [15], then compare that with state-of-the-art methods using two challenging datasets. The error normalized by the inter-pupil distance is used as an evaluation metric, as proposed in [5]. Note that the error is a percentage of the pupil-distance, and we omit the notation % in our reported results for clarity.

In the experiment, our modifications introduce a few adjustable parameters: feature block (normalized by the face size), regularization strength (represented as a vector of which the length equals the number of iterations) and adaptive regression parameter (noted as a three-dimensional vector $[a, b, c]$, which consists of the number of iterations in global regression, part regression and local regression respectively, where $a + b + c = iteration\ time$). In the original algorithm, SDM converges in 4–5 steps. The modifications we adopt would decrease the convergence rate, hence we adopt 1–2 more steps for fair comparison in the experiment.

4.1   Comparison with SDM

In this section, we validate the effect of modifications we proposed using two common facial datasets: LFPW [5] and LFW [26], which correspond to different situations of landmark locating. Experimental configurations are described below.

LFW consists of 13232 images in total. We adopt 10-point annotations and face bounding boxes released by [17]. Among the 13232 images, 10000 are selected randomly as the training data, with the remaining 3232 used for testing. The images in LFW dataset are low-resolution, hence the amounts of landmarks are relatively small.

LFPW are released by Belhumeur et al. [5], containing 1432 face images, 1132 for training and 300 for testing, each of which has 35 fiducial landmark annotations. Since only image URLs are available and some links disappeared as time passed, we failed to download sufficient data for experiments. Thanks to [7], a version of 1035 images is available (each annotated with 68 landmarks), 881 of which are for training and the rest 224 for testing. Since the size of the training data is small, we employ data augmentation to improve generalization ability by initializing each training image multiple times. In this way, the descent direction varies to generate multiple training samples. We enlarge the training data 10 times in our experiment.

In our analysis, feature extracting block size and regularization strength decrease gradually. The adaptive regression parameter follows the global-to-local principle. All the parameter settings are adaptive to the coarse-to-fine criterion in landmark locating. The parameters we adopt are shown in Table 1 (the original SDM in not included). Specifically, for testing the original SDM, we adopt a fixed parameter which gives the best performance: a block size of 0.16 (normalized by face size) for feature extracting.

Table 1. Parameter configuration on two data sets.

Data set	LFW	LFPW
Number of landmarks	10	68
Image size	$100 * 100$	$400 * 400$
Iteration	7	7
Adaptive block	$[0.48, 0.32, 0.16, 0.08,$ $0.04, 0.04, 0.02]$	$[0.24, 0.20, 0.16, 0.12,$ $0.08, 0.04, 0.02]$
Adaptive regression	$[3, 0, 4]$	$[2, 2, 3]$

The two graphs in Fig. 3 illustrate the improvements brought by our three modifications on LFW and LFPW datasets respectively, from which we can observe several similarities. First of all, it is noticed that the original SDM converges faster than others (within 4 iterations), for its average error starts to remain stable earlier in both of the graphs. Also, on both datasets we achieve over 15 % error reduction compared to the original SDM method. Figure 3(a) alone shows that the effect of adaptive block is the most obvious on the LFW dataset, with the error declining from 5.96 to 5.5. In contrast, the curves of Fig. 3(b) indicate that, with the LFPW dataset of 68 landmarks, the rigid regularization applied contributes the most among the three modifications. An explanation is that as the amount of landmarks is large, the initially predicted shape is further from the ground truth, so a stronger global shape constraint is necessary. In this case, more iterations of global regression (instead of part or local regression) should be implemented and strong regularization is preferred in the mean time. Overall, it turns out that with each modification applied, the average error decreases more or less, which demonstrates that all three modifications proposed are effective in improving the performance of the original SDM.

4.2 Comparison with State-of-the-Art

In this section, we compare our method with state-of-the-art on two challenging datasets. These datasets show large variations in face shape, appearance, and

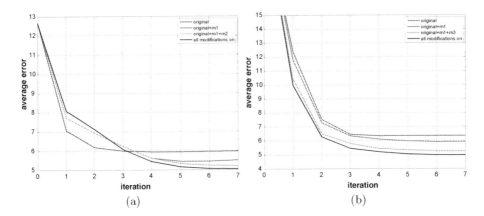

Fig. 3. Curves of average error versus iteration. 3(a) on LFW dataset, 3(b) on LFPW dataset. m1: modification of adaptive feature extracting block; m2: modification of adaptive regression mechanism; m3: modification of rigid regularization.

number of landmarks, which is briefly introduced below. In the experiment, the average of normalized error of all landmarks and images is reported and compared. The adaptive regression parameter is set as $[2, 2, 3]$ and the block sizes (normalized by the length of face size) as $[0.24, 0.20, 0.16, 0.12, 0.08, 0.04, 0.02]$.

Helen (194 landmarks) [6]. It contains $2,330$ high resolution, accurately labeled face images. In our experiment, we follow the setting in [6]: 2000 images for training and 330 images for testing. The most discriminative peculiarities that HENLEN dataset holds are its high resolution and the large number of facial landmarks.

300 Faces in-the-wild (68 landmarks) [7]. It consists of existing datasets including LFPW [5], AFW [27], HELEN [6], XM2VTS [28] and IBUG. All images have been re-annotated by 68 landmarks. This dataset shows the largest variation in pose, expression, illumination, resolution due to cross-database annotation. To compare with state-of-the-art, we follow the setting in [14] which is dividing the training data into two parts in our own experiment. The training set is composed of the whole AFW dataset and the training data from LFPW and HELEN, with 3148 images in total.

The testing set is composed of IBUG and the test sets of HELEN and LFPW, with 689 images in total. The testing set are also divided into two subsets for further comparison: 1. the challenging IBUG subset, consisting of IBUG data; 2. the common subset, consisting of the testing sets from HELEN and LFPW. Face bounding box is offered at the same time[1], produced by their in-house detector. All images are cropped and normalized into a fixed size of $400 * 400$ before operation.

---

[1] http://ibug.doc.ic.ac.uk/resources/300-W.

Results. Recently, a new approach based on boosted-regression called LBF [14] has been proposed. The accuracy appears to be the highest in the literature, outperforming SDM and ESR. Also, its speed is extremely fast. For comprehensive comparison, the performance of LBF is also reported, yet we should note that the aim of our method is to extend the simple, well-understood but efficient facial landmark location algorithm SDM.

From results illustrated in Tables 2 and 3, we find that ESDM outperforms SDM by a large margin on all testing sets, again demonstrating the effectiveness of the modifications we proposed. In general, ESDM achieves comparable result with LBF.

Table 2. Average error of several algorithm on HELEN dataset, results of other methods come from corresponding paper.

Method	HELEN (194 landmarks)
ESDM(Our method)	**5.32**
SDM [15]	5.85
ESR [22]	5.70
RCPR [21]	6.50
LBF [14]	5.41
LBF fast [14]	5.80
STASM [9]	11.1
CompASM [24]	9.10

Table 3. Error on 300-W(68 landmarks) data sets including two subsets, the result of SDM and ESR originate from [14].

Method	Full set	Common subset	Challenging subset
ESDM(our method)	6.44	**4.73**	13.92
SDM [15]	7.52	5.6	15.4
ESR [22]	7.58	5.28	17.0
LBF [14]	**6.32**	4.95	**11.98**
LBF fast [14]	7.37	5.38	15.5

Observing the average error on HELEN dataset reported in Table 2 alone, the performance of our method advances state-of-the-art over all. Regression-based methods [14,15,21,22] significantly outperforms ASM-based [24,29] approaches, proving the advantage of the flexible model of iterative alignment. During our experiment, it is also observed that our approach achieves slightly higher locating accuracy of certain parts than others, such as the contour of lip and the eyes. Table 3 illustrates the performance on 300 Faces in-the-wild dataset including two subsets. Similar to what is done in [15], we experiment on the original SDM algorithm and obtain results of [7.32, 5.40, 15.32] on the fullset, common subset, challenging subset respectively. The rates are comparable with those reported in [15]. Our ESDM method shows superior results over SDM, ESR and LBF fast and is comparable to LBF. It is worth mentioning that on the common test subset, under the condition of identical training data, ESDM performs better than LBF, and we attribute this to the adequacy of training samples. Our method performs less desirably than LBF on the challenging test subset, mainly due to the drawback of the initialization mechanism of our method. In Figs. 4 and 5, some example images and comparison results from the common and the challenging subsets of 300 Faces in-the-wild are displayed.

Fig. 4. Several examples from the Challenging Subset, some results of SDM are shown in the first row, corresponding results of our method are shown in the second line. Compared to SDM our method still achieve good performance on these extremely difficult images with large variation in pose, expression, partial occlusion.

Fig. 5. A comparative result on common subset from 300 Faces in-the-wild.

5   Conclusion and Discussions

In this paper, we propose three modifications on SDM including: (1) Multi-scale HOG feature extraction in a coarse-to-fine manner; (2) Global to local regression of features in cascade; (3) Rigid Regularization applied to obtain more stable prediction results. Extensive experimental results demonstrate that each of the three modifications substantially improves the accuracy and robustness of the traditional SDM method. Furthermore, our method achieves favorable performance over most of the other state-of-the-art methods and comparable results with the LBF method, on several challenging face datasets including LFPW, HELEN and 300 Faces in-the-wild.

Acknowledgement. This work was partially sponsored by National Natural Science Foundation of China (NSFC) under Grant No. 61375031, No. 61471048, and No. 61273217. This work was also supported by the Fundamental Research Funds for the Central Universities, Beijing Higher Education Young Elite Teacher Project, and the Program for New Century Excellent Talents in University.

References

1. Deng, W., Hu, J., Guo, J., Cai, W., Feng, D.: Robust, accurate and efficient face recognition from a single training image: a uniform pursuit approach. Pattern Recogn. **43**, 1748–1762 (2010)
2. Deng, W., Hu, J., Lu, J., Guo, J.: Transform-invariant pca: a unified approach to fully automatic face alignment, representation, and recognition. IEEE Trans. Pattern Anal. Mach. Intell. **36**, 1275–1284 (2014)
3. Deng, W., Hu, J., Guo, J.: Extended src: undersampled face recognition via intraclass variant dictionary. IEEE Trans. Pattern Anal. Mach. Intell. **34**, 1864–1870 (2012)
4. Deng, W., Hu, J., Zhou, X., Guo, J.: Equidistant prototypes embedding for single sample based face recognition with generic learning and incremental learning. Pattern Recogn. **47**, 3738–3749 (2014)
5. Belhumeur, P.N., Jacobs, D.W., Kriegman, D., Kumar, N.: Localizing parts of faces using a consensus of exemplars. In: 2011 IEEE Conference on Computer Vision and Pattern Recognition (CVPR), pp. 545–552. IEEE (2011)
6. Le, V., Brandt, J., Lin, Z., Bourdev, L., Huang, T.S.: Interactive facial feature localization. In: Fitzgibbon, A., Lazebnik, S., Perona, P., Sato, Y., Schmid, C. (eds.) ECCV 2012, Part III. LNCS, vol. 7574, pp. 679–692. Springer, Heidelberg (2012)
7. Sagonas, C., Tzimiropoulos, G., Zafeiriou, S., Pantic, M.: A semi-automatic methodology for facial landmark annotation. In: 2013 IEEE Conference on Computer Vision and Pattern Recognition Workshops (CVPRW), pp. 896–903. IEEE (2013)
8. Cootes, T.F., Edwards, G.J., Taylor, C.J., et al.: Active appearance models. IEEE Trans. Pattern Anal. Mach. Intell. **23**, 681–685 (2001)
9. Milborrow, S., Nicolls, F.: Locating facial features with an extended active shape model. In: Forsyth, D., Torr, P., Zisserman, A. (eds.) ECCV 2008, Part IV. LNCS, vol. 5305, pp. 504–513. Springer, Heidelberg (2008)
10. Saragih, J., Goecke, R.: A nonlinear discriminative approach to aam fitting. In: IEEE 11th International Conference on Computer Vision, ICCV 2007, pp. 1–8. IEEE (2007)
11. Zhou, F., Brandt, J., Lin, Z.: Exemplar-based graph matching for robust facial landmark localization (2013)
12. Cristinacce, D., Cootes, T.: Automatic feature localisation with constrained local models. Pattern Recogn. **41**, 3054–3067 (2008)
13. Saragih, J.M., Lucey, S., Cohn, J.F.: Face alignment through subspace constrained mean-shifts. In: 2009 IEEE 12th International Conference on Computer Vision, pp. 1034–1041. IEEE (2009)
14. Ren, S., Cao, X., Wei, Y., Sun, J.: Face alignment at 3000 fps via regressing local binary features (2014)

15. Xiong, X., De la Torre, F.: Supervised descent method and its applications to face alignment. In: 2013 IEEE Conference on Computer Vision and Pattern Recognition (CVPR),pp. 532–539. IEEE (2013)

16. Sánchez-Lozano, E., De la Torre, F., González-Jiménez, D.: Continuous regression for non-rigid image alignment. In: Fitzgibbon, A., Lazebnik, S., Perona, P., Sato, Y., Schmid, C. (eds.) ECCV 2012, Part VII. LNCS, vol. 7578, pp. 250–263. Springer, Heidelberg (2012)

17. Dantone, M., Gall, J., Fanelli, G., Van Gool, L.: Real-time facial feature detection using conditional regression forests. In: 2012 IEEE Conference on Computer Vision and Pattern Recognition (CVPR), pp. 2578–2585. IEEE (2012)

18. Cao, C., Weng, Y., Lin, S., Zhou, K.: 3d shape regression for real-time facial animation. ACM Trans. Graph. **32**, 41 (2013)

19. Valstar, M., Martinez, B., Binefa, X., Pantic, M.: Facial point detection using boosted regression and graph models. In: 2010 IEEE Conference on Computer Vision and Pattern Recognition (CVPR), pp. 2729–2736. IEEE (2010)

20. Sun, Y., Wang, X., Tang, X.: Deep convolutional network cascade for facial point detection. In: 2013 IEEE Conference on Computer Vision and Pattern Recognition (CVPR), pp. 3476–3483. IEEE (2013)

21. Burgos-Artizzu, X.P., Perona, P., Dollár, P.: Robust face landmark estimation under occlusion. In: ICCV (2013)

22. Cao, X., Wei, Y., Wen, F., Sun, J.: Face alignment by explicit shape regression. In: 2012 IEEE Conference on Computer Vision and Pattern Recognition (CVPR), pp. 2887–2894. IEEE (2012)

23. Dollár, P., Welinder, P., Perona, P.: Cascaded pose regression. In: 2010 IEEE Conference on Computer Vision and Pattern Recognition (CVPR), pp. 1078–1085. IEEE (2010)

24. Efraty, B., Huang, C., Shah, S.K., Kakadiaris, I.A.: Facial landmark detection in uncontrolled conditions. In: 2011 International Joint Conference on Biometrics (IJCB), pp. 1–8. IEEE (2011)

25. Baker, S., Matthews, I.: Lucas-kanade 20 years on: a unifying framework. Int. J. Comput. Vis. **56**, 221–255 (2004)

26. Huang, G.B., Ramesh, M., Berg, T., Learned-Miller, E.: Labeled faces in the wild: A database for studying face recognition in unconstrained environments. Technical report, Technical Report 07–49, University of Massachusetts, Amherst (2007)

27. Zhu, X., Ramanan, D.: Face detection, pose estimation, and landmark localization in the wild. In: 2012 IEEE Conference on Computer Vision and Pattern Recognition (CVPR), pp. 2879–2886. IEEE (2012)

28. Messer, K., Matas, J., Kittler, J., Luettin, J., Maitre, G.: Xm2vtsdb: the extended m2vts database. In: Second International Conference on Audio and Video-Based Biometric Person Authentication, vol. 964, pp. 965–966. Citeseer (1999)

29. Cootes, T.F., Taylor, C.J., Cooper, D.H., Graham, J.: Active shape models-their training and application. Comput. Vis. Image Underst. **61**, 38–59 (1995)

Local Similarity Based Linear Discriminant Analysis for Face Recognition with Single Sample per Person

Fan Liu[1(✉)], Ye Bi[1], Yan Cui[2], and Zhenmin Tang[1]

[1] School of Computer Science and Engineering,
Nanjing University of Science and Technology, Nanjing, China
`fanliu.njust@gmail.com`
[2] Key Laboratory of Broadband Wireless Communication and Sensor Network
Technology, Nanjing University of Posts and Telecommunications, Nanjing, China

Abstract. Fisher linear discriminant analysis (LDA) is one of the most popular projection techniques for feature extraction and has been widely applied in face recognition. However, it cannot be used when encountering the single sample per person problem (SSPP) because the intra-class variations cannot be evaluated. In this paper, we propose a novel method coined local similarity based linear discriminant analysis (LS_LDA) to solve this problem. Motivated by the "divide-conquer" strategy, we first divide the face into local blocks, and classify each local block, and then integrate all the classification results to make final decision. To make LDA feasible for SSPP problem, we further divide each block into overlapped patches and assume that these patches are from the same class. Experimental results on two popular databases show that our method not only generalizes well to SSPP problem but also has strong robustness to expression, illumination, occlusion and time variation.

1 Introduction

Face recognition, as a nonintrusive biometric technology, has been one of the most active research topics in the field of pattern recognition and computer vision, owing to its scientific challenges and useful applications. Nowadays, appearance-based face recognition methods are the mainstream [1] and they usually represent a face image as a high-dimensional vector with plenty of redundant information. Apparently, it is necessary to extract discriminative information from such a high-dimensional vector to form a new low-dimensional vector representation. This process is called feature extraction which is beneficial to both increasing recognition accuracy and reducing computational complexity.

In the last two decades, face recognition systems are considered to be critically dependent on discriminative feature extraction, about which PCA [2,3] and LDA [4] are the two most representative ones. PCA poses the K-L transformation on training face images to find a set of optimal orthogonal bases(also called Eigenfaces) for a low-dimensional subspace and represents a face as a linear combination of Eigenfaces. Though PCA is optimal in the sense of minimum

© Springer International Publishing Switzerland 2015
C.V. Jawahar and S. Shan (Eds.): ACCV 2014 Workshops, Part III, LNCS 9010, pp. 85–95, 2015.
DOI: 10.1007/978-3-319-16634-6_7

reconstruction error, it may not be favorable to classification because it does not incorporate any class specific discriminatory information into feature extraction. On the contrary, LDA takes class discriminatory information into account and seeks to find a set of optimal projection vectors by maximizing the ratio between the between-class and the within-class scatter matrices of the training samples. Existing experimental results demonstrate that LDA generally outperforms PCA in recognition rate when there are enough and representative training samples from per subject [5].

Unfortunately, in some real-world applications, the number of samples of each subject is usually very small. For example, there is only one image available in the scenario of identity card or passport verification, law enforcement, surveillance or access control. This is the so called Single Sample per Person (SSPP) problem which severely challenges existing feature extracting algorithms, especially their robustness to variations such as expression, illumination and disguises. For example, the performance of PCA will degrade seriously [6], while LDA even fails to work because in this case the within-class scatter matrices of the training samples can not be estimated directly.

In this paper, we propose a simple yet effective method, called local similarity based Linear Discriminant Analysis (LS_LDA) to address SSPP problem. The intuitive idea is that to make full use of the single image, we can divide it into many local blocks and analyze them respectively. Then based on the "divide and conquer" [7] strategy, we first classify each local block, and then integrate all the results of classification by majority voting to make a final decision. However, classifying each local block will have to face the difficulty that those local blocks corresponding to the same location but coming from different subjects might be close to each other. Intuitively, if we project these blocks from different persons into a lower-dimensional subspace and keep their respective projection in this subspace as far apart as possible, the classification will have a higher recognition rate. Although LDA is the best choice for this task, it cannot work under SSPP condition because the intra-class scatter matrices cannot be calculated. To make LDA feasible for SSPP problem, we further divide each local block into overlapped patches and propose local similarity assumption. As the patches in a local block have strong similarity, they can be regarded as to be from the same class. Based on this idea, the within-class scatter can be computed by using the overlapped patches in a local block. Finally, the combination of outputs from all local blocks by majority voting further improves the performance. To evaluate the proposed method, we perform a series of experiments on two public datasets including Extended Yale B and AR face databases. Experimental results demonstrate that the proposed method not only outperforms those specially designed methods for SSPP problem, but also has strong robustness to expression, illumination, occlusion and time variation.

The rest of this paper is organized as follows. We start by introducing related work in Sect. 2. Then in Sect. 3, we present the proposed local similarity based linear discriminant analysis. Section 4 demonstrates experiments and results. Finally, we conclude in Sect. 5 by highlighting key points of our work.

2 Related Work

In order to address SSPP problem, many methods have been developed during the last two decades. Shan et al. [8] presented a face-specific subspace method based on PCA which first generates a few virtual samples from single gallery image of per subject and then uses PCA to build a projection subspace for each person. But strong correlation between virtual samples decreases the representativeness of training samples and accordingly limits the performance of this method. In order to make LDA suitable for SSPP problem, Zhang et al. [9] applied SVD decomposition to the only face image of a person and the obtained non-significant SVD basis images were used to estimate the within-class scatter matrix of this person approximately. Generally speaking, the optimal number of non-significant SVD basis images is face-specific and should not be determined equally for all face images as they did. Both methods treat the whole image as a high-dimensional vector and belong to holistic representation. However, some other schemes favor local representation, in which a face image is divided into blocks and vector representation of information is conducted by blocks other than globally. Compared with holistic representation, local representation is proved to be more robust against variations [10]. For example, Chen et al. [11] proposed BlockFLD method which generates multiple training samples for each person by partitioning each face image into a set of same sized blocks and then applies FLD-based methods with these blocks. But the great differences between the appearances of long-distance blocks from one image may go against the compactness of the within-class scatter after projection. Recenlty, Zhu et al. [12] proposed patch based CRC (PCRC) for small sample size (SSS) problem and Kumar et al. [13] proposed patch based n-nearest classifier to improve the stability and generalization ability. However, these patch based methods has a commonality that they just consider each patch independently. Hence, they will lose the local structure information between patches, which is very important for classification. In addition, local feature can also be used to solve SSPP problem. Deng et al. [14,15] propose two representative local feature based methods, which are uniform pursuit approach [14] and linear regression analysis (LRA) [15] respectively.

3 Local Similarity Based Linear Discriminant Analysis for Face Recognition

3.1 Local Structure

To describe the local structure, we illustrate three kinds of neighborhood in Fig. 1. The neighbor pixels on a square of radius R form a squarely symmetric neighbor sets. Suppose there are N pixels in an image. For the i-th pixel in the image, its P neighbor pixels can be denoted by $\Omega_P^i = \{i_j | j = 1, \ldots, P\}$.

For the i-th pixel in the image, we select a $S \times S$ local patch (e.g. S = 3, 5) centered at it. All the S^2 pixels within the patch form a m dimensional local patch vector $\boldsymbol{x}_0^i$, where $m = S^2$. Similarly, the neighbor pixel i_j of the i-th also corresponds to a same sized local patch, whose patch vector is denoted by $\boldsymbol{x}_j^i, j = 1, \ldots, P$. Then, the center patch and neighbor patches determine a local block centered at the i-th pixel. Figure 2 shows an example of a local block containing a central patch and 16 neighbor patches. The size of patch is 3×3 and the size of the block is 7×7. For the pixel on the margin of an image, we use the mirror transform first and then determine its local block.

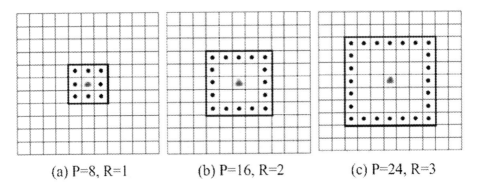

(a) P=8, R=1   (b) P=16, R=2   (c) P=24, R=3

Fig. 1. Squarely symmetric neighbor sets for different R.

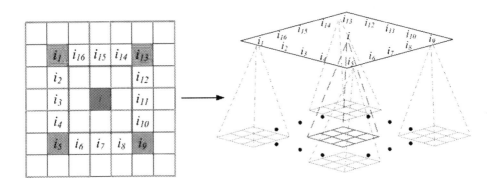

Fig. 2. Illustration of local patches in a local block.

As the patches are overlapped and concentrate in a small block, they are strongly similar. Intuitively, they can be regarded as be from the same class. Therefore, we assume that the overlapped patches in a local block are from the same class.

3.2   Local Similarity Based Linear Discriminant Analysis (LS_LDA)

Suppose there are C training images $\{\boldsymbol{x}_1, \boldsymbol{x}_2, \ldots, \boldsymbol{x}_C\}$ and each of them belongs to a different person. According to the "dividing and conquer" strategy, we first

divide these training images into many local blocks. For example, the only training image $\boldsymbol{x}_k$ from k-th person is divided into a set of N overlapped blocks
$\{\boldsymbol{x}_k^1, \boldsymbol{x}_k^2, \ldots, \boldsymbol{x}_k^N\}$, where the i-th pixel of $\boldsymbol{x}_k$ corresponds to the local block $\boldsymbol{x}_k^i$.
The i-th local blocks from all the C training images form a block training set
$\boldsymbol{B}^i$. Similarly, the probe image $\boldsymbol{y}$ is also decomposed into N overlapped blocks
$\{\boldsymbol{y}^1, \boldsymbol{y}^2, \ldots, \boldsymbol{y}^N\}$ in the same way. Then, we can classify each block respectively.
However, we observe that those local blocks corresponding to the same location
but coming from different subjects might be close to each other because those
individuals are similar-looking.

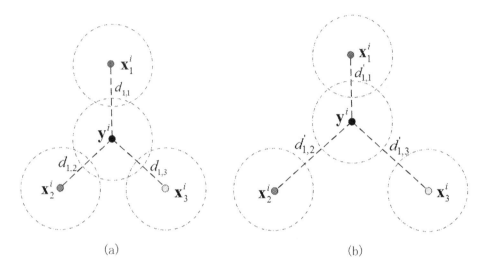

(a)   (b)

Fig. 3. Illustration of the basic idea.

As shown in Fig. 3(a), $\boldsymbol{y}^i$ is supposed to be from Class 1 and $\{\boldsymbol{x}_1^i, \boldsymbol{x}_2^i, \boldsymbol{x}_3^i\}$
are the training images' block from Class 1, Class 2 and Class 3 respectively. It
can be seen that the distance $d_{1,1}$ between $\boldsymbol{y}^i$ and $\boldsymbol{x}_1^i$ is similar with $d_{1,2}$ and
$d_{1,3}$. Hence, these kinds of probe image blocks are easily misclassified. To make
such misclassifications less likely, we aim to learn a mapping to project these
samples into a low-dimensional subspace to enlarge the between-class distance
and shorten the intra-class distance, as shown in Fig. 3(b). Although LDA is
the best choice for this task, it cannot work under SSPP condition because
the intra-class scatter matrices cannot be calculated. To address this problem,
we further divide each block into overlapped patches and take advantage of the
above-mentioned local similarity assumption.

As shown in Fig. 4, the testing block $\boldsymbol{y}^i$ is divided into $P + 1$ overlapped
patches, where $\boldsymbol{y}^i = [\boldsymbol{y}_0^i, \boldsymbol{y}_1^i, \ldots, \boldsymbol{y}_P^i] \in R^{m \times (P+1)}$. Similarly, the training block
$\boldsymbol{x}_k^i$ can also be decomposed into $P+1$ overlapped patches, where $\boldsymbol{x}_k^i = [\boldsymbol{x}_{k,0}^i, \boldsymbol{x}_{k,1}^i,$
$\ldots, \boldsymbol{x}_{k,P}^i] \in R^{m \times (P+1)}$. According to the description of local similarity assump-
tion, these overlapped patches in local block have strong similarity. They can be
theoretically regarded as to be from the same class. Then, the computation of

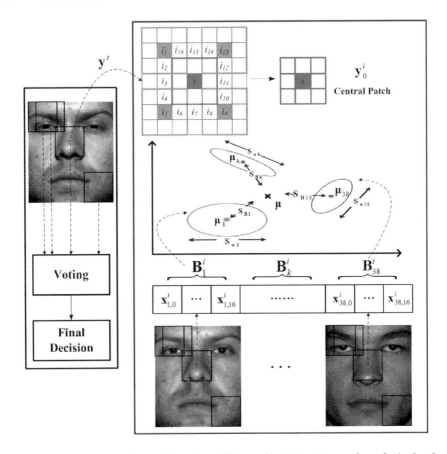

Fig. 4. Diagram of local similarity based linear discriminative and analysis for face recognition.

the within-class scatter and the between-class scatter for the block training set $\boldsymbol{B}^i$ as follows:

$$S_w^i = \sum_{k=1}^{C} \sum_{j=0}^{P} (\boldsymbol{x}_{k,j}^i - \boldsymbol{\mu}_k^i)(\boldsymbol{x}_{k,j}^i - \boldsymbol{\mu}_k^i)^T \tag{1}$$

$$S_b^i = \sum_{k=1}^{C} (P+1)(\boldsymbol{\mu}_k^i - \boldsymbol{\mu}^i)(\boldsymbol{\mu}_k^i - \boldsymbol{\mu}^i)^T \tag{2}$$

where $\boldsymbol{\mu}_k^i$ and $\boldsymbol{\mu}^i$ denote the means of the k-th class and all the classes respectively. Then, LDA can be used to seek the projection matrix $\boldsymbol{W}$ that enlarges the between-class distance and shortens the intra-class distance in the projected subspace. It's equivalent to solve the following optimization problem:

$$J(W) = \frac{|W S_b^i W^T|}{|W S_w^i W_T|} \tag{3}$$

This optimization problem equals to the following generalized eigenvalue problem:

$$S_b^i W = \lambda S_w^i W \tag{4}$$

The solution of (4) can be obtained by applying an eigen-decomposition to the matrix $(S_w^i)^{-1} S_b^i$ if S_w^i is nonsingular, or $(S_b^i)^{-1} S_w^i$ if S_b^i is nonsingular.

After computing the projection matrix W, we project the block training set $\boldsymbol{B}^i$ and testing block $\boldsymbol{y}^i = [\boldsymbol{y}_0^i, \boldsymbol{y}_1^i, \ldots, \boldsymbol{y}_P^i] \in R^{m \times (P+1)}$ into the lower-dimensional subspace and Nearest Neighbor (NN) classifier is used to classify the testing block. In order to decrease the cost of computing, we only focus on the classification of the central patch of the testing block. After classify each block, the classification outputs of all blocks can then be aggregated. Majority voting is used for the final decision making, which means that the test sample is finally classified into the class with the largest number of votes. Fig. 4 shows the whole recognition procedure, which can be summarized as "divide-conquer-aggregate" procedure.

4   Experimental Results

In this section, we use the Extended Yale B [16] and AR [17] databases to evaluate the proposed method and compare it with some popular methods dealing with SSPP problem. These state-of-the-art methods include PCA [3], BlockFLD [11], patch based nearest neighbor (PNN) classifier [13], FLDA_single [13] and patch based CRC (PCRC) [12].

For our proposed method, the neighbor sets are fixed at $P = 8, R = 1$ and the patch size is fixed at 9×9. Since the classification of each local block can be solved independently before combing decisions, we open 4 Matlab workers for parallel computation to improve the efficiency. For BlockFLD and PNN, the patch size is set as 10×10 and overlap is 5 pixels. In addition, histogram equalization is used as illumination preprocessing for BlockFLD, PNN and FLDA_single since it will improve their performance under illumination variation.

4.1   Extended Yale B Database

The Extended Yale B face database [16] contains 38 human subjects under 9 poses and 64 illumination conditions. The 64 images of a subject in a particular pose are acquired at camera frame rate of 30 frames per second, so there is only small change in head pose and facial expression for those 64 images. However, its extreme lighting conditions still make it a challenging task for most face recognition methods. All frontal-face images marked with P00 were used in our experiment. The cropped and normalized 192×168 were captured under various laboratory-controlled lighting conditions [18] and are resized to 80×80 in our experiments. Some sample images of one person are shown in Fig. 5.

To evaluate the performance of our proposed methods to SSPP problem, we use the image under the best illumination condition for training, whose azimuth

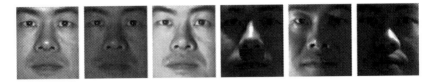

Fig. 5. Samples of a person under different illuminations in Extended Yale B face database.

Table 1. Five Subsets of Extended Yale B.

Subsets	1	2	3	4	5
Lighting angle	$0 \sim 12$	$13 \sim 25$	$26 \sim 50$	$51 \sim 77$	> 77
Number of images	$6 \times 38 = 228$	$12 \times 38 = 456$	$12 \times 38 = 456$	$14 \times 38 = 532$	$19 \times 38 = 722$

and elevation are both 0 degree. The remaining 63 images are used for testing and the average results are reported. The average results of the five subsets of the Extended Yale B are also reported respectively. The details of the five subsets are given in Table 1. The experimental results are shown in Table 2. From the table, we can see that our proposed method achieves the best average result. It also leads to the best results on subset2, subset3 and subset4 respectively with the lighting angles increasing. However, on the subset5 which is under the worst illumination condition, our method is not superior to PNN. This is because the illumination variation is the most drastic on subset5 and more and more test blocks of each face image are regarded as "outlier" blocks that will disturb the final voting result. In addition, we also compared with classical PCA [3] since does not suffer from the SSPP problem. However, it's easy to be subject to gross variations and, thus sensitive to any changes in expression, illumination etc. Therefore, it achieves the best results on the Subset1 that is under the best illumination condition; while the recognition rate drops significantly on Subset2, Subset3, Subset4 and Subset5 whose illumination variation is drastic.

Table 2. Recognition rates (%) on Extended Yale B database for SSPP problem

Method	Subsets1	Subsets2	Subsets3	Subsets4	Subsets5	Average
PCA	$\mathbf{98.7 \pm 3.2}$	94.1 ± 6.7	44.5 ± 22.3	14.1 ± 8.2	17.0 ± 8.0	44.1 ± 36.5
PNN	96.5 ± 3.6	100 ± 0	69.1 ± 15.4	43.4 ± 8.4	$\mathbf{75.5 \pm 14.9}$	73.8 ± 22.8
BlockFLD	80.3 ± 3.2	96.3 ± 4.1	66 ± 7.4	59.2 ± 16.7	53.2 ± 11.2	67.8 ± 19
FLDA_single	96.9 ± 3.1	99.8 ± 0.8	65.8 ± 18.1	32.1 ± 16.5	30.2 ± 10.2	57.1 ± 32.4
PCRC	89 ± 2.6	99.8 ± 0.8	77.6 ± 7.5	58.5 ± 19.2	36.7 ± 12.6	66.3 ± 26.6
LS_LDA	96.9 ± 2.6	100 ± 0	$\mathbf{95 \pm 5}$	$\mathbf{77.3 \pm 14.5}$	67.5 ± 12.1	$\mathbf{83.9 \pm 16.6}$

4.2   AR Database

The AR face database [17] contains over 4,000 color face images of 126 people (70 men and 56 women), including frontal views of faces with different facial expressions, lighting conditions and occlusions. The pictures of 120 individuals (65 men and 55 women) were taken in two sessions (separated by two weeks) and each session contains 13 color images. To demonstrate the performance of our proposed methods for SSPP problem, we use the single image under natural expression and illumination from session 1 for training. The other images of two sessions are used for testing. These 120 individuals are selected to use in our experiment. Some samples of one person are shown in Fig. 6. The images are resized to 32×32 and converted into gray scale.

Fig. 6. Some test samples of a person in AR database. Top row: 8 samples under different expression, illumination and occlusion from session 1; Bottom row: 8 samples from session 2 taken under the same conditions as those in top row.

The classification results on the two sessions are shown in Tables 3 and 4 respectively. One can see from the tables that our method shows superior performance to the other methods. It's not only robust to expression and illumination variations, but also shows great robustness to occlusion. In the experiments of session 1, it leads to at least 10 % improvement to other methods. We also compare it with a special designed method for occlusion (GSRC [19]), which used 7 samples per subject and high resolution images but only achieved a lower recognition rate of 93 % and 79 % in sunglasses and scarves cases. Although there are some blocks with occlusion that will not be classified by LS_LDA correctly,

Table 3. Recognition rates(%) on session 1 of AR database for SSPP problem.

Method	Expression	Illumination	Sunglasses	Scarves	Average
PCA	96.4 ± 1.7	65.3 ± 29.9	91.1 ± 4.2	29.7 ± 7.3	70.6 ± 30.6
PNN	85.8 ± 10.1	91.4 ± 1.9	70.8 ± 12.9	36.9 ± 5.9	71.3 ± 23.4
BlockFLD	72.5 ± 12.0	84.4 ± 3.2	72.5 ± 3.6	41.7 ± 7.5	67.8 ± 17.7
FLDA_single	93.1 ± 3.9	96.1 ± 13	89.4 ± 2.9	43.1 ± 6.5	80.4 ± 22.9
PCRC	84.4 ± 4.9	91.4 ± 2.4	83.6 ± 3.4	63.6 ± 7.9	80.8 ± 11.7
LS_LDA	$\mathbf{96.4 \pm 2.6}$	$\mathbf{98.6 \pm 1.3}$	$\mathbf{96.1 \pm 2.1}$	$\mathbf{82.5 \pm 4.6}$	$\mathbf{93.4 \pm 7.1}$

Table 4. Recognition accuracy(%) on session 2 of AR database for SSPP problem.

Method	Expression	Illumination	Sunglasses	Scarves	Average
PCA	58.9 ± 4.8	35 ± 16.2	50.8 ± 3	18.1 ± 5.7	40.7 ± 18.1
PNN	52.5 ± 11.2	51.9 ± 4.3	34.2 ± 6.0	17.8 ± 2.1	39.1 ± 16.0
BlockFLD	33.1 ± 12.1	41.1 ± 4.6	30.7 ± 5.8	22.5 ± 2.2	31.7 ± 9.3
FLDA_single	57.2 ± 8.0	53.3 ± 1.4	46.1 ± 8.7	20.0 ± 3.6	44.2 ± 16.1
PCRC	43.9 ± 9.4	48.3 ± 3.8	38.3 ± 8.2	32.5 ± 5.2	40.8 ± 8.6
LS_LDA	**66.9 ± 14.4**	**74.2 ± 9.6**	**70 ± 7.6**	**56.9 ± 12.5**	**67 ± 11.7**

there are more blocks without occlusion leading to accurate classification results. These blocks with accurate classification results will suppress the wrong results and achieve the final accurate result. In the experiments of session 2, we can find that the proposed method is also robust to time variation. Compared to other methods, it at least 20 % improvement.

5   Conclusion

In this paper, we propose local similarity based linear discriminant analysis (LS_LDA) to solve SSPP problem. Motivated by the"divide-conquer" strategy, we first divide the face into local blocks, and classify each local block, and then integrate all the classification results to make final decision. To make LDA feasible for the classification of each local block, we further divide each block into overlapped patches and assume that these patches are from the same class. This assumption not only reflects the local structure relationship of the overlapped patches but also makes LDA feasible for SSPP problem. Experimental results on two popular databases show that our method not only generalizes well to SSPP problem but also has strong robustness to expression, illumination, occlusion and time variation. However, the proposed method also relies on the basis that all the training and testing images are well aligned. Therefore, drastic pose variation will decrease their performance. This problem is planned to be solved in our future work.

References

1. Zhao, W., Chellappa, R., Phillips, P.J., Rosenfeld, A.: Face recognition: a literature survey. ACM Comput. Surv. (CSUR) **35**, 399–458 (2003)
2. Kirby, M., Sirovich, L.: Application of the Karhunen-Loeve procedure for the characterization of human faces. IEEE Trans. Pattern Anal. Machine Intell. **12**, 103–108 (1990)
3. Turk, M., Pentland, A.: Eigenfaces for recognition. J. Cogn. Neurosci. **3**, 71–86 (1991)

4. Belhumeur, P., Hespanha, J., Kriegman, D.: Eigenfaces vs. fisherfaces: recognition using class specific linear projection. IEEE Trans. Pattern Anal. Machine Intell. **19**, 711–720 (1997)
5. Martinez, A., Kak, A.: PCA versus IDA. IEEE Trans. Pattern Anal. Machine Intell. **23**, 228–233 (2001)
6. Tan, X., Chen, S., Zhou, Z.H., Zhang, F.: Face recognition from a single image per person: a survey. Pattern Recognit. **39**, 1725–1745 (2006)
7. Stout, Q.F.: Supporting divide-and-conquer algorithms for image processing. J. Parallel Distrib. Syst. **4**, 95–115 (1987)
8. Shan, S., Gao, W., Zhao, D.: Face recognition based on face-specific subspace. Int. J. Imaging Syst. Technol. **13**, 23–32 (2003)
9. Gao, Q.X., Zhang, L., Zhang, D.: Face recognition using FLDA with single training image per person. Appl. Math. Comput. **205**, 726–734 (2008)
10. Martinez, A.: Recognizing imprecisely localized, partially occluded, and expression variant faces from a single sample per class. IEEE Trans. Pattern Anal. Machine Intell. **24**, 748–763 (2002)
11. Chen, S.: Making FLDA applicable to face recognition with one sample per person. Pattern Recognit. **37**, 1553–1555 (2004)
12. Zhu, P., Zhang, L., Hu, Q., Shiu, S.C.K.: Multi-scale patch based collaborative representation for face recognition with margin distribution optimization. In: Fitzgibbon, A., Lazebnik, S., Perona, P., Sato, Y., Schmid, C. (eds.) ECCV 2012, Part I. LNCS, vol. 7572, pp. 822–835. Springer, Heidelberg (2012)
13. Kumar, R., Banerjee, A., Vemuri, B.C., Pfister, H.: Maximizing all margins: pushing face recognition with kernel plurality. In: 2011 International Conference on Computer Vision, IEEE (2011)
14. Deng, W., Hu, J., Guo, J., Cai, W., Feng, D.: Robust, accurate and efficient face recognition from a single training image: a uniform pursuit approach. Pattern Recognit. **43**, 1748–1762 (2010)
15. Deng, W., Hu, J., Zhou, X., Guo, J.: Equidistant prototypes embedding for single sample based face recognition with generic learning and incremental learning. Pattern Recognit. **47**, 3738–3749 (2014)
16. Georghiades, A., Belhumeur, P., Kriegman, D.: From few to many: illumination cone models for face recognition under variable lighting and pose. IEEE Trans. Pattern Anal. Machine Intell. **23**, 643–660 (2001)
17. Martinez, A.M.: The AR face database. CVC Technical Report 24 (1998)
18. Lee, K.C., Ho, J., Kriegman, D.: Acquiring linear subspaces for face recognition under variable lighting. IEEE Trans. Pattern Anal. Machine Intell. **27**, 684–698 (2005)
19. Yang, M., Zhang, L.: Gabor feature based sparse representation for face recognition with Gabor occlusion dictionary. In: Daniilidis, K., Maragos, P., Paragios, N. (eds.) ECCV 2010, Part VI. LNCS, vol. 6316, pp. 448–461. Springer, Heidelberg (2010)

Person Re-identification Using Clustering Ensemble Prototypes

Aparajita Nanda[✉] and Pankaj K. Sa

Department of Computer Science and Engineering,
National Institute of Technology Rourkela, Rourkela, India
512cs102@nitrkl.ac.in

Abstract. This paper presents an appearance-based model to deal with the person re-identification problem. Usually in a crowded scene, it is observed that, the appearances of most people are similar with regard to the combination of attire. In such situation it is a difficult task to distinguish an individual from a group of alike looking individuals and yields an ambiguity in recognition for re-identification. The proper organization of the individuals based on the appearance characteristics leads to recognize the target individual by comparing with a particular group of similar looking individuals. To reconstruct a group of individual according to their appearance is a crucial task for person re-identification. In this work we focus on unsupervised based clustering ensemble approach for discovering prototypes where each prototype represents similar set of gallery image instances. The formation of each prototype depends upon the appearance characteristics of gallery instances. The estimation of k-NN classifier is employed to specify a prototype to a given probe image. The similarity measure computation is performed between the probe and a subset of gallery images, that shares the same prototype with the probe and thus reduces the number of comparisons. Re-identification performance on benchmark datasets are presented using cumulative matching characteristic (CMC) curves.

1 Introduction

Associating individuals across different cameras in a wide coverage space at different instances of time is known as person re-identification. It is a vital task to facilitate cross-camera tracking of people and understanding their global behavior in a wider context. The temporal transition between cameras varies significantly from individual to individual with a great deal of uncertainty. These uncertainty results in images with arbitrary change in pose, variation of illumination, occlusions etc. Figure 1 shows some sample images of individuals captured from two different cameras. It can be seen that there is a significant change in pose and illumination as well. It also demonstrates the difficulty in segmenting the biometric traits like face and iris. This clearly disapproves the use of such traits as prospective candidate for identification. Hence these issues are addressed on a model, that must rely on appearance based features alone. The appearance

© Springer International Publishing Switzerland 2015
C.V. Jawahar and S. Shan (Eds.): ACCV 2014 Workshops, Part III, LNCS 9010, pp. 96–108, 2015.
DOI: 10.1007/978-3-319-16634-6_8

based person re-identification concerns with the establishment of visual correspondence between instances of same individual at different locations and times. Appearance based person re-identification is also considered as non-trivial problem due to visual ambiguities and illumination changes, unknown viewpoint and pose variations, and inter-object occlusions [1].

Fig. 1. Samples from VIPeR Dataset with pose and illumination variations. Top row depicts images of seven different individuals from one angle. Bottom row shows images of same individuals from another angle.

The state-of-art person re-identification methods have majorly focused on two strategies: (i) formulating discriminative feature representations of individuals which are invariant to viewing angle and illuminations [2–4] and (ii) applying learning methods that are capable of making fine distinctions by optimizing the parameters of re-identification model [5]. RankSVM method in [4] aims to find a linear function to weigh the absolute difference of samples through optimization given pairwise relevance constraints. The Probabilistic Relative Distance Comparison (PRDC) [5] shows the probability of a pair of true match having a smaller distance is maximized than that of a wrong matched pair. The requirement of labeled gallery images to discover gallery specific feature importance are described in [3]. The prototype strategy for re-identification problem is introduced in [6] by defining a set of prototypes for each body part and representing them with dissimilarity vectors. A matching strategy is followed between the probe and gallery images with respect to the dissimilarity. However in case of crowded environment where the number individuals appearance increases, the prototypes representation of body parts for each individual is considered as a tedious process for re-identification. In addition the prototype representation and dissimilarity measures remains unsuccessful for individuals wearing same combination of attire. Hence a set of prototypes can be created where each prototype defines a set of image instances, that corresponds to local appearance characteristics shared by different individuals. Most of the existing prototype based approaches [7,8] do follow the simple clustering technique for the prototype formation, whereas these prototypes are not offering a promising representation of features, because for each random initialization of clustering algorithm yields dissimilar prototype labels of representation. Hence the prototypes representation depend upon the selection of random points in clustering algorithm. The better qualitative prototypes representation signifies the qualitative features representation with regard to the commonalities. This motivates the formation of a consensus based prototypes which can be incurred from the ensemble of prototypes representation. The set of consensus prototypes can also termed as clustering ensemble based prototypes that describes the promising representation of features shared by the gallery instances. We depict insights into the optimal prototypes formation of the set of untagged images with given feature sets. The formulation of optimal prototypes representation is assumed as set of consensus prototypes and considered as the best representation of features with respect to commonalities of feature importance of gallery images. The given probe image is classified to a prototype based on its appearance. The similarity measure is computed between the probe and a particular set of gallery images that shares the same prototype.

There exists some commonality in terms of visual features among the instances of gallery representing different individuals. In this work such common features are exploited to form prototypes representing similar instances in the untagged gallery set images. Considering the prototypes as class labels, k-NN classifier assigns a label to each probe image. Similarity measure is computed with a subset of gallery images, that shares the same label with the given probe and hence the number of comparisons are reduced. The resulting scores are listed according to the most similar signature of instances ordered by increasing distance measure. Experimental evaluation of the cluster ensemble based approach is performed on two benchmark datasets.

The paper is structured as follows. In Sect. 2 problem is formulated. Section 3 describes the detail steps of re-identification using cluster ensemble based approach. Experimental evaluation is described in Sect. 4 followed by conclusion in Sect. 5.

2   Problem Formulation

Let $\{Y_i^g\}_{i=1}^n$ be the feature space representing n feature vectors of set of gallery images $\{I_i^g\}_{i=1}^n$. The feature vector of each gallery instance is assumed as signature of the instances. The $\{I_i^g\}_{i=1}^n$ are assigned to $\{P_i\}_{i=1}^K$ prototypes based on the features. For a given probe the objective is to find its corresponding signature in gallery. So for each probe $\{I^{pr}\}$, a prototype P_i is assigned and the matching scores are computed for gallery images $\{I_i^g\}_{i=1}^{n'}$ where $n' \subset n$, the subset instances and the probe shares the same prototype. The gallery and probe images are taken from two different cameras.

3   Cluster Ensemble Based Re-Identification Approach

This section depicts the detail of cluster ensemble based re-identification approach which includes the feature space representation, prototypes formation, classification and similarity measure. The color and texture features are extracted from each image of gallery and represented as feature space. Prototypes are discovered from the feature space based on the appearance characteristics of the gallery instances. Assuming the prototypes as the true label, the feature space is trained by using k-NN classifier. Distance based similarity measure is computed between the subset of gallery images and the given probe image.

3.1   Feature Representation

In case of feature representation, different set of components are extracted from sub parts of an image. Each of the subparts are likely to include sets of local features such as color, texture, interest points or visually discrete features which are able to distinguish an individual from others. This feature sets are considered as the signature of the individual. The principle behind this type of subparts representation is to gain robustness against partial occlusion, pose variations and to roughly captures the body parts. Formally, let $\{I_i^g\}_{i=1}^n$ be the given input of the n untagged gallery set images, where only one image is available for each individual. A d-dimensional feature vector, that is $Feature(I_i^g) = \{y_1,...,y_d\}^T \in R^d$ is extracted from each image instance. Thus $Y = \{Y_i^g\}_{i=1}^n$ represents the feature space of gallery images. Each image of gallery is denoted as an ordered sequence of m parts where $(m \geq 1)$.

$$\{I_i^g\} = \{I_{i,1}^g, \ldots, I_{i,m}^g\} \tag{1}$$

Each part $I_{i,m}^g$ is represented with a set of d' dimensional feature vector $f_{i,m}^{d'}$, $d' \subset d$ and $f_{i,m}^g \in Y$. Where Y denotes the feature space. The feature vectors of all parts are assumed to be represented with same dimensions. In order to roughly capture the head, torso and leg part,the image is partitioned into six equal sized horizontal strips as in [4]. From each strip color features are extracted based on the mixture of color models such as RGB, HS and YCbCr and for texture features 8 Gabor filters [9] and 13 Schmid filters [10] are applied on the luminance channel. The feature vector of each gallery image is integrated to represent the feature space. There exists no single feature, that can be believed to work universally for all instances of images. So the combination of different types of features lead to more discriminant feature space i.e. for individual wearing colorful and bright clothes, the color features yields higher precedence whereas for an individual with high textured clothes, texture features tend to more influencing. To illustrate this, two different images of same individual are considered and the matching rate is computed with regard to different color models, texture features. The matching rates for each feature are determined separately through the average of Euclidean distance measure. Figure 2 shows the matching rate with respect to different types of color and texture features. From Fig. 2 it is observed that, a single feature alone is not able to well perform for all image instances where as the combination of features provide more detailed information.

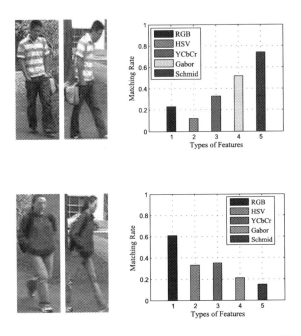

Fig. 2. Matching rate of probe and gallery image on the basis of different color and texture features. RGB, HSV and YCbCr color models are taken for color feature and Gabor and schmid filters are considered for texture feature (Color figure online).

3.2   Cluster Ensemble Based Prototypes Formation

The set of feature vectors is denoted as feature space where each element represents an image instance. The aim of prototype formation is to cluster a set of untagged images with given features into several prototypes representation. Each prototype represents images with similar visual appearance based features such as colors, textures and shapes with colorful shirts, blue jeans, dark jackets or back pack as in Fig. 3. The motivation for prototype formation signifies to distinguish the individuals with similar attire in a crowded environment. Generally a set of prototypes $\{P_i\}_{i=1}^{K}$, is assumed as low-dimensional manifold clusters [11] that group images $\{I_i^g\}_{i=1}^{n}$ with similar appearance based features. In order to formulate the prototypes we construct an ensemble of T prototypes labels with T different random initialization of K-mean clustering algorithm on the feature space Y. Each prototypes label λ^t where $\lambda^t = \bigcup_{a=1}^{K} P_a$ and $P_i \cap P_j = \phi$, is obtained from each random initialization of K mean algorithm where K defines the number of partitions of the input image samples $\{I_i^g\}_{i=1}^{n}$ with respect to their features. We treat the prototype formation problem as a clustering ensemble problem. Cluster ensemble methods have emerged as powerful tools for improving the robustness as well as the accuracy of clusters [12]. The objective of the clustering ensemble task is to search for a combination of multiple prototypes label that provides improved overall prototypes of the given untagged gallery image. Cluster based similarity partitioning algorithm (CSPA) [13] is one of the cluster ensemble technique that can be used for prototypes formation.

Fig. 3. Example of prototype formation on few images of gallery set of VIPeR Dataset. P_1 and P_2 denote the prototypes. P_1 and P_2 represent the images with similar appearances. Based on the feature of the probe image, it only compares with the images belonging to prototype P_1. The green bounding box signifies as true match.

Cluster Based Similarity Partitioning. Clustering algorithm results a prototype label λ^t for the feature space and it signifies a relationship between sample images in the same cluster and can thus be used to establish a measure of pairwise similarity. In order to compute the ensemble prototype formation, for each prototype label λ^t, a co-association matrix is computed. Co-association matrix is a symmetric binary square matrix of size $n \times n$, n being the number of image samples to be classified. The similarity between two sample images is 1 if they are in the same prototype and 0 otherwise.

$$S_{ij} = \frac{1}{T} \sum_{t=1}^{T} I\left(\lambda_i^t, \lambda_j^t\right) \qquad (2)$$

where λ_i^t represents the prototype to which ith sample belongs in prototype label λ^t.

$$I\left(\lambda_i^t, \lambda_j^t\right) = \begin{cases} 1 & (i,j) \in P_a\left(\lambda^t\right) \\ 0 & otherwise \end{cases} \qquad (3)$$

The entry-wise average of T such matrices representing the T sets of groupings yields an overall co-association matrix that is used to re-cluster the sample images, yielding a combined prototype label. The overall similarity matrix is considered as an undirected graph where vertex represents an object and edge weight represents similarities. Given the co-association matrix, a normalized cut algorithm is employed to partition the weighted graph into K clusters. These K clusters are considered as K prototypes. Thus, each untagged probe image is assigned to a prototype P_i. The K value is manually decided by observing the datasets or can be estimated automatically using alternative methods.

Algorithm 1: Formation of cluster ensemble based prototypes

Input   : Ensemble of prototype labels $\{\lambda^t\}_{t=1}^{T}$ where $\lambda^t = \overset{K}{\underset{i=1}{\cup}} P_i$

begin
    for $t = 1$ **to** T **do**
    | Compute the co-association matrix by Eq.3 ;
    end
    Compute the average co-association matrix by Eq.2;
    A normalized cut algorithm is employed to partition the matrix into K
    prototypes;
end
Output: Cluster ensemble prototypes $\{P_i\}_{i=1}^{K}$

3.3   k-NN Classifier

The K prototypes characterized by different appearance characteristics and that are assumed to be the efficient representation of images with similar appearance based features. Moreover each prototype P_i has its own appearance based feature importance which is learned by the k-nearest neighbor. The prototypes that are obtained from the cluster ensemble approach are considered as the class label for the feature set of gallery image. The objective of using k-NN classifier is to assign each untagged probe image $I_i^{p_r}$ to a prototype (class label). So for a given probe image $I_i^{p_r}$ is need to be compared only with a set of gallery images that belongs to the same prototype with the probe image. Thus instead of comparing the probe with all feature vectors of gallery set image, it only compares with the subset of image feature vectors of the prototype that it belongs to and reduces the computational overhead.

Based on the above intuition, we compute the importance of robust prototype assignment of probe according to its ability in discriminating different set of feature vector of image samples. Specifically, we train a k-NN classifier [14] with $\{Y\}$ as inputs and treating the associated prototype labels $\{P_i\}$ as classification outputs. For a given probe image $\{I^{p_r}\}$, we classify it using the learned k-NN classification strategy to obtain its prototype label (class label). Then similarity measure of probe image $\{I^{p_r}\}$ against gallery images $\{I_i^g\}$ of the corresponding prototypes are computed (see Fig. 4).

3.4   Similarity Measure

Given a probe image I^{p_r} is represented as sequence of parts with feature vectors as well, the task is to find the most similar feature vector $x^* \in Y'$, where $Y' \subset Y$, according to similarity measure $D(\cdot, \cdot)$.

$$x^* = \arg\min_{I_i^g} D(I_i^g, I^{p_r}) \tag{4}$$

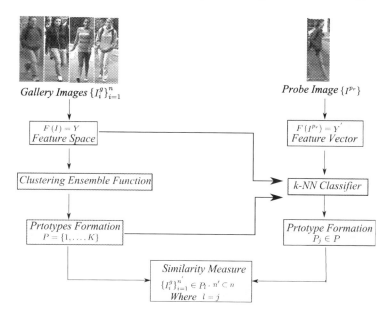

Fig. 4. Overview of cluster ensemble based approach for re-identification

where $D\left(I_i^g, I^{p_r}\right)$ is defined as a function of a similarity measure between sets of feature vectors for sequence of parts. It is the final matching score between a probe I^{p_r} and gallery images I_i^g calculated as the average distance between the image partitions.

$$D\left(I_i^g, I^{p_r}\right) = f\left(dist\left(I_{i,1}^g, I_1^{p_r}\right), \ldots, dist\left(I_{i,m}^g, I_m^{p_r}\right)\right) \tag{5}$$

The measure of $dist\left(\cdot, \cdot\right)$ is the Hausdorff distance d_H [6,15]. Given two set Q and S, d_H is defined as the distance among the minimum distances between all pairs of elements from Q and S. The $dist()$ is defined as the similarity measure of m pairs of parts. For example, $dist()$ can be determined as the additive combination of m distances. The $dist()$ is turned as the kth Hausdroff distance $dist_H$ [16]. The distance $dist_H$ is assumed as kth ranked distance which is the minimum distance between all pair of elements from two sets Q, S.

$$dist_H\left(Q, S\right) = \max\left\{\ h_k\left(Q, S\right),\ h_k\left(S, Q\right)\right\} \tag{6}$$

$$h_k\left(Q, S\right) = kth \min_{q \in Q, s \in S}\left(\|\ q - s\ \|\right) \tag{7}$$

$\|\cdot\|$ denotes the distance metric between the elements of the set. The choice of parameter k is useful for altering robustness to partial occlusions. The value of parameter depends upon the image sub partitions. In our case the images are partitioned into six sub parts and the mixture of color and texture based features are extracted from each part. The partition based distance measure strategy helps to attain robustness to pose variations and experimentally validated in [6].

The result of the similarity measure of the probe is given by the list of the most similar feature vector of the gallery images ordered by increasing dissimilarity. The identity of the probe is determined by finding the gallery images that are most similar to the probe using similarity measure. The similarity measure is computed between probe $\{I^{p_r}\}$ and $\{I^g\}$.

4   Experimental Evaluation

Datasets: Experiments were evaluated on two challenging datasets VIPeR [17], i-LIDS [18] used in former works. The VIPeR dataset consists of 632 pedestrian image pairs taken from two camera views. VIPeR is one of the most promising and challenging dataset with differences in pose, orientation and illumination. It contains only one image for each individual. The i-LIDS dataset contains 476 images of 119 different individuals, captured from disjoint cameras. For i-LIDS the experimental set up of [4] are followed where 208 image pairs from two different camera views are considered. The recognition rates are evaluated with the Cumulative Matching Characteristic (CMC) curves [17]. The CMC curve represents the expectation of finding the correct match in the top rank matches. In other words, a rank recognition rate shows the percentage of the probes that are correctly recognized from the top matches in the gallery images.

Feature representation: Each image was partitioned into six horizontal strips of equal size. Similar to [4,5,19] mixture of color (RGB, HSV and YCbCr) and texture features (8 Gabor filters and 13 Schmid filters) were extracted and forming a 2784-dimensional feature vector for each image. Each feature channel was represented with 16 dimensional feature vector. The Gabor filter used had parameters γ, λ, θ and σ^2 that were set to $(0.3, 0, 4, 2)$, $(0.3, 0, 8, 2)$, $(0.4, 0, 4, 1)$, $(0.4, 0, 4, 1)$, $(0.3, \frac{\pi}{2}, 8, 2)$, $(0.4, \frac{\pi}{2}, 2, 4, 1)$ and $(0.4, \frac{\pi}{2}, 2, 8, 2)$ respectively. The Schmid filters used parameters were set to $(2, 1)$, $(4, 1)$, $(4, 2)$, $(6, 1)$, $(6, 2)$, $(6, 3)$, $(8, 1)$, $(8, 2)$, $(8, 3)$, $(10, 1)$, $(10, 2)$, $(10, 3)$ and $(10, 4)$ respectively.

Comparison Methods: Two different existing ensemble based approaches were demonstrated, that includes the state of art ensemble based Rank SVM [4] and Adaboost based ensemble localized feature [17]. All the three methods were compared using the same image feature set representation with same parameters. The experiments were carried on with 5 random trials and the results were presented by averaging over the trials. The experimented were evaluated for 50 % and 75 % of testing sets of VIPeR and i-LIDS datasets. The impacts of cluster ensemble based approach were evaluated on both the datasets with the above settings of training and testing. The affects of recognition rates on the formation of cluster ensemble based prototypes and cluster based prototypes were also demonstrated. Table 1 represents the comparison of the proposed approach to current approaches on VIPeR dataset and achieves a higher recognition rate in top ranks.

Implementation and Results: In our experiments we follow the same experimental set up of [4]. The number of prototypes depends upon the appearance of

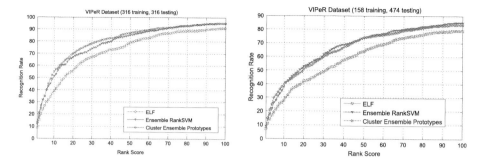

Fig. 5. CMC curves for VIPeR Dataset

individuals in datasets. We manually assume different value of K prototypes for each dataset based on the appearances. For all experiments we fixed $k = 15$ for the k-NN classifiers. Figure 5 presents the CMC curves of VIPeR dataset with 50 % and 25 % of the data used as gallery set images while Fig. 6 show the CMC curves for i-LIDS with 50 % and 75 % of testing sets respectively.

Table 1. The comparison of recognition rates in % with existing approaches on the VIPeR dataset

Ranks	r=1	10	20	50	100
Cluster Ensemble	13	53	67	86	95
ERSVM [4]	12	51	67	85	94
ELF [17]	12	43	60	81	93
PRDC [5]	16	54	70	87	97
LMNN [20]	18	59	75	91	97
MCM [8]	10	32	42	60	72
PS [21]	22	57	71	87	NA

Cluster ensemble based prototypes vs. ensembled rank SVM vs. ELF.
From Fig. 5 it is clearly seen that, by employing the cluster ensemble technique to person re-identification, helps to gain improvement over the ensemble based rank SVM [4] and ensemble localize feature ELF [17]. The performances of distance based learning strategy [17] and the relative ranking scheme degrade, when the individuals with same combination of attire appear in the scene. i.e. images where two individuals wearing combination of white shirt, blue skirt and blue shirt, white skirt, it is difficult to measure the dissimilarity by distance learning or by relative ranking of distance measure between these two images. In contrast the cluster ensemble based approach forms the prototypes from the feature set that best represents the individuals according to their appearance of attire.

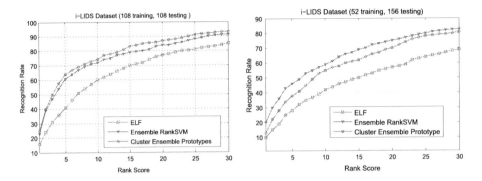

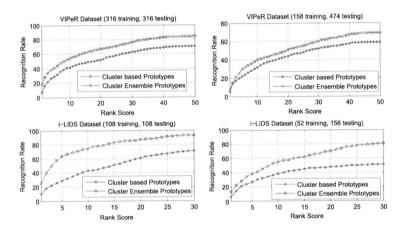

Fig. 6. CMC curves for i-LIDS Dataset

Fig. 7. Recognition rates for VIPeR and i-LIDS datasets by employing cluster based prototypes and cluster ensemble prototypes. The ensemble prototypes are more effective than normal cluster based prototypes. The results are presented by averaging over the five trials.

In case of i-LIDS dataset with 50 % of training and testing sets (Fig. 6), the recognition rates of the cluster ensemble approach seems better than the existing [4,17] approaches. However in Figs. 5 and 6 for 25 % training set, the overall recognition rate decreases due to the less number of training instances, because for a small training data, few number of samples are included for each prototype which affects the recognition rate.

Figure 7 depicts the efficiency of cluster ensemble based prototypes approach over cluster based prototypes. The experimentation were carried out on 50 % and 25 % of testing set for VIPeR and i-LIDS datasets. We followed the same feature set and similarity measure for cluster based prototypes. The CMC curves presented in Fig. 7 is the recognition rate of 5 random trials of K prototypes. For each random trial K-mean algorithm was employed for prototypes formation and the results were presented by averaging over the trials. From Fig. 7 it is observed that, for both of the datasets the cluster ensemble strategy always gives commendable results than the cluster based prototypes.

5 Conclusion

The proposed ensemble based framework for the person re-identification performs well under various challenging conditions. Formation of ensemble based prototypes are able to describe individuals with similar appearance as well as improve the reliability and accuracy under crowded environment. The matching strategy of probe image with a certain group of images, where both shares the same prototype reduces the number of comparisons with gallery images. The proposed approach shows a significant improvement over the existing techniques for re-identification.

The ensemble based prototypes is to be worked out for gaining improvement over recognition rates. The prototype based re-identification will be tested over several other datasets for various challenging condition.

References

1. Doretto, G., Sebastian, T., Tu, P., Rittscher, J.: Appearance-based person re-identification in camera networks: problem overview and current approaches. J. Ambient Intell. Humanized Comput. **2**, 127–151 (2011)
2. Farenzena, M., Bazzani, L., Perina, A., Murino, V., Cristani, M.: Person re-identification by symmetry-driven accumulation of local features. In: IEEE Conference on Computer Vision and Pattern Recognition (2010)
3. Gray, D., Brennan, S., Tao, H.: Evaluating appearance models for recognition, reacquisition and tracking. In: IEEE International Workshop on Performance Evaluation for Tracking and Surveillance 3 (2007)
4. Prosser, B., Zheng, W.S., Gong, S., Xiang, T.: Person re-identification by support vector ranking. In: British Machine Vision Conference, vol. 2 (2010)
5. Zheng, W.S., Gong, S., Xiang, T.: Re-identification by relative distance comparison. IEEE Trans. Pattern Anal. Mach. Intell. **35**, 653–668 (2013)
6. Satta, R., Fumera, G., Roli, F.: Fast person re-identification based on dissimilarity representations. Pattern Recogn. Lett. **33**, 1838–1848 (2012)
7. Satta, R., Fumera, G., Roli, F., Cristani, M., Murino, V.: A multiple component matching framework for person re-identification. In: Maino, G., Foresti, G.L. (eds.) ICIAP 2011, Part II. LNCS, vol. 6979, pp. 140–149. Springer, Heidelberg (2011)
8. Satta, R., Fumera, G., Roli, F.: Exploiting dissimilarity representations for person re-identification. In: Pelillo, M., Hancock, E.R. (eds.) SIMBAD 2011. LNCS, vol. 7005, pp. 275–289. Springer, Heidelberg (2011)
9. Fogel, I., Sagi, D.: Gabor filters as texture discriminator. Biol. Cybern. **61**, 103–114 (1989)
10. Schmid, C.: Constructing models for content-based image retrieval. In: IEEE Computer Society Conference on Computer Vision and Pattern Recognition, vol. 2 (2001)
11. Loy, C.C., Liu, C., Gong, S.: Person re-identification by manifold ranking. In: IEEE International Conference on Image Processing **20** (2013)

12. Topchy, A., Jain, A.K., Punch, W.: Clustering ensembles: Models of consensus and weak partitions. In: IEEE Transactions on Pattern Analysis and Machine Intelligence **27** (2005)

13. Strehl, A., Ghosh, J.: Cluster ensembles-a knowledge reuse framework for combining multiple partitions. J. Mach. Learn. Res. **3**, 583–617 (2003)

14. Joachims, T.: Text categorization with support vector machines: Learning with many relevant features. In: IEEE Computer Society Conference on Computer Vision and Pattern Recognition (1998)

15. Edgar, G.A.: Measure, topology, and fractal geometry. Springer, New York (2008)

16. Wang, J., Zucker, J.D.: Solving multiple-instance problem: A lazy learning approach (2000)

17. Gray, D., Tao, H.: Viewpoint invariant pedestrian recognition with an ensemble of localized features. In: Forsyth, D., Torr, P., Zisserman, A. (eds.) ECCV 2008, Part I. LNCS, vol. 5302, pp. 262–275. Springer, Heidelberg (2008)

18. Zheng, W.S., Gong, S., Xiang, T.: Associating groups of people. British Machine Vision Conference (2009)

19. Hirzer, M., Roth, P.M., Köstinger, M., Bischof, H.: Relaxed pairwise learned metric for person re-identification. In: Fitzgibbon, A., Lazebnik, S., Perona, P., Sato, Y., Schmid, C. (eds.) ECCV 2012, Part VI. LNCS, vol. 7577, pp. 780–793. Springer, Heidelberg (2012)

20. Weinberger, K.Q., Saul, L.K.: Distance metric learning for large margin nearest neighbor classification. J. Mach. Learn. Res. **10**, 207–244 (2009)

21. Hirzer, M., Roth, P.M., Köstinger, M., Bischof, H.: Relaxed pairwise learned metric for person re-identification. In: Fitzgibbon, A., Lazebnik, S., Perona, P., Sato, Y., Schmid, C. (eds.) ECCV 2012, Part VI. LNCS, vol. 7577, pp. 780–793. Springer, Heidelberg (2012)

Automatic Lung Tumor Detection
Based on GLCM Features

Mir Rayat Imtiaz Hossain[✉], Imran Ahmed, and Md. Hasanul Kabir

Department of Computer Science and Engineering,
Islamic University of Technology, Gazipur, Bangladesh
rayat137@iut-dhaka.edu

Abstract. For diagnosis of lung tumors, CT scan of lungs is one of the most common imaging modalities. Manually identifying tumors from hundreds of CT image slices for any patient may prove to be a tedious and time consuming task for the radiologists. Therefore, to assist the physicians we propose an automatic lung tumor detection method based on textural features. The lung parenchyma region is segmented as a preprocessing because the tumors reside within the region. This reduces the search space over which we look for the tumors, thereby increasing computational speed. This also reduces the chance of false identification of tumors. For tumor classification, we used GLCM based textural features. A sliding window is used to search over the lung parenchyma region and extract the features. Chi-Square distance measure is used to classify the tumor. The performance of GLCM features for tumor classification is evaluated with the histogram features.

1 Introduction

Lung cancer is the leading cause of cancer death all over the world [3]. The prevalence of lung cancer is only second to breast cancer in women and prostate cancer in men. About 80 % of the lung cancer patients' present advanced-stage disease (stages III and IV) and are considered inoperable due to loco-regional tumor extension, extra thoracic spread or poor physical condition at the time of diagnosis [4]. If detected at an early stage, life of a cancer patient can be saved.

Manually detecting tumors from CT images of the lungs can be a tedious and time consuming task as the radiologists may have to go thorugh hundreds of slices to predict the tentitive tumorous regions. Automating the process of tumor detcction would assist the physicians to detect tumors tentatively in a much quicker time without having to go through each of the hundred slices manually. However, automating the process is challenging, due to high diversity in appearance of tumor tissue among different patients and in many cases, similarity between tumors and normal tissues because of the low contrast in CT images [1,2].

Computed Tomography is one of the best imaging techniques for soft tissue imaging behind bone structure [5]. A modern multislice CT machine enables the rapid acquisition of precise sets of successive images with very high resolution,

© Springer International Publishing Switzerland 2015
C.V. Jawahar and S. Shan (Eds.): ACCV 2014 Workshops, Part III, LNCS 9010, pp. 109–121, 2015.
DOI: 10.1007/978-3-319-16634-6_9

supporting a more valid diagnosis. CT images help to detect and locate pathological changes. Thus, our purpose is to automate the detection of tumorous
regions from CT images (Fig. 1).

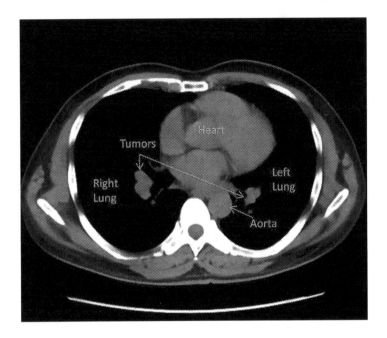

Fig. 1. CT image of lungs.

Because of the ability of the PET-CT scans to determine the stage of tumors
and the high contrast images they produce, many literature works are based
on PET CT images. In their paper, Kanakatte et al. [6] proposed an automatic
lung tumor segmentation from PET CT images using standardized uptake values
(SUV) and connected component analysis. Cherry Ballangan et al. [7] proposed
a tumor-customized downhill method which automatically formulated a tumor-
customized criterion function for improving tumor boundary definition and then
used a monotonic property of the standardized uptake value (SUV) of tumors
to separate the tumor from adjacent regions of increased metabolism. Hui Chui
et al. [8] used neighbourhood grey-tone difference matrix (NGTDM) to calculate contrast features of PET volume in SUV to automatically localize tumor
and then based on analyzing the surrounding CT features of the initial tumor
definition a decision rule was devised. Although, the ability of conventional CT
scan to accurately determine tumors and its stages is limited in comparison to
imaging techniques such PET CT imaging which uses radioactive materials like
Fluro deoxy-glucose (FDG), CT scan is much cheaper and easily obtainable than
PET CT images.

There were several approaches that used semi-automatic click and grow
techniques. Rios Velazquez et al. [9] proposed a semi-automatic click and grow

algorithm for detecting lung lesions from PET CT images. Plajer and Richer [10] proposed a new active contouring algorithm [12] for lung tumor segmentation in CT images where the user needs to specify the initial contours.

Amongst approaches that involved texture analysis Malone et al. [11] proposed a method for identifying diseases in CT image of lungs using 18 textural features and the classification is done using Support Vector Machine (SVM). Wei and Hu [13] used GLCM [17] and GLRLM [19] features for identifying Lobar fissure from Lung CT images using Neural Network. A comparison of using GLCM, GLRLM and wavelet features for segmenting brain tumors from CT images was done by Padma and Sukanesh [14]. Same authors, in their another work [15], used a combination of wavelet transform with GLCM features for brain tumor segmentation. Kadi and Watson [16] proposed a method for differentiating between aggressive and non-aggressive lung tumor using fractal analysis, but the tumorous regions were manually segmented. Although GLCM, GLRLM has been used for brain tumor detection, lobar fissure detection, etc. only a few works has been done to detect lung tumor from CT images using texture analysis.

In this paper, for tumor detection, we are using texture analysis technique. Image texture gives us information about the spatial arrangement of color or intensities in an image or selected region of an image. For texture anlysis, we are using GLCM textural features. We choose GLCM features because it gives us texture pairs with matching second order statistics, which cannot be discriminated by human eye [1]. However, due to textural similarities between normal tissues and tumor, GLCM features alone cannot properly detect tumorous regions. It might yield into lots of false positive values. Therefore, we need to perform our search only within those regions where there is probability of tumor, namely within the lung parenchyma. For this, we segment the lung parenchyma first and then extract the GLCM features. One of the major challenges is to identify tumours that appear to be attached to the chest wall. So segmenting the lung parenchyma properly is an important factor in determining tumor correctly.

One of the major advantages of our approach compared to other approaches is that our method is automatic and does not require the user to specify any seed point or region of interest. Tumorous regions and nodules that are to be detected are found within the lung parenchyma, an area which encompasses only about half of the area of the computed tomography (CT) image slice. Since we are segmenting the lung parenchyma beforehand, the space over which we search for tumors is reduced. So the processing time would be reduced significantly since we would now search only within a specific area. Moreover, the number of false positives would be considerably lower if the lung parenchyma is segmented beforehand.

2   Dataset Description

The CT images of the lungs were collected from Popular Diagnostic Center, Bangladesh. CT scans from 18 patients were collected out of which, 12 were male and 6 were female. All of these patients had tumors and were within an age group of 23 to 77. The slice thickness for the CT scan was 5 mm. The image slices were isotropic and had a resolution of 512×512. For a person with average lung size, an image stack contains more than 200 images.

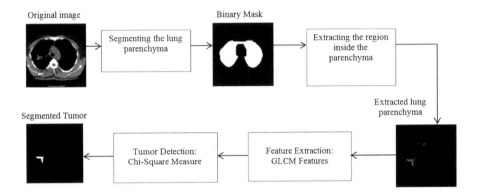

Fig. 2. Overview of our proposed method.

3 Proposed Technique

We proposed a GLCM feature based approach for the detection of lung tumors. First, we segment out the lung parenchyma, using morphological operations, because this is the region where all pathological changes take place. This reduces the search space and also reduces the chance of false detection of tumors because of the textural similarity with some portion of the chest wall and the heart. Once, lung parenchyma is segmented we extract the GLCM features and classify the tumor using a minimum distance classifier. Figure 2 shows the overview of our process illustrating four phases:

- Segmentation of lung parenchyma
- Feature Generation
- Training Phase
- Classification phase

3.1 Segmentation of the Lung Parenchyma

Due to the textural similarities between the tumors, chest walls and the heart GLCM features alone cannot discern them properly. Therefore, proper segmentation of lung parenchyma is important as the tumors reside within these regions. This reduces the space over which we search for the tumor massively as lung parenchyma covers only about half the area of CT image. Moreover, it gets rid of chest walls and the heart and thereby reduces the chance of false positive dections. Figure 3 shows the necessity of segmentation of lung parenchyma segmentation.

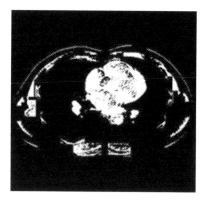 

Fig. 3. Classification using GLCM only without segmenting the lung (Left), Proposed method using GLCM(Right).

From the CT image of lungs it can be observed that the the regions corresponding the lung parenchyma are dark. Therefore, we invert the image and perform global adaptive thresholding [18]. To remove salt and pepper noise that are created due to the thresholding, a 5×5 median filter is used. Once the noise is removed, the connected components are extracted. The largest connected component is removed since it represents the background. The next two largest connected components are kept as they represent the two sides of lung parenchyma. Sometimes due to low contrast in images, the two sides of the lung seem to be attached. In that case, we keep the largest remaining connected component. Since tumors are present within the lung parenchyma they cause holes or intrusions within the large white regions. To fill the holes or intrusions we perform dilation k times and erosion $k + 1$ times where the value of k was emperically determined to be $k = 10$. We ran erosion one time more than dilation to ensure that no part of chest wall or heart falls within the segmented lung parenchyma. Figure 4 shows the result of lung parenchyma segmentation.

In some of the images, the chest wall touches the border of the image. In such cases, the exterior region of the lungs would no longer remain a connected to component. To connect the background, one pixel from each border is padded with zero value. Figure 5 shows the effect of padding zero valued pixels.

3.2   Feature Generation

GLCM Features: Gray Level Co-ocurence Matrix (GLCM) [17] is used to extract the second order statistical texture features. The matrix denoted by $h_{d,\theta}(i,j)$ gives the number of times two pixels with gray level i and j co-occur at a distance d and an angle θ. When divided by the total number of pixels in the image, this estimate $p_{d,\theta}(i,j)$ gives the joint probability of co-ocurence of a pair of pixels.

Four directions are required to describe the texture content in the horizontal $0°$, vertical $90°$, right $45°$ and left-diagonal 135^0 [1] as shown in Fig. 6. A complete

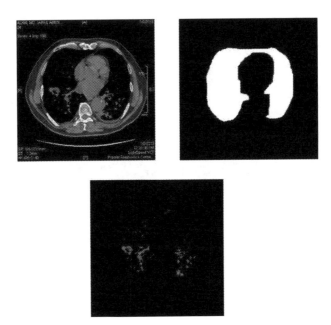

Fig. 4. Original Image (Top Left) , Binary mask corresponding the lung parenchyma (Top Right) , After performing AND Operation (Bottom).

Fig. 5. Binary mask obtained without padding zero (Left), binary Mask obtained after padding zero (Right).

representation of image texture is contained in the co-occurrence matrices calculated in these four directions. Extracting information from these matrices using textural features, which are sensitive to specific elements of texture, provides unique information on the structure of the texture being investigated. Figure 7 shows an example of the calculation of a horizontal co-occurence matrix (at 0°) on a 4×4 image containing four gray levels from 0 to 4.

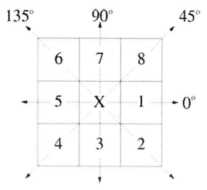

Fig. 6. Eight nearest neighbor pixels used to describe pixel connectivity. Cells 1 and 5 show the horizontal, 4 and 8 the right diagonal, 3 and 7 the vertical and 2 and 6 the left diagonal nearest neighbors.

0	2	2	2
0	3	2	2
1	2	2	2
3	0	3	0

(P_H) Grey-Level

0^0	0	1	2	3
0	0	0	1	4
1	0	0	1	0
2	1	1	10	1
3	4	0	1	0

(Grey-Level along vertical axis)

Fig. 7. Simple example demonstrating the formation of a co-occurrence matrix from an image. Left 4×4 image with four unique grey-levels. Right, the resulting horizontal co-occurrence matrix.

Given an image $f(x, y)$, indicating the pixel intensity value at position (x, y), with a set of G discrete intensity levels, matrix $h_{d,\theta}(i,j)$ is defined in such a way that its (i,j)th entry is equal to the number of times $f(x_1, y_1) = i$ and $f(x_2, y_2) = j$ where (x_1, y_1) and (x_2, y_2) are two pixels such that $(x_2, y_2) = (x_1, y_1) + (d\cos\theta, d\sin\theta)$. This yields a square matrix of dimension equal to the number of intensity levels in the image, for each distance d and orientation θ. The classification of fine textures requires small values of d, whereas coarse textures require large values of d. Reduction of the number of intensity levels (by quantizing the image to fewer levels of intensity) helps increase the speed of computation, with some loss of textural information.

Haralick et al. proposed a set of 14 local features specifically designed for this purpose [17]. In the table we list the features in Table 1. However out of these 14 features we selected 10 features by the method of backward search algorithm [1], because rest of the 4 features are highly correlated. The four features which were not selected are Sum of Squares: Variance, Sum Average, Sum Variance and Maximal Correlation Coefficient. Figure 8 shows the result of using all GLCM features for classification and how selection of 10 featuers make the result better.

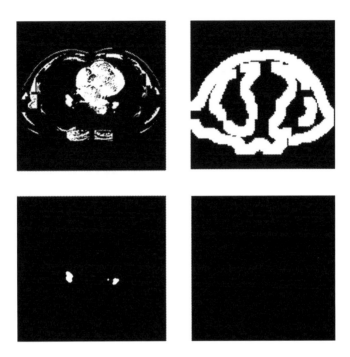

Fig. 8. Using 10 GLCM features (Top Left) without segmenting lung parnechyma, using all the 14 GLCM features (Top Right) without segmenting lung parenchyma, proposed method using 10 GLCM features (Bottom Left), using all 14 GLCM features (Bottom Right).

Table 1. List of GLCM features

Sl No	Features
1	Angular Second Moment
2	Contrast
3	Correlation
4	Sum of Squares: Variance
5	Inverse Difference Moment
6	Sum Average
7	Sum Variance
8	Sum Entropy
9	Entropy
10	Difference Entropy
11	Difference Variance
12	Information measures of Correlation I
13	Information measures of Correlation II
14	Maximal Correlation Coefficient

3.3 Training Phase

For the purpose of training we took 50 patches of tumorous regions as the training set. From these images 10 GLCM features were extracted. In order to generate the GLCM Matrix, Q=16 quantization levels were taken where the 256 grey levels are quantized to 16 grey levels. For each of the training samples, four GLCM matrices were generated for the four directions 0°, 45°, 90° and 135° and a pixel distance of 1 pixel. We calculated the features for each of the four directions and took the average of them. Then the mean feature vector for the training images was calculated. We also applied adaptive thresholding [18] on each of the tumor image to find the maximum and the minimum grey level values of the tumors.

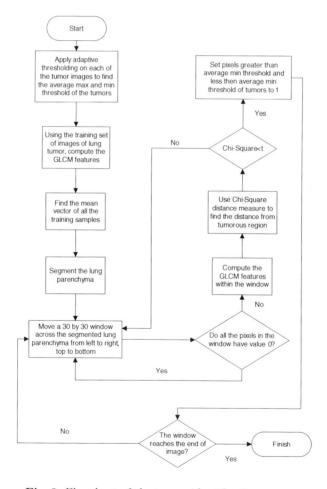

Fig. 9. Flowchart of the tumor identification process.

3.4   Classification Phase

For the purpose of testing 54 images were used, 3 slices from each of the 18 patients. All of these patients had tumors. Classification phase begins by first segmenting out the lung parenchyma of the test image. It is because tumors reside within the lung parenchyma and this reduces the space over which we search for tumor. To search through the segmented lung parenchyma, we have used a sliding window of dimension 30×30 which is slided by 5 pixels in each iteration. If all the pixels within the window are black, the window is simply slided without computing any features within it. Otherwise, the GLCM features within the region are calculated. Then, using chi-square distance measure, the distance from the feature vector of the window to the mean vector is measured. The equation of Chi-Square is given by:

$$\chi^2 = \sum_{j=1}^{J} \frac{(S(j) - M(j))^2}{S(j) + M(j)} \tag{1}$$

where, χ^2 = Chi-square measure, J = No. of features , $M(j)$ = jth feature of mean vector and $S(j)$ = jth feature of the test image patch.

   If the chi-square measure is less than a certain threshold value, t, which is determined emperically, then we can deduce that the window belongs to the tumorous region. However, some part of the window might have non-tumorous pixels. So the pixels which have a gray level intensity within the min and the max threshold value of the tumors are considered to be tumorous.

   Chi-Square methods is preferred over other classifiers like SVM because it is computationally faster and simple, yet produces desirable result for us. The description of our classification process is represented with a flowchart as shown in Fig. 9.

4   Results

We evaluated the performance of our method using GLCM features against the histogram features. For evaluating our results we used Precision, Recall and F-measure. In pattern recoginition and information retrieval, precision (also called positive predictive value) is the fraction of retrieved instances that are relevant, while recall (also known as sensitivity) is the fraction of relevant instances that are retrieved. Both precision and recall are therefore based on an understanding and measure of relevance. Precision can be seen as a measure of exactness or quality, whereas recall is a measure of completeness or quantity. In simple terms, high recall means that an algorithm returned most of the relevant results, while high precision means that an algorithm returned substantially more relevant results than irrelevant. The precision and recall is defined as:

$$Precision = \frac{TP}{TP + FP} \tag{2}$$

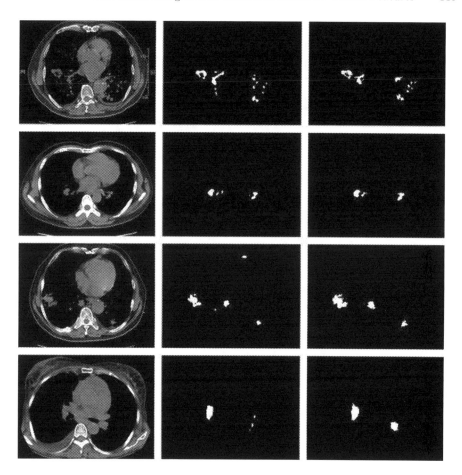

Fig. 10. Original Image (Left), Result of proposed method (center), Ground truth (right).

$$Recall = \frac{TP}{TP + FN} \qquad (3)$$

F-measure is the harmonic mean of precision and recall and is given by:

$$Fmeasure = \frac{2 * precision * recall}{precision + recall} \qquad (4)$$

We implemented our method in MATLAB and executed on Intel Core 2 Duo 2.67 GHz processor with 4.00 GB RAM.

In Fig. 10 we show some of the results that we obtained and compared them with the ground truth. The ground truth were labelled by a radiologist. We can see from Fig. 10 that our method was able to detect tumors properly in most cases, even the ones adhering to the chest walls, heart and pleura.

4.1   Detection Rate of GLCM Features

After we evalating our method using GLCM features with histogram features we found out that GLCM features gives us a precision of 85.5 %, recall of 91 % and f-measure of 88 % compared to a precision of 83.7 %, recall of 85.1 % and a f-measure of 84.4 % for histogram features. The result is shown in Fig. 11.

Although we found that GLCM gives comparativelty better result than the histogram features, it is computationally slower than the histogram feautes. Histogram features took 5.89 s on average. On the other hand GLCM features took 26.57 s on average.

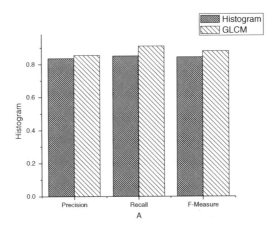

Fig. 11. Result comparison between Histogram and GLCM features.

5   Conclusion

Unlike, previous methods which require the user to specify either the seed points of tumorous regions or the some region of interest, our method can detect tentative tumorous regions automatically. Moreover, since we are segmenting the lung parenchyma initially, our search space is reduced thereby reducing the computational time. Also, morphological operations and GLCM features are computationally fast to compute. Because of the similarity in the texture between tumors and chest walls it is tough to get a desirable result by applying only texture analysis method. Therefore, accurate segmentation of lung parenchyma is an essential task.

References

1. Nailon, W.H.: Texture analysis methods for medical image characterisation. Biomedical Imaging, pp. 75–100. InTech, Vukovar (2010)
2. Castellano, G., Bonilha, L., Li, L.M., Cendes, F.: Texture analysis of medical images. Clin. Radiol. **59**, 1061–1069 (2004)

3. Jemal, A., Siegel, R., Xu, J., Ward, E.: Cancer statistics. CA Cancer J. Clin. **60**, 277–300 (2010)
4. Auperin, A., Le Pechoux, C., Rolland, E., et al.: Meta-analysis of concomitant versus sequential radiochemotherapy in locally advanced non-small-cell lung cancer. J Clin Oncol. **28**, 2181–2190 (2010)
5. von Schulthess, G.K., Steinert, H.C., Hany, T.F.: Integrated PET/CT-3: current applications and future directions. Radiology **238**, 405–422 (2006)
6. Kanakatte, A., Gubbi, J., Mani, N., Kron, T., Binns, D.: A pilot study of automatic lung tumor segmentation from positron emission tomography images using standard uptake values. In: IEEE Symposium on Computational Intelligence In Image and Signal Processing (CIISP), pp. 363–368 (2007)
7. Ballangan, C., Wang, X., Fulham, M., Eberl, S., Yin, Y., Feng, D.: Automated delineation of lung tumors in PET images based on monotonicity and a tumor-customized criterion. IEEE Trans. Inf. Technol. Biomed. **15**, 691–702 (2011)
8. Cui, H., Wang, X., Feng, D.: Automated localization and segmentation of lung tumor from PET-CT thorax volumes based on image feature analysis. In: Conference on Proceedings of IEEE Engineering in Medical and Biology Society, pp. 5384–5387 (2012)
9. Rios Velazquez, E., Aerts, H.J., Gu, Y., Goldgo, D.B., De Ruysscher, D., Dekker, A., Korn, R., Gillies, R.J., Lambin, P.: A semiautomatic CT-based ensemble segmentation of lung tumors: comparison with oncologists' delineations and with the surgical specimen. Radiother. Oncol. **105**, 167–173 (2012)
10. Plajer, I.C., Richter, D.: A new approach to model based active contours in lung tumor segmentation in 3D CT image data. In: Information Technology and Applications in Biomedicine (ITAB), pp. 1–4 (2012)
11. Malone, J., Rossiter, J.M., Prabhu, S., Goddard, P.: Identification of disease in CT of the lung using texture-based image analysis. In: Asilomar Conference on Signals, Systems and Computers, vol. 2, pp. 1620–1624 (2004)
12. Kass, M., Witkin, A., Terzopoulos, D.: Snakes: active contour models. Int. J. Comput. Vis. **1**, 321–331 (1988)
13. Wei, Q., Hu, Y.: A study on using texture analysis methods for identifying lobar fissure regions in isotropic CT images. In; Conference on Proceedings of IEEE Engineering in Medical and Biology Society, pp. 3537–3540 (2009)
14. Padma, A., Sukanesh, R.: Automatic classification and segmentation of brain tumor in CT images using optimal dominant gray level run length texture features. Int. J. Adv. Comput. Sci. Appl. (IJACSA) **2**, 53–59 (2011)
15. Nanthagopal, A.P., Sukanesh, R.: Wavelet statistical texture features-based segmentation and classification of brain computed tomography images. IET Image Process. **7**, 25–32 (2013)
16. Al-Kadi, O.S., Watson, D.: Texture analysis of aggressive and nonaggressive lung tumor CE CT images. IEEE Trans. Biomed. Eng. **55**, 1822–1830 (2008)
17. Haralick, R.M., Shanmugam, K., Dinstein, I.: Textural features for image classification. IEEE Trans. Syst. Man Cybern. **SMC–3**, 610–621 (1973)
18. Gonzalez, R.C., Woods, R.E.: Digital Image Processing, 3rd edn. Pearson Prentice Hall, Upper Saddle River (2008)
19. Tang, X.: Texture information in run-length matrices. IEEE Trans. Image Process. **7**, 1602–1609 (1998)

A Flexible Semi-supervised Feature Extraction Method for Image Classification

Fadi Dornaika[1,2]([✉]) and Youssof El Traboulsi[1]

[1] University of the Basque Country (UPV/EHU), San Sebastian, Spain
fadi.dornaika@ehu.es
[2] IKERBASQUE, Basque Foundation for Science, Bilbao, Spain

Abstract. This paper proposes a novel discriminant semi-supervised feature extraction for generic classification and recognition tasks. The paper has two main contributions. First, we propose a flexible linear semi-supervised feature extraction method that seeks a non-linear subspace that is close to a linear one. The proposed method is based on a criterion that simultaneously exploits the discrimination information provided by the labeled samples, maintains the graph-based smoothness associated with all samples, regularizes the complexity of the linear transform, and minimizes the discrepancy between the unknown linear regression and the unknown non-linear projection. Second, we provide extensive experiments on four benchmark databases in order to study the performance of the proposed method. These experiments demonstrate much improvement over the state-of-the-art algorithms that are either based on label propagation or semi-supervised graph-based embedding.

1 Introduction

Feature extraction with dimensionality reduction is an important step and essential process in embedding data analysis. By computing an adequate representation of data that has a low dimension, more efficient learning and inference [1–4] can be achieved. Although the supervised feature extraction methods had been successfully applied to many pattern recognition applications, they require a full labeling of data samples. It is well-known that it is much easier to collect unlabeled data than labeled samples. The labeling process is often expensive, time consuming, and requires intensive human involvement. As a result, partially labeled datasets are more frequently encountered in real-world problems.

Recently, semi-supervised learning algorithms were developed to effectively utilize limited number of labeled samples and a large amount of unlabeled samples for real-world applications [5,6]. In the past years, many graph-based methods for semi-supervised learning have been developed. The main advantage of graph-based methods is their ability to identify classes of arbitrary distributions. The use of data-driven graphs has led to many progresses in the field of semi-supervised learning (e.g., [7–13]). Toward classification, an excellent subspace should be smooth as well as discriminative. Hence, a graph-theoretic learning

© Springer International Publishing Switzerland 2015
C.V. Jawahar and S. Shan (Eds.): ACCV 2014 Workshops, Part III, LNCS 9010, pp. 122–137, 2015.
DOI: 10.1007/978-3-319-16634-6_10

framework is usually deployed to simultaneously meet the smoothness requirement among nearby points and the discriminative requirement among differently labeled points (e.g., [14]). In addition to the use of partial labelling in
semi-supervised learning, many researchers use pairwise constraints which can
be seen as another form of side information [15].

Despite the success of many graph-based algorithms in dealing with partially labeled problems [16], there are still some problems that are not properly
addressed. Almost all semi-supervised feature extraction techniques can suffer
from one of the following limitations:

1. The non-linear semi-supervised approaches do not have, in general, an implicit
   function that can map unseen data samples. In other words, the non-linear
   methods provide embedding for only the training data. This is the transductive setting, i.e., the test set coincides with the set of unlabeled samples in
   the training dataset. Indeed, solving the out-of-sample extension is still an
   open problem for those techniques adopting non-linear embedding.
2. Almost all proposed semi-supervised approaches target the estimation of a linear transform that maps original data into a low dimensional space. While this
   simplifies the learning processes and gets rid of the out-of-sample problem,
   there is no guarantee that such approaches will be optimal for all datasets.
   The main reason behind this is that the criterion used is already a rigid constraint that contains only the linear mapping. Thus, any coordinate in the
   low-dimensional space is supposed to be a linear combination of the original
   features. Thus, the model has not the flexibility to adapt the linear model to
   a given non-linear model.

In addition to the above limitations, it is not clear what would be the performance of the semi-supervised approaches when minimal labeling is used. In
this paper, we propose an Inductive Flexible Semi Supervised Feature Extraction. The aim is to combine the merits of Flexible Manifold Embedding and the
non-linear graph-based embedding. The proposed method will be flexible since
it estimates a non-linear manifold that is the closest one to a linear embedding.
The non-linear manifold and the mapping are simultaneously estimated. The
dimension of the final embedding obtained by our proposed method is not limited to the number of classes. This allows the application of any kind of classifiers
once the data are embedded in new sub-spaces. Unlike nonlinear dimensionality
reduction approaches which suffer from the out-of-sample problem, our proposed
method has an obvious advantage that the learnt subspace has a direct out-of-
sample extension to novel samples, and are thus easily generalized to the entire
high-dimensional input space.

The paper is structured as follows. In Sect. 2, we briefly review the main
methods for semi-supervised learning including the graph-based label propagation and the semi-supervised embedding methods. In Sect. 3, we introduce the
IFSSFE method. Section 4 states the differences between the proposed method
and the existing ones. Section 5 contains the experimental results obtained with
four public datasets. This section compares the performance of the proposed

method with the that of the competing methods. Finally, in Sect. 6 we present our conclusions. In the sequel, capital bold letters denote matrices and small bold letters denote vectors.

2   Related Work

In order to make the paper self-contained, this section will briefly describe some state-of-the art semi-supervised methods.

2.1   Notation and Preliminaries

We define the training data matrix as $\mathbf{X} = [\mathbf{x}_1, \mathbf{x}_2, ..., \mathbf{x}_l, \mathbf{x}_{l+1}, ..., \mathbf{x}_{l+u}] \in \mathbb{R}^{D \times (l+u)}$, where $\mathbf{x}_i|_{i=1}^{l}$ and $\mathbf{x}_i|_{i=l+1}^{l+u}$ are the labeled and unlabeled samples, respectively, with l and u being the total numbers of labeled and unlabeled samples, D being the feature dimension, and $N = l + u$ being the total number of training samples. Let n_c be the total number of labeled samples in the cth class and represent the labeled samples as $\mathbf{X}_{\mathcal{L}} = [\mathbf{x}_1, \mathbf{x}_2, ..., \mathbf{x}_l] \in \mathbb{R}^{D \times l}$ with the label of $\mathbf{x}_i$ as $y_i \in 1, 2, ..., C$, where C is the total number of classes. Let $\mathbf{S} \in \mathbb{R}^{(l+u) \times (l+u)}$ as the graph similarity matrix with $S(i, j)$ representing the similarity between $\mathbf{x}_i$ and $\mathbf{x}_j$, i.e., $S(i, j) = sim(\mathbf{x}_i, \mathbf{x}_j)$. In a supervised context, one can also consider two similarity matrices $\mathbf{S}_w$ and $\mathbf{S}_b$ that encode the within class and between class graphs, respectively. For each similarity matrix, a Laplacian matrix can be computed. For the similarity matrix $\mathbf{S}$, the Laplacian matrix is given by: $\mathbf{L} = \mathbf{D} - \mathbf{S}$ where $\mathbf{D}$ is a diagonal matrix whose elements are the row (or column since the similarity matrix is symmetric) sums of $\mathbf{S}$ matrix. Similar expression can be found for $\mathbf{L}_b$ and $\mathbf{L}_w$. The normalized Laplacian $\hat{\mathbf{L}}$ is defined by $\hat{\mathbf{L}} = \mathbf{I} - \mathbf{D}^{-1/2} \mathbf{S} \mathbf{D}^{-1/2}$ where $\mathbf{I}$ denotes the identity matrix.

We also define a binary label matrix $\mathbf{Y} \in \mathbb{B}^{N \times C}$ associated with the samples with $Y(i, j) = 1$ if $\mathbf{x}_i$ has label $y_i = j$; $Y(i, j) = 0$, otherwise. In addition to $\mathbf{Y}$, we can define an unknown label matrix denoted by $\mathbf{F} \in \mathbb{R}^{N \times C}$. In a semi-supervised setting, $\mathbf{F} = \begin{pmatrix} \mathbf{F}_{\mathcal{L}} \\ \mathbf{F}_{\mathcal{U}} \end{pmatrix}$ where $\mathbf{F}_{\mathcal{L}} = \mathbf{Y}_{\mathcal{L}}$.

2.2   Graph-Based Label Propagation Methods

In the last decade, the SSL graph-based label propagation methods attracted much attention. All of them impose that samples with high similarity should share similar labels. They differ by the regularization term as well as by the loss function used for fitting label information associated with the labeled samples. All of these methods take as input the weighted graph $\mathbf{S}$ associated with data and the label matrix $\mathbf{Y}$. The state-of-the art label propagation algorithms (can also be called classifiers [17]) can be: Gaussian Fields and Harmonic Functions (**GFHF**) [18], Local and Global Consistency (**LGC**) [19], Laplacian Regularized Least Square (**LapRLS**) [20], Robust Multi-class Graph Transduction (**RMGT**) [21], and Flexible Manifold Embedding (**FME**) [22].

Gaussian Fields and Harmonic Functions. The GFHF algorithm [18] solves the following optimization problem:

$$\min_{\mathbf{F}} \sum_{i,j} ||\mathbf{F}_{i.} - \mathbf{F}_{j.}||^2 S_{ij} = \min_{\mathbf{F}} trace(\mathbf{F}^T \mathbf{L} \mathbf{F}) \;\; s.t. \;\; \mathbf{F}_{\mathcal{L}} = \mathbf{Y}_{\mathcal{L}}$$

Given the graph affinity matrix $\mathbf{S}$ as well as the known labels $\mathbf{Y}_{\mathcal{L}} \in \mathbb{R}^{l \times C}$, the goal is to derive the labels of unlabeled samples, $\mathbf{F}_{\mathcal{U}} \in \mathbb{R}^{u \times C}$. It can be shown that the matrix of unknown labels is given by:

$$\mathbf{F}_{\mathcal{U}} = -\mathbf{L}_{\mathcal{U}\mathcal{U}}^{-1} \mathbf{L}_{\mathcal{U}\mathcal{L}} \mathbf{Y}_{\mathcal{L}} \tag{1}$$

where $\mathbf{L}_{\mathcal{U}\mathcal{U}}$ and $\mathbf{L}_{\mathcal{U}\mathcal{L}}$ are submatrices of the Laplacian matrix $\mathbf{L}$:

$$\mathbf{L} = \begin{pmatrix} \mathbf{L}_{\mathcal{L}\mathcal{L}} & \mathbf{L}_{\mathcal{L}\mathcal{U}} \\ \mathbf{L}_{\mathcal{U}\mathcal{L}} & \mathbf{L}_{\mathcal{U}\mathcal{U}} \end{pmatrix}$$

Local and Global Consistency. The Local and Global Consistency algorithm [19] solves the following optimization problem:

$$\min_{\mathbf{F}} [trace(\mathbf{F}^T \hat{\mathbf{L}} \mathbf{F}) + \mu \, trace((\mathbf{F} - \mathbf{Y})^T (\mathbf{F} - \mathbf{Y}))]$$

which gives the closed-form solution:

$$\mathbf{F} = (\mathbf{I} + \hat{\mathbf{L}}/\mu)^{-1} \mathbf{Y}$$

Robust Multi-class Graph Transduction (RMGT). The RMGT algorithm solves the convex optimization problem $\min_{\mathbf{F}} trace(\mathbf{F}^T \mathbf{L} \mathbf{F})$ s.t. $\mathbf{F}_{\mathcal{L}} = \mathbf{Y}_{\mathcal{L}}$, $\mathbf{F} \mathbf{1}_C = \mathbf{1}_N$, $\mathbf{F}^T \mathbf{1}_N = N\mathbf{\Omega}$, where the vector $\mathbf{\Omega} \in \mathbb{R}^C$ is the class prior probabilities. The solution of this optimization problem is given by:

$$\mathbf{F}_{\mathcal{U}} = -\mathbf{L}_{\mathcal{U}\mathcal{U}}^{-1} \mathbf{L}_{\mathcal{U}\mathcal{L}} \mathbf{Y}_{\mathcal{L}} + \frac{\mathbf{L}_{\mathcal{U}\mathcal{U}}^{-1} \mathbf{1}_u}{\mathbf{1}_u^T \mathbf{L}_{\mathcal{U}\mathcal{U}}^{-1} \mathbf{1}_u} (N\mathbf{\Omega}^T - \mathbf{1}_l^T \mathbf{Y}_{\mathcal{L}} + \mathbf{1}_u^T \mathbf{L}_{\mathcal{U}\mathcal{U}}^{-1} \mathbf{L}_{\mathcal{U}\mathcal{L}} \mathbf{Y}_{\mathcal{L}})$$

Laplacian RLS. The linear LapRLS defines a linear regression function that maps a feature vector $\mathbf{x}$ to its label representation $\mathbf{Y}_{i.}$, i.e., $\mathbf{Y}_{i.} = \mathbf{W}^T \mathbf{x}_i + \mathbf{b}$. The term Laplacian is due to the fact that the regularization term contains the classic Laplacian smoothing criterion. The linear LapRLS estimates the linear transform by optimizes the following criterion:

$$g(\mathbf{W}, \mathbf{b}) = \sum_{i=1}^{l} ||\mathbf{W}^T \mathbf{x}_i + \mathbf{b} - \mathbf{Y}_{i.}||^2 + \lambda_A ||\mathbf{W}||^2 + \lambda_I \, trace(\mathbf{W}^T \mathbf{X} \mathbf{L} \mathbf{X}^T \mathbf{W}) \tag{2}$$

where the two coefficients λ_A and λ_I balance the norm of $\mathbf{W}$, the manifold smoothness and the regression error. The closed-form solution is given by:

$$\mathbf{W} = (\lambda_I \mathbf{X} \mathbf{L} \mathbf{X}^T + \mathbf{X}_{\mathcal{L}} \mathbf{X}_{\mathcal{L}}^T + \lambda_A \mathbf{I})^{-1} \mathbf{X}_{\mathcal{L}} \mathbf{Y}_{\mathcal{L}}$$
$$\mathbf{b} = \mathbf{Y}_{\mathcal{L}}^T \mathbf{1}_l - \mathbf{W}^T \mathbf{X}_{\mathcal{L}} \mathbf{1}_l$$

Flexible Manifold Embedding (FME). Flexible Manifold Embedding can be seen as a flexible variant of non-linear embedding where the embedding is given by the label distribution. FME simultaneously estimates the non-linear embedding of unlabel samples and the linear regression over these non-linear representations. In other words, FME can be seen as a framework that merges LGC and LapRLS in order to solve the out-of-sample extension problem. Compared with LapRLS, FME does not force the prediction labels to lie in the space spanned by all the samples. Therefore, it can be more flexible and it can better cope with the samples which reside on the nonlinear manifold. This framework simultaneously estimates the label matrix as well as a linear mapping by minimizing the following criterion:

$$g(\mathbf{F}, \mathbf{W}, \mathbf{b}) = trace((\mathbf{F} - \mathbf{Y})^T \mathbf{U}(\mathbf{F} - \mathbf{Y})) + trace(\mathbf{F}^T \mathbf{L}\, \mathbf{F})$$
$$+ \mu \left(\|\mathbf{W}\|^2 + \gamma \, \|\mathbf{X}^T \mathbf{W} + \mathbf{1}_N \mathbf{b}^T - \mathbf{F}\|^2 \right)$$

where μ and γ are two balance parameters, and $\mathbf{U}$ is a diagonal matrix whose first l diagonal elements are set to one and the rest $N - l$ are set to zero. As can be seen, the above criterion has four terms: the first is a fitting term over the labeled sample, the second is the smoothing term over all samples, the third is a regularization term, and the fourth term is the regression term. The sought solution $(\mathbf{F}, \mathbf{W}, \mathbf{b})$ is found by minimizing the above criterion. By vanishing the derivative of g with respect to $\mathbf{W}$ and $\mathbf{b}$, a relation between $\mathbf{F}$ and $\mathbf{W}$ can be obtained. Then, by vanishing the derivative with respect to $\mathbf{F}$ a closed form solution can be obtained. This is given by:

$$\mathbf{F} = (\mathbf{U} + \mathbf{L} + \mu\,\gamma\,\mathbf{H}_c - \mu\,\gamma^2\,\mathbf{Q})^{-1}\mathbf{U}\,\mathbf{Y}$$

with $\mathbf{Q} = \mathbf{X}_c^T \mathbf{X}_c\,(\gamma\,\mathbf{X}_c^T \mathbf{X}_c + \mathbf{I})^{-1}$ where $\mathbf{X}_c$ is the centered data matrix and $\mathbf{H}_c = \mathbf{I} - \frac{1}{N}\mathbf{1}_N \mathbf{1}_N^T$ is the centering matrix.

2.3   Graph-Based Embedding Methods

Unlike label propagation techniques that seek label inference, the embedding techniques seek a general coordinate representation where the dimension of the mapped data is not necessarily limited to the number of classes. Two main techniques represent the state-of-the art in semi-supervised graph-based embedding:

Semi-supervised Discriminant Analysis (SDA). Cai et al. extended LDA to SDA [23] by adding a geometrically-based regularization term in the objective function of LDA. The core assumption in SDA is still the manifold smoothness assumption, namely, nearby points will have similar representations in the lower-dimensional space. We define as the data matrix of labeled data $\mathbf{X}_{\mathcal{L}} = [\mathbf{x}_1, \mathbf{x}_2, ..., \mathbf{x}_l]$. LDA can be seen as a particular case of a graph-based embedding.

Semi-supervised Discriminant Embedding (SDE). SDE can be seen as the semi-supervised variant of the Local Embedding (LDE) method [24]. In order to discover both geometrical and discriminant structure of the data manifold,

SDE [25,26] relies on three graphs: the within-class graph G_w (intrinsic graph), the between-class graph G_b (penalty), and the graph defined over the whole set (labeled and unlabeled samples).

3   Inductive Flexible Semi Supervised Feature Extraction (IFSSFE)

In this section, we propose an Inductive Flexible Semi Supervised Feature Extraction (IFSSFE) that can combine the merits of Flexible Manifold Embedding idea and the non-linear graph based embedding. It should be noticed that the dimension of the final embedding is not limited to the number of class. We assume that the non-linear embedding of the seen data samples is given by the matrix $\mathbf{Z} \in \mathbb{R}^{N \times d}$, i.e., the row vector $\mathbf{Z}_{i.}$ is the non-linear representation of the vector $\mathbf{x}_i$. We consider again the within class and between class graphs associated with the labeled data as well as the graph associated the labeled and unlabeled data. The expression of the criteria associated with the non-linear Semi-Supervised Discriminant Embedding will be given by $\min_{\mathbf{Z}} trace(\mathbf{Z}^T \widetilde{\mathbf{L}}_w \mathbf{Z})$ $\max_{\mathbf{Z}} trace(\mathbf{Z}^T \widetilde{\mathbf{L}}_b \mathbf{Z}) \min_{\mathbf{Z}} trace(\mathbf{Z}^T \mathbf{L} \mathbf{Z})$:

By combining the above criteria together with the regression and regularization terms we can define a criterion that should be minimized. This is given by:

$$e(\mathbf{Z}, \mathbf{W}, \mathbf{b}) = trace(\mathbf{Z}^T \mathbf{L} \mathbf{Z}) + trace(\mathbf{Z}^T \widetilde{\mathbf{L}}_w \mathbf{Z}) - \lambda \, trace(\mathbf{Z}^T \widetilde{\mathbf{L}}_b \mathbf{Z}) + \quad (3)$$
$$\mu \left(\|\mathbf{W}\|^2 + \gamma \|\mathbf{X}^T \mathbf{W} + \mathbf{1}_N \mathbf{b}^T - \mathbf{Z}\|^2 \right)$$
$$= trace(\mathbf{Z}^T \mathbf{L}_1 \mathbf{Z}) + \mu \left(\|\mathbf{W}\|^2 + \gamma \|\mathbf{X}^T \mathbf{W} + \mathbf{1}_N \mathbf{b}^T - \mathbf{Z}\|^2 \right) \quad (4)$$

where $\mathbf{L}_1 = \mathbf{L} + \widetilde{\mathbf{L}}_w - \lambda \widetilde{\mathbf{L}}_b$, μ, γ, and λ are three positive balance parameters.

The non-linear embedding as well as the regression should be estimated such that e is minimized. To obtain the optimal solution, we vanish the derivatives of the objective function e with respect to $\mathbf{W}$ and $\mathbf{b}$. We have:

$$\mathbf{b} = \frac{1}{N}(\mathbf{Z}^T \mathbf{1}_N - \mathbf{W}^T \mathbf{X} \mathbf{1}_N) \quad (5)$$

$$\mathbf{W} = \gamma \left(\gamma \mathbf{X}_c \mathbf{X}_c^T + \mathbf{I} \right)^{-1} \mathbf{X}_c \mathbf{Z} = \mathbf{A} \mathbf{Z} \quad (6)$$

where $\mathbf{A} = \gamma \left(\gamma \mathbf{X}_c \mathbf{X}_c^T + \mathbf{I} \right)^{-1} \mathbf{X}_c$. We use the above expression for $\mathbf{W}$ and $\mathbf{b}$ in the regression function $\mathbf{X}^T \mathbf{W} + \mathbf{1}_N \mathbf{b}^T$, we get:

$$\mathbf{X}^T \mathbf{W} + \mathbf{1}_N \mathbf{b}^T = \mathbf{X}^T \mathbf{A} \mathbf{Z} + \frac{1}{N} \mathbf{1}_N \mathbf{1}_N^T \mathbf{Z} - \frac{1}{N} \mathbf{1}_N \mathbf{1}_N^T \mathbf{X}^T \mathbf{A} \mathbf{Z}$$
$$= (\mathbf{I} - \frac{1}{N} \mathbf{1}_N \mathbf{1}_N^T) \mathbf{X}^T \mathbf{A} \mathbf{Z} + \frac{1}{N} \mathbf{1}_N \mathbf{1}_N^T \mathbf{Z}$$
$$= \mathbf{H}_c \mathbf{X}^T \mathbf{A} \mathbf{Z} + \frac{1}{N} \mathbf{1}_N \mathbf{1}_N^T \mathbf{Z} = \mathbf{B} \mathbf{Z}$$

with $\mathbf{B} = \mathbf{H}_c \mathbf{X}^T \mathbf{A} + \frac{1}{N}\mathbf{1}_N \mathbf{1}_N^T$. Thus, the criterion e becomes:

$$
\begin{aligned}
e(\mathbf{Z}, \mathbf{W}, \mathbf{b}) &= trace(\mathbf{Z}^T \mathbf{L}_1 \mathbf{Z}) + \mu\,(trace(\mathbf{Z}^T \mathbf{A}^T \mathbf{A}\,\mathbf{Z}) + \gamma\,trace((\mathbf{B}\,\mathbf{Z} - \mathbf{Z})^T (\mathbf{B}\,\mathbf{Z} - \mathbf{Z})) & (7) \\
&= trace(\mathbf{Z}^T (\mathbf{L}_1 + \mu\,\mathbf{A}^T \mathbf{A} + \mu\gamma(\mathbf{B} - \mathbf{I})^T (\mathbf{B} - \mathbf{I}))\,\mathbf{Z}) & (8) \\
&= trace(\mathbf{Z}^T (\mathbf{L}_1 + \mathbf{E})\,\mathbf{Z}) & (9)
\end{aligned}
$$

where $\mathbf{E} = \mu\,\mathbf{A}^T \mathbf{A} + \mu\gamma(\mathbf{B} - \mathbf{I})^T (\mathbf{B} - \mathbf{I}))$.

Thus, the non-linear embedding $\mathbf{Z}$ is estimated by minimizing the above criterion under a constraint in order to avoid the trivial solution $\mathbf{Z} = \mathbf{0}$.

$$
\mathbf{Z}^\star = \arg\min_{\mathbf{Z}} trace(\mathbf{Z}^T (\mathbf{L}_1 + \mathbf{E})\,\mathbf{Z})\;\; s.t.\;\; \mathbf{Z}^T \mathbf{Z} = \mathbf{I}
$$

Thus $\mathbf{Z}^\star$ is given by the eigenvectors of $\mathbf{L}_1 + \mathbf{E}$ associated with the smallest eigenvalues. Once $\mathbf{Z}^\star$ is estimated the corresponding regression $\mathbf{W}^\star$ and $\mathbf{b}^\star$ are estimated by Eqs. (6) and (5).

Given an unseen sample $\mathbf{x}_{test}$ its embedding (a column vector) is given by $\mathbf{z}_{test} = \mathbf{W}^{\star T} \mathbf{x}_{test} + \mathbf{b}^\star$.

4   Difference Between the Proposed Method and Existing Methods

Obviously, our proposed flexible method has several advantages compared with existing methods. Indeed it can combine the merits of graph-based semi-supervised label propagation and those of graph-based semi-supervised embedding methods. The advantages are as follows. First, unlike the FME method which estimates label distributions, our method estimates a non-linear embedding whose dimension is not limited to the number of classes as it is the case with many frameworks adopting the label propagation algorithm. Second, the proposed method is a kind of a non-linear feature extractor that lends itself nicely to all machine learning tools that can be used in the output space with any dimension in order to infer the class (classification) or the continuous label (regression). Third, the method is still inductive in the sense that it can work with unseen data. Fourth, it inherits the flexibility of FME in the sense that a non-linear embedding and a regression are found such that the non-linear embedding is close to the linear one obtained by regression (see Fig. 1). Thus, the proposed method can better cope with the data sampled from a certain type of nonlinear manifold that is somewhat close to a linear subspace.

5   Performance Evaluation

We test our proposed method on four datasets. In our experiments, we use three face datasets Extended Yale, FacePix and FERET, and one object database (COIL-20).

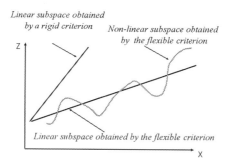

Fig. 1. An illustration of the difference between a rigid linear embedding and the proposed flexible scheme IFSSFE.

5.1 Datasets

– **Extended Yale**[1]: We use the cropped version contains 1774 face images of 28 individuals. The images of the cropped version contain illumination variations and facial expression variations. The image size is 192×168 pixels with 256-bit grey scale. The images are rescaled to 32×32 pixels in our experiments.
– **FERET**[2]: We use a subset of FERET database, which includes 1400 images of 200 distinct subjects, each subject has seven images. The subset involves variations in facial expression, illumination and pose. In our experiment, the facial portion of each original image is cropped automatically based on the location of eyes and resized to 32×32 pixels.
– **FacePix**[3]: This database includes a set of face images with pose angle variations. It is composed of 181 face images (representing yaw angles from −90° to +90° at 1 degree increments) of 30 different subjects, with a total of 5430 images. We used a subset of this dataset in which each person has 18 images.
– **COIL-20**[4]: This dataset (Columbia Object Image Library) consists of 1440 images of 20 objets. Each object has underwent 72 rotations (each object has 72 images). The objects display a wide variety of complex geometry and reflectance characteristics. We used a subset of the database with 18 images for each object (one image for every 20 degree of rotation).

5.2 Semi-supervised Learning and Empirical Setting

We compare our proposed method with GFHF, Class Mass Normalized GFHF (GFHF+CMN), RMGT, LapRLS, SDA, SDE, FME, and LDA. It should be noted that all these methods are semi-supervised except LDA which is supervised. For the embedding methods (LDA, SDA, SDE, and the proposed method),

---

[1] http://vision.ucsd.edu/~leekc/ExtYaleDatabase/ExtYaleB.html.
[2] http://www.itl.nist.gov/iad/humanid/feret/.
[3] http://www.facepix.org/.
[4] http://www.cs.columbia.edu/CAVE/software/softlib/coil-20.php.

any classifier can be used with the obtained mapped data in order to classify the unlabeled and unseen data samples. Since all compared semi-supervised methods used the graph Laplacian $\mathbf{L}$ associated with the training data, the graph was constructed using the classic KNN graph (symmetric KNN) and the RBF (or Gaussian) kernel for the edge weights. The weight associated with each neighboring pair is given by $S(\mathbf{x}_i, \mathbf{x}_j) = exp(-||\mathbf{x}_i - \mathbf{x}_j||^2/t_0)$ where $t_0 \in \mathbb{R}^+$ is the kernel bandwidth parameter. It is set as in many works to the average of squared distances in the training set. The values of neighborhood size was set to 10. For the proposed method, we need to compute the within-class and in between class graph (built on the labeled subset). The weights associated are set to ones or zeros, i.e. the corresponding similarity matrices $\mathbf{S}_b$ and $\mathbf{S}_w$ are binary matrices. It is worthy noting that all compared methods used the same data graph. This makes sure that the difference in performance is due to the embedding method only and not to the data graph.

We randomly select 50 % data as the training dataset and use the remaining 50 % data as the test dataset. Among the training data, we randomly label P samples per class and treat the other training samples as unlabeled data. The above setting is a natural setting to compare different methods. All the training data (labeled and unlabeled samples) are used to learn a subspace (i.e., a projection matrix) for semi-supervised embedding methods or a classifier for the label propagation methods, except that we only use the labeled data for subspace learning in LDA. In all the experiments, PCA is used as a preprocessing step to preserve 98 % energy of the data.

5.3   Method Comparison

For LapRLS, SDA, SDE, FME, two regularization parameters should be tuned. For our proposed method three parameters are used. Each of these parameters is set to a subset of values belonging to $\{10^{-9}, 10^{-6}, 10^{-3}, 1, 10^3, 10^6, 10^9\}$ as in [22], and then we report the top-1 recognition accuracy (best average recognition rate) from the best parameter configuration. Table 1 reports the best mean recognition accuracy (for the four datasets) over ten random splits on the unlabeled data and the test data, which are referred to as Unlabel and Test, respectively. For the embedding methods (LDA, SDA, SDE, IFSSFE), the classification was performed using the Nearest Neighbor classifier.

For the proposed method (IFSSFE), the dimension of the embedding is bounded by the number of training samples N. Thus, for each parameter configuration associated with the criterion and for each split we have a curve for the recognition rate that depicts the rate at several sampled dimensions. Thus, for each parameter configuration, the performance is set to best rate in the mean curve which was obtained by averaging the rate curves over the splits.

Figure 2 illustrates the average recognition rate curves as a function of feature dimensions. These curves were obtained for the test part of data using one labeled sample per class. We recall that FME method does not depend on the dimension, the maximum dimension of SDA method is given by $C - 1$, and the maximum

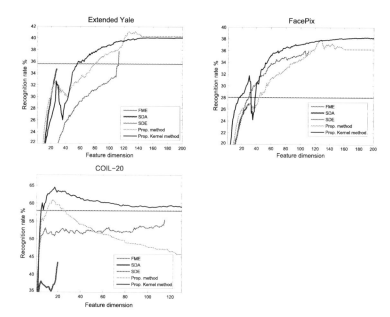

Fig. 2. Recognition accuracy variation as a function of dimensions for Extended Yale, FacePix, and COIL-20 datasets. These curves correspond to the best average curves. One labeled sample per class is used.

dimension of SDE is given by the dimension of input samples. For the proposed method, the maximum dimension is given by the number of training samples.

From the results depicted in Table 1 and Fig. 2, we can draw the following conclusions:

– In general, the proposed method IFSSFE has given the best recognition rate.
– For the non-face dataset (COIL-20), the improvement obtained by the proposed method was very significant compared with the performance of FME and the embedding methods SDA, and SDE. This holds true for all label percentages and for unlabeled and test data.
– For some datasets, the performance obtained with the test part of the data was better than the performance obtained with the unlabeled part. This can be explained by the fact that the captured model has a high generalization capacity.
– The performance of GFHF, GFHF+CMN, and RMGT (direct label propagation methods) was not that good for face datasets.
– More importantly, we can observe that the optimal performance of IFSSFE can be reached with a relatively low dimension. This property makes the proposed method very appealing in practice. Indeed, one needs to find a trade-off between a high recognition rate and a compact representation with a reduced number of dimensions.

Table 1. The best average classification results on ten random splits. GFHF, GFHF+CMN, RMGT, LapRLS, and FME are based on label propagation. LDA, SDA, SDE, and the two proposed method are embedding methods for which nearest neighbor classifier was used after the projection.

Ext Yale	1 labeled sample		2 labeled samples		3 labeled samples	
Method	Unlabel(%)	Test(%)	Unlabel(%)	Test(%)	Unlabel(%)	Test(%)
LDA	36.6	34.8	56.9	55.2	64.4	60.1
GFHF	19.0	-	37.3	-	42.9	-
GFHF+CMN	26.3	-	41.0	-	45.9	-
RMGT	23.0	-	40.5	-	45.6	-
LapRLS	44.9	**41.7**	59.6	56.7	61.3	59.1
SDA	36.6	34.8	57.2	55.0	65.0	61.5
SDE	40.0	37.7	54.5	52.4	50.0	49.0
FME	38.4	35.6	59.9	56.6	64.8	59.1
IFSSFE	**46.3**	41.2	**65.9**	**62.6**	**75.3**	**69.3**

FacePix	1 labeled sample		2 labeled samples		3 labeled samples	
Method	Unlabel(%)	Test(%)	Unlabel(%)	Test(%)	Unlabel(%)	Test(%)
LDA	26.0	26.9	39.1	39.6	48.2	46.6
GFHF	17.3	-	27.9	-	36.6	-
GFHF+CMN	21.6	-	30.3	-	38.0	-
RMGT	18.1	-	29.4	-	38.8	-
LapRLS	31.3	30.0	43.4	40.9	48.4	45.6
SDA	26.0	26.9	42.4	43.0	53.6	50.9
SDE	40.4	36.0	54.4	49.5	57.1	52.2
FME	28.1	28.0	42.7	40.3	50.5	47.7
IFSSFE	**40.6**	**37.8**	**56.6**	**53.6**	**65.8**	**61.2**

FERET	1 labeled sample		2 labeled samples		3 labeled samples	
Method	Unlabel(%)	Test(%)	Unlabel(%)	Test(%)	Unlabel(%)	Test(%)
LDA	21.7	21.0	35.7	36.4	43.9	55.0
GFHF	17.8	-	25.3	-	29.6	-
GFHF+CMN	23.6	-	30.9	-	38.4	-
RMGT	19.2	-	26.4	-	31.1	-
LapRLS	39.0	35.6	50.8	47.9	59.6	60.2
SDA	21.7	21.0	37.7	38.5	46.8	56.2
SDE	24.6	41.2	38.8	**54.6**	42.3	62.1
FME	35.5	27.9	47.2	39.5	54.1	53.0
IFSSFE	**39.7**	**37.4**	**51.3**	50.1	**60.6**	**70.8**

COIL-20	1 labeled sample		2 labeled samples		3 labeled samples	
Method	Unlabel(%)	Test(%)	Unlabel(%)	Test(%)	Unlabel(%)	Test(%)
LDA	43.2	43.5	52.8	59.3	58.5	66.9
GFHF	52.8	-	58.3	-	63.2	-
GFHF+CMN	58.9	-	63.3	-	68.0	-
RMGT	57.1	-	60.4	-	65.1	-
LapRLS	58.9	54.1	64.8	65.7	69.3	71.7
SDA	43.2	43.5	53.3	59.0	60.4	66.9
SDE	56.0	55.5	66.8	65.6	75.4	72.4
FME	62.0	57.9	66.8	64.7	70.7	68.6
IFSSFE	**68.0**	**60.8**	**75.1**	**71.7**	**80.4**	**77.4**

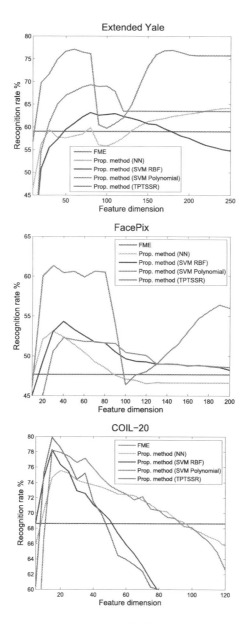

Fig. 3. A performance comparison between FME and the proposed IFSSFE as a function of features for three datasets: Extended Yale, FacePix, and COIL-20. The projection models associated with FME and IFSSFE were optimized on the fixed dimension given by C, namely the number of classes. In each plot, we show the average curve over ten splits. These curves depicts the performance on the test subset in which the number of labeled samples was set to three per class. For the proposed method IFSSFE, we used four classifiers:1-NN, RBF SVM, polynomial SVM (degree 3), and the Two Phase Test Sample Sparse Representation (TPTSSR) classifier.

5.4   Method Performance with Fixed Dimension

In this section, we compare the performance of the FME method with that of our proposed method for which the dimension of features is fixed to the number of classes, C. Note that the FME method is essentially a label propagation algorithm that uses C features. We will show that even in the case where dimensionality of the embedding is fixed to C, IFSSFE is still superior to FME if the balance parameters were optimized at this fixed dimension. This is explained by the fact that the criterion used by the proposed method was the main reason for this obtained superiority. Table 2 illustrates the average performance of FME and IFSSFE in such conditions. Since the proposed IFSSFE is a generic semi-supervised embedding, we used four classifiers: 1-NN, Support Vector Machines (SVM with RBF Kernel), SVM (with polynomial degree equal to 1 and 3), and the Two Phase Test Sample Sparse Representation (TPTSSR) classifier [27]. As can be seen, even when IFSSFE is restricted to work with only C features, its performance is still better than that of FME. We can also observe that the use

Table 2. Comparing the average performance of the FME method and the proposed IFSSFE method obtained at dimension equal to C. For the proposed IFSSFE (a generic semi-supervised embedding), we used four classifiers: 1-NN, RBF SVM, polynomial SVM (degree = 3), and the Two Phase Test Sample Sparse Representation (TPTSSR) classifier.

Ext Yale	1 labeled sample		2 labeled samples		3 labeled samples	
Method	Unlabel(%)	Test(%)	Unlabel(%)	Test(%)	Unlabel(%)	Test(%)
FME	38.4	35.6	59.9	56.6	64.8	59.1
IFSSFE (1-NN)	38.5	36.1	56.2	53.5	65.4	60.8
IFSSFE (RBF SVM)	38.5	36.1	53.0	50.6	59.5	54.9
IFSSFE (Poly. SVM)	44.9	41.0	62.2	57.2	67.3	60.8
IFSSFE (TPTSSR)	49.3	44.7	67.2	62.5	71.6	65.9

FacePix	1 labeled sample		2 labeled samples		3 labeled samples	
Method	Unlabel(%)	Test(%)	Unlabel(%)	Test(%)	Unlabel(%)	Test(%)
FME	28.1	28.0	42.7	40.3	50.5	40.3
IFSSFE (1-NN)	31.6	32.1	47.9	45.9	56.8	53.1
IFSSFE (RBF SVM)	31.6	32.1	48.2	45.8	57.1	53.1
IFSSFE (Poly. SVM)	32.1	29.9	48.4	44.1	56.7	50.6
IFSSFE (TPTSSR)	31.9	30.8	50.9	47.3	61.3	55.7

COIL-20	1 labeled sample		2 labeled samples		3 labeled samples	
Method	Unlabel(%)	Test(%)	Unlabel(%)	Test(%)	Unlabel(%)	Test(%)
FME	62.03	57.9	66.8	64.7	70.7	68.6
IFSSFE (1-NN)	63.6	59.5	72.9	68.7	78.2	75.5
IFSSFE (RBF SVM)	60.2	55.7	68.7	67.7	76.9	76.3
IFSSFE (Poly. SVM)	63.7	60.8	72.6	71.9	79.2	77.9
IFSSFE (TPTSSR)	62.0	57.9	66.8	64.7	70.7	68.6

of other classifiers such as SVM and TPTSSR has enhanced the performance of the IFSSFE with respect to the Nearest Neighbor classifier.

Figure 3 illustrates the average performance of FME and the proposed IFSSFE as a function of feature dimension for three datasets: Extended Yale, FacePix, and COIL-20. The projection models associated with FME and IFSSFE were optimized on the fixed dimension given by C. In each plot, we show the average curve over ten splits. These curves depicts the performance on the test subset in which the number of labeled samples per class was set to three. For the proposed method IFSSFE, we used four classifiers: 1-NN, RBF SVM, polynomial SVM (degree = 3), and the Two Phase Test Sample Sparse Representation (TPTSSR) classifier.

6   Conclusion

This paper presented a novel semi-supervised dimensionality reduction method for classification tasks. We propose an Inductive Flexible Semi Supervised Feature Extraction that retained the merits of Flexible Manifold Embedding and the graph based non-linear embedding. The proposed method simultaneously estimates a non-linear embedding as well as a transform needed for mapping the unseen samples. The proposed method was evaluated on four benchmark databases. We have provided a comparison with several competing methods based on label propagation methods as well as on semi supervised graph-based embedding. Our proposed method outperformed the competing methods in most cases.

Acknowledgment. This work was supported by the project EHU13/40.

References

1. Maaten, L., Postma, E., Herik, J.: Dimensionality reduction: a comparative review. Technical report TiCC TR 2009-005, TiCC, Tilburg University (2009)
2. Saul, L., Weinberger, K., Sha, F., Ham, J., Lee, D.: Spectral methods for dimensionality reduction. In: Chapelle, O., Scholkopf, B., Zien, A. (eds.) Semisupervised Learning, pp. 293–308. MIT Press, Cambridge (2006)
3. Yan, S., Xu, D., Zhang, B., Zhang, H., Yang, Q., Lin, S.: Graph embedding and extension: a general framework for dimensionality reduction. IEEE Trans. Pattern Anal. Mach. Intell. **29**, 40–51 (2007)
4. Zhang, T., Tao, D., Li, X., Yang, J.: Patch alignment for dimensionality reduction. IEEE Trans. Knowl. Data Eng. **21**, 1299–1313 (2009)
5. Chapelle, O., Scholkopf, B., Zien, A.: Semi-Supervised Learning. MIT Press, Cambridge (2006)
6. Silva, T., Zhao, L.: Network-based stochastic semisupervised learning. IEEE Trans. Neural Netw. Learn. Syst. **23**, 451–466 (2012)
7. Camps-Valls, G., Marsheva, T.B., Zhou, D.: Semi-supervised graph-based hyperspectral image classification. IEEE Trans. Geosci. Remote Sens. **45**, 3044–3054 (2007)

8. Huang, H., Li, J., Liu, J.: Enhanced semi-supervised local fisher discriminant analysis for face recognition. Future Gener. Comput. Syst. **28**, 244–253 (2012)

9. Liu, W., He, J., Chang, S.: Large graph construction for scalable semi-supervised learning. In: International Conference on Machine Learning (2010)

10. Pan, F., Wang, J., Lin, X.: Local margin based semi-supervised discriminant embedding for visual recognition. Neurocomputing **74**, 812–819 (2011)

11. Yang, W., Zhang, S., Liang, W.: A graph based subspace semi-supervised learning framework for dimensionality reduction. In: International Conference on Computer Vision (2008)

12. Xu, Z., King, I., Lyu, M.R.T., Rong, J.: Discriminative semi-supervised feature selection via manifold regularization. IEEE Trans. Neural Netw. **21**, 1033–1047 (2010)

13. Zhang, T., Ji, R., Liu, W., Tao, D., Hua, G.: Semi-supervised learning with manifold fitted graphs. In: International Joint Conference on Artificial Intelligence (2013)

14. Liu, W., Tao, D., Liu, J.: Transductive component analysis. In: IEEE International Conference on Data Mining (2008)

15. Cevikalp, H.: Semi-supervised dimensionality reduction using pairwise equivalence constraints. In: International Conference on Computer Vision Theory and Applications (2009)

16. Song, Y., Nie, F., Zhang, C., Xiang, S.: A unified framework for semi-supervised dimensionality reduction. Pattern Recogn. **41**, 2789–2799 (2008)

17. de Sousa, C.A.R., Rezende, S.O., Batista, G.E.A.P.A.: Influence of graph construction on semi-supervised learning. In: Blockeel, H., Kersting, K., Nijssen, S., Železný, F. (eds.) ECML PKDD 2013, Part III. LNCS, vol. 8190, pp. 160–175. Springer, Heidelberg (2013)

18. Zhu, X., Ghahramani, Z., Lafferty, J.: Semi-supervised learning using gaussian fields and harmonic functions. In: International Conference on Machine Learning (2003)

19. Zhou, S., Chellappa, R., Mogghaddam, B.: Visual tracking and recognition using appearance-adaptive models in particle filters. IEEE Trans. Image Process. **13**, 1473–1490 (2004)

20. Belkin, M., Niyogi, P., Sindhwani, V.: Manifold regularization: a geometric framework for learning from labeled and unlabeled examples. J. Mach. Learn. Res. **7**, 2399–2434 (2006)

21. Liu, W., Chang, S.: Robust multi-class transductive learning with graphs. In: Computer Vision and Pattern Recognition (2009)

22. Nie, F., Xu, D., Tsang, I., Zhang, C.: Flexible manifold embedding: a framework for semi-supervised and unsupervised dimension reduction. IEEE Trans. Image Process. **19**, 1921–1932 (2010)

23. Cai, D., He, X., Han, J.: Semi-supervised discriminant analysis. In: IEEE International Conference on Computer Vision (2007)

24. Chen, H., Chang, H., Liu, T.: Local discriminant embedding and its variants. In: IEEE International Conference on Computer Vision and Pattern Recognition (2005)

25. Huang, H., Liu, J., Pan, Y.: Semi-supervised marginal fisher analysis for hyperspectral image classification. In: ISPRS Annals of the Photogrammetry, Remote Sensing and Spatial Information Sciences I-3, pp. 377–382 (2012)

26. Yu, G., Zhang, G., Domeniconi, C., Yu, Z., You, J.: Semi-supervised classification based on random subspace dimensionality reduction. Pattern Recogn. **45**, 1119–1135 (2012)
27. Xu, Y., Zhang, D., Yang, J., Yang, J.Y.: A two-phase test sample sparse representation method for use with face recognition. IEEE Trans. Circuits Syst. Video Technol. **21**, 1255–1262 (2011)

Metric Tensor and Christoffel Symbols Based 3D Object Categorization

Syed Altaf Ganihar[(⊠)], Shreyas Joshi, Shankar Setty, and Uma Mudenagudi

B.V. Bhoomaraddi College of Engineering and Technology, Hubli, India
altafganihar@gmail.com

Abstract. In this paper we propose to address the problem of 3D object categorization. We model 3D object as a piecewise smooth Riemannian manifold and propose metric tensor and Christoffel symbols as a novel set of features. The proposed set of features captures the local and global geometry of 3D objects by exploiting the uniqueness and compatibility of the features. The metric tensor represents a geometrical signature of the 3D object in a Riemannian manifold. To capture global geometry we propose to use combination of metric tensor and Christoffel symbols, as Christoffel symbols measure the deviations in the metric tensor. The categorization of 3D objects is carried out using polynomial kernel SVM classifier. The effectiveness of the proposed framework is demonstrated on 3D objects obtained from different datasets and achieved comparable results.

1 Introduction

In this paper we propose three dimensional (3D) object categorization of a given 3D object using metric tensor and Christoffel symbols [1–4] with the help of kernel based support vector machine (SVM) classifier. With the availability of point cloud data of 3D objects, there is a surge of interest in novel methods for 3D object categorization. Categorization of 3D objects is a challenging problem. To address this issue we propose a set of features based on metric tensor and Christoffel symbols. Metric tensor together with Christoffel symbols captures the unique set of geometric features that are inherent to the 3D object shapes. The physical or intuitive parameter for an object is surface curvature. However they directly do not provide the inherent geometry of the 3D object [1–4]. One of the major challenges lies with the features to consider from the large datasets. Most of the 3D categorization methods use shapes, features and Bag-of-Words extracted from certain projections of the 3D objects. However, we propose to use features extracted from the geometry of 3D objects for categorization. The categorization of 3D objects finds its applications in the areas of content based retrieval, object detection, object recognition, and object tracking.

Humans usually are better in generic than in specific recognition, categorization is considered to be a much harder problem for computers. Since geometric features serves as a key for categorization, this influences the performance in terms of relevance and accuracy of results. Various sets of 3D features considered in literature for categorization and recognition are spin images [5], 3D shape

© Springer International Publishing Switzerland 2015
C.V. Jawahar and S. Shan (Eds.): ACCV 2014 Workshops, Part III, LNCS 9010, pp. 138–151, 2015.
DOI: 10.1007/978-3-319-16634-6_11

context [6], global or local features [7–9], Point Feature Histogram (PFH) [10] and the Viewpoint Feature Histogram (VFH) model [11], Aspect Graph approach [12], spin images combined with other shape and contextual features [13]. In [14] the Global Structure Histogram (GSH) descriptor is presented to represent the point cloud information. The GSH represents objects such that it can generalize over different poses and views, and cope with incomplete data for correct categorization of objects. Use of viewpoints in all possible variations to build the training dataset and unsupervised approach for object categorization could be expensive during learning model building. In [15] authors propose a learning model by using a hypothetical 3D object category. Parts (collection of smaller image patches) of the objects are considered and correspondences between these parts are based on the appearance and geometric consistency. The final model is visualized as parts in a 3D graph based of the learned geometric relations. This approach classifies, localizes and does pose estimation of object in the image. In [16] authors present the work of multi-view part-based model of [15,17], via minimal supervision and detection of objects under arbitrary or unseen poses. The proposed algorithm requires large number of views in training data in order to generalize which is an open issue.

In [18] 3D object categorization is introduced based on Bag-of-Words (BoW) paradigm. The visual vocabulary for bag of words is constructed in a multilevel way considering different seed regions. Hierarchical clustering is followed at each level for different region descriptors to obtain Bag-of-Words histograms for each mesh. Finally, the object categorization is done using one-against-all SVM classifier [19]. If the region descriptors are not properly clustered to obtain 3D visual words and if the vocabulary is so large that it could not distinguish between relevant and irrelevant variations may results into wrong categories. In [20] authors propose a discriminative approach to solve problem of 3D shape categorization. 3D local descriptors are extracted from 3D shapes, quantized using k-means to obtain 3D visual vocabulary and build a BoW representation. A general 3D Spatial Pyramid (3DSP) decomposition with multiple kernels is used to subdivide a cube impressed into surface of 3D shape repeatedly and compute weighted sum of histograms at increasingly fine sub-volumes. In [21] author expresses summary of categorization methods modeled from 2D images to 3D patch based model. In [22] author provides object detection from domestic scenes and based on shape model, categorizes the objects. The shape model is constructed by surface reconstruction method based on Growing Neural Gas (GNG) in combination with a shape distribution based descriptor. However, the shape ambiguity (e.g. bowl and cup) between categories decreases the discrimination because of object similarities under certain perspectives. This requires large number of training data to be provided for improving the discriminative performance. In our approach we address the problem of 3D object categorization by proposing metric tensor and Christoffel symbols as geometric features of the 3D object using point cloud representation. Towards this we make the following contributions:

1. We propose metric tensor and Christoffel symbols to represent basic geometry of 3D object which are intern used for 3D object categorization: we model 3D objects as a set of Riemannian manifolds and compute the features.
2. We propose framework for categorization of 3D objects using the proposed set of features, computed on local basis and captures the global geometry.
3. We demonstrate categorization of 3D objects using models obtained from state of art datasets like SHREC'12 using local patch based classification [23] and Princeton Shape benchmark dataset [24] using a BoW approach.

The rest of the paper is organized as follows. Section 2 describes the motivation and proposed approach. The geometric features are detailed in Sect. 3. We discuss proposed 3D object categorization in Sect. 4. We demonstrate the results on 3D object categorization in Sect. 5. Finally we conclude in Sect. 6.

2   Motivation and Approach

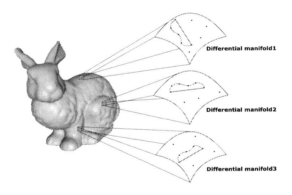

Fig. 1. The 3D object (bunny) exhibits non-uniformity in the distribution of geometrical properties and hence modeled as a set of Riemannian manifolds.

We model the given 3D object in a Euclidean space as a piecewise smooth Riemannian manifold. Let $V(x, y, z)$ be a 3D object in the Euclidean space and is modeled as,

$$V(x, y, z) \mapsto \Psi(\mathcal{M}, g) \tag{1}$$

where $\mathcal{M}, g$ represent the Riemannian manifold.

The piecewise smooth Riemannian manifold constitutes a continuous space and 3D point cloud represented in a Euclidean space is a sampled version of this continuous space. To capture the inherent geometry in Riemannian space, we need to model the 3D object as a piecewise smooth Riemannian manifold to account for the discontinuities in the geometry. A Riemannian manifold is a smooth construct which alone cannot represent the inherent geometry of the

3D object as the 3D object exhibits non-smooth behavior at certain positions due to non-uniform geometry. Consider the 3D model of a cube, the edge of the cube represents the transition in geometric properties and constitutes a non-smooth construct or is a geometric discontinuity. The derivatives for the Riemannian manifold are not defined at these points and contradicts the definition of a smooth Riemannian manifold. This necessitates the need of a piecewise smooth Riemannian manifold (Fig. 1).

When we have a unique mapping of 3D Euclidean space to the Riemannian space, the claim is there exists a unique mapping in the discretized (sampled) version of the 3D Euclidean space to the Riemannian space. 3D point cloud in Euclidean space can be represented by a unique discretized piecewise smooth Riemannian manifold.

Riemannian manifold or Riemannian space $(\mathcal{M}, g)$ [2] is a real smooth differential manifold $\mathcal{M}$ equipped with an inner product g_p on the tangent space at each point p and is given by,

$$p \mapsto g_p(X(p), Y(p)) \tag{2}$$

where the mapping from $p \mapsto g_p$ is a smooth function with $X(p)$ and $Y(p)$ being vector fields in the tangent space of the 3D model at point p. The family g_p of inner products is called Riemannian metric tensor.

The 3D object hence is a piecewise smooth differential manifold in which each tangent space is associated with an inner product which varies smoothly from point to point. The local geometrical properties of the 3D object can hence be inferred from the inner product computed on a local tangent plane for the object.

If $\mathcal{M}$ represents a differential manifold of dimensionality n then a Riemannian metric [2] on the manifold $\mathcal{M}$ is a family of the inner products given by,

$$g_p : T_p\mathcal{M} \times T_p\mathcal{M} \mapsto R | \forall p \in \mathcal{M} \tag{3}$$

where $T_p\mathcal{M}$ represents the tangent space at each point p on the manifold $\mathcal{M}$. The inner product g_p on the manifold $\mathcal{M}$ defines a smooth function from $\mathcal{M} \mapsto R$ [1,2]. The definition of Riemannian metric tensor is dependent on the parametrization technique employed in the tangent plane of the differential manifold. Let f denote the coordinate function defined on the tangent plane of the manifold which defines the map $f : (x, y, z) \mapsto (u, v)$, where (u, v) defines the parameters on the tangent plane of the manifold. The parametrization of the manifold can be obtained by using the inverse map f^{-1} to obtain the basis functions that can be used to compute the metric tensor on the manifold. The parametrization on the manifold is uniquely chosen for all the 3D objects so that the basis functions defined on the tangent plane are velocity vectors.

Given a Riemannian manifold $(\mathcal{M}, g)$ there exists a unique affine connection ∇ on $\mathcal{M}$ that is symmetric and compatible with g [2]. The uniqueness of the affine connection and the compatibility with the metric g is described in Theorem 1.

Theorem 1. *A Riemannian manifold $(\mathcal{M}, g)$ admits precisely one symmetric connection compatible with the metric. This particular connection is called the Riemannian connection or the LeviCivitta connection.*

The affine connection ∇ is called the LeviCivitta connection if

1. It preserves the metric i.e., $\nabla g = 0$
2. It is torsion free. i.e., for any vector fields X and Y we have,

$$\nabla_X Y - \nabla_Y X = [X, Y] \tag{4}$$

$[X, Y]$ is the Lie Bracket [2] of the vector fields X and Y.

The components of the LeviCivitta connection with respect to a system of local co-ordinates are called Christoffel symbols [2].

The Riemannian metric tensor along with Christoffel symbols for a defined parametrization of the tangent space constitute unique set of features for a given 3D object. But the definition of Riemannian metric tensor and Christoffel symbols depends on the choice of a co-ordinate system. To overcome this drawback we define a Cartesian world co-ordinate system in accordance with which we compute the components of metric tensor and Christoffel symbols. The given 3D objects under consideration are coarsely registered according to a predefined Cartesian world co-ordinate system using a ICP (iterative closest point) algorithm for 3D registration [25]. For every category of the 3D objects certain benchmark 3D objects are selected and the rest of the 3D objects are coarsely registered using the ICP algorithm.

In what follows we address the problem of 3D object categorization using a supervised learning approach. The input to the categorization framework is a 3D point cloud obtained either through set of images or from the modeling tools. The categorization is carried out using a supervised learning approach on a kernel based SVM framework. The geometric features for the input point cloud are computed and fed to the SVM. The geometric features used for the learning framework are metric tensor and Christoffel symbols. These set of features are used in a kernel based SVM framework for training and testing of the 3D objects. The testing data fed to the SVM which, after the learning process is able to classify the 3D object on a local patch basis.

3   Features

Riemannian metric tensor and Christoffel symbols effectively portray the inherent geometrical properties for a 3D object due to their compatibility and uniqueness and hence are best suited as features for categorization of 3D objects. The metric tensor represents the geometrical signature of the manifold in the local patch, however this alone may not capture the global geometrical properties since variations in the local patches are not inherently captured in the metric tensor. The Christoffel symbols give the numerical measure for the deviations in the geometrical properties of the manifold in the neighborhood of a patch. Hence variations in the modeled geometrical properties using metric tensor can be effectively captured by Christoffel symbols. The geometrical properties can quantitatively be represented using the combination of metric tensor and Christoffel symbols. The metric tensor and Christoffel symbols vary from point to point for a 3D object and hence can be represented as a mapping from the co-ordinates on the 3D object to the fields defined by metric tensor and Christoffel symbols and is given by,

$$\Phi : V(x, y, z) \mapsto (g, \Gamma) \qquad (5)$$

where $\Psi(x, y, z)$ represents the given 3D object, g and Γ represent the metric tensor and Christoffel symbols. The 3D objects under consideration can be uniquely represented by the pair of fields (g, Γ) defined on them.

3.1   Metric Tensor

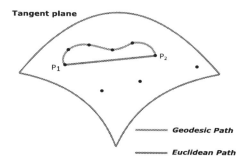

Fig. 2. The shortest distance between the points P_1-P_2 is computed as the geodesic distance (orange) on the manifold and is different from the Euclidean distance (green) between the points due to curvilinear properties of the surface (Color figure online).

Riemannian metric tensor is one of the features used for the classification of the models in the proposed method. The metric g on a manifold is a second order covariant tensor field. The metric tensor $g_{\mu\nu}$ is a symmetric tensor and in 3-dimensions comprises of 6 independent components. The metric tensor gives the quantitative measure for the deviation in the manifold from the Euclidean space. The inner product or the arclength of a curve in the manifold can be computed with the help of the metric tensor and is given by,

$$ds^2 = \sum_{\mu=1}^{3} \sum_{\nu=1}^{3} g_{\mu\nu} dx^\mu dx^\nu \qquad (6)$$

where ds^2 is the arclength of an infinitesimal curve on the manifold and dx^μ, dx^ν are the contravariant tangent vectors in the tangent plane of the manifold.

The deviation in the arclength ds^2 from the Euclidean distance function as shown in Fig. 2, gives a measure of the metric tensor g. To compute the metric tensor we calculate the arclength ds^2 as the geodesic distance between two neighboring points on a local patch as shown in Fig. 3. The geodesic distance is computed on the 3D point cloud by using Algorithm 1.

Data: Pair of points v_1 v_2 on the manifold to compute the geodesic distance.
Result: Geodesic distance between the pair of points.
initialization;
do dist $\leftarrow$ 0;
$k \leftarrow 2$;
$k1 \leftarrow 2$;
$i \leftarrow 1$;
$I_1 \leftarrow 0$;
while $v_2 \notin I_1$ **do**
   | $I_1 \leftarrow$ k-nnsearch(v_1,k);
   | $k \leftarrow k + 1$;
end
while $v_1 \notin I_2$ **do**
   | $I_2 \leftarrow$ k-nnsearch(v_2,$k1$);
   | $k1 \leftarrow k1 + 1$;
end
$I_3 \leftarrow I_1 \cap I_2$;
while $i \neq size(I_3)$ **do**
   | dist $\leftarrow$ dist + EuclideanDistance(I_3[i],I_3[i-1]) ;
   | $i \leftarrow i + 1$;
end

Algorithm 1: Geodesic distance computation on a pair of points on a point cloud.

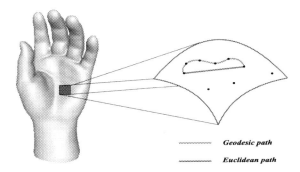

Geodesic path

Euclidean path

Fig. 3. The metric computation for a 3D object is carried out on a local tangent plane by computing the deviation of the geodesic distance from the Euclidean distance.

In matrix notation the relation between the arclength ds and components of the metric tensor g is given by,

$$ds^2 = \begin{pmatrix} dx^1 & dx^2 & dx^3 \end{pmatrix} \begin{pmatrix} g_{11} & g_{12} & g_{13} \\ g_{21} & g_{22} & g_{23} \\ g_{31} & g_{32} & g_{33} \end{pmatrix} \begin{pmatrix} dx^1 \\ dx^2 \\ dx^3 \end{pmatrix} \tag{7}$$

The metric tensor consist of 6 independent components in 3-dimension. To compute the components of metric tensor a minimum of 6 pair of points are required for which the geodesic distances has to be computed. The geodesic distance is computed for 6 pair of points on the tangent plane of the manifold using Algorithm 1. Equation 7 is used to solve for the components of the metric tensor by using the 6 geodesic distances and the contravariant vectors dx_μ and dx_ν in the tangent plane.

3.2   Christoffel Symbols

The Christoffel symbols give a measure of the deviation of the metric tensor as a function of position. The Christoffel symbols in 3-dimensions comprises of 18 independent components. The relation between metric tensor and Christoffel symbols is given by,

$$\Gamma^\sigma_{\mu\nu} = \frac{1}{2} \sum_{\rho=1}^{3} g^{\sigma\rho} \{ \frac{\partial g_{\rho\mu}}{\partial x^\nu} + \frac{\partial g_{\rho\nu}}{\partial x^\mu} - \frac{\partial g_{\mu\nu}}{\partial x^\rho} \} \tag{8}$$

Equation 8 suggests that the computation of Christoffel symbols is dependent on the first derivative of metric tensor. The derivative operator in non-Euclidean space does not preserve the tensorial attributes of Christoffel symbols. They preserve the tensorial attributes under certain non-linear transformations. Equation 8 provides a pseudo-tensor which is utilized as one of the features in the proposed 3D object categorization.

The Christoffel symbols are computed from the metric tensor values for every 12 pair of points belonging to two neighboring tangent planes on the manifold. The computation of the Christoffel symbols for a pair of neighboring tangent planes is as shown in Fig. 4. The Christoffel symbols represent the deviations in the metric tensor from one tangent plane to another due to the phenomenon of parallel transport on a curvilinear surface [3,4].

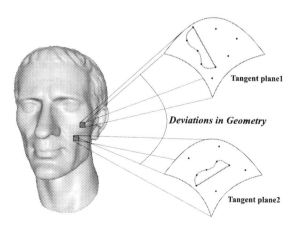

Fig. 4. The computation of Christoffel symbols for a 3D object is carried out on a pair of neighboring local tangent planes by computing the deviations in the metric tensor over the tangent planes.

The explicit geometry of the 3D object is represented by Riemannian curvature tensor which is captured by the combination of metric tensor and Christoffel symbols [2]. The uniqueness and compatibility of the metric tensor and Christoffel symbols as described in Theorem 1 enables us to model the given 3D object uniquely. Therefore metric tensor and Christoffel symbols can be used as geometrical features for the categorization of 3D objects.

4   Categorization of 3D Objects

The categorization of the 3D objects into generic classes is carried out using a SVM framework with the proposed set of features and is shown in Fig. 5. The proposed set of features are computed for a predefined set of models belonging to a particular class of objects and are fed to the SVM for learning.

4.1   Learning Framework

The categorization of 3D objects is carried out using a support vector machine framework [19]. The support vector machine is best suited for the categorization problem as it maps the features for classification into multidimensional vector space and supports non-linear kernels for classification. The features used for the categorization comprise of 24 independent components which are in turn dependent on the geometrical position of the point over which the features are computed. The normalization of the features is carried out by utilizing the positional dependence of the features for the 3D objects. The features for the training dataset are computed for unit scaled models to compensate for the scale dependence of the features. In our case we employ the polynomial kernel for the learning framework in support vector machine as the features, metric tensor and Christoffel symbols exhibit positional dependence.

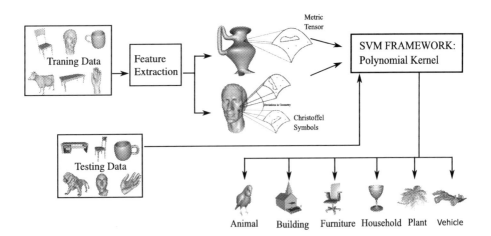

Fig. 5. Overview of the 3D object categorization.

The features, metric tensor and Christoffel symbols for the input 3D model are computed using the proposed method. The features computed are fed to the SVM framework for categorization of the 3D object using local patch based and BoW approach. In local patch based approach, the categorization of 3D objects is carried out into the predefined set of categories on a local patch basis with each patch comprising of 12 points. In BoW approach, k-means clustering is used for the proposed set of features on the training dataset to build the vocabulary. The vocabulary is used to compute the histogram on the proposed set of features. Based on the local patch or BoW approach SVM classifies the 3D objects into predefined set of categories.

5   Results and Discussions

The effectiveness of the proposed categorization framework is demonstrated on models obtained from SHREC'12 [23] and Princeton Shape Benchmark database [24] as shown in Fig. 6. The algorithm is implemented on Intel(R) Core(TM) i7-4700MQ processor @ 2.40 GHz and 8 GB RAM with NVIDIA GeForce GT 755M graphics. The code is written in Matlab and C using point cloud library (PCL). We demonstrate the results for categorization of basic 3D objects in Sect. 5.1 and categorization of 3D objects in Sect. 5.2.

5.1   Categorization of Basic Shapes

The categorization framework is initially demonstrated on basic geometrical models like sphere, cone, cylinder and torus of varied scales. The global geometrical properties for basic shapes like sphere, cone and cylinder can be influenced from the local geometrical properties. The classification of the models is initially carried out on a local patch basis. The decision for the global classification of the model is derived from the results of the classification on the local basis. Table 1 shows the percentage content of each basic geometrical model in the testing models for sphere, cone, cylinder and torus. We obtain an overall classification success rate of 100 % for different scale models for each basic model.

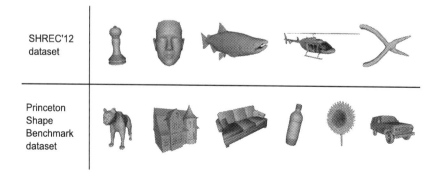

Fig. 6. Samples from the SHREC'12 and Princeton Shape Benchmark dataset

We also obtain an average accuracy of 83.60 % for correct content categorization. The results demonstrates that the local patch exhibits similarity to the global structure for basic shapes and we make the following observations,

1. From the basic geometry it is clear that sphere exhibits more similarity to cone than cylinder and torus and is reinforced in the results.
2. Cone exhibits more similarity to cylinder and sphere than torus as cone is geometrically similar to both sphere and cylinder.
3. Cylinder exhibits more similarity to torus than sphere and cone due to the positive and null curvature regions in cylinder and torus.
4. Torus model is geometrically very similar to a sphere as it comprises of positive Gaussian curvature region in the exterior parts.

Table 1. Result for the categorization of basic models with percentage content.

Testing data set	Sphere content	Cone content	Cylinder content	Torus content
Sphere	89.01 %	8.73 %	0.90 %	1.36 %
Cone	7.31 %	85.19 %	6.50 %	0.99968 %
Cylinder	0.367 %	0.027 %	91.79 %	7.82 %
Torus	22.28 %	7.04 %	2.27 %	68.41 %

5.2   3D Object Categorization

We demonstrate the results for the categorization of the 3D objects using one-against-all testing strategy in SVM learning framework. We have used libSVM [26] for the training and testing of the SVM framework. The training and testing for the SHREC'12 dataset is carried out using local patch based approach with 3rd order polynomial kernel based SVM framework. The SHREC'12 dataset consists of 20 models in each category. We have used 10 randomly selected models for training and the rest 10 models for testing. The overall accuracy for the SHREC'12 dataset is 66.42 % measured as MCC (Mean Correct Classification) [20] for the 5 categories.

The Princeton shape benchmark dataset comprises of 1594 shapes for 6 classes in coarse level two. We have used coarse level two for the evaluation of the proposed framework for categorization. The subsets for training and testing proposed in [27] is used for the experimentation of the proposed categorization framework. The training and testing for the Princeton shape benchmark dataset is carried out using the BoW approach with $K = 200$ for the proposed set of features. The training dataset used for the learning comprises of 807 models and the rest 787 models are used for testing using 15th order polynomial kernel based SVM classifier. The overall accuracy for the Princeton shape benchmark dataset is 67.90 % for 6 categories. The confusion matrix for the SHREC'12 dataset and Princeton shape benchmark dataset is shown in Fig. 7.

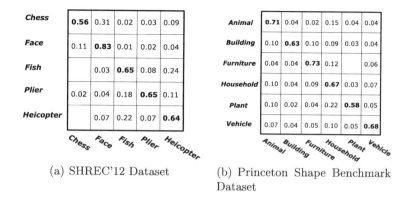

	Chess	Face	Fish	Plier	Heicopter
Chess	**0.56**	0.31	0.02	0.03	0.09
Face	0.11	**0.83**	0.01	0.02	0.04
Fish		0.03	**0.65**	0.08	0.24
Plier	0.02	0.04	0.18	**0.65**	0.11
Heicopter		0.07	0.22	0.07	**0.64**

(a) SHREC'12 Dataset

	Animal	Building	Furniture	Household	Plant	Vehicle
Animal	**0.71**	0.04	0.02	0.15	0.04	0.04
Building	0.10	**0.63**	0.10	0.09	0.03	0.04
Furniture	0.04	0.04	**0.73**	0.12		0.06
Household	0.10	0.04	0.09	**0.67**	0.03	0.07
Plant	0.10	0.02	0.04	0.22	**0.58**	0.05
Vehicle	0.07	0.04	0.05	0.10	0.05	**0.68**

(b) Princeton Shape Benchmark Dataset

Fig. 7. Confusion Matrix for SHREC'12 and Princeton shape benchmark dataset

5.3   Results Comparison

The results for the 3D object categorization framework are carried out using the state of art datasets like SHREC'12 and Princeton shape benchmark dataset used by [20]. We compare our results with the results presented by [20] which uses a similar data set. For BoW approach with K = 200 and L = 0 the authors in [20] have achieved 61.43 % measured as MCC with 3DSP-χ^2 kernel and for K = 1000 and L = 1 have achieved an accuracy of 66.31 % MCC. With K = 200 for the BoW approach we have achieved an accuracy of 67.90 % MCC.

6   Conclusion

In this paper we have addressed the problem of 3D object categorization. We have modeled 3D object as a piecewise smooth Riemannian manifold and propose metric tensor and Christoffel symbols as a novel set of features. The metric tensor along with Christoffel symbols represents the inherent geometry of the 3D object uniquely in a Riemannian manifold and hence used to represent the global geometry for the 3D object. The categorization of 3D objects is carried out using polynomial kernel SVM classifier using proposed set of features with local patch based approach and BoW approach. We have demonstrated 3D objects categorization using SHREC'12 dataset and Princeton shape benchmark dataset.

References

1. Jost, J.: Riemannian Geometry and Geometric Analysis. Springer Universitat Texts. Springer, Heidelberg (2005)
2. Kumaresan, S.: A Course in Differential Geometry and Lie Groups. Texts and Readings in Mathematics. Hindustan Book Agency, New Delhi (2002)
3. Weinberg, S.: Gravitation and Cosmology: Principles and Applications of the General Theory of Relativity. Wiley, New York (1972)

4. Misner, C., Thorne, K., Wheeler, J.: Gravitation. W.H. Freeman and Company, San Francisco (1973)

5. Johnson, A., Hebert, M.: Using spin images for efficient object recognition in cluttered 3d scenes. IEEE Trans. Pattern Anal. Mach. Intell. **21**, 433–449 (1999)

6. Frome, A., Huber, D., Kolluri, R., Bülow, T., Malik, J.: Recognizing objects in range data using regional point descriptors. In: Pajdla, T., Matas, J.G. (eds.) ECCV 2004. LNCS, vol. 3023, pp. 224–237. Springer, Heidelberg (2004)

7. Papageorgiou, C., Poggio, T.: A trainable system for object detection. Int. J. Comput. Vision **38**, 15–33 (2000)

8. Schneiderman, H.: A statistical approach to 3D object detection applied to faces and cars. Ph.D. thesis, Robotics Institute, Carnegie Mellon University, Pittsburgh, PA (2000)

9. Viola, P.A., Jones, M.J.: Rapid object detection using a boosted cascade of simple features. In: CVPR (1), pp. 511–518 (2001)

10. Rusu, R.B., Blodow, N., Beetz, M.: Fast point feature histograms (fpfh) for 3d registration. In: Proceedings of the 2009 IEEE International Conference on Robotics and Automation, ICRA 2009, pp. 1848–1853. IEEE Press, Piscataway (2009)

11. Rusu, R.B., Bradski, G.R., Thibaux, R., Hsu, J.: Fast 3d recognition and pose using the viewpoint feature histogram. In: IROS, pp. 2155–2162. IEEE (2010)

12. Cyr, C.M., Kimia, B.B.: 3d object recognition using shape similarity-based aspect graph. In: ICCV, 254–261 (2001)

13. Golovinskiy, A., Kim, V.G., Funkhouser, T.A.: Shape-based recognition of 3d point clouds in urban environments. In: ICCV, pp. 2154–2161. IEEE (2009)

14. Madry, M., Ek, C.H., Detry, R., Hang, K., Kragic, D.: Improving generalization for 3d object categorization with global structure histograms. In: IROS, pp.1379–1386. IEEE (2012)

15. Savarese, S., Li, F.F.: 3d generic object categorization, localization and pose estimation. In: ICCV, pp. 1–8. IEEE (2007)

16. Savarese, S., Li, F.F.: Multi-view object categorization and pose estimation. In: Cipolla, R., Battiato, S., Farinella, G.M. (eds.) Computer Vision. SCI, vol. 285, pp. 205–231. Springer, Heidelberg (2010)

17. Savarese, S., Li, F.-F.: View synthesis for recognizing unseen poses of object classes. In: Forsyth, D., Torr, P., Zisserman, A. (eds.) ECCV 2008, Part III. LNCS, vol. 5304, pp. 602–615. Springer, Heidelberg (2008)

18. Toldo, R., Castellani, U., Fusiello, A.: A *bag of words* approach for 3D object categorization. In: Gagalowicz, A., Philips, W. (eds.) MIRAGE 2009. LNCS, vol. 5496, pp. 116–127. Springer, Heidelberg (2009)

19. Burges, C.J.: A tutorial on support vector machines for pattern recognition. Data Min. Knowl. Disc. **2**, 121–167 (1998)

20. Lpez-Sastre, R.J., Garca-Fuertes, A., Redondo-Cabrera, C., Acevedo-Rodrguez, F.J., Maldonado-Bascn, S.: Evaluating 3d spatial pyramids for classifying 3d shapes. Comput. Graph. **37**, 473–483 (2013)

21. Pinz, A.: Object categorization. Found. Trends. Comput. Graph. Vis. **1**, 255–353 (2005)

22. Mueller, C.A.: 3D object shape categorization in domestic environments. Technical report, Bonn-Rhein-Sieg University, Sankt Augustin, Germany, February 2012

23. Biasotti, S., Bai, X., Bustos, B., Cerri, A., Giorgi, D., Li, L., Mortara, M., Sipiran, I., Zhang, S., Spagnuolo, M.: In: Spagnuolo, M., Bronstein, M.M., Bronstein, A.M., Ferreira, A. (eds.) 3DOR (Eurographics Association) pp. 101–107

24. Shilane, P., Min, P., Kazhdan, M., Funkhouser, T.: The princeton shape benchmark. In: Proceedings of the Shape Modeling International, SMI 2004, pp. 167–178. IEEE Computer Society, Washington, DC (2004)
25. Rusinkiewicz, S., Levoy, M.: Efficient variants of the ICP algorithm. In: Third International Conference on 3D Digital Imaging and Modeling (3DIM) (2001)
26. Chang, C.C., Lin, C.J.: Libsvm: a library for support vector machines. ACM Trans. Intell. Syst. Technol. **2**, 27:1–27:27 (2011)
27. Knopp, J., Prasad, M., Willems, G., Timofte, R., Van Gool, L.: Hough transform and 3D SURF for robust three dimensional classification. In: Daniilidis, K., Maragos, P., Paragios, N. (eds.) ECCV 2010, Part VI. LNCS, vol. 6316, pp. 589–602. Springer, Heidelberg (2010)

Feature Learning for the Image Retrieval Task

Aakanksha Rana, Joaquin Zepeda$^{(\boxtimes)}$, and Patrick Perez

Technicolor R&I, 975 Avenue des Champs Blancs, CS 17616,
35576 Cesson Sevigne, France
joaquin.zepeda@technicolor.com

Abstract. In this paper we propose a generic framework for the optimization of image feature encoders for image retrieval. Our approach uses a triplet-based objective that compares, for a given query image, the similarity scores of an image with a matching and a non-matching image, penalizing triplets that give a higher score to the non-matching image. We use stochastic gradient descent to address the resulting problem and provide the required gradient expressions for generic encoder parameters, applying the resulting algorithm to learn the power normalization parameters commonly used to condition image features. We also propose a modification to codebook-based feature encoders that consists of weighting the local descriptors as a function of their distance to the assigned codeword before aggregating them as part of the encoding process. Using the VLAD feature encoder, we show experimentally that our proposed optimized power normalization method and local descriptor weighting method yield improvements on a standard dataset.

1 Introduction

Image search methods can be broadly split into two categories. In the first category, *semantic search*, the aim is to retrieve images containing visual concepts. For example, the user might want to find images containing cats. In the second category, *image retrieval*, the search system is given an image of a scene, and the aim is to find all images of the same scene modulo some task-related transformation. Examples of simple transformations include changes in scene illumination, image cropping or scaling. More challenging transformations include drastic changes in background, wide changes in the perspective of the camera, high compression ratios, or picture-of-video-screen artifacts.

Common to both semantic search and image retrieval methods is the need to encode the image into a single, fixed-dimensional feature vector. Many successful image feature encoders have been proposed, and these generally operate on the fixed-dimensional local descriptor vectors extracted from densely [1] or sparsely [2,3] sampled local regions of the image. The feature encoder aggregates these local descriptors to produce a higher dimension image feature vector. Examples of such feature encoders include the bag-of-words encoder [4], the Fisher encoder [5] and the VLAD encoder [6]. All these encoding methods share common parametric post-processing steps where an element-wise power computation and subsequent l_2 normalization are applied. They also depend on specific models of the

© Springer International Publishing Switzerland 2015
C.V. Jawahar and S. Shan (Eds.): ACCV 2014 Workshops, Part III, LNCS 9010, pp. 152–165, 2015.
DOI: 10.1007/978-3-319-16634-6_12

data distribution in the local-descriptor space. For bag-of-words and VLAD, the model is a codebook obtained using K-means, while the Fisher encoding is based on a Gaussian Mixture Model (GMM). In both cases, the model defining the encoding scheme is built in an unsupervised manner using an optimization objective unrelated to the image search task.

For the case of semantic search, recent work has focused on learning the feature encoder parameters to make it better suited to the task at hand. A natural learning objective to use in this situation is the max-margin objective otherwise used to learn support vector machines. Notably, [7] learned the components of the GMM used in the Fisher encoding by optimizing, relative to the GMM mean and variance parameters, the same objective that produces the linear classifier commonly used to carry out semantic search. Approaches based on deep Convolutional Neural Networks (CNNs) [8,9] can also be interpreted as feature learning methods, and these now define the new state-of-the art baseline in semantic search. Indeed Sydorov *et al.* discuss how the Fisher encoder can be interpreted as a deep network, since both consist of alternating layers of linear and non-linear operations.

For the image retrieval task, however, the feature learning literature is lacking. One existing proxy approach is to also use the max-margin objective, and hence features encoders that were learned for the semantic search task [10]. Although the search tasks are not the same, this approach indeed results in improved image retrieval results, since both tasks are based on human visual interpretations of similarity. A second approach instead focuses on learning the local descriptor vectors at the input of the feature encoder. The objective used in this is case engineered to enforce matching, based on the learned local descriptors, of small image blocks centered on the same point in 3-D space, but from images taken from different perspectives [11,12].

One reason why these two approaches circumvent the actual task of image retrieval is the lack of objective functions that are good surrogates for the mean Average Precision (mAP) measure commonly used to evaluate image retrieval systems. Surrogate objectives are necessary because the mAP measure is non-differentiable as it depends on a ranking of the images being searched. The main contribution of this paper is hence to propose a new surrogate objective specifically for the image retrieval task. We show how this objective can be minimized using stochastic gradient descent, and apply the resulting algorithm to select the power-normalization parameters of the VLAD feature encoder. As a second contribution, we also propose a novel method to weight local descriptors for codebook-based image feature encoders that reduces the importance of descriptors too far away from their assigned codeword. We test both contributions independently and jointly and demonstrate improvements on a standard image retrieval performance.

The remainder of this paper is organized as follows: In the next section we describe standard feature encoding methods, focusing on the VLAD encoding that we use in our experiments. In Sect. 3 we described the proposed objective and the resulting learning algorithm, and in Sect. 4 we present the proposed descriptor-weighting method. We present experimental results in Sect. 5 and concluding remarks in Sect. 6.

Notation: We denote scalars, vectors and matrices using, respectively standard, bold, and upper-case bold typeface (e.g., scalar a, vector $\mathbf{a}$ and matrix $\mathbf{A}$). We use $\mathbf{v}_k$ to denote a vector from a sequence $\mathbf{v}_1, \mathbf{v}_2, \ldots, \mathbf{v}_N$, and v_k to denote the k-th coefficient of vector $\mathbf{v}$. We let $[\mathbf{a}_k]_k$ (respectively, $[a_k]_k$) denotes concatenation of the vectors $\mathbf{a}_k$ (scalars a_k) to form a single column vector. Finally, we use $\frac{\partial \mathbf{y}}{\partial \mathbf{x}}$ to denote the Jacobian matrix with (i, j)-th entry $\frac{\partial y_i}{\partial x_j}$.

2   Image Encoding Methods

Image encoders operate on the local descriptors $\mathbf{x} \in \mathbb{R}^d$ extracted from each image. Hence in this work we represent images as a set $\mathcal{I} = \{\mathbf{x}_i \in \mathbb{R}^d\}_i$ of local SIFT descriptors extracted densely [1] or with the Hessian affine region detector [3].

One of the earliest image encoding methods proposed was the bag-of-features encoder (BOF) [4]. The BOF encoder is based on a codebook $\{\mathbf{c}_k \in \mathbb{R}^d\}_{k=1}^L$ obtained by applying K-means to all the local descriptors $\mathcal{T} = \bigcup_t I_t$ of a set of training images. Letting $\mathcal{C}_k$ denote the Voronoi cell $\{\mathbf{x} | \mathbf{x} \in \mathbb{R}^d, k = \text{argmin}_j |\mathbf{x} - \mathbf{c}_j|\}$ associated to codeword $\mathbf{c}_k$, the resulting feature vector for image $\mathcal{I}$ is

$$\mathbf{r}^B = [\# (\mathcal{C}_k \cap \mathcal{I})]_k , \tag{1}$$

where $\#$ yields the number of elements in the set. The Fisher encoder [5] instead relies on a GMM model also trained on $\bigcup_t I_t$. Letting $\beta_i, \mathbf{c}_i, \boldsymbol{\Sigma}_i$ denote, respectively, the i-th GMM component's (i) prior weight, (ii) mean vector, and (iii) covariance matrix (assumed diagonal), the first-order Fisher feature vector is

$$\mathbf{r}^F = \left[\sum_{\mathbf{x} \in \mathcal{I}} \frac{p(k|\mathbf{x})}{\sqrt{\beta_i}} \boldsymbol{\Sigma}_k^{-1} (\mathbf{x} - \mathbf{c}_k) \right]_k . \tag{2}$$

A hybrid combination between BOF and Fisher techniques called VLAD has been proposed [13] that offers a good compromise between the Fisher encoders's performance and the BOF encoder's processing complexity: Similarly to the state-of-the art Fisher aggregator, it encodes residuals $\mathbf{x} - \mathbf{c}_k$, but it hard-assigns each local descriptor to a single cell $\mathcal{C}_k$ instead of using a costly soft-max assignment as in (2). In a later work, [6] further proposed incorporating several conditioning steps that improved the performance of the feature encoder. The resulting complete encoding process begins by first aggregating, on a per-cell basis, the l_2 normalized difference of each local descriptor relative the cell's codeword, subsequently rotating the resulting descriptor using the matrix $\boldsymbol{\Phi}_k$ (obtained by PCA on the training descriptors $\mathcal{C}_k \cap \mathcal{T}$):

$$\mathbf{r}_k^V = \boldsymbol{\Phi}_k \sum_{\mathbf{x} \in \mathcal{I} \cap \mathcal{C}_k} \frac{\mathbf{x} - \mathbf{c}_k}{|\mathbf{x} - \mathbf{c}_k|} \in \mathbb{R}^d, \tag{3}$$

The L sub-vectors thus obtained are then stacked to form a large vector $\mathbf{v}$ that is then power-normalized and l_2 normalized:

$$\mathbf{v} = [\mathbf{r}_k^V]_k \in \mathbb{R}^{dL}, \tag{4}$$

$$\mathbf{p} = [h_{\alpha_j}(v_j)]_j, \tag{5}$$

$$\mathbf{n} = \mathbf{g}(\mathbf{p}). \tag{6}$$

The power normalization function $h_\alpha(x)$ and the l_2 normalization function $\mathbf{n}(\mathbf{v})$ are

$$h_\alpha(x) = \text{sign}(x)|x|^\alpha, \tag{7}$$

$$\mathbf{g}(\mathbf{x}) = \frac{\mathbf{x}}{|\mathbf{x}|_2}. \tag{8}$$

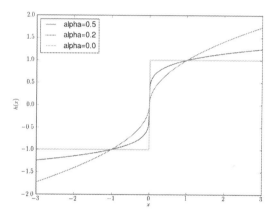

Fig. 1. Plot of $h_\alpha(x)$ for various values of α.

The power normalization function (7) is widely used as a post-processing stage for image features [1,6,14,15]. This post-processing stage is meant to mitigate (respectively, enhance) the contribution of the larger (smaller) coefficients in the vector (*cf.*, Fig. 1). Combining power normalization with the PCA rotation matrices $\boldsymbol{\Phi}_k$ was shown in [6] to yield very good results. In all the approaches using power normalization, the α_j are kept constant for all entries in the vector, $\alpha_j = \alpha, \forall j$. This restriction comes from the fact that α is chosen empricially (often to $\alpha = 0.5$ or $\alpha = 0.2$), and choosing different values for each α_j is difficult. In Sect. 3 we remove this difficulty by applying our proposed feature learning method to the optimization of the α_j.

3   Feature Learning for Image Retrieval

Feature learning has been pursued in the context of image classification [7] or for learning local descriptors akin to parametric variants of the SIFT descriptor [11,12]. Learning features specifically for the image retrieval task, however, has not been pursued previously. In this section we propose an approach to do so, and apply it to the optimization of the parameters of the VLAD feature encoding method described in Sect. 2.

The main difficulty in learning for the image retrieval task lies in the non-smoothness and non-differentiability of the standard performance measures used in this context. These measures are all based on *recall* and *precision* computed over a ground-truth dataset containing known groups of matching images [16,17]: A given query image is used to obtain a ranking $(i_k \in \{1,\dots,N\})_k$ of the N images in the dataset (for example, by an ascending sort of their feature distances relative to the query feature). Given the ground-truth matches $\mathcal{M} = \{i_{k_j}\}_j$ for the query, the recall and precision at rank k are computed using the first k ranked images $\mathcal{F}_k = \{i_1,\dots,i_k\}$ as follows (where $\#$ denotes set cardinality):

$$r(k) = \frac{\#(\mathcal{F}_k \cap \mathcal{M})}{\#\mathcal{M}}, \tag{9}$$

$$p(k) = \frac{\#(\mathcal{F}_k \cap \mathcal{M})}{k}. \tag{10}$$

The *average precision* is then the area under the curve obtained by plotting $p(k)$ versus $r(k)$ for a single query image. A common performance measure is the mean, over all images in the dataset, of the average precision. This mean Average Precision (mAP) measure, and all measures based on recall and precision, are non-differentiable, and it is hence difficult to use them in an optimization framework, motivating the need for an adequate surrogate objective.

3.1   Proposed Objective

We assume that we are given a set of N training images and that for each image i, we are also given labels $\mathcal{M}_i \subset \{1,\dots,N\}$ of images that are a match to image i and labels $\mathcal{N}_i \subset \{1,\dots,N\}$ of images that do not match image i. We assume that some feature encoding scheme has been chosen that is parametrized by a vector $\boldsymbol{\theta}$ and that produces feature vectors $\mathbf{n}_i(\boldsymbol{\theta})$. Our aim is to define a procedure to select good values for the parameters $\boldsymbol{\theta}$ by minimizing the following objective:

$$f(\boldsymbol{\theta}) = \frac{1}{M} \sum_{i,j \in \mathcal{M}_i, k \in \mathcal{N}_i} \phi(\mathbf{n}_i(\boldsymbol{\theta}), \mathbf{n}_j(\boldsymbol{\theta}), \mathbf{n}_k(\boldsymbol{\theta})), \tag{11}$$

where M is the total number of terms in the triple summation and

$$\phi(\boldsymbol{\eta}, \mathbf{a}, \mathbf{b}) = \max(0, \varepsilon - (\boldsymbol{\eta}^{\mathrm{T}}(\mathbf{a} - \mathbf{b}))). \tag{12}$$

The parameter ε enforces some small, non-zero margin that can be held constant (*e.g.*, $\varepsilon = 1e - 2$) or increased gradually during the optimization (*e.g.*, between 0 and $1e - 1$).

An objective based on image triplets similarly to (11) has been previously used in metric learning [18], where the aim is commonly to learn a matrix $\mathbf{W}$

used to compute distances between two given feature vectors $\mathbf{n}_i$ and $\mathbf{n}_j$ using $(\mathbf{n}_i - \mathbf{n}_j)^{\mathrm{T}} \mathbf{W} (\mathbf{n}_i - \mathbf{n}_j)$. Our aim is instead to learn the parameters $\boldsymbol{\theta}$ that define the encoding process. In this work in particular we learn the power normalization parameters α_j in (5).

3.2   Optimization Strategy

Stochastic Gradient Descent (SGD) is a well-established, robust optimization method offering advantages when computational time or memory space is the bottleneck [19], and this is the approach we take to optimize (11). Given the parameter estimate $\boldsymbol{\theta}_t$ at iteration t, SGD substitutes the gradient for the objective

$$\left.\frac{\partial f}{\partial \boldsymbol{\theta}}\right|_{\boldsymbol{\theta}_t} = \frac{1}{M} \sum_{i,j \in \mathcal{M}_i, k \in \mathcal{N}_i} \left.\frac{\partial \phi(\mathbf{n}_i, \mathbf{n}_j, \mathbf{n}_k)}{\partial \boldsymbol{\theta}}\right|_{\boldsymbol{\theta}_t} \tag{13}$$

by an estimate from a single (i, j, k)-triplet drawn at random at time t,

$$\nabla \phi_{i_t j_t k_t}(\boldsymbol{\theta}_t) \triangleq \left.\frac{\partial \phi(\mathbf{n}_{i_t}, \mathbf{n}_{j_t}, \mathbf{n}_{k_t})}{\partial \boldsymbol{\theta}}\right|_{\boldsymbol{\theta}_t}. \tag{14}$$

The resulting SGD update rule is

$$\boldsymbol{\theta}_{t+1} = \boldsymbol{\theta}_t - \gamma_t \cdot \nabla \phi_{i_t j_t k_t}(\boldsymbol{\theta}_t) \tag{15}$$

where γ_t is a learning rate that can be made to decay with t, e.g., $\gamma_t = \gamma/t$, and the parameter γ can be set by cross-validation. SGD is guaranteed to converge to a local minimum under mild decay conditions on γ_t [19].

When the power normalization and l_2 normalization post-processing stages in (5) and (6) are used, the gradient (14) required in (15) can be computed using the chain rule as follows, where we use the notation $\frac{\partial \mathbf{n}}{\partial \mathbf{p}_i} = \left.\frac{\partial \mathbf{n}}{\partial \mathbf{p}}\right|_{\mathbf{p}_i}$:

$$\begin{aligned} \nabla \phi_{i,j,k}(\boldsymbol{\theta}) \triangleq &\left.\frac{\partial \phi}{\partial \boldsymbol{\eta}}\right|_{\mathbf{n}_i} \cdot \frac{\partial \mathbf{n}}{\partial \mathbf{p}_i} \cdot \frac{\partial \mathbf{p}(\mathcal{I}_i)}{\partial \boldsymbol{\theta}} \\ &+ \left.\frac{\partial \phi}{\partial \mathbf{a}}\right|_{\mathbf{n}_j} \cdot \frac{\partial \mathbf{n}}{\partial \mathbf{p}_j} \cdot \frac{\partial \mathbf{p}(\mathcal{I}_j)}{\partial \boldsymbol{\theta}} \\ &+ \left.\frac{\partial \phi}{\partial \mathbf{b}}\right|_{\mathbf{n}_k} \cdot \frac{\partial \mathbf{n}}{\partial \mathbf{p}_k} \cdot \frac{\partial \mathbf{p}(\mathcal{I}_k)}{\partial \boldsymbol{\theta}}. \end{aligned} \tag{16}$$

The partial Jacobians in the above expression are given below, where we use sub-gradients for those expressions relying on the non-differentiable hinge loss:

$$\frac{\partial \phi}{\partial \boldsymbol{\eta}} = \begin{cases} 0, & \text{if } (\boldsymbol{\eta}^{\mathrm{T}}(\mathbf{a} - \mathbf{b})) \geq \varepsilon \\ (\mathbf{b} - \mathbf{a})^{\mathrm{T}}, & \text{otherwise} \end{cases}, \tag{17}$$

$$\frac{\partial \phi}{\partial \mathbf{b}} = -\frac{\partial \phi}{\partial \mathbf{a}} = \begin{cases} 0, & \text{if } (\boldsymbol{\eta}^{\mathrm{T}}(\mathbf{a} - \mathbf{b})) \geq \varepsilon \\ \boldsymbol{\eta}^{\mathrm{T}}, & \text{otherwise} \end{cases}, \tag{18}$$

$$\frac{\partial \mathbf{n}}{\partial \mathbf{p}} = |\mathbf{p}|_2^{-1} (\mathbf{I} - \mathbf{n}\mathbf{n}^{\mathrm{T}}). \tag{19}$$

The above expressions are generic and can be used for any parameter $\boldsymbol{\theta}$ of the feature encoder that one wishes to specialize. In this work we learn the power normalization coefficients α_j in (5) and hence $\boldsymbol{\theta} = \boldsymbol{\alpha}$, and the required Jacobian is

$$\frac{\partial \mathbf{p}}{\partial \boldsymbol{\alpha}} = \text{diag}\left([\log(|v_i|).|v_i|^{\alpha_i}]_i\right). \tag{20}$$

4   Local-Descriptor Pruning

In this section we propose a local-descriptor pruning method applicable to feature encoding methods like BOF, VLAD and Fisher that are based on stacking sub-vectors $\mathbf{r}_k$, where each sub-vector is computed from the local descriptors assigned to a cell $\mathcal{C}_k$. The proposed approach shares some similarities with [20,21].

Unlike the case for low-dimensional sub-spaces, the cells $\mathcal{C}_k$ in high-dimensional local-descriptors spaces are almost always unbounded, meaning that they have infinite volume.[1] Yet only a part of this volume is informative visually. This suggests removing those descriptors that are too far away from the cell center $\mathbf{c}_k$ when constructing the sub-vectors $\mathbf{r}_k$ in (1), (2) and (3). This can be done by restricting the summations in (1), (2) and (3) only to those vectors $\mathbf{x}$ that (i) are in the cell $\mathcal{C}_k$ and (ii) satisfy the following distance-to-$\mathbf{c}_k$ condition:

$$(\mathbf{x} - \mathbf{c}_k)^{\text{T}} \mathbf{M}_k^{-1} (\mathbf{x} - \mathbf{c}_k) \leq \gamma \sigma_k^2. \tag{21}$$

Here γ is determined experimentally by cross-validation and the parameter σ_k is the empirical variance of the distance in (21) computed over those descriptors from the training set that are in the cell. The matrix $\mathbf{M}_k$ can be either

anisotropic: the empirical covariance matrix computed from $\mathcal{T} \cap \mathcal{C}_k$;
axes-aligned: the same as the anisotropic $\mathbf{M}_k$, but with all elements outside the diagonal set to zero;
isotropic: a diagonal matrix $\sigma_k^2 \mathbf{I}$ with σ_k^2 equal to the mean diagonal value of the axes-aligned $\mathbf{M}_k$.

While the anisotropic variant offers the most geometrical modelling flexibility, it also drastically increases the computational cost. The isotropic variant, on the other hand, enjoys practically null computational overhead, but also the least modelling flexibility. The axes-aligned variant offers a compromise between the two approaches.

---

[1] Although l_2 normalization commonly applied to local descriptors limits the effective volume of each cell, one should note that l_2 normalization amounts to a reduction of dimensionality by one dimension, and that l_2-normalized data is still high-dimensional. Yet the question still remains on whether pruning mechanisms other than those proposed herein exist that better take into account the constraints on the data layout.

4.1   Soft-Weight Extension

The pruning carried out by (21) can be implemented by means of 1/0 weights

$$w_k(\mathbf{x}) = [\![(\mathbf{x} - \mathbf{c}_k)^{\mathrm{T}} \mathbf{M}_k^{-1}(\mathbf{x} - \mathbf{c}_k) \leq \gamma \sigma_k^2]\!] \qquad (22)$$

applied to the summation terms in (1), (2) and (3). For example, for (3) the weights would be used as follows:

$$\mathbf{r}_k^V = \boldsymbol{\Phi}_k^{\mathrm{T}} \sum_{\mathbf{x} \in I \cap \mathcal{C}_k} w_k(\mathbf{x}) \frac{\mathbf{x} - \mathbf{c}_k}{|\mathbf{x} - \mathbf{c}_k|} \in \mathbb{R}^d. \qquad (23)$$

A simple extension of the hard-pruning approach corresponding to (22) consists of instead using *exponential weights*

$$w_k(\mathbf{x}) = \exp\left(-\frac{\omega}{\sigma_k^2}(\mathbf{x} - \mathbf{c}_k)^{\mathrm{T}} \mathbf{M}_k^{-1}(\mathbf{x} - \mathbf{c}_k)\right), \qquad (24)$$

where the parameter ω is set experimentally, or *inverse weights*

$$w_k(\mathbf{x}) = \frac{\sigma_k^2}{(\mathbf{x} - \mathbf{c}_k)^{\mathrm{T}} \mathbf{M}_k^{-1}(\mathbf{x} - \mathbf{c}_k)}. \qquad (25)$$

5   Results

Setup: We use SIFT descriptors extracted from local regions computed with the Hessian-affine detector [3] or from a dense-grid using three block sizes (16, 24, 32) with a step size of 3 pixels [1]. When using the Hessian affine detector, we use the RootSIFT variant following [14]. As a training set, we use the Flickr60K dataset [16] composed of 60,000 images extracted randomly from Flickr. This data set is used to learn the codebook, rotation matrices, per-cluster pruning thresholds

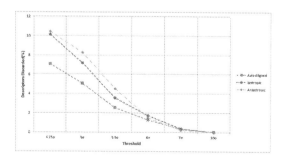

Fig. 2. Percentage of pruned descriptors by anisotropic axes aligned pruning, isotropic pruning, and anisotropic pruning. Holidays dataset with Hessian-Affine SIFT.

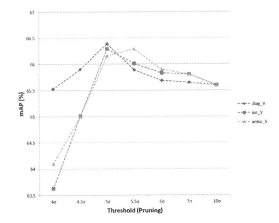

Fig. 3. Impact of Mahalanobis-metric based descriptor pruning on image retrieval performance when using anisotropic axes-aligned pruning (blue), isotropic pruning (red), and anisotropic pruning (green). Holidays dataset with Hessian-Affine SIFT (Color figure online).

and covariance matrices for the computation of the Mahalanobis metrics used for pruning of local descriptors. For testing, we use the INRIA Holidays dataset [16] which contains 1491 high resolution personal photos of 500 locations or objects, where common locations/objects define matching images. The search quality in all the experiments is measured using mAP (mean average precision) using the code provided by the authors [16]. All the experiments have been carried out using the VLAD image encoder and a codebook of size 64 following [6].

Evaluation of pruning methods: In Table 1, we evaluate the pruning approaches discussed in Sect. 4. Each variant is specified by a choice of weight type (hard, exponential or inverse), metric type (isotropic, anisotropic or axes-aligned), and local feature (dense or Hessian affine). The best result overall is obtained using axes-aligned exponential weighting (74.28 % and 67.02 % for dense and Hessian affine detections, respectively). The choice of the weighting parameter for exponential pruning is empirically set to $\omega = 1.55$. For completeness, we provide plots, for the case of hard-pruning, depicting the percentage of local descriptors removed (Fig. 2) and the resulting mAP score (Fig. 3) as a function of $\sqrt{\gamma}\sigma_k$. The values plotted in Fig. 2 are averaged over all cells $\mathcal{C}_k$.

Evaluation of α learning: In Fig. 4, we provide a plot of the cost in (11) as a function of the number of SGD iterations (15) using a dataset of $M = 8,000$ image triplets. The cost drops from 0.0401 to less than 0.0385. The resulting mAP is given in Table 2, where we present results both for the case where α is learned with and without exponential weighting of the local descriptors. The combined effect of exponential weighting and α learning is a gain of 1.86 mAP points.

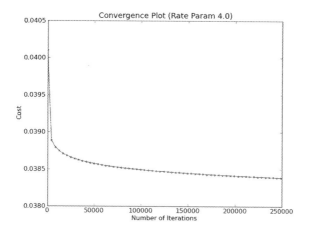

Fig. 4. Convergence plot for the α_j learning procedure.

Table 1. Summary of feature pruning results for all combinations of detectors-dense or Hessian-affine, metrics - isotropic (Iso), anisotropic (Aniso), and axes-alinged (Ax-align) and weighting schemes - hard, exponential and inverse. Underlines indicate best-in-row and bold best overall. The baseline results are for the system in [6].

Descriptors	mAP (%)				
	Baseline	Weights	Iso	Aniso	Ax-align
Hessian Affine	65.60	hard	66.29	66.29	<u>66.40</u>
		inverse	66.40	66.39	<u>66.55</u>
		exponential	66.45	66.40	**67.02**
Dense	72.71	hard	73.34	73.37	<u>73.56</u>
		inverse	73.45	73.45	<u>73.60</u>
		exponential	73.69	73.61	**74.28**

In Figs. 5 and 6 we provide two examples of top-ten ranked results for two different query images using our proposed modifications. We also provide examples of query images that resulted in improved (Fig. 7) and worsened (Fig. 8) ranking.

Table 2. Summary of best results (with dense detection) when using *(i)* only exponential weighting, *(ii)* learned α_j parameters without exponential weighting, and *(iii)* combined exponential weighting and learning of the α_j parameters. The baseline results are for the system in [6].

Baseline	Exp. weighting only	Learned α_j only	Exp. weighting and learn α_j
72.71	74.28	74.30	**74.57**

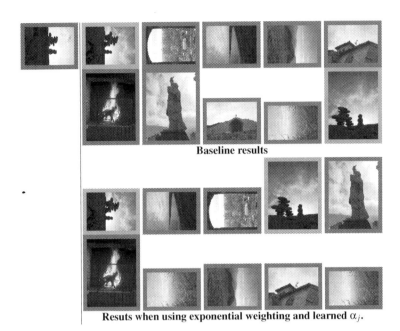

Fig. 5. Top-ten ranked results. *Top:* Baseline. *Bottom:* With exponential weighting/α_j learning.

Fig. 6. Top-ten ranked results. *Top:* Baseline. *Bottom:* With exponential weighting/α learning.

Fig. 7. Query images that result in improved ranking when using α_j learning with exponential weighting.

Fig. 8. Query images that result in degraded ranking when using α_j learning with exponential weighting.

6   Conclusions

In this paper we proposed learning the power normalization parameters commonly applied to image feature encoders using an image-triplet-based objective that penalizes erroneous ranking in the image retrieval task. The proposed feature learning approach is applicable to other parameters of the feature encoder.

We also propose, for the case of codebook-based feature encoders, weighting local descriptors based on their distance from the assigned codeword. We evaluate both methods experimentally and show that they provide improved results on a standard dataset.

Acknowledgement. This work was partially supported by the FP7 European integrated project AXES.

References

1. Chatfield, K., Lempitsky, V., Vedaldi, A., Zisserman, A.: The devil is in the details: an evaluation of recent feature encoding methods. In: Proceedings of the British Machine Vision Conference, pp. 76.1–76.12. British Machine Vision Association (2011)
2. Lowe, D.G.: Distinctive image features from scale-invariant keypoints. Int. J. Comput. Vision **60**, 91–110 (2004)
3. Mikolajczyk, K., Tuytelaars, T., Schmid, C., Zisserman, A., Matas, J., Schaffalitzky, F., Kadir, T., Gool, L.V.: A comparison of affine region detectors. Int. J. Comput. Vision **65**, 43–72 (2005)
4. Sivic, J., Zisserman, A.: Video Google: a text retrieval approach to object matching in videos. In: International Conference on Computer Vision, pp. 2–9 (2003)
5. Perronnin, F., Sánchez, J., Mensink, T.: Improving the fisher kernel for large-scale image classification. In: Daniilidis, K., Maragos, P., Paragios, N. (eds.) ECCV 2010, Part IV. LNCS, vol. 6314, pp. 143–156. Springer, Heidelberg (2010)
6. Delhumeau, J., Gosselin, P.H., Jégou, H., Pérez, P.: Revisiting the VLAD image representation. In: Proceedings of ACM International Conference on Multimedia, vol. 21, pp. 653–656 (2013)
7. Sydorov, V., Sakurada, M., Lampert, C.: Deep fisher kernels - end to end learning of the fisher kernel GMM parameters. In: Computer Vision and Pattern Recognition (2014)
8. Krizhevsky, A., Sutskever, I., Hinton, G.E.: ImageNet classification with deep convolutional neural networks. In: Proceedings of Neural Information Processing Systems, pp. 1–9 (2012)
9. Oquab, M., Bottou, L.: Learning and transferring mid-level image representations using convolutional neural networks. In: Proceedings of Computer Vision and Pattern Recognition (2014)
10. Mensink, T., Verbeek, J., Perronnin, F., Csurka, G.: Metric learning for large scale image classification: generalizing to new classes at Near-Zero cost. Pattern Anal. Mach. Intell. **34**, 1704–1716 (2012)
11. Brown, M., Hua, G., Winder, S.: Discriminative learning of local image descriptors. IEEE Trans. Pattern Anal. Mach. Intell. **33**, 43–57 (2011)
12. Simonyan, K., Vedaldi, A., Zisserman, A.: Descriptor learning using convex optimisation. In: Fitzgibbon, A., Lazebnik, S., Perona, P., Sato, Y., Schmid, C. (eds.) ECCV 2012, Part I. LNCS, vol. 7572, pp. 243–256. Springer, Heidelberg (2012)
13. Jegou, H., Douze, M., Schmid, C., Perez, P.: Aggregating local descriptors into a compact image representation. In: Proceedings of Computer Vision and Pattern Recognition, pp. 3304–3311 (2010)

14. Arandjelovic, R., Zisserman, A.: Three things everyone should know to improve object retrieval. In: Proceedings of Computer Vision and Pattern Recognition (2012)
15. Jegou, H., Perronnin, F., Douze, M., Jorge, S., Patrick, P., Schmid, C.: Aggregating local image descriptors into compact codes. IEEE Trans. Pattern Anal. Mach. Intell., 1–12 (2011)
16. Jegou, H.: INRIA Holidays dataset (2014)
17. Nistér, D., Stewénius, H.: Scalable recognition with a vocabulary tree (2006)
18. Chechik, G., Shalit, U.: Large scale online learning of image similarity through ranking. J. Mach. Learn. Res. **11**, 1109–1135 (2010)
19. Bottou, L.: Stochastic gradient descent tricks. In: Montavon, G., Orr, G.B., Müller, K.-R. (eds.) Neural Networks: Tricks of the Trade, 2nd edn. LNCS, vol. 7700, 2nd edn, pp. 421–436. Springer, Heidelberg (2012)
20. Avila, S., Thome, N., Cord, M., Valle, E., de A. Araujo, A.: BOSSA: extended bow formalism for image classification. In: 2011 18th IEEE International Conference on Image Processing (ICIP), pp. 2909–2912 (2011)
21. Avila, S., Thome, N., Cord, M., Valle, E., De A. AraúJo, A.: Pooling in image representation: the visual codeword point of view. Comput. Vis. Image Underst. **117**, 453–465 (2013)

Evaluation of Smile Detection Methods with Images in Real-World Scenarios

Zhoucong Cui, Shuo Zhang, Jiani Hu, and Weihong Deng[✉]

Beijing University of Posts and Telecommunications, Beijing, China
whdeng@bupt.edu.cn

Abstract. Discriminative methods such as SVM, have been validated extremely efficient in pattern recognition issues. We present a systematic study on smile detection with different SVM classifiers. We experimented with linear SVM classifier, RBF kernel SVM classifier and a recently-proposed local linear SVM (LL-SVM) classifier. In this paper, we focus on smile detection in face images captured in real-world scenarios, such as those in GENKI4K database. In the meantime, illumination normalization, alignment and feature representation methods are also taken into consideration. Compared with the commonly used pixel-based representation, we find that local-feature-based methods achieve not only higher detection performance but also better robustness against misalignment. Almost all the illumination normalization methods have no effect on the detection accuracy. Among all the SVM classifiers, the novel LL-SVM is verified to find a balance between accuracy and efficiency. And among all the features including pixel value intensity, Gabor, LBP and HOG features, we find that HOG features are the most appropriate features to detect smiling faces, which, combined with RBF kernel SVM, achieve an accuracy of 93.25 % on GENKI4K database.

1 Introduction

Facial expression recognition has been an active topic for the last two decades. Common methods such as feature-point-based expression classifiers [1–3], 3D face modeling [4,5] and dynamic analysis of video sequences [6–8] have achieved great success, while few have specially focused on smile detection. Shinohara and Otsu [9] used a Fisher weight map as an assist to gain an accuracy of 97.9 % on 96 testing face images. The result is desirable, but the data is limited. Another smile detector presented in [10] achieves an accuracy of 96.1 % on 4928 testing face images. However, the face images used in these studies are captured in limited conditions, and are mainly frontal. Many of the testing images were collected by asking subjects to deliberately pose certain expressions, thus exaggerating the effect of the expressions, which would seldom occur in our daily life. In this paper, we focus on face images collected in real-world scenarios, and present a systematic study on smile detection with different SVM classifiers using various features. To measure the real-world performance on smile detection, we choose the GENKI4K database and some of the image examples are shown in Fig. 1.

© Springer International Publishing Switzerland 2015
C.V. Jawahar and S. Shan (Eds.): ACCV 2014 Workshops, Part III, LNCS 9010, pp. 166–179, 2015.
DOI: 10.1007/978-3-319-16634-6_13

Fig. 1. Examples of real-life faces from the GENKI4K database. The top two rows are smile faces and the rest are nonsmile faces.

As regards classifiers, support vector machine has been universally accepted as an outstanding classifier for both of its efficiency and accuracy. Most SVM-based works adopt RBF kernel SVM, but one point deserves mentioning is that as the size of the training data increases, RBF is much more time-consuming compared to linear SVM.

Another classifier, Orthogonal Coordinate Coding (OCC) SVM, seems to strike a balance between accuracy and efficiency, which we will discuss later.

This paper is organized as follows. In Sect. 2, we will give a brief introduction to the features we used. Then some SVM classifiers will be discussed in Sect. 3, including the improved OCC. We also provide a sensitivity analysis in Sects. 4 and 5 will demonstrate the conclusions.

2   Features

This section introduces three well-known local features for image description, which have been evaluated in previous research on face recognition [11–13]. We would like to evaluate and compare their performance on smile detection tasks.

2.1   Local Binary Pattern

The original LBP operator, introduced by Ojala et al. [14], is a powerful means of texture description. The operator labels the pixels of an image by thresholding the 3×3 neighbourhood of each pixel with the center value and considering the result as a binary number. Then the histogram of the labels are used as a texture descriptor.

Later research extended the operator to use neighbourhood of different sizes [15]. Using circular neighbourhood and bilinearly interpolating the pixel values allow any radius and number of pixels in the neighbourhood.

Rotation invariance was later taken into consideration in LBP-based representation due to the circular property of the pattern. The experiment done by Ojala et al. (Ojala et al., 2002) indicates that only a particular subset of local binary patterns (those containing at most two bitwise transitions from 0 to 1 or vice versa) are typically present in most of the pixels contained in real images, and these patterns are referred to as uniform [15].

Concretely, the LBP operator is employed to obtain the LBP value of each pixel of the image in the uniform mode. Then the whole image is divided into same-sized small blocks of which a histogram is calculated, containing information about the distribution of the local texture, such as edges, spots and flat areas. Finally, regional histograms are concatenated to build a global description of the face image.

2.2   Histogram of Oriented Gradient

The basic idea behind the Histogram of Oriented Gradient descriptors is that local appearance and shapes within an image can be described by the distribution of intensity gradients or edge directions. The implementation of these descriptors can be achieved by dividing the image into small connected regions, called cells, and for each cell compiling a histogram of gradient directions or edge orientations for the pixels within the cell. The combination of these histograms then forms a descriptor. To improve accuracy, the local histograms can be contrast-normalized by calculating a measure of the intensity across a larger region of the image, called a block, and then using this value to normalize all cells within the block. The normalization results in better invariance to changes in illumination or shadowing.

With normalized pixel values, we calculate the gradient of each pixel point. Then we divide the face image into small cells, for example, 5*5, and calculate the histogram of each cell to form the cell descriptor. And we combine several cells to a block, and concatenate each cell descriptor into a block descriptor. Finally, block descriptors are concatenated to build a global description of the face.

2.3   Gabor Features

A Gabor filter can be seen as a sinusoidal plane of a particular frequency and orientation, modulated by a Gaussian envelop [16,17]. A 2-D Gabor function g(x, y) and its Fourier transform G(u, v) are defined as

$$g(x,y) = \frac{1}{2\pi\sigma_x\sigma_y} exp[-\frac{1}{2}(\frac{x^2}{\sigma_x^2} + (\frac{y^2}{\sigma_y^2}) + 2\pi jWx] \tag{1}$$

where $j = \sqrt{-1}$, and W is the frequency of the modulated sinusoid

$$G(u, v) = exp[-\frac{1}{2}[\frac{u - W^2}{\sigma_u^2} + \frac{v^2}{\sigma_v^2}] \qquad (2)$$

where $\sigma_u = 1/2\pi\sigma_x, \sigma_v = 1/2\pi\sigma_y$.

A self-similar filter dictionary can be obtained by associating an appropriate scale factor α and a rotation parameter θ with the mother wavelet g(x,y). M and N represent the scales and orientations of the Gabor wavelets, respectively.

$$g_{mn}(x, y) = \alpha(x', y'), 0 \leq m \leq M - 1, 0 \leq n \leq N - 1 \qquad (3)$$

where $x' = \alpha^{-m}(xcos\theta + ysin\theta), y' = \alpha^{-m}(-xsin\theta + ycos\theta)$ and $\theta = n\pi/K$, while K is the total number of orientations.

3   SVM

As a supervised learning model, the well-known support vector machine has already gained its dominance in classification issues. An interesting property of SVM is that it is an approximate implementation of the Structural Risk Minimisation (SRM) induction principle that aims at minimising a bound on the generalisation error of a model, rather than minimising the mean square error over the data set.

The original SVM was proposed to solve binary classification problems by finding the optimal separating hyperplane. It has achieved great success in linear separable problems. However, most practical problems are not linearly separable. To cope with this problem, previous research has found solutions by using kernel tricks, which maps the data to a higher dimension. The kernel function effectively solves the nonlinear problems, but in the meantime, it also brings limitations in practice: First, the complexity is highly dependent on the size of the training data, which means as the size of the training set grows, it takes increasingly long time to test a smile. Another limitation is the lack of theoretic support on how to choose kernel functions and setting parameters for specific problems.

Recently, a novel locally linear SVM has been proposed, which has smooth decision boundary and bounded curvature.

The standard linear SVM classifier is adopted as our first classifier, which takes the form

$$H(x) = \sum_{i=1}^{m} \alpha_i y^{(i)} < x^{(i)}, x > +b \qquad (4)$$

where $H(x) = 0$ is the boundary plane, $x^{((i))}$ is the support vector and $y^{(i)}$, α_i is the responding label and weight respectively. b is the bias value.

As for kernel SVM, it takes the form

$$H(x) = \sum_{j=1} y_j \alpha_j K(x, x^{(j)}) + b \qquad (5)$$

where $K(x, x_j)$ stands for the kernel we choose. We choose an SVM classifier with square kernel as our second classifier, which is $K(x, z) = (x^T z)^2$. For our third classifier, we adopt an RBF kernel, which is

$$K(x, z) = e^{-\frac{||x-z||^2}{2\sigma^2}} \tag{6}$$

To introduce the orthogonal coordinate coding, first we take another look at the standard linear SVM classifier function:

$$H(x) = w(x)^T x + b(x) \tag{7}$$

The thought behind locally linear SVM is that w is adapted with different input x, which could be described as

$$H(x) \approx \sum (c_v w(v)^T x + c_v b) = < W^T, c_x x^T > + c_x^2 b \tag{8}$$

where v belongs to the coding anchor points and $c_v(x)$ is the related coefficient. Imagine we select m anchor points, then W^T would be an m by n matrix and each row $w(v)^T$ is the corresponding anchor vector.

To preserve local characteristics, large amounts of anchor points are needed, which plays a significant role in the performance of locally linear SVM. As a result, if the input data is sparse, it could lead to many holes in the feature space. To solve this problem, Ziming Zhang [18] proposed a new LL-SVM exploiting an orthogonal coordinate coding (OCC) scheme, which encodes data using orthogonal anchor planes rather than anchor points.

Given an input matrix $X = [x^1, x^2, ...]$, if we select N orthogonal vectors to code X, the algorithm is presented as follows

4   Experiment

In this section we analyze the effect of illumination normalization, face image scales, face alignment and various controlled misalignments on smile detection.

Algorithm 1. The OCC coding algorithm.

Require: X: data matrix, $X = [x^1, x^2, ...]$
Ensure: find the coding vector $\hat{c}_x$
  1: Calculate the SVD of X . $X = U\Sigma V$, where U and V are two unitary matrices. Σ is a rectangular diagonal matrix, of which the elements are all non-negative real numbers and set in descending order;
  2: We select a few top eigenvectors u_i as our orthogonal basis vectors to form the basic matrix $\hat{U}$. Consider one input x, we code it as $\hat{c}_i = \frac{x^T u_i}{\sigma_i}$, where σ_i is the corresponding singular value. Then we define a generator matrix $G = \hat{U}\Sigma^{-1}$

  and $\Sigma = \begin{bmatrix} \sigma_1 & ... & 0 \\ ... & ... & ... \\ 0 & ... & \sigma_n \end{bmatrix}$. As a result, given a vector x, we code it as $\hat{c}_x = G^T x$;

  3: Normalize the $\hat{c}_x$ as $\hat{c}_x = \hat{c}_x = \frac{\hat{c}_x}{||\hat{c}_x||_1}$;

4.1 Data

By far, almost all the expression recognition studies have been carried out on facial expression databases that were collected under tightly controlled conditions containing limited number of subjects [19,20]. Hence, it leads to a lack of diversity in illumination conditions and individual differences which is unavoidable in our real world. To address this problem, we carried out our experiments on the publicly-available GENKI4K database. The face images were taken not by laboratory scientists, but by common people from all over the world taking photographs of themselves for personal purpose. The database consists of 4000 face images, of which 2162 are smile faces, and 1838 are nonsmile faces as shown in Fig. 1. As we can see, the images vary significantly in illumination conditions as well as poses and resolutions.

In our experiment, the images were first converted to gray scale. As for scaling and alignment, one set of images are normalized to into canonical faces of three different sizes: 24×24, 36×36 and 48×48 pixels, which is done using the face detector provided by Open CV 1.0. Another set of images are normalized to the same three levels but based on the manually labeled eye positions. Figure 2 illustrates some examples of the aligned face images and the unaligned ones.

4.2 Procedure

As regards feature extraction in the training and testing processes, we choose pixel value intensity, LBP, HOG, and Gabor features. Three classifiers, including linear SVM, OCC SVM and RBF kernel SVM, were applied in the experiments. We divide the face images into 4 similar sets and each contains similar number of smile faces and nonsmile faces. All of the sets are applied to a fourfold cross-validation, which means when one set is used as the test data, the other three

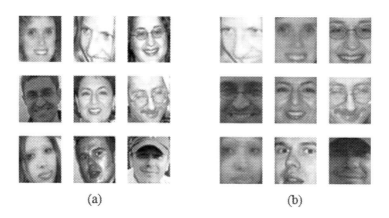

(a)   (b)

Fig. 2. (a) Face images that are detected directly by the detector provided by Open CV 1.0. (b) Face images that are aligned and cropped based on the manually labeled eyes positions

Table 1. Test variables and their conditions.

Variable	Conditions
Features	Pixel value intensity
	LBP features
	Gabor features
	HOG features
Classifier	Linear SVM
	OCC
	RBF SVM
Alignment	None
	Manual
Input image size	24×24
	36×36
	48×48

sets are used as training data. The procedure repeats four times for each set. Test variables are shown in Table 1.

A. Illumination Normalization

Face images from Figs. 1 and 2 witness a wide range of illumination conditions. To alleviate the impact of illumination conditions, we adopt several illumination normalization methods proposed in previous research for our experiment:

(1) Gaussian disposition: It first filters out the uncontrolled illumination changes by dividing the intensity value of each pixel in the original image by the one after performing Gaussian Smooth [21].
(2) Histogram equalization (HE): HE is a simple and widely-used technique for normalizing illumination effects.
(3) Discrete cosine transform (DCT): The DCT-based normalization [22] sets a number of DCT coefficients corresponding to low frequencies as zero to achieve illumination invariance.

We apply each of these methods to grayscale face images and then conduct face detection with the four features and three SVM classifiers mentioned above. The results can be seen in Table 2. Abnormally, it seems all the illumination normalization methods fail to work except the case of HE method employed with OCC and RBF SVM classifiers using LBP feature. All the other methods with the four features, however, bring the recognition rate down. It might be caused by the complexity of illumination conditions in the real world. Another possible explanation could be the characteristics of a smile face mainly rely on the structure of the face itself and the state of the organs, especially the mouth. Since these illumination normalization methods do not work as expected, the following experiments omit the illumination normalization step.

Table 2. Experiment results of different illumination normalization methods.

Illumination normalization	Features	Linear (%)	OCC (%)	RBF SVM (%)
None	Pixel	89.70	91.80	91.50
	LBP	92.28	92.47	92.38
	Gabor	91.13	92.03	92.08
	HOG	93.05	93.15	93.25
Gaussian	Pixel	84.70	87.35	87.35
	LBP	89.45	89.25	89.25
	Gabor	89.20	89.42	90.10
	HOG	91.70	92.03	92.13
HE	Pixel	88.50	91.30	90.65
	LBP	92.15	**92.70**	**92.40**
	Gabor	89.95	90.35	91.13
	HOG	91.15	92.48	92.00
DCT	Pixel	88.10	90.52	90.57
	LBP	91.32	91.47	91.52
	Gabor	90.56	91.05	90.89
	HOG	91.03	92.30	92.40

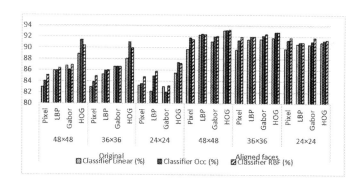

Fig. 3. Classification rates with different classifiers and face sizes

B. Scales and Alignment

The alignment method is conducted as follows: Firstly, we manually find the eye locations of each face image. Then we rotate the images to make the two eyes in the same horizontal line. Finally, we scale and crop the face images to ensure the two eyes to be at fixed positions.

All the methods achieve better classification rate with manually aligned faces than with unaligned ones, as can be seen from Table 3. Besides, it can be noticed that with unaligned face images, small-size images (24×24) are inferior to the larger ones in terms of classification rate. In contrast, as the size varies, the rate with aligned faces seems much more robust.

Table 3. Classification rates with different classifiers and face sizes.

	Image size	Feature	Classifier		
			Linear (%)	OCC (%)	RBF (%)
Original faces	48 × 48	Pixel	82.98	84.06	85.09
		LBP	85.93	85.96	86.39
		Gabor	86.73	86.07	86.91
		HOG	88.87	91.40	90.40
	36 × 36	Pixel	82.98	83.91	84.91
		LBP	85.22	85.91	86.01
		Gabor	86.63	86.65	86.65
		HOG	88.07	91.00	89.95
	24 × 24	Pixel	83.26	83.59	84.75
		LBP	82.19	84.94	85.80
		Gabor	83.04	82.05	83.22
		HOG	85.44	87.36	87.20
Aligned faces	48 × 48	Pixel	89.70	91.80	91.50
		LBP	92.28	92.47	92.38
		Gabor	91.13	92.03	92.08
		HOG	93.05	93.15	93.25
	36 × 36	Pixel	89.63	91.36	91.95
		LBP	91.50	92.00	92.03
		Gabor	91.52	92.13	92.48
		HOG	91.80	92.78	92.80
	24 × 24	Pixel	89.83	91.35	91.80
		LBP	90.63	90.93	90.93
		Gabor	90.55	91.08	83.22
		HOG	90.98	91.28	91.43

The classification rates for different face image sizes are illustrated in Table 3 and Fig. 3. Almost all the best classification rates are achieved with the image size 48 × 48, but compared with the size of 36 × 36, there is no statistically significant difference. When the image size goes down to 24 × 24, almost all the classification rates have a more noticeable drop.

It can also be observed that, regardless of the image size and the feature chosen, the OCC and RBF SVM classifiers perform almost equally well, much better than the linear SVM classifier does.

As for features, all the other features outperform the pixel value intensity. HOG feature performs the best almost under any condition, especially with the original faces of 48 × 48 and 36 × 36 sizes. The classification rates with LBP and Gabor features achieve similar results. However, with the size of 24 × 24, the advantage of the other features over pixel value intensity becomes less obvious.

One possible reason is that when the image size gets smaller, local texture could not be represented by the LBP and Gabor features appropriately.

C. Sensitivity Analysis

In this section, we give an analysis on the sensitivity of smile classifiers to various detection and alignment inaccuracies as Baluja and Rowley did in [23]. We use the manually aligned faces and add translations in horizontal and vertical directions from -3 to $+3$ pixels, in-plane rotation of the faces from $-45°$ to $+45°$, and Gaussian noise with the variance from 1 to 7.

As expected, we observe significant advantages of the three other features over pixel value intensity for they are more robust against image rotation, translation and Gaussian noise.

From the sensitivity analysis against rotation shown in Fig. 4, we can see that Gabor feature shows the lowest sensitivity against rotation, since its recognition rate remains the highest among the four features under the same conditions.

As for sensitivity against translation, which is shown in Fig. 5, HOG feature turns out to be the best among all the features.

As shown in Fig. 6, with the variance of Gaussian noise increasing from 1 to 7, the average classification rate with pixel value intensity decrease from 90.7 % to 73.1 %, LBP feature from 91.3 % to 78.9 % and HOG feature from 92.43 % to 79.38 %. Moreover, we observe that with the variance of Gaussian noise grows, the OCC and RBF SVM classifiers depict increasingly notable advantage over the linear SVM classifier.

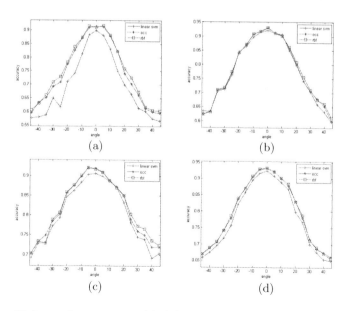

Fig. 4. Sensitivity against rotation. (a)–(d) illustrate the sensitivity against rotation with features of pixel intensity, LBP, Gabor and HOG respectively.

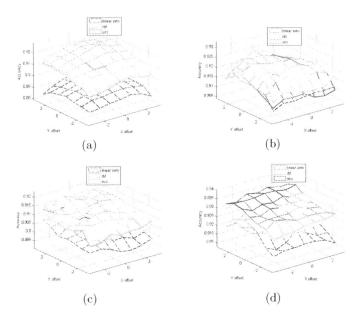

(a)   (b)

(c)   (d)

Fig. 5. Sensitivity of the classifiers against translation. (a)–(d) illustrate the sensitivity against translation with features of pixel intensity, LBP, Gabor and HOG respectively.

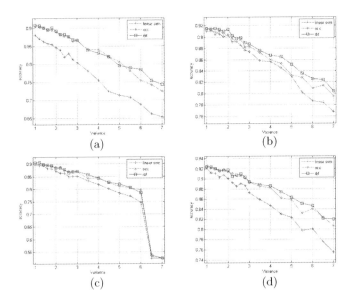

(a)   (b)

(c)   (d)

Fig. 6. Sensitivity of classifiers against Gaussian noise (a)–(d) illustrate the sensitivity against Gaussian noise with features of pixel intensity, LBP, Gabor and HOG respectively.

Besides, in almost all the sensitivity analyses above, the OCC classifier perform equally well with RBF SVM classifier, outperforming the linear SVM classifier.

5   Summary and Conclusions

Database: Most of the existing facial expression databases were collected under strictly controlled conditions containing limited number of subjects, which lack diversity in illumination conditions and individual differences. To address this problem, we select the publicly-available GENKI4K database which was built with world-wide self-portrait photos under various conditions.

Illumination and Size: As the results elucidated, there is no method found to fit the real-world scenarios, for the tested illumination normalization methods do not work properly to increase the smile detection rates and sometimes even prove to lower the rates. It might be caused by the diversity of the real-world environment. The size of input face images exerts limited impact on the smile detection rates. Almost all the best classification rates are achieved with the image size of 48×48, but the rates given by the size of 36×36 present no statistically difference. In general, the size of the images does not make much difference, but if it is too small, it may affect some of the features such as LBP and Gabor.

Alignment: As expected, the recognition rate is improved obviously with the alignment method. Under the same conditions including size, features and classifiers, the aligned face images all gain a better result. Surprisingly, under the same settings except for the alignment, on average, our result is 5.88 % higher than Shan and Caifeng's method in [24]. Our alignment method proves to be very effective. Besides, it is very simple and easy to conduct.

Features and Classifiers: Among the features of pixel value intensity, LBP, Gabor and HOG features, HOG feature can best represent a smiling face. In all the cases tested, it always gains the highest accuracy. All the other three features perform better than pixel value intensity, but one point deserves mentioning is that when the size of face images is small, LBP and Gabor features tend to be inferior to pixel value intensity. As for the classifiers, in terms of accuracy, OCC and RBF SVM classifiers perform almost equally well and both outperform the linear SVM with pixel value intensity. And when it comes to time consumption, OCC SVM classifiers are ten times faster than the RBF kernel. Compared with linear SVM and RBF SVM, it seems that OCC SVM balances well between accuracy and efficiency.

Acknowledgement. This work was partially sponsored by National Natural Science Foundation of China (NSFC) under Grant No. 61375031, No. 61471048, and No. 61273217. This work was also supported by the Fundamental Research Funds for the Central Universities, Beijing Higher Education Young Elite Teacher Project, and the Program for New Century Excellent Talents in University.

References

1. Pantic, M., Rothkrantz, L.J.: Facial action recognition for facial expression analysis from static face images. IEEE Trans. Syst. Man Cybern. Part B Cybern. **34**, 1449–1461 (2004)
2. Tian, Y.L., Kanade, T., Cohn, J.F.: Recognizing action units for facial expression analysis. IEEE Trans. Pattern Anal. Mach. Intell. **23**, 97–115 (2001)
3. Kapoor, A., Qi, Y., Picard, R.W.: Fully automatic upper facial action recognition. In: IEEE International Workshop on Analysis and Modeling of Faces and Gestures, 2003, AMFG 2003, pp. 195–202. IEEE (2003)
4. Wen, Z., Huang, T.S.: Capturing subtle facial motions in 3D face tracking. In: Proceedings of the Ninth IEEE International Conference on Computer Vision, 2003, pp. 1343–1350. IEEE (2003)
5. Sebe, N., Sun, Y., Bakker, E., Lew, M.S., Cohen, I., Huang, T.S.: Towards authentic emotion recognition. In: IEEE International Conference on Systems, Man and Cybernetics, 2004, vol. 1, pp. 623–628. IEEE (2004)
6. Cohn, J.F., Schmidt, K.L.: The timing of facial motion in posed and spontaneous smiles. Int. J. Wavelets Multiresolut. Inf. Process. **2**, 121–132 (2004)
7. Cohen, I., Sebe, N., Garg, A., Chen, L.S., Huang, T.S.: Facial expression recognition from video sequences: temporal and static modeling. Comput. Vis. Image Underst. **91**, 160–187 (2003)
8. Zhang, Y., Ji, Q.: Active and dynamic information fusion for facial expression understanding from image sequences. IEEE Trans. Pattern Anal. Mach. Intell. **27**, 699–714 (2005)
9. Shinohara, Y., Otsu, N.: Facial expression recognition using fisher weight maps. In: Proceedings of the Sixth IEEE International Conference on Automatic Face and Gesture Recognition, 2004, pp. 499–504. IEEE (2004)
10. Deniz, O., Castrillon, M., Lorenzo, J., Anton, L., Bueno, G.: Smile detection for user interfaces. In: Bebis, G., et al. (eds.) ISVC 2008, Part II. LNCS, vol. 5359, pp. 602–611. Springer, Heidelberg (2008)
11. Deng, W., Hu, J., Guo, J.: Extended SRC: undersampled face recognition via intraclass variant dictionary. IEEE Trans. Pattern Anal. Mach. Intell. **34**, 1864–1870 (2012)
12. Deng, W., Hu, J., Lu, J., Guo, J.: Transform-invariant PCA: a unified approach to fully automatic face alignment, representation, and recognition. IEEE Trans. Pattern Anal. Mach. Intell. **36**, 1275–1284 (2014)
13. Deng, W., Hu, J., Guo, J.: Linear ranking analysis. In: 2013 IEEE Conference on Computer Vision and Pattern Recognition (CVPR). IEEE (2014)
14. Ojala, T., Pietikainen, M., Maenpaa, T.: Multiresolution gray-scale and rotation invariant texture classification with local binary patterns. IEEE Trans. Pattern Anal. Mach. Intell. **24**, 971–987 (2002)
15. Ahonen, T., Hadid, A., Pietikäinen, M.: Face recognition with local binary patterns. In: Pajdla, T., Matas, J.G. (eds.) ECCV 2004. LNCS, vol. 3021, pp. 469–481. Springer, Heidelberg (2004)
16. Deng, W., Hu, J., Guo, J., Cai, W., Feng, D.: Emulating biological strategies for uncontrolled face recognition. Pattern Recogn. **43**, 2210–2223 (2010)
17. Deng, W., Hu, J., Guo, J., Cai, W., Feng, D.: Robust, accurate and efficient face recognition from a single training image: a uniform pursuit approach. Pattern Recogn. **43**, 1748–1762 (2010)

18. Zhang, Z., Ladicky, L., Torr, P., Saffari, A.: Learning anchor planes for classification. In: Advances in Neural Information Processing Systems, pp. 1611–1619 (2011)
19. Kanade, T., Cohn, J.F., Tian, Y.: Comprehensive database for facial expression analysis. In: Proceedings of the Fourth IEEE International Conference on Automatic Face and Gesture Recognition, 2000, pp. 46–53. IEEE (2000)
20. Sim, T., Baker, S., Bsat, M.: The CMU pose, illumination, and expression database. IEEE Trans. Pattern Anal. Mach. Intell. **25**, 1615–1618 (2003)
21. Wang, H., Li, S.Z., Wang, Y.: Generalized quotient image. In: Proceedings of the 2004 IEEE Computer Society Conference on Computer Vision and Pattern Recognition, 2004, CVPR 2004, vol. 2, p. II-498. IEEE (2004)
22. Chen, W., Er, M.J., Wu, S.: Illumination compensation and normalization for robust face recognition using discrete cosine transform in logarithm domain. IEEE Trans. Syst. Man Cybern. B Cybern. **36**, 458–466 (2006)
23. Baluja, S., Rowley, H.A.: Boosting sex identification performance. Int. J. Comput. Vision **71**, 111–119 (2007)
24. Shan, C.: Smile detection by boosting pixel differences. IEEE Trans. Image Process. **21**, 431–436 (2012)

Everything is in the Face? Represent Faces with Object Bank

Xin Liu[1,2], Shiguang Shan[1(⊠)], Shaoxin Li[1,2], and Alexander G. Hauptmann[3]

[1] Key Lab of Intelligent Information Processing of Chinese Academy of Sciences
(CAS), Institute of Computing Technology, CAS, Beijing 100190, China
{xin.liu,shiguang.shan,shaoxin.li}@vipl.ict.ac.cn
[2] University of Chinese Academy of Sciences, Beijing 100049, China
[3] School of Computer Science, Carnegie Mellon University,
Pittsburgh, PA 15213, USA
alex@cs.cmu.edu

Abstract. Object Bank (OB) [1] has been recently proposed as an object-level image representation for high-level visual recognition. OB represents an image from its responses to many pre-trained object filters. While OB has been validated in general image recognition tasks, it might seem ridiculous to represent a face with OB. However, in this paper, we study this anti-intuitive potential and show how OB can well represent faces amazingly, which seems a proof of the saying that "Everything is in the face". With OB representation, we achieve results better than many low-level features and even competitive to state-of-the-art methods on LFW dataset under unsupervised setting. We then show how we can achieve state of the art results by combining OB with some low-level feature (e.g. Gabor).

1 Introduction

To achieve accurate automatic face recognition, faces must be modeled appropriately. In the past decades, faces are widely represented by computer vision researchers with hand-crafted local features, e.g., Gabor [2], Local Binary Pattern (LBP) [3] and its high dimensional variant [4], patterns of oriented edge magnitudes(POEM) [5], Local Quantized Pattern (LQP) [6], Scale-Invariant Feature Transform (SIFT) [7], and Histogram of Oriented Gradients (HOG) [8], etc. However, designing an effective local descriptor demands considerable domain specific knowledge and a great deal of efforts.

Recently, data-driven representation learning is becoming popular and reports promising accuracy. In [9,10], filters are learned to maximize the discriminative power for face recognition. In [11], a multi-layer face representation framework is constructed based on large-scale random filter searching. In [12–14], codebook learning technologies are utilized for robust face representation design. More recently, faces are represented with mid-level or high-level semantic information. For instance, the attributes and simile classifier [15] represents faces by the mid-level face attributes and so-called simile feature. And the Tom-versus-Pete classifier [16] encodes faces with high-level semantic information by the

© Springer International Publishing Switzerland 2015
C.V. Jawahar and S. Shan (Eds.): ACCV 2014 Workshops, Part III, LNCS 9010, pp. 180–193, 2015.
DOI: 10.1007/978-3-319-16634-6_14

output scores of a large number of person-pair classifiers. In the last two years, deep learning methods are applied to face recognition to learn hierarchical face representation and report state-of-the-art results [17,18].

On the other hand, general object recognition has also been studied extensively in the past decade [19,20] and has achieved significant progress recently [21]. Actually, in terms of representation, general object recognition roughly shares similar conceptual paradigms to those for face recognition, i.e., low-level local features [22–24], mid-level features [25], high-level semantic features [1], and their combination via hierarchical deep models [21]. Especially, many low-level features, such as SIFT and HOG, have been applied successfully to both face recognition [7,8] and general object recognition [22,26]. However, it is unclear whether the same mid-level features and high-level features can be general enough to boost both face recognition and general object recognition.

Intuitively, it seems ridiculous to apply mid-level attribute or high-level semantic features for general object representation to face recognition. This intuition is further strengthened by the findings in the cognitive neuroscience literature that human vision system seems to have specific vision cortices for face recognition, separated from those for general object recognition. There is a lot of evidence that the primary locus for human face perception is found on the fusiform gyrus (so called fusiform face area, abbr. as FFA) of the extra-striate visual cortex [27,28]. In other words, perception mechanism of face recognition should be different from general object recognition. At least, the high-level semantic features for general object recognition are probably not sharable to face recognition, as there is indeed neither truck nor ship on human faces. Thus, computationally, it will seem silly to represent a face with general object filters like OB [1].

However, in this paper, we reach an anti-intuitive observation that faces can be computationally represented by high-level general "object filters", which seems to validate the saying that "everything is in the face". Specifically, we model faces by encoding the responses of OB filters operated on the input faces. With this OB representation, we surprisingly achieve results better than many low-level features and even competitive to state-of-the-art methods on LFW dataset [29] under unsupervised setting. We also investigate the influence of pooling methods, the scales of object filters, and importance of various objects. Finally, we show how we can achieve state-of-the-art results by combining OB with some low-level features (e.g. Gabor).

Our finding, i.e., faces can be represented by high-level object-based models, seems incredible at first glance. Nevertheless, we argue that it might be actually consistent with some cognitive observations, which supports the argument that faces are not processed specifically different from general object recognition [30]. It has been suggested that rather than being a true face module, the FFA may be responsible for performing either subordinate or expert-level categorization of generic objects, as suggested from both behavioral studies [28,31] and neuroimaging studies [32]. However, we must admit that it is still too early to make any determinate decision on what our findings suggest.

The rest of this paper is organized as follows: Sect. 2 briefly reviews OB and presents its application to face recognition. Then, in Sect. 3, we evaluate its accuracy on LFW restricted face recognition evaluation under unsupervised setting. The following section then describes how performance can be further improved by combining OB with other low-level features. Finally, we conclude the work and discuss future work in the last section.

2   Object Bank Face: Face Representation with Object Bank

In this section, we firstly briefly review the OB method [1]. Then, we describe how we apply OB to represent faces for face recognition.

2.1   Object Bank

OB is a high-level image representation which explicitly encodes the appearance and spatial location information of object in the image. OB represents image by collecting its responses to a large number of pre-trained object filters.

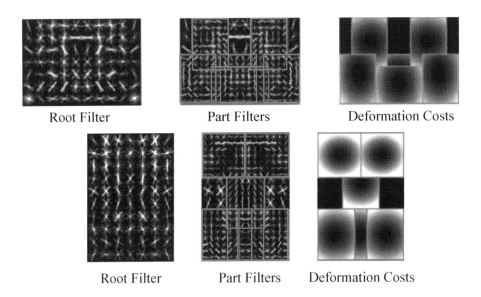

Root Filter   Part Filters   Deformation Costs

Root Filter   Part Filters   Deformation Costs

Fig. 1. An example DPM of bicycle trained on ImageNet dataset [20]. Two components are demonstrated, each component consists of a coarse root filter (1), multiple part filters (2) and a spatial model indicates the relative position of each part (3). The upper row shows the first component, which captures sideways views of bicycles. The lower row shows the second component, which captures frontal and near frontal views of bicycles.

In the OB framework, Deformable Part Model (DPM) [26] is employed as the object detector/filter. DPM is a widely used method for learning multi-view object detectors, and it is able to discriminatively learn a part-based object detector model using training samples only with the bounding box labels. This is done by treating the object parts as latent variables and learning by latent SVM method [26]. Each DPM is defined by a coarse root filter with multiple part filters. Figure 1 shows an example DPM of bicycle.

In Fig. 2, we demonstrate the feature extraction procedure of OB, including two main steps: convolution and pooling.

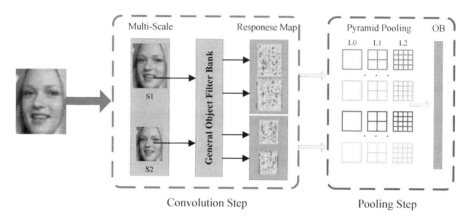

Fig. 2. Illustration of the OB. The convolution step is encircled by blue dotted box and the pooling step is encircled by dark red dotted box.

In the convolution step, pre-trained object filters are convolved with a multi-scale image pyramid of the input image. The response maps of all object filters in different scales are collected all together. Note that each object filter has two sets of filters with respect to different viewpoints, i.e., frontal and profile. So, in each scale, there are two response maps for each object filter. The final number of response map is $NumObject * NumScale * 2$. The value of each location on the response map indicates how likely an image patch contains a specific object.

In the pooling step, OB encodes both the appearance and spatial location information of objects. OB employs a 3-level pyramid pooling strategy. Each response map is divided into $1 * 1$ (L0), $2 * 2$ (L1) and $4 * 4$ (L2) grids. Two kinds of pooling operators are considered in this work. The first is the max-pooling, which computes the maximum response value in each grid. The second is the mean-pooling, which computes the mean response value in each grid. After pooling, the dimension of OB will be $NumObject * NumScale * 2 * 21$.

2.2   Object Bank Face (OBF)

Here we simply apply OB to face representation and name it as Object Bank Face (OBF). OBF can be seen as a high-level face representation, which is driven by models learned from the general object classes rather than the specific face classes.

Clearly, face images do not contain any general object, but face images can still form meaningful responses to OB filters. The response value reflects how likely a face patch "looks alike" the specific object in the HOG [23] feature space, as the object filters in OB do convolution on the HOG feature maps [26]. One can refer to [33] for more interesting discovery on how object detectors see this world in the HOG feature space.

3   Experimental Evaluation

In this section, we validate the OBF on face recognition task. Firstly, we introduce the evaluation dataset LFW [29] and setting of parameters. Then, we evaluate OBF on LFW and analyze the influence of different OBF setting in face recognition. Lastly, we discuss the novelty and insight of OBF.

3.1   Dataset and Setting of Parameters

In this subsection, we briefly introduce the LFW dataset, as well as the setting of parameters in our experiments.

Evaluation Dataset. LFW is a very challenging real-world face verification benchmark. The LFW dataset contains 13,233 face images of 5,749 persons. The face images in LFW have large variations in pose, expression, illumination, occlusion, etc. In Fig. 3, we show some example images of LFW.

Face Normalization. In this work, we utilize the LFW-a images [34] provided by Lior Wolf *et al.* and crop the face image to $120 * 96$ pixels with fixed eye center locations as $(29, 51)$ and $(66, 51)$;

OBF Details. We use the 208 pre-trained object filters provided by Li-Jia Li *et al.* [35], which are trained on ImageNet dataset [20]. Six detection scales and

Fig. 3. Example images in LFW which show the variations of face images in pose, expression, illumination and occlusion.

3-Level pyramid pooling are used as suggested in [1]. The dimension of the final OBF representation is $208*4*2*21 = 52,416$. Without additional specification, we take max-pooling as default pooling operator. In this work, whitening PCA (WPCA) is employed to reduce the dimensionality of OBF feature to 1,000, and finally cosine similarity is exploited for face matching.

3.2   Evaluation of OBF on LFW

We follow the LFW restricted protocol, which splits the LFW View 2 dataset into 10 subsets with each subset containing 300 positive pairs and 300 negative pairs. In this work, we follow the unsupervised setting. As requested by the protocol, we perform 10-fold cross-validation, among which 9 folds are used to train the WPCA model and the rest one is used for testing. The evaluation results are shown in Table 1. Note that, we use "(sqrt)" behind a specific method to denote additional signed square root operation of features. These results are given because many previous works have shown the effectiveness of sqrt operator especially for histogram-like features.

Table 1. Comparisons between OBF and state-of-the-art methods on LFW dataset under unsupervised setting.

Method	Mean accuracy rate(%)
LHS [36]	73.40 ± 0.40
LARK [37]	78.90
MRF-MLBP [38]	80.08 ± 0.13
POEM [5]	82.71 ± 0.59
OCLBP [39]	82.78 ± 0.41
High-dim LBP [4]	84.08
High-dim LE [4]	84.58
I-LQP [6]	86.20 ± 0.46
SFRD [14]	84.81
OBF	**83.80 ± 0.42**
OBF (sqrt)	83.75 ± 0.46

As can be seen from the table, the OBF surprisingly achieves 83.80 % mean accuracy on LFW View 2 under unsupervised setting, which outperforms many specifically designed face representation methods such as LARK [37], POEM [5] and OCLBP [39]. The performance of OBF is even competitive to state-of-the-art methods, which seems a proof of the saying that "Everything is in the face". In terms of face recognition, it seems that who you are is determined by what objects are in your face.

3.3   Analysis of OBF

In this section, we analyze the role of each component in OBF. We demonstrate the sensitivity of object filters on face and the influence of different OBF setting on face recognition performance. We eventually provide a good understanding of OBF.

Sensitivity of Object Filters to Face. In order to visualize the sensitivity of object filters to face, we do statistic of the mean response maps of each object filters on face. In this experiment, we use all the 13,233 face images from LFW dataset View 2 [29]. We randomly choose eight object filters and demonstrate the mean response maps in Fig. 4. Interesting phenomenon can be explored: big response value appears mostly around the eyes, nose and mouth. "Everything is in the face" has intuitive explanations.

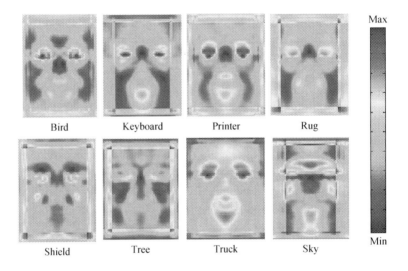

Fig. 4. Mean response maps of selected object filters on face.

Comparisons of Different Pooling Operator. As described earlier, the pooling process in OB encodes the spatial location and semantic meaning of objects in the input image. In this experiment, we analyze the performance of different pooling operator, i.e., max-pooling and mean-pooling, on face verification. We conduct the experiment on LFW View 2 under unsupervised setting. We reduce the original features to 1,000 dimensions using WPCA and take the cosine similarity. As it is shown in Table 2, the max-pooling outperforms mean-pooling. The same conclusion is also presented in [1], where mean-pooling outperforms max-pooling when applying OB for scene classification task.

Role of Scale. In OBF, there are totally six detection scales. In this experiment, we evaluate the pooled features of each scale independently. The results are

Table 2. Comparisons of the pooling operators of OBF on LFW dataset under unsupervised setting.

Pooling strategy	Mean accuracy rate(%)
OBF + Mean-Pooling	80.35 ± 0.55
OBF + Mean-Pooling (sqrt)	80.32 ± 0.54
OBF + Max-Pooling	83.80 ± 0.42
OBF + Max-Pooling (sqrt)	83.75 ± 0.46

Table 3. Comparisons of the performance of each scale of OBF on LFW dataset under unsupervised setting. The left column presents the feature map size, and the object filter kernel size is around $10 * 8$ or $8 * 10$.

Feature map size in pixel	Mean accuracy rate(%)
S1: $61 * 48$	75.92 ± 0.78
S2: $42 * 33$	76.82 ± 0.79
S3: $29 * 23$	78.78 ± 0.76
S4: $20 * 16$	79.52 ± 0.70
S5: $14 * 11$	80.33 ± 0.65
S6: $9 * 7$	79.78 ± 0.75

demonstrated in Table 3. One can see that individual scale performs different with each other, and small size scales delivers better performance. Furthermore, in this experiment, we also study single object performance in different scale. The same conclusion is also hold, and the single object filter performs best on small size scale S5 or S6. This phenomenon is not hold when applying OB for scene classification task as reported in [1], where middle size scale performs better. An intuitive interpretation is that in natural image, object often occupies middle size of the input image. While for face image, it shares common visual pattern with general object in bigger face regions.

Role of Object. In this experiment, we analyze the effectiveness of different types of objects on face verification performance. It is an interesting problem that which object contributes more to the performance. The test is also taken on LFW View 2 under unsupervised setting. We reduce the dimension of original feature using WPCA and take the cosine similarity. We sort the object filters according to the 10 folder mean accuracy on LFW View 2. In Fig. 5, we report the best 20 objects and worst 20 objects together with their mean accuracies.

In this experiment, we also evaluate the performance of using different number of object filters in OBF. For computation simplicity, we conduct the experiment on LFW View 1 under unsupervised setting. The LFW View 1 has a training set of 1,000 positive pairs and 1,000 negative pairs, and a test set of 500 positive pairs and 500 negative pairs. We reduce the dimension of original feature using WPCA. Dimension is fixed to the minimum number of principal

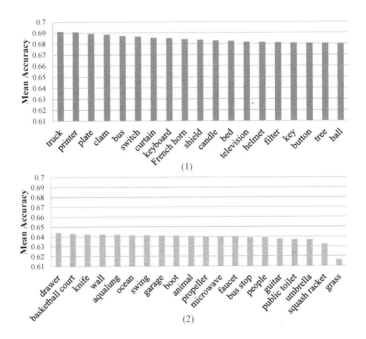

Fig. 5. Performance by using OBF generated by individual object on LFW View 2 under unsupervised setting. (1) The best 20 objects; (2) The worst 20 objects.

components to preserve 80 % of the training set variance. By adding one object at a time, we get the performance curve in Fig. 6. The classification accuracy increases along with the number of object filters progressively, and we believe more powerful and discriminative face representation can be constructed with the growth of number of object filters.

3.4   Discussion

Above experiments and observations suggest that general object filters can indeed achieve very promising accuracy for face recognition. This observation seems anti-intuitive, as we are not so silly to believe there are indeed trucks or printer on face. So, if we were not wrong in implementing, the effectiveness has to be hidden somewhere in the mechanism. To explain this in principal, we argue that, each object is actually composed of a large number of low-level features or mid-level attributes. In other words, an object filter models the co-occurrence of plenty of low-level filters and/or mid-level filter. Therefore, the effectiveness of OB for face recognition might essentially come from its modeling of co-occurrence of many lower-level features. We can imagine that, a truck, in a suitable scale, can be like human eye, especially in the sense of several horizontal edges.

In terms of cognitive neuroscience, as mentioned in the introduction part, it has long been controversial whether face processing is a dedicated procedure. Our findings suggest that, at least, in the low-level or even mid-level, the perception

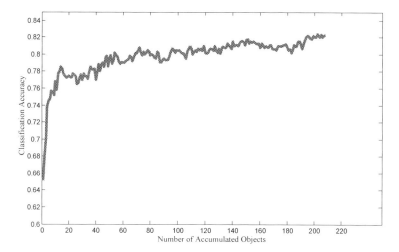

Fig. 6. Classification performance on LFW View 1 by using OBF corresponding to accumulative objects under unsupervised setting. X axis is the number of objects. Y axis represents the average classification accuracy.

mechanism might be similar (if not the same) for face and general object processing. There is just no need to distinguish the horizontal edges on a truck or those on a face. What matters is how the low-level features are combined together in some co-occurrence mode, in the higher level of perceptual organization.

4   Combination with Low-Level Features

As a high-level representation, OB is intuitively complementary for low-level feature based face representation. In the following, we firstly present the experimental setting. Then, we evaluate the proposed method on LFW.

We combine the OBF with LBP Face, SIFT Face or Gabor Face and evaluate on LFW View 2 under unsupervised setting. We reduce the dimension of original feature using WPCA and take the cosine similarity. In this experiment, we take score-level fusion with equal weight by simply summating the corresponding similarity matrices. The setting of parameters for low-level feature extraction is presented as follows.

LBP Face Details. We take the uniform LBP setting as suggested in [3], and we set the block size as $8 * 8$ and the radius as 2.

SIFT Face Details. We take the dense SIFT feature, which compute SIFT descriptor [22] on dense grid on the face image. In this experiment, we set the grid size as $4 * 4$.

Gabor Face Details. In the Gabor face, 40 Gabor wavelets with 5 scales and 8 orientations are utilized, the parameters are set as suggested in [40]. We take a $4 * 4$ mean pooling on the 40 Gabor magnitude images to reduce the feature dimensionality.

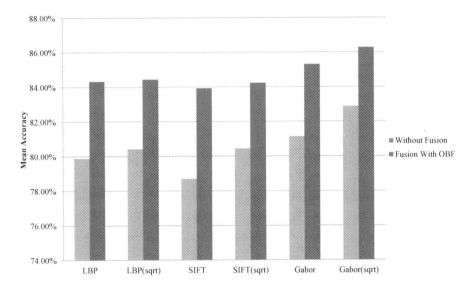

Fig. 7. Score-level fusion of OBF and low-level feature based face representation on LFW dataset under unsupervised setting.

The results are shown in Fig. 7. As it is shown, the fusing of OBF and low-level feature significantly improves the performance. Specially, we achieve 86.27 % mean accuracy by fusing OBF and Gabor (sqrt), which is the state-of-the-art performance on LFW under unsupervised setting. We can conclude that OBF is an efficient complementary representation to improve face representation robustness.

5   Conclusions and Future Work

Face representation is a fundamental problem in face recognition research. In this work, we apply general object filters, i.e., OB, to face representation. The experimental evaluations on LFW demonstrate that face representation can well leverage the high-level models of general objects. Furthermore, by combining with low-level features, state-of-the-art performance can even be achieved on LFW.

The observation of this work naturally raises the conjecture whether midlevel features, e.g., attributes, for general object recognition, can better facilitate face recognition. So, in the future, we will study how general attribute can be exploited for face recognition.

Acknowledgement. The work is supported by Natural Science Foundation of China under contract No. 61222211.

References

1. Li, L.J., Su, H., Lim, Y., Fei-Fei, L.: Object bank: an object-level image representation for high-level visual recognition. Int. J. Comput. Vis. **107**, 1–20 (2013)
2. Liu, C., Wechsler, H.: Gabor feature based classification using the enhanced fisher linear discriminant model for face recognition. IEEE Trans. Image Process. **11**, 467–476 (2002)
3. Ahonen, T., Hadid, A., Pietikainen, M.: Face description with local binary patterns: application to face recognition. IEEE Trans. Pattern Anal. Mach. Intell. **28**, 2037–2041 (2006)
4. Chen, D., Cao, X., Wen, F., Sun, J.: Blessing of dimensionality: high-dimensional feature and its efficient compression for face verification. In: Computer Vision and Pattern Recognition (CVPR), pp. 3025–3032. IEEE (2013)
5. Vu, N.S., Caplier, A.: Enhanced patterns of oriented edge magnitudes for face recognition and image matching. IEEE Trans. Image Process. **21**, 1352–1365 (2012)
6. Hussain, S.U., Napoléon, T., Jurie, F., et al.: Face recognition using local quantized patterns. In: British Machive Vision Conference (BMVC) (2012)
7. Bicego, M., Lagorio, A., Grosso, E., Tistarelli, M.: On the use of sift features for face authentication. In: CVPRW, pp. 35–35. IEEE (2006)
8. Albiol, A., Monzo, D., Martin, A., Sastre, J., Albiol, A.: Face recognition using HOG-EBGM. Pattern Recogn. Lett. **29**, 1537–1543 (2008)
9. Kumar, R., Banerjee, A., Vemuri, B.C., Pfister, H.: Trainable convolution filters and their application to face recognition. IEEE Trans. Pattern Anal. Mach. Intell. **34**, 1423–1436 (2012)
10. Lei, Z., Yi, D., Li, S.Z.: Discriminant image filter learning for face recognition with local binary pattern like representation. In: Computer Vision and Pattern Recognition (CVPR), pp. 2512–2517. IEEE (2012)
11. Cox, D., Pinto, N.: Beyond simple features: a large-scale feature search approach to unconstrained face recognition. In: 2011 IEEE International Conference on Automatic Face and Gesture Recognition and Workshops (FG), pp. 8–15. IEEE (2011)
12. Xie, S., Shan, S., Chen, X., Meng, X., Gao, W.: Learned local gabor patterns for face representation and recognition. Sig. Process. **89**, 2333–2344 (2009)
13. Cao, Z., Yin, Q., Tang, X., Sun, J.: Face recognition with learning-based descriptor. In: Computer Vision and Pattern Recognition (CVPR), pp. 2707–2714. IEEE (2010)
14. Cui, Z., Li, W., Xu, D., Shan, S., Chen, X.: Fusing robust face region descriptors via multiple metric learning for face recognition in the wild. In: Computer Vision and Pattern Recognition (CVPR), pp. 3554–3561. IEEE (2013)
15. Kumar, N., Berg, A.C., Belhumeur, P.N., Nayar, S.K.: Attribute and simile classifiers for face verification. In: Computer Vision and Pattern Recognition (CVPR), pp. 365–372. IEEE (2009)
16. Berg, T., Belhumeur, P.N.: Tom-vs-pete classifiers and identity-preserving alignment for face verification. In: British Machine Vision Conference (BMVC), vol. 1, p. 5 (2012)
17. Huang, G.B., Lee, H., Learned-Miller, E.: Learning hierarchical representations for face verification with convolutional deep belief networks. In: Computer Vision and Pattern Recognition (CVPR), pp. 2518–2525. IEEE (2012)
18. Taigman, Y., Yang, M., Ranzato, M., Wolf, L.: Deepface: closing the gap to human-level performance in face verification. In: Computer Vision and Pattern Recognition (CVPR) (2014)

19. Fei-Fei, L., Fergus, R., Perona, P.: Learning generative visual models from few training examples: an incremental bayesian approach tested on 101 object categories. Comput. Vis. Image Underst. **106**, 59–70 (2007)

20. Deng, J., Dong, W., Socher, R., Li, L.J., Li, K., Fei-Fei, L.: Imagenet: a large-scale hierarchical image database. In: Computer Vision and Pattern Recognition (CVPR), pp. 248–255. IEEE (2009)

21. Krizhevsky, A., Sutskever, I., Hinton, G.E.: Imagenet classification with deep convolutional neural networks. In: NIPS, vol. 1, p. 4 (2012)

22. Lowe, D.G.: Distinctive image features from scale-invariant keypoints. Int. J. Comput. Vision **60**, 91–110 (2004)

23. Dalal, N., Triggs., B.: Histograms of oriented gradients for human detection. In: Computer Vision and Pattern Recognition (CVPR) (2005)

24. Bay, H., Tuytelaars, T., Van Gool, L.: SURF: speeded up robust features. In: Leonardis, A., Bischof, H., Pinz, A. (eds.) ECCV 2006, Part I. LNCS, vol. 3951, pp. 404–417. Springer, Heidelberg (2006)

25. Fernando, B., Fromont, E., Tuytelaars, T.: Mining mid-level features for image classification. Int. J. Comput. Vision **108**, 186–203 (2014)

26. Felzenszwalb, P.F., Girshick, R.B., McAllester, D., Ramanan, D.: Object detection with discriminatively trained part-based models. IEEE Trans. Pattern Anal. Mach. Intell. **32**, 1627–1645 (2010)

27. Kanwisher, N., McDermott, J., Chun, M.M.: The fusiform face area: a module in human extrastriate cortex specialized for face perception. J. Neurosci. **17**, 4302–4311 (1997)

28. Grill-Spector, K., Knouf, N., Kanwisher, N.: The fusiform face area subserves face perception, not generic within-category identification. Nat. Neurosci. **7**, 555–562 (2004)

29. Huang, G.B., Ramesh, M., Berg, T., Learned-Miller, E.: Labeled faces in the wild: A database for studying face recognition in unconstrained environments. Technical report, Technical report 07–49, University of Massachusetts, Amherst (2007)

30. Sinha, P., Balas, B., Ostrovsky, Y., Russell, R.: Face recognition by humans: nineteen results all computer vision researchers should know about. Proc. IEEE **94**, 1948–1962 (2006)

31. Gauthier, I., Tarr, M.J.: Becoming a greeble expert: exploring mechanisms for face recognition. Vis. Res. **37**, 1673–1682 (1997)

32. Gauthier, I., Anderson, A.W., Tarr, M.J., Skudlarski, P., Gore, J.C.: Levels of categorization in visual recognition studied using functional magnetic resonance imaging. Curr. Biol. **7**, 645–651 (1997)

33. Vondrick, C., Khosla, A., Malisiewicz, T., Torralba, A.: HOGgles: visualizing object detection features. In: International Conference on Computer vision (ICCV) (2013)

34. Taigman, Y., Wolf, L., Hassner, T.: Multiple one-shots for utilizing class label information. In: British Machine Vision Conference (BMVC), pp. 1–12 (2009)

35. Vision Lab, S.U.: http://vision.stanford.edu/projects/objectbank/

36. Sharma, G., ul Hussain, S., Jurie, F.: Local higher-order statistics (LHS) for texture categorization and facial analysis. In: Fitzgibbon, A., Lazebnik, S., Perona, P., Sato, Y., Schmid, C. (eds.) ECCV 2012, Part VII. LNCS, vol. 7578, pp. 1–12. Springer, Heidelberg (2012)

37. Seo, H.J., Milanfar, P.: Face verification using the lark representation. IEEE Trans. Inf. Forensics Secur. **6**, 1275–1286 (2011)

38. Arashloo, S.R., Kittler., J.: Efficient processing of mrfs for unconstrained-pose face recognition. Biometrics: Theory, Applications and Systems 2013 (2013)

39. Barkan, O., Weill, J., Wolf, L., Aronowitz, H.: Fast high dimensional vector multi-plication face recognition. In: International Conference on Computer vision (ICCV) (2013)
40. Li, Y., Shan, S., Zhang, H., Lao, S., Chen, X.: Fusing magnitude and phase features for robust face recognition. In: Lee, K.M., Matsushita, Y., Rehg, J.M., Hu, Z. (eds.) ACCV 2012, Part II. LNCS, vol. 7725, pp. 601–612. Springer, Heidelberg (2013)

Quasi Cosine Similarity Metric Learning

Xiang Wu, Zhi-Guo Shi$^{(\boxtimes)}$, and Lei Liu

School of Computer and Communication Engineering, University of Science
and Technology Beijing, No. 30 Xueyuan Road, Haidian District, Beijing, China
szg@ustb.edu.cn

Abstract. It is vital to select an appropriate distance metric for many
learning algorithm. Cosine distance is an efficient metric for measuring
the similarity of descriptors in classification task. However, the cosine
similarity metric learning (CSML) [3] is not widely used due to the complexity of its formulation and time consuming. In this paper, a Quasi
Cosine Similarity Metric Learning (QCSML) is proposed to make it easy.
The normalization and Lagrange multipliers are employed to convert
cosine distance into simple formulation, which is convex and its derivation is easy to calculate. The complexity of the QCSML algorithm is
$O(t \times p \times d)$ (The parameters t, p, d represent the number of iterations,
the dimensionality of descriptors and the compressed features.), while
the complexity of CSML is $O(r \times b \times g \times s \times d \times m)$ (From the paper
[3], r is the number of iterations used to optimize the projection matrix,
b is the number of values tested in cross validation process, g is the
number of steps in the Conjugate Gradient method, s is the number of
training data, d and m are the dimensions of projection matrix.). The
experimental results of our method on UCI datasets for classification
task and LFW dataset for face verification problem are better than the
state-of-the-art methods. For classification task, the proposed approach
is employed on Iris, Ionosphere and Wine dataset and the classification
accuracy and the time consuming are much better than the compared
methods. Moreover, our approach obtains 92.33 % accuracy for face verification on unrestricted setting of LFW dataset, which outperforms the
state-of-the-art algorithms.

1 Introduction

An appropriate distance measure (or metric) is fundamental to many supervised
and unsupervised learning algorithm such as k-means, kernel method, the nearest neighborhood classification and so on. Besides, it is important for varieties
of application such as image retrieval or face recognition to choose a proper distance metric to measure the similarity or dissimilarity between different images.
Therefore, to apply an appropriate distance metric for practical applications, lots
of distance metric learning algorithm methods are proposed to find the special
latent relevance between different samples.

However, choosing a proper distance metric is highly problem-specific and
ultimately dictates the success of the actual learning algorithm. Many existing

C.V. Jawahar and S. Shan (Eds.): ACCV 2014 Workshops, Part III, LNCS 9010, pp. 194–205, 2015.
DOI: 10.1007/978-3-319-16634-6_15

algorithms for metric learning have been shown to perform well in different application, but most of them do not perform well in high dimensional input. The high dimensional descriptor exists in a wide range of application such as image retrieval [2], face recognition [3,4] and natural language processing. In these occasions, it is necessary to compress the high dimensional descriptors into low dimensional ones due to high computation and storage. So the distance metric learning is not only used to make data separately but also to compress the high dimensional vectors.

Recently, Mahalanobis distance metric has been widely applied in many aspects as a metric learning measure [1,5,6]. Xing et al. [1] applied semidefinite programming (SDP) objective function to learn a Mahalanobis distance metric for clustering. They minimize the sum of Euclidean distance between similarity labeled inputs and maintained a lower bound on the distance between different ones. Davis et al. [5] use information-theoretic regularization term for Euclidean distance. Moreover, Qi et al. [6] formulate a sparse Mahalanobis matrix which reflects the intrinsic nature of sparsity. They impose a sparse prior and show the obtained l_1-penalized Log-Determinant optimization problem for sparse metric can be minimized by a block coordinate descent algorithm [7], which is faster than SDP method widely used in metric learning.

Moreover, cosine similarity is an efficient distance metric to comparing the difference between vectors and it is an effective alternative to Euclidean distance in metric learning problem. Nguyen proposed a cosine similarity metric learning (CSML) [3] which can improve the generalization ability of an existing metric significantly in most cases. But it is not useful for high dimensional descriptors because of the highly memory used and computing burden for gradient descent method. In paper [4], Cao proposed a similarity metric learning (SML) which combines Euclidean distance and cosine similarity as the metric learning objective function. The formulation of similarity metric learning is convex and Cao optimized the dual formulation to obtain the global solution instead.

In this paper, due to the high computation and memory used, we proposed a Quasi Cosine Similarity Metric Learning (QCSML) for high dimensional vectors. There are two main contributions of our method. One is that we have introduced a novel solution for cosine similarity metric learning problem. The other is that QCSML is efficient for high dimensional vectors which are usually as the descriptors for face recognition, image classification and image retrieval. QCSML is not only discriminative for classification tasks but also used for dimensionality reduction.

The paper is organized as following. In Sect. 2, we review distance metric and similarity metric for classification. In Sect. 3, we formulate the detail of Quasi Cosine Similarity Metric Learning. The learning algorithm and gradient descent optimization method is introduced. Section 4 evaluates the proposed Quasi Cosine Similarity Metric Learning algorithm on UCI datasets for classification and LFW datasets for face recognition. Finally, the conclusion is given in Sect. 5.

2   Preliminary

In this section, we briefly review general distance metric learning [1], cosine similarity metric learning [3] and similarity metric learning [4].

2.1   Metric Learning

Given a set of points $X = \{x_1, \ldots, x_n\}, x_i \in \mathbb{R}^d$, we can define a positive definite matrix $A \in \mathbb{R}^{d \times d}$ which represents the Mahalanobis distance.

$$d_A^2(x_i, x_j) = (x_i - x_j)^T A(x_i - x_j) \tag{1}$$

The goal of metric learning is to adapt the metric function to the problem using information from the training datasets. Because of positive definite characteristics, the matrix A can be decomposed as $A = W^T W, W \in \mathbb{R}^{p \times d}$, therefore, the metric function can be shown as

$$d_W^2(x_i, x_j) = (x_i - x_j)^T W^T W(x_i - x_j) = \|Wx_i - Wx_j\|_2^2 \tag{2}$$

Here, we assume we have known the prior knowledge about the relationship constraining the similarity or dissimilarity between pairs of points.

$$\begin{aligned} S : (x_i, x_j) \in S \quad & \text{if } x_i \text{ and } x_j \text{ are similar} \\ D : (x_i, x_j) \in D \quad & \text{if } x_i \text{ and } x_j \text{ are dissimilar} \end{aligned} \tag{3}$$

This gives the optimization problem

$$\begin{aligned} \min \ & \textstyle\sum_{(x_i, x_j) \in S} \|Wx_i - Wx_j\|_2^2 \\ \text{s.t.} \ & \textstyle\sum_{(x_i, x_j) \in D} \|Wx_i - Wx_j\|_2^2 \geq 1 \end{aligned} \tag{4}$$

Then in paper [1], Xing introduced the efficient algorithm using the Newton-Raphson method to optimize the objective function.

2.2   Cosine Similarity Metric Learning

Compared with distance metric, cosine similarity between two vectors can be defined as

$$d_W^2(x_i, x_j) = \frac{(Wx_i)^T Wx_j}{\|Wx_i\|\|Wx_j\|} \tag{5}$$

Given the similar sets S and dissimilar sets D, the objective function can be shown as

$$\max \sum_{(x_i, x_j) \in S} d_W^2(x_i, x_j) - \alpha \sum_{(x_i, x_j) \in D} d_W^2(x_i, x_j) \tag{6}$$

where d_W^2 is defined as Eq. (5)

In this optimization problem, it is difficult to calculate the derivation of objective function which is used for gradient descent method. In paper [3], Nguyen gives the gradient as

$$
\begin{aligned}
\frac{\partial d_W^2(x_i, x_j)}{\partial W} &= \frac{\partial(\frac{u(W)}{v(W)})}{\partial W} \\
&= \frac{1}{v(W)} \frac{\partial u(W)}{\partial W} - \frac{u(W)}{v(W)^2} \frac{\partial v(W)}{\partial W}
\end{aligned}
\tag{7}
$$

where

$$
\begin{cases}
\frac{\partial u(W)}{\partial W} = W(x_i x_j^T + x_j x_i^T) \\
\frac{\partial v(W)}{\partial W} = \frac{\|W x_j\|}{\|W x_i\|} W x_i x_i^T - \frac{\|W x_i\|}{\|W x_j\|} W x_j x_j^T
\end{cases}
\tag{8}
$$

As is shown in Eq. (8), the complexity of the gradient is too high to compute if the dimensionality of descriptors x_i is large. For example, if the descriptors are high dimensional Fisher Vector (FV) which is about 67586-d in the paper [8], the cosine similarity metric learning will be inefficient and ineffective for dimensionality reduction and data classification.

3   Quasi Cosine Similarity Metric Learning

In this section, we first give the objective function of Quasi Cosine Similarity Metric Learning (QCSML), and introduce the hinge-loss to represent the objective function. The stochastic gradient descent (SGD) is used to optimize the objective function.

3.1   Problem Formulation

We begin this section with some notation definitions. Our goal is to learn the cosine similarity in Eq. (5) from a set of feature space $X = \{x_1, \ldots, x_n\}, x_i \in \mathbb{R}^d$. Obviously, the gradient of the cosine similarity function in Eqs. (7) and (8) is complex and if the dimension of feaature d is large, the optimization processing will take up lots of memory which personal computer cannot afford.

As is shown in Eq. (5), the cosine similarity metric can be written as

$$
d_W^2(x_i, x_j) = \frac{(W x_i)^T W x_j}{\|W x_i\|\|W x_j\|} \geq \frac{(W x_i)^T W x_j}{(\|W\|\|x_i\|)(\|W\|\|x_j\|)}
\tag{9}
$$

because of the inequality $\|ab\| \leq \|a\|\|b\|$. Therefore, the cosine similarity metric can be written as

$$
d_W^2(x_i, x_j) \geq \frac{(W x_i)^T W x_j}{\|W\|^2 \|x_i\|\|x_j\|}
\tag{10}
$$

Here, to simplify the cosine similarity metric in Eq. (10), we can give the prior knowledge about the projection matrix W which is represented as

$$
\|W\|_F^2 = \mathrm{Tr}(WW^T) = \sum_{i=1}^{p} \sum_{j=1}^{d} W_{ij}^2 = 1
\tag{11}
$$

where $\|\ \|_F$ is the Frobenius norm of matrix. And then we normalize the descriptors $\{x_i\}$ in feature space which means $\|x_i\|_2 = 1, x_i \in \mathbb{R}^d, i = 1, \ldots, n$.

With these constrains, the cosine similarity metric can be written as

$$d_W^2(x_i, x_j) \geq \tilde{d}_W^2(x_i, x_j) = (Wx_i)^T Wx_j \tag{12}$$

Then we define the label y_{ij} represents the similarity or dissimilarity between a pair of two vectors (x_i, x_j). Therefore, we can denote a threshold $b \in \mathbb{R}$ that the pair is similar or different if the cosine similarity $d_W^2(x_i, x_j)$ is upon or below the threshold b. Therefore, these constrains can be defined as

$$y_{ij}(\tilde{d}_W^2(x_i, x_j) - b) > 1 \tag{13}$$

where $y_{ij} = 1$ if x_i and x_j are similar, which means $(x_i, x_j) \in S$, and $y_{ij} = -1$ otherwise.

According the constrains in Eq. (13), the quasi cosine similarity metric learning problem can be defined as

$$\begin{aligned} &\min \sum_{i,j} \max \left[1 - y_{ij}(\tilde{d}_W^2(x_i, x_j) - b), 0\right] \\ &\text{s.t.} \quad \|x_i\|_2 = 1, x_i \in S \cup D, i = 1, \ldots, n \\ &\qquad\quad \text{Tr}(WW^T) = 1 \end{aligned} \tag{14}$$

The objective function is to make the margin between the positive and negative pairs to be large, since it is hinge-loss. The hinge loss function is used for max-margin classification problem, mostly notably for support vector machine (SVM).

Due to the only equality constraint and the normalization of the input vectors $\{x_i\}$, the Lagrange multipliers method can be used to convert the QCSML problem into an unconstrained problem and the converted objective function can be written as

$$\min \sum_{i,j} \max \left[1 - y_{ij}(\tilde{d}_W^2(x_i, x_j) - b), 0\right] + \lambda(\text{Tr}(WW^T) - 1) \tag{15}$$

where λ is the Lagrange multipliers and $\lambda > 0$. Because of the $W^T W$ is positive definite matrix, the objective function is convex. The constrain $\text{Tr}(WW^T) = \|W\|_F^2 = 1$ can also be treated as the regularization to prevent overfitting and in this aspect, λ can be considered as the trade-off parameters.

3.2   Algorithm and Complexity

Due to the convexity of objective function, we can use the gradient descent method to get the global value of Quasi Cosine Similarity Metric Learning.

Instead of Conjugate Gradient method [3], we employ stochastic gradient descent (SGD) method [9] to optimize the objective function due to large dataset. The projection matrix W can be updated as following

$$W_{t+1} = \begin{cases} W_t & \text{if} \quad y_{ij}(d_W^2(x_i, x_j) - b) > 1) \\ W_t - \alpha \frac{\partial f(W)}{\partial W} & \text{otherwise} \end{cases} \tag{16}$$

Algorithm 1. Quasi Cosine Similarity Metric Learning Optimization

---

Input   : Train data: $(X, y), X = \{x_i\} \subseteq \mathbb{R}^d, \|x_i\| = 1, y_i \in \{-1, 1\}$
Output: The parameters:$\Theta = \{W, b\}$

begin
   Parameters initialization;

   /* Initialize W */
   Set W_{init} =PCA_whitening(X);

   /* Initialize b */
   (X_s, y_s) =sample(X,y);
   $\phi_s = W X_s$;
   score$_s$ = $\tilde{d}_w^2(x_i, x_j)$ $\forall x_i, x_j \in X_s$;
   b_{init} = accuracy_best(score$_s$, y_s);

   /* SGD iteration   */
   for $t = 1$ **to** n **do**
      switch y_t **do**
         case *positive*
            score = $\tilde{d}_W^2(x_i, x_j)$ $\forall x_i, x_j \in X_t$;
            if score $< b + 1$ **then**
               $W_{t+1} = W_t - \alpha \frac{\partial \tilde{d}_W^2}{\partial W}$;
               $b_{t+1} = b_t - \gamma_b$;
            break;
         case *negative*
            score = $\tilde{d}_W^2(x_i, x_j)$ $\forall x_i, x_j \in X_t$;
            if score $> b - 1$ **then**
               $W_{t+1} = W_t + \alpha \frac{\partial \tilde{d}_W^2}{\partial W}$;
               $b_{t+1} = b_t + \gamma_b$;
            break;
   return $\Theta = \{W, b\}$;

---

where α is learning rate, t is the number of iterations and $f(W)$ represents the objective function. The gradient of objective function is written as

$$\frac{\partial f(W)}{\partial W} = y_{ij} W (x_i x_j^T + x_j x_i^T) + \lambda W \tag{17}$$

where λ is the parameter which is set by us. The threshold b can be updated by

$$b_{t+1} = \begin{cases} b_t & \text{if} \quad y_{ij}(\tilde{d}_W^2(x_i, x_j) - b) > 1) \\ b_t - \gamma_b y_{ij} & \text{otherwise} \end{cases} \tag{18}$$

where γ_b is the threshold bias.

We can use SGD to update the parameters $\Theta = (W, b)$. The detail of Quasi Cosine Similarity Metric Learning is given in Algorithm 1. The choice of parameters α and λ is important for the optimization. The learning rate α controls the speed of gradient descent for the optimization processing. With the SGD method for optimization, the learning rate α should be smaller than L-BFGS [10] or conjugate gradient [3] because we calculate the gradient only use one sample (x_i, x_j) from data set at every iteration. Moreover, the trade-off parameter λ is also important for optimization if it is suitable or not. Although the objective function is convex, we also need to give a good initialization for projection matrix W due to the SGD method. K-means and PCA-whitening are both choices for initialization and the effectiveness will be performed in Sect. 4.

According to the Algorithm 1, the complexity of computing the gradients of objective function is $O(p \times d)$ where d is the dimensionality of descriptors x_i and p is the dimension of reduction by projection matrix W. Therefore, the complexity of QCSML algorithm is $O(t \times p \times d)$ where t is the number of iterations to update the projection matrix W by SGD. It is faster than CSML which the complexity is $O(r \times b \times g \times s \times d \times m)$, where r is the number of iterations used to optimize the projection matrix, b is the number of values tested in cross validation process, g is the number of steps in the Conjugate Gradient method, s is the number of training data, d and m are the dimensions of projection matrix. And the experiments of QCSML for time consuming are in next section.

4 Experiments

In this section, we employ the proposed Quasi Cosine Similarity Metric Learning (QCSML) on various benchmark UCI datasets and LFW datasets [12] for face recognition.

4.1 UCI Datasets Classification

We evaluate the algorithm on three UCI datasets: Iris[1], Ionosphere[2] and Wine[3]. And we deal with three UCI benchmark as following:

1. Iris dataset: This dataset has 150 instances for 3 classes (50 in each of 3 classes). The dimension of descriptors is 4 and we use 120 instances for training (40 in each class) and 30 instances for testing.
2. Ionosphere dataset: This dataset has 351 instances for binary classification task, which dimension of each feature is 34. We use 200 instances for training data and the rest for testing.
3. Wine dataset: This dataset has 178 instances for 3 classes and the number of attributes is 13. We consider 150 instances as training data and the rest as testing data.

The proposed QCSML method is compared with the following algorithm for two aspects: classification performance and computational costs.

---

[1] https://archive.ics.uci.edu/ml/datasets/Iris.
[2] https://archive.ics.uci.edu/ml/datasets/Ionosphere.
[3] https://archive.ics.uci.edu/ml/datasets/Wine.

Table 1. Classification error rates(%) for different distances across various UCI benchmark datasets.

Algorithm	Iris	Ionosphere	Wine
Euclidean	4.00	14.86	4.5
InvCov	8.67	17.71	43.82
LMNN [11]	3.34	14.29	2.25
ITML [5]	3.00	17.14	3.94
SDML(Identity Matrix) [6]	2.00	13.71	0.5618
SDML(Inverse Covariance) [6]	2.00	12	**0**
QCSML(Random Projection)	4.34	13.49	5.81
QCSML(K-means)	**1.04**	7.99	**0**
QCSML(PCA-whitening)	2.22	**6.93**	**0**

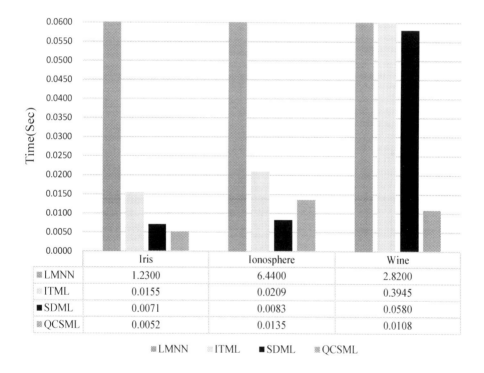

	Iris	Ionosphere	Wine
LMNN	1.2300	6.4400	2.8200
ITML	0.0155	0.0209	0.3945
SDML	0.0071	0.0083	0.0580
QCSML	0.0052	0.0135	0.0108

▓ LMNN   ░ ITML   ■ SDML   ▒ QCSML

Fig. 1. Training time used different distance metric learning on different datasets

1. **Euclidean:** The squared Euclidean distance $\|x_i - x_j\|_2^2$ as a baseline algorithm for classification.
2. **InvCov:** A Mahalanobis distance parameterized by the inverse of sample covariance. It is equivalent to performing PCA over the input data and then computing the squared Euclidean distance in the transformed space.

3. **LMNN:** Large margin nearest neighbor method is proposed by [11]. This method trains the classifier to separate different classes by a large margin.
4. **ITML:** Information-theoretic metric learning is proposed by [5]. It formulates to learn the Mahalanobis matrix by optimizing the differential relative entropy loss function.
5. **SDML:** Sparse distance metric learning is proposed by [6]. It formulates the loss function by log-determinant divergence with a prior knowledge M_0 and the L1-norm regularization for sparsity.

The experimental result is illustrated in Table 1. We can see that the proposed QCSML has the smaller error rates across the datasets compared with other distance metric learning. On the other hand, the PCA-whitening initialization method performs better than other initialization.

Finally, we compare the computational costs of these metric learning algorithms. Figure 1 proves the computing efficiency of the proposed QCSML algorithm. We find our method is faster than LMNN, ITML and SDML in most cases. Moreover, with the different gradient descent method employing, our algorithm will be faster if the input data is large and high dimensional because SDML and ITML algorithm need to compute the gradient by all the training data every iteration, while QCSML only need one sample per each iteration.

4.2   LFW Dataset Face Recognition

In this section, we show the performance of the proposed QCSML metric learning on LFW dataset[4] in detail.

LFW dataset contains 13233 images of 5749 people for face verification. For evaluation, the face data is divided in 10 folds which contain different identities and 600 face pairs for evaluation. There are two evaluation setting about LFW training and testing: restricted and unrestricted. In restricted setting, the predefine image pairs is fixed by author (each fold contains 5400 pairs for training and 600 pairs for testing). And in unrestricted setting, the identities of people within each fold for training is allowed to be much larger.

For face verification, it is also important to extract robust descriptors for representing the images. In this paper, we employ Fisher Vector (FV) [13] which is widely used in image classification [14], image retrieval [15] and face recognition [8]. We extract dense SIFT for each aligned image and learn Gaussian Mixture Model (GMM) parameters by EM algorithm. Then the local descriptors are encoded into Fisher Vectors via GMM parameters.

The Receiving Operating Characteristic Equal Error Rate (ROC-EER) measure is used for evaluations. In the restricted setting, we compare the proposed QCSML method with Combined B/G sample based method [16], LDML [17], DML-eig combined method [18], LBP-CSML [3], SML [4] and Fisher Vector Face [8]. The face verification results are shown in Table 2. Compared with the compressed FV after PCA-whitening, our QCSML method improve the accuracy

---

[4] http://vis-www.cs.umass.edu/lfw/.

Table 2. Comparison of QCSML method with other state-of-the-art methods in restricted setting of LFW.

Method	Dimension	Accuracy(%)
Combined B/G sample based methods, aligned [16]	-	86.83 ± 0.34
LDML, funneled [17]	-	79.27 ± 0.60
DML-eig combined, funneled & aligned [18]	-	85.65 ± 0.56
LBP+CSML, aligned [3]	200	85.57 ± 0.52
Sub-SML, funneled & aligned [4]	300	86.73 ± 0.53
FV+PCA-Whitening funneled & aligned [8]	128	$78.60 \pm N/A$
Fisher Vector Faces, funneled & aligned [8]	128	$\mathbf{87.47 \pm 1.49}$
FV+QCSML, aligned	256	87.10 ± 1.25
FV+QCSML, aligned	128	$\mathbf{87.47 \pm 1.99}$
FV+QCSML, aligned	64	85.20 ± 1.39
FV+QCSML, aligned	32	84.53 ± 1.74

Table 3. Comparison of QCSML method with other state-of-the-art methods in unrestricted setting of LFW.

Method	Accuracy(%)
LDML-MKNN, funneled [17]	87.50 ± 0.40
PLDA combined, funneled & aligned [19]	90.07 ± 0.51
Joint Bayesian combined [20]	90.90 ± 1.48
Sub-SML combined, funneled & aligned [4]	90.75 ± 0.64
Fisher Vector Faces, funneled & aligned [8]	$\mathbf{93.03 \pm 1.05}$
FV+QCSML, aligned	$\mathbf{92.33 \pm 1.12}$

about 9 % and it is the same performance as the Large Margin Dimensionality Reduction (LMDR) which employed metric similarity distance [8]. Besides, the proposed QCSML obtains 87.47 % verification rate, which mostly outperforms other state-of-the-art method in the restricted setting.

Moreover, we evaluate the proposed QCSML method in unrestricted setting of LFW. The results of our method performance are shown in Table 3 and Fig. 2. Our method achieves 92.33 % accuracy, closely matching the Fisher Vector Face [8], which achieves 93.03 %. According to the Table 3, it is obvious that our method obtains 92.33 % verification rate and outperforms most state-of-the-art methods such as LDML-MKNN [17], PLDA [19], joint Baysian [20] and Sub-SML [4]. Although our method cannot obtain higher verification rate than Fisher Vector Faces [8], we can find our method performs better at false accept rate (false positive rate) 1 % point than them in Fig. 2(b), which means our method has more valuable for practical systems because the threshold is often selected when the false accept rate is at 0.1 % or 1 % point instead of equal error rate.

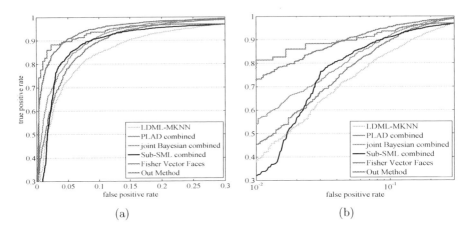

Fig. 2. ROC curves of our method and the state-of-the-art techniques in LFW-unrestricted setting. The left is shown in Linear-Axis and the right is in Log-Axis.

5   Conclusion

In this paper, we proposed the Quasi Cosine Similarity Metric Learning (QCSML) method for classification and face verification tasks. We employ normalization and Lagrange multipliers to convert the cosine similarity metric into a new formulation and it makes the computation faster for high dimensional features and the complexity of QCSML is $O(t \times p \times d)$ which precedes CSML method. In practice, our QCSML performs considerably better on both UCI classification datasets and LFW dataset. In the future, we plan to investigate the optimization processing to make the method more effective and efficient and extend our QCSML to other applications.

Acknowledgement. This work was jointly supported by Beijing Natural Science Foundation under Grant No. 4122049, Beijing Higher Education Young Elite Teacher (No.YETP0381), and the Fundamental Research Funds for the Central Universities (FRF-JX-12-002).

References

1. Xing, E.P., Ng, A.Y., Jordan, M.I., Russell, S.: Distance metric learning with application to clustering with side-information. In: Advances in Neural Information Processing Systems, pp. 521–528 (2003)
2. Guillaumin, M., Mensink, T., Verbeek, J., Schmid, C.: Tagprop: Discriminative metric learning in nearest neighbor models for image auto-annotation. In: 2009 IEEE 12th International Conference on Computer Vision (ICCV), pp. 309–316. IEEE (2009)
3. Nguyen, H.V., Bai, L.: Cosine similarity metric learning for face verification. In: Kimmel, R., Klette, R., Sugimoto, A. (eds.) ACCV 2010, Part II. LNCS, vol. 6493, pp. 709–720. Springer, Heidelberg (2011)

4. Cao, Q., Ying, Y., Li, P.: Similarity metric learning for face recognition. In: 2013 IEEE International Conference on Computer Vision (ICCV), pp. 2408–2415. IEEE (2013)

5. Davis, J.V., Kulis, B., Jain, P., Sra, S., Dhillon, I.S.: Information-theoretic metric learning. In: Proceedings of the 24th International Conference on Machine Learning, pp. 209–216. ACM (2007)

6. Qi, G.-J., Tang, J., Zha, Z.-J., Chua, T.-S., Zhang, H.-J.: An efficient sparse metric learning in high-dimensional space via l 1-penalized log-determinant regularization. In: Proceedings of the 26th Annual International Conference on Machine Learning, pp. 841–848. ACM (2009)

7. Friedman, J., Hastie, T., Tibshirani, R.: Sparse inverse covariance estimation with the graphical lasso. Biostatistics 9, 432–441 (2008)

8. Simonyan, K., Parkhi, O.M., Vedaldi, A., Zisserman, A.: Fisher vector faces in the wild. In: Proceedings of the BMVC, vol. 1, p. 7 (2013)

9. Bottou, L., Bousquet, O.: The Tradeoffs of Large Scale Learning. Advances in Neural Information Processing Systems, p. 2. MIT Press, Cambridge (2007)

10. Zhu, C., Byrd, R.H., Lu, P., Nocedal, J.: Algorithm 778: L-BFGS-B: Fortran subroutines for large-scale bound-constrained optimization. ACM Trans. Math. Softw. (TOMS) 23, 550–560 (1997)

11. Weinberger, K., Blitzer, J., Saul, L.: Distance metric learning for large margin nearest neighbor classification. In: Advances in Neural Information Processing Systems 18, p. 1473 (2006)

12. Huang, G.B., Ramesh, M., Berg, T., Learned-Miller, E.: Labeled faces in the wild: A database for studying face recognition in unconstrained environments. Technical report, 07–49, University of Massachusetts, Amherst (2007)

13. Perronnin, F., Dance, C.: Fisher kernels on visual vocabularies for image categorization. In: IEEE Conference on Computer Vision, Pattern Recognition, CVPR 2007, pp. 1–8. IEEE (2007)

14. Perronnin, F., Sánchez, J., Mensink, T.: Improving the fisher kernel for large-scale image classification. In: Daniilidis, K., Maragos, P., Paragios, N. (eds.) ECCV 2010, Part IV. LNCS, vol. 6314, pp. 143–156. Springer, Heidelberg (2010)

15. Perronnin, F., Liu, Y., Sánchez, J., Poirier, H.: Large-scale image retrieval with compressed fisher vectors. In: 2010 IEEE Conference on Computer Vision, Pattern Recognition (CVPR), pp. 3384–3391. IEEE (2010)

16. Wolf, L., Hassner, T., Taigman, Y.: Similarity scores based on background samples. In: Zha, H., Taniguchi, R., Maybank, S. (eds.) ACCV 2009, Part II. LNCS, vol. 5995, pp. 88–97. Springer, Heidelberg (2010)

17. Guillaumin, M., Verbeek, J., Schmid, C.: Is that you? Metric learning approaches for face identification. In: 2009 IEEE 12th International Conference on Computer Vision, pp. 498–505 (2009)

18. Ying, Y., Li, P.: Distance metric learning with eigenvalue optimization. J. Mach. Learn. Res. 13, 1–26 (2012)

19. Li, P., Fu, Y., Mohammed, U., Elder, J.H., Prince, S.J.D.: Probabilistic models for inference about identity. IEEE Trans. Pattern Anal. Mach. Intell. 34, 144–157 (2012)

20. Chen, D., Cao, X., Wang, L., Wen, F., Sun, J.: Bayesian face revisited: a joint formulation. In: Fitzgibbon, A., Lazebnik, S., Perona, P., Sato, Y., Schmid, C. (eds.) ECCV 2012, Part III. LNCS, vol. 7574, pp. 566–579. Springer, Heidelberg (2012)

Hand Gesture Recognition Based on the Parallel Edge Finger Feature and Angular Projection

Yimin Zhou[1,2]([✉]), Guolai Jiang[1], Guoqing Xu[1], and Yaorong Lin[3]

[1] Shenzhen Institutes of Advanced Technology,
Chinese Academy of Sciences, Shenzhen, China
{ym.zhou,gl.jiang,gq.xu}@siat.ac.cn
[2] The Chinese University of Hong Kong, Hong Kong, China
[3] School of Electronic and Information Engineering,
South China University of Technology, Guangzhou, China
eeyrlin@scut.edu.cn

Abstract. In this paper, a novel high-level hand feature extraction method is proposed by the aid of finger parallel edge feature and angular projection. The finger is modelled as a cylindrical object and the finger images can be extracted from the convolution with a specific operator as salient hand edge images. Hand center, hand orientation and wrist location are then determined via the analysis of finger image and hand silhouette. The angular projection of the finger images with origin point on the wrist is calculated, and five fingers are located by analyzing the angular projection. The proposed algorithm can detect extensional fingers as well as flexional fingers. It is robust to the hand rotation, side movements of fingers and disturbance from the arm. Experimental results demonstrate that the proposed method can directly estimate simple hand poses in real-time.

1 Introduction

Computer vision based hand gesture recognition has become a hot topic recently, due to its natural human-computer interaction characteristics. Hand gestures are usually composed of different hand postures and their motions. Estimating hand posture in real-time is the necessary step for hand gesture recognition. However, human hand is an articulated object with over 20 degrees of freedom (DOF) [8], and many self-occlusions can occur in its projection results. Moreover, hand motion is often too fast comparing with current computer processing speed. Therefore, real-time hand posture estimation is still a challenging work.

In recent years, many 3D sensors, such as binocular cameras, Kinect and leap motion, have been applied for hand gesture recognition with good performance. However, hand gesture recognition based on monocular camera has quite a limitation, since 3D sensors are not always available in many systems, i.e., Google Glasses.

Feature extraction is a crucial module in hand posture estimation system. Different kinds of features with low-level precision are available for detecting

© Springer International Publishing Switzerland 2015
C.V. Jawahar and S. Shan (Eds.): ACCV 2014 Workshops, Part III, LNCS 9010, pp. 206–217, 2015.
DOI: 10.1007/978-3-319-16634-6_16

and recognizing hand postures. Skin color segmentation is wildly used for hand localization [10,15]. Silhouette, the outline of segmented hand area, can be used for matching predefined hand postures [16]. Edge is another common feature for model based matching [17]. Histogram of oriented gradients has been implemented in [14]. Combinations of multiple features can be taken to improve robustness of the feature extraction [5].

However, systems using low-level features can only recognize several predefined postures. High-level features (fingers, fingertips, joint locations etc.) from hand image should be extracted for various hand postures estimation, since they can provide very compact representation of the input hand postures in real-time.

In [12,13], fingertips were found by fingertip masks considered the characteristics of fingertips, and the location of fingertips can be achieved based on feature matching. However, the objects which share the similar shape of fingertips, will result in a misjudgment. In [1], fingertips were located by generalized Hough Transform and probabilistic models. The detected edge segments of the black and white images is computed by Hough Transform for fingertip detection. But the light and brightness would seriously damage the quality of the images and detection.

In [3], palm and fingers were detected by skin-colored blob and ridge features. In [6,9], high-level hand features were extracted by analyzing hand contour. Hand postures can be recognized through the geometric features and external shape of the Palm and fingers [2]. It proposes a prediction model for showing the hand postures. Though the measurement error would be large due to the complexity of the hand gestures and varieties of hands.

Since flexional fingers will not appear on the hand contour, little attention has been paid on extracting high-level features of these fingers directly. In [7], a finger detection method using grayscale morphology and blob analysis is introduced, which can be used for flexional finger detection.

In this paper, a novel method for extracting and locating all the fingers based on a digital camera is proposed, where the hand postures are viewed from the orthogonal direction to the palm. Firstly, the salient hand edge image is extracted, which includes hand contour and boundaries of flexional fingers. Considering that each finger has two main "parallel" edges on its two sides, a finger image is extracted from convolution result of salient hand edge image with a specific operator G. Then hand orientation and wrist location are computed from the finger image and hand image. Finally, fingers and their angles are extracted from the local maxima of angular projection.

The remainder of the paper is organized as follows. Section 2 introduces a novel finger extraction and modelling approach. The high-level hand feature extraction is described in Sect. 3. Simulation experiments and results are shown in Sect. 4. Conclusions and future works are given in Sect. 5.

2   The Procedure of Finger Extraction

2.1   Description of the Finger Model

The most notable parts of a hand are the fingers. Fingers are cylindrical objects with nearly constant diameter from root to fingertip. The proposed finger model is shown in Fig. 1(A). Here, the boundary of the ith finger, C_{fi}, is considered as the composition of arc edges (C_{ti}, fingertip or joint) and a pair of parallel edges ($C_{ei} + C'_{ei}$, finger body):

$$C_{fi} = (C_{ei} + C'_{ei}) + \sum C_{ti} \tag{1}$$

The finger center line (FCL) C_{FCLi} is introduced, which can be regarded as the center line of the finger parallel edges. Fingertip/joint O_{ti} locates at the end point of FCL, and it is the center of the arc curve C_{ti} as well. The distance between the parallelled finger edges are known as $2 \times d$, where d is the radius of all fingers.

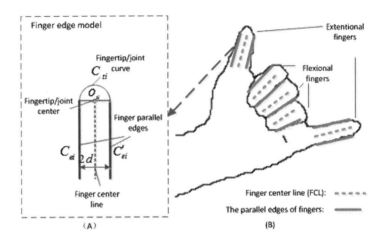

Fig. 1. Finger edge model (A) and the parallel edges (B)

As it is shown in Fig. 1(B), all fingers are represented by the areas near FCLs, which can be produced from the parallelled finger edges.

2.2 Salient Hand Edge Image Extraction

The salient hand edge image $I_{edge}(x, y)$ is mainly made up of hand contour which includes boundaries of extensional fingers and arm/palm and flexional fingers. As shown in Fig. 2: (a) is the grayscale hand image $I_{gray}(x, y)$; (b) is the threshold image $I_{black}(x, y)$; (c) is the canny edge image $I_{canny}(x, y)$; (d) is the contour image $I_{contour}(x, y)$; (e) is the obtained salient hand edge image $I_{edge}(x, y)$. Then the extraction of $I_{edge}(x, y)$ is summarized as follows:

1. Extract grayscale hand image $I_{gray}(x, y)$ and hand contour image $I_{contour}(x, y)$ from source color image by using skin color segmentation method in [10].
2. Extract canny edge image $I_{canny}(x, y)$ from $I_{gray}(x, y)$ [4].

3. In most hand postures, the boundaries of flexional fingers are dark edges. Applysing the threshold Th_{black} (predefined) to gray-level hand image for extraction, then the obtained $I_{black}(x,y)$ is:

$$I_{black}(x,y) = \begin{cases} 1 & (I_{gray}(x,y) < Th_{black}) \\ 0 & (I_{gray}(x,y) \geq Th_{black}) \end{cases} \tag{2}$$

The boundaries of the flexional fingers are extracted by the interlapped area of $I_{canny}(x,y) \cap I_{black}(x,y)$.
4. Then the salient hand edge image $I_{edge}(x,y)$ is extracted by:

$$I_{edge}(x,y) = I_{black}(x,y) \cap I_{canny}(x,y) \cup I_{contour}(x,y) \tag{3}$$

The whole procedure is demonstrated in Fig. 2.

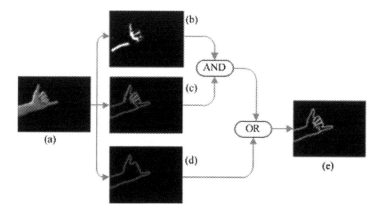

Fig. 2. Diagram of extracting hand edge procedure

2.3   Finger Extraction Using Parallel Edge Feature

In order to extract fingers from their parallel edges, a rotation invariant operator G is introduced, which is given by:

$$G(x,y) = \begin{cases} -1 & 0 \leq D(x,y) \leq r_1 \\ 1 & r_2 < D(x,y) \leq r_3 \\ 0 & \text{else} \end{cases} \tag{4}$$

$$D(x,y) = \sqrt{(x-x_0)^2 + (y-y_0)^2} \tag{5}$$

where (x_0, y_0) denotes the center point of G; $D(x,y)$ is the distance from the center point; r_1, r_2 and $r_3 (r_1 < r_2 < r_3)$ are the defined parameters shown in Fig. 3.

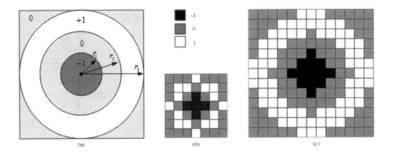

Fig. 3. (a) Continuous Operator G; (b) Discrete operator G (N = 1); (c) Discrete operator G (N = 2);

For each edge point in $I_{edge}(x, y)$, if the information of edge orientation is not considered, it can be regarded as a rough estimation of its correlative points on FCL. Here we simply assume $r_1 = r_2 - r_1 = r_3 - r_2 = N$, and $N = 0.5d$. As shown in Fig. 3, the size of continuous G operator is defined as $(6N \times 6N)$, while the size of discrete G is $((6N + 1)(6N + 1))$, without loss of generality.

The finger image $I_f(x, y)$ can be extracted by the following steps, shown in Fig. 4. Figure 4(a) is the salient hand edge image $I_{edge}(x, y)$ (see Fig. 2(e)) obtained from the steps in Sect. 2.2; Fig. 4(b) is the convolution result $I_w(x, y)$ and Fig. 4(c) is the finger image $I_f(x, y)$. The steps are described in details:

1. Decide the size of G with $N = 0.5d$. When the dealt hand image is too large, it should be adapted to proper size firstly;
2. Calculate the convolution $I_w(x, y) = I_{edge}(x, y) \times G$;
3. Extract rough finger image $I_{f'}(x, y)$ from $I_w(x, y)$ by,

$$I_{f'(x,y)} = \begin{cases} 1 & I_{w(x,y)} > T_h \\ 0 & I_{w(x,y)} \leq T_h \end{cases}$$

where T_h is the proper threshold in $(\sqrt{5}d, 2\sqrt{5}d)$. Due to space limit, the details of T_h definition is not given here.

4. Apply morphological dilation operation to $I_{f'}(x, y)$ to get the finger image $I_f(x, y)$.

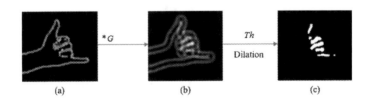

Fig. 4. Diagram of extracting fingers

It should be noted that in the extracted finger images $I_f(x, y)$, noise pixels are mainly generated from edges of palm print near palm center, while faulty detection is often caused by incomplete extraction of flexional finger boundaries.

3 High-Level Hand Feature Extraction

3.1 The Definition of the Hand Center

The whole hand model is depicted in Fig. 5. Here, palm center P_{hc} is regarded as the hand center since it is insensitive to various hand postures. P_{hc} is also assumed as a stable point which has the maximum distance to the closest boundary edge during estimation. Morphological erosion operation has been taken to remove the extensional fingers in the hand area [12]. Thus, the center of the hand region is defined as the mass center of the remaining region.

However, the width of a arm is quite close to wrist width, which makes it difficult to be removed through erosion operation. Large remaining arm area may cause error for the hand center determination. As a result, arm should be separated initially during this method application. In our approach, based on the finger image extracted in Sect. 2, finger center P_{fc} could be regarded as the mass center of the finger area. A circle centered on P_{fc} is drawn to remove the arm, which is the large dashed circle in Fig. 5. The diameter of the circle is set to $5R_p$, where R_p denotes the radius of the palm.

3.2 Hand Orientation Determination

Hand orientation is another important basic feature for hand posture recognition. It can be estimated by calculating the direction of principal axes of the silhouettes, or determined by the arm [12]. However, these methods are not always effective because the silhouettes of some postures ("fist") don't have obvious principal axes, while arm may not always appear.

In this paper, since the fingers are always located on one side of the hand palm, the vector from the hand center P_{hc} to the middle finger center P_{fc} will give a more reliable estimation of the hand orientation. The hand orientation $\vec{N}_p$ is determined by:

$$\vec{N}_p = \frac{\overrightarrow{P_{hc}P_{fc}}}{\left|\overrightarrow{P_{hc}P_{fc}}\right|} \tag{6}$$

where $\vec{\cdot}$ is the vector denotation, $|\cdot|$ is the norm operation. The wrist position P_w could be determined through the palm radius R_p calculation,

$$P_w = P_{hc} - R_p\vec{N}_p \tag{7}$$

The diagram of finding the hand center P_{fc}, hand orientation $\vec{N}_p$ and wrist position P_w is shown in Fig. 5.

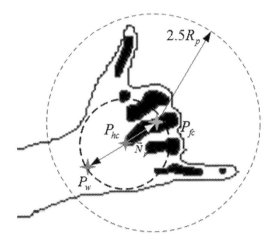

Fig. 5. The diagram of the parameter detrmination for hand model

3.3   Angular Projection for Finger Localization

(A) Finger angle determination
According to the kinematic hand model [8], if there is no finger crossing, all the fingers can be regarded as radials starting from the wrist. The angular projection method described in [11] for face recognition is a feasible choice for finger angle determination via angle distribution characteristics of fingers pixels.

Here, the finger image $I_f(x, y)$ (shown in Fig. 4) is transformed into polar image $I_f(r, \theta)$ (see Fig. 5), with the origin point on the wrist center point P_w. Different from normal angular projection methods, considering the noise pixels in $I_f(x, y)$ located near palm center, the points with larger radius vector r should have higher probability of being a point on fingers. The adjusted angular projection in each direction is calculated as follows:

$$p_n = \int_{r1}^{r2} \int_{\theta_n}^{\theta_{n+1}} rI_f(r, \theta)d\theta dr, \quad n = 0, 1, 2, ..., M - 1 \qquad (8)$$

where $[r_1, r_2]$ is the range of radius vector; $[\theta_n, \theta_{n+1}]$ is the nth subdivision of polar angle; M is the number of subdivisions of polar angle. Then the normalized P_n is given by:

$$p_{n\_norm} = \frac{p_n}{\sum\limits_{n=0}^{M-1} p_n}, \quad n = 0, 1, 2, ..., M - 1 \qquad (9)$$

(B) Finger location
The polar angle subdivisions of the five fingers can be determined by the largest five local maxima of $\{p_{n\_norm}\}$. In the angle subdivision of each finger, the point on the hand contour with farthest distance D_i to P_w is determined as

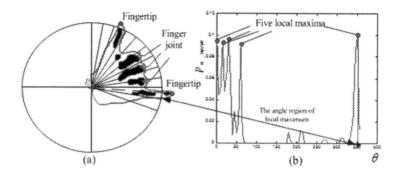

Fig. 6. (a): Angular projection of finger image; (b): The value of P_n in different angles

fingertip (for extensional fingers) or finger joint (for flexional fingers). The status of the fingers (extensional or flexional) could be determined by: $D_i > \alpha_i R_p$ and $D_i < \alpha_i R_p$, where R_p is the palm radius. α_i is the constant, determined by the finger length and varied with different fingers, which is depicted in Fig. 6.

4 Experiments and Result Analysis

The proposed hand extraction algorithm is used in a real-time hand posture estimation system for matching and tracking. The experiments are running with a PC (CPU: Pentium Core 2 Duo E7200, RAM: 2 GB) and a monocular web camera. The system could process real-time video of resolution 320*240 pixels and frame rate 15fps. The tested hand stays at a position at roughly $0.5m$ distance in front of the camera, keeping the palm in the direction facing the camera with freely moved hand. A video is attached for the parts of the experiments. Figure 7 shows several examples of hand estimation generated from hand features in the video. The left column is the input frames, where the red marks and blue marks indicate the estimated hand center and hand wrist center. The output hand postures are shown in the right column, where the white segments denote flexional fingers and green segments denote extensional fingers.

Table 1. Correct rate for detecting fingers

Total frames	Extensional fingers	Flexional fingers
5300	96.2 %	75.2 %

Table 1 lists the numerical results of correct rate for the detected fingers. It can be seen that the recognition for the extensional fingers are quite accurate. As for the flexional finger, it still has 75.2 % recognition rate, which is not bad as well. Figure 8 shows the hand orientation estimation result for a gesture sequence in the experiments. The blue solid line denotes the real hand orientation, determined by the marks of the palm center and the wrist. The red dashed line (close

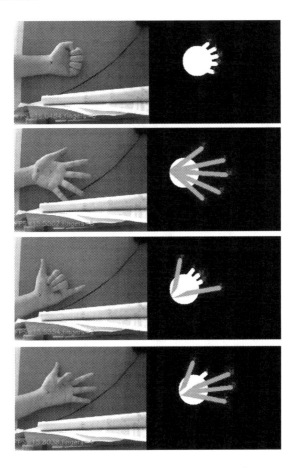

Fig. 7. Real-time 2D hand posture estimation directly using features extracted in this paper (Color figure online)

to blue line) denotes the estimated hand orientation with the proposed method. The black small dashed line is the estimation error, which is calculated by:

$$Error = \sqrt{\frac{1}{N} \sum_{i=1}^{N} (y_i - \hat{y}_i)^2} \tag{10}$$

where y_i is the actual hand orientation or final angle; $\hat{y}_i$ is the estimation values.

In Fig. 9, it shows the middle finger angle estimation result for another gesture sequence in the experiments. The blue solid line denotes the real middle finger angle determined by the marks on the middle finger and the wrist. The red dashed line denotes the estimated middle finger angle via the proposed method, and the black dashed line is the estimation error.

Experiment results demonstrate that the proposed method can give a robust estimation for the features of the hand location, hand orientation and extensional

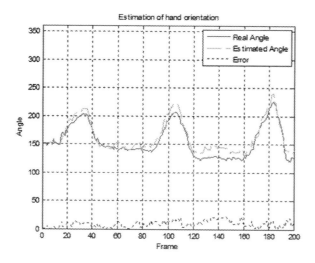

Fig. 8. Estimation of hand orientation (Color figure online)

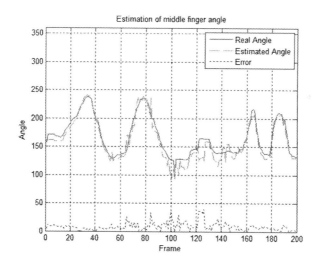

Fig. 9. Estimation of middle finger angle (Color figure online)

fingers. In this case, it takes the whole fingers into account and not only the finger tips. As for the flexional fingers, the error is mainly from the bad hand edge estimation or flexional thumb finger, since they do not fulfill the assumption that made in the angular projection. Although there are errors in the hand orientation and finger angle estimation, the results for real-time hand posture recognition has satisfied performance.

5   Summary and Future Work

Feature extraction is a crucial module in computer vision based hand posture recognition system. The implementation of this module has a considerable effect on the robustness and processing speed of the system. High level hand features are desirable because they can provide very compact representation of the input hand in real-time operational mode.

In this paper, a high level hand feature extraction method has been proposed for recognizing hands with whose palm facing to the camera. Hand edges have been extracted for detecting fingers via the parallel edge feature. Angular projection centered on wrist point is then used for getting the angle and length of each finger. A simple 2D hand model can be produced directly from these features. The proposed method can make robust estimation for features of hand location, hand orientation and extensional fingers. Hence, the estimated hand model can be used for controlling mechanical hands or interacting with Google Glasses.

However, if the salient hand edge is not well detected, false or miss detection for flexional fingers could occur. Future work will focus on improving salient hand edge detection algorithm with tracking modules.

Acknowledgment. This work is partially supported under the Shenzhen Science and Technology Innovation Commission Project Grant Ref. JCYJ20120615125931560 and partially supported by Introduced Innovative R&D Team of Guangdong Province (201001D0104648280).

References

1. Barrho, J., et al.: Finger localization and classification in images based on generalized hough transform and probabilistic models. In: The 9th International Conference on Control, Automation, Robotics and Vision, pp. 1–6 (2006)
2. Bhuyan, M.K., et al.: Hand pose recognition using geometric features. In: National Conference on Communications (NCC), pp. 1–5 (2011)
3. Bretzner, L., et al.: Hand gesture recognition using multi-scale colour features, hierarchical models and particle filtering. In: 5th IEEE International Conference on Proceedings (2002)
4. Canny, J.: A computational approach to edge detection. IEEE Trans. Pattern Anal. Mach. Intell. **8**, 679–698 (1986)
5. Cao, C., et al.: Real-time multi-hand posture recognition. In: International Conference on Computer Design and Applications (ICCDA) (2010)
6. Chang, C.-C., et al.: Feature alignment approach for hand posture recognition based on curvature scale space. Neurocomputing **71**, 1947–1953 (2008)
7. Dung Duc, N., et al.: Finger extraction from scene with grayscale morphology and BLOB analysis. In: IEEE International Conference on Robotics and Biomimetics, pp. 324–329 (2008)
8. Erol, A., et al.: Vision-based hand pose estimation: a review. Comput. Vis. Image Underst. **108**, 52–73 (2007)
9. Feng, Z., et al.: Features extraction from hand images based on new detection operators. Pattern Recogn. **44**, 1089–1105 (2011)

10. Jones, M.J., Rehg, J.M.: Statistical color models with application to skin detection. In: Proceedings of the IEEE Computer Society Conference on Computer Vision and Pattern Recognition, vol. 1, pp. 274–280 (1999)
11. Kim, H.I., et al.: Rotation-invariant face detection using angular projections. Electron. Lett. **40**, 726–727 (2004)
12. Oka, K., et al.: Real-time fingertip tracking and gesture recognition. IEEE Comput. Graphics Appl. **22**, 64–71 (2002)
13. Parker, J.R., Baumback, M.: Finger recognition for hand pose determination. In: IEEE International Conference on Systems, Man and Cybernetics, pp. 2492–2497 (2009)
14. Romero, J., et al.: Monocular real-time 3D articulated hand pose estimation. In: 9th IEEE-RAS International Conference on Humanoids (2009)
15. Sha, L., et al.: Hand posture recognition in video using multiple cues. In: IEEE International Conference on Multimedia and Expo (2009)
16. Shimada, N., et al.: Real-time 3-D hand posture estimation based on 2-D appearance retrieval using monocular camera. In: The Proceedings of the IEEE ICCV Workshop on Recognition, Analysis, and Tracking of Faces and Gestures in Real-Time Systems (RATFG-RTS 2001) (2001)
17. Ying, W., et al.: Capturing natural hand articulation. In: 8th IEEE International Conference on Computer Vision, vol. 2, pp. 426–432 (2001)

Image Retrieval by Using Non-subsampled Shearlet Transform and Krawtchouk Moment Invariants

Cheng Wan[✉] and Yiquan Wu

College of Electronic and Information Engineering,
Nanjing University of Aeronautics and Astronautics,
29 Yudao St., Nanjing 210016, China
wanch@nuaa.edu.cn

Abstract. In this paper, we use non-subsampled shearlet transform (NSST) and Krawtchouk Moment Invariants (KMI) to realize image retrieval based on texture and shape features. Shearlet is a new sparse representation tool of multidimensional function, which provides a simple and efficient mathematical framework.We decompose the images by NSST. The directional subband coefficients are modeled by Generalized Gaussian Distribution (GGD). The distribution parameters are used to build texture feature vectors which are measured by Kullback–Leibler distance (KLD). Meanwhile, low-order KMI are employed to extract shape features which are measured by Euclidean distance (ED). Finally, the image retrieval is achieved based on weighted distance measurement. Experimental results show the proposed retrieval system can obtain the highest retrieval rate comparing with the methods based on DWT, Contourlet, NSCT and DT-CWT.

1 Introduction

Retrieving similar image from an image database is a challenging task. Text based image retrieval systems (TBIR) were used from the 1970's. Searching images based on the content of the image is called Content Based Image Retrieval (CBIR) which uses different features of the image to search similar images from an image database. The CBIR appeared in the 1990's and it is suitable for both large and small size databases [1–3]. CBIR adopts feature extraction and similarity measurement for image retrieval process. Image content features such as colour, texture and shape, which are analyzed and extracted automatically by computer achieves the effective retrieval.

Texture can efficiently reflect the structural, directional, granularity, or regularity differences of diverse regions in a visual image. Therefore, texture features become one of the most efficient and effective models used in CBIR systems. Discrete wavelet transform can represent the texture features into some subband properties such as the subband energy, which was studied sufficiently at the early stage [4–6]. Later, Gabor wavelets and complex wavelets were developed to remove

© Springer International Publishing Switzerland 2015
C.V. Jawahar and S. Shan (Eds.): ACCV 2014 Workshops, Part III, LNCS 9010, pp. 218–232, 2015.
DOI: 10.1007/978-3-319-16634-6_17

the limitation of the wavelet transform to capture more directional information in subbands. Contourlet transform [7,8] is a new image analysis tool, which is anisotropic and has good directional selectivity. The sub-sampling operation is employed in the implementation of Contourlet transform. So it is not shift-invariant and sub-band spectrum aliasing phenomenon takes place, which weakens the directional selectivity of Contourlet transform. To overcome this disadvantage, Cunha et al. [9] proposed non-subsampled Contourlet transform (NSCT) which is an improved version of Contourlet transform]. But the computational efficiency of NSCT is low. In 2005, Labate et al. [10] proposed a new multi-scale geometric analysis tools: shearlet which is optimally sparse in representing images. The decomposition of shearlet is similar to Contourlet transform, but an important advantage of shearlet over Contourlet transform is that there are no restrictions on the number of directions for the shearing. In addition, the inverse shearlet transform only requires a summation of the shearing filters rather than inverting a directional filter banks, thus the implementation of shearlet is more efficient computationally. Due to these advantages with the shearlet, it has been applied in many image processing fields, such as image denoising, edge detection, and image fusion. The shearlet transform also uses the subsampling operation, so it is not shift-invariant and it will produce pseudo-Gibbs phenomenon around the singular point. The non-subsampled shearlet transform [11] by cascading of non-subsampled pyramid filter banks and shearing filter banks has all the advantages of the shearlet transform and does not need the implementation of the sub-sampling operation. In wavelet domain, the detail subband coefficients of a texture image usually exhibit a striking non-Gaussian behavior, and their marginal distribution can be well characterized by the generalized Gaussian distribution (GGD) [12]. We use GGD to model the subbands of NSST and use maximum likelihood (ML) estimation to estimate the parameters of GGD.

On the other hand, image retrieval based on shape content remains a more difficult task than that based on other visual features [13–15]. Shape is one of the most basic and meaningful characteristics, shape descriptors should be invariant to translation, rotation and scaling changes of the object on the basis of distinguishing different objects. Moment has been widely used in pattern recognition applications to describe the geometrical characteristics of different objects, its calculation is related to all of the relevant pixels of the image or region, it can describe the global information, and Krawtchouk moment [16] proposed by Yap is a set of discrete orthogonal moments, there is no need for spatial normalization, hence the error in the computed Krawtchouk moments due to discretization is nonexistent, and it has smaller redundancy than other moments, at the same time Krawtchouk moment invariants is invariant to rotation, scaling, and translation of the image. Krawtchouk moment invariants as global shape features achieve image retrieval. We extract Krawtchouk moment invariants as global feature of the image, and only use low-order moments that can describe the shape information. High-order moments need great amount of calculation and only describe the details of an image.

In this paper, we decompose the image with NSST and compute the subband GGD parameters to represent the texture features. Kullback–Leibler distance (KLD) is used to measure the distance between texture features. Meanwhile, low-order Krawtchouk moment invariants are calculated to be the shape features which are measured by Euclidean distance (ED). Finally, the weighted distance of KLD and ED is derived to measure the similarity between the query image and the database images. The effectiveness of the proposed retrieval scheme has been demonstrated by the experiments.

2   Proposed Image Retrieval Method

The proposed image retrieval system works in two stages to collect the image information. The first stage measures the distribution of the textures in the image. The second stage measures the shape information from the image. Thus the collected information helps in enhanced retrieval. For each image in the database, its texture and shape feature vectors have been extracted and stored. The texture and shape features of the query image are calculated. To obtain the texture feature, the image is decomposed by NSST and the subband coefficients of NSST is modeled by Generalized Gaussian Density (GGD). The parameters of GGD are the texture feature of image. Meanwhile, we figure out the Krawtchouk moment invariants (KMI) of the image as the shape feature. Feature dissimilarity between the query image and the database images is measured by weighted distance of Kullback–Leibler distance (KLD) and Euclidean distance (ED). KLD is used to compute the distance between texture features and ED is for measuring the similarity of shape features. The system then outputs the most relevant images to the user in the ascending order of the weighted distance. The block diagram is given in Fig. 1 and the detailed explanation of every block is given in the following sections.

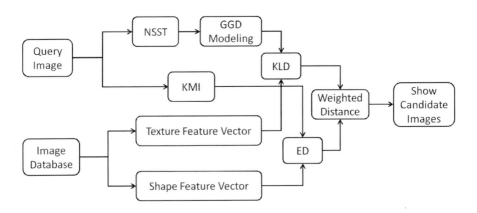

Fig. 1. Block diagram of the proposed image retrieval system

3   Texture Feature Extraction

3.1   Non-subsampled Shearlet Transform (NSST)

In dimension n = 2, the affine systems of ST is defined as follows:

$$\{\psi_{j,l,k}(x) = |\det \mathbf{A}|^{j/2}\psi(\mathbf{B}^l\mathbf{A}^j x - k), j, l \in \mathbb{Z}, k \in \mathbb{Z}^2\} \tag{1}$$

where ψ is a collection of basis function and satisfies $\psi \in L^2(\mathbb{R}^2)$, $\mathbf{A}$ denotes the anisotropy matrix for multi-scale partitions, $\mathbf{B}$ is a shear matrix for directional analysis. j, l, k are scale, direction and shift parameter respectively. $\mathbf{A}$, $\mathbf{B}$ are both 2×2 invertible matrices and $\det |\mathbf{B}| = 1$. For each $a > 0$ and $b \in \mathbf{R}$, the matrices of $\mathbf{A}$ and $\mathbf{B}$ are represented as:

$$\mathbf{A} = \begin{pmatrix} a & 0 \\ 0 & \sqrt{a} \end{pmatrix}, \mathbf{B} = \begin{pmatrix} 1 & b \\ 0 & 1 \end{pmatrix} \tag{2}$$

These two matrices are important roles in the process of shearlet transform. The former dominates the scaling of shearlet, and the latter controls the orientation of shearlet. When a = 4, s = 1, (2) is written as follows:

$$\mathbf{A} = \begin{pmatrix} 4 & 0 \\ 0 & 2 \end{pmatrix}, \mathbf{B} = \begin{pmatrix} 1 & 1 \\ 0 & 1 \end{pmatrix} \tag{3}$$

For any $\forall \xi = (\xi_1, \xi_2) \in \widehat{\mathbb{R}}^2, \xi_1 \neq 0$, $\psi^{(0)}$ for ST can be described as:

$$\widehat{\psi}^{(0)}(\xi) = \widehat{\psi}^{(0)}(\xi_1, \xi_2) = \widehat{\psi}_1(\xi_1)\widehat{\psi}_2(\frac{\xi_2}{\xi_1}) \tag{4}$$

Here, $\widehat{\psi}$ is the Fourier transform of ψ, $\text{supp}\widehat{\psi}_1 \subset [-\frac{1}{2}, -\frac{1}{16}] \cup [\frac{1}{16}, \frac{1}{2}]\text{supp}\widehat{\psi}_2 \subset [-1, 1]$.
   Assume that

$$\sum_{j \geq 0} |\widehat{\psi}_1(2^{-2j}\omega)|^2 = 1, \quad |\omega| \geq \frac{1}{8} \tag{5}$$

$$\sum_{l=-2^j}^{2^j-1} |\widehat{\psi}_2(2^{-2j}\omega - l)| = 1, \quad |\omega| \leq 1, j \geq 0 \tag{6}$$

we can see $\psi_{j,l,k}(x)$ has the frequency support as follows.

$$\text{supp}\widehat{\psi}_{j,l,k}^{(0)} \subset \{(\xi_1, \xi_2) : \xi_1 \in [-2^{2j-1}, -2^{2j-4}] \cup [2^{2j-4}, 2^{2j-1}], |\frac{\xi_2}{\xi_1} + l2^{-j}| \leq 2^{-j}\} \tag{7}$$

Hence, $\widehat{\psi}_{j,l,k}$ is supported on a pair of trapeziform zones, whose sizes all approximate to $2^{2j} \times 2^j$.
   The shearlet transform has the following main properties: well localizing, parabolic scaling, highly directional sensitivity, spatially localizing and optimally sparse. The non-subsampled shearlet transform (NSST), which combined the

non-subsampled Laplacian pyramid transform with several different combinations of the shearing filters, is the shift-invariant version of the shearlet transform. The NSST differs from the shearlet transform in that the NSST eliminates the down-samplers and up-samplers. The NSST is a fully shift-invariant, multi-scale, and multi-direction expansion. Consequently, introduction of NSST into image retrieval could make use of the good characters of NSST in effectively extracting features from original images.

3.2 GGD Probability Density Function (PDF)

The PDF of a zero-mean generalized Gaussian distribution (GGD) is defined as:

$$p(x; \sigma^2, \beta) = \frac{\beta \eta(\sigma, \beta)}{2\Gamma(1/\beta)} \exp\{-[\eta(\sigma, \beta) |x|]^\beta\} \tag{8}$$

where σ^2 and β denote the variance and the shape parameter of the distribution respectively. Γ is the gamma function given by $\Gamma(x) = \int_0^\infty t^{x-1} e^{-t} dt, x > 0$, and

$$\log \Gamma(x) = -\gamma x - \log(x) + \sum_{k=1}^\infty [\frac{x}{k} - \log(1 + \frac{x}{k})] \tag{9}$$

where $\gamma = 0.577$ denotes the Euler constant. In addition, $\eta(\sigma, \beta)$ is given by

$$\eta(\sigma, \beta) = \frac{1}{\sigma}[\frac{\Gamma(3/\beta)}{\Gamma(1/\beta)}]^{1/2} \tag{10}$$

Here, the smaller β corresponds to the sharper distribution, while the bigger β represents the flatter distribution.

3.3 GGD Modeling of NSST Direction Subband Coefficients

Given N samples $\mathbf{x} = (x_1, x_2, \cdots, x_N)$ with zero mean PDF of GGD, The ML method gives the estimates of parameters σ and β as follows.

$$\hat{\sigma} = \left[\frac{\Gamma(3/\beta)}{\Gamma(1/\beta)}\right]^{1/2} \left[\frac{\hat{\beta}}{N} \sum_{i=1}^N |x_i|^{\hat{\beta}}\right]^{1/\hat{\beta}} \tag{11}$$

$$1 + \frac{\Psi(1/\hat{\beta})}{\hat{\beta}} - \frac{\sum_{i=1}^N |x_i|^{\hat{\beta}} \log |x_i|}{\sum_{i=1}^N |x_i|^{\hat{\beta}}} + \frac{\log(\frac{\hat{\beta}}{N} \sum_{i=1}^N |x_i|^{\hat{\beta}})}{\hat{\beta}} = 0 \tag{12}$$

where the digamma function is given by $\Psi(x) = \frac{d \log(\Gamma(x))}{dx}$. For the transcendental equation (12), $\hat{\beta}$ can be solved numerically, and then $\hat{\sigma}$ is obtained from (11).

Therefore, it is sufficient to obtain the estimation of $\hat{\beta}$. [6] used the Newton-Raphson method with an initial value $E[\|x\|]/\sigma$ from the moment estimation to estimate $\hat{\beta}$.

In this paper, we use a zero mean GGD model to characterize the NSST subband coefficients. The GGD parameters, σ^2 and β, are extracted as image texture features using ML method. Each subband corresponds one pair of $\{\sigma^2, \beta\}$. Assume NSST gives n subbands, we can derive $2n$ parameters of texture feature. The texture feature vector is described as: $\mathbf{V}_t = [\sigma_1^2, \beta_1, \sigma_2^2, \beta_2, \cdots, \sigma_n^2, \beta_n]$.

4   Shape Feature Extraction

4.1   Krawtchouk Moment Invariants (KMI)

The Krawtchouk moments are a set of orthogonal moments formed by using the discrete classical Krawtchouk polynomials as the basis function set. Because the basis polynomial is discrete, there is no need for spatial normalization, hence, the error in the computed Krawtchouk moments due to discretization is nonexistent. The n-th order Krawtchouk polynomial [16] is defined as follows.

$$K_n(x; p, N) = \sum_{k=0}^{n} a_{k,n,p} x^k =_2F_1(-n, -x; -N; \frac{1}{p}) \tag{13}$$

where $x, n = 0, 1, 2, \cdots, N, N > 0, p \in (0, 1)$. The hypergeometric function $_2F_1(a, b; c; z) = \sum_{k=0}^{n} \frac{(a)_k (b)_k}{(c)_k} \frac{z^k}{k!}$. $(a)_k$ is the Pochhammer symbol given by

$$(a)_k = a(a + 1) \cdots (a + k - 1). \tag{14}$$

In order to achieve numerical stability, a set of weighted Krawtchouk polynomials $\bar{K}_n(x; p, N)$ is defined by

$$\bar{K}_n(x; p, N) = K_n(x; p, N) \sqrt{\frac{w(x; p, N)}{\rho(n; p, N)}} \tag{15}$$

where $w(x; p, N) = \binom{N}{x} p^x (1 - p)^{N-x}$, $\rho(n; p, N) = (-1)^n (\frac{1-p}{p})^n \frac{n!}{(-N)_n}$.

The Krawtchouk moments of order $(n + m)$ for an $M_1 \times M_2$ image with intensity function $f(x)$ is defined as

$$Q_{nm} = \sum_{x=0}^{M_1-1} \sum_{y=0}^{M_2-1} \bar{K}_n(x) \bar{K}_m(y) \tilde{f}(x, y) \tag{16}$$

where $\tilde{f}(x, y) = [w(x; p, M_1) w(y; q, M_2)]^{-(1/2)} f(x, y)$, $\bar{K}_n(x) = \bar{K}_n(x; p, M_1 - 1)$, $\bar{K}_m(y) = \bar{K}_m(y; q, M_2 - 1)$.

The set of Krawtchouk moment invariants (KMI) is defined as follows.

$$\tilde{Q}_{nm} = [\rho(n)\rho(m)]^{-\frac{1}{2}} \sum_{i=0}^{n} \sum_{j=0}^{m} a_{i,n,p} a_{j,m,q} \tilde{v}_{ij} \tag{17}$$

where $\rho(n) = \rho(n; p, M_1)$, $\rho(m) = \rho(m; q, M_2)$,

$$\tilde{v}_{ij} = \sum_{i=0}^{n} \sum_{j=0}^{m} \binom{n}{i} \binom{m}{q} \left(\frac{N^2}{2}\right)^{(i+j)/2+1} \left(\frac{N}{2}\right)^{(n+m-i-j)} v_{ij}. \tag{18}$$

v_{ij} is the standard set of geometric moment invariants.

4.2   Shape Feature Extraction by KMI

The Krawtchouk moment invariants are rotation, scale and translation invariant, hence in this paper shape features of images are extracted with KMI. The shape feature vector is represented as: $\mathbf{f} = [\tilde{Q}_{00}, \tilde{Q}_{01}, \tilde{Q}_{10}, \cdots, \tilde{Q}_{nm}]$. Normalize the Krawtchouk moment invariants:

$$\bar{Q}_{ij} = \frac{\tilde{Q}_{ij} - \mu_{\mathbf{f}}}{\sigma_{\mathbf{f}}} \tag{19}$$

where $\mu_{\mathbf{f}}$ and $\sigma_{\mathbf{f}}$ are mean and standard deviation of $\mathbf{f}$ respectively. Then the normalized shape feature vector is written as:

$$\mathbf{V}_s = [\bar{Q}_{00}, \bar{Q}_{01}, \bar{Q}_{10}, \cdots, \bar{Q}_{nm}] \tag{20}$$

5   Similarity Measurement

5.1   Distance Between Texture Vectors

The PDF of each NSST subband coefficients can be described by parameter σ^2 and β in the GGD model. Here, the distance between the GGD parameter pairs is calculated with the Kullback–Leibler distance (KLD):

$$D(p(.; \sigma^2, \beta) \| p(.; \sigma'^2, \beta')) = log\left(\frac{\beta \sigma'^2 \Gamma(1/\beta')}{\beta' \sigma^2 \Gamma(1/\beta)}\right) + \left(\frac{\sigma^2}{\sigma'^2}\right)^{\beta'} \frac{\Gamma((\beta'+1)/\beta)}{\Gamma(1/\beta)} - \frac{1}{\beta} \tag{21}$$

The similarity measurement between two NSST subbands can be figured out effectively by the PDF parameters. Meanwhile, the NSST coefficients in different subbands are independent, therefore, the overall distance between two images is the sum of all the KLDs across the NSST subbands. The texture feature distance between the query image and the database image is represented as follows.

$$D(\mathbf{V}_t, \mathbf{V}_t') = \sum_{i=1}^{S} D(p(.; \sigma_i^2, \beta_i) \| p(.; \sigma_i'^2, \beta_i')) \tag{22}$$

where S is the number of the subbands. There is no need for normalization on texture feature vectors in this method of similarity measurement.

5.2 Distance Between Shape Vectors

The similarity measurement between the shape feature vectors is selected to be Euclidean distance, which is the most common distance measurement and is defined as follows.

$$D(\mathbf{V}_s, \mathbf{V}_s') = \left(\sum_{i=1}^{K} (V_{si} - V_{si}')^2 \right)^{\frac{1}{2}} \tag{23}$$

where $\mathbf{V}_s$ is the shape feature vector of the query image, $\mathbf{V}_s'$ is the shape feature vector of the image in the database, and K is the number of vector elements.

5.3 Weighted Distance Measurement

The final distance between the query image $\mathbf{I}$ and the database image $\mathbf{I}'$ is defined by the weighted distance formula as follows.

$$D(\mathbf{I}, \mathbf{I}') = wD(\mathbf{V}_t, \mathbf{V}_t') + (1 - w)D(\mathbf{V}_s, \mathbf{V}_s') \tag{24}$$

where w is the weighted coefficient and $0 \leq w \leq 1$. We use a minimum distance criterion and sort the database images for each query.

6   Experimental Studies

By extracting the texture features and shape features, and measuring the weighted distance developed above, we proposed a NSST-KML-based retrieval system. The retrieval task is to search the top N images that are similar to one query image within a large database of total M unlabeled images. In the proposed system, each image is decomposed into multiple directional subbands. Then, we use the GGD to model the subband coefficients. Parameters $\{\sigma^2, \beta\}$ are derived with ML method, which are the extracted image texture features. The Kullback–Leibler distance (KLD) between the query image and each database image is measured. Meanwhile, the low-order Krawtchouk moment invariants are extracted to be the image shape features. Calculate the Euclidean distance (ED) between the query image and each database image. Finally, the weighted distance of KLD and ED is measured. The top N database images that have the smallest weighted distance are retrieved.

To validate the performance of the proposed retrieval system, we use 40 classes obtained from VisTex database. A test database of 640 texture images is constructed by dividing each 512×512 image into 16 non-overlapping 128×128 subimages. In retrieval experiments, the query image is taken randomly from 640 subimages, and relevant candidate images are the other 15 subimages from the same class. We show 3 sets of experiments and analyze the results. First, we compare the mean-variance-based method, GGD-based method and GGD-KMI-based method across 5 transform: DWT, Contourlet, NSCT, DT-CWT and NSST. Secondly, we change the decomposition levels of NSST to evaluate the retrieval rate with the method mentioned above. Thirdly, we show the image retrieval result interfaces derived from the methods combining 5 different transforms with KMI respectively.

6.1 Comparison Results from Different Transforms

In this experiment, the images are all decomposed into 3 levels by DWT, Contourlet, NSCT, DT-CWT and NSST. For the proposed method, 6, 6, 10 directions are decomposed in the scales from coarser to finer for NSST, and the pyramid filter of NSST is set as maxflat. The shape feature vector built based on Krawtchouk moment invariants is constructed as: $\mathbf{V}_s = [\bar{Q}_{02}, \bar{Q}_{11}, \bar{Q}_{03}, \bar{Q}_{12}]$. The weighted coefficient w is set to be 0.7. The selection of parameters is according to the experimental results.

As shown in Table 1, we computed the overall recognition rate with 7 methods on the all 5 transforms. Overall recognition rate denotes the average *recall* for all the images in the database. $\{\mu\}$&ED denotes only the mean of each subband is extracted to measure the similarity by Euclidean distance. σ is the standard deviation of the subband. $\{\mu, \sigma\}$&ED+KMI is the method combining $\{\mu, \sigma\}$&ED with KMI. GGD&KLD represents the retrieval system just extracts the texture features, which are modeled by GGD and measure the distance between the feature vectors with KL distance. GGD&ED describes the similarity measurement of GGD parameters is obtained by ED. GGD&KLD+KMI is the method combining GGD&KLD with KMI.

From Table 1, we can see that:

(1) Method $\{\mu, \sigma\}$&ED+K is better than $\{\mu\}$&ED, $\{\sigma\}$&ED and $\{\mu, \sigma\}$&ED. The retrieval rate of $\{\mu\}$&ED or $\{\sigma\}$&ED is lower than $\{\mu, \sigma\}$&ED, since more features bring more information of images, which improves the efficiency of retrieval system. One the other hand, KMI provide information of image shape and global features, hence the hybrid method is better than the single feature method.

Table 1. Overall recognition rate(%) with 7 methods across 5 different transforms

Transform	Method						
	$\{\mu\}$&ED	$\{\sigma\}$&ED	$\{\mu, \sigma\}$&ED	$\{\mu, \sigma\}$&ED +KMI	GGD&ED	GGD&KLD	GGD&KLD +KMI
DWT	70.3320	69.1211	73.3887	75.8984	63.2813	80.1758	80.8594
Coutourlet	74.4727	71.9824	75.2344	78.6523	68.4570	80.2051	82.6270
NSCT	74.3164	73.7012	76.9141	79.9609	74.5313	81.3477	83.6621
DT-CWT	75.1367	74.2188	78.9160	81.5723	71.2146	80.8691	83.1738
NSST	76.2305	75.2051	78.2617	81.0449	75.9668	82.6172	84.5410

(2) Method GGD&KLD is better than GGD&ED. To GGD parameters $\{\sigma^2, \beta\}$, KLD is the more appropriate similarity measurement than ED.

(3) GGD&KLD+KMI method is better than the second best method GGD& KLD based on any transform. The reason is as same as the first summary. Texture features combining with shape features can describe an image more accurately. In addition, in order to extract shape features, we just use four low-order KMI, and the measurement of feature vectors is ED, hence the time cost on shape feature extraction and similarity measurement is quite low. However that brings better results.

(4) Methods based on NSST show the best results comparing with the methods on the other transforms. That demonstrates the good performance of NSST in comparison with other transforms.

6.2   Results from Different Decomposition Levels of NSST

We also evaluate the performance of proposed method based on different decomposition levels of NSST. Table 2 shows the overall recognition rate with 7 methods on NSST of one scale with 6 directions, 2 scales with 6 directions respectively and three scales with 6, 6, 10 directions. NSST of three decomposition levels gives the best results than that of one and two decomposition levels. The proposed method still derives best results comparing with the other methods.

Table 2. Overall recognition rate(%) with 7 methods based on different decomposition levels of NSST

Scale	Method						
Direction	$\{\mu\}$&ED	$\{\sigma\}$&ED	$\{\mu, \sigma\}$&ED	$\{\mu, \sigma\}$&ED +KMI	GGD&ED	GGD&KLD	GGD&KLD +KMI
1 [6]	59.1602	58.8867	66.0742	71.2500	65.2930	71.2207	74.7949
2 [6 6]	70.8008	69.7754	74.1992	77.7637	73.2617	79.5215	82.2754
3 [6 6 10]	76.2305	75.2051	78.2617	81.0449	75.9668	82.6172	84.5410

6.3   Image Retrieval Examples

Four sets of comparison examples of GGD&KLD+KMI method based on DWT, Coutourlet, NSCT, DT-CWT and NSST are shown in Figs. 2, 3, 4 and 5 respectively. Examples of retrieval results are computed from 640 texture images based on the VisTex database. In each example, the query image is on the top left corner and all other images are ranked in the order of similarity with the query image from left to right, top to bottom. The query images are Fabric, Food, Leaves and Tile respectively. The proposed method of GGD&KLD+KMI based on NSST shows better retrieval performance than other methods.

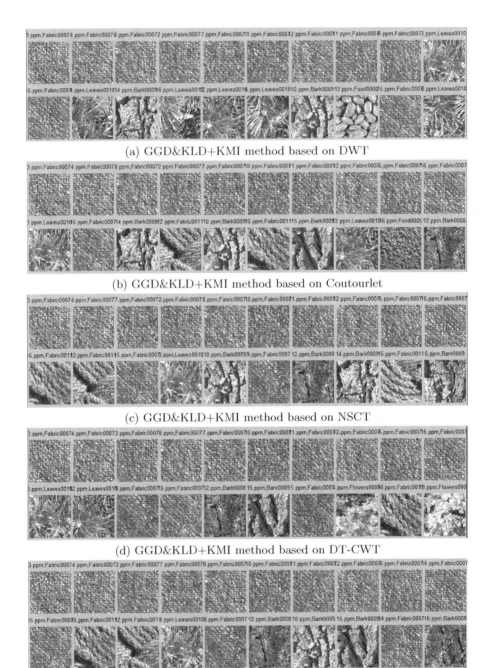

(a) GGD&KLD+KMI method based on DWT

(b) GGD&KLD+KMI method based on Coutourlet

(c) GGD&KLD+KMI method based on NSCT

(d) GGD&KLD+KMI method based on DT-CWT

(e) GGD&KLD+KMI method based on NSST

Fig. 2. Image retrieval example 1

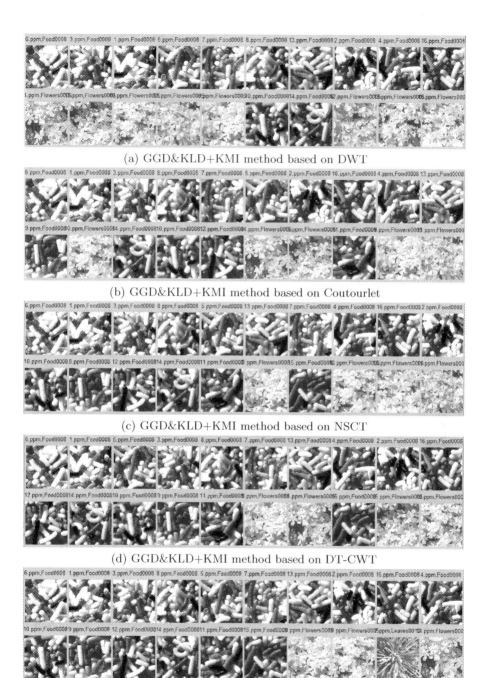

(a) GGD&KLD+KMI method based on DWT

(b) GGD&KLD+KMI method based on Coutourlet

(c) GGD&KLD+KMI method based on NSCT

(d) GGD&KLD+KMI method based on DT-CWT

(e) GGD&KLD+KMI method based on NSST

Fig. 3. Image retrieval example 2

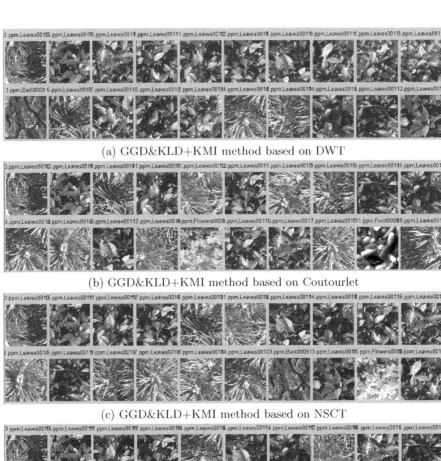

(a) GGD&KLD+KMI method based on DWT

(b) GGD&KLD+KMI method based on Coutourlet

(c) GGD&KLD+KMI method based on NSCT

(d) GGD&KLD+KMI method based on DT-CWT

(e) GGD&KLD+KMI method based on NSST

Fig. 4. Image retrieval example 3

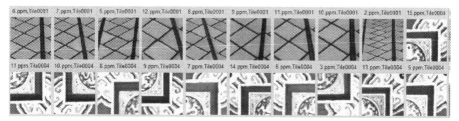

(a) GGD&KLD+KMI method based on DWT

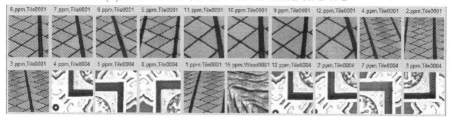

(b) GGD&KLD+KMI method based on Coutourlet

(c) GGD&KLD+KMI method based on NSCT

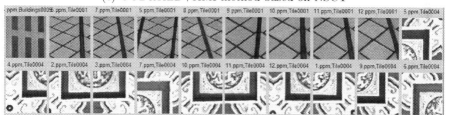

(d) GGD&KLD+KMI method based on DT-CWT

(e) GGD&KLD+KMI method based on NSST

Fig. 5. Image retrieval example 4

7   Conclusion

In this paper, we proposed a new image retrieval system which is a hybrid method of NSST and KMI. The image was decomposed by NSST and GGD was used to fit the subband coefficients of NSST. The parameters of GGD were regarded as texture features which were measured by Kullback–Leibler distance. On the other hand, low-order Krawtchouk moment invariants were calculated to be the shape features which were measured by Euclidean distance. At last, the weighted distance of two kinds of distance was computed to measure the similarity between the query image and the database images. Three sets of experiments based on 7 methods and 5 different transforms showed the proposed method can obtain higher retrieval rate than other methods.

References

1. Rui, Y., Huang, T.S., Chang, S.F.: Image retrieval: current techniques, promising directions and open issues. J. Vis. Commun. Image Represent. **10**, 39–62 (1999)
2. Smeulders, A.W.M., Worring, M., Santini, S., Gupta, A., Jain, R.: Content-based image retrieval at the end of the early years. IEEE Trans. Pattern Anal. Mach. Intell. **22**, 1349–1380 (2000)
3. Shandilya, S.K., Singhai, N.: A survey on: content based image retrieval systems. Int. J. Comput. Appl. **4**, 22–26 (2010). Published By Foundation of Computer Science
4. Do, M.N., Vetterli, M.: Wavelet-based texture retrieval using generalized gaussian density and kullback-leibler distance. IEEE Trans. Image Process. **11**, 146–158 (2002)
5. Laine, A., Fan, J.: Texture classification by wavelet packet signatures. IEEE Trans. Pattern Recognit. Mach. Intell. **15**, 1186–1191 (1993)
6. Chang, T., Kuo, C.J.: Texture analysis and classification with tree-structure wavelet transform. IEEE Trans. Image Process. **2**, 429–441 (1993)
7. Do, M.N.: Contoulets and sparse image expansions. Proc. SPIE Appl. Signal Image Process. X **5207**, 560–570 (2003)
8. Do, M.N., Vetterli, M.: The contourlet transform: an efficient directional multiresolution image representation. IEEE Trans. Image Process. **14**, 2091–2106 (2005)
9. Cunha, A.L., Zhou, J.P., Do, M.N.: The nonsubsampled contourlet transform: theory, design and application. IEEE Trans. Image Process. **15**, 3089–3101 (2006)
10. Labate, D., Lim, W., Kutyniok, G.: Sparse multidimensional representation using shearlets. SPIE Proc. **5914**, 254–262 (2005)
11. Easley, G., Labate, D., Lim, W.: Sparse directional image representations using the discrete shearlet transform. Appl. Comput. Harmonic Anal. **25**, 25–46 (2008)
12. Varanasi, M.K., Aazhang, B.: Parametric generalized gaussian density estimation. J. Acoust. Soc. Am. **86**, 1404–1415 (1989)
13. Zhang, D.S.: Image retrieval based on shape. Monash University, Australia (2002)
14. Ding, X.F., Wu, H., Zhang, H.J.: Review on shape matching. Acta Automatica Sin. **27**, 678–694 (2001)
15. Zhang, D.S., Lu, G.J.: Review of shape representation and description techniques. Pattern Recogn. **37**, 1–19 (2004)
16. Yap, P.T., Paramesran, R., Ong, S.H.: Image analysis by krawtchouk moments. IEEE Trans. Image Process. **12**, 1367–1377 (2003)

Curve Matching from the View of Manifold for Sign Language Recognition

Yushun Lin[1,2](✉), Xiujuan Chai[1], Yu Zhou[3], and Xilin Chen[1]

[1] Key Laboratory of Intelligent Information Processing of Chinese Academy of
Sciences (CAS), Institute of Computing Technology, CAS, Beijing 100190, China
[2] University of Chinese Academy of Sciences, Beijing 100049, China
yushun.lin@vipl.ict.ac.cn
[3] Institute of Information Engineering, CAS, Beijing 100093, China

Abstract. Sign language recognition is a challenging task due to the
complex action variations and the large vocabulary set. Generally, sign
language conveys meaning through multichannel information like trajectory, hand posture and facial expression simultaneously. Obviously,
trajectories of sign words play an important role for sign language recognition. Although the multichannel features are helpful for sign representation, this paper only focuses on the trajectory aspect. A method of
curve matching based on manifold analysis is proposed to recognize isolated sign language word with 3D trajectory captured by Kinect. From
the view of manifold, the main structure of the curve is found by the
intrinsic linear segments, which are characterized by some geometric fea-
tures.Then the matching between curves is transformed into the matching between two sets of sequential linear segments. The performance
of the proposed curve matching strategy is evaluated on two different
sign language datasets. Our method achieves a top-1 recognition rate
of 78.3 % and 61.4 % in a 370 daily words dataset and a large dataset
containing 1000 vocabularies.

1 Introduction

Sign language is one of the most common communication means for hearing-
impaired community. With the rising public concern on deaf-mutes, sign language recognition has become a hot research topic nowadays. At the same time,
it's also a complex and challenging problem for dealing with multichannel information, including trajectory, hand posture and even facial expression.

There exist several common approaches for sign language recognition. The
early approaches often applied artificial neural networks (ANN). A method for
Japanese sign language recognition using recurrent neural network was proposed
by Murakami et al. (1991) [1]. They developed a system to recognize a finger
alphabet of 42 symbols, with the input of data gloves. Huang et al. (1995) [2]
presented an isolated sign recognition system using a Hopfield ANN, which can
recognize 15 different gestures accurately. After that, Kim et al. (1996) [3] trained
a Fuzzy Min Max ANN with x, y, z coordinates and angles provided by data

© Springer International Publishing Switzerland 2015
C.V. Jawahar and S. Shan (Eds.): ACCV 2014 Workshops, Part III, LNCS 9010, pp. 233–246, 2015.
DOI: 10.1007/978-3-319-16634-6_18

gloves to recognize 25 isolated gestures with a success rate of 85 %. Another popular approach is Hidden Markov Model, as sign language recognition is a kind of temporal pattern recognition, just like speech recognition. Grobel et al. (1997) [4] presented an HMM based isolated sign (gesture) recognition system, which achieved a recognition rate of 94 % for 262 signs. They used cotton gloves with several colour-markings in fingers, palm and back of the hand to get both trajectory and hand shape features. Starner et al. (1998) [5] proposed an HMM based system for recognizing sentence-level continuous American Sign Language (ASL) in a lexicon of 40 words and demonstrated that HMMs presents a strong technique for recognizing sign language.

However, in this paper we tried to focus on the problem of sign language recognition with only trajectory information. Then the problem of sign language recognition can be transformed to curve matching in 3D space. In this field, many work has been done on 2D curve matching. Mokhtarian et al. (1986) [6] introduced a scale-based description at varying levels of details to recognize planer curves. An affine-invariant method was proposed by Zuliani et al. (2004) [7]. Efrat et al. (2007) [8] introduced continuous dynamic time warping measures for calculating the curves' similarity. As to 3D curve matching, Pajdla et al. (1995) [9] presented a method for this problem using semi-differential invariants, without computation of high order derivatives. Kishon et al. (1990) [10] proposed a method, which transformed the 3D curves into 1D numerical strings of rotation and translation invariant to match. Most of work above focuses on finding out the longest corresponding parts of two curves, while in sign language recognition application we concern more about the feature representation and the matching, i.e. the quantization of the distance between two curves.

To realize the curve matching, our solution is inspired by the view of manifold. Each sign curve is regarded as a manifold, then the curve matching problem is reformulated as the distance calculating between two corresponding manifolds. Maximum Linear Segment (MLS) by hierarchical divisive clustering is introduced to explore the structure information of curves. To well describe the sign, a hand-elbow co-occurrence feature is proposed, in which, the hand is dominant and the elbow is subordinative. With a novel designed distance measurement, the final matching between two sets of sequential linear segments is achieved with dynamic time warping (DTW) matching.

The remaining part of this paper is organized as follows. Section 2 gives an overview on the manifold analysis based curve matching method. Section 3 is the detail for each key module, including preprocessing, Maximum Linear Segment, the hand-elbow co-occurrence feature and the matching of sequential linear segments. Section 4 is about the experiments and the analysis. Section 5 is the conclusion.

2   Method Overview

To solve the problem of sign language recognition with only trajectory information, we propose the manifold analysis based curve matching method.

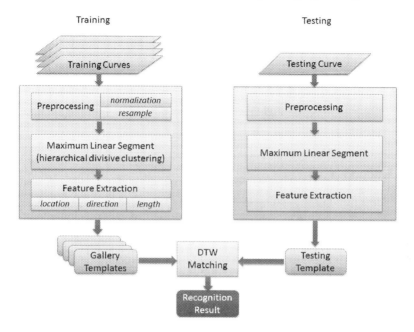

Fig. 1. Flowchart of the proposed method.

The proposed method mainly consists of three parts, i.e., preprocessing, Maximum Linear segment (MLS), hand-elbow co-occurrence feature extraction and matching of sequential linear segments. Figure 1 gives the illustration for the main flowchart of the proposed method.

The preprocessing step attempts to normalize the data and remove the noise. In consideration of various scales from different signers, the trajectory curves should be normalized by signers' size at the first step of preprocessing. Then it must be re-sampled in order to remove the effect of noise, especially those brought by inconsistent velocities.

Maximum Linear Segments are further designed to model the locally linear structure of the curves by using hierarchical divisive clustering (HDC). Then the intrinsic structure of the curves can be discovered. Some geometric features are extracted to characterize each linear segment and to well describe the sign curve, a novel hand-elbow co-occurrence feature is proposed. The final matching between sequential linear segments is obtained by DTW matching. It is worth mentioning that in the case of two-hands words, the hands are dealt separately and their DTW distances are combined with an equal weight for the final decision.

3   Manifold Analysis Based Curve Matching

3.1   Preprocessing

The target of preprocessing consists of two aspects, scale normalization and data resampling. On one hand, the raw data of trajectory obtained from the device are

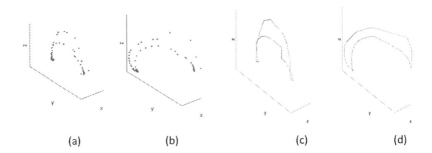

Fig. 2. Some exemplars of resampling. (a) and (b) are raw curves. (c) and (d) are re-sampled ones corresponding to (a) and (b) respectively.

3D locations of joints, which should not be used directly. Since different signers have different body sizes, the scales of trajectory curves played by them may also be different. In order to avoid the effect of various scales, data of trajectory should be normalized from the very beginning. In this work, the signers are required to stand in front of the device with their hands down at the beginning of recording. The locations of both hands and the head in the first frame are selected as reference points for normalization.

On the other hand, resampling is essential for two reasons. First of all, it helps to remove the noise caused by inconsistent velocities. Because there is no explicit rules about velocity in standard Chinese Sign Language, velocity of the trajectory seems to be a random factor. It means that different signers may play a same word in different velocities, and even one signer may play the same word in different velocities in two times. Resampling operation can effectively remove the difference of velocities and keep the curve shape information well. The other reason is for the convenience of division of Maximum Linear Segment, which requires sufficient sample points. There exists many algorithms for resampling [11], and the $1 algorithm proposed by Wobbrock, J.O., et al. (2007) [12] is used in our method because of its less time cost and simpleness. The main idea of this algorithm is the equidistant linear interpolation along the curve. The interval length of interpolation is the length of the curve divided by the number of re-sampled points. Some exemplars of resampling are shown in Fig. 2, from which, it can be seen that the curve is re-sampled evenly while maintaining the original shape.

3.2   Maximum Linear Segment

After preprocessing, normalized curves that consist of a certain number of even points are obtained. Each curve can be regarded as a simple manifold of one dimension. In mathematics, manifold is a topological space that resembles Euclidean space near each point locally. Similar to Maximum Linear Patch (MLP) introduced by [13], we introduce a concept of Maximum Linear Segment (MLS), to represent the structure information of curves. The maximum

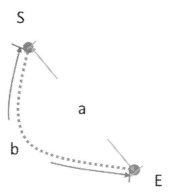

Fig. 3. The illustration of Euclidean distance a and geodesic distance b between point S and point E on the blue curve (Color figure online).

here means that the partition of segments made by MLS has maximum linearity for current step of division. In order to quantify the linear degree of curves, the definition of non-linearity degree is introduced. The non-linearity degree of a curve is defined as the geodesic distance along the curve divided by Euclidean distance between the starting point and end point. An illustration of geodesic distance and Euclidean distance between two points on a curve is given in Fig. 3. Denote $d(p)$ as the non-linearity degree of segment $S(p)$ with n points from p_1 to p_n. Then $d(p) = \sum_{i=1}^{n-1} ||p_{i+1} - p_i|| / ||p_n - p_1||$. Just as the name implies, the larger non-linearity degree, the less linear it is.

The main idea of MLS is to regard the whole curve as a manifold, then split it into certain segments according to the non-linearity degree. Different partitions represent different distributions of local linear structure of curves. MLS is able to keep the local linear property of segments as much as possible.

Let's denote X as the whole curve with N points, and M is the number of segments that the curve will be divided into. Then problem can be formulated as looking for a set of segments $X^{(i)} (i = 1, 2, ...M)$ that the non-linearity degree of each division step is minimum and satisfy the following equations:

$$X = \bigcup_{i=1}^{M} X^{(i)} (i = 1, 2, ...M). \tag{1}$$

$$X^{(i)} \cap X^{(j)} = \emptyset (i \neq j, i, j = 1, 2, ...M). \tag{2}$$

$$X^{(i)} = \left\{ x_1^{(i)}, x_2^{(i)}, ..., x_{Ni}^{(i)} \right\} \left(\sum_{i=1}^{M} Ni = N \right). \tag{3}$$

To achieve the target, we use hierarchical divisive clustering (HDC) to get the segments. The algorithm for MLS by HDC is described in Algorithm 1.

The specific value of M is a tuning parameter. In our experiment, we find that 72 is a proper value. Finally, we can get a set of Maximum Linear Segments

---

1 Given a curve C and a fixed number M ;
2 Define the set of segments as S and the number of segments in S as n ;
3 Set S empty ;
4 Add C to S ;
5 n=1 ;
6 **while** *n is not equal to M* **do**
7 | Find the segment S_m with maximum non-linearity degree in S ;
8 | **for** *Each point p in S_m* **do**
9 | | Divide S_m from p into two segments: $S_1(p)$, $S_2(p)$ and calculate their non-linearity degree $d_1(p)$, $d_2(p)$;
10 | **end**
11 | Find the point p_m that minimize $max(d_1(p), d_2(p))$;
12 | Add $S_1(p_m)$, $S_2(p_m)$ to S ;
13 | Remove S_m from S ;
14 | n=n+1 ;
15 **end**
16 Output the final set of segments S ;

Algorithm 1. MLS by hierarchical divisive clustering (HDC).

divided from a curve, which are the elementary units for the following matching steps. An example of HDC with 4 steps is shown in Fig. 4. Figure 5 gives the MLS result of a real sign trajectory curve.

3.3   Matching of Sequential Segments

Since we have gotten the Maximum Linear Segments for each curve, the problem of calculating the distance between two curves is transformed to the calculation of distance between two sequences of segments. In order to accomplish the target, there are two problems needed to be solved.

Measurement of Two Matched Segments. The first problem is feature extraction to measure two matched segments. Considering that the geometric information is most important in trajectory data, the location, direction and length are extracted to characterize each segment. Then based on this kind of feature representation, the distance between two matched segments should be defined.

From the view of manifold, the segment is a linear subspace, which can be abstracted to the line segment. The location of a segment is defined as the midpoint of the line segment. Just as its name implies, the direction of a segment is defined by the difference between the end point and the starting point. The length of segment is defined as the Euclidean distance between the starting point and end point of the line segment. Denote the two matched segments as S_1, S_2, the distance between them is defined as follows. (see Eq. 4)

$$d(S_1, S_2) = (d_e + d_{len})/\left[(d_p + 1)/2\right].\tag{4}$$

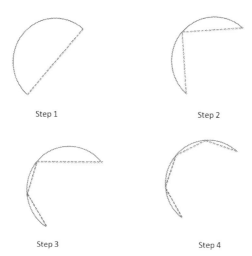

Step 1   Step 2

Step 3   Step 4

Fig. 4. An example of a HDC with 4 steps. The segments of the curve are represented by dash lines.

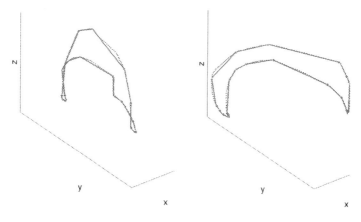

Fig. 5. The MLS of a real sign trajectory. The green lines represent the Maximum Linear Segments. In this example, M is set to 20 for easy understanding (Color figure online).

where d_e is the Euclidean distance between the locations of two segments, d_{len} is the length difference between two segments, that is $d_{len} = |len(S_1) - len(S_2)|$ where $len(S)$ represents the length of segment S. $d_p = \cos\theta$ is the cosine of the angle between two segments. (see Fig. 6) Actually, the plane in Fig. 6 is just simplified for visualization. Two segments S_1 and S_2 are the vectors in 3D space. Thus the angle between these two 3D vectors can be calculated.

Measurement of Segment Sequences. The other problem is the matching between two sets of sequential linear segments. Although the amount of segments we get from each curve is set to be the same, it doesn't means that the

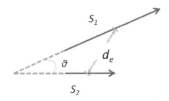

Fig. 6. Illustrations of features. S_1 and S_2 are two matched segments (blue line segments) (Color figure online).

segments should be matched simply just by their indexes. In spite that the sign curves belonging to the same vocabulary will have roughly similar structure, corresponding segments may appear with different indexes due to the random variance when playing a sign. The signer cannot repeat the same word with exactly the same trajectory.

However, it's true that the matching of segment sequences for two curves should obey the rules of sequential matching. Let's denote $X_1^{(i)}$ as the ith segment of curve X_1, and denote $X_2^{(j)}$ in a similar way. The sequential matching means that there will not exist a match between $X_1^{(i)}$ and $X_2^{(j)}$ if there already exists a match between $X_1^{(m)}$ and $X_2^{(n)}$, where $(i - m) * (j - n) < 0$. To meet the requirement of order preservation, DTW is applied in our solution. In sequential analysis, DTW is used for measuring similarity between two temporal sequences which may vary in time. It has been widely used in many applications, including handwriting recognition [14] and signature verification [15]. The most well-known one should be automatic speech recognition [16], in which DTW is used to cope with different speaking speeds. Generally speaking, DTW can be applied to the analysis for any temporal data which can be turned into a linear sequence, including the trajectory data in our problem. It can determine the matches that minimize the distance between two segment sequences under the condition of sequential matching.

Since we have measurement for two segments, the DTW distance between two segment sequences can be calculated to evaluate the similarity between two curves. The formulation of DTW distance between two segment sequences Seq_1 and Seq_2 is given in Eq. 5.

$$D_{dtw}(Seq_1, Seq_2) = \frac{1}{Card(I)} \sum_{(i,j) \in I} d(S_{1i}, S_{2j}). \tag{5}$$

where S_{1i} denotes the ith segment in Seq_1 and S_{2j} is the jth segment in Seq_2. I is the matching set in the warping path, which minimize the DTW distance between Seq_1 and Seq_2. $Card(I)$ denotes the cardinality of set I.

3.4   Hand-Elbow Co-occurrence Feature

It is obvious that the hands' location features describe the majority of trajectory information in a sign word. At the same time, the elbow can also provide some

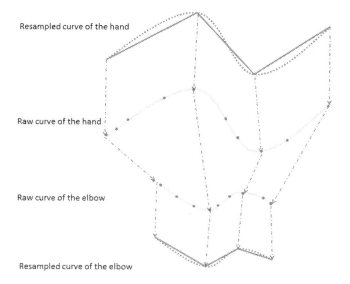

Fig. 7. An illustration for the procedure of building the hand and elbow mapping relations. The blue points represent the raw data points on raw curves while the green lines represent the segments on re-sampled curves. The red dash lines indicates the mapping relations (Color figure online).

discriminative information for trajectory curve recognition. The elbow movement can be regarded as a subordinative motion to hand. In order to avoid the noise effect, a novel hand-elbow co-occurrence feature is proposed.

Considering that the hand movement is dominant, the Maximum Linear Segments are obtained from hand curve and set as the reference to get the corresponding elbow segments. In other words, there is a one-to-one correspondence between raw data points of hands and elbows. Unfortunately, this one-to-one correspondence will disappear after resampling and MLS. Our solution is mapping the hand segment points of re-sampled curves back to the nearest raw hand curves. Then we can get corresponding segment points of the raw elbow curves because of the one-to-one correlation. In the end, mapping the segment points of the raw elbow curves to the re-sampled elbow curves. The correspondence between hands and elbows' segments is built. The procedure for building the mapping relations is illustrated in Fig. 7.

4 Experiments

4.1 Datasets and Experimental Setting

To evaluate the performance of the proposed method, our experiments are conducted on two Chinese Sign Language datasets collected by ourselves.

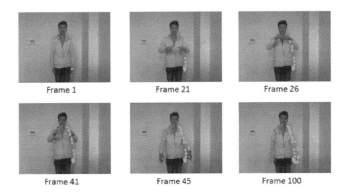

Fig. 8. An example of key frames from one of the words in Dataset B.

Dataset A. Dataset A has 370 vocabularies that are widely used in daily life. Each word is played 5 times. So there are 5×370 sign videos totally. All of the data is played by one female deaf student.

Dataset B. Dataset B has 1000 different vocabularies. Each word is played 3 times, in other words there are 3×1000 words in all. All the words are played by a male deaf student. Figure 8 shows an example of the key frames for a word in this dataset.

The data is captured by Kinect, which can provide color image and depth map simultaneously. With the public windows SDK, the joint locations can be obtained in real-time. For the convenience to get abundant data, Kinect has been used in gesture recognition [17], 3D body scanning [18] and sign language recognition [19,20]. Among all 20 joints provided, 5 joints including head, both hands, and both elbows are used in our implementation. It is obvious that the locations of both hands are essential for our problem of sign language recognition. The locations of both elbows are good supplement for hands. The location of the head is used for normalization in preprocessing.

Since we have 5 groups of data in Dataset A and 3 groups of data in Dataset B, the leave-one-out cross-validation strategy is adopted in our experiments.

4.2 Baseline Method

Hidden Markov Model (HMM) is a statistical learning method for modeling a system, that is assumed to be a Markov process with hidden states. It is a classic method in temporal pattern recognition, such as speech, gesture and handwriting recognition [21], which is similar to our problem of trajectory recognition in sign language recognition. Therefore HMM is regarded as a baseline method in this paper. On the other hand, we also compare to the curve matching method presented by Chai, X., et al. (2013) [22], which matches the location of trajectory points (denoted as PLM), for a sanity check.

To give a concrete comparison with our method, we carried out the HMM experiments in two versions with different features. One is based on the simple

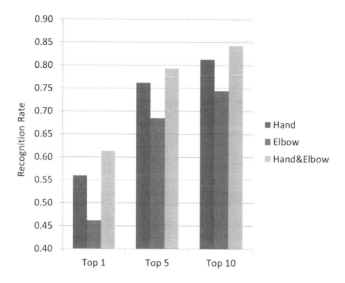

Fig. 9. The comparison among the recognition rates by using elbows' feature, hands' feature and hand-elbow co-occurrence feature on Dataset B.

Table 1. The recognition rates for all four methods in Dataset A.

Dataset A	HMM_N	HMM_P	PLM	Our method
Top 1	0.792	0.753	0.810	0.783
Top 5	0.901	0.879	0.919	0.902
Top 10	0.928	0.914	0.946	0.929

normalized skeleton points with head reference, which is denoted as HMM_N. And the other adopts the pairwise features as the input (denoted as HMM_P). The pairwise feature is also generated by hands, elbows and the head, which is introduced by Wang, J., et al. (2012) [23].

4.3   Experimental Results and Analysis

Location feature of elbows is considered to be a supplement to the location feature of hands. We have shown the comparison among the recognition rates of elbows' feature, hands' feature and both hands and elbows' feature on Dataset B in Fig. 9. From this figure, it can be seen that the hand-elbow co-occurrence feature can enhance the recognition rate significantly by considering the elbow feature simultaneously.

The recognition rates of HMM with normalized feature (HMM_N), HMM with pairwise feature (HMM_P), method proposed in [22] (PLM) and our method on both Dataset A (see Table 1) and Dataset B (see Table 2) are shown in the following tables. Figure 10 gives the top-1 recognition rates for all three methods on the two datasets.

Table 2. The recognition rates for all four methods in Dataset B.

Dataset B	HMM_N	HMM_P	PLM	Our method
Top 1	0.336	0.597	0.591	0.614
Top 5	0.563	0.813	0.770	0.793
Top 10	0.655	0.868	0.819	0.843

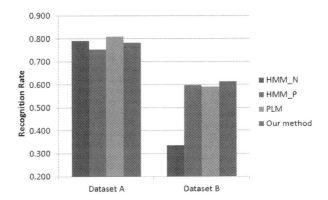

Fig. 10. The comparison of top-1 recognition rates among HMM_N, HMM_P, PLM and the proposed method.

Judging from these results, in Dataset A, a dataset of smaller scale, our method is comparable with the other methods. In Dataset B, a more persuasive dataset with large vocabulary, our method achieves much higher recognition rate than the other methods. The HMM_N performs well on Dataset A, which is of smaller scale. But its performance decreases significantly on challenging Dataset B. However, our method of manifold analysis based curve matching performs much better on Dataset B, for the reason that it characterizes the local linear structure of curves better.

5   Conclusion

In this paper, we propose a novel 3D curve matching method for sign language recognition. The method is inspired from the view of manifold and divides the trajectory curve into Maximum Linear Segment (MLS), which describes the intrinsic structure of the curve. Based on the definition of distance between two matched segments, the similarity between two curves is also evaluated through the matching between sequential linear segments. Furthermore, a novel hand-elbow co-occurrence feature is proposed to enhance the representation ability for sign curve by considering the hand and elbow movement simultaneously. The experiments show that our method performs better than the baseline methods on a large dataset containing 1000 vocabularies. However, it can be seen that the trajectory information may be not enough for sign language recognition, especially when confronted with the large vocabularies. A directly future work is to explore the fusion feature by combining the motion trajectory and the hand shape for robust sign language recognition.

Acknowledgements. This work was partially supported by the Microsoft Research Asia, and Natural Science Foundation of China under contract Nos. 61303170, 61472398, and the FiDiPro program of Tekes.

References

1. Murakami, K., Taguchi, H.: Gesture recognition using recurrent neural networks. In: Proceedings of the SIGCHI Conference on Human Factors in Computing Systems, CHI 1991, pp. 237–242. ACM, New York (1991)
2. Huang, C.L., Huang, W.Y., Lien, C.C.: Sign language recognition using 3-d hopfield neural network. In: Proceedings of the International Conference on Image Processing, vol. 2, pp. 611–614 (1995)
3. Kim, J.S., Jang, W., Bien, Z.: A dynamic gesture recognition system for the korean sign language (ksl). IEEE Trans. Syst. Man Cybern. Part B: Cybern. **26**, 354–359 (1996)
4. Grobel, K., Assan, M.: Isolated sign language recognition using hidden markov models. In: 1997 IEEE International Conference on Systems, Man, and Cybernetics, Computational Cybernetics and Simulation, vol. 1, pp. 162–167 (1997)
5. Starner, T., Weaver, J., Pentland, A.: Real-time american sign language recognition using desk and wearable computer based video. IEEE Trans. Pattern Anal. Mach. Intell. **20**, 1371–1375 (1998)
6. Mokhtarian, F., Mackworth, A.: Scale-based description and recognition of planar curves and two-dimensional shapes. IEEE Trans. Pattern Anal. Mach. Intell. **PAMI–8**, 34–43 (1986)
7. Zuliani, M., Bhagavathy, S., Manjunath, B., Kenney, C.: Affine-invariant curve matching. In: 2004 International Conference on Image Processing, ICIP 2004, vol. 5, pp. 3041–3044 (2004)
8. Efrat, A., Fan, Q., Venkatasubramanian, S.: Curve matching, time warping, and light fields: New algorithms for computing similarity between curves. J. Math. Imaging Vis. **27**, 203–216 (2007)
9. Pajdla, T., Gool, L.V.: Matching of 3-d curves using semi-differential invariants. In: Proceedings of the Fifth International Conference on Computer Vision, pp. 390–395 (1995)
10. Kishon, E., Hastie, T., Wolfson, H.: 3-d curve matching using splines. In: Faugeras, O. (ed.) ECCV 1990. LNCS, vol. 427, pp. 589–591. Springer, Heidelberg (1990)
11. Shahraray, B., Anderson, D.: Uniform resampling of digitized contours. IEEE Trans. Pattern Anal. Mach. Intell. **PAMI–7**, 674–681 (1985)
12. Wobbrock, J.O., Wilson, A.D., Li, Y.: Gestures without libraries, toolkits or training: A \$1 recognizer for user interface prototypes. In: Proceedings of the 20th Annual ACM Symposium on User Interface Software and Technology, UIST 2007, pp. 159–168. ACM, New York (2007)
13. Wang, R., Shan, S., Chen, X., Chen, J., Gao, W.: Maximal linear embedding for dimensionality reduction. IEEE Trans. Pattern Anal. Mach. Intell. **33**, 1776–1792 (2011)

14. Bahlmann, C., Burkhardt, H.: The writer independent online handwriting recognition system frog on hand and cluster generative statistical dynamic time warping. IEEE Trans. Pattern Anal. Mach. Intell. **26**, 299–310 (2004)
15. Martens, R., Claesen, L.: On-line signature verification by dynamic time-warping. In: Proceedings of the 13th International Conference on Pattern Recognition, vol. 3, pp. 38–42 (1996)
16. Sakoe, H., Chiba, S.: Dynamic programming algorithm optimization for spoken word recognition. IEEE Trans. Acoust. Speech Signal Process. **26**, 43–49 (1978)
17. Ren, Z., Meng, J., Yuan, J., Zhang, Z.: Robust hand gesture recognition with kinect sensor. In: Proceedings of the 19th ACM International Conference on Multimedia, MM 2011, pp. 759–760. ACM, New York (2011)
18. Tong, J., Zhou, J., Liu, L., Pan, Z., Yan, H.: Scanning 3d full human bodies using kinects. IEEE Trans. Visual Comput. Graphics **18**, 643–650 (2012)
19. Zafrulla, Z., Brashear, H., Starner, T., Hamilton, H., Presti, P.: American sign language recognition with the kinect. In: Proceedings of the 13th International Conference on Multimodal Interfaces, ICMI 2011, pp. 279–286, ACM, New York (2011)
20. Sun, C., Zhang, T., Bao, B.K., Xu, C., Mei, T.: Discriminative exemplar coding for sign language recognition with kinect. IEEE Trans. Cybern. **43**, 1418–1428 (2013)
21. Al-Hajj Mohamad, R., Likforman-Sulem, L., Mokbel, C.: Combining slanted-frame classifiers for improved hmm-based arabic handwriting recognition. IEEE Trans. Pattern Anal. Mach. Intell. **31**, 1165–1177 (2009)
22. Chai, X., Li, G., Lin, Y., Xu, Z., Tang, Y., Chen, X., Zhou, M.: Sign language recognition and translation with kinect. In: IEEE Conference on AFGR (2013)
23. Wang, J., Liu, Z., Wu, Y., Yuan, J.: Mining actionlet ensemble for action recognition with depth cameras. In: 2012 IEEE Conference on Computer Vision and Pattern Recognition (CVPR), pp. 1290–1297 (2012)

Learning Partially Shared Dictionaries for Domain Adaptation

Viresh Ranjan[1(✉)], Gaurav Harit[2], and C.V. Jawahar[1]

[1] CVIT, IIIT Hyderabad, Hyderabad, India
vireshranjan@gmail.com
[2] IIT Jodhpur, Jodhpur, India

Abstract. Real world applicability of many computer vision solutions is constrained by the mismatch between the training and test domains. This mismatch might arise because of factors such as change in pose, lighting conditions, quality of imaging devices, intra-class variations inherent in object categories etc. In this work, we present a dictionary learning based approach to tackle the problem of domain mismatch. In our approach, we jointly learn dictionaries for the source and the target domains. The dictionaries are partially shared, i.e. some elements are common across both the dictionaries. These shared elements can represent the information which is common across both the domains. The dictionaries also have some elements to represent the domain specific information. Using these dictionaries, we separate the domain specific information and the information which is common across the domains. We use the latter for training cross-domain classifiers i.e., we build classifiers that work well on a new target domain while using labeled examples only in the source domain. We conduct cross-domain object recognition experiments on popular benchmark datasets and show improvement in results over the existing state of art domain adaptation approaches.

1 Introduction

Visual object recognition schemes popularly use feature descriptor such as SIFT [1], HOG [2] followed by a classification strategy such as SVMs [3]. They train on a set of annotated training set images and evaluate on a set of similar images for quantifying the performance. However, such object recognition schemes may perform badly in the case of large variations between the source domain and the target domain [4]. Variations between the source and target domain might arise from changes in pose, illumination or intra-class variations inherent in object categories. In Fig. 1, we show sample images of the categories chair and bottle from three different domains, namely Amazon, DSLR and Webcam [5]. The domain Amazon is visually very different from the other two domain, the reason being large intra-class variations. The difference between the domains DSLR and Webcam arises because of change in pose, camera quality and lighting conditions.

To tackle the issue of variations across the source and target domains, various domain adaptation (DA) techniques have been proposed in the natural language

© Springer International Publishing Switzerland 2015
C.V. Jawahar and S. Shan (Eds.): ACCV 2014 Workshops, Part III, LNCS 9010, pp. 247–261, 2015.
DOI: 10.1007/978-3-319-16634-6_19

processing as well as computer vision communities. In Fig. 2, we present the overall idea behind a general DA approach. The figure depicts the idea that a classifier trained on the source domain may need further adaptation in order to perform well on the target domain. In the natural language processing community, DA techniques have been applied for tasks such as sentiment classification, parts of speech tagging etc. Blitzer *et al.* [6] present a DA technique to modify discriminative classifiers trained on the source domain to classify samples from the target domain. The primary aspect of their work is identifying the *pivot* features, i.e. those features which occur frequently and behave similarly across the two domains. Hal Daume [7] presents a feature augmentation approach where source, target and a common domain representation are obtained by replicating the original feature. Jiang and Zhai [8] present an instance weighting approach where they prune misleading examples from the source domain and give more weight to the labeled examples from the target domain.

Fig. 1. Sample images of categories "bottle" and "chair" from the domains Amazon, DSLR and Webcam [5]. Images from Amazon are visually very different in comparison to the other two domains. Visual mismatch between DSLR and Webcam is relatively less and arises from factors such as changes in pose, image resolution and lighting conditions.

In recent years, there has been a surge of interest in the visual domain adaptation task. Several DA strategies have been proposed which adapt either the feature representation or the classifier. These strategies are semi-supervised or unsupervised depending on whether some labeled data from the target domain is available or not. Utilizing labeled examples from source as well as target domains, Saenko *et al.* [5] learn a transformation to map vectors from one domain to another. This transformation tries to bring closer the intra-class vectors from the two domains and push the inter-class vectors farther apart. In a similar feature transformation based approach, Kulis *et al.* [9] do an extension of this work in which the vectors in the two domains can have different dimensions. Unlike the previous two works, the feature transformation based approach presented by Fernando *et al.* [10] is completely unsupervised. They model the source subspace by the eigenvectors obtained by doing PCA over the source domain and similarly for the target subspace. They align the source subspace with the target subspace by learning a transformation matrix. The source and the target domain samples are then projected to their corresponding aligned subspace. Gopalan *et al.* [11] present a feature augmentation based DA approach where the source and the target subspaces are modeled as points on a Grassmann manifold. They sample points along the geodesic between the source subspace and the target subspace to obtain intermediate subspaces. The data points are projected along all the intermediate subspaces to obtain a domain independent representation. Gong *et al.* [12] propose a geodesic flow kernel based approach and instead of sampling finite number of subspaces along the geodesic from source subspace to target subspace, they integrate over infinite number of intermediate subspaces. Jhuo *et al.* [13] present a semi-supervised DA approach based on low-rank approximation. The samples from the source domain are mapped to an intermediate representation where the transformed source samples can be expressed as a linear combination of target samples. The authors consider single source domain as well as multiple source domain scenarios in this work. Apart from the feature adaptation based DA techniques, classifier adaptation based DA techniques have also been proposed. Yang *et al.* [14] present a classifier adaptation based DA approach where they adapt a source domain SVM classifier by using few labels from the target domain.

Recently, sparse representation has been used for various visual DA tasks such as object recognition, face recognition [15,16] and action recognition [17]. Zheng *et al.* [17] propose a dictionary learning approach for doing cross-domain action recognition. Given correspondence between videos from two domains, i.e. videos of same action shot from two different views, they learn two separate dictionaries while forcing the sparse representation for corresponding video frames from the two domains to be same. Using this view independent representation, action model learned from the source view video can be directly applied on the target view video. Ni *et al.* [16] present a dictionary learning based DA approach when correspondence information across domains is not available. Given a dictionary in one domain, say source, they iteratively modify the dictionary to be suitable for the target domain. They store all the intermediate dictionaries and use all of them to obtain a view independent representation of images from both the domains. Shekhar *et al.* [15] present a dictionary learning based approach where they map samples from both the domains to a low dimensional subspace and learn a common dictionary by minimizing reconstruction error for the projected samples in the low dimensional subspace.

In our current work, we present a dictionary learning approach for learning partially shared dictionaries across different domains. We learn separate dictionaries for the source and target domains. These dictionaries have some shared atoms which represent the common information which is present in both the domains. The dictionaries also have some domain specific atoms to represent the

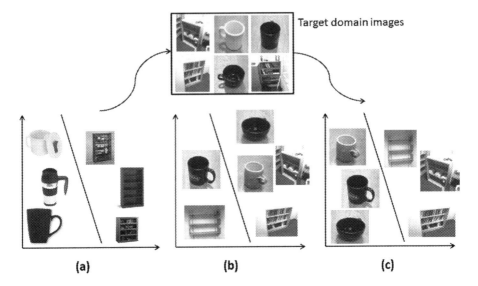

Fig. 2. Overall idea behind a Domain Adaptation approach is shown. Source and target domains are Amazon and Webcam respectively. The two object categories are mug and bookcase. (a) shows a classifier which perfectly separates the two object categories in the source domain. (b) shows the same classifier misclassifies images from the target domain. (c) shows the scenario after domain adaptation, the classifier now correctly classifies the target domain images. The target domain images aid the DA strategy. These examples can be labeled or unlabeled depending on whether the DA approach is semi-supervised or unsupervised.

domain specific information. We show the effectiveness of our dictionary learning strategy by using it for the cross-domain classification task. The domain specific information can cause confusion while doing cross-domain classification. Hence, we ignore the domain specific information and use the representation obtained from the common dictionary elements for training cross-domain classifier. The highlights of our approach are

1. We present a strategy for jointly learning partially shared dictionaries across domains.
2. We design the dictionaries to have two types of elements, i.e. domain specific elements and domain independent elements. As the name suggests, the domain specific atoms represent the domain specific information whereas the domain independent elements capture the information common to both the domains.
3. A form of selective block sparsity arises naturally from the partially shared dictionary learning formulation. More specifically, depending on the underlying domain of the signal, a specific block of sparse coefficients is forced to consist only of zeros. A simple strategy for obtaining sparse representation in presence of selective block sparsity is given.

4. Our dictionary learning approach can be seen as making few modifications over an existing dictionary learning approach [18]. However, using this simple approach, we obtain comparable results to the state of the art visual DA approaches.

2 Domain Adaptation Using Partially Shared Dictionaries

2.1 Method Overview

A dictionary learned from the source domain might not be suitable for representing signals from the target domain. Using such a dictionary for representing both the domains might result in a scenario where the sparse representation obtained for the same class signals from the two domains are very different. Clearly, such a representation will lead to poor cross-domain classification performance. Hence, while designing dictionaries in the presence of domain mismatch, further steps are required to accommodate signals from the new domains. We have presented a partially shared dictionary learning strategy to tackle the issue of domain mismatch. Our strategy is based on the idea that there could be some commonalities between the source and target domains. The same set of dictionary atoms can be used to represent this common information. Also, the signals from a domain will have certain domain specific aspects. This domain specific information can be represented well by dictionary atoms which are exclusive to the particular domain. Using the common atoms for sparse decomposition will lead to similar representation for the same class signals from both the domains. Hence, such a representation is more suited for the cross-domain classification task. The overview of our approach is shown in Fig. 3. As shown in the figure, some atoms are shared across the dictionaries from the source and the target domains. Apart from these common atoms, the dictionaries also have some domain specific atoms.

2.2 Sparse Representation of Signals

A signal $y \in R^n$ can be sparsely represented using a dictionary $D \in R^{n \times K}$, consisting of K atoms or prototype signals. The atoms of D can be pre-defined using discrete cosine transform basis [19], wavelets [20] or they can be learned from the available signals. The learned dictionaries have been shown to perform better than pre-defined dictionaries for tasks such as reconstruction [21]. For learning dictionaries from the data, several efficient dictionary learning strategies such as K-SVD [22] and MOD [18] have been proposed in the past. These dictionary learning techniques solve the following optimization problem

$$\min_{D,A} \|Y - DA\|_F^2 \quad \text{subject to} \quad \forall i, \quad \|a^i\|_0 \leq T_0. \tag{1}$$

Here the signals are arranged along the columns of Y and the columns of A, i.e. a^i, contain the corresponding sparse representation. The dictionary learning

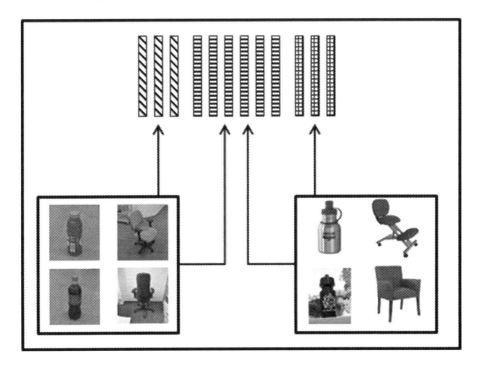

Fig. 3. Overview of partially shared dictionary learning. Dictionary for each domain consists of two types of atoms, domain specific atoms and atoms shared across the domains. Shared atoms are learned using samples from both the domains whereas domain specific atoms are learned using samples from the corresponding domain.

techniques solve this problem by alternating between solving for A, i.e. sparse coding step and updating D, i.e. dictionary update step. In [18], the dictionary update consists of updating all the dictionary elements while keeping the sparse representation unchanged. The dictionary learning approach given in [22], however, updates a single dictionary atom at a time. The sparse coefficients also change during the update so that the number of nonzero coefficients further reduces or remains the same. The sparse decomposition problem with the l_0 penalty is NP hard and greedy algorithms are used to solve this. When D is fixed, sparse representation a^i can be obtained using greedy pursuit algorithms such as OMP [23]. Sparse decomposition can also be done by relaxing the l_0 penalty and using a l_1 penalty in its place [24].

2.3 Partially Shared Dictionary Learning

Dictionary learned from one visual domain might not be suitable for representing signals from another visual domain. Hence, we propose a dictionary learning strategy which jointly learns a dictionary which is suitable for both the source as well as target visual domains. We believe that examples from any domain can be represented effectively using a dictionary which has some domain specific atoms as well as some domain independent atoms, i.e. which are common across domains. This assumption is supported by the observation that instances of same category across different domains generally have some similarity between them. Hence, we represent the source domain dictionary D_s and the target domain dictionary D_t as

$$D_s = [D_{src} \ D_c] ; \quad D_t = [D_{tgt} \ D_c], \tag{2}$$

where D_{src}, D_{tgt} are source and target domain specific atoms and D_c are the common atoms across the two domains. Also, we represent the combined dictionary D as

$$D = [D_{src} \ D_c \ D_{tgt}]. \tag{3}$$

The objective for jointly learning D is given as given as

$$\min_{D,A,B} \ \|[Y_s \ Y_t] - D[A \ B]\|_F^2,$$
$$\text{subject to} \quad a_{tgt}^i = [0 \ 0 \ .... \ 0]^T, \ b_{src}^j = [0 \ 0 \ .... \ 0]^T, \tag{4}$$
$$\|a^i\|_0 \le T_0, \ \|b^j\|_0 \le T_0,$$

where $a^i = \begin{bmatrix} a_{src}^i \\ a_{com}^i \\ a_{tgt}^i \end{bmatrix}, b^j = \begin{bmatrix} b_{src}^j \\ b_{com}^j \\ b_{tgt}^j \end{bmatrix}$, index i corresponds to the $i-th$ source domain sample, i.e. $i-th$ column of Y_s, similarly index j corresponds to the $j-th$ target domain sample. Both the sparse coefficient vectors a^i and b^j can be seen as a concatenation of three blocks of coefficient vectors. Depending upon the underlying domain of the corresponding signal, one of these three blocks, i.e. a_{tgt}^i or b_{src}^j, is forced to have all elements as zero. The equality constraints thus give rise to a specific form of block sparsity [25], which we call selective block sparsity. The above optimization problem allows for jointly learning the source as well as the target domain dictionaries. The equality constraint $a_{tgt}^i = [0 \ 0 \ .... \ 0]^T$ makes sure that the dictionary atoms D_{tgt} are used only for representing the target domain signals Y_t. Hence, D_{tgt} captures only the target domain information. Similarly, the equality constraint $b_{src}^j = [0 \ 0 \ .... \ 0]^T$ makes sure that D_{src} captures only the source domain information. The block of sparse coefficients a_{com}^i and b_{com}^j correspond to the common dictionary atoms D_c. As both a_{com}^i and b_{com}^j can have non-zero terms, the dictionary atoms D_c are used while representing signals from both the domains, hence, these atoms capture the common information across the source and target domains.

To effectively solve the optimization problem given in Eq. 4, we rewrite it as

$$\min_{D_s,D_t,A_s,B_t} \ \|Y_s - D_s A_s\|_F^2 \ + \ \|Y_t - D_t B_t\|_F^2,$$
$$\text{subject to} \quad \|a_s^i\|_0 \le T_0, \ \|b_t^j\|_0 \le T_0, \tag{5}$$

where $a_s^i = \begin{bmatrix} a_{src}^i \\ a_{com}^i \end{bmatrix}, b_t^j = \begin{bmatrix} b_{tgt}^j \\ b_{com}^j \end{bmatrix}$, a_s^i is the $i-th$ column of A_s, b_t^j is the $j-th$ column of B_t. We would like to point out here that to learn D_s and D_t using

---

Algorithm 1. Partially Shared Dictionary Learning(PSDL)

Input: source domain vectors Y_s, target domain vectors Y_t, n
Output: D, A_s, B_t
   initialize:D_s, $P = [R \ Q]$
   for i= 1 **to** n **do**
      $A_s \leftarrow OMP(Y_s, D_s)$
      $\begin{bmatrix} B_{tgt} \\ B_{com} \end{bmatrix} = B_t \leftarrow OMP(Y_t, D_s P)$
      $D_s \leftarrow (Y_s A_s^T + Y_t B_t^T P^T)(A_s A_s^T + P B_t B_t^T P^T)^{-1}$
      $R \leftarrow (D_s^T D_s)^{-1} D_s^T E_t B_{tgt}^T (B_{tgt} B_{tgt}^T)^{-1}$
   end for

Eq. 5, one might be tempted to use MOD and alternate between sparse coding and dictionary update by taking derivative of the energy term with respect to D_s and D_t. However, such an approach would not ensure the structure we desire to be present in the dictionaries D_s and D_t, as presented in Eq. 2. We take a short digression to describe how to solve Eq. 4 in case the dictionary structure given in Eq. 2 is not present. In such a scenario, we can rewrite the optimization problem given in Eqs. 4 as 5. For dictionary learning, we can use MOD. Obtaining a_s^i and b_t^j is straightforward and these can be obtained via OMP. If a_s^i and b_t^j are available, a^i and b^j can be obtained trivially by concatenating a vector of zeros at appropriate position.

Now we get back to our original dictionary learning formulation. To maintain the desired structure in the two dictionaries D_s and D_t, we further couple the two dictionaries D_s and D_t using the following relation between the two

$$D_t = D_s P, \tag{6}$$

where P is a square matrix. Since we want some elements to be common among D_s and D_t, we fix a set of the columns of P, i.e. Q, to have a single element as 1 and remaining elements as 0. The location of 1 in each column of Q corresponds to location of common atoms D_c. Hence P can be represented as

$$P = [R \ Q]. \tag{7}$$

Using Eqs. 5 and 7 can be rewritten as

$$\min_{D_s, R, A_s, B_t} \quad \|Y_s - D_s A_s\|_F^2 \quad + \quad \|Y_t - D_s Q B_{com} - D_s R B_{tgt}\|_F^2, \tag{8}$$
$$\text{subject to} \quad \|a_s^i\|_0 \leq T_0, \|b_t^j\|_0 \leq T_0,$$

where $B_t = \begin{bmatrix} B_{tgt} \\ B_{com} \end{bmatrix}$.

To solve this optimization problem, we alternate between updating D_s and R followed by sparse coding step. We set the first order derivative with respect to D_s equal to zero and obtain the following closed form expression for D_s

$$D_s = (Y_s A_s^T + Y_t B_t^T P^T)(A_s A_s^T + P B_t B_t^T P^T)^{-1}. \tag{9}$$

Similarly, the update for R is done using the following closed form expression.

$$R = (D_s^T D_s)^{-1} D_s^T E_t B_{tgt}^T (B_{tgt} B_{tgt}^T)^{-1}, \tag{10}$$

where $E_t = Y_t - D_s Q B_{com}$. In the sparse coding step, D_s and R is kept fixed and OMP is used to obtain the sparse representation. We summarize our partially shared dictionary learning (PSDL) strategy in Algorithm 1.

2.4   Cross-Domain Classification Using PSDL

Using unlabeled data from the source and the target domains, the dictionary D is learned as described in the previous section. The dictionary atoms which are common across the two domains, i.e. D_c, are then used to obtain sparse representations for signals from both the domains. The sparse decomposition using D_c maps signals from both the domains to a common subspace. The sparse representation of samples from source and target domain, thus obtained, are used directly for doing cross-domain classification. The coefficients corresponding to D_{src} and D_{tgt} are ignored while doing cross-domain classification.

The dictionary atom subsets D_{src} and D_{tgt} represent the domain specific information, hence, using their coefficients also for the cross-domain classification task will create confusion for the classifier. By using just the coefficients corresponding to D_c, we effectively extract only the common information which is shared across the source and target domains. This results in similar sparse representation for same class signals across the two domains. Clearly, such a representation is better suited for the crosss-domain classification task.

As stated before, we use just the coefficients corresponding to D_c for representing signals from the source and the target domains. The classifiers are trained using the sparse representation for plenty of labeled data from the source domain as well as a small amount of labeled data from the target domain. We use SVMs for the cross-domain classification task, as in [16].

3   Results and Discussions

We validate our approach by conducting object recognition experiments in a cross-domain setting on benchmark datasets. We conduct the experiments using the same experimental setup as in [12,16]. Our dictionary learning approach PSDL is unsupervised, and does not use any label information from the source or the target domains. We compare our dictionary learning based DA approach with various baseline approaches as well as a recently proposed dictionary learning based DA approach [16]. We also compare our approach with other DA techniques [11,12].

3.1   Dataset and Representation

We conduct object recognition experiments on 4 datasets, i.e. Amazon(images downloaded from online merchants), Webcam(images taken by a low resolution

Table 1. Classification accuracies for PSDL is compared with baseline approaches as well as the DA approaches given in [11,12,16]. For baseline approaches we learn source and target domain dictionaries using [18]. The acronyms A, C, D, W represent the domains Amazon, Caltech, DSLR and Webcam respectively. In the notation C → A, C is the source domain and A is the target domain. Similar notation is followed for the other dataset pairs. Experiments are done in semi-supervised setting.

Method	C → A	C → D	A → C	A → W	W → C	W → A	D → A	D → W
SVM$_S$	40.6	45.7	36.5	26.5	23.2	30.6	35.3	69.7
SVM$_{ST}$	46.4	52.1	38.9	38.7	31.0	40.2	42.3	74.1
MOD$_{source}$	44.9	50.5	39.2	46.6	27.3	38.5	37.6	67.2
MOD$_{target}$	49.2	53.6	39.4	50.7	34.2	44.4	44.3	72.0
SGF [11]	40.2	36.6	37.7	37.9	29.2	38.2	39.2	69.5
GFK [12]	46.1	55.5	39.6	56.9	32.8	46.2	46.2	80.2
Ni *et al.* [16]	50.0	57.1	41.5	57.8	40.6	51.5	50.3	87.8
PSDL	53.9	59.4	41.8	57.7	37.0	46.8	48.9	83.3

webcam), DSLR(images taken by a digital SLR camera) and Caltech(images taken from the Caltech-256 [26] dataset). The first three datasets were introduced in [5] whereas the fourth one was first studied by [12]. Each of the dataset are considered as a separate domain. Datasets consist of images pertaining to the following 10 classes BACKPACK, TOURING-BIKE, CALCULATOR, HEAD-PHONES, COMPUTER-KEYBOARD, LAPTOP, COMPUTER-MONITOR, COMPUTER-MOUSE, COFFEE-MUG, VIDEO-PROJECTOR. There are atleast 8 images and a maximum of 151 images per category in each domain. The datasets consist of a total of 2533 images.

Scale invariant interest points were detected in the images using the SURF detector [27]. A 64 dimensional SURF descriptor was used to describe the patch

Table 2. Classification accuracies for PSDL is compared with baseline approaches as well as the DA approaches given in [11,12,16]. The acronyms A, C, D, W represent the domains Amazon, Caltech, DSLR and Webcam respectively. In the notation C → A, C is the source domain and A is the target domain. Similar notation is followed for the other dataset pairs. Experiments are done in unsupervised setting.

Method	C → A	C → D	A → C	A → W	W → C	W → A	D → A	D → W
MOD$_{source}$	39.8	42.1	37.0	36.2	19.8	26.8	30.1	55.3
MOD$_{target}$	44.4	44.0	36.8	38.2	30.5	35.4	34.5	69.5
SGF [11]	36.8	32.6	35.3	31.0	21.7	27.5	32.0	66.0
GFK [12]	40.4	41.1	37.9	35.7	29.3	35.5	36.1	79.1
Ni *et al.* [16]	45.4	42.3	40.4	37.9	36.3	38.3	39.1	86.2
PSDL	47.6	48.5	39.8	38.9	31.8	36.0	37.9	79.1

around the interest points. A codebook consisting of 800 visual words was constructed by clustering random descriptors from the AMAZON dataset, using k-means clustering. A histogram representation was obtained for each of the images by obtaining the count of each of the visual words in the image. All the histograms were z-score normalized to have zero mean and unit deviation along each dimension.

3.2   Experiments

For experiments, two domains are picked from the datasets. We use one of them as the source domain and the other is used as the target domain. Goal of the experiments is to classify target domain data points. We conduct experiments in unsupervised setting as well as semi-supervised setting. In the unsupervised setting, labeled examples are present only in the source domain. In the semi-supervised setting, along with the labeled source examples, we also sample few labeled examples from the target domain. When WEBCAM or DSLR are the source domains, we sample 8 labeled points from them. In case AMAZON or CALTECH are the source domains, 20 labeled examples are sampled. In semi-supervised setting, 3 labeled examples are sampled from the target domain. For dictionary learning using PSDL, we utilize unlabeled samples from both the domains. The optimal parameters for PSDL(number of dictionary atoms) and for the SVM classifier are obtained by empirically searching over the parameter space. Sparse representation is obtained using Orthogonal Matching Pursuit(OMP) [23]. Following the previous works [11,12,16] , all the experiments are repeated 20 times and the mean classification accuracy over the 20 trials is reported in each case.

In Table 1, we compare our approach with baseline approaches as well as other popular DA approaches. This experiment is done in a semi-supervised setting, i.e. we use few labeled examples from the target domain along with the labeled examples from the source domain. SVM$_S$, SVM$_{ST}$ and MOD based dictionary learning approaches are taken as the baseline in this experiment. SVM$_S$ refers to SVM classifiers learned using only the source examples, SVM$_{ST}$ refers to

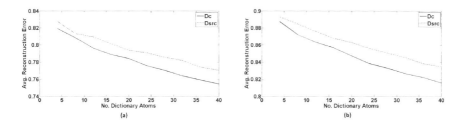

Fig. 4. Average reconstruction error for target dataset is plotted as a function of number of dictionary atoms. Dc represents the shared dictionary and $Dsrc$ the source domain dictionary. The shared dictionary Dc does a better reconstruction of the target domain samples in comparison to $Dsrc$. For (a), the source and target domains are Amazon and Caltech respectively and for (b), Caltech and Amazon.

Fig. 5. Test images from the DSLR domain on the left and two adjacent rows of corresponding nearest neighbors from the Amazon domain. The top row adjacent to each query image is obtained using our dictionary learning approach, bottom row adjacent to each query corresponds to nearest neighhbors obtained with original features.

SVM classifiers learned using source as well as target domain examples. We use MOD for learning dictionaries from the source as well as the target domains. For MOD_{source}, sparse decomposition of signals from both the domains is done using the dictionary learned from just the source domain. Similarly, for MOD_{target}, the dictionary learned from the target domain is used for sparse representation. We also present the results given in the dictionary learning based DA approach given in [16]. We also compare our results with two other DA approaches [11,12]. Our dictionary learning approach as well as [16], always outperform the baseline approaches as well as the DA approaches given in [11,12]. For the dataset pair WEBCAM and DSLR, all the approaches perform better compared to the other dataset pairs. The reason for this high classification accuracy is the high similarity across these two domains, i.e. these datasets consist of images of the same object instances obtained using different imaging devices. On the other hand, all the approaches tend to show low accuracy for some dataset pairs, for example AMAZON and CALTECH. This can be explained by the large variations across these two domains. We observe that for the first four dataset pairs, our method performs almost as good or better than [16]. For the remaining dataset pairs, [16] outperforms our method. In Table 2, results are reported for experiments in unsupervised setting. For the unsupervised setting, PSDL as well as [16] outperform the baseline approaches as well as [11,12]. Like semi-supervised setting, here also PSDL lags behind [16] for the last four cases. For these cases, training domain has 8 labeled examples whereas for the first four cases, it has 20 examples. In the objective given in Eq. 8, the target reconstruction term may dominate over the source term when the source has fewer samples than the target domain. For tasks 5–8 (in Tables 1 and 2), there are only 8 samples in the source domain. We believe this leads to a common dictionary which represent the target domain well but not the source domain.

In Fig. 4, we show the comparison between the average reconstruction error obtained by dictionaries D_c and D_{src} while representing the target domain. Irrespective of the dictionary size, D_c results in less reconstruction error, thus showing that it is more suited for representing the target domain. In Fig. 5, we show three test images from the target domain and corresponding top five nearest neighbors from the source domain. We provide nearest neighbors for two cases, i.e. results obtained via our dictionary learning based approach and no adaptation case using original features. In all the three examples, improvement because of our approach is clearly observable.

4 Conclusion

We present a partially shared dictionary learning approach. Our approach allows dictionaries from the source and the target domains to share some atoms. We use the shared atoms to represent signals from both the domains. This results in similar sparse representation for same class signals across the two domains. Our results show that such a representation is better suited for cross-domain visual object recognition task.

We show the effectiveness of our approach by performing cross-domain object recognition on the benchmark datasets. We also compare our approach with various existing approaches and show improvement in results over the existing state of art approaches. For our future work, we plan to learn discriminative shared dictionaries by utilizing available label information.

References

1. Lowe, D.G.: Object recognition from local scale-invariant features. In: IEEE International Conference on Computer Vision. IEEE (1999)
2. Dalal, N., Triggs, B.: Histograms of oriented gradients for human detection. In: IEEE Computer Society Conference on Computer Vision and Pattern Recognition, CVPR. IEEE (2005)
3. Perronnin, F., Sánchez, J., Mensink, T.: Improving the fisher kernel for large-scale image classification. In: Daniilidis, K., Maragos, P., Paragios, N. (eds.) ECCV 2010, Part IV. LNCS, vol. 6314, pp. 143–156. Springer, Heidelberg (2010)
4. Torralba, A., Efros, A.A.: Unbiased look at dataset bias. In: IEEE Conference on Computer Vision and Pattern Recognition (CVPR). IEEE (2011)
5. Saenko, K., Kulis, B., Fritz, M., Darrell, T.: Adapting visual category models to new domains. In: Daniilidis, K., Maragos, P., Paragios, N. (eds.) ECCV 2010, Part IV. LNCS, vol. 6314, pp. 213–226. Springer, Heidelberg (2010)
6. Blitzer, J., McDonald, R., Pereira, F.: Domain adaptation with structural correspondence learning. In: Proceedings of the Conference on Empirical Methods in Natural Language Processing, Association for Computational Linguistics (2006)
7. Daumé III, H.: Frustratingly easy domain adaptation. In: ACL (2007)
8. Jiang, J., Zhai, C.: Instance weighting for domain adaptation in NLP. In: ACL (2007)
9. Kulis, B., Saenko, K., Darrell, T.: What you saw is not what you get: Domain adaptation using asymmetric kernel transforms. In: IEEE Conference on Computer Vision and Pattern Recognition (CVPR). IEEE (2011)
10. Fernando, B., Habrard, A., Sebban, M., Tuytelaars, T., et al.: Unsupervised visual domain adaptation using subspace alignment. In: Proceedings of ICCV (2013)
11. Gopalan, R., Li, R., Chellappa, R.: Domain adaptation for object recognition: an unsupervised approach. In: IEEE International Conference on Computer Vision (ICCV). IEEE (2011)
12. Gong, B., Shi, Y., Sha, F., Grauman, K.: Geodesic flow kernel for unsupervised domain adaptation. In: IEEE Conference on Computer Vision and Pattern Recognition (CVPR). IEEE (2012)
13. Jhuo, I.H., Liu, D., Lee, D., Chang, S.F.: Robust visual domain adaptation with low-rank reconstruction. In: IEEE Conference on Computer Vision and Pattern Recognition (CVPR). IEEE (2012)
14. Yang, J., Yan, R., Hauptmann, A.G.: Cross-domain video concept detection using adaptive svms. In: Proceedings of the 15th International Conference on Multimedia. ACM (2007)
15. Shekhar, S., Patel, V.M., Nguyen, H.V., Chellappa, R.: Generalized domain-adaptive dictionaries. In: IEEE Conference on Computer Vision and Pattern Recognition (CVPR). IEEE (2013)
16. Ni, J., Qiu, Q., Chellappa, R.: Subspace interpolation via dictionary learning for unsupervised domain adaptation. In: IEEE Conference on Computer Vision and Pattern Recognition (CVPR). IEEE (2013)
17. Zheng, J., Jiang, Z., Phillips, P.J., Chellappa, R.: Cross-view action recognition via a transferable dictionary pair. In: BMVC (2012)
18. Engan, K., Aase, S.O., Husoy, J.H.: Method of optimal directions for frame design. In: 1999 IEEE International Conference on Acoustics, Speech, and Signal Processing, Proceedings. IEEE (1999)

19. Ahmed, N., Natarajan, T., Rao, K.R.: Discrete cosine transform. IEEE Trans. Comput. **23**, 90–93 (1974)
20. Mallat, S.: A Wavelet Tour of Signal Processing. Academic Press, San Diego (1999)
21. Elad, M., Aharon, M.: Image denoising via learned dictionaries and sparse representation. In: IEEE Computer Society Conference on Computer Vision and Pattern Recognition. IEEE (2006)
22. Aharon, M., Elad, M., Bruckstein, A.: K-SVD: an algorithm for designing overcomplete dictionaries for sparse representation. IEEE Trans. Sig. Process. **54**, 4311–4322 (2006)
23. Tropp, J.A., Gilbert, A.C.: Signal recovery from random measurements via orthogonal matching pursuit. IEEE Trans. Inf. Theory **53**, 4655–4666 (2007)
24. Donoho, D.L.: Compressed sensing. IEEE Trans. Inf. Theory **52**, 1289–1306 (2006)
25. Eldar, Y.C., Kuppinger, P., Bolcskei, H.: Block-sparse signals: Uncertainty relations and efficient recovery. IEEE Trans. Sig. Process. **58**, 3042–3054 (2010)
26. Griffin, G., Holub, A., Perona, P.: Caltech-256 object category dataset (2007)
27. Bay, H., Ess, A., Tuytelaars, T., Van Gool, L.: Speeded-up robust features (surf). (Computer vision and image understanding)

Learning Discriminative Hidden Structural Parts for Visual Tracking

Longyin Wen[1]($\boxtimes$), Zhaowei Cai[1], Dawei Du[2], Zhen Lei[1], and Stan Z. Li[1]

[1] CBSR & NLPR, Institute of Automation, Chinese Academy of Sciences,
95 Zhongguancun Donglu, Beijing 100190, China
{lywen,zwcai,zlei,szli}@nlpr.ia.ac.cn
[2] School of Computer and Control Engineering, University of Chinese
Academy of Sciences, No.3, Zhongguancun Nanyitiao, Beijing 100049, China
dawei.du@vipl.ict.ac.cn
http://www.cbsr.ia.ac.cn

Abstract. Part-based visual tracking is attractive in recent years due to its robustness to occlusion and non-rigid motion. However, how to automatically generate the discriminative structural parts and consider their interactions jointly to construct a more robust tracker still remains unsolved. This paper proposes a discriminative structural part learning method while integrating the structure information, to address the visual tracking problem. Particulary, the state (e.g. position, width and height) of each part is regarded as a hidden variable and inferred automatically by considering the inner structure information of the target and the appearance difference between the target and the background. The inner structure information considering the relationship between neighboring parts, is integrated using a graph model based on a dynamically constructed pair-wise Markov Random Field. Finally, we adopt Metropolis-Hastings algorithm integrated with the online Support Vector Machine to complete the hidden variable inference task. The experimental results on various challenging sequences demonstrate the favorable performance of the proposed tracker over the state-of-the-art ones.

1 Introduction

Visual tracking is one of the most important and challenging problems in computer vision field. Traditionally, the majority of existing methods always focus on modeling the holistic appearance of the target within a bounding box, and they have achieved good performance in some conditions [1–8]. Intuitively however, they ignore the local information of the target, which greatly limits their application in the scenario where the target is partially occluded or the global appearance of the target changes a lot.

Recently, methods [9–18] with part-based appearance representation are popular in visual tracking task to deal with some situations that the holistic appearance based methods fail. The part based tracker combines multiple parts with local information to achieve stronger representation ability, and is able to explicitly model the target structure variations. However, the obvious shortcomings

© Springer International Publishing Switzerland 2015
C.V. Jawahar and S. Shan (Eds.): ACCV 2014 Workshops, Part III, LNCS 9010, pp. 262–276, 2015.
DOI: 10.1007/978-3-319-16634-6_20

of these previous part based trackers are: (1) the number of parts are assigned before empirically, which is hard to obtain better discriminative ability with suitable parts; (2) no relationships between neighboring generated parts are considered, which loses the structural information of the target. Therefore, their methods lack strong discriminative ability and fail to combine the parts into a whole to complete the tracking task.

In this paper, we propose a discriminative Hidden Structural Part Tracker (HSPT), which tracks arbitrary objects without any assumptions on the scenarios. The proposed method learns the discriminative parts automatically by integrating the structure and discriminative information. In the learning step, both the appearance of the parts and the relationships between them are considered. Since the discriminative parts are not located in the fixed location to the target center, we regard the state of them as hidden variables in the objective function. Then, the objective is optimized by the Metropolis-Hastings (MH) algorithm [19] integrated with the online Support Vector Machine (SVM) method [20] iteratively. In order to achieve more robust performance in complex environments, the bounding box based appearance of the target is also incorporated in our tracker. The contributions of this paper are concluded as follows:

- We propose a hidden discriminative structural parts learning based tracker, which simultaneously learns multiple structural parts of the target to represent the target better to enhance its robustness.
- The MH algorithm and the online SVM are interestingly combined to infer the optimal state of the discriminative parts, and this optimization method handles varying number of parts well.
- The structural supporting between parts are naturally integrated through the dynamically constructed pair-wise MRF model.
- Extensive tracking experiments are various publicly available challenge sequences demonstrate the favorable performance against the state-of-the-art methods.

2   Related Works

The part based model has been developed recently in visual tracking task. Shahed et al. proposed a part-based tracker HABT [10], which generated the target parts by manually labeling in the first frame. And the appearance model of the target is assumed to be fixed in the tracker, which limits its performance in the complex environment. Another tracker Frag [9] utilized the regularly partitioned parts to model the appearance of the target, but the appearance model of each part is also fixed. In [18], Yao et al. introduced an online part-based tracking with latent structured learning. However, the proposed method uses fix number of parts to represent the target object, which makes it insufficient to describe the appearance variations when large deformation happens.

There also exist some other part based appearance representation methods. Kwon et al. [11] proposed the BHMC tracker, which generates the parts based on SURF-like key points without global structure constraints and the appearance of the parts were updated roughly. Wang et al. [14] proposed a discriminative appearance model based on superpixels, called SPT, in which the probabilities of superpixels belonging to the foreground were utilized to discriminate the target from the background. However, the relation between superpixels was not incorporated, which makes the tracker easily affected by the similar backgrounds. Godec et al. [12] extended the hough forest to the online domain and integrated the voting method for tracking, regardless of the structure information. Cehovin et al. [13] proposed a coupled-layer visual tracker, which combined the global and local appearance of the target together in part generation. However, the ignorance of the relationships between parts makes it unstable when clutter backgrounds, occlusions or non-rigid motion happen. Cai et al. [17] designed a dynamic graph based method which works well in non-rigid motion, but the parts are generated only based on color feature and some background parts will be easily misclassified as foreground.

3   Overview of Proposed Method

Generally, the tracking task is formulated as a Markovian state transition process, where the current target state is determined by its previous state. Let $p(O_t|Z_t)$ be the appearance model and $p(Z_t|Z_{t-1})$ be the motion model of the target at time t. The state of the target is represented as $Z_t = (\ell_t, s_t)$. ℓ_t is the position of the target in the 2D image plane and s_t is the size of the target consisting of the width and height. The motion model $p(Z_t|Z_{t-1})$ and the appearance model $p(O_t|Z_t)$ are described as follows.

Motion Model. Similar to [2], we assume the position and size of the target varies independently in the motion model, that is:

$$p(Z_t|Z_{t-1}) = p(\ell_t, s_t|\ell_{t-1}, s_{t-1}) = p(\ell_t|\ell_{t-1})p(s_t|s_{t-1}), \qquad (1)$$

where the target position transition probability $p(\ell_t|\ell_{t-1}) = 1$, if $\|\ell_t - \ell_{t-1}\|_2 < R_s$; otherwise, it equals to zero. R_s is the predefined searching radius. The target scale transition probability is similarly handled as [2].

Appearance Model. The appearance of our tracker consists of two parts, the learned discriminative structural parts model $\mathcal{A}^{(0)}$ and the bounding box based appearance model $\mathcal{A}^{(1)}$. The learned structural parts focus on the local variations and the bounding box based appearance focuses on the holistic variations of the target. Intuitively, the combination of them can achieve more robust performance. The appearance model of our tracker is formulated as follows:

$$p(O_t|Z_t) = \left(p^{(0)}(O_t|Z_t)\right)^{\lambda_b} \cdot \left(p^{(1)}(O_t|Z_t)\right)^{(1-\lambda_b)}, \qquad (2)$$

where λ_b is a predefined balance parameter, and $p^{(0)}(O_t|Z_t)$ and $p^{(1)}(O_t|Z_t)$ are the probabilities of the target candidate given out by $\mathcal{A}^{(0)}$ and $\mathcal{A}^{(1)}$, respectively. To model the appearance of the target, the online SVM [20] is adopted for each part.

For the discriminative structural parts model $\mathcal{A}^{(0)}$, the probability of the candidate, including the appearance likelihood and the deformation likelihood of the parts, is calculated as

$$p^{(0)}(O_t|Z_t) = \prod_{i,j \in \mathcal{N}} \phi(Z_{i,t}, Z_{j,t}) \cdot \prod_{i=1}^{n} p(O_{i,t}|Z_{i,t}), \tag{3}$$

where n is the number of parts, $p(O_{i,t}|Z_{i,t})$ is the probability of the part i applauding the candidate to be positive, $O_{i,t}$ is the observation of part i, $\mathcal{N}$ is the neighboring system of the parts and $\phi(Z_{i,t}, Z_{j,t})$ is the pairwise interaction potentials between the learned part i and part j.

In our model, the state of part i at time t is defined as $Z_{i,t} = (x_{i,t}, y_{i,t}, w_{i,t}, h_{i,t}, \Delta x_{i,t}, \Delta y_{i,t})$, where $(x_{i,t}, y_{i,t})$, $w_{i,t}$ and $h_{i,t}$ are the position, width and height of part i at time t respectively. $(\Delta x_{i,t}, \Delta y_{i,t})$ is the spatial offset of the part relative to the target center. The interaction potential term is expressed by means of Gibbs distribution:

$$\phi(Z_{i,t}, Z_{j,t}) \propto \exp\left(-\lambda_\phi \|v_t(i,j) - \tilde{v}(i,j)\|_2\right), \tag{4}$$

where $\lambda_\phi = 0.2$ is the scaler parameter in the experiments and $v_t(i,j) = \ell_t(i) - \ell_t(j)$ represents the vector pointing from the position $\ell_t(i)$ of part i to the location $\ell_t(j)$ of part j at time t, which encodes the supporting between neighboring parts (structural information of the target). $\tilde{v}(i,j)$ is the learnt relative position of the part i and j. Here, we model the relationship between different parts, rather than modeling the exclusions between close targets as in [21].

Let $\Phi_p(O_{i,t})$ represent the HOG feature [22] of the part observation, and $\omega_{i,p}^t$ is the SVM parameter corresponding to part i at time t. The likelihood of its appearance is calculated as

$$p(O_{i,t}|Z_{i,t}) \propto \exp\left(\omega_{i,p}^{(t)} \cdot \Phi_p(O_{i,t})\right), \tag{5}$$

In order to reduce the influence of some badly learned parts, we only utilize η percent high confident parts to score the candidates. Then the probability of the candidate (3) can be rewritten as

$$p^{(0)}(O_t|Z_t) \propto \prod_{i,j \in \mathcal{N}} \phi(Z_{i,t}, Z_{j,t}) \cdot \exp\left(\sum_{i \in \mathcal{I}} \omega_{i,p}^{(t)} \cdot \Phi_p(O_{i,t})\right), \tag{6}$$

where $\mathcal{I}$ is the index set of the selected η high confident parts.

For the bounding box based appearance model $\mathcal{A}^{(1)}$, the probability of the candidate is presented as:

$$p^{(1)}(O_t|Z_t) \propto \exp(\omega_b^{(t)} \cdot \Phi_b(O_t)), \tag{7}$$

where the SVM parameter $\omega_b^{(t)}$ and the HOG feature $\Phi_b(O_t)$ are determined based on the whole target.

4   Learning the Discriminative Parts

In the tracking task, the appearance of the target changes dramatically. In order to adapt the part models to the target appearance variations, some of the target parts should be added or deleted, and their state should be determined to represent the target optimally. Therefore, we design a reasonable objective function to perform the parts learning, whose goal consists of three aspects: (1) maximize the margin between the target and the background; (2) retain the structure information of the target; (3) cover the most of the target foreground area. The state of each part is treated as a hidden variable, which will be inferred based on the acquired observation information.

As discussed above, our objective in terms of optimization is to find the optimized SVM parameter $\omega_{i,p}$ and parts state $Z_{i,t}$. In this section, the MH algorithm and the online SVM are integrated into an unified optimization framework to complete the inference task. The details will be presented as follows.

4.1   Objective

For the local parts to be learned, we expect that they acquire better representation ability to ensure robust tracking performance. The objective for the i-th part learning is formulated as

$$G(\omega_{i,p}, Z_i; \mathcal{X}) = \alpha \cdot \rho \cdot \omega_{i,p} \cdot \Phi_p(\mathcal{X}_{Z_i}) + \beta \cdot \mathcal{R}(\mathcal{F}, Z_i), \qquad (8)$$

where the first term is to separate the target parts from the background parts by maximizing the margin between them, and the second term encourages the learned part to cover more target foreground area. $\mathcal{F}$ is foreground area, $\rho \in \{-1, 1\}$ is the binary label of the updating sample $\mathcal{X}$ indicating the foreground and background, and $\mathcal{X}_{Z_i}$ represents the part observations of the updating sample with the part state Z_i. $\mathcal{R}(\mathcal{F}, Z_i)$ means the overlap ratio between Z_i and $\mathcal{F}$. We set the balancing parameters $\alpha = 0.7$, $\beta = 0.2$ in all of our experiments.

Naturally, we infer the optimal state of each part jointly and integrate the structure information in the inference process. The objective for the target is proposed as

$$G(\omega_p, \tilde{Z}; \mathcal{X}) = \alpha \cdot \rho \cdot \omega_p \cdot \Phi_p(\mathcal{X}_{\tilde{Z}}) + \beta \cdot \mathcal{R}(\mathcal{F}, \tilde{Z}), \qquad (9)$$

where $\tilde{Z} = (Z_1, \cdots, Z_n)$ is the combination of the parts state and $\omega_p = (\omega_{1,p}^{(t)}, \cdots, \omega_{n,p}^{(t)})$ is the concatenated SVM weight of each part. $\mathcal{R}(\mathcal{F}, \tilde{Z}) = \sum_{i=1}^{n} \mathcal{R}(\mathcal{F}, Z_i)$ represents the coverage ratio of the true target area.

4.2   Optimal Parts Inference

The MH algorithm [19] has been applied in the multiple target tracking task [21,23], where the authors use it for particle filter sampling to identify the state of each target precisely. However, in our paper, we do not aim at identifying each part all the time. Instead, we utilize the MH algorithm to clean up useless parts and discover new parts adaptively according to the target appearance variation. Firstly, we need convert our objective (9) into the probability form:

$$p(\tilde{Z}, \omega_p | \mathcal{X}) \propto \exp\left(\zeta \cdot G(\omega_p, \tilde{Z}; \mathcal{X})\right), \tag{10}$$

where ζ is the scale factor. Then maximizing the objective (9) is equivalent to solve the maximum posterior probability problem:

$$\{\tilde{Z}, \omega_p\} = \arg\max_{\tilde{Z}, \omega_p} p(\tilde{Z}, \omega_p | \mathcal{X}). \tag{11}$$

Due to the dependance between the hidden variable $\tilde{Z}$ and ω_p, it is difficult to optimize them simultaneously. Hence, we decompose the inference task of the two hidden variables in (10) into a two stage iterative optimization problem. In each pass r, we solve the objective by dividing it into two steps to iteratively update $\{\tilde{Z}, \omega_p\}$ using the following procedure.

Optimize ω_p. Given the optimized parts state $\tilde{Z}^{(r)}$, (11) is equivalent to the following optimization problem:

$$\omega_p^{(r)} = \arg\max_{\omega_p} p(\tilde{Z}^{(r)}, \omega_p | \mathcal{X}) = \arg\max_{\omega_p} \left\{\alpha \cdot \rho \cdot \omega_p \cdot \Phi_p(\mathcal{X}_{\tilde{Z}^{(r)}})\right\}. \tag{12}$$

Then in analogy to classical SVM, we train the parameter $\omega_p^{(r)}$ by solving the following optimization problem:

$$\omega_p^{(r)} = \arg\min_{\omega_p} \left\{\frac{1}{2}\|\omega_p\|^2 + \gamma \sum_{i=1}^{m} \max\left(0, 1 - \rho^{(i)} \cdot f_{\omega_p}(\mathcal{X}^i)\right)\right\}, \tag{13}$$

where $\{(\mathcal{X}^1, \rho^1), \cdots, (\mathcal{X}^m, \rho^m)\}$ is the collected sample pool, $\rho^i \in \{-1, 1\}$ is the label of the ith collected sample $\mathcal{X}^i$, m is the number of samples. We set parameter $\gamma = 5$ in our experiments. The score of the sample is calculated as $f_{\omega_p}(\mathcal{X}_{\tilde{Z}^{(r)}}) = \omega_p \cdot \Phi_p(\mathcal{X}_{\tilde{Z}^{(r)}})$, where $\Phi_p(\mathcal{X}_{\tilde{Z}^{(r)}})$ is the concatenated HOG feature of parts.

Optimize $\tilde{Z}$. With the determined $\omega_p^{(r)}$, we sample a proposal part state $\tilde{Z}^{(r)\prime}$ according to the previous parts state $\tilde{Z}^{(r)}$, and calculate the acceptance ratio in the MH algorithm based on the optimized model parameter $\omega_p^{(r)}$ to get the optimized parts state $\tilde{Z}^{(r+1)}$. Therefore, the MAP solution of the parts state in the constructed Markov Chain of MH algorithm is utilized to get the optimized state $\tilde{Z}$.

To that end, five moves are defined for states change of each part. *Birth*, indicates the move of adding the candidate parts in the sampler; *Death*, indicates the move of removing the newly added candidate parts in the sampler. The reversible pair focuses on the newly added candidates generated by SLIC sampling [24] (i.e. the external rectangle region of the generated superpixel is used as the candidate) and the deleted candidates, if the target is changing the pose so that some old parts will disappear and some new parts will be generated. *Stay*, indicates the move of adding the disappeared parts in previous iterations in the sampler; *Leave*, indicates the move of removing the learned parts in previous iterations in the sampler. The reversible pair determine the state of the learned parts in previous sampling iterations when the target undergoes heavy occlusion so that some old parts are missed temporarily and appear again then. *Update*, indicates the move of updating the parts in the sampler, which deals with the dynamically updated appearance of the target as a self-reversible pair.

For easy description, we omit the iteration mark r in the following. We define two sets in the optimization process: (1) $T^\star = \{T_1^\star, \cdots, T_n^\star\}$ is the learned parts set and its corresponding state set is $Z^\star = \{Z_1^\star, \cdots, Z_n^\star\}$; (2) $T^+ = \{T_1^+, \cdots, T_m^+\}$ is the birth candidate set and its corresponding state set is $Z^+ = \{Z_1^+, \cdots, Z_m^+\}$. The notation $Z_i^{(\cdot)}$ ($Z_i^\star$ or Z_i^+) is the current part state and $Z_i^{(\cdot)\prime}$ is the proposal state of the part.

Let $\mathcal{N}_{i,t}$ be the neighbors of part i at time t. Let $C = \{C_b, C_d, C_s, C_l, C_u\}$ represent the prior probability of each move type and we set $C = \{0.3, 0.1, 0.1, 0.01, 3.0\}$ in the experiments empirically. The proposal distribution $q = \{q_b, q_d, q_s, q_l, q_u\}$ and the acceptance ratio are calculated as follows.

Birth: Select the part T_i^+ from the birth candidate parts set T^+ with the uniform distribution. The birth proposal distribution can be calculated as $q_b(Z_i^{+\prime}; Z_i^+) = \frac{C_b}{m}$, if $(T^{\star\prime}, Z^{\star\prime}) = (T^\star \cup \{T_i^+\}, Z^\star \cup \{Z_i^+\})$, and otherwise it equals to zero. Then the acceptance ratio is presented as

$$\alpha_b = \min\left(1, p(\mathcal{X}|Z_i^{+\prime}, \omega_p) \cdot p(Z_i^{+\prime}) \cdot \frac{q_d(Z_i^+; Z_i^{+\prime})}{q_b(Z_i^{+\prime}; Z_i^+)}\right), \tag{14}$$

where $p(Z_i^{+\prime})$ represents the birth transition probability.

Death: Select the part $T_i^\star$ as the death part with the uniform distribution from the newly added candidate set $T^+ \cap T^\star$. The death proposal distribution is defined as $q_d(Z_i^{\star\prime}; Z_i^\star) = \frac{C_d}{|T^\star \cap T^+|}$, if $(T^{\star\prime}, Z^{\star\prime}) = (T^\star \backslash \{T_i^+\}, Z^\star \backslash \{Z_i^+\})$, and otherwise it equals to zero. Then the acceptance ratio is presented as

$$\alpha_d = \min\left(1, \frac{1}{p(\mathcal{X}|Z_i^{\star\prime}, \omega_p)} \cdot \frac{1}{p(Z_i^{\star\prime})} \cdot \frac{q_b(Z_i^\star; Z_i^{\star\prime})}{q_d(Z_i^{\star\prime}; Z_i^\star)}\right), \tag{15}$$

Stay: Select the part $T_i^\star$ to be the stay part. We introduce a set $T^{(d)} = T^\star \backslash T_i^\star$, $T^\star$ is the union set of the parts set in the previous iterations. Then the stay proposal distribution is presented as $q_s(Z_i^{\star\prime}; Z_i^\star) = \frac{C_s}{|T^{(d)}|} \cdot J(Z_i^{\star\prime})$, if $|T^{(d)}| \neq 0$, and otherwise it equals to zero. The function $J(Z_i^{\star\prime})$ represents the probability of part $T_i^\star$ staying at the state $Z_i^{\star\prime}$, and it is modeled as a normal density centered at the disappearing point. Z_i^l is the state of the part $T_i^\star$ at the disappearing point in the previous iterations. Then the acceptance ratio is presented as

$$\alpha_s = \min\left(1, p(\mathcal{X}|Z_i^{\star\prime}, \omega_p) \cdot \prod_{j \in \mathcal{N}_t} \phi(Z_i^{\star\prime}, Z_{j,t}) \cdot p(Z_i^{\star\prime}|Z_i^\star) \cdot \frac{q_l(Z_i^\star; Z_i^{\star\prime})}{q_s(Z_i^{\star\prime}; Z_i^\star)}\right), \tag{16}$$

where $\mathcal{N}_t$ is the part neighboring system at t, $p(Z_i^{*\prime}|Z_i^l)$ is the state transition probability.

Leave: Select the part T_i^* to be the leave part. The leave proposal distribution is defined as $q_l(Z_i^{*\prime}; Z_i^*) = \frac{C_l}{|T^*|}$, if $(T^{*\prime}, Z^{*\prime}) = (T^* \backslash \{T_i^*\}, Z^* \backslash \{Z_i^*\})$. Otherwise, it equals to zero. Then the acceptance ratio is presented as

$$\alpha_l = \min\left(1, \frac{1}{p(\mathcal{X}|Z_i^{*\prime}, \omega_p)} \cdot \frac{1}{\prod_{j \in \mathcal{N}_t} \phi(Z_i^{*\prime}, Z_{j,t})} \cdot \frac{1}{p(Z_i^{*\prime}|Z_i^*)} \cdot \frac{q_s(Z_i^*; Z_i^{*\prime})}{q_l(Z_i^{*\prime}; Z_i^*)}\right), \quad (17)$$

Update: Select the part T_i^* to be the update part. The appearance of the part will be updated if the move is accepted. The proposal distribution of this move type is defined as $q_u(Z_i^{*\prime}; Z_i^*) = \frac{1}{|T^*|}$, if $T_i^* \in T^*$. Otherwise it equals to zero. Then the acceptance ratio is presented as

$$\alpha_u = \min\left(1, \frac{p(\mathcal{X}|Z_i^{*\prime}, \omega_p)}{p(\mathcal{X}|Z_i^*, \omega_p)} \cdot \frac{\prod_{j \in \mathcal{N}_t} \phi(Z_i^{*\prime}, Z_{j,t})}{\prod_{j \in \mathcal{N}_t} \phi(Z_i^*, Z_{j,t})}\right), \quad (18)$$

In the above acceptance ratio calculation Eqs. (14–18), the birth transition probability $p(Z_i^{(\cdot)}) = \exp(-\lambda_o \cdot \mathcal{R}(Z_i^{(\cdot)}, Z_i^*))$. This term penalizes the overlap ratio between the candidate part T_i^+ and the existing learned parts to avoid adding redundant parts. We set $\lambda_o = 2$ in our experiments. The probability $p(\mathcal{X}|Z_i^{(\cdot)}, \omega_p) \propto \exp(G(\mathcal{X}; \omega_p, Z_i^{(\cdot)}))$, where $\mathcal{X}$ is the current updating sample and $Z_i^{(\cdot)}$ (Z_i^*, $Z_i^{*\prime}$, Z_i^+ or $Z_i^{+\prime}$) is the part state.

In addition, the target parts interact with each other, especially for the neighboring ones, so it is inappropriate to assume the independence between target parts. We should integrate the relationships between parts in optimization process rather than optimize the objective in (8) for each part individually. Motivated by [21], we propose to utilize the dynamically constructed pairwise MRF to model the supporting between different parts in the part learning process. We set all pairs of parts as the neighbors in the graph. Thus, the part transition probability in (16) and (17) is presented as follows:

$$p(\tilde{Z}'|\tilde{Z}) \propto \prod_i p(Z_i'|Z_i) \prod_{ij \in \mathcal{N}} \phi(Z_i', Z_j'), \quad (19)$$

where $\tilde{Z} = (Z_1, \cdots, Z_n)$ is the combination state of multiple parts, $\phi(Z_i', Z_j')$ is the pairwise interaction potentials between parts similarly defined as (4), $p(Z_i'|Z_i)$ is the part transition model, and $\mathcal{N}$ is the neighbor system of the parts. Z_i' is the proposal state of part i and Z_i is the state of part i currently. The part transition is modeled as a normal density centered at the previous state, which is presented as $Z_i'|Z_i = Z_i + \Delta Z_i$, where $\Delta Z_i \sim [\mathcal{N}(0, \sigma_x^2), \mathcal{N}(0, \sigma_y^2), \mathcal{N}(0, \sigma_w^2), \mathcal{N}(0, \sigma_h^2), \mathcal{N}(0, \sigma_{\Delta x}^2), \mathcal{N}(0, \sigma_{\Delta y}^2)]$.

Figure 1 is an example to illustrate how the discriminative parts are automatically learned over time in *shirt* sequence. The shirt is crinkled in frame ♯0026 and ♯0040, where the target bounding box contains considerable background in

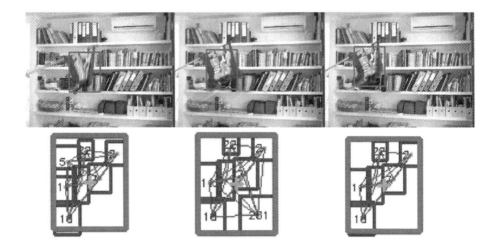

Fig. 1. The first row is the tracking results of our tracker in the sequence *shirt* and the second row presents the learned discriminative parts and the corresponding structure. The nodes in the graph represent each learned part and the lines represent the spatial relationships between neighboring parts. The green cross represents the center of the target (Colour figure online).

the right bottom corner. In contrast, the target bounding box in frame ♯0038 contains only foreground. In this case, the proposed part-based method adaptively generates a part 261 to cover the new foreground in frame ♯0038, and deletes it in frame ♯0040. In this way, our part based model can adapt to the variations of the target better than the bounding box based model. The final optimization scheme is summarized in Algorithm 1.

5   Experimental Results

5.1   Parameters

The parameters in our experiment are detailed in the following. In the learning phase, we run $P_n = 400$ iterations to complete the part state inference task and $P_{n_0} = 100$ of them are burn-in in the MH algorithm. Generally, the algorithm will converge after about 300 iterations. The motion model parameters we used are $\sigma_x^2 = 3$, $\sigma_y^2 = 3$, $\sigma_w^2 = 0.2$, $\sigma_h^2 = 0.2$, $\sigma_{\Delta x}^2 = 0.1$ and $\sigma_{\Delta y}^2 = 0.1$. The cell size of HOG is set as 8×8 pixels. A block consists of 4 cells and the strides of the cell are set 4 in both x and y directions. The linear kernel is exploited in the online learning SVM model. The target area is divided into about 15 or 20 superpixels in the SLIC algorithm. Meanwhile, in the tracking phase, the searching radius $R_s \in [20, 60]$. The balance parameter λ_b in (2) is set in the interval $[0, 0.5]$.

Algorithm 1. Discriminative Hidden Structural Part Tracker

1: Initialize the target state $\hat{Z}_1$.

2: **for** $t = 2$ to N **do**

3:     Get the foreground $\mathcal{F}$ and collect the birth candidate parts based on the opti-mized target state $\hat{Z}_t = \hat{Z}_{t-1}$, and get the initial joint parts state $\tilde{Z}^{(1)}$.

4:     Set the sample set $\Gamma = \emptyset$ in the MH algorithm.

5:     **for** $r = 1$ to P_n **do**

6:         Get optimal $\omega_p^{(r)}$ based on the current state $\tilde{Z}^{(r)}$.

7:         Generate the proposal joint state $\tilde{Z}^{(r)\prime}$ based on $\tilde{Z}^{(r)}$:

   – Choose a move type according to the move prior probability C.
   – Select a part i according to the move proposal distribution q and compute the acceptance ratio α for $\tilde{Z}^{(r)\prime}$ in (14), (15), (16), (17), (18).
   – Accept the proposal state, if $\alpha \geq 1$, and add it to Γ, $\tilde{Z}^{(r+1)} = \tilde{Z}^{(r)\prime}$; otherwise generate a uniform random number $u \in [0,1]$. If $u < \alpha$, accept the proposal state and add it to Γ, $\tilde{Z}^{(r+1)} = \tilde{Z}^{(r)\prime}$; otherwise reject the proposal state and add the previous state to Γ, $\tilde{Z}^{(r+1)} = \tilde{Z}^{(r)}$.
   – $\tilde{Z}^{(r)} = \tilde{Z}^{(r+1)}$.

8:     **end for**

9:     Discard the first P_{n_0} burn-in samples in Γ.

10:     Get the optimized parts Z^* by the MAP solution of Γ in (11).

11:     Update the SVM model and the MRF graph model of the parts with $\{Z^*, \omega_p\}$ in $\mathcal{A}^{(0)}$, and update the SVM model with $\{\hat{Z}_t, \omega_b\}$ in $\mathcal{A}^{(1)}$.

12: **end for**

5.2   Effectiveness of $\mathcal{A}^{(0)}$ and $\mathcal{A}^{(1)}$

Firstly, we chose four representative sequences to demonstrate the behavior of $\mathcal{A}^{(1)}$ and $\mathcal{A}^{(0)}$. The results shown in Table 1 demonstrate the performance of HSPT is improved mainly due to the local discriminative parts learning rather than the features or the classifiers adopted.

Table 1. Comparison results of $\mathcal{A}^{(1)}$, $\mathcal{A}^{(0)}$, and HSPT.

Seq.	AECP Metric			PASCAL VOC Metric		
	$\mathcal{A}^{(1)}$	$\mathcal{A}^{(0)}$	HSPT	$\mathcal{A}^{(1)}$	$\mathcal{A}^{(0)}$	HSPT
tiger2	26.7	14.7	8.39	80	215	280
shirt	30.2	22.5	8.34	680	920	1310
pedestrian	105	14.9	3.76	110	310	355
car	29.7	10.3	7.78	770	870	895

As presented in Table 1, the combined HSPT outperforms the individual $\mathcal{A}^{(1)}$ and $\mathcal{A}^{(0)}$ in all tested sequences. $\mathcal{A}^{(1)}$ focuses on holistic appearance and $\mathcal{A}^{(0)}$ focuses on inner structure of the target. $\mathcal{A}^{(0)}$ is superior over $\mathcal{A}^{(1)}$ because the local discriminative structure model represents the target better than appearance only model. Especially in the sequences *shirt* and *pedestrian* where nonrigid deformations and illumination variations frequently happen, the targets were still well tracked by $\mathcal{A}^{(0)}$ even when several parts undergo changes in location and appearance. In contrast, $\mathcal{A}^{(1)}$ was affected more seriously by these challenges. $\mathcal{A}^{(1)}$ focuses on the holistic appearance, which can enhance the stability of the tracker in the complex situations. HSPT inherits the advantages both from $\mathcal{A}^{(1)}$ and $\mathcal{A}^{(0)}$ and thus performed best against the other evaluated trackers on the evaluated sequences.

5.3   Comparison with Other Trackers

Then, we compare our tracker with some state-of-the-art methods, including bounding box based methods (MIL [2], VTD [7], $\ell1$ [4], TLD [5]), and part based methods (Frag [9], HABT [10], BHMC [11], SPT [14]). All the codes are provided by the authors on their websites. Ten challenging sequences (nine of them are publicly available [2,5,7,25,26] and the other one is collected by ourself) are utilized in the experiment. These sequences cover most of the challenging situations in tracking task: non-rigid motion, in-plane and out-of-plane rotation, large illumination changes, heavy occlusions and complex background (see Fig. 3).

Table 2. Comparison results based on the AECP metric.

Seqences	MIL	$\ell1$	TLD	VTD	Frag	HABT	BHMC	SPT	HSPT
football	12.7	26.3	13.1	**6.25**	9.92	70.0	70.9	170	4.39
tiger1	**8.35**	60.4	25.9	22.3	29.3	16.9	-	16.3	8.29
tiger2	5.91	47.3	25.3	31.5	39.3	53.7	-	-	**8.39**
david	15.6	77.3	4.49	31.7	55.9	44.7	-	144	**5.44**
shirt	32.1	68.4	70.8	27.5	19.9	79.5	**9.47**	36.6	8.34
pedestrian	64.3	128	36.7	86.0	56.9	80.8	-	27.1	3.76
stone	9.07	11.6	9.69	26.1	92.9	127	40.3	50.3	6.14
carchase	39.9	8.92	3.77	81.1	12.6	21.0	-	153	**3.95**
car	80.3	23.3	11.1	51.8	28.6	26.3	-	153	7.78
portman	42.5	60.6	30.7	45.1	54.5	39.4	**23.3**	73.4	17.6

Table 3. The successfully tracked frames based on the PASCAL VOC metric.

Seqences	Frames	MIL	$\ell1$	TLD	VTD	Frag	HABT	BHMC	SPT	HSPT
football	**362**	272	206	272	357	302	206	55	30	362
tiger1	**354**	279	80	150	189	155	209	-	65	279
tiger2	**365**	315	15	60	85	35	25	-	-	**280**
david	**462**	328	124	422	65	159	253	-	30	462
shirt	**1365**	760	55	5	760	890	10	330	595	1310
pedestrian	**355**	55	85	75	110	80	55	-	85	355
stone	**593**	135	384	339	384	95	5	40	115	473
carchase	**424**	91	283	419	66	368	318	-	61	424
car	**945**	105	765	880	575	650	480	-	25	895
portman	**301**	94	69	118	168	74	148	30	74	271

The proposed tracker is implemented in C++ and it runs about 0.1 fps on the Intel 3.0 GHz PC platform. We present the tracking results in this section and more results as well as demos can be found in supplementary materials. Two metrics are utilized to quantify the performance, namely the Average Error Center Location in Pixels (AECP) metric ($\downarrow$) and the PASCAL VOC object detection metric [27] ($\uparrow$). The symbol $\uparrow$ means methods with higher scores perform better, and $\downarrow$ indicate methods with lower scores perform better. The quantitative comparison results are shown in Tables 2 and 3.

Heavy Occlusion. The target in sequences *carchase*, *stone*, *car*, *tiger1* and *tiger2* undergoes heavy occlusion multiple times. As shown in Fig. 3, most of the trackers drift away when heavy occlusion happens, and some of them can not recapture the target after occlusion. In the sequence *car*, TLD with detection module works relatively well. Some other trackers such as Frag and $\ell 1$ who have intuitive robustness to occlusion also track the car well. However, our tracker still outperforms other methods due to the consideration of the relationships between parts which alleviates the influence of some badly learned parts.

Large Illumination Variations. The frequent large illumination variations in *tiger1*, *tiger2* and *pedestrian* sequences challenge the performance of the trackers. Since the appearance features are easily affected by illumination variations, most of the previously proposed trackers fail to track the target in these sequences. For example, when the light is shining in the sequence *tiger1* and *tiger2*, Frag fails to track the tiger, and when the woman is under the shadow in the *pedestrian* sequence, VTD and SPT shrink to those parts that are not shadowed. In contrast, our tracker optimally partitions the target into several parts, and the target can be located with the help of those less affected parts. The combined appearance features of different parts, the structure information between parts, and the structure information between the part and the target center make our tracker outperforms other trackers.

Pose Changes. The inner structure changes caused by pose changes usually make the bounding box based trackers fail to track the target. Nevertheless, part based trackers including Frag and BHMC work relatively well because they focus on the parts instead of the holistic bounding box template, which can be demonstrated in the sequence *portman* and *shirt* in Fig. 3, Tables 2 and 3. Especially in the sequence *shirt*, the bounding box based trackers such as $\ell 1$, VTD and MIL fail because of the error accumulation when the non-rigid motion happens. Since the combination of part appearance is less influenced than the bounding box based appearance under pose changes and several correctly learned parts are good enough to locate the target, our tracker still works very well in these sequences.

Complex Background. In the *football* and *stone* sequences, many similar objects confuse the trackers a lot. As shown in Figs. 2 and 3, TLD and HABT frequently skip to other similar objects. The similar appearance between the target and the background in the sequences *football*, *stone*, *tiger1* and *tiger2* makes it hard to precisely track the target. Through combining the target holistic appearance, the

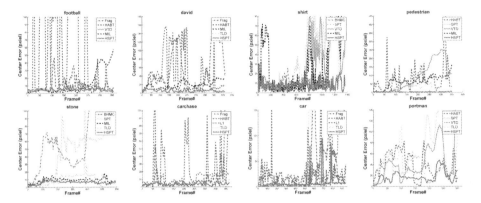

Fig. 2. Tracking results of MIL [2], VTD [7], ℓ1 [4], TLD [5], Frag [9], HABT [10], BHMC [11], SPT [14] and the proposed HSPT tracker. The results of five trackers with relatively better performance are displayed.

Fig. 3. Tracking results of different trackers. Only the trackers with relatively better performance are displayed.

detailed appearance of parts and the structure information between different parts, our tracker can discriminate the target from the complex background. While some specific trackers outperform ours in some specific sequences, but comprehensively speaking, our tracker performs the best.

6   Conclusion

In this paper, a novel online learned discriminative part-based tracker is proposed. The appearance of the target is described by the combination of multiple learned discriminative structural parts. In the parts learning phase, we utilize the MH algorithm based optimization framework integrated with the online SVM to infer the optimal parts state. We introduce the dynamically constructed pairwise MRF to model the interaction between neighboring parts. The experiments demonstrate the superiority of the proposed method. In the future, we will optimize the codes to make the tracker run in real-time.

Acknowledgement. This work was supported by the Chinese National Natural Science Foundation Projects #61105023, #61103156, #61105037, #61203267, #61375037, #61473291, National Science and Technology Support Program Project #2013BAK02 B01, Chinese Academy of Sciences Project No. KGZD-EW-102-2, and AuthenMetric R&D Funds.

References

1. Lim, J., Ross, D.A., Lin, R.S., Yang, M.H.: Incremental learning for visual tracking. In: NIPS (2004)
2. Babenko, B., Yang, M.H., Belongie, S.J.: Visual tracking with online multiple instance learning. In: CVPR, pp. 983–990 (2009)
3. Wen, L., Cai, Z., Lei, Z., Yi, D., Li, S.Z.: Robust online learned spatio-temporal context model for visual tracking. IEEE Trans. Image Process. **23**, 785–796 (2014)
4. Mei, X., Ling, H.: Robust visual tracking using $\ell 1$ minimization. In: ICCV, pp. 1436–1443 (2009)
5. Kalal, Z., Matas, J., Mikolajczyk, K.: P-N learning: bootstrapping binary classifiers by structural constraints. In: CVPR, pp. 49–56 (2010)
6. Wen, L., Cai, Z., Yang, M., Lei, Z., Yi, D., Li, S.Z.: Online multiple instance joint model for visual tracking. In: IEEE Ninth International Conference on Advanced Video and Signal-Based Surveillance, pp. 319–324 (2012)
7. Kwon, J., Lee, K.M.: Visual tracking decomposition. In: CVPR, pp. 1269–1276 (2010)
8. Wen, L., Cai, Z., Lei, Z., Yi, D., Li, S.Z.: Online spatio-temporal structural context learning for visual tracking. In: Fitzgibbon, A., Lazebnik, S., Perona, P., Sato, Y., Schmid, C. (eds.) ECCV 2012, Part IV. LNCS, vol. 7575, pp. 716–729. Springer, Heidelberg (2012)
9. Adam, A., Rivlin, E., Shimshoni, I.: Robust fragments-based tracking using the integral histogram. In: CVPR, pp. 798–805 (2006)
10. Shahed, S.M.N., Ho, J., Yang, M.H.: Visual tracking with histograms and articulating blocks. In: CVPR (2008)
11. Kwon, J., Lee, K.M.: Tracking of a non-rigid object via patch-based dynamic appearance modeling and adaptive basin hopping monte carlo sampling. In: CVPR (2009)
12. Godec, M., Roth, P.M., Bischof, H.: Hough-based tracking of non-rigid objects. In: ICCV, pp. 81–88 (2011)

13. Cehovin, L., Kristan, M., Leonardis, A.: An adaptive coupled-layer visual model for robust visual tracking. In: ICCV, pp. 1363–1370 (2011)
14. Wang, S., Lu, H., Yang, F., Yang, M.H.: Superpixel tracking. In: ICCV, pp. 1323–1330 (2011)
15. Shu, G., Dehghan, A., Oreifej, O., Hand, E., Shah, M.: Part-based multiple-person tracking with partial occlusion handling. In: CVPR, pp. 1815–1821 (2012)
16. Zhong, W., Lu, H., Yang, M.H.: Robust object tracking via sparsity-based collaborative model. In: CVPR, pp. 1838–1845 (2012)
17. Cai, Z., Wen, L., Yang, J., Lei, Z., Li, S.Z.: Structured visual tracking with dynamic graph. In: Lee, K.M., Matsushita, Y., Rehg, J.M., Hu, Z. (eds.) ACCV 2012, Part III. LNCS, vol. 7726, pp. 86–97. Springer, Heidelberg (2013)
18. Yao, R., Shi, Q., Shen, C., Zhang, Y., van den Hengel, A.: Part-based visual tracking with online latent structural learning. In: CVPR, pp. 2363–2370 (2013)
19. Hastings, W.: Monte carlo sampling methods using markov chains and their applications. Biometrika **57**, 97–109 (1970)
20. Bordes, A., Ertekin, S., Weston, J., Bottou, L.: Fast kernel classifiers with online and active learning. J. Mach. Learn. Res. **6**, 1579–1619 (2005)
21. Khan, Z., Balch, T.R., Dellaert, F.: MCMC-based particle filtering for tracking a variable number of interacting targets. IEEE Trans. Pattern Anal. Mach. Intell. **27**, 1805–1918 (2005)
22. Dalal, N., Triggs, B.: Histograms of oriented gradients for human detection. In: CVPR, vol. 1, pp. 886–893 (2005)
23. Yang, M., Liu, Y., Wen, L., You, Z., Li, S.Z.: A probabilistic framework for multitarget tracking with mutual occlusions, pp. 1298–1305 (2014)
24. Achanta, R., Shaji, A., Smith, K., Lucchi, A., Fua, P., Ssstrunk, S.: SLIC Superpixels. Technical report (2010)
25. Oron, S., Bar-Hillel, A., Levi, D., Avidan, S.: Locally orderless tracking. In: CVPR, pp. 1940–1947 (2012)
26. Jia, X., Lu, H., Yang, M.H.: Visual tracking via adaptive structural local sparse appearance model. In: CVPR, pp. 1822–1829 (2012)
27. Everingham, M., Gool, L.J.V., Williams, C.K.I., Winn, J.M., Zisserman, A.: The pascal visual object classes (voc) challenge. Int. J. Comput. Vis. **88**, 303–338 (2010)

Image Based Visibility Estimation During Day and Night

Sami Varjo$^{(\boxtimes)}$ and Jari Hannuksela

The Center for Machine Vision Research, Department of Computer Science
and Engineering, University of Oulu, P.O. Box 4500, 90014 Oulu, Finland
{sami.varjo,jari.hannuksela}@ee.oulu.fi

Abstract. The meteorological visibility estimation is an important task,
for example, in road traffic control and aviation safety, but its reliable
automation is difficult. The conventional light scattering measurements
are limited into a small space and the extrapolated values are often erroneous. The current meteorological visibility estimates relying on a single
camera work only with data captured in day light. We propose a new
method based on feature vectors that are projections of the scene images
with lighting normalization. The proposed method was combined with
the high dynamic range imaging to improve night time image quality.
Visibility classification accuracy (F1) of 85.5 % was achieved for data
containing both day and night images. The results show that the approach can compete with commercial visibility measurement devices.

1 Introduction

Meteorological visibility is an important measure in many fields such as road traffic safety, flight control and aviation safety, as well as coastal and marine activities. Visibility is usually reduced by different sized particles or aerosols, such as
fog, rain, or snow [1]. In addition to precipitation, visibility can be reduced by
air pollution which levels can be monitored using visibility estimation [2]. Traditionally, visibility is estimated by a human observer who estimates the visibility
range from known distant objects like forest line, stars or other light sources.
Manual observations are not only costly and time consuming but also biased by
the individuals.

The visibility degradation can be modeled using physical models that are
based on light absorption and scattering in the air. One commonly applied model
is based on the Beer-Lambert law that states that the light intensity drops exponentially as a function of the traveled distance and the medium absorption coefficient. The non-imaging based measurements with lasers are typically carried out
in very small volumes and the results are extrapolated to cover distances up to
20 km away from the actual measurement location. Local weather phenomenon
at some distance cannot be captured by these methods.

Cameras enable monitoring affordably panoramic scenes at all directions.
Tan [3] has applied random Markov fields on single images to improve the image
quality in low visibility images, but no metric visibility measurements were

© Springer International Publishing Switzerland 2015
C.V. Jawahar and S. Shan (Eds.): ACCV 2014 Workshops, Part III, LNCS 9010, pp. 277–289, 2015.
DOI: 10.1007/978-3-319-16634-6_21

attempted. The Beer-Lambert law has been also applied with images, which allows learning of simple regression models for visibility estimation [4,5].

The observations are usually required to be carried out continuously. So far the large illumination changes have not been addressed and the image based estimation methods are usually shown to work only at day light or during night but not both. Du et al. [6] have used two cameras at different distances to capture images of the same target during night and transmission based estimation was obtained. The transmission measurements are limited on a line of sight and multiple cameras are required. Also distant light sources and scattering models have been used to estimate the visibility during night [7]. That approach has limited usability during the day time as the sky luminance is often dominant over the man-made light sources. We suggest an approach that does not rely on external light sources and use a single camera to estimate the visibility 24 h a day.

We show that commonly applied gradient and light attenuation based regression models fail in continuous monitoring. We propose a new method where Retinex filtering is used to create light intensity invariant images that are used for projection based features. The visibility estimation is based on machine learning with support vector machines. Further, HDR techniques have not been previously utilized in metric visibility estimation and we show that it is beneficial when combined with the proposed feature classification. The proposed method is targeted for scenes with a visible horizon, like at air ports or city skylines.

2   Related Work

Human observations are still widely used in visibility estimation as no extra equipment is required. The estimation is based on known distant objects whose contrast against the background is assessed and, if the visibility vary in different directions, usually only the smallest visibility is reported. Manual observations are relatively expensive and may be biased by the observer.

The visibility sensors do not directly measure how far one can see but instead the clarity of the air is measured. Locality of the non-imaging sensors is a major problem that would require multiple measuring sites, which usually is not cost efficient. Cameras can be utilized for estimating visibility, but large lighting changes limit the current methods to day time imaging. We discuss these issues in the following chapters in more detail.

2.1   Non-imaging Based Estimation

Non-imaging based automatic measuring systems rely either on light scattering or transmission. Light scattering measurements are carried out in a small volume, for example, a cubic decimeter and the estimates are extrapolated to cover ranges up to 20 km. Transmission meters measure the intensity drop in a light beam on a given path. While not as point-like as the scattering methods, these are also considered local measurement devices. Both measuring methods are considered too expensive to be utilized in numbers for more comprehensive coverage.

The extrapolated visibility measurements do not always correspond to the true situation at some distance. For example, by visual inspection the Vaisala FD12P forward scattering device used in this study, produced estimates that were wrong in about 23 % of the 6650 captured images. Figure 1 shows few examples of images and visibility estimates from the device where the measurements are clearly misleading. Too high visibility estimates announced by aviation traffic control are problematic at airports where reliable information is required 24 h a day. This is one of the main motivations behind this work.

Fig. 1. Examples of cases where the light scattering equipment failed: (left,day) visibility measure 8180 m and (right,night) 13 km. The proper visibilities are less than 1000 m and less than 5 Km respectively. The clearly false measurements cause problems for example in aviation during day and especially at night.

2.2   Image Based Models

The existing visibility measures are typically based on contrast or gradients computed from the captured images. Multivariate linear regression, for example, can be used to learn visibility models from edges obtained using Sobel-filters [4]. The model coefficient of determination values (R^2) ranged between 0.780 and 0.845 for day time images.

Babari et al. also used gradients to estimate a physical model described by the Beer-Lambert law [5]. They reported the average errors of 30 % (R^2 between 0.89 and 0.95) for images taken during day with luminance between 10 and 8000 cd/m^2. Their approach uses Sobel filtering to extract gradient magnitudes from an image I and the response is weighted by an estimate of the Lambertianess $W_{x,y}$ of the pixels in the scene for a contrast measure:

$$G = \sqrt{\left(\begin{bmatrix} +1 & 0 & -1 \\ +2 & 0 & -2 \\ +1 & 0 & -1 \end{bmatrix} * I\right)^2 + \left(\begin{bmatrix} +1 & +2 & +1 \\ 0 & 0 & 0 \\ -1 & -2 & -1 \end{bmatrix} * I\right)^2} \cdot W_{x,y} \qquad (1)$$

$$W_{x,y} = corr(I_{x,y}^{set}, I_{sky}^{set}),$$

where $[*]$ denotes convolution, and $[\cdot]$ point wise multiplication at pixels x, y. The $W_{x,y}$ is calculated as the correlation of pixel intensities respect to the sky illumination over time. The final measure $\hat{E}$ is formed by integrating the response G over the image. The obtained responses are used to form a regression model with parameters A and B that relates the measure to the Beer-Lambert law:

$$\hat{E} = A + B \log\left(V_{met}\right). \tag{2}$$

The inverse of the model can be used to convert the measured response to metric visibility V_{met}. Figure 2 shows a fit to data containing both day and night images from Matilda database [5].

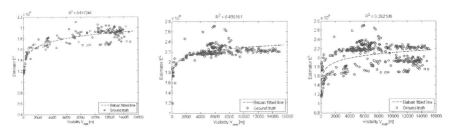

Fig. 2. Model (2) fitted on day and night data from the Matilda database and combined data. Only the day data fits the model with a reasonable coefficient of determination, and weighting with a surface Lambertianess estimate improves the fit as shown in [5]. However, data is separated so that no static weighting can remove all the lighting covariance from the data and the day and night data is clearly separated to sub groups.

The data clearly separates into two groups and a single model is clearly not enough. The changing sky luminance cannot be handled with the regression based models. The ambient luminance can change over a day radically at the northern and southern hemisphere making the reliable estimation of the Lambertian weights $W_{x,y}$ almost impossible in these scenarios, at least the estimation should be adaptive instead of static. Further, as the weighting is based on the illumination of the sky this cannot be done during night.

Other approach is to use machine learning to categorize the samples into visibility classes. Yin et al. [8] used support vector regression to learn visibility from regions of interest where the features were formed from the mean local contrast of 4×4 sub blocks resulting in 16-dimensional features for each image. A mean success ratio calculated from the reported results show 81 % performance for day time images. In this case poor visibility was most often misinterpreted, which is the most important class to be classified correctly.

3   Proposed Method

The main problem with the existing image based visibility estimation is the covariance of the extracted response with the ambient illumination. The estimation of the ambient illumination based on the sky is not reliable as clouds change and reflect light and at night there is only light from moon and ground reflections.

We suggest forming a new, lighting invariant, descriptor for scene visibility classification. The features are formed by projecting Retinex filtered images on the horizontal direction corresponding to different distances in the view as shown in Fig. 3. While we propose to use HDR imaging to capture the images, the approach is shown to work also with conventional low dynamic range images. The features are further processed for robustness by taking the absolute values of their gradients which are normalized. The visibility estimation is then finally based on support vector machine based classification. In the following, the steps are described in more detail.

3.1   HDR Imaging for Visibility Estimation

Luminance of the night sky is typically in range of millicandelas per square meter while street lamps are capable of producing some tens of kilocandelas per square meter. Therefore night scenes with urban surroundings have a dynamic range of six to seven decades, which is on the lower limit of an adapted human eye dynamic contrast ratio. Digital cameras commonly use analog to digital converters with 8 to 16 bit precision enabling at most contrast ratios of 1:65536, which is not enough for representing the low light scenes as can be seen from Fig. 4. Short exposure times enable sufficient capture of the street lights vicinity but unilluminated vegetation is poorly shown. Long exposure times capture even clouds and the stars from the night sky but the lit streets are severely over exposed.

High dynamic range (HDR) imaging in combination with tone mapping techniques is typically used to produce visually pleasing images to overcome the limited camera sensor and display dynamic ranges [9]. It is also possible to enhance an image quality by producing HDR images without using physical models [10,11].

HDR images are commonly constructed by solving the camera response function (CRF) that describes the relation of recorded intensity and the exposure time to the scene radiance. Inverse CRF is then used to linearize the collected data which is combined for a radiance map [12]. Further the radiance map is used with tone mapping to represent the color information correctly for the human viewer using low dynamic range displays or print outs [13]. It has been shown that a dense set of images produces an improved signal to noise ratio for HDR radiance map compared with a minimal number of images [14]. The dense image sampling strategy is selected to suite the needs of this application.

There are no real-time requirements for the system and it is enough to have an updated estimate every ten minutes. To capture stacks of images, exposure times between 25 μs and 23.6 s were used yielding maximum total exposure time of 35.4 s. This enables the reconstruction of HDR images under a large range of lighting conditions. In this work, we are more interested in the normalization of the lighting changes and the visibility than the esthetics of the images.

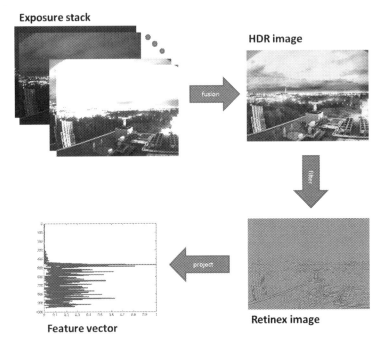

Fig. 3. An exposure stack of images is captured for a high dynamic range image, which is Retinex filtered and the rows of the image are projected on the y-axis for the feature vector.

Fig. 4. Six subsequent images from an exposure time stack captured during a night. The exposure times are 72.9 ms, 218.7 ms, 656.1 ms, 1.968 s, 5.905 s, and 11.810 s.

An adaptive method was utilized to stack the HDR images. As the sensor behaves here linearly (see Section Data, Fig. 7), the fusion of the images I_{HDR} in an exposure stack is an integration of the intensity values I_i over the pixels weighted with the exposure times t_i:

$$I_{HDR} = \sum_{i}^{N} I_i \cdot t_i. \tag{3}$$

The number of used images N, was selected by estimating the contrast in result image I_{HDR} after each summing step and selecting the result image with the highest obtained contrast at the end. Figure 5 shows an example of the combined images for a scene during a night and a day. The night image corresponds to the inputs shown also in the Fig. 4. The HDR image clearly contains more information than the single input images.

Fig. 5. HDR images obtained in a night and during a day. The images are tone mapped for viewing using contrast limited adaptive histogram equalization (CLAHE) [15].

3.2   Retinex Filtering

Multiscale retinex (MSR) filtering was originally developed to improve the visual quality of both 8-bit images and wide dynamic range images [10]. MSR aims for creating images that match lightness and color perception of human vision. While the filtering is aimed for color images, it handles each color channel separately and thus can be used also on gray scale images. The response R_{MSR} for MSR is a weighted sum of single scale retinex filters R_i where the Gaussian filter F_i scale c_i is varied:

$$R_{MSR} = \sum_{n=1}^{N} w_n R_n, \tag{4}$$

$$R_i(x, y) = log I(x, y) - log \left[F_i(x, y) * I(x, y) \right], \tag{5}$$

$$F_i(x, y) = K e^{-r^2/c_i^2}. \tag{6}$$

The parameters suggested in [10] were applied here; $N = 3$, $c_i = 15, 80, 250$ and $w_i = 1/3$. For the $r = sqrt(x^2 + y^2)$ 9×9 local window was applied. The applied filter does not produce here visually stunning results, but the differences between day and night time images are very small in visual inspection (see Fig. 6).

3.3   Feature Extraction

The retinex filtering can be characterized as high pass filtering in logarithmic space, which makes the result much less susceptible towards the lighting changes. The scene distance from the camera increases towards to upper part of the image. According to (2) the contrast or here the high frequencies are reduced with the increased distance in case of visibility degradation.

The projection of the frequency response in the image on y-axis can be considered a good descriptor for the visibility when the horizon is captured levelly in the images. In this work, no image rectification was used but the camera was only positioned carefully to capture horizon images. In practice, the feature can be formed by summing up the retinex filter response row wise over the image. The robustness is further increased by taking a gradient of the y-projection. Finally, the magnitude vector of the obtained gradient is normalized in range [0, 1]. The features were orthogonalized using principal component analysis (PCA). The best results were obtained containing all the components of the PCA transformation indicating that all the feature vector components contain relevant visibility information that is useful in classification.

The Fig. 6 exhibits examples of retinex filtered images and the corresponding feature vectors. The difference between night and day images is visually compared small. The corresponding good visibility and poor visibility feature vectors appear very similar.

3.4   SVM Training

The feature vectors extracted from the images were used to learn the required visibility classes. Here support vector machines with RBF kernels were utilized to learn the five classes named in Table 1. The applied learning strategy for the given class was one-against-all. In the multiclass classification the distance from the decision boundary of each class was used as the classification score and the largest distance was selected as a winner.

The learning parameters for soft margin C and the kernel Gaussian variance σ were optimized using a grid search and the performance was tested using leave one out cross validation for Matilda data and 5-fold cross validation for HDR data. F1-measure, that is the harmonic mean of precision and recall, was used as an accuracy measure.

4   Data

Images from two sources were used and all images top part of the sky was cropped out. With the Matilda database 31 pixel rows, containing time stamps,

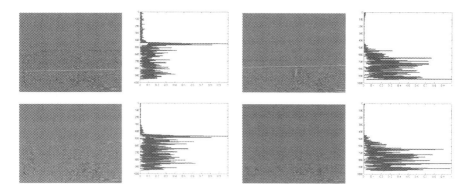

Fig. 6. Retinex filtered images in good (left) and poor visibility (right) during day (up) and night (bottom) and the corresponding feature vectors.

were cropped out leaving 100 pixel rows sky and 348 rows ground. With the HDRvisMe data 40 pixel rows of the sky were left leaving 511 pixel rows of ground. The Matilda database was utilized for the experiments with 8-bit images [5]. The data consist of CCTV camera images and meteorological optical range and sky luminance values measured with meteorological instruments. The similar division into visibility classes as in [5] was utilized, with the distinction that also images having visibility above 15 Km were included. Table 1 contains the number of samples in each of the class. Images with luminance below 10 cd/m^2 were treated as night images. It should be noted that in [5] only day images and four first classes were utilized.

For HDR data set (HDRvisMe)[1] a 14 bit gray scale camera (AVT, Prosilica GT1290) was used for capturing the images. It was housed in a weatherproof casing with a heater element to prevent water condensation in the casing and on the case window. The setup was installed on a roof about 30 m from the ground. The camera was secured so that image registration due to camera movement was not required. For comparison also here the data was divided to day, night, and dusk&dawn sets based on the capture time. Data was manually labeled to the visibility classes in Table 1.

Vaisala FD12P light scattering measurement device was used to collect the reference visibility data. The manufacturer of FD12P gives the error tolerance of ±10 % for the range 10–10000 m and ±20 % for range 10–50 m and instrument consistency of 4 %. It should be noted that the device has the sampling volume of 0.1 l.

Lack of the color filters in the camera removes the need for separate camera response calibration for color channels. The color filters would also reduce slightly the incident light hitting the sensor. The images were captured using manual settings so that exact parameters, such as gain and exposure time, were used. Also all the in-build data manipulations, including gamma correction, were

---

[1] The data and Matlab implementation are available upon a request from the authors.

Table 1. Visibility classes and number of samples.

Database		<400 m	400–1000 m	1–5 km	5–15 km	>15 km
Matilda	Day	17	6	26	109	49
	Night	9	3	44	148	26
	Total	26	9	70	257	75
HDRvisMe	Day	48	58	208	725	1765
	Night	15	68	393	955	1339
	Dawn and dusk	10	20	104	317	625
	Total	73	146	705	1997	3729

disabled. This enabled the near linear behavior of the camera sensor, which was ensured by inspecting the camera response curve. This was constructed by selecting uniform intensity patches from sets of images with different exposure times, which were further reduced to median values for each of the exposure time settings. These values were converted to relative intensity values using exposure times as presented in Fig. 7. In addition the method from Mitsunaga and Nayar [16] was used to fit polynomials with different degrees. The best fit was a straight line, proving the linear response of the imaging setup.

5   Results and Discussion

The proposed method reached 86.5 % F1-accuracy for the combined day and night images in the Matilda database with leave one out cross validation (soft margin $C = 3.5$ and $\sigma_{RBF} = 1.05$). With separation to day and night, the accuracies were 85.2 % and 82.6 % respectively. One would expect that the classification would work better with separated data sets, but apparently the data contains a limited number of samples, especially the second class for visibility range 400–1000 m. This can also be seen in the confusion matrix in Fig. 8 as large variation with the second class.

A more comprehensive data set was captured for HDR testing with over 10 times more images available than in Matilda data set. Here for the combined day and night data, the 5-fold cross validation yielded 85.5 % F1-accuracy. With the HDR data, the confusion matrix is less scattered around the true classes and the result is more reliable than with the low dynamic range image set. Data separated to day and night sets resulted in 86.2 % and 85.0 % classification accuracy respectively. The utilized training parameters here for SVMs were for soft margin $C = 5.3593$ and for the RBF-kernel shape $\sigma_{RBF} = 1.5625$ with all classes.

Classifiers were tested also by using only the day data for training and night data for testing and vice versa. Here the mean accuracy was 71 %, showing that some lighting invariance has been obtained for the descriptor. The slight drop indicates that the invariance is not perfect, but with samples of all possible lighting conditions the changing illumination can be handled with the descriptor and SVMs.

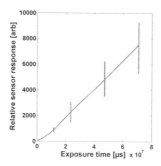

Fig. 7. The camera response function for the utilized camera.

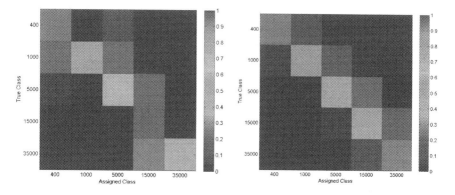

Fig. 8. Confusion matrices for all Matilda data (left) and for the HDR data (right).

It should be noted that the HDR data was manually annotated to the visibility categories as the visibility sensor was realized to produce correct estimates for about 77 % of input images. However, it is difficult also for human to estimate visibilities around 10 Km in the wild and here annotation was done based on images only. Therefor, some errors may exist in annotations. The confusion matrix with HDR data supports this assumption as the first two classes are classified more robustly than the problematic classes 1–5 km and 5–15 km, while the clear weather was easier to classify for both the human and the proposed system.

6   Conclusions

We show the first time that also night time images can be used for estimating the visibility using a single camera. High dynamic range imaging can be used to capture images under varying lighting conditions and combined with Retinex filtering the illumination changes become almost negligible. The projection of filtered images on the horizontal direction captures the visibility degradation as function of distance which can be used for visibility classification.

The proposed features were used to train support vector machines for five visibility classes using data from two different sources. The image sets resulted in 85 % accuracy measured using F1-score. The confusion matrices show the HDR data set to be more reliable than the CCTV-based image set. While the sample size differences in the data sets leave room for speculation, HDR imaging can be considered a good way to capture images in varying lighting conditions.

It was shown that the method reached 85 % classification accuracy, which can be safely stated to be at least on the same level or better than the reference commercial light scattering instrument, that produced visually correct results in 77 % of the cases. The confusion matrices show that the proposed classification approach tends to misclassify some of the samples to the neighboring classes, which can be the indication of a data annotation problem, while the commercial instrument failed some times grossly. Also the classifiers were trained using the same training parameters for all the classes and it can be expected that some improvement can be obtained using more advanced machine learning approach.

The main criticism for the method, and in fact for most of the image based approaches, can be considered to be the need for training data. It can be very time consuming to gather the required training data that contains all the possible weather conditions. For future work, one might consider if computational methods can be used to generate simulated training data, so that only a few good visibility images in varying lighting conditions would be enough for the training.

Acknowledgment. The Finnish meteorological institute is acknowledged for both the image data and the reference data used in the HDR experiments. Nicolas Hautière and Éric Dumont are thanked for the Matilda database. Eliska Nyrönen is acknowledged for the implementation of the regression model based method described in [5].

References

1. Nayar, S., Narasimhan, S.: Vision in bad weather. In: The Proceedings of the Seventh IEEE International Conference on Computer Vision, vol. 2, pp. 820–827 (1999)
2. Hyslop, N.P.: Impaired visibility: the air pollution people see. Atmos. Environ. **43**, 182–195 (2009)
3. Tan, R.: Visibility in bad weather from a single image. In: IEEE Conference on Computer Vision and Pattern Recognition, pp. 1–8 (2008)
4. Graves, N., Newsam, S.: Visibility cameras: where and how to look. In: Proceedings of the 1st ACM International Workshop on Multimedia Analysis for Ecological Data, pp. 7–12 (2012)
5. Babari, R., Hautière, N., Dumont, È., Paparoditis, N., Misener, J.: Visibility monitoring using conventional roadside cameras emerging applications. Transp. Res. Part C Emerg. Technol. **22**, 17–28 (2012)
6. Du, K., Wang, K., Shi, P., Wang, Y.: Quantification of atmospheric visibility with dual digital cameras during daytime and nighttime. Atmos. Meas. Tech. **6**, 2121–2130 (2013)

7. Narasimhan, S., Nayar, S.: Shedding light on the weather. In: Proceedings of the IEEE Computer Society Conference on Computer Vision and Pattern Recognition, vol. 1, pp. I-665–I-672 (2003)

8. Yin, X.-C., He, T.-T., Hao, H.-W., Xu, X., Cao, X.-Z., Li, Q.: Learning based visibility measuring with images. In: Lu, B.-L., Zhang, L., Kwok, J. (eds.) ICONIP 2011, Part III. LNCS, vol. 7064, pp. 711–718. Springer, Heidelberg (2011)

9. Cadík, M., Wimmer, M., Neumann, L., Artusi, A.: Evaluation of HDR tone mapping methods using essential perceptual attributes. Comput. Graph. **32**, 330–349 (2008)

10. Daniel, J., Jobson, ZuR, Woodell, G.A.: A multiscale retinex for building the gap between color images and the human observation of scenes. IEEE Trans. Image Process. **6**, 965–976 (1997)

11. Mertens, T., Kautz, J., Van Reeth, F.: Exposure fusion: a simple and practical alternative to high dynamic range photography. Comput. Graph. Forum **28**, 161–171 (2009)

12. Debevec, P.E., Malik, J.: Recovering high dynamic range radiance maps from photographs. In: Proceedings of the 24th Annual Conference on Computer Graphics and Interactive Techniques. SIGGRAPH 1997, pp. 369–378 (1997)

13. Kirk, A.G., O'Brien, J.F.: Perceptually based tone mapping for low-light conditions. ACM Trans. Graph. **30**(4), 42:1–42:10 (2011). SIGGRAPH 2011

14. Barakat, N., Darcie, T., Hone, A.: The tradeoff between snr and exposure-set size in hdr imaging. In: 15th IEEE International Conference on Image Processing 2008, pp. 1848–1851 (2008)

15. Zuiderveld, K.: Contrast limited adaptive histograph equalization. Graphics Gems IV, pp. 474–485. Academic Press, Cambridge (1994)

16. Mitsunaga, T., Nayar, S.: Radiometric self calibration. IEEE Conference on Computer Vision and Pattern Recognition. **1**, 374–380 (1999)

Symmetric Feature Extraction
for Pose Neutralization

S.G. Charan[✉]

M S Ramaiah Institute of Technology, Bangalore, India
charansg333@gmail.com

Abstract. This paper proposes a method to neutralize the pose of facial databases. *Efficient use of the feature extractors* and its properties leads to the pose neutralization. Feature extractors discussed here are few transforms like Discrete Cosine Transform (DCT) and Discrete Fourier Transform (DFT). *Symmetric behavior* of transforms is the basis of the proposed method. *Modulo* based approach in extracting the features was found to provide better results than the conventional techniques for pose neutralization. Experiments are conducted on various benchmark facial databases mainly pose variant FERET and FEI which show the promising performance of the proposed method in neutralizing pose.

1 Introduction

Face Recognition (FR) is a vast and challenging domain. Identification or verification of the person from the digital image is face recognition. FR finds many applications in several fields such as security, authentication systems and surveillance. Extensive research work is going on due the large number of applications of this field.

Two important stages of FR system is training and testing. Training stage involves the learning of the faces of various personalities. Learning involves preprocessing, feature extraction and feature selection. Recognition stage involves all those processes of training stage and in addition it has classification stage.

All the practical FR systems are affected by the variance in illumination, pose, expression and background [1]. There are algorithms which specifically target few of the above mentioned variables [2–6]. Pose is the most challenging discrepancy in FR. View-specific eigenface [7,8] analysis as extraction method and neural network to recognize the personality was used by Fu Jie Huang in an attempt to neutralize pose [9]. Pose variant FR was also done by using a single example view by creating virtual views by MIT labs [10]. Facial symmetry [11] was used to neutralize pose in 2D FR [12]. Local regression was an effective solution to develop frontal pose from non-frontal pose and this method also tried to neutralize pose [13]. Texture based pose invariant techniques is discussed in Ref. [14].

Electronic supplementary material The online version of this chapter (doi:10. 1007/978-3-319-16634-6_22) contains supplementary material, which is available to authorized users.

© Springer International Publishing Switzerland 2015
C.V. Jawahar and S. Shan (Eds.): ACCV 2014 Workshops, Part III, LNCS 9010, pp. 290–305, 2015.
DOI: 10.1007/978-3-319-16634-6_22

Research in the field of FR is going on in both 2D and 3D field [15]. Spatial domain feature extraction, template matching is one of the very efficient methods for recognition [16]. There are many works in the principal component analysis method [17]. The extraction [18–20] methodology used is not so efficient to extract the required features. We try to improve the research gap here by effectively utilizing the available extractors. This paper proposes a novel method of extracting the features utilizing the *symmetric property* of the extractors.

The rest of the paper is organized as follows. In Sect. 2, we explain the problem and the proposed solution in brief. In Sect. 3, we discuss in detail analysis of the proposed extraction. Section 4 completely deals with the experiments conducted and analysis concluded. Section 5 summarizes about limitations of the proposed methodology of extraction and the future work to be done. The paper is concluded in Sect. 6.

2   Problem Definition and Solution

Pose neutralization is a challenging task in Face recognition. System must be able to recognize the person given his different poses. Pose variant FR is tackled using proposed methodology as mentioned below.

2.1   Symmetric Extractors for Pose Neutralization

The *symmetric properties* of the extractors is used so as the pose is neutralized. Face has vertical symmetry by its nature, the feature extractors which follow the principle of symmetry are capable of extracting the facial features more efficiently. *Modelling the transforms so as to make them symmetric is the proposed methodology. Modulo technique avails in achieving the symmetry.*

3   Proposed Extraction Technique

The extractors that we use here are few transforms like DCT and DFT. These two transforms are modelled so as to exhibit the symmetry property. We analyze DCT in first subsection and then DFT in second subsection. Symmetry behavior of these transforms can be seen when the proposed modulo technique is used. We discuss both in depth analysis and as well as a general proof for both the transforms.

3.1   DCT as Symmetric Extractor

Discrete Cosine Transforms [21–23] respond in a peculiar fashion for the symmetric input data. This peculiar behavior is very useful and is the basis of this paper. DCT is analyzed for an assumed particular set of data and as well as a general approach is given so as to figure out this property. The complete analysis of DCT equations are discussed in this section.

DCT is given by Eq. 1. This is for two dimensional approach. We are trying to find out a method which can make system invariant to original and mirrored sequences (symmetrical sequences). Symmetry property is exhibited by 1D DCT also and is given by Eq. 2, it is just a modification of the Eq. 1. The whole symmetry property of DCT is due to *cosine function* and is proved below.

$$B_{pq} = \alpha_p \alpha_q \sum_{m=0}^{M-1} \sum_{n=0}^{N-1} A_{mn} \cos \frac{\pi(2m+1)p}{2M} \cos \frac{\pi(2n+1)q}{2N} \tag{1}$$

$$0 \le p \le M - 1 \quad and \quad 0 \le q \le N - 1$$

$$\alpha_p = \left\{ \begin{array}{ll} \frac{1}{\sqrt{M}} & p = 0 \\ \sqrt{\frac{2}{M}} & 1 \le p \le M - 1 \end{array} \right\} \quad and \quad \alpha_q = \left\{ \begin{array}{ll} \frac{1}{\sqrt{N}} & q = 0 \\ \sqrt{\frac{2}{N}} & 1 \le q \le N - 1 \end{array} \right\}$$

$M : number\ of\ elements\ in\ the\ sequence\ counted\ \ column\ wise$

$N : \ number\ of\ elements\ in\ the\ sequence\ counted\ row\ wise$

$A_{mn} : \ Value\ of\ the\ element\ at\ the\ m^{th}\ column,\ n^{th}\ row.$

$$B_{p0} = \alpha_p \sum_{m=0}^{M-1} A_{m0} \cos \left[\frac{\pi(2m+1)p}{2M} \right] \tag{2}$$

General Proof. The whole analysis for various cases is performed mathematically to explain how DCT behaves to mirrored input and normal sequence and is solved completely in supplementary material. Let us prove it generally using the main DCT expression given by Eq. 3.

$$\begin{array}{l} B_{p0} \\ {\scriptstyle (original)} \end{array} = \alpha_p \sum_{j=0}^{M-1} x(j) cos \left[\frac{\pi(2j+1)p}{2M} \right] \tag{3}$$

$$\begin{array}{l} B_{p0} \\ {\scriptstyle (mirror)} \end{array} = \alpha_p \sum_{j=0}^{M-1} z(j) cos \left[\frac{\pi(2j+1)p}{2M} \right]$$

$$= \alpha_p \sum_{j=0}^{M-1} x(M-j-1) cos \left[\frac{\pi(2j+1)p}{2M} \right]$$

$$Substituting \quad s = M - j - 1$$

$$\begin{aligned}
B_{p0}_{(mirror)} &= \alpha_p \sum_{s=M-1}^{0} x(s)\cos\left[\frac{\pi(2(M-s-1)+1)p}{2M}\right] \\
&= \alpha_p \sum_{s=M-1}^{0} x(s)\cos\left[\frac{\pi(2M-2s-2+1)p}{2M}\right] \\
&= \alpha_p \sum_{s=0}^{M-1} x(s)\cos\left[p\pi - \frac{\pi(2s+1)p}{2M}\right] \\
&= \alpha_p \sum_{j=0}^{j-1} x(j)\cos\left[p\pi - \frac{\pi(2j+1)p}{2M}\right]
\end{aligned}$$

$$B_{p0}_{(mirror)} = \begin{cases} +\alpha_p \sum_{j=0}^{j-1} x(j)\cos\left[\frac{\pi(2j+1)p}{2M}\right] & p = even \\ -\alpha_p \sum_{j=0}^{j-1} x(j)\cos\left[\frac{\pi(2j+1)p}{2M}\right] & p = odd \end{cases}$$

$$B_{p0}_{(mirror)} = \begin{cases} + B_{p0}_{(original)} & p = even \\ - B_{p0}_{(original)} & p = odd \end{cases} \tag{4}$$

Important conclusion drawn from the above general proof is stated as in Eq. 4 and we can say this exists because of *cosine function*.

$$\therefore \left| B_{p0}_{mirror} \right| = \left| B_{p0}_{original} \right| \tag{5}$$

Equation 5 shows that using the modulo operation, the DCT output of original and mirrored sequence can be same. This proposed method to make the system blind to the original and mirrored sequence helps in pose invariance. System using the proposed methodology cannot differentiate between 2 different poses of the same person.

3.2   DFT as Symmetric Extractor

Discrete Fourier Transform [23,24] can also aid in pose invariance by exhibiting property of symmetry. Here also a general approach and the in depth analysis is done. Our aim is to prove that using some methodology the DFT output of original and mirrored sequence can be made equal. DFT is given by Eq. 6.

$$X(k) = \sum_{j=1}^{M} x(j) w_M^{(j-1)(k-1)} \tag{6}$$

$$Twiddle\ Factor = w_M = e^{(-2\pi i)/M} \quad and \quad 1 \le k \le M$$

General Proof. Various cases are considered and are analyzed mathematically in supplementary material. Here we generalize to draw an important conclusion. Equation 7 represents DFT expression for original sequence where as the next equation is for the mirrored sequence.

$$\underset{original}{X(k)} = \sum_{j=1}^{M} x(j) \left| \left(\frac{-2\pi i(j-1)(k-1)}{M} \right) \right.$$ (7)

$$\underset{mirror}{X(k)} = \sum_{j=1}^{M} z(j) \left| \left(\frac{-2\pi i(j-1)(k-1)}{M} \right) \right.$$

$$= \left\{ \sum_{j=1}^{M} z(j) \left| \left(\frac{-2\pi i(j-1)(k-1)}{M} \right) - \left(\frac{-2\pi i(M-1)(k-1)}{M} \right) \right. \right\}$$

$$\times \left| \left(\frac{-2\pi i(M-1)(k-1)}{M} \right) \right.$$

$$= \left\{ \sum_{j=1}^{M} z(j) \left| \frac{2\pi i(k-1)}{M}(M-j) \right. \right\} \times \left| \left(\frac{-2\pi i(M-1)(k-1)}{M} \right) \right.$$

$$= \left\{ \sum_{j=1}^{M} x(M-j+1) \left| \frac{2\pi i(k-1)}{M}(M-j) \right. \right\} \times \left| \left(\frac{-2\pi i(M-1)(k-1)}{M} \right) \right.$$

$$Substituting \ \ q = M - j + 1$$

$$\underset{mirror}{X(k)} = \left\{ \sum_{q=M}^{1} x(q) \left| \frac{2\pi i(k-1)}{M}(M-(M-q+1)) \right. \right\} \times \left| \left(\frac{-2\pi i(M-1)(k-1)}{M} \right) \right.$$

$$= \left\{ \sum_{q=1}^{M} x(q) \left| \frac{2\pi i(k-1)}{M}(q-1) \right. \right\} \times \left| \left(\frac{-2\pi i(M-1)(k-1)}{M} \right) \right.$$

$$= \left\{ \sum_{j=1}^{M} x(j) \left| \frac{2\pi i(k-1)(j-1)}{M} \right. \right\} \times \left| \left(\frac{-2\pi i(M-1)(k-1)}{M} \right) \right.$$

$$\left| \underset{mirror}{X(k)} \right| = \sum_{j=1}^{M} x(j) \left| \frac{2\pi i(k-1)(j-1)}{M} \right.$$

$$\left| X(k) \atop original \right| = \left| \sum_{j=1}^{M} x(j) \frac{-2\pi i(k-1)(j-1)}{M} \right|$$

$$\therefore \left| X(k) \atop mirror \right| = \left| X(k) \atop original \right| \tag{8}$$

$$Since \ \left| \frac{2\pi i(k-1)(j-1)}{M} \right| = \left| \frac{-2\pi i(k-1)(j-1)}{M} \right|$$

Thus it is proved that the modulo value of both the DFT outputs of original and mirrored sequences are equal and is shown in Eq. 8 . This symmetry property is due to the exponential term which is varied with 'j '.

This method of applying the modulo over the DFT resultant matrix is the proposed method which makes the system invariant to normal and mirror sequences.

4   Experiments

This section we shall look at in depth analysis and behavior of DCT and DFT. We here in this paper emphasize only on showing variance of Recognition rate by use of symmetry property of transforms, so we didn't try to add any other morphological techniques or classifiers. We also apply this to few benchmark datasets and prove that the proposed method is working as explained.

4.1   Analysis of Proposed Technique

An input data sequence and its mirror sequence is used and the extraction techniques are applied. The results due to DCT and DFT are discussed in following two subsections.

DCT as Extractor. Input sequence details and the corresponding output values are tabulated in Table 1. Graphs are plotted showing the symmetry of the DCT output (Fig. 1). We just assume a sequence of 10 elements (numbers). DCT is applied to the original sequence and the results are tabulated.

The proposed method result is also tabulated in the same Table 1. On careful observation of the two results, we see that system by using the proposed method becomes invariant to the mirrored and original sequence.

DFT as Extractor. Input sequence details and the corresponding output values are tabulated in Table 2. This is the same input taken to analyse DCT also. Graphs are plotted showing the symmetry of the DFT output (Fig. 2). *We prefer to express solution in modulo and angle form since expressing it in complex co-ordinate form doesn't show the symmetry.*

The proposed method result is also tabulated in Table 2. System by using the proposed method becomes invariant to the mirrored and original sequence as shown in Table 2.

4.2   Face Recognition as an Application

Experiments have been performed on various standard benchmark databases to exemplify the robustness of the proposed methodologies. The proposed FR system is designed to handle various pose present in the benchmark database. If the total number of test images including all classes is 'b', and if 'a' number of images are correctly identified, then the Recognition Rate (RR) is calculated as defined in Eq. 9.

$$Recognition\ Rate = \frac{a}{b} \tag{9}$$

The RR is used as a metric to demonstrate the performance of the experiments performed. We have made use of 4 benchmark face databases. These databases are: ORL, FERET, Head Pose and FEI. Experiments on ORL and Head Pose dataset is included in supplementary material.

Customized databases are created for each of the databases for the following four reasons. First, it covers almost all the expected changes such as pose, expression and illumination variations. Second, to reduce the burden of resource utilization. Large number of images may not be required as many images are redundant, in the sense that the customized database contains the variations possessed by all those images. Third, the various databases contain different number of images per subject, so it will be tedious job to fix the training set in order to have uniform contribution of RR from each subject. Fourth, to compare our results with the standard proven results (Fig. 4).

All the experiments have been performed on a computer with specifications Intel Core^{TM}i7 2.2 GHz CPU and 8 GB RAM on MATLAB$^{\circledR}$ 2012b [25]. Let the total number of subjects in the customized database be T_{sub} and total images

Table 1. This explains numerically how proposed technique is suitable for flipping invariance. Original data, its DCT result, DCT result of the mirrored sequence input and also final optimal proposed solution can be seen here. Taking modulo for the results obtained, we make the proposed solution optimal for flipping invariance. The related graphs are shown in Fig. 1.

Original Data	DCT Result Original	DCT Result Original	**Proposed Method**
12	217.88	217.88	**21.88**
25	−113.82	113.82	**113.82**
40	3.14	3.14	**3.14**
48	−17.80	17.80	**17.80**
62	6.99	6.99	**6.99**
77	−12.96	12.96	**12.96**
88	2.53	2.53	**2.53**
92	−0.69	0.69	**0.69**
110	0.51	0.51	**0.51**
135	1.52	−1.52	**1.52**

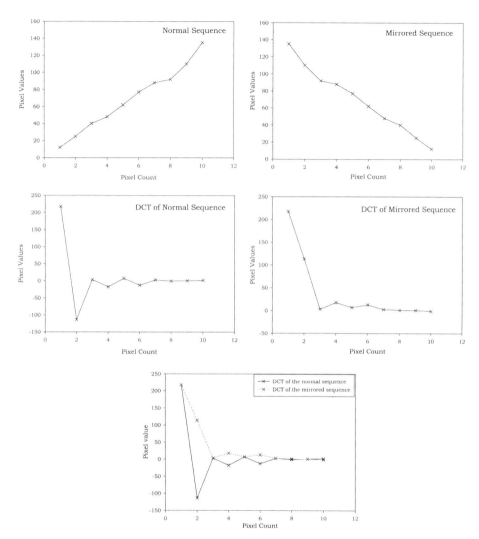

Fig. 1. Figure depicts the response of DCT to original and mirrored sequence. The original (normal) sequence and its corresponding mirrored sequence is seen in first row. DCT output results of normal and mirrored sequence is shown in second row. We cannot make out the symmetry by looking at 2 different output plots, therefore we have plotted both the DCT results in a single figure in third row. Considering the zero pixel value line as base we can see the symmetry of the plots, which is exploited here and is the proposed method. The Pixel values of all the graphs are tabulated in Table 1.

per subject be T_{img}. Let the total number of *random* images taken for training be T_{tr} per subject. Then the total number of testing images are all the remaining ones given by $T_{img} - T_{tr}$ per subject. Each experiment is conducted for 20 trials and the results are averaged to *reduce the randomness introduced due to different training sets.*

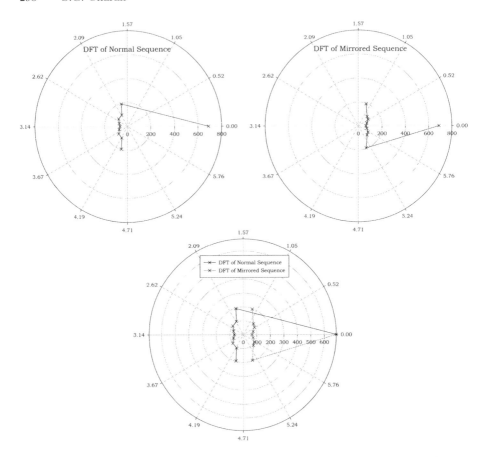

Fig. 2. The input sequence of Table 2 is applied with DFT. The result obtained from DFT for original and mirrored sequence are complex and we use polar plots to show symmetry. First row shows these results. To make out the symmetry in the result obtained, we combine both plots into a single plot and is shown in second row. One can see that all the corresponding points of both the output sequences lie on the same radius circle (modulo circle). This is the proposed method where we choose the radius or the modulo which makes the system flip data invariant.

Color FERET database FERET database [26–29] which stands for Facial Recognition Technology database mainly accounts for pose, scale and costume variations. This is the challenging database to illustrate various poses. Images in this database consists of different sizes. We have chosen the 'smaller' type images whose size is 256×384. The customized database consists of total of 35 classes and 20 images per each class, accounting a total of 700 images. Each class includes 2 sets of fa, fb, pl, hl, pr, hr, qr, ql images and 4 random images totalling 20 images [30].

This experiment is performed to explain the importance of the proposed FR system using proposed methodology. The image is converted to gray scale.

Table 2. This explains numerically how proposed technique is suitable for flipping invariance for DFT. Neglecting the angle and only choosing the magnitude, we make the proposed solution optimal for flipping invariance. The graphs are in Fig. 2.

	DFT Output Mirrored		DFT Output Original		Proposed Method
Original	Modulo	Angle (radians)	Modulo	Angle (radians)	Modulo
12	689.00	0	689.00	0	**689.00**
25	195.89	−1.22	195.89	1.84	**195.89**
40	107.42	−0.79	107.42	2.05	**107.42**
48	98.43	−0.57	98.43	2.46	**98.43**
62	76.27	−0.29	76.27	2.81	**76.27**
77	65.00	0	65.00	3.14	**65.00**
88	76.27	0.29	76.27	−2.81	**76.27**
92	98.43	0.57	98.43	−2.46	**98.43**
110	107.42	0.79	107.42	−2.05	**107.42**
135	195.89	1.22	195.89	−1.85	**195.89**

This image is reduced by a factor of 2 using Gaussian reduction technique. This converts the image from original size of 256×384 to 64×96. Now extraction is carried out using conventional DCT, DFT (next time the extraction is via proposed DCT, DFT techniques). The feature matrix is then reshaped to one dimension using raster scan [31]. Then the feature vector is sent to well known evolutionary algorithm Binary Particle Swarm Optimisation (BPSO) [31–38] for feature selection (to reduce the number of features in computation). The number of selected features after this stage comes out to be around 60 % of the input features. This is stored in feature gallery. Then the feature vector of the testing image is extracted in the same way as that of the training images. The extracted

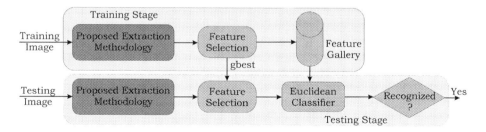

Fig. 3. Block diagram of proposed FR system. Our aim is to show the pose invariance, so we start directly with extraction. Extraction by DCT and DFT. Here we use both conventional method and proposed method to obtain the RR. After extraction its feature selection using BPSO. Finally the classification. Our main intention is to show the proposed methodology is better than the conventional for pose invariance,hence the image intensity correction, normalisation and all other precursors are left out.

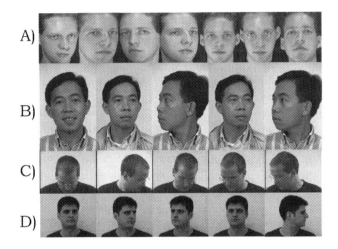

Fig. 4. Sample images of (A) ORL dataset (B) FERET dataset (C) HP dataset (D) FEI dataset.

Table 3. Comparison of RR for different ratios of FERET.

Ratio	DCT + DFT [30]	Proposed DCT method	Proposed DFT method
8:12	80.23	**83.55**	**83.97**
12:8	88.07	**90.04**	**89.96**

testing gallery is multiplied with the g_{best} of the BPSO (obtained during the training time). This becomes the test image gallery. A simple Euclidean distance is used for the classification. This process is clearly shown in Fig. 3.

The whole process is repeated but this time using the proposed methodology of extraction. Here we choose the modulus value of the extracted vector before it moves on to next pipelines.

Figures 5 and 6 show the proposed techniques. The image and its conventional DCT is shown and further due to proposed methodology, the new plot of only

Table 4. Cross correlation for two extreme poses is noted for conventional and proposed method. Higher cross-correlation is better. This experiment is done using 2 extreme profiles of FERET.

Method	Cross-Correlation Value
Conventional DCT	0.8723
Conventional DFT real part	0.9564
Conventional DFT imaginary part	−0.0106
Proposed Modulo DCT	**0.9798**
Proposed Modulo DFT	**0.9851**

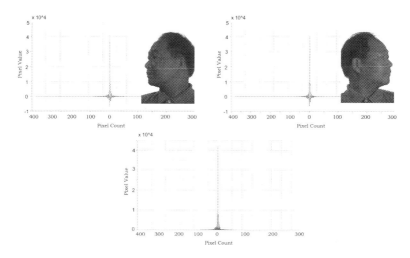

Fig. 5. Conventional DCT outputs of 2 extreme poses is shown in first row. This will have both negative and positive components along pixel values. Proposed modulo technique produces only the positive part which is exactly same for both the profile images and shown in second row. This means that for two different symmetric poses the system produces same features indicating pose invariance.

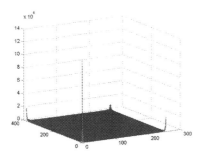

Fig. 6. Proposed DFT Output for an facial image of FERET dataset. The value on 'y' axis strats from zero indicating we have chosen only magnitude i.e., modulo radius and neglected the angle. DFT basically gives the frequency spectrum as shown in this figure.

positive values are shown. DFT gives both radius and angle, but due to selection of radius Fig. 6 is obtained.

Proposed application of modulus to the outputs of DCT and DFT has made the system invariant to the symmetric poses (if one pose is considered as original image, then other is its mirrored image). The above proved concept of modulo application is thus the reason for the pose neutralisation.

Consider cross correlation as a metric to measure the factor of pose neutralization. Cross-correlation is taken between feature vectors of symmetric poses of the same personality. This cross correlation is measured for the extracted vectors

Table 5. Results for different ratio for all the databases. This includes RR from common conventional approach and the proposed approach. Both DFT, DCT results are tabulated in a single table so as it is very easy to compare.

	DCT based		DFT based	
FERET	Normal	**Proposed Method**	Normal	**Proposed Method**
Ratio	Avg. RR (%)	Avg.RR (%)	Avg. RR (%)	Avg.RR (%)
4:16	56.91	**65.78**	22.58	**69.72**
6:14	67.83	**76.14**	25.67	**78.43**
8:12	75.47	**83.55**	27.07	**83.97**
10:10	78.85	**85.89**	29.80	**86.43**
12:8	84.53	**90.04**	31.79	**89.96**

	DCT based		DFT based	
FEI	Normal	**Proposed Method**	Normal	**Proposed Method**
Ratio	Avg. RR (%)	Avg.RR (%)	Avg. RR (%)	Avg.RR (%)
4:6	84.72	**88.57**	35.00	**84.81**
5:5	88.40	**91.83**	34.46	**87.89**
6:4	91.50	**92.29**	38.79	**90.15**
7:3	92.67	**95.14**	40.28	**92.24**
8:2	92.15	**95.15**	39.43	**92.29**
9:1	92.86	**97.15**	38.57	**93.72**

from conventional and proposed methods for both DFT and DCT. These are tabulated in Table 4. From this table it is very much clear that the proposed technique increases the correlation between the symmetric poses and proposed method is the possible explanation for pose invariance. Final result is '0.9851' not '1'. 'One' would have been the result if only mirror images are fed. Cross correlation between the full frontal pose and extreme left profile image will increase from 0.6236 to 0.8942. This shows that not only mirror images are recognized but also other poses. The similar center line features are enhanced.

Comparison of RR of FERET with others is done and is shown in Table 3. **FEI database.** FEI dataset [39] is also one of the benchmark facial database with pose variance like FERET database. This dataset contains 10 different pose of facial images. We have about 35 subjects in this class. As this has pose variant images, the proposed methodology increases RR which is as shown in Table 5. The size of the image initially is 640×480. We apply three stages of Gaussian reduction to decrease the size and then the process as shown in Fig. 3.

5   Limitations and Future Work

This paper studies the symmetric property which helps for pose neutralization. There can be various properties of the extractors which has to be explored for various applications. We only discussed two transforms here, study has to be done on *various extractors and their properties*.

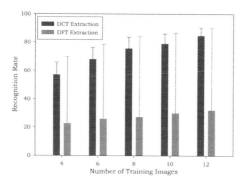

Fig. 7. The variation of use of proposed method over conventional method for FERET dataset is seen. The filled bars are the RR obtained for the conventional methods. The line extension is the increase provided by the proposed method. Proposed method increases RR to a greater extent when DFT is used.

We have not concentrated on the initial pre-processing steps which will increase the recognition rates, as our concern was to prove the efficiency of the proposed method. Various pre-processing can be applied to obtain high recognition rates (Fig. 7).

Future work is in extending this algorithm for various databases and real-time video applications. There are various properties like this symmetry property which can help in eliminating various discrepancies. *Efficient use of properties has to be done to improve the productivity of extractors* (Fig. 5).

6   Conclusion

Symmetry behavior of DCT and DFT are analyzed. General approach and specific approach both has been dealt here. We model the symmetric behavior of DCT and DFT for pose neutralization application.

High recognition rates are achieved for pose variant FERET and FEI dataset by using the proposed methodology and these results outperform the conventional methods.

Efficient use of properties DCT and DFT is achieved here. Application of *modulus* over DCT and DFT outputs proved to neutralize pose.

We here try to fill the research gap in the applications of various properties of feature extractors for efficient extraction.

References

1. Zhao, P.P.W., Chellappa, R., Rosenfeld, A.: Face recognition: a literature survey. ACM Comput. Surv. **35**, 399–458 (2003)
2. Zhang, X., Gao, Y.: Face recognition across pose: a review. Pattern Recogn. **42**, 2876–2896 (2009)

3. De Vel, O., Aeberhard, S.: Line-based face recognition under varying pose. IEEE Trans. Pattern Anal. Mach. Intell. **21**, 1081–1088 (1999)
4. Belhumeur, P.N., Hespanha, J.P., Kriegman, D.J.: Eigenfaces vs. fisherfaces: recognition using class specific linear projection. IEEE Trans. Pattern Anal. Mach. Intell. **19**, 711–720 (1997)
5. Bartlett, M.S., Movellan, J.R., Sejnowski, T.J.: Face recognition by independent component analysis. IEEE Trans. Neural Netw. **13**, 1450–1464 (2002)
6. Wiskott, L., Fellous, J.-M., Kuiger, N., Von Der Malsburg, C.: Face recognition by elastic bunch graph matching. IEEE Trans. Pattern Anal. Mach. Intell. **19**, 775–779 (1997)
7. Turk, Matthew, Pentland, Alex: Eigenfaces for recognition. J. Cogn. Neurosci. **3**(1), 71–86 (1991)
8. Agarwal, N.J.M., Agrawal, H., Kumar, M.: Face recognition using principle component analysis, eigenface and neural network. In: Proceedings of the 2010 International Conference on Signal Acquisition and Processing, ICSAP 2010, pp. 310–314. IEEE Computer Society (2010)
9. Huang, F.J.: Pose invariant face recognition. In: Fourth IEEE International Conference on Automatic Face and Gesture Recognition (2000)
10. Beymer, D., Poggio, T.: Face recognition from one example view. In: Fifth International Conference Computer Vision (1995)
11. Troje, N.F., Blthoff, H.H.: How is bilateral symmetry of human faces used for recognition of novel views? Vis. Res. **38**(1), 79–89 (1998)
12. Gonzlez-Jimnez, D., Alba-Castro, J.L.: Toward pose-invariant 2-d face recognition through point distribution models and facial symmetry. IEEE Trans. Inf. Forensics Secur. **2**(3), 413–429 (2007)
13. Gonzlez-Jimnez, D., Alba-Castro, J.L.: Locally linear regression for pose-invariant face recognition. IEEE Trans. Image Process. **16**(7), 1716–1725 (2007)
14. Troje, N.F., Blthoff, H.H.: Face recognition under varying poses: the role of texture and shape. Vis. Res. **36**(12), 1761–1771 (1996)
15. Abate, A.F., Nappi, M., Riccio, D., Sabatino, G.: 2D and 3D face recognition: a survey. Pattern Recogn. Lett. **28**, 1885–1906 (2007)
16. Brunelli, R., Poggio, T.: Face recognition: features versus templates. IEEE Trans. Pattern Anal. Mach. Intell. **15**, 1042–1052 (1993)
17. Chen, S., Zhu, Y.: Subpattern-based principal component analysis. Pattern Recogn. **37**, 1081–1083 (2004)
18. Nixon, M., Aquado, A.S.: Feature Extraction & Image Processing, 2nd edn. Academic Press, New York (2008)
19. Chen, S., Zhang, D., Zhou, Z.H.: Enhanced (pc) 2 a for face recognition with one training image per person. Pattern Recogn. Lett. **25**, 1173–1181 (2004a)
20. Chen, S., Liu, J., Zhou, Z.H.: Making flda applicable to face recognition with one sample per person. Pattern Recogn. **37**, 1553–1555 (2004b)
21. W.L. Scott II, "Block-level discrete cosine transform coefficients for autonomic face recognition. Ph.D. thesis (2003)
22. Hafed, Z.M., Levine, M.D.: Face recognition using the discrete cosine transform. Int. J. Comput. Vis. **43**, 167–188 (2001)
23. Samra, A.S., Allah, S.E.T.G., Ibrahim, R.M.: Face recognition using wavelet transform, fast fourier transform and discrete cosine transform. In: Proceedings 46th IEEE International Midwest Symposium Circuits and Systems (MWSCAS 2003), vol. 1, pp. 272–275 (2003)
24. Bracewell, R.N.: The Fourier Transform and its Applications. McGraw-Hill, New York (1986)

25. MATLAB, www.mathworks.com

26. Color FERET. http://www.nist.gov/itl/iad/ig/colorferet.cfm

27. Phillips, P.J., Moon, H., Rizvi, S.A., Rauss, P.J.: The feret evaluation methodology for face-recognition algorithms. IEEE Trans. Pattern Anal. Mach. Intell. **22**, 1090–1104 (2000)

28. Phillips, P.J., Wechslerb, H., Huang, J., Rauss, P.J.: The feret database and evaluation procedure for face recognition algorithms. Image Vis. Comput. **16**, 295–306 (1998)

29. Zhu, Y., Liu, J., Chen, S.: Semi-random subspace method for face recognition. Image Vis. Comput. **27**, 1358–1370 (2009)

30. Deepa, G., Keerthi, R., Meghana, N., Manikantan, K.: Face recognition using spectrum-based feature extraction. Appl. Soft Comput. **12**, 2913–2923 (2012)

31. Ramadan, R.M., Abdel-Kader, R.F.: Face recognition using particle swarm optimization-based selected features. Int. J. Signal Process. Image Process. Pattern Recogn. **2**, 51–65 (2009)

32. Engelbrecht, A.P.: Fundamentals of Computational Swarm Intelligence. Wiley, Chichester (2005)

33. Fan, X., Verma, B.: Face recognition: a new feature selection and classification technique. In: Proceedings of the 7th Asia-Pacific Conference on Complex Systems (2004)

34. Kanan,H.R., Faez, K., Hosseinzadeh, M.: Face recognition system using ant colony optimization-based selected features. In: Proceedings IEEE Symposium Computational Intelligence in Security and Defense Applications (CISDA 2007) pp. 57–62 (2007)

35. Eberhart, R., Kennedy, J.: A new optimizer using particles swarm theory. In: Proceedings 6th International Symposium Micro Machine and Human Science pp. 39–43 (1995)

36. Kennedy, J., Eberhart, R.: A discrete binary version of the particle swarm algorithm. Proc. IEEE Int. Conf. Syst. Man Cybern. **5**, 4104–4108 (1997)

37. Liu, C., Wechsler, H.: Evolutionary pursuit and its application to face recognition. IEEE Trans. Pattern Anal. Mach. Intell. **22**, 570–582 (2000)

38. Narendra, P., Fukunage, K.: A branch and bound algorithm for feature subset selection. IEEE Trans. Comput. **6**, 917–922 (1977)

39. FEI Database. http://fei.edu.br/cet/facedatabase.html

3D Laplacian Pyramid Signature

Kaimo Hu and Yi Fang$^{(\boxtimes)}$

Electrical and Computer Engineering, New York University in Abu Dhabi,
Abu Dhabi, UAE
{kaimo.hu,yfang}@nyu.edu

Abstract. We introduce a simple and effective point descriptor, called
3D Laplacian Pyramid Signature (3DLPS), by extending and adapting
the Laplacian Pyramid defined in 2D images to 3D shapes. The signature is represented as a high-dimensional feature vector recording the
magnitudes of mean curvatures, which are captured through sequentially
applying Laplacian of Gaussian (LOG) operators on each vertex of 3D
shapes. We show that 3DLPS organizes the intrinsic geometry information concisely, while possessing high sensitivity and specificity. Compared
with existing point signatures, 3DLPS is robust and easy to compute, yet
captures enough information embedded in the shape. We describe how
3DLPS may potentially benefit the applications involved in shape analysis, and especially demonstrate how to incorporate it in point correspondence detection, best view selection and automatic mesh segmentation.
Experiments across a collection of shapes have verified its effectiveness.

1 Introduction

Signatures play an important role in shape analysis. On one hand, they are
concise presentations of the model and are easily commeasurable; on the other
hand, a good signature preserves much geometric information of the shape.

Current signatures defined on 3D shapes can be classified into *shape signature*
(also referred to as shape descriptor) and *point signature*. Shape signatures are
usually defined on a whole 3D model or a partial model, while point signatures
are defined on each vertex of the mesh and capture the local or global shape
characteristics from the perspective of that vertex. Since point signatures possess
more local information, and sometimes can be easily converted to shape signature
by calculating their statistic distributions [1], they have wider applications than
shape signatures, such as point clouds registration, mesh saliency detection and
mesh segmentation.

Early work on designing point signatures mainly focused on characterising
the spatial geometric information of a shape [2–5]. While performing well in
capturing the spacial geometrical attributes of models ranging from local to
global, they are incompetent to describe the intrinsic frequency domain information embedded in a 3D model, which is sometimes essential for model analysis
and understanding. Recently, some new point signatures have been proposed to
reflect the frequency domain information [6–9]. However, they usually suffer from
heavy calculations (see the related work in Sect. 2 for detail).

© Springer International Publishing Switzerland 2015
C.V. Jawahar and S. Shan (Eds.): ACCV 2014 Workshops, Part III, LNCS 9010, pp. 306–321, 2015.
DOI: 10.1007/978-3-319-16634-6_23

Inspired by the successful applications of Laplacian Pyramid operator applied in image processing [10], we introduce a simple yet effective point descriptor: 3D Laplacian Pyramid Signature (3DLPS), based on the evaluation of Laplacian Pyramid after applying a series of Laplacian of Gaussian (LOG) operators at each vertex on a 3D model. Every vertex on the meshed surface is then characterized with a high-dimensional feature vector that measures the magnitudes of mean curvatures, which are captured from these sequentially applied LOG operators. We show that 3DLPS has the following desirable properties: (1) it organizes the intrinsic frequency information of a shape in an efficient, multi-scale way; (2) it shows high sensitivity and specificity; and (3) it is insensitive to articulated objects. Due to these nice properties, 3DLPS has the potential to benefit many applications involved in shape analysis, including point correspondence detection, best view selection and automatic mesh segmentation. We have illustrated these across a collection of shapes. In summary, our main contributions are:

– The 3D Laplacian Pyramid Signature is introduced to characterize the geometric information embedded in the shape in a concise way.
– The desirable properties of 3DLPS are explained, which ensure the effectiveness of potential applications in shape analysis.
– The applications of point correspondence detection, best view selection and automatic mesh segmentation based on 3DLPS are evaluated.

2   Related Work

Signatures have been extensively studied across areas as diverse as computer vision, structure biology and others. In this paper, we mainly review the related works that are designed as point signatures in the context of shape analysis. Furthermore, as is related to 3DLPS, we also briefly review the applications of the Laplacian operator and pyramid defined on 2D images and 3D shapes.

Point signatures on 3D shapes. Early work on designing point signatures is based on spatial domain. To make signatures robust against rigid transformation, a common strategy is to summarize the shape distribution in neighborhoods of a point [2]. For instance, the spin image method [3] constructs a 2D histogram that encodes the density of oriented points, and the shape context [4,5,11] captures the distribution over distances and angles of all other points on the shape according to the current point. While these signatures are widely used in spacial domain, they do not catch the frequency domain information, which are sometimes essential for shape analysis and understanding.

Recent efforts to track the problem of a robust shape signature are diffusion based approaches [6,7,12,13]. This process provides a natural notion of scale to describe the shape around a point. Rustamov [6] proposed the global point signature (GPS), which shows that the eigenfunctions of the Laplace-Beltrami nicely characterize the geometric features for points. Sun et al. [7] defined the heat kernel signature (HKS) by restricting the well-known heat kernel to the temporal domain. Dey et al. [8] merged the concept of HKS with the persistent homology, and designed a pose-oblivious algorithm for partial shape matching. Aubry et al. [9] proposed the Wave Kernel Signature (WKS), which represents the average probability of measuring a quantum mechanical particle at a specific location. All these methods somewhat catches the frequency domain information through a multi-scale way. However, they suffer from a computational challenge of estimating an eigen-decomposition of a huge Laplacian matrix. On the contrary, 3DLPS does not require the calculation of eigenvalues and eigenvectors, and thus is more robust and easier to compute, while still capturing the frequency domain information for shape analysis.

Another category of approaches for defining multi-scale point signatures is to capture the features of points at shapes that are resulted from geometry processing operations. For example, Li and Guskov [14] first obtained a series of increasingly smoothed version of a given shape, and then constructed point signatures for features found at each smoothed version of the shape. Manay et al.'s integral invariant signature [15] also falls into this category. 3DLPS is somewhat similar to this class of signatures. However, we use the Laplacian Pyramid operators, which concentrate more on the frequency geometric information.

Applications of Laplacian-beltrami and Pyramid. The discrete versions of Laplace and Laplace-Beltrami operators, both referred to as Laplacians, are widely used in image processing [16] and geometry processing [17]. To mention a few, Taubin [18] used the graph Laplacian for surface fairing; Ni et al. [19] used different weight of Laplacians to control the number of critical points; Dong et al. [20] described an approach to the quadrangulation of manifold meshes using Laplacian eigenfunctions; Reuter et al. [21] investigated the discrete Laplace-Beltrami operators for shape segmentation.

Pyramid is a type of multi-scale signal representation developed by the computer vision and image processing communities, in which a signal or an image is subject to repeated smoothing and subsampling [22]. Burt and Adelson [10] applied the Laplacian operator and Gaussian filter in the smoothing step, and presented the Laplacian Pyramid operator for compact image coding and edge enhancement. Kobbelt et al. [23] and Guskov et al. [24] introduced the ideas of pyramid to 3D shape processing for mesh modeling, smoothing and sampling.

Though both the Laplacian operator and the ideas of Pyramid have been successfully applied in image and mesh processing, we found little work has been addressed on deriving point signatures based on Laplacian Pyramid. Since Laplacian Pyramid can be regarded as a spectral decomposition and compression of 3D shapes, it is promising to apply it for capturing the frequency domain information embedded in the shape, which would provide an alternative perspective for shape analysis. This is the main motivation of our work.

3   3D Laplacian Pyramid Signature

Similar to real-world digital images, 3D shapes are in general both scale-variant and highly nonstationary in space. Inspired by the successful applications of

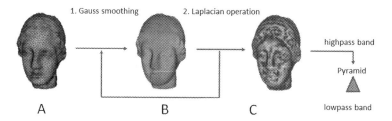

Fig. 1. Pipeline of 3DLPS extraction. The color ramps in (B) and (C) indicate the magnitudes of δ-coordinates od the models before and after the corresponding operations.

Laplacian Pyramid decomposition in images, we define the Laplacian Pyramid operator on 3D surface shapes. Since the Laplacian operator filters the high frequency information in frequency domain, and the Pyramid decomposition tracks 3D shapes in different frequencies, 3DLPS is supposed to be capable of capturing the intrinsic properties of a point on different frequencies with respect to the shape. In this section, we first give the definition and construction algorithm of 3DLPS, and then explain its desirable attributes for shape analysis.

3.1 Definition and Construction

Laplacian Matrix on Mesh. Let $\mathscr{M} = (V, E, F)$ be a triangular mesh with n vertices. For each vertex v_i, the δ-coordinates is defined as the difference between the absolute coordinates of v_i and the weighted average of its immediate neighbors in $\mathscr{M}$ [17],

$$\delta_i = (\delta_i^{(x)}, \delta_i^{(y)}, \delta_i^{(z)}) = v_i - \frac{\sum_{j \in N(i)} w_{ij} v_j}{\sum_{j \in N(i)} w_{ij}}, \tag{1}$$

where $N(i) = \{j | (i, j) \in E\}$. To better approximate the mean-curvature normals using δ-coordinates, we employ the "cotangent weights" $w_{ij} = (\cot \alpha_{ij} + \cot \beta_{ij})/(2 * |\Omega_i|)$ as proposed in [25], where $|\Omega_i|$ is the size of the Voronoi cell of v_i, and α_{ij}, β_{ij} are the two angles opposite to edge (i, j). Finally, the normalized Laplacian Matrix L is defined as

$$L_{ij} = \begin{cases} 1 & \text{if } i = j \\ -w_{ij}/\sum_{j \in N(i)} w_{ij} & \text{if } (i, j) \in E. \\ 0 & \text{otherwise} \end{cases} \tag{2}$$

Laplacian Pyramid on Vertices. Similar to the Laplacian Pyramid decomposition applied in images, we decompose the Cartesian coordinates on 3D shapes into "highpass" bands and a "lowpass" band. This is achieved by iteratively applying the Laplacian Matrix on the Cartesian coordinates of $\mathscr{M}$. Let $A = D - L$, where D is the unit diagonal matrix, then the decomposition of the Cartesian coordinates on 3D shapes can be represented as

$$D \cdot \mathbf{C} = A \cdot \mathbf{C} + L \cdot \mathbf{C}, \tag{3}$$

where $\mathbf{C}$ is a $n \times 3$ matrix representing the Cartesian coordinates of the vertices. Here $A \cdot \mathbf{C}$ is the "lowpass" band, and $L \cdot \mathbf{C}$ is the "highpass" band at the current iteration. At each iteration, the "highpass" band $L \cdot \mathbf{C}$ is concatenated to form the vertex coordinates differences, while the "lowpass" band $A \cdot \mathbf{C}$ is passed down as the input of the next iteration (as shown in Fig. 1).

Laplacian of Gaussian on 3D Surfaces. Triangular meshes are discrete approximations of continuous surfaces defined on $\mathbb{R}^3$. Hence, directly applying the Laplacian Pyramid operator may cause high noise due to the mesh irregularity. To reduce its sensibility to the discrete approximation as well as the noise, we apply the Gaussian smoothing filter each time before performing Laplacian operator. For Vertex v_i, the Cartesian coordinate after Gaussian smoothing is

$$C(v_i) = \sum_{v_j \in V, ||v_i - v_j|| < 3\sigma} \frac{1}{2\pi\sigma^2} \exp(-\frac{||v_i - v_j||^2}{2\sigma^2}), \qquad (4)$$

where σ is the standard deviation of Gaussian distribution, and $||v_i - v_j||$ represents the distance between v_i and v_j. We use the Euclidean distance in our experiments because it gives better results than geodesic distance and is easy to compute [26]. Note that if $||v_i - v_j|| \geq 3\sigma$, the values become very small in Gaussian distribution, we only consider a small neighborhood in which all the vertices has Euclidean distances smaller than 3σ with v_i.

Algorithm 1. Laplacian on 3D surface mesh

Input: m: the maximum iterations;
      $\mathcal{M}$: the input surface mesh.
Output: the concatenated "highpass" bands $\mathcal{C}$.

Initialize $\mathcal{C}$ as a null matrix;
for $i \leftarrow 1$ *to* m **do**
    | Apply Gaussian smoothing on $\mathcal{M}$ using Equ. 4;
    | Decompose the Cartesian coordinates $\mathbf{C}$ on $\mathcal{M}$ into "highpass" band $L \cdot \mathbf{C}$
    | and "lowpass" band $A \cdot \mathbf{C}$, as formulated in Equ. 3;
    | $\mathcal{C} \leftarrow [\mathcal{C}, L \cdot \mathbf{C}]$;   ▷ Concatenate "highpass" band
    | $\mathcal{M} \leftarrow A \cdot \mathbf{C}$.   ▷ Update $\mathcal{M}$ as "lowpass" band
end

The Laplacian Pyramid operation is finally shown in Algorithm 1. Given a 3D mesh $\mathcal{M}$, we repeatedly apply Gaussian smoothing and Laplacian decomposition. After each decomposition, we concatenate the "highpass" band to the output, and update the 3D surface mesh as the "lowpass" band. The final output is the concatenation of v's δ-coordinates after all iterations.

Construction of 3D Point Signature. The concatenation of δ-coordinates itself is able to characterize the mean curvatures of vertices on the surfaces. However, it contains redundant information. To make the 3DLPS efficient for shape analysis, we covert the concatenation of δ-coordinates into a compact and concise representation. Suppose the concatenation of differential vertex coordinates $\mathscr{C}$ has the form $\mathscr{C} = \{\delta_1, \delta_2, ...\delta_m\}$, then we define the final 3DLPS as $S = \{|\delta_1|, |\delta_2|, ...|\delta_m|\}$, where $|\delta|$ is the length of the vector δ.

3.2 Attributes of 3D Laplacian Pyramid Signature

Inherited from the properties of Laplacian Pyramid operator defined on 2D images, 3DLPS exhibits some nice attributes that are desirable for 3D shape analysis. In the following, we will explain them in detail.

Intrinsic and multi-frequency. By intrinsic and multi-frequency, we mean that 3DLPS captures the intrinsic information, and represents them in a multi-frequency way.

Laplacian operator is a smoothing schema, where the high frequency information tends to be smoothed out first, and the low frequency information will be retained for longer time. It is well known from differential geometry that [17]

$$\lim_{|\gamma| \to 0} \frac{1}{|\gamma|} \int_{v \in \gamma} (v_i - v) dl(v) = -H(v_i)\mathbf{n}_i, \tag{5}$$

where $H(v_i)$ is the mean curvature at v_i and $\mathbf{n}_i$ is the surface normal at v_i. Therefore, the magnitude approximates a quantity proportional to the local mean curvature [18]. This means that at each iteration in Algorithm 1, the "highpass" band $L \cdot \mathbf{C}$ encapsulates the local mean curvature normal at each vertex.

We apply the "cotangent weights" schema to approximate the mean-curvature normal. These geometry-dependent weights lead to $L \cdot \mathbf{C}$ with normal components only, rather than encoding the tangential components, which may be non-zero on planar 1-rings. This makes our 3DLPS approximately captures the δ−differences along the normal component, and only captures the local frequency geometry information rather than mesh regularities.

With the increase of iterations, the high frequencies will be smoothed out sequentially, and the lower frequencies will be decomposed into $L \cdot \mathbf{C}$. Therefore, our 3DLPS represents the decreasing frequency information of the vertex. Figure 2 illustrates the affections of Laplacian Pyramid operations on the homer model. We see that at the beginning, the high frequency parts (nose, fingers and mouth) are smoothed out. With the increase of iterations, the arms and legs shrink tenderly. Since the belly is in low frequency, it changes placidly during the Laplacian operations. Our 3DlPS captures all these information concisely.

Sensitivity and specificity. By sensitivity and specificity, we mean 3DLPS shows high discrimination among different points, and share high similarity among analogous points with respect to the frequencies embedded in the shape.

The sensitivity and specificity of the 3DLPS can be deduced from the attributes of intrinsic and multi-frequency. Since 3DLPS captures the features at different frequencies, if two points are embedded with different frequency information (e.g. the points on nose and the points on belly in Fig. 2), their 3DLPSs

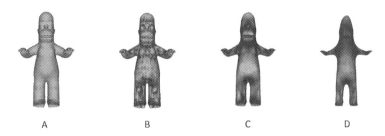

Fig. 2. Affections of Laplacian Pyramid operations on homer. (A) The oiginal model; (B) magnitudes of the 1st δ-coordinates; (C): magnitudes of the 15th δ-coordinates; (D) magnitudes of the 30th δ-coordinates.

will be distinctive from each other. On the contrary, if two points are embedded with similar frequency information (e.g. the points on the left leg and the corresponding points on the right leg in Fig. 2), their 3DLPS will be similar. We can intuitively verify that from the color ramp depicted in Fig. 2.

Figure 3 gives a more intuitive illustration. P_1 lies on the ear, where there are plenty of high frequency information, thus its 3DLPS has large values in high frequency components. Since P_2 lies on the flat areas, all the components in the 3DLPS are stationary and low; the 3DLPS of P_3 are between those of P_1 and P_2, since the legs of the giraffe have higher frequency than the belly.

Insensitive to articulated deformation. By insensitive to articulated deformation, we mean that for two models, in which one is articulated deformed from another, their 3DLPS signatures are differs only in the joint parts, but remains similar in other regions.

Intuitively, the articulated deformation only changes the geometries near the joint parts, so our 3DLPSs on the two models only differ from each other in these joint parts. Though in theory, the 3DLPS of a point P is affected by its k-ring neighbors, where k is the iteration times, in practice the far distance neighbors affect little of the 3DLPSs in large scales. Figure 4 shows the corresponding points

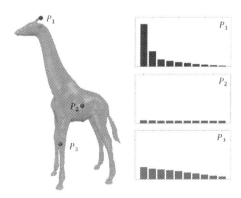

Fig. 3. 3DLPS of different points on the giraffe model.

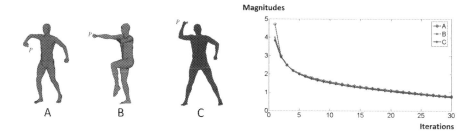

Fig. 4. 3DLPSs of P on different poses of similar men. Due to the different local geometries, the first few components of the 3DLPSs are a little different.

of similar men with different poses and their according 3DLPSs. Since P is far away from joints, their 3DLPSs are pretty similar in high frequency components.

4 Applications

Since 3DLPS concisely captures the frequency information of points, it has the potential to benefit applications involved in shape analysis from the perspective of frequency domain. As a brief illustration, we will show how 3DLPS is successfully applied in point correspondence detection, best view selection and mesh segmentation. Other applications may also potentially benefit from 3DLPS.

4.1 Point Correspondence Detection

The general solution of correspondence detection is to define the descriptors on surfaces or points, and then match the descriptors between them. Since 3DLPS reflects the multi-frequency information embedded in the shape, it can be incorporated into the general solution for point correspondence detection.

However, the original 3DLPS is sensitive to local geometries in high frequencies, whose magnitudes usually dominate the feature vectors (See the curves in Fig. 4). To solve this problem, we define a new signature based on 3DLPS, called *Context 3DLPS*, which is more robust and adaptive, yet rotation-invariant. We apply it as the new tool for robust point correspondence detection.

Definition 1 (Context 3DLPS). Given a vertex v, we average over all vertices in its i-th neighbor ring, and denote it as $\bar{S}_i$ (the 0th neighbor vertex is v itself), then the Context 3DLPS of v is defined as $S(v, nr) = (\bar{S}_0, \bar{S}_1, ..., \bar{S}_{nr})$, where nr is the number of neighbor rings.

The similarities of Context 3DLPSs are measured by χ^2 distance in our experiments, which is defined as

$$Sim(h_1, h_2) = \sum_{i=1}^{m} \frac{(h_1(i) - h_2(i))^2}{h_1(i) + h_2(i)}, \tag{6}$$

where h_1 and h_2 are the Context 3DLPS of v_1 and v_2 respectively, and m is the dimension of the Context 3DLPS.

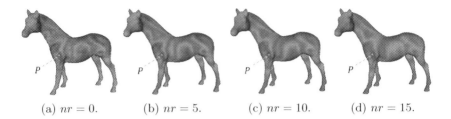

(a) $nr = 0$.   (b) $nr = 5$.   (c) $nr = 10$.   (d) $nr = 15$.

Fig. 5. Illustration of Context 3DLPS with various parameter nr. The color ramp depicts the differences of Context 3DLPS between their according locations and P, in which hot color means high differences. When nr is too small, the Context 3DLPS only captures the very local context information, which makes its specificity low (dissimilar points are wrongly regarded as similar).

For Context 3DLPS, the larger nr is, the larger context of information around v we considered, yet the larger memory storage and computation are required. In our experiments, we found $nr = 10$ is a good tradeoff between context information amount and computational cost, as shown in Fig. 5.

To detect the corresponding points of P, we first calculate the Context 3DLPS for each vertex, and then measure the similarities between P and all other vertices using Eq. 6. The top k points with the smallest χ^2 distances are regarded as the candidates. To make the candidates distinct from each other and reduce the number of candidates, we merge the candidates within the top k points as long as their distances are smaller than a user-specified threshold ϵ (in experiments, we set ϵ as 10 times the average edge length of the mesh), and set the new merged candidate as the one whose χ^2 distance is the smallest to P. Finally, Users can manually select the best correspondence point, or apply some algorithms to further refine the candidates [27–29]. As our main goal is to demonstrate the power of 3DLPS, we do not involve the matching algorithms here.

Figure 6 shows the results of corresponding point candidates detected by our algorithm. In this experiment, we first designate points P_1, P_2 and P_3 on model 1, and then calculate the top 3 most similar points on model 1 and model 2 respectively. In Fig. 6 (a), we verify that our method is capable of detecting the symmetric points. For instance, Q_{21} and Q_{31} are detected as the most similar points to P_2 and P_3 respectively. In Fig. 6 (b), we verify that our method is insensitive to articulated deformation. Though model 2 is a deformation of model 1, our algorithm can still detect Q_{21} and Q_{31} as the most similar points to P_2 and P_3, followed by their symmetric points Q_{22} and Q_{32}. We use color ramp to illustrate the χ^2 distance between the designated points and other vertices, in which cool colors indicate small distances.

4.2 Best View Selection

Automatic generation of best views for 3D models has drawn much attention [30], due to its applications in 3D model browsing, automatic camera replacement, 3D scene generation and view-based 3D object recognition [31,32].

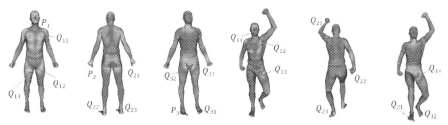

(a) P_1, P_2, P_3 and their most similar points on model 1.

(b) The most similar points of P_1, P_2 and P_3 on model 2.

Fig. 6. Illustration of our methods to detect the corresponding points. P_i is the ith user designated point, and Q_{ij} indicates the jth similar point to P_i. We designate P_1, P_2 and P_3 on model 1, and detect the top 3 similar points on both model 1 and model 2. Note that model 2 is an articulated deformation of model 1.

Among all the existing algorithms, one category aims at maximizing the amount of features visible from the viewpoint. These methods usually associate a goodness measure to a number of candidate views, where the goodness measure is a function of some objectives related to the geometrical properties of the object [26]. 3DLPS can be naturally incorporated as the geometrical properties for best view selection. In addition, since it exhibits various frequency information in 3DLPS, it provides an adaptive mechanism for visualizing different frequency information according to user specifications.

The framework of our algorithm is the same as the mesh saliency [26], in which the only difference is we replace the mesh saliency of v by the function $f(v)$ defined on v based on 3DLPS. For a viewpoint P, let $F(P)$ be the set of vertices visible from P, we compute the visible information from P as

$$U(P) = \sum_{v \in F(P)} f(v). \tag{7}$$

Then the viewpoint with maximum visible information P_m is

$$P_m = \underset{P}{\operatorname{argmax}}\ U(P). \tag{8}$$

Similar to [26], we use the gradient-descent-based optimization heuristic to select good viewpoints for efficiency.

Figure 7 intuitively shows how magnitudes change with the variations of frequency in 3DLPSs. If we define $f(v)$ as the magnitude of the first component of v's 3DLPS, then the best view is selected as the left view in Fig. 7(a). Note that this view is similar to the view based on maximizing the pure curvatures, since both of them tend to exhibit the high frequency information. If $f(v)$ is defined as the 30th component of v's 3DLPS, our algorithm tend to select the views that exhibit the low frequency information, as shown in the right view of Fig. 7(b).

By defining different $f(v)$ based on 3DLPS, our algorithm is adaptive to generate different best views that maximize different information. For instance,

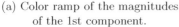

(a) Color ramp of the magnitudes   (b) Color ramp of the magnitudes
of the 1st component.   of the 30th component.

Fig. 7. (a): different views of the magnitudes of the 1st component of the 3DLPSs.
The left view is determined as the best view by our algorithm; (b) different views of
the magnitudes of the 30th component of the 3DLPSs. The right view is determined
as the best view by our algorithm.

if we define $f(v)$ as the 5th component of v's 3DLPS, the best view selected by
our algorithm is as Fig. 8(e).

A more reasonable strategy is to exhibit different frequency information
approximately equally in the best view. To achieve this goal, we define

$$f(v) = \sum_{i=1}^{m} rank(v, i), \tag{9}$$

where m is the length of 3DLPS, and $rank(v, i)$ indicates the index of v sorted
according to the ith components of all vertices in ascending order. This function
normalizes the low frequency information and high frequency information into a
commeasurable metric space. Thus, by maximizing $f(v)$s across all the vertices
with all the specified frequencies, the best view is supposed to exhibit various
frequency information equally, as shown in Fig. 8(f).

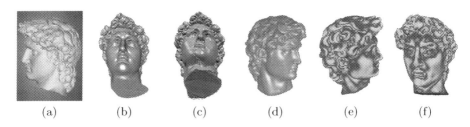

(a)   (b)   (c)   (d)   (e)   (f)

Fig. 8. Comparison of the best views of David head. (a): Result of [26]; (b): Result
of [33]; (c): Results of [34]; (d): Result of [35]; (e): Our result by maximizing the
magnitudes of the 15th component in 3DLPSs; (f): Our result by defining $f(v)$ as
Equ. 9. Note in (e) and (f), the pictures are mapped to color ramps, in which cool
color means small value. To make our best views visibly pleasant, we manually rotated
our best views to the correct upright direction.

Figure 8 also gives a brief comparison with some of the state-of-the-art approaches. Note that the result of [26] is similar to our result in Fig. 8(f). However,

our result exhibits more information on different frequencies (refer to the color ramp in Fig. 7 for comparison). Actually, the method proposed in [26] is exactly the Difference of Gaussian (DOG) defined on 3D, which is an approximation of Laplacian of Gaussian (LOG) utilized in our method. Since we separate different frequencies and define the signature as a vector, rather than summing them up to form a scale value [26], our method is more adaptive to exhibit different frequency information.

4.3 Automatic Mesh Segmentation

By observation, we found in many models, functional parts are usually different in frequency domain, and the boundaries between them often coincide with the regions where the frequency magnitudes change rapidly (see Fig. 9(c)). Since 3DLPS provides a good measure for frequency differences, it provides a natural way for segmentation of this kind of objects.

Inspired by this observation, we propose a label-swop based mesh segmentation method, as shown in Algorithm 2. In this algorithm, we first initialize the vertices with different labels, and then segment the mesh using a label-swop mechanism, as shown in lines 5–17 in Algorithm 2. We first estimate the probability density function of each vertex (shown in line 8–9) using Gaussian weights, and then swop the label of each vertex to its exemplar that possesses the locally highest estimated probability density function value. Finally, the segments are merged sequentially until the user specified segments number is achieved.

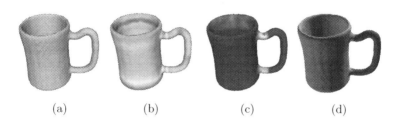

(a)   (b)   (c)   (d)

Fig. 9. Illustration of our label-swop based mesh segmentation algorithm. (a): the original model; (b): the color ramp mappings of 3DLPSs ($h = 15$); (c): the local variations of low frequency components in (b) ($nr = 3$); (d): the final segmentation result ($n = 2$).

Figure 9 illustrates how our label-swop algorithm performs on the cup model. In this experiments, we set $n = 2$, $h = 15$ and $nr = 3$. As shown in Fig. 9(c), the local variations of the specified components for all the vertices clearly marked the boundaries between the body and the handle of the cup. This leads to consistent segmentations with human intuition.

We provide a comparison of our method with some of the state-of-the-art methods. We conclude that for the models whose ground-truth segmentations

---

Algorithm 2. Label-swop based mesh segmentation

Input: n: the number of segments specified by users;
       h: the to be used index of component in 3DLPS;
       nr: the parameter to determine local regions.
Output: L: label array that represents the result.

for $i \leftarrow 1$ *to* $size(V)$ **do**
   | $L(v_i) \leftarrow i$; $\triangleright$ initialize vertices with different labels
   | $M(v_i) \leftarrow$ the hth component of the 3DLPS of v_i;
end
for $i \leftarrow 1$ *to* $size(V)$ **do**
   | $N(v_i) \leftarrow \max\limits_{v_j \in N(v_i, nr)} ||M(v_i) - M(v_j)||$; $\triangleright$ $N(v_i, nr)$ is v_i's nr-ring neighbors
end
for $i \leftarrow 1$ *to* $size(V)$ **do**
   | $D(v_i) \leftarrow \sum_{j \in N(v_i, nr)} N(v_j) \exp(-\frac{||v_i - v_j||^2}{2\sigma^2})$;
end
foreach *Vertex v in the mesh* **do**
   | $v_{final} \leftarrow v$;
   | **while** $\exists v_j \in N(v_i, 1)$ *such that* $D(v_j) > D(v_{final})$ **do**
   | | $v_{final} \leftarrow v_j$; $\triangleright$ swop v_j as the new exemplar
   | **end**
   | $L(v) \leftarrow L(v_{final})$;
end
$m \leftarrow$ the number of different labels in L;
while $m > n$ **do**
   | **foreach** *adjacent segments S_i and S_j* **do**
   | | $v_{min}(i, j) \leftarrow$ the vertex v whose $D(v)$ is the minimum along the
   | | boundaries of S_i and S_j;
   | **end**
   | $v_{join} \leftarrow \operatorname*{argmax}_{v} D(v_{min}(i, j))(1 \leq i, j \leq m, i \neq j)$;
   | join the segments adjacent to v_{join} by updating L;
end

---

coincide with the low frequency variations, such as ants, octopuses, tables and so on, our algorithm performs consistently better than existing methods (see Fig. 10 for details).

However, our segmentation algorithm suffers from two main limitations: (1) users have to specify the number of segments in advance; (2) for the models whose semantic segmentations do not coincide with low frequency variations, such as human faces, our algorithm may fail. As demonstrated in [1,29,42–44] that no one automatic segmentation algorithm is better than the others for all types of objects, we are confident that our algorithm provides an alternative means of segmentation for some types of models.

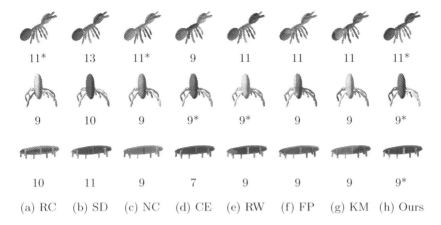

11*	13	11*	9	11	11	11	11*
9	10	9	9*	9*	9	9	9*
10	11	9	7	9	9	9	9*
(a) RC	(b) SD	(c) NC	(d) CE	(e) RW	(f) FP	(g) KM	(h) Ours

Fig. 10. Comparison with other segmentation methods. The number below each image indicates the number of the according segments, and the "*" following it means the according segmentation approximately coincides with the ground-truth manual segmentation. The methods used from left to right columns are (a): Randomized cuts (RC) [36]; (b): Shape diameter (SD) [37]; (c) Normalized cuts (NC) [36]; (d): Core extraction (CE) [38]; (e) Random walks (RW) [39]; (f) Fitting primitives (FP) [40]; (g) K-Means (KM) [41]; (h) Our method (Ours), respectively.

5   Discussion and Conclusion

In this paper, we have proposed a simple yet efficient point descriptor, called 3D Laplacian Pyramid signature, which reveals the frequency domain information of points embedded in the shape. Our 3DLPS is a multi-scale representation of the points and thus can be applied to plenty of applications involved in shape analysis, such as point correspondence detection, best view selection and automatic mesh segmentation. Compared with existing point signatures such as GPS, HKS and WKS, our proposed 3DLPS is more simpler to calculate, yet efficient enough for plenty of applications, since it inherits the nice properties of Laplacian Pyramid defined in 2D images.

However, 3DLPS still suffers from some limitations. For example, given two isometric deformed models, the 3DLPSs of vertices on the joint regions may be a little different, since 3DLPS simply catches the magnitudes of changes induced by Laplacian operations. This means that 3DLPS is not completely invariant under isometric transformations. Another limitation is that 3DLPS may suffer from noises of high frequency information (see Fig. 5). In this case, we have to introduce Context 3DLPS for some applications such as robust point correspondence detection. We will further investigate how to solve these problems.

References

1. Fang, Y., Sun, M., Ramani, K.: Temperature distribution descriptor for robust 3d shape retrieval. In: 4th Workshop on Non-Rigid Shape Analysis and Deformable Image Alignment (NORDIA 2011) (2011)

2. Chua, C.S., Jarvis, R.: Point signatures: a new representation for 3d object recognition. Int. J. Comput. Vis. **25**, 63–85 (1997)
3. Johnson, A.: Spin-images: a representation for 3-d surface matching. Ph.D. thesis, Robotics Institute, Carnegie Mellon University, Pittsburgh, PA (1997)
4. Belongie, S., Malik, J., Puzicha, J.: Shape context: a new descriptor for shape matching and object recognition. In: NIPS, pp. 831–837 (2000)
5. Kokkinos, I., Bronstein, M.M., Litman, R., Bronstein, A.M.: Intrinsic shape context descriptors for deformable shapes. In: CVPR, pp. 159–166 (2012)
6. Rustamov, R.M.: Laplace-beltrami eigenfunctions for deformation invariant shape representation. In: Symposium on Geometry Processing, pp. 225–233 (2007)
7. Sun, J., Ovsjanikov, M., Guibas, L.: A concise and provably informative multi-scale signature based on heat diffusion. In: Proceedings of the Symposium on Geometry Processing, SGP 2009, pp. 1383–1392 (2009)
8. Dey, T.K., Li, K., Luo, C., Ranjan, P., Safa, I., Wang, Y.: Persistent heat signature for pose-oblivious matching of incomplete models. Comput. Graph. Forum **29**, 1545–1554 (2010)
9. Aubry, M., Schlickewei, U., Cremers, D.: The wave kernel signature: a quantum mechanical approach to shape analysis. In: ICCV Workshops, pp. 1626–1633 (2011)
10. Burt, P.J.: The laplacian pyramid as a compact image code. IEEE Trans. Commun. **31**, 532–540 (1983)
11. Gal, R., Shamir, A., Cohen-Or, D.: Pose-oblivious shape signature. IEEE Trans. Vis. Comput. Graph. **13**, 261–271 (2007)
12. Bronstein, A.M., Bronstein, M.M., Kimmel, R., Mahmoudi, M., Sapiro, G.: A gromov-hausdorff framework with diffusion geometry for topologically-robust non-rigid shape matching. Int. J. Comput. Vision **89**, 266–286 (2010)
13. Bronstein, A.M., Bronstein, M.M., Guibas, L.J., Ovsjanikov, M.: Shape google: Geometric words and expressions for invariant shape retrieval. ACM Trans. Graph. **30**, 1:20–1:20 (2011)
14. Li, X., Guskov, I.: Multiscale features for approximate alignment of point-based surfaces. In: Symposium on Geometry Processing. Volume 255 of ACM International Conference Proceeding Series, pp. 217–226. Eurographics Association (2005)
15. Manay, S., Hong, B.-W., Yezzi, A.J., Soatto, S.: Integral invariant signatures. In: Pajdla, T., Matas, J.G. (eds.) ECCV 2004. LNCS, vol. 3024, pp. 87–99. Springer, Heidelberg (2004)
16. Paris, S., Hasinoff, S.W., Kautz, J.: Local laplacian filters: edge-aware image processing with a laplacian pyramid. ACM Trans. Graph. **30**, 68 (2011)
17. Sorkine, O.: Laplacian Mesh Processing. Ph.D. thesis, School of Computer Science, Tel Aviv University (2006)
18. Taubin, G.: A signal processing approach to fair surface design. In: Proceedings of the 22nd Annual Conference on Computer Graphics and Interactive Techniques, SIGGRAPH 1995, pp. 351–358. ACM New York, NY, USA (1995)
19. Ni, X., Garland, M., Hart, J.C.: Fair morse functions for extracting the topological structure of a surface mesh. ACM Trans. Graph. **23**, 613–622 (2004)
20. Dong, S., Bremer, P.T., Garland, M., Pascucci, V., Hart, J.C.: Spectral surface quadrangulation. ACM Trans. Graph. **25**, 1057–1066 (2006)
21. Reuter, M., Biasotti, S., Giorgi, D., Patan, G., Spagnuolo, M.: Discrete laplacecbeltrami operators for shape analysis and segmentation. Comput. Graph. **33**, 381–390 (2009)
22. Meer, P., Baugher, E.S., Rosenfeld, A.: Frequency domain analysis and synthesis of image pyramid generating kernels. IEEE Trans. Pattern Anal. Mach. Intell. **9**, 512–522 (1987)

23. Kobbelt, L., Campagna, S., Vorsatz, J., Seidel, H.P.: Interactive multi-resolution modeling on arbitrary meshes. In: Proceedings of the 25th Annual Conference on Computer Graphics and Interactive Techniques, SIGGRAPH 1998, pp. 105–114. ACM New York, NY, USA (1998)

24. Guskov, I., Sweldens, W., Schröder, P.: Multiresolution signal processing for meshes. In: Computer Graphics Proceedings (SIGGRAPH 99), pp. 325–334 (1999)

25. Pinkall, U., Polthier, K.: Computing discrete minimal surfaces and their conjugates. Exp. Math. **2**, 15–36 (1993)

26. Lee, C.H., Varshney, A., Jacobs, D.W.: Mesh saliency. ACM Trans. Graph. **24**, 659–666 (2005)

27. Dubrovina, A., Kimmel, R.: Matching shapes by eigendecomposition of the Laplace-Beltrami operator. In: 3DPVT (2010)

28. Funkhouser, T., Shilane, P.: Partial matching of 3D shapes with priority-driven search. In: Symposium on Geometry Processing (2006)

29. Sun, M., Fang, Y., Ramani, K.: Center-shift: an approach towards automatic robust mesh segmentation (arms), pp. 2145–2152. In: CVPR (2012)

30. Hu, K.M., Wang, B., Yuan, B., Yong, J.H.: Automatic generation of canonical views for cad models. In: CAD/Graphics, pp. 17–24 (2011)

31. Dutagaci, H., Cheung, C.P., Godil, A.: A benchmark for best view selection of 3d objects. In: Proceedings of the ACM Workshop on 3D Object Retrieval, 3DOR 2010, pp. 45–50 (2010)

32. Secord, A., Lu, J., Finkelstein, A., Singh, M., Nealen, A.: Perceptual models of viewpoint preference. ACM Trans. Graph. **30**, 109 (2011)

33. Mortara, M., Spagnuolo, M.: Semantics-driven best view of 3d shapes. Comput. Graph. **33**, 280–290 (2009)

34. Yamauchi, H., Saleem, W., Yoshizawa, S., Karni, Z., Belyaev, A.G., Seidel, H.P.: Towards stable and salient multi-view representation of 3d shapes. In: SMI, vol. 40. IEEE Computer Society (2006)

35. Leifman, G., Shtrom, E., Tal, A.: Surface regions of interest for viewpoint selection. In: CVPR, pp. 414–421. IEEE (2012)

36. Golovinskiy, A., Funkhouser, T.: Randomized cuts for 3D mesh analysis. ACM Transa. Graph. (Proc. SIGGRAPH ASIA) **27**, 145–157 (2008)

37. Shapira, L., Shamir, A., Cohen-Or, D.: Consistent mesh partitioning and skeletonisation using the shape diameter function. Vis. Comput. **24**, 249–259 (2008)

38. Katz, S., Leifman, G., Tal, A.: Mesh segmentation using feature point and core extraction. Vis. Comput. **21**, 649–658 (2005)

39. Kun Lai, Y., Min Hu, S., Martin, R.R., Rosin, P.L.: Fast mesh segmentation using random walks. In. ACM Symposium on Solid and Physical Modeling (2008)

40. Attene, M., Falcidieno, B., Spagnuolo, M.: Hierarchical mesh segmentation based on fitting primitives. Vis. Comput. **22**, 181–193 (2006)

41. Shlafman, S., Tal, A., Katz, S.: Metamorphosis of polyhedral surfaces using decomposition. In: Computer Graphics Forum, pp. 219–228 (2002)

42. Chen, X., Golovinskiy, A., Funkhouser, T.: A benchmark for 3D mesh segmentation. ACM Trans. Graph. (Proc. SIGGRAPH) **28**, 1–10 (2009)

43. Fang, Y., Liu, Y.S., Ramani, K.: Three dimensional shape comparison of flexible proteins using the local-diameter descriptor. BMC Struct. Biol. **9**, 29 (2009)

44. Fang, Y., Sun, M., Kim, M., Ramani, K.: Heat-mapping: a robust approach toward perceptually consistent mesh segmentation, pp. 2145–2152. In: CVPR (2011)

Commonality Preserving Multiple Instance Clustering Based on Diverse Density

Takayuki Fukui[✉] and Toshikazu Wada

Graduate School of Systems Engineering,
Wakayama University, Wakayama, Japan
`fukui@vrl.sys.wakayama-u.ac.jp`

Abstract. Image-set clustering is a problem decomposing a given image set into disjoint subsets satisfying specified criteria. For single vector image representations, proximity or similarity criterion is widely applied, i.e., proximal or similar images form a cluster. Recent trend of the image description, however, is the local feature based, i.e., an image is described by multiple local features, e.g., SIFT, SURF, and so on. In this description, which criterion should be employed for the clustering? As an answer to this question, this paper presents an image-set clustering method based on commonality, that is, images preserving strong commonality (coherent local features) form a cluster. In this criterion, image variations that do not affect common features are harmless. In the case of face images, hair-style changes and partial occlusions by glasses may not affect the cluster formation. We defined four commonality measures based on Diverse Density, that are used in agglomerative clustering. Through comparative experiments, we confirmed that two of our methods perform better than other methods examined in the experiments.

1 Introduction

Image-set clustering is a problem dividing a given image set into disjoint image subsets (clusters). Images belonging to a cluster should satisfy some criteria, e.g. mutual similarity or proximity. The image-set clustering can be utilized for many applications. For example, the clustering can be used as a chunking process of training images for accelerating the learning process of large-scale image classification systems. Also, the cluster information helps image annotation, labeling, and arranging the pictures to create photo albums.

For single vector image descriptions, similarity or proximity criterion is widely applied, i.e., mutually similar images (proximal image vectors) form a cluster. However, recent trend of the image description is the local feature based, i.e., an image is described by multiple local features, e.g. SIFT [1], SURF [2], and so on. Such image descriptions have some advantages over single-vector description, i.e., robustness against occlusions and geometric transformations. This description principle can be generalized from local features to any other features. That is, an image can be described by arbitrary number of vectors depending on the

© Springer International Publishing Switzerland 2015
C.V. Jawahar and S. Shan (Eds.): ACCV 2014 Workshops, Part III, LNCS 9010, pp. 322–335, 2015.
DOI: 10.1007/978-3-319-16634-6_24

amount of intrinsic information in the image. In this description, which criterion should be used for the clustering?

Of course, Bag of Features (BoF) [3] is a useful description that enables us to use many algorithms developed for the single-vector image description. If we employ BoF description, we can keep relying on similarity or proximity based clustering algorithms. BoF vector description, however, requires large scale feature-vector clustering to create a code book (visual words), which accumulates the occurrence of local features. This clustering consumes significant computational time and memory. Also, since the performance depends on the code book size, we have to run the vector clustering many times by changing the codebook size. Because of these reasons, we stick on the question above, i.e., which criterion should be used for image-set clustering for multiple-vector image description.

This paper responds to the question that *commonality* of the feature vectors can be a criterion of image-set clustering. The term "commonality" in this paper means the number and/or strength of commonly existing features across the image set. Basically, similarity or dissimilarity is defined between two images, but commonality measure is naturally defined on an image set. From this viewpoint, it is clear that commonality and similarity are essentially different criteria.

Based on this idea, we propose an image-set clustering method, i.e., images preserving strong commonality (coherent feature vectors) form a cluster. This clustering method has the robustness against common feature preserving image variations. In the case of face images, hair-style changes and partial occlusions by glasses rarely affect the clustering result.

We first define four commonality measures based on Diverse Density (DD) [4,5]. These measures are used in agglomerative (bottom-up) clustering that iteratively merges the image-set pair having the maximum commonality measure to create hierarchical image-set clusters. Thus, we get four clustering methods. Through comparative experiments conducted on "the ORL database of face" consisting of 400 face images and a part of "Nister's image dataset", we confirmed that two of our methods are almost competitive and perform better than other methods including k-means++ [6] and Ward's clustering [7] applied to BoF vectors and Hausdorff clustering applied to multiple-vector image descriptions.

2   Related Works

Clustering methods can be classified into two types "partitional" and "hierarchical" methods [8]. Partitional clustering divides given dataset into several subsets without checking all possible subset systems. On the contrary, hierarchical clustering builds possible cluster hierarchy, which is represented by tree-shaped data structure known as *dendrogram*. Of course the partition algorithm is more efficient than hierarchical one, because all possible subsets are not examined. But, this drawback of hierarchical clustering can be relaxed by introducing stop conditions.

K-means clustering [9,10] and k-medoid clustering [11] are the popular partition clustering methods under the assumption that the number of clusters is known. In contrast to the k-means clustering designed for vector data, k-medoids chooses data points as cluster centers (medoids or exemplars) so as to work in any metric space. For vector data, k-means clustering is one of the most popular method, but it depends on the initial locations of the cluster centers. For solving this problem, k-means++ [6] (k-means clustering with cluster center initialization) has been proposed.

Hierarchical clustering can further be classified into agglomerative (bottom-up) and divisive (top-down) methods [12]. Both methods require *linkage metric*: agglomerative clustering iteratively merges cluster pairs having minimum linkage metric, and divisive clustering extracts the subcluster pairs having maximum linkage metric by separating a cluster. The computational cost of divisive clustering is more expensive than agglomerative clustering, because of the bigger number of separation cases.

Based on the discussion above, we can notice that agglomerative clustering is one of the simplest clustering method. It only requires linkage metric, and the computational cost is not so big. Because of this simplicity, we employ this algorithm and focus on the linkage metric design.

Most linkage metric is obtained by extending the between-image distance to between-image-set distance. This framework is represented by Lance-Williams dissimilarity updating formula [13]. This formula is defined as

$$d(i \cup j, k) = \alpha_i d(i,k) + \alpha_j d(j,k) + \beta d(i,j) + \gamma \left| d(i,k) - d(j,k) \right|, \qquad (1)$$

where i, j and k represent image clusters, α_i, α_j, β, and γ are the parameters that define agglomerative criterion, and $| \cdot |$ represents the absolute value. Each image clusters have centers $\boldsymbol{g}_i$, $\boldsymbol{g}_j$ and $\boldsymbol{g}_k$, and the between-class dissimilarity is defined based on the distance between cluster centers. Just by assigning these parameters, dissimilarity definition, and cluster center update formula, wide varieties of agglomerative clustering algorithms can be described [14]. For example, Ward's clustering method [7], which merges the cluster pair so as to minimize the variance of the merged cluster, can be described by the following settings.

$$\alpha_i = \frac{|i| + |k|}{|i| + |j| + |k|}, \ \beta = -\frac{|k|}{|i| + |j| + |k|}, \ \gamma = 0, \qquad (2)$$

$$\boldsymbol{g}_{i \cup j} = \frac{|i|\boldsymbol{g}_i + |j|\boldsymbol{g}_j}{|i| + |j|}, \ d(i,j) = \frac{|i||j|}{|i| + |j|} \left\| \boldsymbol{g}_i - \boldsymbol{g}_j \right\|^2, \qquad (3)$$

where $| \cdot |$ and $\| \cdot \|$ represents the number of elements in the cluster and Euclidian norm, respectively.

The other approach is to employ between-set distance in agglomerative clustering. The most popular one is Hausdorff distance. This measure is the greatest of all the distances from a data point in one set to the closest data point in the other set. This metric is often used in the template matching scenario, because the metric is inverse proportional to the resemblance of two dataset distributions [15]. For the image clustering, however, Hausdorff distance is not suitable because of its excessive sensitivity to the outliers. For example, if a feature vector leaps into one set at the furthest point from the other set, the Hausdorff distance between them changes. That is, only one vector may drastically change the distance between two sets.

Same as the linkage metric, similarity measure can be defined and used in agglomerative clustering. This approach has wider variations than metric approach. This is because the similarity measure has lesser axioms than metrics and bigger degree of freedom. Furthermore, just by providing similarity values between any pairs instead of defining explicit form of similarity function, we can perform clustering like affinity propagation [16].

A question arose here, how do we define the linkage metric for multi-vector representation of images? Of course, when we employ Bag of Features (BoF) [3] image description, we can keep using standard clustering methods, such as k-means++ or Ward's method. The other possibility is the Hausdorff clustering between two sets of feature vectors. The method, however, is not suitable for linkage metric, because one image may contain wide variety of vectors and the Hausdorff distance is very sensitive to the outliers, as mentioned above.

As discussed above, no effective clustering method for multiple-vector image description has been proposed so far.

In the following sections, we propose commonality preserving clustering in Sect. 2, comparative experiments of image-set clustering conducted over "the ORL database of face" and a part of "Nister's image dataset" are shown and discussed in Sect. 4, and we conclude the discussion in Sect. 5.

3 Commonality Preserving Image-Set Clustering

In this section, we define a commonality measure between two image sets for agglomerative clustering. This commonality measure is defined based on Diverse Density (DD) $DD(\boldsymbol{x})$, which represents how the feature $\boldsymbol{x}$ commonly appears in positive bags (images) and never appears in negative bags. By integrating the value $DD(\boldsymbol{x})$ in the whole feature space, we can define the commonality measure. In some cases, it may be necessary to find the maxima of $DD(\boldsymbol{x})$ for common feature extraction from positive bags. For such computation, EM-DD has been proposed. This is an accelerated hill-climbing method of $DD(\boldsymbol{x})$ in the feature space. In this section, we shortly introduce both DD and EM-DD, and define four commonality measures based on them.

3.1 Diverse Density

In the field of MIL [4,5], common feature extraction has been regarded as one of the essential problems, which can be formalized as the local maxima search problem of DD in the feature space. Compared with other sophisticated MIL methods, DD is very sensitive to the incorrect labeling. That is, if a positive feature leaps into a negative bag (image) or a feature is removed from a positive bag (image), the value of DD may change drastically. This sensitivity is unwanted property for many MIL applications that use manually labeled data. But for commonality preserving clustering, this sensitivity is welcome, because the task is not extracting features from incorrectly labeled data, but gathering images preserving strong commonality.

Based on the following definitions and terminology, we define the DD.

Bag $\mathcal{B}$: A set of instances. This corresponds to an image in our problem.

Label $+, -$: We assign positive labels to those bags where we want to found out the commonality. Also, negative labels are assigned to those bags where the commonality never be expected. These are denoted by $\mathcal{B}_i^+, (i = 1, \dots, m)$ and $\mathcal{B}_i^-, (i = 1, \dots, n)$, respectively.

Instance $\boldsymbol{B}_{ij}^+, \boldsymbol{B}_{ij}^-$: An element belonging to a bag. This corresponds to a local feature vector. Positive and negative instances are denoted by $\boldsymbol{B}_{ij}^+ \in \mathcal{B}_i^+$ and $\boldsymbol{B}_{ij}^- \in \mathcal{B}_i^-$, respectively.

First, the following function represents a potential generated by an instance $\boldsymbol{B}_{ij}$ at a point $\boldsymbol{x}$ in feature space.

$$P(\boldsymbol{x}|\boldsymbol{B}_{ij}) \equiv \exp\left(-\frac{\|\boldsymbol{B}_{ij} - \boldsymbol{x}\|^2}{\sigma^2}\right). \tag{4}$$

The maximum and the minimum values of this potential are 1 and 0, respectively.

The following function represents the integrated potential $P(\boldsymbol{x}|\mathcal{B}_i^+)$ generated by instances in a positive bag $\mathcal{B}_i^+$.

$$P(\boldsymbol{x}|\mathcal{B}_i^+) \equiv 1 - \prod_{\boldsymbol{B}_{ij}^+ \in \mathcal{B}_i^+} \left(1 - P\left(\boldsymbol{x}|\boldsymbol{B}_{ij}^+\right)\right). \tag{5}$$

Subtraction of an individual potential from 1 can be regarded as the similar meaning to negation and the product can be regarded as logical AND. Under this interpretation, Eq. (5) can be regarded as integration by logical OR of the individual potentials in the positive bag by applying De Morgan's laws.

For negative bag, integrated potential from a negative bag $\mathcal{B}_i^-$ can be defined as follows.

$$P(\boldsymbol{x}|\mathcal{B}_i^-) \equiv \prod_{\boldsymbol{B}_{ij}^- \in \mathcal{B}_i^-} \left(1 - P\left(\boldsymbol{x}|\boldsymbol{B}_{ij}^-\right)\right). \tag{6}$$

Same as the interpretation of Eq. (5), this integration can be regarded as logical NOR.

The potentials generated by positive and negative bags are further integrated by their product to obtain Diverse Density $DD(\boldsymbol{x})$.

$$DD(\boldsymbol{x}) \equiv \prod_i^m P\left(\boldsymbol{x}|\mathcal{B}_i^+\right) \prod_i^n P\left(\boldsymbol{x}|\mathcal{B}_i^-\right). \tag{7}$$

At a local maximum of $DD(\boldsymbol{x})$ having enough value in the feature space, the point $\boldsymbol{x}$ can be regarded as a common local feature among the positive bags and does not contain similar features in all negative bags.

By integrating the $DD(\boldsymbol{x})$ value in whole feature space, we can measure the commonality of the local features in a positive image set.

3.2 EM-DD

$DD(\boldsymbol{x})$ can be regarded as a commonality measure at a point $\boldsymbol{x}$ in the feature space. We sometimes need the points that locally maximize $DD(\boldsymbol{x})$ for determining the features specifying the given cluster (positive and negative image sets). EM-DD [17] estimates a local maximum of DD in the feature space by hill climbing iterations, in which an accelerated approximation of DD like EM-algorithm [18] is employed. EM-DD algorithm iteratively finds local maxima in the feature space starting from all positive instances. The DD is defined in Eq. (7), but the computation using all positive and negative instances is cumbersome. For avoiding this, EM-DD approximates DD value by using proximal instances, each of which is the 1-nearest instance picked up from a positive or negative bag. Since this selection process is similar with expectation process, and the hill climbing can be regarded as maximization process, this algorithm is called EM-DD.

3.3 Commonality Between Two Clusters

In this section, we discuss how to define the commonality measure between two image sets. The measure should represent how many and/or strong common local features are preserved after merging two image sets. According to this principle, the measure $\mathcal{C}$ can be easily defined as follows.

$$\mathcal{C}_{\mathbb{N}}(\mathbb{A} \cup \mathbb{B}) \equiv \int_{\boldsymbol{x} \in \mathcal{F}} DD(\boldsymbol{x}) d\boldsymbol{x}, \tag{8}$$

where $\mathbb{A}$ and $\mathbb{B}$ represent positive image sets, $\mathbb{N}$ negative image set, and $\mathcal{F}$ feature space. This commonality $\mathcal{C}_{\mathbb{N}}(\mathbb{A} \cup \mathbb{B})$ is obtained by integrating $DD(\boldsymbol{x})$ over feature space $\mathcal{F}$. However, this computation is practically impossible, because the feature space $\mathcal{F}$ is infinitely vast.

For avoiding this endless computation, the computation of $DD(\boldsymbol{x})$ can be restricted on sampling points. In this case, one positive image set is assigned as the positive image set and the other is used to define the sampling point in the feature space. So as to produce a commonality measure that is independent of number of sampling points, we compute the sample mean of $DD(\boldsymbol{x})$. By assigning the image set $\mathbb{A}$ to produce sampling point set $\mathcal{S}_{\mathbb{A}}$ in feature space and $\mathbb{B}$ as positive image set, we get

$$\mathcal{C}_{\mathbb{N}}^{\mathcal{S}_{\mathbb{A}}}(\mathbb{B}) \equiv \frac{1}{|\mathcal{S}_{\mathbb{A}}|} \sum_{\boldsymbol{x} \in \mathcal{S}_{\mathbb{A}}} DD(\boldsymbol{x}), \tag{9}$$

where $\mathcal{S}_{\mathbb{A}}$ represents the feature set extracted from all images in the image set $\mathbb{A}$. This computation is possible, but the resulted value has asymmetric property depending on the assignment as shown in Eq. (10).

$$\mathcal{C}_{\mathbb{N}}^{\mathcal{S}_{\mathbb{A}}}(\mathbb{B}) \neq \mathcal{C}_{\mathbb{N}}^{\mathcal{S}_{\mathbb{B}}}(\mathbb{A}). \tag{10}$$

For guaranteeing the symmetric property, we first define a commonality measure by taking arithmetic mean as

$$\mathcal{C}_{\mathbb{N}}^{1}(\mathbb{A}, \mathbb{B}) \equiv \frac{1}{2}(\mathcal{C}_{\mathbb{N}}^{\mathcal{S}_{\mathbb{B}}}(\mathbb{A}) + \mathcal{C}_{\mathbb{N}}^{\mathcal{S}_{\mathbb{A}}}(\mathbb{B})). \tag{11}$$

We can also define other commonality measure by taking geometric mean as

$$C_\mathbb{N}^2(\mathbb{A}, \mathbb{B}) \equiv \sqrt{C_\mathbb{N}^{\mathcal{S}_\mathbb{B}}(\mathbb{A}) * C_\mathbb{N}^{\mathcal{S}_\mathbb{A}}(\mathbb{B})}. \tag{12}$$

The above definitions require excessive sampling points in the feature space for big image data sets. For avoiding this problem, we can reduce the number of sampling points by applying EM-DD to image set $\mathbb{A}$ to produce reduced sampling point set $\mathcal{M}_\mathbb{A}$. By using this as a point set, we can define other commonality measure as

$$C_\mathbb{N}^{\mathcal{M}_\mathbb{A}}(\mathbb{B}) \equiv \frac{1}{|\mathcal{M}_\mathbb{A}|} \sum_{x \in \mathcal{M}_\mathbb{A}} DD(x). \tag{13}$$

Same as the discussion above, we propose the following commonality measures:

$$C_\mathbb{N}^3(\mathbb{A}, \mathbb{B}) \equiv \frac{1}{2}(C_\mathbb{N}^{\mathcal{M}_\mathbb{B}}(\mathbb{A}) + C_\mathbb{N}^{\mathcal{M}_\mathbb{A}}(\mathbb{B})), \tag{14}$$

$$C_\mathbb{N}^4(\mathbb{A}, \mathbb{B}) \equiv \sqrt{C_\mathbb{N}^{\mathcal{M}_\mathbb{B}}(\mathbb{A}) * C_\mathbb{N}^{\mathcal{M}_\mathbb{A}}(\mathbb{B})}. \tag{15}$$

As discussed above, we propose four commonality measures, $C_\mathbb{N}^1(\mathbb{A}, \mathbb{B})$, $C_\mathbb{N}^2(\mathbb{A}, \mathbb{B})$, $C_\mathbb{N}^3(\mathbb{A}, \mathbb{B})$, $C_\mathbb{N}^4(\mathbb{A}, \mathbb{B})$ in this paper. These commonality measures are used in the Algorithm1.

Algorithm 1. Agglomerative clustering

Initialize: Create singleton clusters, each of which consists of a single image.
Step1: Merge the cluster pair having maximum commonality measure among all clus-
ter pairs.
Step2: If the number of cluster is greater than one, go to Step 1.
End:

---

4   Experiments

We conducted comparative experiments on image-set clustering. We first introduce the normalized mutual information (NMI) [19] for evaluating the accuracy
of the clustering results. Next, we describe the experimental settings and parameter tuning of different methods (Only the resulted parameters are shown in
this paper because of the page limitation). Finally, we compared NMI scores by
changing number of clusters, and examine images in resulted clusters.

4.1   Evaluation Criterion

We employ NMI for measuring the clustering accuracy. Let $\mathbb{X}$ be the set of
clusters obtained by a clustering algorithm and $\mathbb{Y}$ obtained from the ground
truth. Then, the NMI measure between $\mathbb{X}$ and $\mathbb{Y}$ is defined as

$$NMI(\mathbb{X}, \mathbb{Y}) = \frac{I(\mathbb{X}; \mathbb{Y})}{(H(\mathbb{X}) + H(\mathbb{Y}))/2}, \tag{16}$$

where
$$I(\mathbb{X}; \mathbb{Y}) = H(\mathbb{X}, \mathbb{Y}) - H(\mathbb{X}|\mathbb{Y}) - H(\mathbb{Y}|\mathbb{X}), \qquad (17)$$

is the mutual information between $\mathbb{X}$ and $\mathbb{Y}$, $H(\mathbb{X}, \mathbb{Y})$ and $H(\mathbb{X}|\mathbb{Y})$ are the joint and conditional entropies of $\mathbb{X}$ and $\mathbb{Y}$, respectively. When $\mathbb{X}$ and $\mathbb{Y}$ are the same set of clusters, $NMI(\mathbb{X}, \mathbb{Y})$ is 1. Conversely, When $NMI(\mathbb{X}, \mathbb{Y})$ is nearly 0, $\mathbb{X}$ and $\mathbb{Y}$ are different set of clusters. By using $NMI(\mathbb{X}, \mathbb{Y})$, we evaluate the clustering accuracy.

4.2   Experimental Settings

The experimental settings are shown below.

Methods to be Compared

In the experiments, we compare the following clustering methods. K-means++ and Ward's clustering are applied to BoF vectors.

Agglomerative Clustering with $\mathcal{C}_{\mathbb{N}}^1(\mathbb{A}, \mathbb{B})$: $AM\_ALL(\sigma)$, where σ represents the parameter in Eq. (4).

Agglomerative Clustering with $\mathcal{C}_{\mathbb{N}}^2(\mathbb{A}, \mathbb{B})$: $GM\_ALL(\sigma)$, where σ represents the parameter in Eq. (4).

Agglomerative Clustering with $\mathcal{C}_{\mathbb{N}}^3(\mathbb{A}, \mathbb{B})$: $AM\_EM(\sigma)$, where σ represents the parameter in Eq. (4).

Agglomerative Clustering with $\mathcal{C}_{\mathbb{N}}^4(\mathbb{A}, \mathbb{B})$: $GM\_EM(\sigma)$, where σ represents the parameter in Eq. (4).

k-means++ Clustering: $KMPP(num)$, where num represents the code book size.

Ward's Clustering: $WARD(num)$, where num represents the code book size.

Hausdorff Clustering: $HAUS$. For both between-image (feature set) distance and between-image-set distance, Hausdorff distance is used.

Image Dataset

In the experiments, we use "the ORL database of face" and "Nister's image dataset", because we have to know the ground truth for the evaluation. The ORL database consists of 400 face images, ten different images of each of 40 distinct subjects. Nister's image dataset consists of 10200 images, four different images taken under different conditions of each 2550 objects. For the Nister's dataset, top 400 images (100 objects) are used.

Local Features

The local features used in the experiments are 64D SURF features extracted by using OpenCV 2.4.2 library.

Note that we didn't use negative images in the experiments, because of the fairness, i.e., k-means++, Ward's Clustering, and Hausdorff Clustering do not use negative images.

4.3   Parameter Tuning

Except for Hausdorff clustering, all methods have parameters: proposed methods have parameter σ, k-means++ and Ward's Clustering are applied to BoF vectors that depend on the code book size *num*. These parameters are tuned so as to maximize $NMI(\mathbb{X}, \mathbb{Y})$ for given dataset.

In this paper, detailed parameter tuning results are omitted, and only the resulted parameters are shown. For ORL database, $AM\_ALL(0.01)$, GM_ALL (0.05), $AM\_EM(0.005)$, $GM\_EM(0.01)$, $KMPP(50)$ and $WARD(500)$ were the best in examined parameters. For Nister's dataset, $AM\_ALL(0.05)$, $GM\_$ ALL (0.05), $AM\_EM(0.005)$, $GM\_EM(0.01)$, $KMPP(30)$ and $WARD(50)$ were the best in examined parameters.

4.4   Comparative Results

We compared NMI scores of different clustering methods with tuned parameters described above. Figure 1 shows the graphs of NMI scores on ORL database. The horizontal and vertical axes represent the number of clusters and NMI score, respectively. From this figure, $GM\_ALL(0.05)$ and $AM\_ALL(0.01)$ are competitive and provide the better results than other methods. The best NMI score is obtained at #cluster $= 58$. The third best method is $GM\_EM(0.01)$, and the fourth best is $WARD(500)$. Following these methods, NMI scores of $AM\_EM(0.005)$, $KMPP(50)$, and $HAUS$ degenerates in this order.

Figure 2 shows the graphs of NMI scores on Nister's image dataset. From this figure, $AM\_ALL(0.05)$ and $GM\_ALL(0.05)$ are almost competitive and

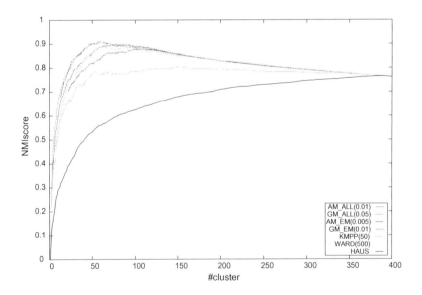

Fig. 1. NMI graphs of different clustering methods for the ORL database of faces using best parameters: Horizontal axis is the number of clusters, vertical is the NMI score.

Table 1. The NMI scores of different clustering methods on different image datasets. For the ORL database, NMI scores are measured at #cluster = 40. For Nister's image dataset, NMI score is measured at #cluster = 100.

Dataset (#clusters)	Clustering method						
	AM_ALL	GM_ALL	AM_EM	GM_EM	KMPP	WARD	HAUS
ORL(40)	.874106	**.877920**	.789015	.830801	.728934	.840150	.496100
Nister(100)	**.917861**	.911005	.841304	.879985	.823335	.844039	.729781

perform better than other methods. The third best method is $GM\_EM(0.01)$. $WARD(50)$ and $AM\_EM(0.005)$ are almost competitive. Following these methods, NMI scores of $KMPP(30)$, and $HAUS$ degenerates in this order.

Table 1 summarizes the best NMI scores for different clustering methods on different image datasets. From this table, we can notice that AM_ALL and GM_ALL are almost competitive and perform better than other examined methods. Following these methods, GM_EM and WARD are performing better than AM_EM and k-means++, and Hausdorff clustering is the worst.

This means that agglomerative clustering based on Eqs. (11) and (12) have advantages over other clustering methods. Ward's method and k-means++ have been widely applied to many clustering tasks and proven as standard clustering methods. But, the experimental results demonstrate that some commonality based clustering methods perform better than these method applied to BoF image representation.

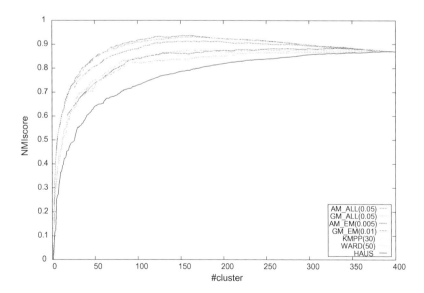

Fig. 2. NMI graphs of different clustering methods for Nister's image dataset using best parameters: Horizontal axis is the number of clusters, vertical is the NMI score.

This implies that multiple-vector representation of images is better than BoF representation for some tasks. If this is true, there is a chance to renew Pattern Recognition or Image Retrieval algorithms for multiple-vector representation without using BoF.

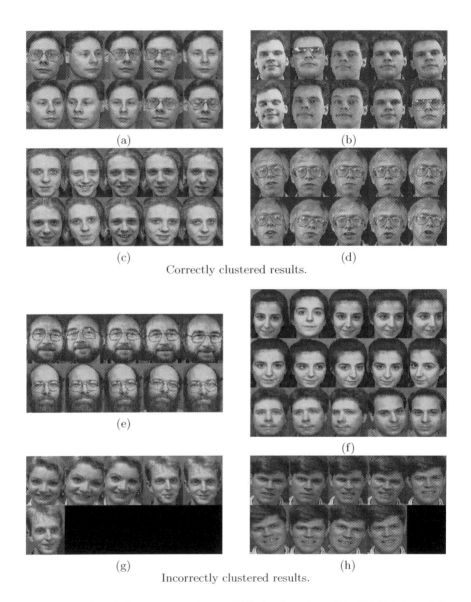

(a)   (b)

(c)   (d)

Correctly clustered results.

(e)

(f)

(g)   (h)

Incorrectly clustered results.

Fig. 3. Examples of clustering results on ORL database by $GM\_ALL(0.05)$ at #cluster $= 58$, where NMI score becomes maximum. Correct clusterings, in other words cluster has all the images of the same subject in the dataset, are from (a) to (d), and incorrect clustering are from (e) to (h).

4.5 Examples of Clustered Images

Figure 3 shows examples of clustered images of ORL database by using GM_ALL (0.05) at #cluster = 58, where the NMI score becomes maximum. The correct clustering results (a)~(d) demonstrate that wearing glasses, small rotation, scale change and local shading do not affect the clustering results. The incorrect results (e)~(h) show that if there exists similar images of other subjects, that may cause mis-clustering.

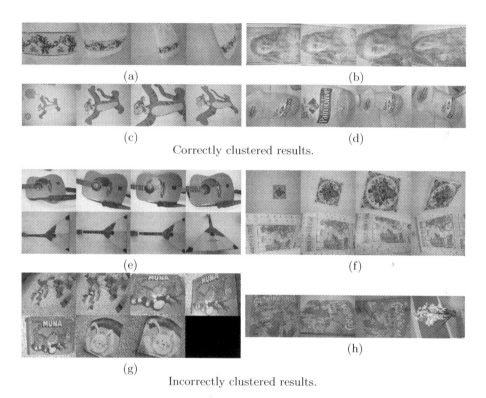

(a)   (b)

(c)   (d)

Correctly clustered results.

(e)   (f)

(g)

(h)

Incorrectly clustered results.

Fig. 4. Examples of clustering results on Nister's dataset by $GM\_ALL(0.05)$ at #cluster = 100. Correct clusterings, in other words cluster has all the images of the same subject in the dataset, are from (a) to (d), and incorrect clustering are from (e) to (h).

Figure 4 shows examples of clustered images of Nister's dataset by using $GM\_ALL(0.05)$ at #cluster = 100. The correct clustering results (a)~(d) demonstrate that rotations, homography, and partial occlusions do not affect the clustering results. The incorrect results (e)~(h) show that if there exists common local features, e.g. frets and strings found in guitar and balalaika and background carpet textures may cause mis-clustering.

As shown above, objects having distinctive features can be clustered correctly, but objects including common local features may be clustered incorrectly. If we can properly assign negative samples, this problem can be solved or relaxed.

5   Conclusion

This paper proposes commonality preserving clustering method for local feature representation of images. Four commonality measures are proposed based on Diverse Density and examined through extensive experiments. As a result, agglomerative clustering methods using two commonality measures defined by Eqs. (11) and (12) are almost competitive and perform better than examined methods including k-means++, Ward's method, and Hausdorff clustering.

Essentially, similarity or dissimilarity is defined between two images, but commonality measure is naturally defined on an image set. From this viewpoint, it is clear that commonality and similarity are essentially different and the idea of commonality preserving clustering has not been proposed so far.

The commonality measure represents the number and strength of commonly existing features across the image set. The advantages of commonality preserving clustering are (1) it can be applied directly to the multiple-vector representation of images without using BoF (code book is not necessary), (2) we can guarantee the integrity of clusters in terms of common features, (3) some commonality measure produce better clustering results than other methods.

In the experiments, we didn't use negative images for guaranteeing the fairness. But, by properly assigning negative images, we can emphasize the distinctive features and enlarge the difference between clusters. Also, we can create models representing clusters by using EM-DD and the resulted models can be utilized in the classifier. These tasks should be done in the future works.

Acknowledgement. This work was supported by "R&D Program for Implementation of Anti-Crime and Anti-Terrorism Technologies for a Safe and Secure Society", Funds for integrated promotion of social system reform and research and development of the Ministry of Education, Culture, Sports, Science and Technology, the Japanese Government.

References

1. Lowe, D.G.: Distinctive image features from scale-invariant keypoints. IJCV **60**, 91–110 (2004)
2. Bay, H., Ess, A., Tiytelaars, T., Gool, L.J.V.: Surf: speeded up robust features. CVIU **110**, 346–359 (2008)
3. Fei-Fei, L.: A bayesian hierarchical model for learning natural scene categories. In: CVPR, pp. 524–531 (2005)
4. Maron, O., Lozano-Pérez, T.: A framework for multiple-instance learning. In: Advances in Neural Information Processing Systems, pp. 570–576. MIT Press (1998)
5. Maron, O., Ratan, A.L.: Multiple-instance learning for natural scene classification. In: The Fifteenth International Conference on Machine Learning, pp. 341–349. Morgan Kaufmann (1998)
6. Arthur, D., Vassilvitskii, S.: k-means++: the advantages of careful seeding. In: SODA 2007: Proceedings of the Eighteenth Annual ACM-SIAM Symposium on Discrete Algorithms, pp. 1027–1035. Society for Industrial and Applied Mathematics, Philadelphia (2007)

7. Ward, J.: Hierarchical grouping to optimize an objective function. J. Am. Stat. Assoc. **58**, 236–244 (1963)
8. Berkhin, P.: A survey of clustering data mining techniques. In: Kogan, J., Nicholas, C., Teboulle, M. (eds.) Grouping Multidimensional Data. Springer, Berlin (2006)
9. Forgy, E.: Cluster analysis of multivariate data: efficiency versus interpretability of classification. Biometrics **21**, 768–769 (1965)
10. MacQueen, J.: Some methods for classification and analysis of multivariate observations. In: Cam, L.M.L., Neyman, J., eds.: Proceedings of the Fifth Berkeley Symposium on Mathematical Statistics and Probabilitym, vol. 1, pp. 281–297. University of California Press (1967)
11. Kaufman, L., Rousseeuw, P.J.: Finding Groups in Data: An Introduction to Cluster Analysis. Wiley, New York (1990)
12. Jain, A., Dubes, R.: Algorithms for Clustering Data. Prentice Hall, Upper Saddle River (1988)
13. Lance, G.N., Williams, W.T.: A general theory of classificatory sorting strategies 1. hierarchical systems. Comput. J. **9**, 373–380 (1967)
14. Murtagh, F., Contreras, P.: Methods of hierarchical clustering. CoRR abs/1105.0121 (2011)
15. Huttenlocher, D., Klanderman, G.A., Kl, G.A., Rucklidge, W.J.: Comparing images using the hausdorff distance. IEEE Trans. Pattern Anal. Mach. Intell. **15**, 850–863 (1993)
16. Frey, B.J., Dueck, D.: Clustering by passing messages between data points. Science **315**, 972–976 (2007)
17. Zhang, Q., Goldman, S.A.: Em-dd: An Improved Multiple-instance Learning Technique. MIT Press, Cambridge (2001)
18. Dempster, A.P., Laird, N.M., Rubin, D.B.: Maximum likelihood from incomplete data via the em algorithm. J. Roy. Stat. Soc. Ser. B **39**, 1–38 (1977)
19. Witten, I.H., Frank, E., Holmes, G.: Data Mining : Practical Machine Learning Tools and Techniques. The Morgan Kaufmann series in data management systems. Morgan Kaufmann, Amsterdam (2011)

Third International Workshop on Intelligent Mobile and Egocentric Vision (IMEV2014)

Interactive RGB-D SLAM on Mobile Devices

Nicholas Brunetto, Nicola Fioraio$^{(\boxtimes)}$, and Luigi Di Stefano

CVLab - Department of Computer Science and Engineering, University of Bologna,
Viale Risorgimento, 2 - 40135 Bologna, Italy
`nicholas.brunetto@studio.unibo.it`,
`{nicola.fioraio,luigi.distefano}@unibo.it`

Abstract. In this paper we present a new RGB-D SLAM system specifically designed for mobile platforms. Though the basic approach has already been proposed, many relevant changes are required to suit a user-centered mobile environment. In particular, our implementation tackles the strict memory constraints and limited computational power of a typical tablet device, thus delivering interactive usability without hindering effectiveness. Real-time 3D reconstruction is achieved by projecting measurements from aligned RGB-D keyframes, so to provide the user with instant feedback. We analyze quantitatively the accuracy vs. speed trade-off of diverse variants of the proposed pipeline, we estimate the amount of memory required to run the application and we also provide qualitative results dealing with reconstructions of indoor environments.

1 Introduction

In the past decades, the problem of Simultaneous Localization And Mapping (SLAM) has been mainly addressed by either expensive 3D sensors, *e.g.* laser scanners, or monocular RGB cameras. Laser scanners feature high precision and have enabled impressive results in the SLAM realm [1,2]. However, their size and expensiveness limit the range of addressable applications. On the other hand, monocular SLAM has reached a considerable maturity [3,4] but still mandates massive parallelization by GPGPU to attain dense 3D reconstruction [5].

The availability of consumer-grade RGB-D cameras capable of delivering color and depth information in real-time, such as the Microsoft Kinect and Asus Xtion Pro Live, has fostered a new approach in solving the SLAM problem. Nowadays, a newer generation of devices is appearing, *e.g.* the Structure sensor [6], the Creative Senz3D [7] and Google's Project Tango [8], which opens up new possibilities to introduce 3D sensing into the mobile world. However, most RGB-D SLAM systems have not been conceived for mobile platforms and typically rely on massive GPGPU processing to reach real-time execution on desktop environments [9].

Very recently, though, the SlamDunk [10] RGB-D SLAM algorithm has been proposed, which features a few threads, low memory consumption and can achieve real-time operation without any GPGPU acceleration. As SlamDunk is quite accurate and robust, it qualifies itself as a suitable reference design towards the creation of an RGB-D SLAM framework for mobile platforms. Hence, rather

© Springer International Publishing Switzerland 2015
C.V. Jawahar and S. Shan (Eds.): ACCV 2014 Workshops, Part III, LNCS 9010, pp. 339–351, 2015.
DOI: 10.1007/978-3-319-16634-6_25

than investigating on a possible brand-new pipeline, in this paper we focus on the adaptation and implementation of the key steps of the SlamDunk algorithm on an Android-operated tablet, underlining the difficulties faced and the solutions adopted to overcome them. Accordingly, our contribution may pave the way for further research on the novel topic of interactive RGB-D SLAM on mobile devices, as countless exciting applications and new scenarios may be envisioned should this technology trend grow and reach maturity.

The remainder of this paper is organized as follows. The next section reviews related publications, including SlamDunk and similar approaches, so to better motivate the adoption of the former as a SLAM reference design "to go mobile" as well as to highlight the key steps that will be adapted to enable the actual mobile implementation. Then, in Sect. 3 we briefly summarize the SlamDunk pipeline and in Sect. 4 present the software architecture deployed in the final SLAM mobile application. Section 5 discusses the adaptations made to the original algorithm, chiefly needed to improve efficiency and better meet the computational requirement of a mobile platform. Experimental findings are reported in Sect. 6, while in Sect. 7 we draw concluding remarks and point out the major issues to be addressed by future work.

2   Related Work

In the field of RGB-D SLAM, most state-of-the-art proposals can now produce high quality meshes in real-time. However, they usually require hardware acceleration by massive GPGPU processing together with high-end desktop platforms [9,11,12]. As such, these systems can hardly provide reference design guidelines towards rendering interactive SLAM feasible on resource limited platforms such off-the-shelf mobile devices. Among those not relying on GPGPU acceleration, RGB-D Mapping [13] has been one of the first proposals aimed at exploiting RGB-D sensing for SLAM. Their camera tracking is based on pairwise matching of image features, the system performing also global pose graph optimization to handle camera drift. To constrain nodes and make the optimization effective, they look for possible loop closures by matching image features within a subset of previous keyframes and perform a global pose graph relaxation accordingly. Therefore, as the explored space gets wider and the graph size increases, more time is spent in finding loop closures and optimizing poses. A similar approach is deployed by RGB-D SLAM [14,15], where, moreover, near real-time processing (less than 15 FPS) is achieved by using the GPU for extraction of SIFT visual features. Nonetheless, the speed usually drops after gathering many frames due to the increasing complexity of the global optimization routine.

SlamDunk [10] is a very recent and particularly "lightweight" approach to RGB-D SLAM. Unlike previous work, it builds a Local Map against which the camera is robustly tracked and avoids full-graph optimization to preserve real-time operation even along large trajectories. Moreover, the Local Map enables implicit loop closure handling and allows for efficient operation as the camera moves back and forth within an already mapped portion of the workspace.

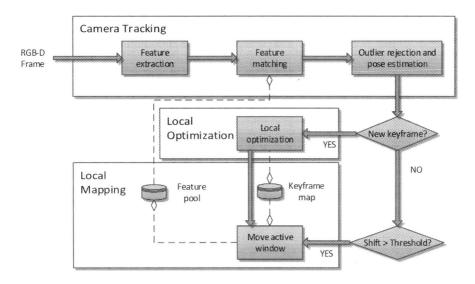

Fig. 1. The SlamDunk pipeline comprises three main modules: Local Mapping, Camera Tracking and Local Optimization.

As SlamDunk is computationally rather inexpensive and can achieve real-time performance (*i.e.* more than 30 FPS) without any GPGPU acceleration, it appears as an interesting approach for mobile RGB-D SLAM.

As regards SLAM systems specifically proposed for a mobile platform, we point out that they usually rely on the RGB camera of the device and the internal sensors, *e.g.* accelerometer and gyroscope, while we target RGB-D data. Klein and Murray [16] showed how to run PTAM [4] on an Apple iPhone 3G. However, while tracking is typically carried out in 30 ms, the bundle adjustment step may take seconds to complete. Lee *et al.* [17] process frames off-line on a remote server, so the computation is not performed on the device. Pan *et al.* [18] produces 3D reconstructions of very large scenes on a phone. The models are suitable for AR, but the system does not run in real-time and the user can perceive the result only after seconds of processing. More recently, Tanskanen *et al.* [19] achieved real-time tracking on a smartphone by leveraging on inertial sensing and GPU processing. Also, the system is designed for object capturing with model reconstruction running at about 0.3–0.5 Hz. In this paper we demonstrate a much higher frame rate without exploiting any other sensors than the RGB-D camera.

3   SlamDunk System Overview

As depicted in Fig. 1, the SlamDunk algorithm [10] can be decoupled in three main modules: Local Mapping, Camera Tracking and Local Optimization. Local Mapping models the camera path as a collection of RGB-D keyframes and stores their poses within a quad-tree data structure. The algorithm does not consider the full path for tracking but, instead, selects a subset of spatially adjacent keyframes by retrieving an *Active Keyframe Window* from the quad-tree.

Local visual features associated with such keyframes, *e.g.* SURF [20] or SIFT [21] descriptors, are gathered and stored into a local *Feature Pool*, which is a KD-Tree forest [22] enabling efficient feature matching.

The Camera Tracking module is the first to execute, taking the RGB-D frame coming from the sensor as input and returning the estimated camera pose. Purposely, visual features are extracted from the RGB image and matched into the *feature pool* to find correspondences between the current frame and the local map. Using the associated depth measurements, matching pixels are back-projected in the 3D space leading to 3D correspondences. Accordingly, a full 6DOF pose can be robustly estimated by running a standard Absolute Orientation algorithm [23] within a RANSAC framework. Camera pose is represented as a 4×4 matrix in the following format:

$$T = \begin{pmatrix} R & t \\ \mathbf{0}_3^\top & 1 \end{pmatrix}, \tag{1}$$

where R represents a 3×3 rotation matrix and t is a translation vector. Points are projected onto the image plane by means of the camera matrix

$$K = \begin{pmatrix} f_x & 0 & c_x \\ 0 & f_y & c_y \\ 0 & 0 & 1 \end{pmatrix}, \tag{2}$$

where f_x, f_y are the focal lengths and c_x, c_y the principal points. Then, given a pixel position $m = (u, v)$ and its associated depth measurement D, its 3D back-projection is computed as

$$p = K^{-1} \cdot \begin{pmatrix} u \cdot D \\ v \cdot D \\ D \end{pmatrix}. \tag{3}$$

Then, as detailed in [10], if the currently tracked frame exhibits a limited overlap with respect to the Local Map, it is promoted as a new keyframe, thereby triggering the Local Optimization module and updating the Local Map itself, which, accordingly, gets centered around the newly spawned keyframe.

The Local Optimization module is in charge of optimizing poses across a pose graph associated with the keyframes to minimize the reconstruction error. The cost function is expressed as a sum of squared residuals, where each residual draws from a successful match:

$$r_{ij} = p_i - T_i^{-1} T_j q_j. \tag{4}$$

Here, (p_i, q_j) is the 3D point pair derived from a 2D feature match, while T_i and T_j are the associated camera poses. The optimization considers only the keyframe poses within a certain distance along the graph, in order to preserve local consistency while being able to scale smoothly to large workspaces. Indeed, a global pose graph optimization, though usually reaching a more accurate solution, would grow in complexity and time requirements with the number of keyframes.

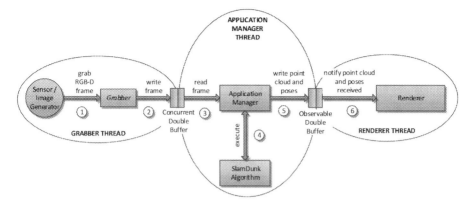

Fig. 2. The software architecture of the mobile application, divided into three main execution modules: Grabber, Application Manager and Renderer.

4   Software Architecture

A global system overview of the proposed framework is shown in Fig. 2. The application consists of three main threads, namely, the *Grabber* (see Sect. 4.1), the *Application Manager* (see Sect. 4.2) and the *Renderer* (see Sect. 4.3). The next sub-sections describe these components, focusing on their specific implementation on an Android mobile platform.

4.1   Grabber

The Grabber entity is used to abstract from the actual hardware device that would gather RGB-D frames. In this way, the other modules can execute independently and the specific kind of sensor will not influence execution of the SLAM algorithm. This entity is also used for testing purposes to emulate the behavior of an RGB-D camera. Different implementations of the actual Grabber may be easily included in the future, thus allowing for easy expansion of the range of sensors supported by the application.

The Grabber entity executes independently of the other components as a single thread which takes the data coming from the specific RGB-D sensor (step 1 in Fig. 2) and inserts them into a double buffer, called *Concurrent Double Buffer* (step 2 in Fig. 2). The choice of a double buffer is used in a multi-threaded environment to handle concurrency issues: the read operation on the buffer would block the associated thread if the data were not yet available, while the write operation frees the threads waiting for the data, if any.

4.2   Application Manager

The Application Manager is the most resource and time consuming thread, as it endlessly collects new data and runs the core SLAM algorithm, *i.e.* the adapted

SlamDunk pipeline. First, RGB-D data are read from the input buffer previously filled by the Grabber (step 3 in Fig. 2). Then, the SlamDunk algorithm is executed (step 4 in Fig. 2). In case of tracking failure, the frame is discarded, otherwise the estimated camera pose is retrieved, together with updated keyframe poses had an optimization been run.

Should tracking be successful and the current frame promoted to keyframe, each pixel (u, v) on the RGB image is then expressed in 3D homogeneous coordinates and its associated depth value D is applied as:

$$
\begin{pmatrix} u \cdot D \\ v \cdot D \\ D \\ 1 \end{pmatrix}, \tag{5}
$$

without including the effect of the camera parameters. This is the only projection executed by the Application Manager, thus avoiding the multiplication with the inverse of the camera matrix and then with the pose estimated by the SLAM algorithm. Instead, we extend the inverse of the camera matrix to 4×4 as

$$
\mathrm{K}_e^{-1} = \begin{pmatrix} \mathrm{K}^{-1} & \overline{\mathbf{0}}_3 \\ \overline{\mathbf{0}}_3^{\top} & 1 \end{pmatrix} \tag{6}
$$

and produce the affine transformation which is given as input to the Renderer together with the partially projected points:

$$
\mathrm{T}_f = \mathrm{T} \cdot \mathrm{K}_e^{-1}. \tag{7}
$$

The point cloud generated by the projection, together with the model matrix and the other matrices representing updated poses, are inserted into the *Observable Double Buffer* (step 5 in Fig. 2). This buffer follows the Observer design pattern: the entity that observes the buffer, in this case the Renderer, is notified whenever the data changes. Accordingly, updates are promptly reported to the user interface as soon as new data become available.

4.3   Renderer

The final module in our mobile application is the Renderer, which manages the user interface. The Renderer is notified when a new keyframe has been detected or existing poses have been updated in the Observable Double Buffer (step 5 in Fig. 2). Then, it reads the points and the affine transformations (step 6 of Fig. 2) to show them properly on the screen of the mobile device. The Renderer draws the new point cloud as well as updates those that have their poses updated because of a local optimization.

To improve the overall efficiency, the actual multiplication between the points and the corresponding affine matrix has been delegated to the graphical pipeline. In our specific case, the rendering part has been developed using *OpenGL ES 2* [24].

5   SlamDunk on a Mobile Platform

The SlamDunk pipeline proposed in [10] has been modified to better match the capabilities of modern mobile devices. The main changes refer to the feature detection, description and matching approaches. Indeed, the visual features algorithms deployed by the desktop version, *i.e.* SURF [20] and SIFT [21], cannot be run at interactive frame-rate on a modern mobile platform. Moreover, the descriptors generated by these algorithms are quite long (up to 128 floating-point numbers), so that a nearest neighbor search based on calculation of the Euclidean distance does not allow for sufficiently fast matching, further limiting speed on a mobile platform.

To overcome these limitations, several other feature detectors and descriptors have been considered. We list below the most promising variants and, in Sect. 6, compare them in terms of both accuracy and speed:

- ORB keypoint detector and feature descriptor [25];
- ORB keypoint detector and BRISK feature descriptor [26];
- Upright-SURF (U-SURF) optimized keypoint detector and BRISK feature descriptor.

The implementation of the above methods have been taken from the OpenCV library [27]. However, we have further optimized the U-SURF detector for a mobile platform by means of ARM NEON parallel instructions. It is noteworthy to point out that the feature descriptors chosen, namely ORB and BRISK, are binary descriptors that allow for replacing the Euclidean distance with the faster Hamming distance during nearest-neighbor search.

Fig. 3. The user interface of the developed Android application.

6   Experimental Results

In this section we present quantitative and qualitative results obtained by running the mobile version of SlamDunk with the different algorithms considered for feature detection and description. We also present an estimate of the amount of memory used by the application. Figure 3 shows the user interface of the developed Android application. The reference mobile platform used in all the experiments is a Samsung Galaxy Tab Pro 10.1 tablet.

As for quantitative experiments, we have processed several sequences from the publicly available TUM RGB-D benchmark dataset [14], which includes color and depth streams acquired by a Microsoft Kinect and an Asus Xtion Pro Live, together with ground truth data obtained by a motion capture system tracking camera movements. Results are shown in Table 1. Clearly, the *U-SURF + BRISK* variant provides overall the best accuracy. We ascribe this to the quality of SURF keypoints, which, in particular, show higher repeatability than ORB features across viewpoint changes. Indeed, between the other two variants, *i.e. ORB + ORB* and *ORB + BRISK*, the difference is far less evident. These two variations can be considered equivalent in terms of accuracy, though they differ significantly in terms of timing performance, as discussed below.

As for the efficiency of feature algorithms, we have measured separately the mean time spent for detection and description (see Table 2). As expected, the ORB detector is much faster than U-SURF, due to the simpler pipeline of the algorithm. On the other hand, the BRISK feature descriptor is considerably more efficient than ORB. Indeed, it is also worth noticing that the combination *ORB + BRISK* is faster than *ORB + ORB*, the former taking an average time of about 35 ms compared to the 100 ms required by the latter. The combination *U-SURF + BRISK* runs at an approximate speed of about 84 ms. However, as shown above, it has the advantage of a notably higher precision. Therefore, we consider *U-SURF + BRISK* as the preferred feature detection and description

Table 1. RMSE of the absolute trajectory error (meters) for several sequences of the TUM RGB-D benchmark.

Sequence	ORB + ORB	ORB + BRISK	U-SURF + BRISK
fr1/floor	0.058	0.055	**0.051**
fr1/desk	0.049	0.052	**0.042**
fr1/room	0.270	0.278	**0.140**
fr3/structure_texture_near	0.092	0.047	**0.025**
fr3/structure_texture_far	0.052	0.045	**0.028**
fr3/nostructure_texture_near_with_loop	0.046	0.057	**0.030**
fr3/nostructure_texture_far	0.178	0.139	**0.083**
fr3/long_office_household	0.058	0.063	**0.041**

Table 2. Average execution times (milliseconds) for the different feature combinations.

Algorithms	Detection	Description
ORB + ORB	**28**	72
ORB + BRISK	**28**	**7**
U-SURF + BRISK	70	14

Table 3. Execution times of the other SlamDunk modules.

Operation	Time of Execution
Feature Matching	20–40 ms
Robust Pose Estimation	1–10 ms
Local Optimization	20 ms–2 s or more

approach for our current implementation of a SLAM system in a mobile environment, while *ORB + BRISK* turns out a more favorable choice than *ORB + ORB* due to faster speed and equivalent accuracy.

In Table 3 we highlight the efficiency of the other steps of the SlamDunk algorithm in terms of maximum and minimum execution times. Though the actual times vary depending on the observed scene, it is clear that graph optimization is by far the most time consuming step, especially when a high number of feature matches constrains the poses. However, a local optimization is run only when a new portion of environment starts being explored and a new keyframe is spawn; thus we are able to run the application at an average speed of about 10 frames per second using the *ORB + BRISK* variant, while *U-SURF + BRISK* can work at an average speed of about 6–7 frames per second.

The final reconstructions can also be qualitatively inspected. Figures 5 and 6 show two reconstructions from the TUM RGB-D benchmark, while Fig. 4 is an indoor environment explored by connecting an Asus Xtion Pro Live camera to our Galaxy Tab Pro 10.1 tablet. These reconstructions have been attained using the *U-SURF + BRISK* variant of the algorithm.

With regard to the memory footprint, we report now an estimate of the amount of memory required by the SlamDunk algorithm as well as the visualization module. Every keyframe stored in the map contains a list of features, each one having a length set by the algorithm used for the extraction. Also, for each feature we keep the 3D point representing the projected keypoint in camera space, expressed by 3 floating-point values, and an integer index used to find the corresponding keyframe during the feature matching operation. Considering both the ORB and BRISK algorithms, which define, respectively, a 32- and 64-byte descriptor, the memory needed by 100 keyframes is computed as follows:

$$M_{ORB} = (32\,\text{B} + 3 \cdot 4\,\text{B} + 4\,\text{B}) \cdot 100 \cdot 500 = 2400000\,\text{B} \approx 2.3\,\text{MB}, \quad (8)$$

$$M_{BRISK} = (64\,\text{B} + 3 \cdot 4\,\text{B} + 4\,\text{B}) \cdot 100 \cdot 500 = 4000000\,\text{B} \approx 3.8\,\text{MB}, \quad (9)$$

Fig. 4. 3D reconstruction of a kitchen using the Asus Xtion Pro Live camera.

where we have considered an average number of 500 features extracted from each keyframe. It is worth to notice that the SIFT and SURF descriptors used by the desktop version have a much higher memory consumption, being the feature vector composed of 64 or 128 floating-point numbers.

As for the pose graph, for each node a quaternion and a translation vector are stored as 7 double-precision floating point numbers, coupled with a unique integer index; considering 100 keyframe, it amounts to

$$M_{poses} = (7 \cdot 8\,\mathrm{B} + 4\,\mathrm{B}) \cdot 100 = 6000\,\mathrm{B} \approx 6\,\mathrm{KB}. \qquad (10)$$

Fig. 5. 3D reconstruction of a wooden floor from dataset fr1/floor.

Fig. 6. 3D reconstruction of some objects from dataset fr3/structure_texture_far.

For each feature match a constraint is created and the corresponding 3D points are kept in double-precision format. Assuming an average number of 200 matches per keyframe during the tracking phase, we get

$$M_{matches} = 2 \cdot 3 \cdot 8\,\mathrm{B} \cdot 200 \cdot 100 = 960000\,\mathrm{B} \approx 938\,\mathrm{KB}. \tag{11}$$

The optimization routine allocates space for the Hessian matrix as a collection of 6×6 double-precision blocks, one for each keyframe and one for each pair of keyframes connected by an edge in the graph. If each keyframe is connected, on average, to 10 other keyframes in the local map at the end of the tracking phase, we reach a memory usage of

$$M_{hessians} = 6 \cdot 6 \cdot 8\,\mathrm{B} \cdot (1 + 10) \cdot 100 = 316800\,\mathrm{B} \approx 310\,\mathrm{KB}. \tag{12}$$

We are now able to estimate the memory usage of the entire SlamDunk algorithm for both ORB and BRISK features:

$$M_{SlamDunkORB} = 2.3\,\mathrm{MB} + 6\,\mathrm{KB} + 938\,\mathrm{KB} + 310\,\mathrm{KB} \approx 3.5\,\mathrm{MB}. \tag{13}$$

$$M_{SlamDunkBRISK} = 3.8\,\mathrm{MB} + 6\,\mathrm{KB} + 938\,\mathrm{KB} + 310\,\mathrm{KB} \approx 5.0\,\mathrm{MB}. \tag{14}$$

These values highlight the low memory requirement of the SlamDunk framework and, also, they are suitable for a mobile environment, which often shows limited memory availability. As for the storage and visualization of point clouds, for each 3D point its position and its color have to be saved, the former as three single-precision floating-point values, the latter as three bytes accounting for the RGB components. Considering 100 keyframes at VGA resolution, we get:

$$M_{PointClouds} = 640 \cdot 480 \cdot (3 + 3 \cdot 4\,\text{B}) \cdot 100 = 460800000\,\text{B} \approx 439\,\text{MB}. \quad (15)$$

Although this value appears very high, usually not all the image pixels need to be projected into the 3D space. Indeed, sub-sampling the depth images by a factor of two can reduce the memory occupancy by 75 %, *i.e.* $\approx$110 MB.

7   Conclusions

We have presented an implementation of a state-of-the-art RGB-D SLAM algorithm on a mobile platform. The original pipeline has been modified and adapted to better cope with limited memory and hardware resources, leading to an application running at interactive frame-rate on an Android tablet. We have investigated on diverse feature detection and description approaches, thereby determining that the combination Upright-SURF and BRISK provides the preferred trade-off between reconstruction quality and responsiveness. Though our mobile SLAM application can run at about 6–7 Hz, or even faster when deploying ORB keypoints (10 Hz), the timing performance of the local optimization step should be addressed in the future. The main challenge here is to keep a bounded maximum complexity, *e.g.* by marginalization of 3D point matches in terms of pose-to-pose constraints, while locally refining the trajectory and reducing the overall reconstruction error. Finally, we hope that our work may foster further research on the new topic of mobile RGB-D SLAM, so to devise soon more efficient solutions that could deliver real-time operation together with high reconstruction quality.

References

1. Borrmann, D., Elseberg, J., Lingemann, K., Nüchter, A., Hertzberg, J.: Globally consistent 3d mapping with scan matching. J. Robot. Auton. Syst. **56**, 130–142 (2008)
2. Montemerlo, M., Thrun, S.: Large-scale robotic 3-d mapping of urban structures. In: International Symposium on Experimental Robotics (ISER), Singapore (2004)
3. Davison, A., Reid, I.D., Molton, N.D., Stasse, O.: MonoSLAM: real-time single camera SLAM. IEEE Trans. Pattern Anal. Mach. Intell. (PAMI) **29**, 1052–1067 (2007)
4. Klein, G., Murray, D.: Parallel tracking and mapping for small ar workspaces. In: International Symposium on Mixed and Augmented Reality (ISMAR), pp. 225–234 (2007)
5. Newcombe, R., Lovegrove, S., Davison, A.: DTAM: dense tracking and mapping in real-time. In: International Conference on Computer Vision (ICCV), pp. 2320–2327 (2011)
6. Occipital Inc.: The Structure sensor (2014). http://structure.io/
7. Creative Technology Ltd.: The Creative Senz3D sensor (2014). http://us.creative.com/p/web-cameras/creative-senz3d

8. Google Inc.: Project Tango (2014). https://www.google.com/atap/projecttango/
9. Kim, D., Davison, A., Kohli, P., Shotton, J., Hodges, S., Fitzgibbon, A.: KinectFusion: Real-time dense surface mapping and tracking. In: 10th IEEE International Symposium on Mixed and Augmented Reality (ISMAR), pp. 127–136, Washington, DC, USA (2011)
10. Fioraio, N., Di Stefano, L.: SlamDunk: affordable real-time RGB-D SLAM. In: ECCV Workshop on Consumer Depth Cameras for Computer Vision, Zurich, Switzerland (2014)
11. Bylow, E., Sturm, J., Kerl, C., Kahl, F., Cremers, D.: Real-time camera tracking and 3d reconstruction using signed distance functions. In: Robotics: Science and Systems (RSS), Berlin, Germany (2013)
12. Whelan, T., Kaess, M., Leonard, J., Mcdonald, J.: Deformation-based loop closure for large scale dense RGB-D SLAM. In: International Conference on Intelligent Robot Systems (IROS), Tokyo, Japan (2013)
13. Henry, P., Krainin, M., Herbst, E., Ren, X., Fox, D.: RGB-D mapping: Using kinect-style depth cameras for dense 3D modeling of indoor environments. Int. J. Robot. Res. **31**, 647–663 (2012)
14. Sturm, J., Engelhard, N., Endres, F., Burgard, W., Cremers, D.: A benchmark for the evaluation of RGB-D SLAM systems. In: International Conference on Intelligent Robot Systems (IROS) (2012)
15. Endres, F., Hess, J., Sturm, J., Cremers, D., Burgard, W.: 3D mapping with an RGB-D camera. IEEE Transactions on Robotics (2013)
16. Klein, G., Murray, D.: Parallel tracking and mapping on a camera phone. In: International Symposium on Mixed and Augmented Reality (ISMAR), pp. 83–86 Orlando, Florida, USA (2009)
17. Lee, W., Kim, K., Woo, W.: Mobile phone-based 3D modeling framework for instant interaction. In: International Conference on Computer Vision Workshops (ICCV Workshops) (2009)
18. Pan, Q., Arth, C., Rosten, E., Reitmayr, G., Drummond, T.: Rapid scene reconstruction on mobile phones from panoramic images. In: International Symposium on Mixed and Augmented Reality (ISMAR) (2011)
19. Tanskanen, P., Kolev, K., Meier, L., Camposeco, F., Saurer, O., Pollefeys, M.: Live metric 3D reconstruction on mobile phones. In: International Conference on Computer Vision (ICCV), Sydney, Australia (2013)
20. Bay, H., Ess, A., Tuytelaars, T., Gool, L.V.: Speeded-up robust features (SURF). Comput. Vis. Image Underst. **110**, 346–359 (2008)
21. Lowe, D.G.: Distinctive image features from scale-invariant keypoints. Int. J. Comput. Vis. (IJCV) **60**, 91–119 (2004)
22. Muja, M., Lowe, D.G.: Fast approximate nearest neighbors with automatic algorithm configuration. In: International Conference on Computer Vision Theory and Applications (VISAPP) (2009)
23. Arun, K.S., Huang, T.S., Blostein, S.D.: Least-squares fitting of two 3-d point sets. IEEE Trans Pattern Anal. Mach. Intell. (PAMI) **9**, 698–700 (1987)
24. Segal, M., Akeley, K.: The opengl graphics system: a specification (2004). http://www.opengl.org/documentation/specs/version2.0/glspec20.pdf
25. Rublee, E., Rabaud, V., Konolige, K., Bradski, G.: ORB: an efficient alternative to SIFT or SURF. In: International Conference on Computer Vision (ICCV), Barcelona, Spain (2011)
26. Leutenegger, S., Chli, M., Siegwart, R.: BRISK: binary robust invariant scalable keypoints. In: International Conference on Computer Vision (ICCV) (2011)
27. Bradski, G.: The OpenCV library. Dr. Dobb's J. Softw. Tools **25**, 120–126 (2000)

3D Reconstruction with Automatic Foreground Segmentation from Multi-view Images Acquired from a Mobile Device

Ping-Cheng Kuo[✉], Chao-An Chen, Hsing-Chun Chang,
Te-Feng Su, and Shang-Hong Lai

Department of Computer Science, National Tsing Hua University,
Hsinchu, Taiwan
vul3kuo@hotmail.com
http://cv.cs.nthu.edu.tw/index.php

Abstract. We propose a novel foreground object segmentation algorithm for a silhouette-based 3D reconstruction system. Our system requires several multi-view images as input to reconstruct a complete 3D model. The proposed foreground segmentation algorithm is based on graph-cut optimization with the energy function developed for planar background assumption. We parallelize parts of our program with GPU programming. The 3D reconstruction system consists of camera calibration, foreground segmentation, visual hull reconstruction, surface reconstruction, and texture mapping. The proposed 3D reconstruction process is accelerated with GPU implementation. In the experimental result, we demonstrate the improved accuracy by using the proposed segmentation method and show the reconstructed 3D models computed from several image sets.

1 Introduction

Three-dimensional reconstruction from multi-view images is a challenging problem in computer vision. In general, most previous 3D reconstruction systems are based on point correspondences between multi-view images [1–3]. In this paper, we develop a 3D object reconstruction system based on visual hull, which does not require any point correspondences on the object. In this work, we use a handheld camera to capture images surrounding the target object freely from different views. Compared to other handheld devices (e.g. Kinect [4]), our system has the advantages of low cost and ease of use. An example of input images are shown in Fig. 1, and the flow chart of our system is depicted in Fig. 2.

Inspired by GraphCut algorithm, many papers apply GrphCut algorithm into image segmentation such as Boykov and Jolly's work [5], Pham et al. [6], and Rother et al. [7]. The image segmentation method proposed in this paper is aspired by [6,7]. We assume that the target object be placed roughly in the middle of each image. So firstly, we define a fixed bounding box in the center region to reduce user interaction. Then we compute the homography between the

© Springer International Publishing Switzerland 2015
C.V. Jawahar and S. Shan (Eds.): ACCV 2014 Workshops, Part III, LNCS 9010, pp. 352–365, 2015.
DOI: 10.1007/978-3-319-16634-6_26

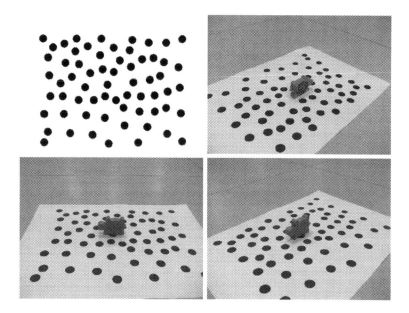

Fig. 1. The left top image is the pattern image. The others are input images to our system. The target object is placed in the middle of each image. These images are taken from various viewpoints around the target object.

pattern image and input image to obtain an initial guess of the bounding box. Finally we use GraphCut optimization to label the region in the bounding box.

The segmentation methods in [6,7] can be applied for images of natural environment because pixels with color similar to those of the reference background will be labeled as foreground as long as there is a little difference between them. However, this property may not be appropriate for our dataset. With illumination change and different shadow of different view point, the background in the bounding box may contain some difference to the reference background outside the box. In this work, we propose to combine the color distribution constraint citeconfspscvprspsPhamTN11 and the multi-view homography analysis (MHA) to solve this problem. We speed up the entire system toward segmentation and image-based visual hulls via GPU programming. For segmentation, we use GraphCut to label each pixel with foreground or background. We take each working item to roughly compute each pixels initial label for GraphCut. In image-based visual hull, each working item converts the 2D pixels to 3D points and find local neighbors for each 3D point to compute its normal vector. The 3D reconstruction process is quite time consuming, and it can be sped-up by taking advantage of parallel computing. The remainder of this paper is organized as follows. Section 2 introduces our 3D reconstruction system. In Sect. 3, we are going to describe our proposed segmentation method in this paper. And we describe the GPU acceleration of the proposed 3D reconstruction system in

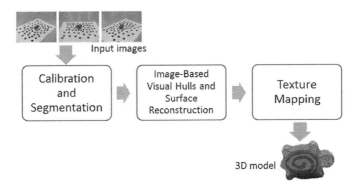

Fig. 2. The flow chart of our system. Our system only needs some pictures as input images to reconstruct the 3D model. It consists of camera calibration, foreground segmentation, image-based visual hulls, surface reconstruction, and texture mapping.

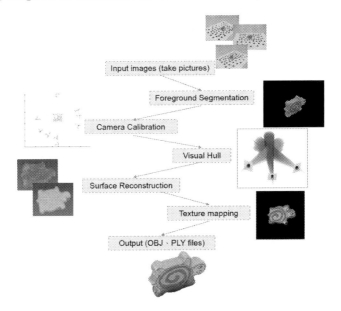

Fig. 3. The flow chart of our 3d reconstruction system.

Sect. 4. The experimental results are given in Sect. 5, followed by conclusion in Sect. 6 (Fig. 3).

2   Proposed 3D Reconstruction System

2.1   Foreground Segmentation

In the beginning of our 3D reconstruction system, we need to extract the foreground region (the object we would like to reconstruct) from the background which has a preset dot pattern. The dot pattern is used for camera calibration in step Sect. 2.2. One of the challenges in foreground segmentation step is to remove the background which has dot pattern. Most segmentation methods lead to unfavorable segmentation results in this case, since they tend to be regard the black dot as the foreground. As a result, we propose a novel approach for image segmentation, which can produce a better result by solving the foreground segmentation for multiple images of the scene simultaneously in a graphcut framework. In this paper, we focus on the foreground segmentation algorithm, which will be described in details in Sect. 3.

2.2   Camera Calibration

Our calibration utilizes a dot pattern to compute the associated camera projection matrix [8]. It is based on using a multi-view stereo algorithm to calibrate an image sequence, which is a dot pattern on the planar in calibration. We print the dot pattern paper provided from the website, and then place the object on the top of the dot pattern paper. The pattern is designed for detecting the dots from a different view and then finding the corresponding dots in each different view. We can take images around the object on the pattern to obtain camera intrinsic and extrinsic parameters. Once both camera parameters are obtained, we can compute the camera projection matrix. Thus the homography matrices can be estimated by the corresponding dots from every two images.

2.3   Visual Hull

In this step, we will obtain a point cloud of 3D object and the normal vectors for all data points by applying the image-based visual hulls (IBVH) [9] after we compute the camera projection matrix and a set of silhouettes of foreground images. The IBVH algorithm [9] is an efficient image-based approach to computing and shading visual hulls from silhouette image data. In our system, we use the IBVH algorithm [9] to obtain the point cloud of 3D object, and we perform surface reconstruction from the point cloud in step Sect. 2.4.

2.4   Surface Reconstruction

The mesh of a 3D model is reconstructed in this step. After obtaining a point cloud for a 3D object in step Sect. 2.3, we apply the Poisson surface reconstruction [10] to reconstruct the mesh of a 3D object. Poisson surface reconstruction formulates surface reconstruction as the solution to a Poisson equation, and creates watertight surfaces from oriented point sets finally.

2.5   Texture Mapping

The last step is texture mapping. The texture mapping technique used in this work is multiresolution spline [11]. This method combines two or more images into a larger image mosaic, and we map this image mosaic to the mesh generated in step Sect. 2.4.

3   Proposed Foreground Segmentation Method

We define a fixed bounding box (see Fig. 4(a)) and users are requested to place the target object in the middle of images. Inspired by [6], we state this problem as an energy minimization problem (see Fig. 4(b)). Then we add a multi-view homography constraint term into the energy function. Unlike the iterated distribution matching in [6], we use homography relationship between the input image and pattern image to compute a good initial guess. Hence we can obtain the segmentation result efficiently by using GraphCut to minimize the energy function.

3.1   The Color Distribution Method (CDM)

According to the three conditions and the method in [6], we implement their energy function and find that their method is sensitive to the small color distance. Even if there only exists small color difference between the pixel color values in the bounding box and the reference background, their method may determine the pixel as foreground (see Fig. 5), which is not appropriate. With the illumination change and different shadow due to different viewpoint, the background in the bounding box may exist some difference to the reference background outside the box (see Fig. 6).

Inspired by Pham et al.'s work [6], we state the problem as the minimization of an energy function, and apply the following function as the energy function.

$$E_1\left(L\right) = \underbrace{B\left(P_1\left(L\right), H\right)}_{F_1(L)} - \underbrace{B\left(P_0\left(L\right), H\right)}_{F_2(L)} + \lambda S\left(L\right) \tag{1}$$

There exists a segmentation L that separates the image region into background $R_0^L = \{p|L_p = 0\}$ and $R_1^L = \{p|L_p = 1\}$. Where $P_i\left(L\right)$ is the distribution of the region R_i^L and $B\left(f, g\right)$ is the Bhattacharyya distance which expresses the similarity of the two distributions. The terms $F_1\left(L\right)$ and $F_2\left(L\right)$, which evaluate global similarities between distributions, cannot be express purely in the form of data and smoothness terms. It follows that cannot be solved directly by Graph-Cut. Therefore, Pham et al. [6] estimated an upper bound of $F_1 + F_2$ and it can be expressed purely by data terms. $S\left(L\right)$ is a smoothness term that enforces smoothness property to L.

$$S\left(L\right) = \sum_{(p,q)\in N} \delta_{L_p \neq L_q} \left(\frac{1}{1 + \|I_p - I_q\|^2} + \frac{\epsilon}{\|p - q\|}\right) \tag{2}$$

3.2   Multi-view Homography Analysis (MHA)

To decrease the error resulting from shadow or illumination changes, we apply multi-view homography analysis with the homography matrix, H, obtained from

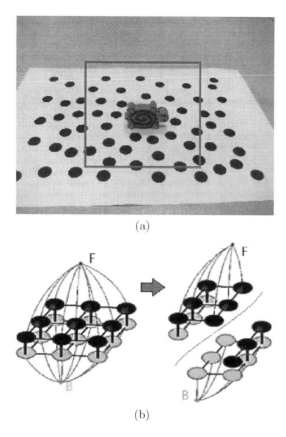

(a)

(b)

Fig. 4. In (a), the red frame is the fixed bounding box (i.e., each of our image has the same bounding box). (b) shows the energy minimization problem. Each pixel is a node and the sum of all edge weights is the total energy. We can minimize the energy by using GraphCut algorithm (Color figure online).

calibration to integrate more reliable information into the data term of the energy function. Let F^i be the i-th frame, N be the number of neighbor frames which are included into the multi-view analysis for F^i. In our experiment, we set N to 6. Let F^n be a neighbor frame of F^i, and the set of all neighboring frames are denoted by:

$$F^{i,N} = \left\{ F^n \mid i - \frac{N}{2} < n < i + \frac{N}{2}; n \neq i \right\} \qquad (3)$$

If n is less than 0, we take neighbor frames from the tail frame. For each pixel p in F^i, there is a likelihood, W_p^i, which ranges from 0 to 1.

$$W_{p^i}^i = \lambda V_{p^i}^{i,C} + (1 - \lambda) V_{p^i}^{i,G} \qquad (4)$$

The higher value $W_{p^i}^i$ is, the more probable it is a foreground pixel. Otherwise, the pixel is more probable to be background. $W_{p^i}^i$ is determined by the color distortion $V_{p^i}^{i,C}$ and gain information $V_{p^i}^{i,G}$. λ is used to control the importance of these two terms. $V_{p^i}^{i,C}$ is the color distortion of p on the i-th frame, which is related with the angle between two color vectors.

$$V_{p^i}^{i,C} = 1 - exp\left(-\left(P_{p^i}^{i,C}\right)^2 * \alpha\right) \tag{5}$$

$$P_{p^i}^{i,C} = \frac{\sum_{n\epsilon N} f_{cd}\left(R_{p^i}^i, R_{p^{i+n}}^{i+n}\right)}{N} \tag{6}$$

$$f_{cd}(u, v) = \cos^{-1}\left(\frac{u \cdot v}{|u|\,|v|}\right) \tag{7}$$

Here, R_b^a means the color vector $[r, g, b]$ of pixel b in the a-th frame. p^{i+n} means the corresponding pixel of p^i on the neighbor frame via homography matrix, H_{i+n}

$$P^{i+n} = H_{i+n} * p^i \tag{8}$$

If $R_{p^i}^i$ and $R_{p^{i+n}}^{i+n}$ are similar (e.g. dark green and light green), the angle will be smaller, hence $V_{p^i}^{i,C}$ will be more capable to overcome the problem of illumination variation. $V_{p^i}^{i,G}$ is the gain information of p^i on the i-th frame, which is related with illumination as follows:

$$V_{p^i}^{i,G} = 1 - exp\left(-\left(P_{p^i}^{i,G}\right)^2 * \beta\right) \tag{9}$$

$$P_{p^i}^{i,G} = \frac{\sum_{n\epsilon N} f_{gain}\left(I_{p^i}^i, I_{p^{i+n}}^{i+n}\right)}{N} \tag{10}$$

$$f_{gain}(I_a, I_b) = \frac{|I_a - I_b|}{I_b} \tag{11}$$

Here, I_b^a means the intensity of pixel b in the a-th frame. The gain value will be high when the brightness difference between two pixels is large. Since the gain value for background shadow is still relatively small with foreground, it can be used to overcome the shadow problem. $V_{p^i}^{i,C}$ and $V_{p^i}^{i,G}$ are the color distortion and gain information in F^i. Exponential function is used to convert the range to $(0, 1]$. To expand the difference between the foreground and background values, we take the square of $P_{p^i}^{i,C}$ and then multiply it by a constant α. Similarly, we take the square of $P_{p^i}^{i,G}$ and multiply it by a constant β.

3.3 Initial Guess and Energy Minimization

We add the MHA information into the data term of the energy function referred to [6] and it can be denoted as follows:

$$Q(L, L^*, \alpha)$$
$$= \sum_{p \in R_1^L} m_p(1) + W_p + (sign(F(L^*))\alpha + 1) \sum_{p \in R_0^L} m_p(0) \qquad (12)$$

where $m_p(1)$ and $m_p(0)$ are data terms given for each p:

$$m_p(1) = \frac{\delta_{L_p^*=0}}{2A(R_1^{L^*})} \sum_{z \in Z} K_z(I_p) \sqrt{\frac{H(z)}{P_1^{L^*}(z)}}$$
$$+ \frac{\delta_{L_p^*=0}}{A(R_0^{L^*})} \left(\sum_{z \in Z} K_z(I_p) \sqrt{\frac{H(z)}{P_1^{L^*}(z)}} + F(L^*) \right)$$
$$m_p(0) = \frac{F(L^*)}{A(R_0^{L^*})} \qquad (13)$$

Here, δ is the Kronecker delta function. When $R_1^{L^*} = \emptyset$(empty), we consider $A(R_1^{L^*})$ and $P_1^{L^*}(z)$ as 1.

We use homography projection to warp each image onto pattern image and use the color distance to obtain an initial guess. Unlike the iterated distribution matching in [6], we use homography relationship between the input image and pattern image to compute a better initial guess, hence we can obtain the segmentation result quickly by using GraphCut.

Fig. 5. The results of [6]. This method can discriminate small color difference between foreground and background (image source: [6]) (Color figure online).

3.4 Model Reconstruction

When we have camera projection matrix and a set of silhouettes of foreground images, we can obtain a point cloud for the 3D object by applying the image-based visual hulls (IBVH) [9], which recovers the space of 3D viewing cones by intersecting them from different views decided by the known camera information.

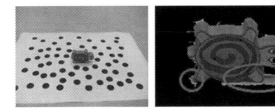

Fig. 6. The method in [6] is too sensitive to consider the shadow region to be foreground.

We set a virtual desired view that is above the model on the pattern in our system, and the remaining images are reference views. There are dense rays from the desired view that intersect with other reference views rays, and then project onto each reference view to get the set of intersecting lines. Finally, we project the intersecting lines on each reference view to the 3D space to get the model crave by the silhouette of foreground object. The algorithm can be speded up by using epipolar constraint [9]. Thus, we can obtain a set of vertices in 3D coordinates. The vertices are point cloud of the 3D model which is segmented by silhouette of reference views. We then compute the normal of each 3D vertex by finding the neighbor vertices. It is based on using the viewing direction of the ray, the segmented information provided by the reference image and the neighbors of the ray on the desired view. While we have point cloud and the normal vectors for all data points, we can apply Poisson surface reconstruction technique [10] and texture mapping algorithm to create the reconstructed 3D model.

4   GPU Acceleration

In the 3D reconstruction system, we speed up the foreground segmentation and image-based visual hulls components via parallelization with GPU programming. We are going to illustrate each part in details in the following.

4.1   Speed up Foreground Segmentation

In foreground segmentation, in order to determine each node's label, we apply GraphCut to solve this problem. We assume each node is a pixel that limit to two possible labels: foreground pixel, background pixel. It is necessary to give an initial label as input to the GraphCut procedure.

We take each GPU working item to compute each image view's color distribution and each pixel's color distance. Each GPU working item represents a pixel in the whole image. We set the homography matrix as global parameters and every GPU working item can compute the color distance from pattern view to other view through the homography matrix. Once obtaining each pixel's color distance, we can roughly decide the initial label from the distance.

4.2   Speed up Image-Based Visual Hull

In the image-based visual hulls, we parallelize two parts in our 3D reconstruction system to speed up the computation. For every reference view and the desired view on the ray intersection, we can obtain the visual hull vertices, which come from the 2D image on a pixel required by the known camera projection information, and they are converted to 3D spatial coordinates. Thus, we first assign each pixel to GPU working item to calculate their 3D coordinates by known projection matrix.

Second, we parallelize the procedure of finding each voxel's neighbors for calculating a voxel's normal, which is required in the Poisson surface reconstruction. Against all of the 3D points we look for the neighbor-ray, and for every ray we select a closest point. Once we obtain a certain number of neighbors of a 3D point, the normal vector of 3D point can be determined by the set of neighboring points.

The GPU acceleration of the above components in our 3D reconstruction system can speed up the computational performance and the result will be shown in Sect. 5.3.

5   Expermemtal Results

In this section, we demonstrate the performance of the proposed segmentation method and compare the segmentation results with the GrabCut algorithm [7] implemented by OpenCV. Besides, we show our reconstruction system by using three image datasets. There are 14 images of resolution 608×456 in each dataset. We evaluate the results of segmentation and 3D model reconstruction.

5.1   Evaluation of Segmentation

We manually label the ground truth and employ error rate as a measurement to evaluate the results

$$errorrate = \frac{No.of\ misclassified\ pixels}{No.pixels\ in\ inference\ region}$$

Figure 7 shows the segmentation results of our proposed method and prior work on several datasets. For each dataset, we report the average error rate in Table 1. According to Table 1, the experimental results show that the average error rate of GrabCut [7] is about 10 %, and the average error rate of the Color Distribution Method (CDM) [6] is much less than the error rate of GrabCut [7]. Our proposed method outperforms the CDM [6] since we include the Multi-view Homography Analysis (MHA) to improve better results. Furthermore, the average error rate of our proposed method is less than 2 % in average from the three datasets, and these errors has quite small influence to the 3D reconstruction results.

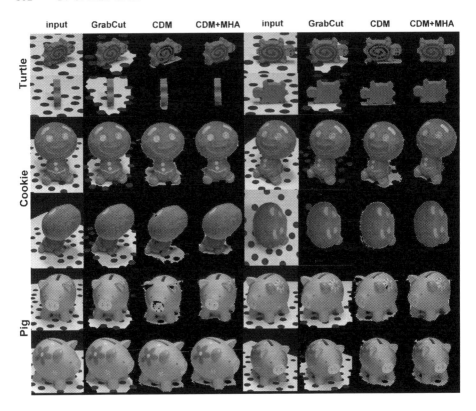

Fig. 7. Comparison of segmentation results. Each row shows results from a different dataset. The columns show (1, 5) input image, (2, 6) the results of OpenCV GrabCut, (3, 7) the results of CDM, and (4, 8) the results of CDM with MHA.

5.2 Evaluation of Reconstructed 3D Model

We use the coverage rate to evaluate 3D models:

$$Coverage\ rate = \frac{P_{GT} \cap P_i}{P_{GT}}$$

where P_i is the plane pixels projected by 3D model reconstructed from segmentation results of i-th method. The coverage rate is the number of pixels overlapped by P_{GT} and P_i divided by the number of all pixels on P_{GT}. The results are reported in Table 2. Our 3D reconstruction results are depicted in Fig. 8.

5.3 Performance by GPU Acceleration

We have developed a set of parallel algorithms to speed up the original sequential version of the 3D reconstruction system, which includes foreground segmentation and image-based visual hulls. In GPU version, each part is speded up by about 4–7 times. The sequential version of the 3D reconstruction process is time

Table 1. Comparison of the average error rate

Dataset	GrabCut [7] (%)	CDM [6] (%)	CDM with MHA (%)
Turtle	10.99	2.99	0.31
Cookie	9.54	4.31	1.22
Pig	10.73	4.10	2.95

Table 2. Comparison of the coverage rates

Method	Coverage rate (%)
GrabCut [7]	84.99
CDM [6]	88.17
CDM with MHA	93.50

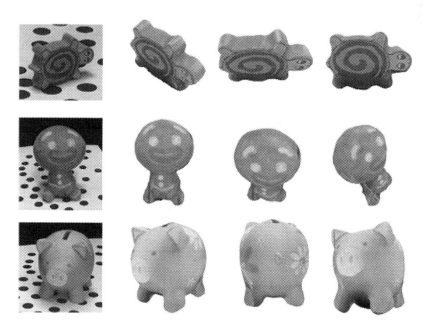

Fig. 8. Our results of 3D model

consuming, taking about 40 to 60 s, and for the parallel version it takes only 20 to 40 s. The GPU acceleration was implemented on AMD Radeon ^{TM} HD 7850 Graphics with OpenCL, and the CPU version was implemented on a Intel i5-3570 CPU with 8 GB RAM. The results are shown in Table 3.

In addition, we also run our system on InFocus M320 mobile phone. Its CPU version uses MediaTek MT6592, 1.7 GHz, 8 core, and the GPU version uses Mali-450. As Fig. 9, we port our system to Android 4.2.2 to evaluate its performance. The kernel which includes segmentation and image-based visual hulls part in the

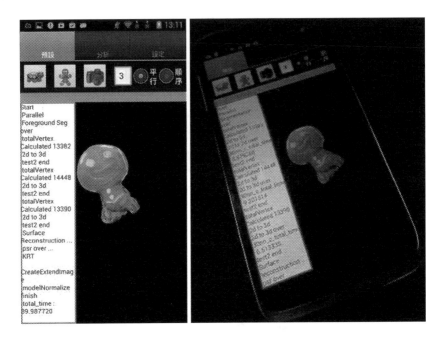

Fig. 9. Our system's srceenshot on mobile phone

Table 3. Comparison of the execution time by the parallel algorithm by GPU and sequential version on CPU on PC platform

Module	GPU (s)	CPU (s)	Speedup ratio
Segmentation	2.208	17.413	7.89
Visual hulls (2D to 3D)	0.009	0.037	4.11
Visual hulls (find neighbor)	0.646	3.983	6.15
Execution time	24.757	43.504	1.76

Table 4. Comparison of the execution time between the parallel algorithm with GPU speedup and the sequential version on ImFocus M320 mobile phone

	Sequential (s)	Parallel (s)	Speedup ratio
Kernel	153.3481	26.34496	5.820777
Total	211.7985	92.04223	2.301102

sequential version takes about 153.3481 s for a 3D reconstruction task, while the parallel version only takes about 26.34496 s. In other words, comparing with the sequential C program, the kernel with GPU acceleration can achieve about 5.8 times speedup in this case (Table 4).

6   Conclusion

We propose a novel approach for image segmentation from multi-view images, which is included in an automatic 3D reconstruction system. For each view, we compute the homography with the first view, and we analyze the multi-view information about the color distortion and gain value. We use these multi-view information to improve the segmentation results of CDM [6]. From our experimental results, we can see the reconstructed 3D model is satisfactory from its visual appearance. In addition, the 3D reconstruction system is quite efficient by using GPU implementation to speed up the computation.

References

1. Hartley, R., Zisserman, A.: Multiple View Geometry in Computer Vision, 2nd edn. Cambridge University Press, New York (2003)
2. Laurentini, A.: The visual hull concept for silhouette-based image understanding. IEEE Trans. Pattern Anal. Mach. Intell. **16**, 150–162 (1994)
3. Pollefeys, M., Koch, R., Gool, L.J.V.: Self-calibration and metric reconstruction inspite of varying and unknown intrinsic camera parameters. Int. J. Comput. Vis. **32**, 7–25 (1999)
4. Wang, R., Choi, J., Medioni, G.G.: Accurate full body scanning from a single fixed 3d camera. In: 3DIMPVT, pp. 432–439. IEEE (2012)
5. Boykov, Y., Jolly, M.P.: Interactive graph cuts for optimal boundary and region segmentation of objects in n-d images. In: ICCV, pp. 105–112 (2001)
6. Pham, V.Q., Takahashi, K., Naemura, T.: Foreground-background segmentation using iterated distribution matching. In: CVPR, pp. 2113–2120. IEEE (2011)
7. Rother, C., Kolmogorov, V., Blake, A.: "GrabCut": interactive foreground extraction using iterated graph cuts. ACM Trans. Graph. **23**, 309–314 (2004)
8. Vogiatzis, G., Hernandez, C.: Automatic camera pose estimation from dot pattern (2010). http://jabref.sourceforge.net/help/LabelPatterns.php
9. Matusik, W., Buehler, C., Raskar, R., Gortler, S.J., McMillan, L.: Image-based visual hulls. In: SIGGRAPH, pp. 369–374 (2000)
10. Kazhdan, M., Hoppe, H.: Screened poisson surface reconstruction. ACM Trans. Graph. **32**, 29:1–29:13 (2013)
11. Burt, P.J., Adelson, E.H.: A multiresolution spline with application to image mosaics. ACM Trans. Graph. **2**, 217–236 (1983)

Accelerating Local Feature Extraction Using Two Stage Feature Selection and Partial Gradient Computation

Keundong Lee[(⊠)], Seungjae Lee, and Weon-Geun Oh

ETRI, Daejeon, Republic of Korea
`zacurr@etri.re.kr`

Abstract. In this paper, we present a fast local feature extraction method, which is our contribution to ongoing MPEG standardization of compact descriptor for visual search (CDVS). To reduce time complexity of feature extraction, two-stage feature selection, which is based on the feature selection method of CDVS Test Model (TM), and partial gradient computation are introduced. The proposed method is examined on SIFT and compared to SIFT and SURF extractor with the previous feature selection method. In addition, the proposed method is compared to various feature extraction methods of the current CDVS TM 11 in CDVS evaluation framework. Experimental results show that the proposed method significantly reduces the time complexity while maintaining the matching and retrieval performance of previous work. For its efficiency, the proposed method has been integrated into CDVS TM since 107th MPEG meeting. This method will be also useful for feature extraction on mobile devices, where the use of computational resource is limited.

1 Introduction

Local features such as SIFT [14] and SURF [3] are widely used in visual search, object recognition and image classification for their robustness against scale, rotation, illumination changes and affine transformation.

However, these local features are not feasible in mobile visual search or large-scale image retrieval due to large size and high computational complexity. To handle this problem, MPEG-7 is working on the standardization of compact descriptors for visual search (CDVS) [11]. They study on feature selection, feature compression and searching method for mobile visual search. In feature selection, it selects more helpful features for correct match based on probabilistic model [7] instead of using all the detected features to reduce computational burden in descriptor extraction and achieve compactness of descriptor. However, the complexity of feature extraction is still high.

In this paper, the efficient feature extraction with considering feature selection is proposed. To reduce computational complexity, two-stage feature selection and partial gradient computation are introduced. Experimental results verify the effectiveness of the proposed method.

© Springer International Publishing Switzerland 2015
C.V. Jawahar and S. Shan (Eds.): ACCV 2014 Workshops, Part III, LNCS 9010, pp. 366–380, 2015.
DOI: 10.1007/978-3-319-16634-6_27

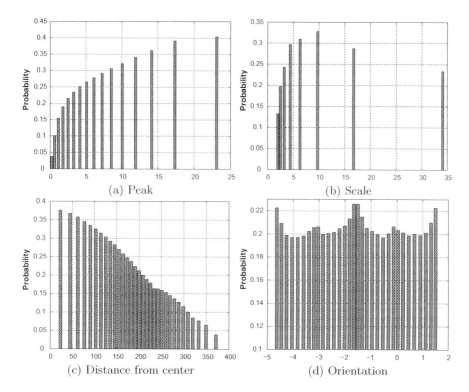

Fig. 1. Conditional probabilities of correct match on SIFT given the peak, scale, distance from center and orientation. (see Footnote 1)

This paper is organized as follows: Sect. 2 describes the feature selection method of CDVS TM. Section 3 introduces the proposed method in detail. Experimental results on uncompressed SIFT framework and CDVS evalation framework are presented in Sect. 4. Finally, we conclude in Sect. 5.

2   Feature Selection for Compact Descriptor

Feature selection is the method to select more important feature among the detected features. G. Francini et al. [7] proposed feature selection based on the characteristics of detector to generate compact descriptor with specific size such as 512 B, 1 KB, and this method has been adopted in CDVS TM. They estimated conditional probability of correct match as a function of four characteristics of the feature: scale, orientation, peak response (e.g., Difference-of-Gaussian (DoG) response), and distance from the center of image as shown in Fig. 1.[1] For each

---

[1] This figure was reproduced by the author with the permission of G. Francini et al. [7] using the program and dataset which they provided during the MPEG meetings. Note that this figure is not just taken from [7], the larger datasets [4,8,17,18] were used to obtain the results compared to that of [7].

 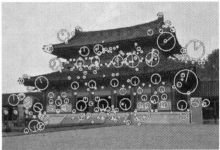

(a) Before feature selection   (b) After feature selection

Fig. 2. An example of feature selection [7] with DoG detector [14].

feature, the relevance, the probability of correct match, is assigned by multiplying four probabilities.

And then, only a small number of features with high relevance are selected to satisfy its target descriptor length. An exemplar result of feature selection is shown in Fig. 2.

Consequently, they also reduced time complexity of descriptor extraction by selecting features before descriptor extraction. However, there are still rooms to be improved in view of time complexity.

In this paper, we propose more efficient feature extraction method with feature selection and show that improvement in time complexity can be achieved.

3   Proposed Method

We propose a fast feature extraction method based on SIFT descriptor which significantly reduces the time complexity while maintaining the matching and retrieval accuracy of [7]. This was achieved by two-stage feature selection, and partial gradient computation. Two methods are described in the following subsections.

3.1   Two-Stage Feature Selection

In [7], keypoints are detected for input image, and orientation is computed for each keypoint, after that the relevance of feature is assigned by its peak, scale, location and orientation, and then, N most relevant features are selected to extract descriptor. Even though time complexity is reduced by extracting descriptor only for the selected features, it is still computational burden that orientation is computed for every keypoint.

To handle this problem, we divide the feature selection method into two stages: upright feature selection and feature selection. We introduce the following definition for better understanding:

– Upright Feature: Keypoint having detectors output such as peak (e.g., DoG response), scale and location (coordinate) except orientation
– Feature: Keypoint having peak, scale, location and orientation after orientation assignment

As shown in Fig. 3, upright feature selection stage is inserted before orientation assignment. In this step, first, the relevance of upright feature is computed by its peak, scale, location (Eq. (1)), and then, top N upright features with high relevance are selected to compute orientation. With this approach, time complexity can be reduced by selecting relevant upright features in advance of orientation assignment.

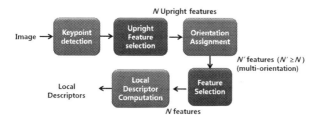

Fig. 3. The proposed feature extraction pipeline based on two-stage feature selection.

The relevance of i_{th} upright feature, $r(u_i)$ is presented by the following equation:

$$r(u_i) = c(p_{u_i})c(s_{u_i})c(l_{u_i}) \tag{1}$$

where p, s, l are peak, scale, and location of upright feature, respectively, and $c(x)$ is the probability of correct match corresponding to x as in Fig. 1.

In orientation assignment stage, N upright features produce N' features. N' is usually bigger than N because there are features with different orientations at the same location and scale. After computing orientation, the relevancies of features are updated for each feature by multiplying the relevance computed in upright feature selection stage and the relevance corresponding to its orientation (Eq. (2)).

$$r(f_{i,j}) = r(u_i)c(o_{i,j}) \tag{2}$$

where j is positive integer, and $f_{i,j}$ is feature u_i with orientation $o_{i,j}$, and $r(f_{i,j})$ is the relevance of $f_{i,j}$.

Finally, N features are selected to extract descriptors. The detail of procedure is presented in Algorithm 1. As we expected from the low variance of probabilities according to orientation shown in Fig. 1, the selected features with the proposed method is very similar to the features selected by [7].

To verify the effect of two-stage feature selection in computational complexity, the average number of upright features, for which orientation is computed, is compared to that of the previous feature selection [7] for 250 images of VGA resolution according to the target number of selected features in Table 1.

Algorithm 1. Two-stage Feature Selection.

Input: $\mathbb{U}$, N ($\mathbb{U}$: detected upright features , N: target number of selected features)
Output: selected features $\mathbb{F}$
1: **for each** upright feature $u_i \in \mathbb{U}$ **do**
2:     $r(u_i) \leftarrow$ relevance of upright feature in Eq. (1)
3: **end for**
4: sort the relevancies of upright features in descending order
5: $\mathbb{U}_s \leftarrow$ select top N upright features with high relevance in $\mathbb{U}$
6: **for each** selected upright feature $u_i \in \mathbb{U}_s$ **do**
7:     compute orientations $\mathbb{O}_i$
8:     **for each** orientation $o_{i,j} \in \mathbb{O}_i$ **do**
9:         $f_{i,j} \leftarrow$ feature u_i with orientation $o_{i,j}$
10:        $r(f_{i,j}) \leftarrow$ relevance of feature in Eq. (2)
11:    **end for**
12: **end for**
13: sort the relevancies of features in descending order
14: $\mathbb{F} \leftarrow$ select top N features with high relevance

Table 1. Comparison of the average number of upright features in orientation assignment stage.

Target # of Selected Features	Feature Selection [7]	Two-stage Feature Selection	Computational Improvement
150	913.8	150.4	83.54 %
300	913.8	300.3	67.14 %
500	913.8	492.1	46.15 %
650	913.8	621.4	32.00 %

When the target number of selected feature is 150, orientation is computed only for the 16.46 % of keypoints in the proposed method which means large computational saving can be achieved.

3.2 Efficient Partial Gradient Computation

In this subsection, we present an efficient partial gradient computation method. In SIFT, Gaussian smoothed image patch around each keypoint are extracted, and gradient magnitude and angle of each pixel in this patch are used to compute orientation or descriptor. The patches used in orientation and descriptor computation are called orientation patch and description patch, respectively. Orientation patch belongs to descriptor patch.

Considering the complexity of gradient computation, previous SIFT implementations are inefficient. In OpenCV [16], gradients are directly computed for orientation and description patches, so there are unnecessary multiple calculations at the same pixel, because orientation patch belongs to description patch, and usually this patch is overlapped with other features patch. In Vlfeat [19], gradient map is used, which computes the gradients in advance for all pixels of scale space images to avoid multiple calculations at the same pixel. However, there are unused areas in gradient map for descriptor computation.

To tackle this problem, proposed method combines two approaches for efficient gradient computation. Gradients are calculated only for the pixels used in orientation and descriptor computation, and updated in the gradient map to avoid multiple calculations. The procedure is as follows:

1. Initialize gradient map to -1
2. When the gradients are required, read the values of gradient magnitude in the map
   (a) If the value is -1, compute gradients and update the map
   (b) Otherwise, use the values

It is obvious that multiple calculations at the same pixel will not occur, because once gradient magnitude is computed, it will have non-negative value, which means that it cannot be equal to -1. Its effectiveness is enhanced with two-stage feature selection, which significantly reduces the regions where gradients computation is required for orientation computation as shown in Fig. 4(e).

To show the efficiency of partial gradient computation and compare the two-stage feature selection with [7] in terms of computational complexity, we compared five different implementations of SIFT extractor:

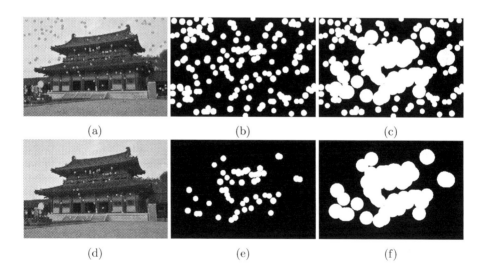

(a)   (b)   (c)

(d)   (e)   (f)

Fig. 4. Partial gradient computation applied to (a)-(c) the previous feature selection [7] and (d)-(f) the two-stage feature selection in 2nd layer of 1st octave. (a) 159 detected upright features by DoG detector. (b) Orientation patches. (c) Orientation and description patches. (d) 52 pre-selected upright features with two-stage feature selection scheme. (e) Orientation patches (f) Orientation and description patches.

- FS+GM: Feature selection [7] and Gradient map
- FS+DGC: Feature selection [7] and Direct gradient computation
- TFS+DGC: Two-stage Feature selection and Direct gradient computation
- FS+PGC: Feature selection [7] and Partial gradient computation
- Proposed: Two-stage Feature selection and Partial gradient computation

For each implementation, the average number of pixels, for which gradients are computed, is counted for 250 images of VGA resolution according to the target number of selected features, and the relative computational complexity with respect to FS+GM is summarized in Table 2. Comparing FS+DGC and FS+PGC, there are many unnecessary multiple calculation at the same pixel in direct gradient computation. Partial gradient computation greatly reduced unnecessary calculation in this case. Comparing FS+PGC and proposed method, the complexity is further reduced when the two-stage feature selection is combined with partial gradient computation. This is visualized in Fig. 4. FS+PGC and proposed method case is shown in Fig. 4(a)-(c), and (d)-(f), respectively.

Table 2. Comparison of complexity in gradient computation.

Target # of Selected Features	FS+GM	FS+DGC	FS+PGC	TFS+DGC	Proposed
150	1	0.48	0.28	0.26	0.11
300	1	0.67	0.33	0.50	0.20
500	1	0.90	0.39	0.79	0.31
650	1	1.05	0.44	0.98	0.38
Average	1	0.78	0.36	0.63	0.25

For simple comparison, only one layer of scale space is considered in this example. By DoG detector, 159 upright features are detected in this layer as shown in Fig. 4(a). Corresponding to these upright features, the white regions in Fig. 4(b) represent orientation patches where gradients are calculated. When 300 features are selected in whole layer by [7], the gradients in description patch are computed and updated for the 61 selected features in this layer as shown in Fig. 4(c). When two-stage feature selection is combined, computational complexity can be further reduced as shown in Fig. 4(d)-(f).

4   Experimental Results

The proposed method is evaluated on uncompressed SIFT framework, and CDVS framework (where descriptors are compressed). The experimental results are presented in Sects. 4.1 and 4.2, respectively. MPEG CDVS datasets [9] were used in both experiments as described in Table 3. These datasets can be divided into two datasets: 2D planar object (books, CDs, and paintings) dataset and 3D non-planar object (buildings or common 3D objects) dataset.

Table 3. CDVS dataset used in the experiments.

Name	Categoty	# of Images	# of Matching Pairs	# of Non-matching Pairs
2D	Graphics	2500	3000	30000
	Paintings	455	364	3640
	Video frames	500	400	4000
3D	Buildings	14935	4005	48675
	Objects	10200	2550	25500

4.1   Evaluation on SIFT Framework

To evaluate the proposed method, we compared the time complexity and matching performance of the proposed method with that of [7] on the SIFT framework.
Target number of selected features is varied from 150 to 650 to examine the effect
of the proposed method. Moreover, the feature selected SURF based on [7] was
examined for comparison. Vlfeat 9.14 [19] and OpenCV 2.4.6 [16] were used to
extract SIFT and SURF respectively and several training sets in [4,8,17,18] were
used to estimate the conditional probabilities of correct match for each descriptor as a function of scale, orientation, peak, and distance from image center. As
peak threshold of SURF influences the number of features, time complexity, and
matching performance, SURF with various peak threshold were compared.

Time Complexity. Time complexity was measured in terms of average extraction time including feature detection and descriptor extraction time. For fair
comparison, image loading and resizing time was excluded because different
libraries were used for SIFT and SURF, respectively. Average extraction time
was measured with i7-2600 @ 3.4 GHz PC on 250 randomly selected images with
a resolution of VGA. All of the results were measured on single core not to
consider speed-up by parallel processing of different implementation.

The results are shown in Fig. 5(a). The extraction time is increased according
to the target number of selected features. The SIFT with the proposed method
is more than 1.7 times faster compared to SIFT with [7], and 1.4 times faster on
average compared to SURF based on [7] with various peak response threshold.
When compared to SIFT with all features, the speed-up factor is up-to 3.8 and on
average 2.96. While SURF with [7] extracts 150 features, SIFT with proposed
method can extract about 500 features. These results tell us that the proposed
method significantly improved the time complexity.

Matching Performance. To verify the influence of the proposed method on
matching performance, receiver operating characteristic (ROC) curve and true
positive rate (TPR) at 1 % of false positive rate (FPR) were compared. Nearest
neighbor distance ratio matching scheme based on Euclidean distance was used
with the ratio of 0.85, and geometrical consistency was checked by DISTRAT [13]
with the Chi-square percentile of 95.

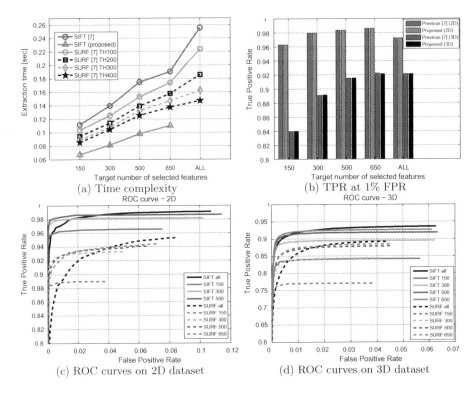

(a) Time complexity

(b) TPR at 1% FPR

(c) ROC curves on 2D dataset

(d) ROC curves on 3D dataset

Fig. 5. Performance comparisons.

Table 4. The influence of peak response threshold of SURF.

	Threshold	# of Features	Extraction Time (sec)	TPR in 3D @ 1 % FPR
SIFT	0	1115.5	0.2556	0.921
SURF	100	1575.5	0.2238	0.872
	200	1229.0	0.1858	0.862
	300	1025.3	0.1624	0.844
	400	886.3	0.1474	0.831

As shown in Table 4, peak response threshold of SURF influences the number of features, and affects the time complexity and matching performance. Considering the trade-off between complexity and performance, 200 was used in the following experiments for the threshold of SURF.

In Fig. 5(b), the proposed method and [7] on SIFT are compared in terms of TPR at 1 % FPR for 2D and 3D dataset according to the target number of selected features. The performance of the proposed method is almost the same as that of [7]. It is noticeable that the proposed method significantly improved the time complexity without affecting the matching performance.

ROC curves of feature selected SIFT and SURF are compared in Fig. 5(c-d) These curves were drawn by changing the threshold of the number of the inlier matching pair in DISTRAT [13]. From the results, it is shown that the matching performance of 300 selected features for 2D and 500 selected features for 3D is comparable to that of all features both on SIFT and SURF, and even outperform at low FPR. This is because the irrelevant features from background or clutters degrade the matching performance while increasing descriptor length.

Moreover, it is noticeable that SIFT with 150 features for 2D dataset and 300 features for 3D dataset outperform SURF with all features. This tells us that SIFT with the proposed scheme for 2D dataset is 2.8 times (2.3 times for 3D) faster in extraction time than SURF while using 25 % (40 % for 3D) of descriptor length and still outperforming in the matching performance compared to SURF.

4.2 Evaluation on MPEG CDVS Framework

The proposed method has been integrated into CDVS TM as an fast mode of feature extractor for its efficiency since 107th meeting. In current CDVS TM11 [10,11], there are four different modes on feature extraction. A Low-degree Polynomial (ALP) detector [6] and SIFT descriptor are baseline of feature extraction, and its three variants:

- ALP low memory mode [1]: ALP detector with block-based processing and spatial domain filtering
- ALP fast mode [12]: fast extraction mode of ALP detector (Proposed method)
- ALP BF [5]: ALP detector with block-based processing and frequency domain filtering

ALP detector approximates the LoG filtering by polynomials to find keypoints of image.

In Table 5, four different feature extractors are compared. For ALP, ALP low memory, ALP BF cases, the previous feature selection method [7] is used. However, features are selected after SIFT descriptor extraction in ALP low memory, ALP BF for their block-based processing to acheive low memory usage.

Table 5. Comparison of TM11 feature extractors (TFS: Two-stage feature selection, GM: Gradient map. PGC: Partial gradient computation, DGC: Direct gradient computation.).

	ALP	ALP Low Memory	ALP Fast Mode	ALP BF
Feature selection	[7]	[7]	TFS	[7]
Gradient Computation	GM	GM	PGC	DGC
Average extraction time (sec)	0.1385	0.2689	0.0915	0.1374
Speed-up Factor	1	0.5149	1.5130	1.0076
Max Memory usage (512 B)	28.2 MB	21.0 MB	24.3 MB	21.2 MB
Max Memory usage (16 KB)	28.3 MB	22.2 MB	25.7 MB	21.0 MB

For gradient computation, ALP and ALP low memory mode use gradient map, which computes the gradients in advance for all pixels, and ALP BF compute gradient directly for each detected keypoint where multiple calculations at the same pixel can exist.

In fast mode of ALP, the proposed method was integrated into ALP detector and SIFT descriptor. Because the proposed method speed up the computation after keypoints are detected, it could be easily integrated into ALP detector framework.

In this section, we compare the time complexity, memory usage, and mathcing and retrieval accuracy of the proposed method (ALP fast mode) with three other modes of feature extractor on MPEG CDVS framework.

Experimental Setup. Evaluation procedure follows the evaluation framework for CDVS [9]. Descriptors are extracted at six operating point (512 B, 1 KB, 2 KB, 4 KB, 8 KB, and 16 KB), and the target number of selected features are 250 (512 B, 1 KB, 2 KB), 300 (4 KB), 500 (8 KB), and 650 (16 KB), respectively. For feature selection, the conditional probabilities of correct match in TM11 were used. MPEG CDVS datasets [9] were used in matching and retrieval experiments as described in Table 3. In retrieval experiment, 7814 images were queried to retrieve relevant images in a dataset of 1 million images including distractor set collected from Flicker.

Time Complexity. Time complexity was measured for each operating point of CDVS. The other condition is the same of Sect. 4.1 Time Complexity, but extraction time measures whole processing time including image loading, resizing and descriptor encoding in this case, because all of TM11 feature extractors are built on the same framework.

The results are shown in Fig. 6(a), and average extraction time for six operating points and speed-up factor compared to ALP are shown in Table 5.

For ALP BF and ALP low memory cases, the extraction times over six operating points are nearly constant because that features are selected after SIFT

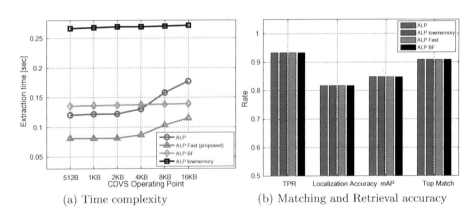

(a) Time complexity   (b) Matching and Retrieval accuracy

Fig. 6. Comparison of feature extractors of CDVS TM11

extraction for their block-based processing to achieve low memory usage. For ALP and the proposed method (ALP Fast mode), the extraction time is increased according to the target number of selected features (250 features are selected at 512 B, 1 KB, 2 KB).

The proposed method outperforms the others in terms of extraction time at every operating point. Especially, the proposed method at 16 KB, where 650 features are selected, is faster than other methods at 512 B, where only 250 features are selected. On average, the proposed method is more than 1.5 times faster compared to ALP and ALP BF, and 2.94 times faster compared to ALP low memory mode. From these results, it is verified that the proposed method significantly improved the time complexity of feature extraction in CDVS TM. Moreover, in [2], they examined ALP, ALP low memory and ALP fast mode (proposed method) on ARM based device (LG Nexus5) and reported the proposed method is not only faster than the other methods, but also consumes the lowest energy (about $0.75 \sim 0.95$ J per image) because of its shorter extraction time.

Memory Usage. In this section, memory usages of the proposed method (ALP fast mode), and other ALP variants of TM11 in feature extraction are examined by profiling with their binaries. All the binaries of four different feature extractor were compiled in 64 bit release mode, and Visual Studio 2010 performance wizard in CPU sampling method was used as profiling tool. While extracting feature of 250 randomly selected VGA images at 512 B and 16 KB operating point, the actual memory used by the extraction binaries were measured by Working Set Peak memory counter on i7-2600 @ 3.4 GHz / 16 GB RAM PC. The definition of Working Set Peak is as follows from [15]:

– Working Set: The current number of bytes in the working set of the process. The working set is the set of memory pages touched recently by the threads in the process. It includes both shared and private data.
– Working Set Peak: The maximum size, in bytes, in the working set of the process at any point in time.

Profiling results are shown in Fig. 7. In this figure, both of extraction speed and memory usage can be compared. Memory usage of ALP BF and ALP Low memory was lower than the other methods at 512 B, 16 KB operating point both. This is because two methods are based on block-wise processing. The maximum of working set peak in this figure is presented in Table 5. The proposed method uses 3.1 MB and 4.7 MB more than ALP BF at 512 B and 16 KB, respectively. Even though, the proposed method is based on whole-image processing, the memory usage difference with that of feature extractor based on block-wise processing is not huge. It is because two-stage feature selection is helpful to keep less features in feature extraction process.

Matching and Retrieval Performance. Matching and retrieval performance of the proposed method, and other feature extractors of TM11 are compared on MPEG CDVS evaluation framework. For pair-wise matching performance,

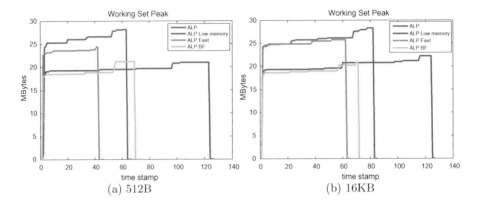

Fig. 7. Memory profiling results.

true positive rate (TPR) at a fixed false positive rate (1 %), and localization accuracy were measured. Mean average precision (mAP) and success rate for top match were measured for retrieval performance. All of measures were measured for each dataset at each operating point, and their average values are compared.

Experimental results are shown in Fig. 6(b). It shows that the performance is similar for all methods.

5   Conclusion

In this paper, a fast feature extraction method based on two-stage feature selection and efficient partial gradient computation is proposed. SIFT with the proposed method is compared to SIFT and SURF with [7]. With the proposed scheme, feature selected SIFT is not only much faster than SURF, but also outperforms in terms of matching accuracy.

Also, the proposed method applied to ALP detector is compared to other feature extraction methods of TM11 on CDVS evaluation framework. Experimental results show that the proposed method significantly reduces the time complexity while maintaining the matching and retrieval performance of TM11. Moreover, the proposed method are compared to ALP and ALP low memory on ARM-based device (LG Nexus5) in [2]. From the results of [2], the proposed method is not only faster than the other methods, but also consumes the lowest energy (about $0.75 \sim 0.95$ J per image) because of its shorter extraction time.

From the results, we can conclude that the proposed method significantly reduces the time complexity of feature extraction without affecting the matching and retrieval accuracy, and this method will be useful for fast extraction of the high accuracy feature on mobile devices, where the use of computational resource is limited.

For further work, the proposed method can be easily extended to SURF framework, and time complexity can be further reduced when combined with parallel processing or GPU implementation. Memory usage can be also improved considering block-based processing.

Acknowledgement. This work was supported by the ICT R& D program of MSIP/ IITP. [2014(R2012030111), Development of The Smart Mobile Search Technology based on UVD(Unified Visual Descriptor)]

References

1. Balestri, M., Francini, G., Lepsøy, S., Lee, K.D., Na, S.I., Lee, S.J.: CDVS: ETRI and TI's response to CE1 - an invariant low memory implementation of the ALP detector with a simplified usage interface. In: 107th MPEG Meeting, M31987 (2014)
2. Ballocca, G., Mosca, A., Fiandrotti, A., Mattelliano, M.: CDVS: TM10 extraction evaluation on ARM architectures. In: 109th MPEG Meeting, M34086 (2014)
3. Bay, H., Ess, A., Tuytelaars, T., Van, G.L.: Speeded-up robust features (SURF). Comput. Vis. Image Underst. **110**, 346–359 (2008)
4. California Institute of Technology: Pasadena Buildings 2010 dataset. http://www. vision.caltech.edu/archive.html (accessed 2014)
5. Chen, J., Duan, L.Y., Huang, T., Gao, W., Kot, A.C., Balestri, M., Francini, G., Lepsøy, S.: CDVS CE1: a low complexity detector ALP BFLoG. In: 108th MPEG Meeting, M33159 (2014)
6. Francini, G., Balestri, M., Lepsøy, S.: CDVS: telecom Italia's response to CE1 - interest point detection. In: 106th MPEG Meeting, M31369 (2013)
7. Francini, G., Lepsøy, S., Balestri, M.: Selection of local features for visual search. Signal Process. Image Commun. **28**, 311–322 (2013)
8. Jegou, H., Douze, M., Schmid, C.: Hamming embedding and weak geometric consistency for large scale image search. In: Forsyth, D., Torr, P., Zisserman, A. (eds.) ECCV 2008, Part I. LNCS, vol. 5302, pp. 304–317. Springer, Heidelberg (2008)
9. ISO/IEC JTC1/SC29/WG11: Evaluation framework for compact descriptors for visual search. In: 97th MPEG Meeting, N12202 (2011)
10. ISO/IEC JTC1/SC29/WG11: Study text of ISO/IEC DIS 15938–13 compact descriptors for visual search. In: 109th MPEG Meeting, N14681 (2014)
11. ISO/IEC JTC1/SC29/WG11: Test Model 11: compact descriptors for visual search. In: 109th MPEG Meeting, N14682 (2014)
12. Lee, K.D., Na, S.I., Lee, S.J., Balestri, M., Francini, G., Lepsøy, S.: CDVS: ETRI and TI's response to CE1 - a fast feature extraction based on ALP detector. In: 107th MPEG Meeting, M31991 (2014)
13. Lepsøy, S., Francini, G., Cordara, G., de Gusmao, P.P.B.: Statistical modelling of outliers for fast visual search. In: IEEE International Conference on Multimedia and Expo (ICME), pp. 1–6 (2011)
14. Lowe, D.G.: Distinctive image features from scale-invariant keypoints. Int. J. Comput. Vision **60**, 91–110 (2004)
15. Microsoft: Counters in Process performance object. http://msdn.microsoft.com/ en-us/library/ms804621.aspx (accessed 2014)
16. OpenCV Library: http://opencv.org/ (accessed 2014)

17. Telecom Italia: 201 Books, InternetArchive and DistractorPairs dataset. http://pacific.tilab.com/ (accessed 2013)
18. University of Oxford: The Oxford Buildings Dataset. http://www.robots.ox.ac.uk/~vgg/data/oxbuildings/ (accessed 2014)
19. VLFeat Library. http://www.vlfeat.org/ (accessed 2014)

Hybrid Feature and Template Based Tracking for Augmented Reality Application

Gede Putra Kusuma Negara$^{(\boxtimes)}$, Fong Wee Teck, and Li Yiqun

Visual Computing Department, Institute for Infocomm Research,
1 Fusionopolis Way, Singapore 138632, Singapore
{igpknegara,wtfong,yqli}@i2r.a-star.edu.sg

Abstract. Visual tracking is the core technology that enables the vision-based augmented reality application. Recent contributions in visual tracking are dominated by template-based tracking approaches such as ESM due to its accuracy in estimating the camera pose. However, it is shown that the template-based tracking approach is less robust against large inter-frames displacements and image variations than the feature-based tracking. Therefore, we propose to combine the feature-based and template-based tracking into a hybrid tracking model to improve the overall tracking performance. The feature-based tracking is performed prior to the template-based tracking. The feature-based tracking estimates pose changes between frames using the tracked feature-points. The template-based tracking is then used to refine the estimated pose. As a result, the hybrid tracking approach is robust against large inter-frames displacements and image variations. It also accurately estimates the camera pose. Furthermore, we will show that the pose adjustment performed by the feature-based tracking reduces the number of iterations necessary for the ESM to refine the estimated pose.

1 Introduction

The vision-based augmented reality (AR) application has been popularized by the rise of the smart-phones. Visual tracking is the core technology that enables the vision-based augmentation. Visual tracking approaches can be roughly categorized into three main groups: feature-based, template-based and hybrid tracking approaches. The feature-based tracking approaches track a set of local features across image sequence. Tracking local features in image sequence can be done by detection or by frame-to-frame tracking. The local features can be detected from salient regions of a reference image. The detected features can then be matched to the features of the input image. Features of both images that provide the best matching scores are considered as the matching pairs.

Electronic supplementary material The online version of this chapter (doi:10.1007/978-3-319-16634-6_28) contains supplementary material, which is available to authorized users. Videos can also be accessed at http://www.springerimages.com/videos/978-3-319-16633-9.

C.V. Jawahar and S. Shan (Eds.): ACCV 2014 Workshops, Part III, LNCS 9010, pp. 381–395, 2015.
DOI: 10.1007/978-3-319-16634-6_28

The camera pose is then estimated from the pairs of features using Levenberg-Marquardt algorithm [1] or RANSAC homography [2]. Popular choice of feature extraction methods include SIFT [3] and SURF [4]. The feature locations can also be tracked between frames using a feature tracker such as Kanade-Lucas-Tomasi (KLT) feature tracker [5]. The KLT uses spatial intensity information of the image to direct the search for the feature location that yields the best match. The KLT has been known to efficiently and robustly track the feature locations in an image sequence. The feature-based tracking approaches rely heavily on feature detection and cannot be applied to texture-less objects that do not contain reliable features to track. The camera pose estimated from a set of feature pairs is usually not so accurate. However, the feature-based tracking approach is usually robust to handle large inter-frames displacements and image variations.

The estimated camera pose can be refined through an iterative numerical method. As the projective geometry of the pinhole image formation process is non-linear, second-order numerical methods are typically used to minimize the mean squared errors between the values predicted by a pose model and those obtained from measurements. For feature-based tracking, standard methods, such as the Levenberg-Marquardt, perform well. However, its accuracy depends on the number of feature points available. This can be partially addressed by using all the available pixel intensity information directly. The Efficient Second-order Minimization (ESM) [6] is one such method, which uses the current pose to warp the current image back to the reference image, so as to minimize the resultant pixel intensity errors. As this method depends on the image gradients, it can work well for well-textured images. Fong et al. [7] added sub-grids to exclude sub-regions within the larger reference image with little textures from computation. Furthermore, the sub-grids are also used to estimate the change in illumination, as well as to handle occlusion. The template-based tracking approaches make use of image intensity information to estimate the camera pose. The pose is estimated by adjusting parameters of a pose model that minimizes an error measure based on image brightness. The pose estimated by the template-based tracking approach is usually more accurate than the one estimated by the feature-based tracking. However, the template-based approach is easier to lose track for large inter-frames displacements and image variations.

The feature-based and template-based tracking approaches are complementary in nature. Therefore, it is logical to combine both feature-based and template-based tracking approaches into a hybrid tracking approach to obtain performance gain. Their specific strengths can be exploited to improve the overall tracking performance. Combining the feature-based and template-based tracking approaches into a hybrid tracking model is not a new idea. A hybrid tracking for augmented reality application has been proposed by Ladikos et al. [8]. They adopted the extended version of the ESM as the template-based tracking and Harris points as the feature-based tracking. It is observed that the template-based tracking works well for small inter-frames displacements and feature-based tracking can deal with larger inter-frames displacements. Therefore, they implemented an adaptive switching strategy between the template-based and feature-based tracking depending on the scene condition. The template-based tracking is used as the default tracking. While, the feature-based tracking is designed to act as a backup to recover the pose in the event that the template-based tracking fails. They avoided running the template-based and feature-based tracking at the same time, because they believed that the combined approach would increase the computational burden and the inter-frames displacements.

Our idea of combining the feature-based and template-based tracking approaches is quite the opposite of theirs. In this contribution, we propose a hybrid tracking approach that combines the feature-based and template-based tracking approaches, where the feature-based tracking is performed prior to the template-based tracking.The feature-based tracking estimates pose changes between frames using the tracked feature-points. The template-based tracking then refines the estimated pose. The feature-based tracking is used as a coarse estimate of the pose for large inter-frames displacements. In order words, the feature-based tracking has reduced the inter-frames displacements for the template-based tracking to handle. As a result, the hybrid tracking approach is robust against large inter-frames displacements and image variations. It also accurately estimates the camera pose. One may expect that it will result in slower processing speed due to larger computation burden. But, according to our experiments, the hybrid tracking is faster than the template-based tracking alone. The pose adjustment performed by the feature-based tracking has reduced the number of iterations necessary for the template-based tracking to refine the estimated pose. Therefore, the hybrid tracking approach is benefited by the strength of both feature-based and template-based tracking approaches; also it is faster than the template-based tracking approach.

2   Combining Feature-Based and Template-Based Tracking

2.1   Overview

Figure 1 shows the overview of the proposed hybrid tracking approach. It consists of three different parts: initialization, feature-based tracking and template-based tracking. The initialization is only done once at the starting point of a tracking process. The initialization process starts with keypoints detection and features extraction. The detected keypoints are used to recognize the object contained in the image. The recognition process is repeated until a match is found (1). The initial pose is then estimated from the keypoints pairs between the input and reference images. A set of keypoints are selected for tracking and also the template image is generated from the input image using the initial estimated pose.

The location of the keypoints in the subsequent input images are tracked using Kanade-Lucas-Tomasi (KLT) feature tracker [5]. The average displacement of the keypoints between successive images is calculated to judge the movement level. If large movement is detected, the pose will be updated by the RANSAC Homography [2] using the keypoints pairs; otherwise, this step will be skipped

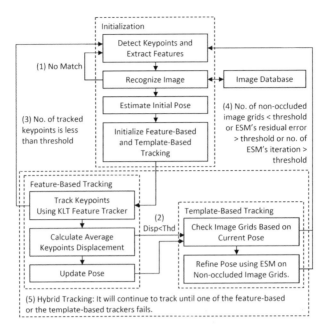

Fig. 1. Overview of the proposed hybrid tracking approach.

(2). The feature-based tracking is stopped when the number of tracked keypoints falls below a certain threshold (3). The re-initialization will be invoked when a tracking failure is detected.

The template-based tracking is performed to refine the estimated pose. It divides the template image into sub-grids of smaller patches. Each sub-grid is checked individually for occlusion in the current image based on the estimated pose. A sub-grid is occluded when its average pixel error is above a defined threshold. If there are enough non-occluded sub-grids available, the pose is refined using the ESM performed on the pixels of the non-occluded sub-grids. The template-based tracking is stopped if the number of non-occluded sub-grids is lower than threshold, the residual error of the ESM is larger than threshold, or the number of ESM's iteration is larger than threshold (4).

The hybrid tracking approach combines the template-based and feature-based tracking. It will stop tracking when any of the feature-based or the template-based trackers fails (5). The details of each part will be described in the followings.

2.2   Initialization

The initialization process is started by detecting keypoints and extracting feature descriptors on the input image. Here, we adopt the Fast-Hessian keypoints detector and SURF descriptor proposed by Bay et al. [4] because of its speed and robustness. Object recognition is then performed based on the detected features to obtain the identity of the object contained in the image. We employ the appearance-based object recognition method based on weighted longest increasing subsequence proposed by Kusuma et al. [9]. The features are matched to the database of images features indexed by the k-d tree data structure [10]. The matching features are then subjected to a geometric validation method based on the longest increasing subsequence [11]. The identity of the object is defined by the class in the database that has the highest similarity score to the input image.

Once the identity of the object has been known, the tracking process can then be initialized. The objective of the tracking process is to estimate a pose matrix that positions the reference coordinate frame to the current coordinate frame. The pose matrix is composed by a rotation matrix $\mathbf{R} \in SO(3)$ and a translation vector $\mathbf{T} \in \mathbb{R}^3$. The only available information for the pose estimation are a reference image of the target object $\mathbf{I}_0$ and an image of the current scene $\mathbf{I}_t$. Assuming a planar target, the reference and current images are related by a homography. The pose matrix can then be estimated by decomposing the homography matrix using the SVD-based homography decomposition [12]. The homography decomposition requires the intrinsic parameters of the camera, which can be obtained through camera calibration [13].

The initial homography is estimated from a set of feature pairs using RANSAC homography [2]. The SURF features of the input image are matched to the features of the selected reference image. The matched features are sorted according to the matching distances. Only, the top 100 feature pairs are used to calculate the initial homography. Let us define $W(\mathbf{H})$ as a warping function of an image based on a homography $\mathbf{H}$ and $\mathbf{H}_{ij}$ is a homography that transform the ith image to the jth image. The initial homography $\mathbf{H}_{01}$ will transform the reference image $\mathbf{I}_0$ to the initial image $\mathbf{I}_1$ by

$$\mathbf{I}_1 = W(\mathbf{H}_{01})\mathbf{I}_0. \tag{1}$$

The feature-based tracking requires a set of initial keypoints to start the tracking. The initial keypoints are selected from the reference keypoints that are consistent with the initial homography.

Meanwhile, the template-based tracking requires an initial estimate of the pose and an image template to start the tracking. There are two ways to set the template image: set template image from a reference image or generate template image from the initial image. The reference image is an image of the target object that is prepared prior to the tracking process. Since the reference image is readily available, therefore the accuracy of the initial pose estimation is less critical for the template image set from a reference image. However, it requires more memory to store the reference image. This approach also requires that the reference image to be similar to the expected image quality and conditions of the target object in the scene. These two requirements are impractical for AR application on the phone. The phone has limited memory space and the camera quality between phones varies. Furthermore, the illumination variation alters the appearance of the target object.

In this contribution, the template image is generated from the initial image. The initial image is warped to the template using the inverse of the initial pose, such as

$$\mathbf{I}_0^* = W\left(\mathbf{H}_{01}^{-1}\right)\mathbf{I}_1. \tag{2}$$

This approach does not require storage space. Furthermore, it is easier to track the template because the appearance of the template image is similar to the appearance of the target object in the scene. However, the accuracy of the initial pose estimation is now more critical. It also adds to the processing load to generate the template image. The initial pose estimation and the template image generation can be done in a different thread to avoid jittery display. Hence, a more expensive method can be used to accurately estimate the initial pose.

The accuracy of the initial pose can be improved by implementing a loop of pose tuning process. A set of SURF features is extracted from the generated template image. These features are matched to the features of the reference image. The homography between the template and reference images is estimated from the keypoint pairs similar to the approach described above. This homography is then used to update the initial homography. This tuning process is performed until an acceptable level of pose accuracy or the maximum number of loop is achieved. The pose accuracy is measured from the number of consistent keypoints to the estimated homography. Since the initialization is done in a separate thread, the user will not really notice it. It will just increase the latency to start tracking, which is usually only a fraction of a second.

2.3   Feature-Based Tracking

The locations of the initial keypoints are tracked throughout the image sequence using Kanade-Lucas-Tomasi (KLT) feature tracker [5]. The KLT has been known to efficiently and robustly track the feature locations in an image sequence. It uses spatial intensity information of the image to direct the search for the feature location that yields the best match. In the current implementation, we employ the pyramidal implementation of the Lucas-Kanade feature tracker [14]. It calculates the optical flow for sparse features using the iterative version of the Lucas-Kanade method in pyramids.

The average displacement of the keypoints between successive images is calculated to judge the movement level. If large movement is detected, the pose will be updated using the between-frames homography. The between-frames homography is estimated from the between-frames keypoint pairs using RANSAC Homography. The current homography is then defined by multiplying the between-frames homography to the previous homography, such as

$$\mathbf{H}_{0t} = \mathbf{H}_{t-1t} * \mathbf{H}_{t-2t-1} * ...\mathbf{H}_{12} * \mathbf{H}_{01}. \tag{3}$$

The homography will only be updated by the feature-based tracking if the average displacement of the keypoints between successive images is larger than threshold. The feature-based tracking may lose some of the keypoints during tracking due to occlusions or image variations. The set of tracked keypoints is constantly updated to the last set of trackable keypoints. The feature-based tracking is stopped when the number of tracked keypoints falls below threshold. The re-initialization will be invoked when the feature-based tracking failure is detected.

2.4   Template-Based Tracking

The template-based tracking is performed to fine-tune the estimated homography. The homography is refined using the efficient second order minimization (ESM) such that to minimize the pixel error between the current and template images. The ESM is developed as an iterative method for second-order minimization of image errors. Compared to the widely-used iterative methods, such as Gauss-Newton and Levenberg-Marquardt, the ESM is shown to have a higher convergence rate [6]. In general, ESM requires a model of the transformation of the image of a surface due to camera motion. The current image is transformed to match the template image using the current camera pose. With a suitable parameterization of small motion about the current camera pose, the ESM can iteratively converge to the camera pose that gives the minimal image error between the template and warped images. For the tracking of planar surfaces, homography is used to correctly model the perspective transformations due to camera motion. As a large number of pixels are used in an efficient manner, the end result is highly accurate and jitter-free pose estimation.

In this contribution, we adopt the modified version of ESM proposed by Fong et al. [7] as the template-based tracking. The template image is divided into sub-grids. The average image gradient within each sub-grid is computed, and only those sub-grids where the gradient is above 10 grey levels per pixel are used in tracking. The sub-grids filtering are performed based in the image gradient as it is used to construct the Jacobian matrices used in the ESM. Experimental observation shows that image regions with low gradients do not contribute additional information for ESM convergence, and in certain cases causes convergence towards the wrong minima.

In the current implementation, the template image is resized to 160×120 pixels and then divided into 12×9 sub-grids with the size of 12×12 pixels per grid. Some of the remaining pixels around the image borders are ignored. The formulation of ESM tracking in terms of sub-grids improves the tolerance to illumination changes and partial occlusion. For illumination changes, both the mean and standard deviation of the pixel intensities within each warped sub-grid is adjusted to match those of the corresponding template sub-grid. As the initial pose is close to the actual pose, the compensation required for illumination changes can be directly computed using the warped and template sub-grids. As both the transformation and illumination models are accurate, the occlusion of a sub-grid can be simply detected when its average pixel error is above a predefined threshold, which is set to 25 in the current implementation. The average pixel error is defined as the mean of absolute differences between corresponding pixels in gray scale.

The ESM pose estimation is then performed based on a set of pixels from the non-occluded sub-grids. The average pixel error of the set of pixels is calculated based on the ESM's estimated pose. The template-based tracking failure is detected when the average pixel error is larger than 10. The tracking will also be stopped when the number of non-occluded sub-grids is less than 12 grids or the number of ESM's iterations is larger than 10. The modified ESM [7] is observed to converge within five iterations.

2.5 Hybrid Tracking

The feature-based and template-based tracking are combined to form a hybrid tracking. Based on our observation, the feature-based tracking is more robust against large inter-frames displacements and image variations than the template-based tracking. On the other hand, the template-based tracking is more accurately estimate the pose than the feature-based tracking. Therefore, the feature-based tracking is performed prior to the template-based tracking. The feature-based tracking updates the homography based on the between-frames keypoint pairs, such as

$$\mathbf{H}_{0t^*} = \mathbf{H}_{t-1t^*} * \mathbf{H}_{0t-1} \ , \tag{4}$$

where $\mathbf{H}_{0t-1}$ is the homography up to the previous frame, $\mathbf{H}_{t-1t^*}$ is the between-frames homography estimated by the feature-based tracking, and $\mathbf{H}_{0t^*}$ is the updated homography. The hybrid tracking approach also checks the average displacement of the keypoints between successive images in order to reduce the computation burden. The homography will only be updated by the feature-based tracking if the average displacement is larger than threshold.

The template-based tracking is then performed to refine the estimated homography, such as

$$\mathbf{H}_{0t} = ESM\left(\mathbf{H}_{0t^*}\right) \ , \tag{5}$$

where $ESM\left(\mathbf{H}\right)$ indicates the ESM iterative refining process. It is also observed that refining the homography from $\mathbf{H}_{0t^*}$ requires less number of iteration than directly from $\mathbf{H}_{0t-1}$. Thus, the hybrid tracking approach is not only combining the strength of both feature-based and template-based tracking approaches; it is also reducing the number of ESM iteration necessary to refine the pose.

3   Experiments

3.1   Methodology

To evaluate the performance of the proposed hybrid tracking approach, we perform experiments on public benchmarking datasets presented by Lieberknecht et al. [15]. We perform our experiments on the normal textured targets: a car (Isetta) and a cityscape (Philadelphia), as shown in Fig. 2. There are five image sequences of different motion patterns for each target. The motion patterns include "Angle", "Range", "Fast Far", "Fast Close", and "Illumination". Therefore, there are ten different image sequences for the experiments and each image sequence contains 1200 image frames.

The image frame is resized to 320×240 pixels for the feature-based tracking and 160×120 pixels for the template-based tracking. The feature-based, template-based and hybrid tracking are performed separately on the image

Fig. 2. The target objects: Isetta (car) and Philadelphia (cityscape).

sequences. Their performances are measured by the ratio of the successfully tracked images in the sequence. The root mean square (RMS) of the pixel distance is also defined for each image in the sequence based on four reference points. The RMS of the pixel distance (err) for an image frame (i) in the sequence is computed as:

$$err_i = \sqrt{\frac{1}{4} \sum_{j=1}^{4} \left\| \mathbf{x}_j - \mathbf{x}_j^* \right\|^2} , \tag{6}$$

where $\mathbf{x}_j$ and $\mathbf{x}_j^*$ are the reference point in the current frame and the ground truth of the reference point respectively. All frames with $err_i > 10$ pixels are considered to be unsuccessfully tracked. Hence, the ratio of the successfully tracked images is calculated based on the filtered results.

Additional experiments are also performed to compare between the performances of the template-based tracking based on a template image set from a reference image and a template image generated from the initial image. The first approach is adopted by Ladikos et al. [8] where a reference image is used as a template image. These additional experiments are also performed on the same datasets and evaluated using the same performance measure as described above.

3.2   Experimental Results

Table 1 shows the ratio of successfully tracked images for the feature-based, template-based and hybrid tracking approaches without err_i thresholding. These results are based on the performances of maintaining tracking throughout the image sequence without considering their accuracies. The results show that the feature-based tracking outperforms the template-based tracking on all image sequences. They also show that the hybrid-based tracking is benefited by the performance of the feature-based tracking. The results of the hybrid tracking are slightly lower than the results of the feature-based tracking. Overall, the feature-based and hybrid tracking achieved much higher ratio of successfully tracked images (without err_i thresholding) than the template-based tracking.

Meanwhile, Table 2 shows the ratio of successfully tracked images for feature-based, template-based and hybrid tracking approaches with err_i thresholding.

Table 1. Ratio of successfully tracked images (without err_i thresholding): feature-based, template-based, and hybrid tracking.

Target	Image sequence	Feature-based tracking	Template-based tracking	Hybrid tracking
Isetta	Angle	100.00 %	79.67 %	100.00 %
	Range	99.92 %	55.33 %	99.92 %
	Fast Far	78.00 %	7.33 %	78.00 %
	Fast Close	93.42 %	81.00 %	93.25 %
	Illumination	99.75 %	96.58 %	99.75 %
Philadelphia	Angle	100.00 %	71.50 %	100.00 %
	Range	99.58 %	62.17 %	91.83 %
	Fast Far	94.58 %	14.67 %	74.92 %
	Fast Close	81.08 %	49.83 %	81.83 %
	Illumination	99.92 %	88.17 %	99.92 %
Average		94.63 %	60.63 %	91.94 %

Table 2. Ratio of successfully tracked images (with $err_i < 10$ pixels): feature-based tracking, template-based tracking, hybrid tracking, and baseline ESM.

Target	Image sequence	Feature-based tracking	Template-based tracking	Hybrid tracking	Baseline ESM
Isetta	Angle	0.40 %	43.40 %	99.10 %	95.42 %
	Range	0.80 %	17.60 %	97.00 %	77.75 %
	Fast Far	5.30 %	5.40 %	54.30 %	7.50 %
	Fast Close	8.10 %	29.10 %	38.40 %	67.08 %
	Illumination	9.80 %	91.50 %	99.60 %	76.75 %
Philadelphia	Angle	12.80 %	35.90 %	96.00 %	99.58 %
	Range	4.90 %	34.70 %	66.40 %	99.00 %
	Fast Far	5.20 %	6.20 %	56.00 %	15.67 %
	Fast Close	14.60 %	36.60 %	40.90 %	86.75 %
	Illumination	13.20 %	50.70 %	99.90 %	90.67 %
Average		7.51 %	35.11 %	74.76 %	71.62 %

These results are based on the performances of maintaining tracking throughout the image sequence as well as their accuracies in estimating the camera pose. Image frames with err_i more than 10 pixels are considered to be unsuccessfully tracked and removed from the performance calculation. We have also added the baseline ESM performance extracted from [15] in the last column.

From the results in Tables 1 and 2, we can clearly see significant drops in the ratio of successfully tracked images for the feature-based tracking. These indicate that the pose estimated by the feature-based tracking is not accurate, even though it can maintain its tracking throughout the image sequence.

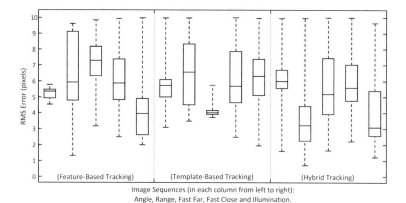

Fig. 3. Box-plots of RMS error (err_i) for Isetta dataset.

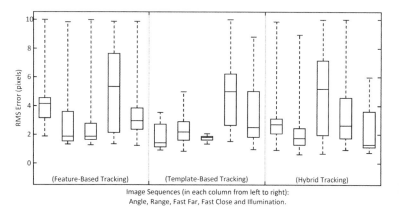

Fig. 4. Box-plots of RMS error (err_i) for Philadelphia dataset.

On the other hand, smaller drops are observed for the feature-based and hybrid tracking approaches. They accurately estimate the camera pose in the majority of the tracked images. It also shown in Table 2 that the ratios of successfully tracked images for hybrid tracking are much higher than the feature-based and template-based tracking approaches on all image sequences. These results prove that the hybrid tracking has combined the strengths of the feature-based and template-based tracking approaches.

The performance of our hybrid tracking is better than the baseline ESM in 6 out of 10 image sequences. On average, our hybrid tracking approach achieved 3.14 % higher ratio of successfully tracked images. Unfortunately, our template-based tracking is not as good as the baseline ESM. This opens up a room for further improvement.

We also show the box-plots of the err_i for Isetta and Philadelphia datasets in Figs. 3 and 4 respectively. Please note that these box-plots are based on the

Table 3. Ratio of successfully tracked images (with $err_i < 10$ pixels) for template-based and hybrid tracking: template image set from reference image vs. template image generated from initial image.

Target	Image sequence	Template-based tracking		Hybrid tracking	
		Reference image	Initial image	Reference image	Initial image
Isetta	Angle	20.10 %	43.40 %	64.70 %	99.10 %
	Range	20.80 %	17.60 %	12.20 %	97.00 %
	Fast Far	5.00 %	5.40 %	5.20 %	54.30 %
	Fast Close	20.20 %	29.10 %	29.20 %	38.40 %
	Illumination	49.40 %	91.50 %	82.60 %	99.60 %
Philadelphia	Angle	8.20 %	35.90 %	29.00 %	96.00 %
	Range	7.80 %	34.70 %	7.90 %	66.40 %
	Fast Far	2.20 %	6.20 %	5.20 %	56.00 %
	Fast Close	18.10 %	36.60 %	46.30 %	40.90 %
	Illumination	16.10 %	50.70 %	24.20 %	99.90 %
Average		16.79 %	35.11 %	30.65 %	74.76 %

ratio of successfully tracked images with err_i thresholding shown in Table 2. Figures 3 and 4 show that, out of the successfully tracked images, the template-based tracking estimates the camera pose more accurately than the feature-based tracking. The distributions of err_i values for the hybrid tracking is within the range of the err_i values for the template-based and feature-based tracking, even though the number of tracked images for the hybrid tracking is much higher than the template-based and feature-based tracking.

We have also compared the tracking performance of the template image set from a reference image and template image generated from the initial image for the template-based and hybrid tracking. Their ratios of successfully tracked images with err_i thresholding are shown in Table 3. Please refer to Sect. 2.2 for the details of the template image initialization. These results clearly show that template image generated from the initial image is easier to track than the template image set from the reference image. The template image generated from the initial image is similar to the appearance of the target object in the current scene. Also, generating template image from the initial image is more practical for mobile augmented reality application since it does not require storage space to store the reference images.

All of the experiments are performed on a personal computer powered by Intel® Core™ i7-2600 3.40 GHz processor, 3.16 GB of RAM, and Windows XP operating system. The entire codes are written in C++ programming language. The average frame rates for the feature-based, template-based and hybrid tracking approaches are 64.0 fps, 46.5 fps and 59.0 fps respectively. The feature-based tracking is faster than the template-based tracking. There is a slight drop in

Table 4. Average number of ESM iteration for template-based and hybrid tracking.

Target	Image sequence	Template-based tracking	Hybrid tracking
Isetta	Angle	3.84	2.58
	Range	3.46	2.96
	Fast Far	3.84	2.88
	Fast Close	4.16	3.00
	Illumination	3.78	2.02
Philadelphia	Angle	3.44	2.02
	Range	3.24	2.02
	Fast Far	3.78	2.02
	Fast Close	3.54	2.02
	Illumination	3.52	2.02
Mean of average		3.66	2.35

frame rates for the hybrid tracking compared to the feature-based tracking. However, the hybrid tracking is faster than the template-based tracking alone. It proves that the pose adjustment performed by the feature-based tracking has reduced the number of iteration necessary for the template-based tracking to refine the estimated pose.

To back-up our claim, we also measured the average number of ESM iteration for the template-based and hybrid tracking for all image sequences. Their average numbers of ESM iteration are shown in Table 4. It clearly shows that the hybrid tracking requires less average number of ESM iteration than the template-based tracking for all image sequences.

The obvious drawback of the proposed hybrid tracking compared to the feature-based or template-based tracking is its higher memory footprint. The memory requirements for feature-based, template-based and hybrid tracking during run-time are summarized in Table 5. It is shown that the feature-based tracking requires slightly more memory than the template-based tracking; and the hybrid tracking requires memory slightly less than the combined memory requirements for the feature-based and template-based tracking. The memory requirement for the hybrid tracking is mainly for storing the template and current frames required by the template-based tracking and the image pyramids of two successive frames required by the feature-based tracking. This memory requirement is still very low compared to the memory space available in a smartphone. Current smart-phones available in the market are usually equipped with at least 1 GB of RAM. As a demo, we have created a sample mobile AR application using the proposed hybrid tracking method. It can run on iPad Air at around 60 fps.[1]

---

[1] Demo video can be found at http://scholar-milk.i2r.a-star.edu.sg/demo/imev14_videos.html.

Table 5. Run-time memory requirements for feature-based, template-based and hybrid tracking.

Methods	Run-time memory
Feature-based tracking	1.85 MB
Template-based tracking	1.66 MB
Hybrid tracking	3.25 MB

4   Conclusion

We have presented in this paper a hybrid tracking approach that combines the feature-based and template-based tracking approaches. The feature-based tracking is performed prior to the template-based tracking. The feature-based tracking estimates the pose changes between successive frames using feature points; while the template-based tracking refines the estimated pose. It has been shown that the proposed hybrid tracking has combined the strength of both feature-based and template-based tracking approaches. It is robust against large inter-frames displacements and image variations and also accurately estimates the camera pose. The pose adjustment performed by the feature-based tracking has also been shown to reduce the number of iteration required by the template-based tracking to refine the estimated pose. Therefore, the hybrid tracking is faster than the template-based tracking alone.

References

1. Lourakis, M.I.A.: homest: A c/c++ library for robust, non-linear homography estimation, July 2006. http://www.ics.forth.gr/~lourakis/homest/. Accessed on 17 Dec 2011
2. Hartley, R., Zisserman, A.: Multiple View Geometry in Computer Vision. Cambridge University Press, New York (2003)
3. Lowe, D.G.: Distinctive image features from scale-invariant keypoints. Int. J. Comput. Vis. **60**, 91–110 (2004)
4. Bay, H., Tuytelaars, T., Van Gool, L.: SURF: speeded up robust features. In: Leonardis, A., Bischof, H., Pinz, A. (eds.) ECCV 2006, Part I. LNCS, vol. 3951, pp. 404–417. Springer, Heidelberg (2006)
5. Shi, J., Tomasi, C.: Good features to track. In: IEEE Conference on Computer Vision and Pattern Recognition, pp. 593–600 (1994)
6. Benhimane, S., Malis, E.: Homography-based 2d visual tracking and servoing. Int. J. Robot. Res. **26**, 661–676 (2007)
7. Fong, W.T., Ong, S.K., Nee, A.Y.C.: Computer vision centric hybrid tracking for augmented reality in outdoor urban environments. In: Proceedings of the International Conference on Virtual Reality Continuum and its Applications in Industry, VRCAI 2009, pp. 185–190. ACM, New York (2009)
8. Ladikos, A., Benhimane, S., Navab, N.: A real-time tracking system combining template-based and feature-based approaches. In: VISAPP, pp. 325–332 (2007)

9. Kusuma, G.P., Szabo, A., Li, Y., Lee, J.A.: Appearance-based object recognition using weighted longest increasing subsequence. In: Proceedings of the International Conference on Pattern Recognition, Tsukuba, Japan, pp. 3668–3671 (2012)
10. Bentley, J.L.: Multidimensional binary search trees used for associative searching. Commun. ACM **18**, 509–517 (1975)
11. Fredman, M.L.: On computing the length of longest increasing subsequences. Discrete Math. **11**, 29–35 (1975)
12. Ma, Y., Soatto, S., Kosecka, J., Sastry, S.S.: An Invitation to 3-D Vision: From Images to Geometric Models. Springer, New York (2003)
13. Zhang, Z.: A flexible new technique for camera calibration. IEEE Trans. Pattern Anal. Mach. Intell. **22**, 1330–1334 (2000)
14. Bouguet, J.Y.: Pyramidal implementation of the lucas kanade feature tracker. Intel Corporation, Microprocessor Research Labs (2000)
15. Lieberknecht, S., Benhimane, S., Meier, P., Navab, N.: A dataset and evaluation methodology for template-based tracking algorithms. In: Proceedings of IEEE International Symposium on Mixed and Augmented Reality, ISMAR 2009, pp. 145–151. IEEE Computer Society, Washington, DC (2009)

A Mobile Augmented Reality Framework for Post-stroke Patient Rehabilitation

Sujay Babruwad$^{(\boxtimes)}$, Rahul Avaghan, and Uma Mudenagudi

B.V. Bhoomaraddi College of Engineering and Technology, Hubli, India
sujaybabruwad@gmail.com

Abstract. In this paper, we put forward a novel framework based on mobile augmented reality (AR) to enhance the post stroke patients participation in the rehabilitation process. The exercises performed in the rehabilitation centers are monotonous and thus requires maximum effort and time from both the patients and the occupational therapists. We propose to combine these tedious activities with the interactive mobile augmented reality technologies. We call this framework Cogni-Care. In this paper, we introduce the underlying architecture of the system that eases the work of the stakeholders involved in the process of stroke recovery. We also present two exercises to improve the fine motor skills, AR-Ball exercise and AR-Maze exercise, as examples and perform the initial usability study.

1 Introduction

Paralysis of one side of the body is called Hemiplegia and is caused by injury or disease to the brain motor centers [5]. Hemiplegia most commonly occurs in brain stroke. Depending on the severity and location of the lesion, brain strokes can cause a variety of locomotor disorders. Spinal cord injury, example Brown-Sequard syndrome, diseases affecting the brain, or severe brain injury are the other causes of hemiplegia. Symptoms other than weakness include decreased movement control, spasticity and decreased endurance [6]. Premature babies show much higher incidence of hemiplegia than full term babies. There is also a high incidence of hemiplegia during pregnancy and experts believe that this may be related to either a traumatic delivery, use of forceps or some event which causes brain injury. Trauma, bleeding, brain infections and cancers are the most common causes of hemiplegia in adults [6]. Uncontrolled diabetes, hypertension or smokers have a higher chance of occurrence of stroke [7]. Facial palsy is a condition where weakness on one side of the face may occur and is caused by viral infection, cancer or stroke [6].

In many of the instances the cause for hemiplegia is not known, but it appears that the brain is deprived of oxygen and this result in the death of neurons. Hemiplegia is diagnosed by clinical examination and investigation by a health professional, such as a physiotherapist or doctor. Radiological studies like a CT scan or magnetic resonance imaging of the brain should be used to confirm injury

C.V. Jawahar and S. Shan (Eds.): ACCV 2014 Workshops, Part III, LNCS 9010, pp. 396–406, 2015.
DOI: 10.1007/978-3-319-16634-6_29

in the brain and spinal cord, but alone cannot be used to identify movement disorders [6].

Rehabilitation is the most important treatment of patients having hemiplegia. The main aim of rehabilitation in hemiplegic patients is to regain maximum function and improvement of quality of life. Improvement in the quality of life is achieved by both physical and occupational therapies [14]. Most of the rehabilitation activities are carried out in the rehabilitation centres located in hospitals. Stroke patients with restricted physical mobility will have problems in getting access to such facilities. In addition, these rehabilitation methods need a lot of equipments and man power in the form of trained therapists [15]. The studies carried out in [1,9] have shown that for effective motor rehabilitation the important factors are duration, capacity and intensity of the exercise sessions. But, with the lack of proper equipments and qualified therapists, efficient treatment of the stroke patients is challenging. These activities concentrate on the patients motor functionality and hence are repetitive, trivial and boring. For example, moving the hand back and forth frequently. This causes the patients to quickly lose interest in the rehabilitation process. Therefore, for a successful post-stroke recovery the patients have to be motivated to perform these exercises [15]. In this paper, we propose a novel framework that uses mobile augmented reality to make these activities motivating and entertaining.

1.1 Related Work

Rehabilitation applications are specifically designed to recover certain handicapped functions of the individuals with disabilities. In the past two decades, a lot of research has been conducted making use of computer applications to help brain stroke patients recover basic motor skills. In this paper we focus on augmented reality and hence we will discuss the state-of-the-art research that has been carried out in augmented and virtual reality domain.

In the study conducted by G.Burdea [15], the author shows that virtual reality (VR) rehabilitation systems can make the dull occupational therapy exercises interesting and entertaining. In the virtual reality system the patients are exposed to pre-programmed tasks designed under the guidance of the occupational therapist. To keep up the patients engaged in these computer generated exercises, audio-visual motivational messages are triggered frequently in the application. A significant improvement is seen in the patients doing these exercises [4]. But the major drawback of this virtual reality application is that it requires sophisticated settings and a dedicated place to perform these exercises. Virtual Reality also separates the patient from the real world, which makes it difficult to correlate the exercises done during the actual occupational therapy sessions.

On the other hand, with the development of the augmented reality technologies some of these problems have been addressed. Augmented reality blends the virtual objects seamlessly in the real world by capturing and processing the real world scene using trackers. The authors of the paper [2,8] have integrated a training environment with AR in the process of repetitive grasp and release tasks.

The major criticism of this system was that the patient required therapist intervention to assist in wearing the required equipments. The authors of the paper [3], have proposed a framework to take the advantage of the AR technology using 2D cameras and fiducial markers. This system addresses the training of daily activities, but due to the use of the fiducial markers therapist intervention is necessitated and is to be performed in a confined environment.

To the best of our knowledge, this is the first framework to make use of the android mobile devices in the rehabilitation of the brain stroke patients. We propose the intuitive augmented reality application that can help the patients practice the fundamental motor movements. These movements are central to all daily life activities, such as hand stretching and fine finger movements. An active participation in such interactive activities is expected to develop a positive psychological feeling to the rehabilitation experience. Developing these exercises on the mobile platform will allow the patients to perform them in their free time, at home or in the garden, without the intervention of the occupational therapists. This on the other hand, would help the qualified occupational therapists to cater to more number of stroke patients. This is due to decrease in the number of patient-therapist sessions per patient in hospitals, as they can perform the same exercises on their mobile devices which in the earlier systems had to be performed in the scheduled session under the guidance of the therapist.

The key contributions of the paper are:

1. We propose a portable, low cost, interactive framework that will help in the speedy recovery of the brain stroke patients during their rehabilitation process.
2. We propose a hand exercise, AR- Ball, which addresses the fundamental hand exercise for the wrist, elbow and shoulder movements, that covers a wide range of stroke patients with different weaknesses.
3. We propose a finger movement exercise to address the fine motor skills in the form of an interactive AR-Maze solving game.
4. We provide the initial experimental results of the usability study performed on 12 healthy subjects.

The proposed system does not require any equipment other than the mobile device. This makes the feasibility and usability better than the currently available augmented reality rehabilitation systems.

The rest of the paper is organized as follows. Section 2 explains the core architecture with the implementation and design details of Cogni-Care Framework. Section 3 describes the AR-Ball exercise and the AR-Maze exercise, which are part of our framework. Section 4 presents the experimental setup and the results of the usability study we conducted on 12 healthy subjects. Finally in the Sect. 5, we summarize this paper's contents and provide outlook for future work.

2   Cogni-Care Framework

Our proposed framework is called Cogni-Care. It is developed to study the use of AR technique in the rehabilitation of the brain stroke patients. We also present two exercises namely, AR-Ball exercise and AR-Maze exercise, developed using augmented reality technologies. The AR-Ball exercise focuses on the hand movement involving wrist, elbow and shoulder exercise. The AR-Maze exercise focuses only on the finger movement. In this section, we will explain the fundamental architecture and technology used in implementing this framework.

The Cogni-Care framework consists of patient module and the therapist module (see Fig. 1). The patient module consists of the front end application. It involves the mobile application in which 3D models are rendered using the AR technique. The patient first logs-in the system. Depending upon the patients profile, the exercises are selected. The patient selects one of these exercises and performs the necessary movement related to the task. The fine motor movements can be recorded for future analysis. The inbuilt accelerometers and the gyroscopes can be used to track the special movements of the patient. The selected augmented exercises are rendered to the patients mobile screen through the AR tracker system. The second module is for the therapists. In this module the therapists can design the exercises for the patients. The therapists can also control and modify the exercises remotely. In this paper, we will describe in detail only the patient module. In this framework we have used the Qualcomm Vuforia SDK [10] for the AR rendering mechanism. The example exercises are designed

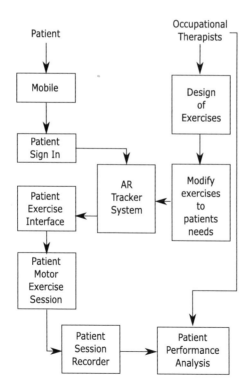

Fig. 1. Cogni-Care architecture.

with the Unity3D game engine [11] and are written in C# using the MonoDevelop IDE [12]. The application is deployed on the android mobile device for the preliminary usability study. The Vuforia SDK is used to capture the camera frames and search for the user defined target image. The user defined target image is the real world captured scene that the patient selects as the reference base to perform the tasks. To create an intuitive game like environment, we used Unity3D game engine for the game development and to detect the collisions involved in performing the tasks [13].

3   AR-Ball and AR-Maze Exercises

3.1   AR-Ball Exercise

Fig. 2. The patient selects the target on which he wants to perform the task. On selecting the target the system will generate and render the 3D ball in the real world scene. The patient has to move his hand towards the ball to collect it.

The objective of the ball exercise is to perform the hand related exercises (see Fig. 2). The patient selects the reference base on which he wants to perform the task which is called the user defined target(target). On selecting the target the system will generate and render the 3D ball in the real world scene. To collect the ball, the patient has to move his hand, holding the mobile, towards the ball. As soon as the ball is touched, sound is triggered to show that the ball is collected. In addition, to have a game like experience scores are updated for collecting each ball. In this way we address the wrist, elbow and shoulder movements. The steps involved in the AR-Ball exercise are as follows:

1. The patient registers in the mobile application and selects the AR-Ball exercise.
2. The patient selects the target on which he wants to perform the task.

3. Once the camera recognizes the selected target the 3D balls are rendered in the real world scene.
4. Now the patient moves the mobile towards the balls.
5. Once the patient collects the ball, a sound is played and the score is updated.
6. This procedure is repeated again until all the balls are collected.
7. The time taken to perform the task is recorded and displayed.

3.2 AR-Maze Exercise

The main objective of this exercise is to specifically address the finger motor skills of the brain stroke patients. This is done by creating a Maze structure, in which the patient solves the maze by guiding the augmented character. The movement of the character is controlled by tapping the onscreen Graphical User Interface (GUI) by using the fingers (see Figs. 3 and 4). The steps involved in the AR-Maze exercise are as follows:

Fig. 3. The patient selects the target on which he wants to perform the task. On selecting the target the system will render the maze arena in the real world scene. The user taps his fingers on the mobile to control the chracter movement.

1. The patient registers in the mobile application and selects the AR-Maze exercise.
2. The patient selects the target on which he wants to perform the task.
3. Once the camera recognizes the selected target the system renders the maze in the real world scene.
4. The patient taps on up, down, right and left GUI to control the movement of the character.
5. Once the patient solves the maze, a cheering sound effect is played to show that he/she has successfully completed the level.
6. This procedure is followed to complete the other maze level.
7. The time taken to perform this session is recorded and displayed.

Fig. 4. This is the AR-Maze exercise with a different maze arena.

4   Cogni-Care Usability Study

In this section we present the outcomes of the usability study we conducted on this system. The main motive of this pilot study is to verify that our framework is interesting and motivating tool for rehabilitation. The study is conducted on 12 healthy subjects/participants (5 female and 7 male) randomly selected to perform these exercises. The mobile devices used in this study are Samsung Galaxy S3, Xolo A500S, HTC Explorer and Samsung Google Nexus S. Each subject is randomly given one of the mobile devices. The study is conducted in a classroom setting and all the 12 subjects are new to this system. The participants did not take part in any medical check-up prior to participating in the study. They self reported a normal eyesight and sense of touch. Each participant was tested two times for both exercises in which he/she undertook five consequent sessions of treatment per exercise. After the completion of the session, we requested every subject to fill in a questionnaire designed to evaluate the system. The subjects scored each question a numerical score of [1–10]. Here [1–3] represents that the subject strongly disagrees; [8–10] means that he/she strongly agrees with what have been asked and [4–7] means that they are not sure.

The following questions were asked to each subject:

1. The instructions on the mobile are clear.
2. I have used mobile phones before this time.
3. I have tried AR applications on mobile phones before this time.

4. I can perceive the depth in the scene.
5. I can easily control the game character with my fingers.
6. I was able to complete the task successfully.
7. I enjoyed playing the exercise.
8. Next time do you require any external assistance in performing this exercise.
9. I felt some hand pain in performing the exercises.
10. I felt that I was doing some sort of exercise.
11. I can perform the same exercise with my other non-dominant hand.

4.1   Results Discussion

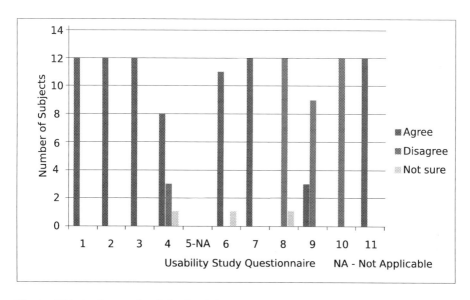

Fig. 5. This is the result of the Usability Study conducted on the AR-Ball exercise. The horizontal axis depicts the various questions used to analyse the system and the vertical axis represents the number of subjects in the study.

We can clearly see that from the Fig. 5 for the AR-Ball exercise and Fig. 6 for the AR-Maze exercise that most of the participants were motivated to perform the exercises. From the analysis of the scores of the questionnaire, we found that all the 12 subjects were interested to carry out the exercises. The AR-Ball exercise addresses the hand movement that involves the wrist, elbow and the shoulder movements. But, the AR-Maze exercise addresses the only the finger movements. Almost all the subjects agreed that they did not realize that they were doing some kind of a rehabilitation related exercises. Most of the subjects found the mobile instructions self-explanatory. In addition, all the subjects confirmed that this framework is easy to use and can be performed at home, garden and rehabilitation center easily. This corroborates that the framework can be used without any supervision.

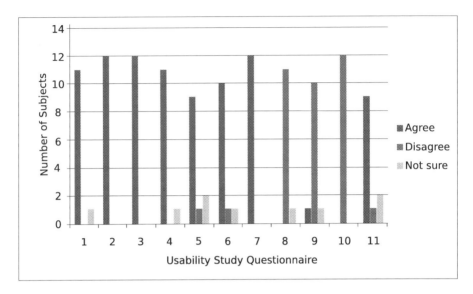

Fig. 6. This is the result of the Usability Study conducted on the AR-Maze exercise. The horizontal axis depicts the various questions used to analyse the system and the vertical axis represents the number of subjects in the study.

Fig. 7. Future work includes development of new gaming exercises to facilitate a variety of therapy solutions to cater the individual patients specific needs.

Some subjects raised a difficulty in the depth perception during the AR-ball exercise. This is because they pointed the mobile camera at a distant place as the reference target image. This caused the difficulty in perceiving the depth of the 3D objects with reference to the mobile. On the other hand, this was not observed in the AR-Maze exercise, as the subjects did not require the depth at which the objects were rendered. In the AR-Maze exercise one person complained of hand fatigue, while the others finished the exercises without any difficulty. However, three subjects suffered hand fatigue in the AR-Ball exercise.

5   Conclusions and Future Work

In this paper, we have proposed an Cogni-Care framework that takes the advantage of the augmented reality technology for the brain stroke patients rehabilitation. We described the architecture of the framework and two exercises, AR-Ball and AR-Maze exercises. Here beautiful and attractive 3D models are overlaid on top of the real world, so that the patients enjoy the boring and repetitive tasks. The usability study has shown that the implemented framework achieves these objectives completely. The potential ease of use and ability to blend motivating tasks in 3D are obvious advantages of this framework. In addition, the patient can work independently anywhere with no assistance from the family or occupational therapist (Fig. 7).

In the future, we will register the patients activity on the server for the regular analysis by the occupational therapists. We will automate the tracking of patients hand and finger movement. For this we will use gyroscope and accelerometer inbuilt in the mobiles to track the hand trajectory and speed of the hand movement while performing the tasks. These additional results will help the therapists to gauge the overall improvement of the patient. Future work also includes development of new gaming exercises to facilitate a variety of therapy solutions to cater the individual patient's specific needs. This will enable the therapist to moderate the amount of hand movement exercises necessary to keep the rehabilitation advancing.

References

1. Kwakkel, G., Wagenaar, R.C., Koelman, T.W., Lankhorst, G.J., Koetsier, J.C.: Effects of intensity of rehabilitation after stroke. Stroke **28**, 1550–1556 (1997)
2. Sucar, L.E., Leder, R.S., Reinkensmeyer, D., Hernández, J., Sanchez, I., Saucedo, P.: Gesture therapy: a vision-based system for arm rehabilitation after stroke. BIOSTEC 2008. CCIS, vol. 25, pp. 531–540. Springer, Heidelberg (2009)
3. Correa, O.A.G.D., Ficheman, I.K., do Nascimento, M., de Deus Lopes, R.: Computer assisted music therapy: a case study of an augmented reality musical system for children with cerebral palsy rehabilitation. In: Ninth IEEE International Conference on Advanced Learning Technologies, pp. 15–17, July 2009
4. Sveistrup, H.: Motor rehabilitation using virtual reality. Neuro Eng. Rehabil. **1**, 10 (2004)
5. Hemi-Kids: Hemiplegic Cerebral Palsy (1997). http://www.hemikids.org/hemi plegia.htm. Accessed 25 Dec 2013
6. Wikipedia Hemiplegia. http://en.wikipedia.org/wiki/Hemiplegia. Accessed 20 Dec 2013
7. World Heart Federation. Cardiovascular disease risk factors: diabetes (1978). http://www.world-heart-federation.org/cardiovascular-health/cardiovascular-dis ease-risk-factors/diabetes/. Accessed 12 Dec 2013

8. Ong, S.K., Shen, Y., Zhang, J., Nee, A.Y.C.: Augmented reality in assistive technology and rehabilitation engineering. In: Furht, B. (ed.) Handbook of Augmented Reality, pp. 603–630. Springer, New York (2011)
9. Nudo, R., Wise, B., SiFuentes, F., Milliken, G.: Neural substrates for the effects of rehabilitative training on motor recovery after ischemicinfarct. Science **272**, 1791–1794 (1996)
10. Qualcomm Vuforia. http://www.qualcomm.com/Vuforia. Accessed 20 Jan 2014
11. Unity3D. http://unity3d.com/. Accessed 11 Jan 2014
12. MonoDevelop. http://monodevelop.com/. Accessed 15 Jan 2014
13. Unity3D Collision Detection. http://docs.unity3d.com/ScriptReference/Collision.html. Accessed 12 Mar 2014
14. American Heart Association - American Stroke Association : Post-Stroke Rehabilitation. http://www.strokeassociation.org/. Accessed 16 Dec 2013
15. Burdea, G.: Virtual rehabilitation benefits and challenges. In: Review Paper in Proceedings of the International Medical Informatics Association, Yearbook of Medical Informatics, Heidelberg, Germany (2004)

Estimation of 3-D Foot Parameters Using Hand-Held RGB-D Camera

Yang-Sheng Chen[1], Yu-Chun Chen[1], Peng-Yuan Kao[1],
Sheng-Wen Shih[2], and Yi-Ping Hung[1(✉)]

[1] Department of CSIE, National Taiwan University, Taipei, Taiwan
hung@csie.ntu.edu.tw
[2] Department of CSIE, National Chi Nan University, Nantou, Taiwan

Abstract. Most people choose shoes mainly based on their foot sizes. However, a foot size only reflects the foot length which does not consider the foot width. Therefore, some people use both width and length of their feet to select shoes, but those two parameters cannot fully characterize the 3-D shape of a foot and are certainly not enough for selecting a pair of comfortable shoes. In general, the ball-girth is also required for shoe selection in addition to the width and the length of a foot. In this paper, we propose a foot measurement system which consists of a low cost Intel Creative Senz3D RGB-D camera, an A4-size reference pattern, and a desktop computer. The reference pattern is used to provide video-rate camera pose estimation. Therefore, the acquired 3-D data can be converted into a common reference coordinate system to form a set of complete foot surface data. Also, we proposed a markerless ball-girth estimation method which uses the lengthes of two toes gaps to infer the joint locations of the big/little toes and the metatarsals. Results from real experiments show that the proposed method is accurate enough to provide three major foot parameters for shoe selection.

1 Introduction

Shoes were invented to help people get to their goal by protecting and comforting their feet and joints. They have long been a necessity in human being life. They keep evolving according to requirements from fashion, rehabilitation, and sports activities. Nowadays, there are so many choices out there for shoes that choosing a suitable one is usually not easy. In general, people will use their foot sizes to screen out unfitted ones. However, the foot size only reflects the length of one's foot, which is not enough to determine a pair of suitable shoes without repetitive try-on. Therefore, selecting a pair of well-fitted shoes can be very time consuming. Also, the requirement of trying on shoes makes online shoe shopping not convenient. The solution to make choosing a pair of well-fitted shoes easier can be revealed by inspecting how shoe producers design their products. They produce each product for prospects with a specific foot shape which is parameterized with a set of foot parameters. Techniques for measuring those foot parameters can help people choose well-fitted shoes more conveniently and are strongly demanded.

© Springer International Publishing Switzerland 2015
C.V. Jawahar and S. Shan (Eds.): ACCV 2014 Workshops, Part III, LNCS 9010, pp. 407–418, 2015.
DOI: 10.1007/978-3-319-16634-6_30

Foot measurement methods can be classified into two categories according to their purposes. The first category includes methods for shoe last design [1,2], which require accurate 3-D foot scanners and are out of the scope of this paper. The second category includes methods for measuring foot parameters for assisting shoes selection. Most of the methods in the second category use a 2-D footprint to estimate the length and the width of a foot [3–7], which are not good enough because a 2-D footprint only provide partial information of a 3-D foot. In 2006, Witana *et al.* proposed a method to measure foot parameters from 3-D foot scanning data [8]. However, their measurement method requires to manually mark 10 anatomical landmarks on the foot surface which requires a foot expert to identify the landmarks correctly.

To help consumers choosing comfortable sports shoes, Asics [9] developed a foot measurement system which uses lasers and cameras to scan the foot surface. To improve the system robustness, their foot measurement system also requires to manually mark three anatomical landmarks on the foot surface. Ildiko Gal Company also provides a 3-D foot scanner consists of a single RGB camera and a turn table. Instead of projecting structure light patterns on the foot surface, they ask users to wear a sock with mesh patterns. Detailed information about their proprietary foot measurement method is unknown.

Shoefitr [10], on the other hand, developed an interesting method for recommending shoes to customers. Their method does not require any hardware device but to ask a user to provide the brand name and the type of the shoes that he/she is wearing. They use the size parameters of the old shoes to search for suitable new shoes. However, when the input brand name of the old shoes is not in their database, their method will fail to provide useful feedback.

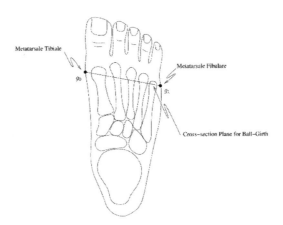

Fig. 1. The ball-girth's cross-section plane.

In this paper, we describe a system, which consists of a low cost Intel Creative Senz3D RGB-D camera, an A4-size printed reference pattern, and a desktop computer, to estimate three foot parameters for shoe selection. The three foot parameters include two basic and one advanced parameters. The two basic parameters are the length and the width of a foot. The advanced parameter is the ball-girth which is defined to be the circumference of the cross-section curve of the foot whose cross-section plane is perpendicular to the ground plane and passing through the Metatarsale Tibiale and the Metatarsale Fibulare as shown in Fig. 1. The two basic parameters can be used to screen out shoes which are either too short or too narrow. However, those two 2-D parameters do not guarantee that the selected shoes are comfortable. Therefore, the foot size is represented with a 3-D vector which includes the length, the width, and the ball-girth.

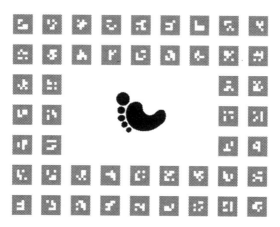

Fig. 2. The reference pattern consists of a set of AR codes.

The reference pattern is an array of AR codes (see Fig. 2). Each AR code is a square pattern encoding a number which can be easily recognized with ArToolKit [11]. This reference pattern can facilitate real-time camera pose tracking. Therefore, we can use a hand-held camera to scan user's foot from different directions. The depth data acquired with the RGB-D camera are transformed to the coordinate system of the reference pattern for 3-D data integration. The proposed method does not require manually marked anatomical landmarks. Therefore, it is more flexible than existing methods and can be operated by users without foot measurement knowledge.

The remainder of this paper is organized as follows. Section 2 introduces the foot measurement system which includes the camera pose estimation, definition of the foot coordinate system, estimation of the foot width/length, estimation of the girth cutting points, and estimation of the ball-girth. Section 3 shows the experimental results of the proposed system. Section 4 concludes this paper.

2   Hand-Held Foot Measurement System

Figure 3 shows the schematic diagram of using a hand-held RGB-D camera to obtain 3-D surface data of the foot. Notably, since the AR codes captured in the RGB image can be easily identified as shown in Fig. 4, the 3-D coordinates of the four corners of the AR codes can be used to estimate the camera pose with the method described in Sect. 2.1. Thanks to the depth information provided by the RGB-D camera, the pose estimation can be accomplished in video rate. Therefore, one can transform the acquired 3-D data into a common reference coordinate system so as to perform 3-D data fusion. However, since the RGB-D camera generates about 300,000 3-D data points per $1/30$ s, fusing such large amount of 3-D data will slow down the system reaction time considerably. In fact, we only need a few key RGB-D frames acquired at a few specific locations/orientations to reconstruct the 3-D surface of the foot. Therefore, we propose to use the video-rate camera pose estimation results to guide the user moving the camera toward one of the m pre-specified locations. When the system detects that the camera pose is close enough to the i-th pre-specified camera pose, $1 \leq i \leq m$, an RGB-D image is automatically captured. Then, the system will guide the user to move the camera to the $(i + 1)$-th location. After all m RGB-D images are captured. The system automatically perform foot parameter estimation procedure described in Sects. 2.2–2.5.

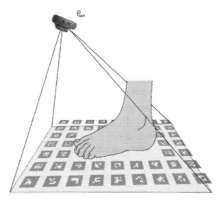

Fig. 3. The Schematic diagram of the proposed foot measurement system using a hand-held camera.

Fig. 4. The identified AR code fiducial marks.

2.1   Camera Pose Estimation

In this work, we adopt the camera model proposed by Tsai [11]. The relationship between a 3-D point, denoted as $\mathbf{p}_c = [x\ y\ z]^\top$, in the camera coordinate system (CCS) and its corresponding image point, denoted as $\mathbf{q}_c = [u\ v]^\top$, is given by

$$\begin{bmatrix} u \\ v \\ 1 \end{bmatrix} \sim \mathbf{A} \begin{bmatrix} x \\ y \\ z \end{bmatrix} \tag{1}$$

where

$$\mathbf{A} = \begin{bmatrix} \alpha_u & 0 & u_o \\ 0 & \alpha_v & v_o \\ 0 & 0 & 1 \end{bmatrix} \tag{2}$$

is the calibration matrix which contains the intrinsic parameters, i.e., α_u, α_v, u_0, and v_0, of the Intel Creative Senz3D RGB-D camera all provided by the manufacturer. Nonlinear lens distortion is not considered in this work.

The extrinsic parameters of the camera consist of a rotation matrix $\mathbf{R}$ and a translation vector $\mathbf{t}$ transforming coordinates from the reference coordinate system (RCS) to the CCS. Let $\mathbf{E} = [\mathbf{R}\ \ \mathbf{t}]$. Then, the complete camera projection matrix can be represented as follows.

$$\mathbf{C} = \mathbf{A} \cdot \mathbf{E} \tag{3}$$

Camera pose estimation is equivalent to calibrating the extrinsic parameters which are the coordinate transformation matrix from the RCS to the CCS. The RCS is defined with a calibration pattern placed on the foot standing area. The calibration pattern is printed on a planar object and is composed of a set of AR code patterns. For each recognized AR code, we can extract the image coordinates of its four exterior corners of the square AR code pattern. Therefore, we can obtain n image points which are denoted by $\mathbf{q}_{ci}$'s, where $1 \leq i \leq n$. By using the image positions of the recognized AR codes, we can compute an ROI which contains the whole calibration pattern. Then, for each image point inside the ROI, its 3-D coordinates are computed using the depth map. If the 3-D coordinates are valid (a valid depth value is greater than zero), then the 3-D coordinates are recorded. We then use the RANSAC algorithm to find a reliable plane equation in the CCS. For each $\mathbf{q}_{ci}$, we back project the image point to form a 3-D ray as follows

$$\mathbf{r}_{ci} = \mathbf{A}^{-1} \cdot \mathbf{q}_{ci}. \tag{4}$$

The 3-D coordinates of the calibration points in the CCS can be computed by finding the intersection points $\mathbf{p}_{ri}$'s of the 3-D ray $\mathbf{r}_{ci}$'s and the plane π. Since the calibration pattern is designed and printed by ourselves, the coordinates of the corners of every AR code pattern are known. Therefore, we have the corresponding 3-D coordinates of the n image points which are denoted by $\mathbf{p}_{ti}$s, where $1 \leq i \leq n$. Therefore, for each 3-D point in the CCS, i.e., $\mathbf{p}_{ri}$, we can determine its corresponding point in the RCS, i.e., $\mathbf{p}_{ti}$. The coordinate transformation matrix from the RCS to the CCS can be estimated by using the least-square fitting algorithm proposed by Arun et al. [12] which completes the camera pose estimation process.

2.2 Foot Coordinate System

After capturing m RGB-D images around the m specified locations, the depth maps are converted to point clouds. In the first step, we perform noise removal to delete points belonging to the background. The remaining points are called the foot point cloud Pt_{ft}. In order to measure the correct width and length of the foot, we need to orthogonally project all points in Pt_{ft} onto the x-y plane of the RCS which is the surface of the foot standing area. Then, we perform a 2-D principal component analysis (PCA) to find the long-axis and the short-axis of the 2-D foot print (see Fig. 5). The long-axis and the short-axis of the foot print are used to define the y-axis and the x-axis of a new 3-D coordinate system, called the foot coordinate system (FCS).

Fig. 5. The Foot axes computed with PCA of the 2-D foot print points.

2.3 Estimation of the Foot Width and the Foot Length

By transforming the 2-D points in the foot print area to the FCS, we have Pt'_{ft}. The width and the length of the foot is estimated by evaluating the width and length of the bounding box of Pt'_{ft}. Figure 6 shows a schematic diagram of the foot width and the foot height estimation method. The x-axis and the y-axis of the FCS shown in Fig. 6 are denoted by X_{ft} and Y_{ft}, respectively. Also, the leftmost, the rightmost, the topmost, and the bottommost points of the foot in the FCS is marked with Pt_{W0}, Pt_{W1}, Pt_{H0} and Pt_{H1}, respectively.

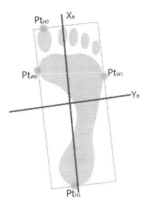

Fig. 6. The evaluation of the width and the length of the footprint.

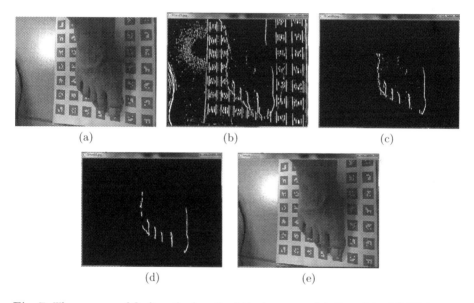

Fig. 7. The process of finding the length of the toe gaps: (a) the original RGB image, (b) the Canny edge detection result, (c) clear edge pixels outside ROI, (d) noise removal result, and (e) estimated endpoints of toes gaps.

2.4   Estimation of the Girth Cut Points

According to the definition of the ball-girth, before we can compute it, we need to determine the girth cut points g_0 and g_1, which are the joints of the metatarsals and the big/little toes, respectively. The main challenge to find those two points is that the foot bone positions are not visible. Therefore, existing methods require an operator to manually mark those two points. In this work, we propose to infer the girth cut points by using the length of the toe gap next to the big/little toe. Since the toe gap can be very narrow, it is unreliable to estimate the toe gap length. Therefore, we determine the toe gap length solely with RGB image. First,

Canny edge detection algorithm is applied to the intensity of the RGB image (see Fig. 7(b)). Then, all the edge pixels outside the ROI of the foot stand area are cleared (see Fig. 7(c)). Perform connected component analysis and remove those blobs smaller than an empirically determined threshold value (see Fig. 7(d)). Finally, we can determine the end points of the toes gaps as shown in Fig. 7(e). The end points of the big/little toe gaps are denoted as V_0 and V_1 (see Fig. 8), respectively.

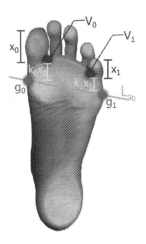

Fig. 8. The estimation of the girth cutting points.

Let x_0 and x_1 denotes the length of the toe gaps next to the big toe and the little toe, respectively. Points V_0 and V_1 are offseted toward the heel by $k_0 x_0$ and $k_1 x_1$ where k_0 and k_1 are the self-defined constants. Finally, a straight line connecting the two offset points is computed. The intersection points of the footprint contour and the straight line are the girth cutting points (see Fig. 8).

2.5 Estimation of the Ball-Girth

To compute the ball-girth value defined in Sect. 1, we transform Pt_{ft} into the FCS. Assume that we have known the stable girth cut points $\mathbf{g}_0$ and $\mathbf{g}_1$, we need to find a plane equation that includes those two cut points and is perpendicular to the x-y plane of the FCS (see Fig. 9). The normal vector of the cutting plane is given by

$$\mathbf{g}_{pl} = \begin{bmatrix} g_x & g_y & g_z \end{bmatrix} = \mathbf{g} \times \mathbf{z} \tag{5}$$

where $\mathbf{g} = \mathbf{g}_0 - \mathbf{g}_1$ and $\mathbf{z}$ is the z-axis vector of the FCS. With the plane normal vector, we have the following ball-girth's plane equation parameterized by using $\mathbf{g}_{pl}$ and $\mathbf{g}_0$

$$G_1 = \begin{bmatrix} \mathbf{g}_{pl} & c \end{bmatrix} \cdot \begin{bmatrix} x \\ y \\ z \\ 1 \end{bmatrix} = g_x\, x + g_y\, y + g_z\, z + c \tag{6}$$

where $c = \mathbf{g}_0 \cdot \mathbf{g}_{pl}^T$. The point-to-plane distance between each point in Pt_{ft} and the cutting plane is computed and if it is smaller than a specified threshold δ, the point is projected to the Since the 3-D data projected onto the cutting plane is still noisy, the contour points are smoothed and connected by using the snake algorithm [13,14]. The length of the snake curve is computed as the ball-girth value.

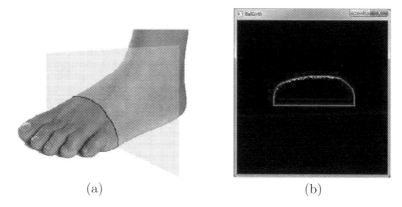

(a)   (b)

Fig. 9. The Estimation of the ball-grith: (a) the cross-section plane computed with girth cutting points, and (b) the extracted cross-section curve for determining the ball-girth.

In summary, the point clouds from multiple cameras are transformed into the RCS. Points which belong to the background objects are removed to yield an integrated point cloud of the foot surface. Then, we find the rotation invariant FCS using the PCA technique. After obtaining the FCS, we can compute the width and the length of the foot. For computing the girth value, we need to process the image to find the toe position and use the toe position to infer the girth cut points $\mathbf{g}_0$ and $\mathbf{g}_1$. Finally, we can calculate the ball-girth value using the above-mentioned method.

3   Experiment

To test the proposed method, we invite five male adults to measure their left and right foot parameters. Each subject was asked to repeat the measurement for 10 times. The following tables list the standard deviations of the estimations of all foot parameters. The results show that the repeatabilities of the foot length and the foot width estimations are good enough for selecting shoes. Their standard deviations are less than 3.5 mm. Conversely, the girth estimations are slightly noisier than the first two parameters. However, the variations of the estimated values are still within the 6 mm which are good enough for shoe selection (Tables 1 and 2).

Table 1. The standard deviations of the estimated left foot parameters.

Subject	Length(mm)	Width(mm)	Girth(mm)
Subject1	2.17	1.45	2.76
Subject2	2.10	2.42	3.10
Subject3	3.50	2.17	4.25
Subject4	2.85	3.11	3.09
Subject5	2.25	2.14	3.04

Table 2. The standard deviations of the estimated right foot parameters.

Subject	Length(mm)	Width(mm)	Girth(mm)
Subject1	1.18	1.32	2.16
Subject2	2.51	2.15	3.57
Subject3	1.93	1.81	3.03
Subject4	2.59	2.04	5.69
Subject5	2.34	1.94	3.42

With the computed foot parameters, we can search for a suitable shoes from a database provided by a shoe producer. The user can then choose a shoe to perform further analysis. First, the 3-D model of the selected shoe is retrieved from the database. Second, the 3-D shoe model is projected to the x-y plane to compute its long and short axes using the PCA method. Third, align the heels, the long axes and the short axes of both the shoe and user's footprint. Fourth, adjust the foot orientation to maximize the overlapping area. Finally, signed distances between the foot surface point cloud and the shoe surfaces are evaluated and are converted to a subjective loose/tight index using a predetermined lookup table. Figure 10 shows the degree of fitness of three shoes using the pseudo color technique. Figure 10(a) shows that the shoe perfectly match user's

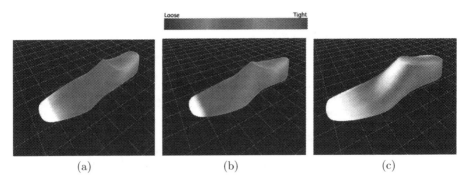

(a)   (b)   (c)

Fig. 10. Estimation of shoe fitness: (a) A suitable shoe, (b) a slightly larger one, and (c) an oversize shoe for the same foot.

foot. If we replace the shoe with a slightly larger one, then the computed pseudo colors shift toward blue as shown in Fig. 10(b). Also, Fig. 10(c) shows that the shoe is too large for the user.

Results of the two experiments show that the 3-D foot reconstruction results using a hand-held RGB-D camera is accurate enough for providing satisfactory foot measurements.

4   Conclusion

In this paper, we propose a method to reconstruct the 3-D foot surface using a hand-held RGB-D camera. A reference pattern consists of an array of AR codes is used to provide registration information so that real-time camera pose estimation can be achieved. The estimated camera pose is used to guide the users to move the camera to a sequence of predetermined poses. When the camera pose is close enough to a predetermined one, the system will acquire an RGB-D image automatically. Upon the completion of image acquisition, the system will automatically evaluate three foot parameters for finding suitable shoes. Results from real experiments show that the repeatability of the proposed method is good enough and can be used to recommend well-fitted shoes.

References

1. Wang, C.S., Chang, T.R., Lin, M.C.: A systematic approach in shoe last design for human feet. In: IEEE International Conference on Industrial Engineering and Engineering Management, pp. 1204–1208. IEEE (2008)
2. Zhao, J., Xiong, S., Bu, Y., Goonetilleke, R.S.: Computerized girth determination for custom footwear manufacture. Comput. Ind. Eng. **54**, 359–373 (2008)
3. Kouchi, M.: Analysis of foot shape variation based on the medial axis of foot outline. Ergonomics **38**, 1911–1920 (1995)
4. Kouchi, M.: Foot dimensions and foot shape: differences due to growth, generation and ethnic origin. Anthropol. Sci. **106**, 161–188 (1998)
5. Oda, T., Sato, N., Nakano, I., Kaneko, Y., Ota, T.: System and method for assisting shoe selection. United States Patent 7089152 (2006)
6. Nguyen, P., Hong, B.: Method and system for sizing feet and fitting shoes. United States Patent 20040073407 (2004)
7. White, J.P.: Foot measurement and footwear sizing system. United States Patent 5128880 (1992)
8. Witana, C.P., Xiong, S., Zhao, J., Goonetilleke, R.S.: Foot measurements from 3-dimensional scans : a comparison and evaluation of different methods. Int. J. Ind. Ergon. **36**, 789–807 (2006)
9. Asics: http://www.asics.com.tw/
10. Shoefitr: http://shoefitr.com/
11. Tsai, R.Y.: A versatile camera calibration technique for high-accuracy 3d machine vision metrology using off-the-shelf tv cameras and lenses. IEEE J. Robot. Autom. **RA–3**, 323–344 (1987)
12. Arun, K.S., Huang, T.S., Blostein, S.D.: Least-squares fitting of two 3-d point sets. IEEE Trans. Pattern Anal. Mach. Intell. **PAMI–9**, 698–700 (1987)

13. Kass, M., Witkin, A., Terzopoulous, D.: Snakes: active contour models. Int. J. Comput. Visi. **1**, 321–331 (1988)
14. Cohen, L.D.: On active contour models and balloons. CVGIP: Image Underst. **53**, 211–218 (1991)

A Wearable Face Recognition System on Google Glass for Assisting Social Interactions

Bappaditya Mandal[(✉)], Shue-Ching Chia, Liyuan Li, Vijay Chandrasekhar, Cheston Tan, and Joo-Hwee Lim

Visual Computing Department, Institute for Infocomm Research,
Connexis, Singapore
{bmandal,scchia,lyli,vijay,cheston-tan,joohwee}@i2r.a-star.edu.sg

Abstract. In this paper, we present a wearable face recognition (FR) system on Google Glass (GG) to assist users in social interactions. FR is the first step towards face-to-face social interactions. We propose a wearable system on GG, which acts as a social interaction assistant, the application includes face detection, eye localization, face recognition and a user interface for personal information display. To be useful in natural social interaction scenarios, the system should be robust to changes in face pose, scale and lighting conditions. OpenCV face detection is implemented in GG. We exploit both OpenCV and ISG (Integration of Sketch and Graph patterns) eye detectors to locate a pair of eyes on the face, between them the former is stable for frontal view faces and the latter performs better for oblique view faces. We extend the eigenfeature regularization and extraction (ERE) face recognition approach by introducing subclass discriminant analysis (SDA) to perform within-subclass discriminant analysis for face feature extraction. The new approach improves the accuracy of FR over varying face pose, expression and lighting conditions. A simple user interface (UI) is designed to present relevant personal information of the recognized person to assist in the social interaction. A standalone independent system on GG and a Client-Server (CS) system via Bluetooth to connect GG with a smart phone are implemented, for different levels of privacy protection. The performance on database created using GG is evaluated and comparisons with baseline approaches are performed. Numerous experimental studies show that our proposed system on GG can perform better real-time FR as compared to other methods.

1 Introduction

Knowing the identity of a person is the first step in a face-to-face social interaction. When meeting an unknown or unfamiliar person, recognizing who he/she is and knowing the essential personal information about him/her, such as name, job and working company, can be very helpful in engaging with the person for an interaction or conversation. Sometimes it is embarrassing when we are unable to recall somebody's name with whom we have met and/or interacted for a sufficiently long time in the near past. Hence, it has become an interesting research topic on developing wearable systems that can aid visual memory to an individual, especially

© Springer International Publishing Switzerland 2015
C.V. Jawahar and S. Shan (Eds.): ACCV 2014 Workshops, Part III, LNCS 9010, pp. 419–433, 2015.
DOI: 10.1007/978-3-319-16634-6_31

in remembering names or recognizing people with whom we meet/recall or have interests [1 3]. Such wearable system will also be very helpful to persons with difficulty in remembering and recognizing faces, such as some form of prosopagnosia [4] which causes difficulty in distinguishing facial features and differentiating people in their social lives [5–8].

With the emergence of popular wearable devices like Google Glass, it has become possible to develop wearable FR system for users as an online assistant for social interactions. We propose a real-time wearable FR system on GG that can recognize persons with various face poses under natural lighting conditions and provide essential personal information for social interactions. This helps to log interaction events automatically, remember names of the person and hence, acts as a memory aid and information service to an individual. One important capability of wearable FR for social interactions is the ability to recognize faces of various poses, scales and under varying lighting conditions. For example, before a formal face-to-face engagement, you may not have been able to capture a frontal view of the person from your view point. Next is that the user would prefer to perform FR locally, *e.g.* solely on GG or just connecting to the user's smartphone, because of the fact that the face and personal information are highly private. There are a few existing FR systems with GG, like NameTag [9], but they are not meant to be used as an online assistant for social interactions.

In this paper, we first enhance the face detection from OpenCV [10] for non-frontal views of faces. Next, we integrate two eye detectors, OpenCV [10] and ISG [11] eye detectors, for better performance on FPV videos. We improve the eigenfeature regularization and extraction (ERE) [12,13] face recognition approach by introducing subclass discriminant analysis (SDA) [14] to perform whole space subclass discriminant analysis (WSSDA) for face feature learning and face recognition. We implement two architectures on GG, *i.e.* one local scheme solely on GG and another Client-Server scheme using Bluetooth to connect GG to the user's smartphone, for the tradeoff of computational burden on GG and privacy protection levels. Details of evaluations and comparisons with baseline approaches are presented.

The remainder of the paper is organized as follows. The next subsection discusses related work. The system configuration and modules on face detection, eye localization, face recognition, and implementations on GG are described in Sect. 2. The evaluation of performance for eye localization, face recognition and computational efficiency are presented in Sect. 3. Finally, conclusions and future work are discussed in Sect. 4.

1.1 Related Work

Face recognition (FR) is perhaps one of the most well studied computer vision research problems. Researchers from diverse areas have been studying problems associated with FR for over four decades [15]. Recently, good progress has been made in recognizing frontal face images with even lighting conditions and tolerance to large variations in expressions. However, the performance drops to a very large extent for changes with pose, uneven lighting conditions and ageing [16–18]. A systematic independent evaluation of recent face recognition algorithms from commercial and academic institutions can be found in the face recognition vendor test (FRVT) 2013 report [19].

There are a few systems proposed for wearable FR [1–3]. Krishna *et al.* [1], developed an iCare Interaction Assistant device for helping visually impaired individuals for social interactions. Their evaluations are limited to only 10 subjects' face images captured under tightly controlled and calibrated face images using classical subspace methodologies like principal component analysis (PCA) and linear discriminant analysis (LDA). Utsumi, *et al.* proposed a coarse-to-fine scheme for FR based on simple image matching [2]. In [3], a simple HMM model is used to capture engagement faces from online FPV videos. However, none of them has been implemented on GG.

Another eye wearable system for improving social lives of prosopagnosics is developed by Wang *et al.* [8]. This system enables prosopagnosic patients to identify the people they come across. Due to their high processing requirements, their modules run on a smartphone. The camera and display units are placed in the wearable eye glass (Vuzix STAR 1200XL third generation augmented reality device). For this system also, the performance evaluations are limited to 20 subjects using local binary pattern (LBP) features.

We propose and implement a FR system on GG with improved face detection, eye localization and face recognition for various face poses and lighting conditions naturally observed from FPV videos in social interactions.

2   The Proposed System

In this section, system configuration and implementation of the main modules are described.

2.1   System Overview

Figure 1 shows the block diagram of our wearable FR system on GG for social interaction assistance. The input FPV image is captured by the camera in GG and then face detection is performed using OpenCV face detector. A pair of eyes are located by fusing the results from OpenCV eye detector [10] and ISG (integration of sketch and graph patterns) eye detector [11]. Using the detected eye coordinates, faces are aligned, normalized and cropped following the CSU Face Identification Evaluation System [20]. Normalized face image is used for FR. According to the FR result, the person is identified and his/her personal information is retrieved from the database. The most relevant personal information is displayed on the screen of GG to assist the user in social interaction.

Training for FR is performed using within-subclass based statistical learning method extended from [12,13]. Low dimensional face discriminative features are extracted and stored in the database. Any incoming novel image is first converted into features and then compared against those stored in the database and a matching ID is obtained using minimum distance measure.

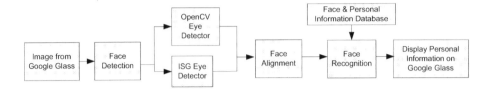

Fig. 1. Our algorithm design flow in Google Glass for face recognition

2.2   Face Detection and Eye Localization

The original API (application protocol interface) provided in GG allows us to capture images at 30 frames per second (fps). However, to reduce the computation burden on GG, we reduce the capturing rate to 5 fps. Firstly, OpenCV face detector [21] is applied to find faces in the incoming images. It is very robust in finding frontal faces, as it is trained with a huge number of training samples. However, it cannot detect faces with pose variations. In a ten-minute video recording of an interaction involving 3 people, the OpenCV detector can only detect about 40 % of the faces in the poses of looking at the GG (frontal faces). At all other times it fails due to non-frontal view faces, motion blurry, severe shadows and other poor lighting conditions. Therefore, we train a new face detector specially for the non-frontal view faces using the OpenCV face detection algorithm, *i.e.*, the algorithm based on Harr features and Adaboost classifier [10]. Both face detectors are applied to find faces in the input image frame.

OpenCV eye detector [10] performs well to locate eyes in the front-view face images, even with closed eyes. But it often fails to locate the pair of eyes in an oblique-view face (poses other than frontal views) or face of smaller scale (people who are at far distances). To alleviate these problems, we use the ISG eye detector developed for human-robot-interaction in [11]. It is re-trained for detecting and locating a pair of open eyes in both frontal-view and oblique-view faces. It performs much better to detect and locate a pair of open eyes in oblique-view faces than the OpenCV eye detector. Also, using this eye detector, the correct detection rate is much higher for faces of small scales in the image compared to OpenCV eye detector. However, it might fail to find the correct locations of closed eyes. Through the integration of both eye detectors, we are able to achieve high success rate of eye localization in the face images of FPV for both frontal and non-frontal faces at various scales (sizes).

Using the eye coordinates, faces are aligned, eye coordinates are placed at fixed distances, cropped and re-sized to 67×75 pixels. We use the face normalization technique described in [20]. The images collected using GG are often blurry in nature as the person wearing the GG moves his/her head quite frequently. Also, sometimes the images are out of camera focus. The face and eye detectors also serve as filters to remove images with large motion blur or poor image quality.

2.3   Within-Subclass Subspace Learning for Face Recognition

To make natural social interactions online viable, the wearable system should be able to recognize person's faces with various poses and lighting conditions. Especially, recognizing a person's face and providing the related personal information, even before the person is engaged in the interaction process. In such situations, often the person has not directly faced the user yet, so no frontal view face image might be available to the user's FPV observations.

Traditional discriminant analysis employs between-class and within-class scatter information for face pattern classification [15]. When applied to uncontrolled illumination, expression and multi-pose FR, it may lose crucial discriminant information in individual's face images [22–25]. In this system, we propose to introduce the subclass discriminant analysis (SDA) [14] into existing subspace learning approach for FR.

In training stage, for each person enrolled in the database, seven face images of different poses, *e.g.* looking front, up, down, left and right, are collected. All these face images are normalized and preprocessed following the CSU Face Identification Evaluation System [20]. The training face images of each person are clustered into subclasses using mixture of Gaussians representation as done in [14]. Then, we compute the within-subclass scatter matrix. Eigenfeature regularization scheme [22] is applied to regularize features obtained from whole space within-subclass scatter matrix. On these regularized features, total-subclass and between-subclass scatter matrixes (depending on the clusters for each person and the number of people in the database [22,26]) are computed. Finally, only those features are used for which the corresponding eigenvalues (variances) are largest.

This kind of regularized features are reported to perform better for both the tasks of FR, which are face identification (FI) [22] and face verification (FV) [12,27] as compared to other subspace methods (like Eigenfaces [28], Fisherfaces [29], Bayesian FR [30,31] and other variants of Fisherfaces [32–35]) and local features [36]. Particularly, we selected this method because the features obtained are optimal as they are extracted after the whole space discriminant analysis [22]. Also, this method achieves good recognition rates with very small number of features [22,23]. The proposed method is named as whole space subclass discriminant analysis (WSSDA) for FR. The proposed method has been evaluated on the challenging YouTube face (YTF) database following the existing protocol for FR [37]. Figure 2 shows the average receiver operating characteristics (ROC) curves that plots the true acceptance rate (TAR) against the false acceptance rate (FAR) following the 10-fold cross-validation pairwise tests protocol suggested for the YTF database [37]. It can be seen that our method performs better among the popular baseline approaches like mixture discriminant analysis (MDA) [38], SDA [14], mixture subclass discriminant analysis (MSDA) [39] and ERE [22,23] for FV task.

After training, only the gallery features and transformation matrix are stored in the system. When more people have to be enrolled in the database, the incoming face images are transformed using the above generated training module (transformation matrix) and only the gallery features are stored.

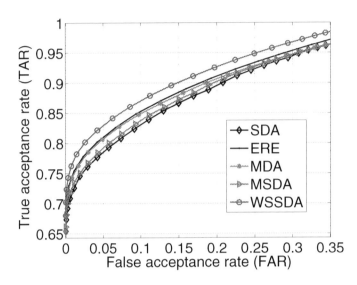

Fig. 2. Average of 10-folds of cross-validation ROC curves plotting true acceptance rate against the false acceptance rate on YouTube database [37] (best viewed in color)(Color figure online)

In recognition stage, any incoming face image vector is converted into a feature vector using the transformation matrix learned by WSSDA method. The feature vector is used to perform recognition by matching it with the gallery features. We use cosine distance measures with 1-nearest neighbor (NN) as the best match for each of the faces in a frame.

2.4   User-Interface Design

The resolution of GG's display is of 640×360 pixels. Keeping the small display resolution in mind, we design a simple user interface with large fonts of brief description of personal information. Another consideration for designing a simple interface is to cause less distraction to the natural social interactions. The app interface has three main components: (i) an option menu to trigger the start of recognition, (ii) a guided viewfinder for positioning of the portrait face and (iii) a three-page display showing the personal information of the recognized person, as shown in Fig. 3. App navigation and control is performed via touch gesture: (1) menu options are activated and canceled by tapping and swiping down respectively and (2) navigation is done by swiping to the left or right.

During test, the guided viewfinder will guide the user to capture a good face image of the targeted person. Once a face image is captured resulting in a successful face recognition, related information is displayed on the screen of GG. The displayed personal information include name, job title, company and portrait of the recognized person in the first page. Information on education, age, hobby, city, and food are displayed in second and third pages of the GG screen.

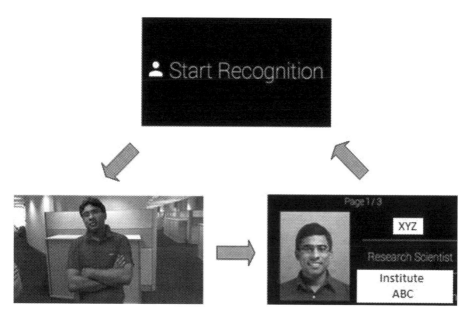

Fig. 3. Face recognition process flow in Google Glass (best viewed in zoomed version)

2.5 Implementation

GG runs on a dual core System on Chip (SoC) with 1GB RAM. The image captured by GG camera has resolution of 640×360 pixels. To investigate the tradeoff between computational complexity and privacy protection level, we designed and implemented two types of architectures for the system:

1. Local architecture solely on GG and
2. Client-Server (CS) architecture via Bluetooth to connect GG and a smartphone.

For the *Local architecture* GG, the whole process runs entirely on GG as a standalone system. The images captured using GG camera are processed directly on GG CPU and recognition results are shown on GG screen. For the *Client-Server architecture* (CS), GG serves as the client in charge of image acquisition and result display, and a smartphone serves as a server performing face recognition. Images captured on GG are compressed to 95 % JPEG quality and cropped to 360 × 240 pixels before sending over to the smartphone. Upon receiving each image, the smartphone proceeds to perform face recognition. Recognition results are then sent back to GG. Finally, GG displays the results on its screen.

For coding development, we are using Eclipse IDE with Android Development Toolkit (ADT). The smartphone is HTC One M8. The target SDKs on Glass and smartphone are Glass Development Kit Preview (GDK) 4.4.2 and Android 4.4.2 respectively. Face recognition algorithms are written in native C/C++ codes and the Java/C++ interfacing is done via Java Native Interface (JNI).

Android codes for the user-interface and Bluetooth communication are written in Java. OpenCV 2.4.8 is used for image capturing as well as processing.

3   System Evaluation

In this section, we present experimental evaluations on eye localization, recognition and computational costs for wearable FR on FPV videos from Google Glass.

3.1   Eye Localization

To be adaptive to head pose and lighting changes in FPV videos for eye localization, two eye detectors are employed in this system. The two detectors perform differently for front-view and oblique-view faces. The benefit of integration is also evaluated.

From FPV videos captured using GG, we randomly select 2675 detected faces. Among them, we further randomly select 100 frontal faces and 118 non-frontal faces. Both the sets have large changes in scales (near and far away faces). We perform the eye detection study using three methodologies: OpenCV, ISG and Fusion. For all the cases, if both eyes are successfully detected then it is considered to be a success, otherwise it is a failure. In our evaluation, it is found out that for both frontal and non-frontal views, the two eye detectors perform complementary roles. So a fusion of these two detectors gave us an improved eye detection results in FPV videos. As shown in Fig. 4, our Fusion system achieves over 90 % accurate rate for frontal view cases and over 70 % accurate rate for non-frontal view cases. This significantly increases the chance to recognize a person using GG before initiating an social interaction engagement.

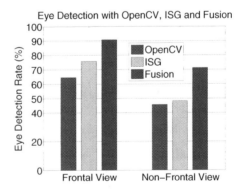

Fig. 4. Eye Detection Rate (%) using OpenCV, ISG and Fusion computed using Google Glass database (best viewed in color)(Color figure online)

Fig. 5. Left, original image captured by Google Glass. Right, extracted normalized face images (red boxes: both eye coordinates are detected, blue box: either of the eyes is not detected). (Best viewed in color.)(Color figure online)

3.2   FPV Database Description

To evaluate the performance of face recognition on FPV images [40], we use the face images collected using two wearable devices: Google Glass and head mounted webcam connecting to a tablet. This database contains faces of persons observed from FPV in natural social interactions, where people are involved in group meetings, indoor social interactions, business networking and all other activities in indoor office environment. There are large changes in poses, expressions, illuminations and jitters because of head and/or camera movement. It is collected between Sep 2012 to Aug 2014 comprising of 7075 images of 88 people (average 80.4 images per person). Out of which 46 people are collected using Logitech C190 webcam and rest 42 people by using first version of the Google Glass [41]. The database is composed of 9 females and 79 males across 9 races. One sample image captured by GG and the extracted and normalized face images are shown in Fig. 5. The red box is shown in the face image where both the eye coordinates are successfully detected and blue box shows a face in which either one of the eye coordinates is not detected.

3.3   Evaluation of Face Recognition on FPV Database

We evaluate the proposed face recognition algorithm WSSDA on the database built from FPV videos. We randomly select images of 42 people for training and images of the remaining 46 people are used for testing. We test the performance of face recognition for two application scenarios. In the first case, only one frontal view face image for each person is stored in the gallery database, rest all images in the probe database (termed as G1 for each of the compared methods in Fig. 6). This is similar to the commercial database of personal information containing

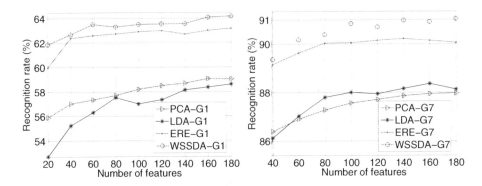

Fig. 6. Recognition rate vs Number of features used in the matching on wearable device database for two scenarios: Left (G1), 1 image per person in the gallery, rest all as probe images. Right (G7), 7 images per person in the gallery, rest all as probe images (best viewed in color)(Color figure online)

only one mug shot image for each person. As mentioned previously and presented in the recent state-of-the-art wearable FR devices [1,2], that keeping one mug shot image in the gallery may not be suitable for wearable FR for natural social interactions. However, in this work we perform experiments for such challenging scenarios.

In the second case, we select 7 images of different poses for each person and use them to form the gallery database and the remaining images are used as probe images (termed as G7 for each of the compared methods in Fig. 6). Two methods, *i.e.* LDA and PCA are selected as baseline methods for comparison as they are recommended for wearable FR in [1]. The comparison with previous method of eigenfeature regularization and extraction (ERE) on within-class subspace [22] is also performed.

Figure 6 shows the plots of recognition rates (%) against the number of features used in the matching for two application scenarios: G1 (left) and G7 (right). It can be seen that the first scheme with only one mug shot image per person in the gallery database cannot generate satisfied result for wearable FR. In the second scheme, when using dimension of 100 features, the accuracy rate is close to 91%. For any kind of wearable device like GG, it is really very important to achieve good recognition rates while using small number of features. Evidently, a little more efforts would be required to build the gallery database as compared to the first simpler scheme in real-world applications.

3.4   Evaluations on Computational Efficiency and Battery Life

Computational Efficiency. The computational efficiency is crucial for real-time apps on GG and other mobile devices. If we run the whole FR process fully on GG, it turns hot in a short span of time (4–6 minutes, also depending on the number of recognitions performed) and the processing frame rate drops significantly. We performed an analysis of computational cost for FR on both GG

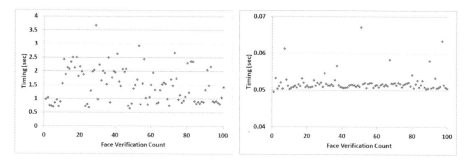

Fig. 7. Timings taken for 100 consecutive successful face recognitions on Google Glass (Left) and Smartphone (Right) with Client-Server Bluetooth architecture

alone and smartphone Client-Server via Bluetooth, and then, selected practical strategies of task arrangement for both the implementations.

The timings taken for FR on GG is shown in Fig. 7, left. The timings are taken for 100 successful recognitions performed consecutively. It can be seen that the timings are quite long and unstable when solely running on GG, however, on the Client-Server Bluetooth architecture, timings are shorter and much more stable as compared to GG, as shown in Fig. 7, right.

The statistics and comparisons are shown in Table 1. The descriptions of the various timing breakdowns are as follows:

(a) Face Recognition refers to performance of the face recognition algorithm.
(b) Bluetooth Image Transmission refers to the timing needed to transfer the input image from GG to mobile device (not required for Local architecture).
(c) Total refers to the total round trip execution timing when GG acquires the input image till when the recognition result is achieved.

During our evaluation, we found that the Local architecture placed a huge strain on the Glass hardware, resulting in GG heating up quickly and performance degrades significantly. To minimize this heating effect, we stopped all recognition operations and allow GG to cool down for about 1 min, after each set of 10 recognitions.

Table 1. Performance comparison between the Local architecture and Client-server Bluetooth architecture for 100 successful recognitions.

	Local architecture		Client-server bluetooth	
	Average (sec)	Stand. Deviation	Average (sec)	Stand. Deviation
Face recognition	1.4973	0.6409	0.0520	0.0025
Bluetooth image transmission	-	-	0.2808	0.0897
Total	1.4973	0.6409	0.3328	0.0922

Table 2. Battery life comparison between the Local architecture and Client-server Bluetooth architecture for 100 consecutive recognitions.

Number of recognitions	% Battery remaining		
	Local (GG)	Client-server bluetooth	
		Google glass	Mobile device
0	100	100	100
10	92	97	100
20	87	94	100
30	80	91	100
40	74	88	99
50	67	85	99
60	60	82	99
70	52	79	99
80	45	76	98
90	39	73	98
100	33	70	97

From Fig. 7 and Table 1, it can be seen that the performance of the GG Client-Server architecture is much more consistent as compared to the Local GG architecture. The time fluctuations is mainly due to the heating up of the GG, when used as a standalone device. For Local mode, the time taken from capturing raw image to final recognition ID takes about 1.5 seconds, whereas for CS architecture, the total time taken is only 0.33 seconds. So our system can be applied as a real-time FR on the GG in real-world condition.

Battery Life. For battery life evaluation, we recorded the battery consumption for performing 100 consecutive recognitions on both Local and Client-Server architectures in Table 2. Both the GG and smartphone were unplugged and not connected to any power outlet.

From Table 2, it is evident that the battery life drops significantly when GG performs FR as a standalone device. However, the battery life drops only a little when FR is performed on smartphone operated in Client-Server Bluetooth mode.

4   Conclusions and Future Work

In this paper, we have described a wearable FR system on GG for assisting people in social interactions. Our proposed system works in two modes of operations: local (standalone) and client-server Bluetooth (with a mobile phone) architectures. Numerous existing methodologies achieving high accuracy rates might not be suitable for wearable devices because of their limited hardware computing and power resources. We propose a system that includes multiple face and eye detections, regularized subspace based methods for training and testing of individuals in an unconstrained environment. Our system is able to achieve 91 % accuracy rates with 7 face images in the gallery database. Out of the two modes of operations, local standalone system takes 1.4973 seconds for each recognition and client-server Bluetooth architecture takes only 0.3328 seconds, which is a real time implementation of FR on GG.

In future, we plan to improve the accuracy of FR, considering outdoor business meeting, social interactions and other group meeting scenarios. We intend to optimize our code and reduce the time taken for recognition in both local and client-server Bluetooth modes of operations. The optimization may also help in reducing of the build up of heat on GG, leading to better performance and usability for natural social interactions.

References

1. Krishna, S., Little, G., Black, J., Panchanathan, S.: A wearable face recognition system for individuals with visual impairments. In: ACM SIGACCESS Conference on Computer and Accessbility, pp. 106–113 (2005)
2. Utsumi, Y., Kato, Y., Kunze, K., Iwamura, M., Kise, K.: Who are you?: a wearable face recognition system to support human memory. In: ACM Proceedings of the 4th Augmented Human International Conference, pp. 150–153 (2013)
3. Singletary, B.A., Starner, T.E.: Symbiotic interfaces for wearable face recognition. In: In HCII2001 Workshop on Wearable Computing (2001)
4. H.F., E., Haan, D., Campbell, R.: A fifteen year follow-up of a case of developmental prosopagnosia. Cortex, a Journal Devoted to the Study of the Nervous System and Behaviour 27 (1991) 489–509
5. Bate, S.: Face recognition and its disorders. Palgrave Macmillan, New York (2013)
6. Kennerknecht, I., Grueter, T., Welling, B., Wentzek, S., Horst, J., Edwards, S., Grueter, M.: First report of prevalence of non-syndromic hereditary prosopagnosia (hpa). Am. J. Med. Genet. **140**, 1617–1622 (2006)
7. Grter, T., Grter, M., Carbon, C.: Neural and genetic foundations of face recognition and prosopagnosia. J. Neuropsychol. **2**, 79–97 (2008)
8. Wang, X., Zhao, X., Prakash, V., Shi, W., Gnawali, O.: Computerized-eyewear based face recognition system for improving social lives of prosopagnosics. In: Proceedings of the 7th International Conference on Pervasive Computing Technologies for Healthcare, pp. 77–80 (2013)
9. Nametag (2014). http://www.nametag.ws/
10. OpenCV: Open source computer vision (2014). http://opencv.org/
11. Yu, X., Han, W., Li, L., Shi, J.Y., Wang, G.: An eye detection and localization system for natural human and robot interaction without face detection. In: Groß, R., Alboul, L., Melhuish, C., Witkowski, M., Prescott, T.J., Penders, J. (eds.) TAROS 2011. LNCS, vol. 6856, pp. 54–65. Springer, Heidelberg (2011)
12. Mandal, B., Jiang, X., Eng, H.L., Kot, A.: Prediction of eigenvalues and regularization of eigenfeatures for human face verification. Pattern Recogn. Lett. **31**, 717–724 (2010)
13. Jiang, X.D., Mandal, B., Kot, A.: Face recognition based on discriminant evaluation in the whole space. In: IEEE 32nd International Conference on Acoustics, Speech and Signal Processing (ICASSP 2007), pp. 245–248. Honolulu, Hawaii, USA (2007)

14. Zhu, M., Martinez, A.: Subclass discriminant analysis. IEEE PAMI **28**, 1274–1286 (2006)
15. Zhao, W., Chellappa, R., Phillips, P.J., Rosenfeld, A.: Face recognition: a literature survey. ACM Comput. Surv. **35**, 399–458 (2003)
16. Phillips, P.J.: Face & ocular challenges. Presentation: http://www.cse.nd.edu/BTAS_10/BTAS_Jonathon_Phillips_Sep_2010_FINAL.pdf (2010)
17. Rowden, L., Klare, B., Klontz, J., Jain, A.K.: Video-to-video face matching: establishing a baseline for unconstrained face recognition. In: IEEE BTAS (2013)
18. Jiang, X.D., Mandal, B., Kot, A.: Complete discriminant evaluation and feature extraction in kernel space for face recognition. Mach. Vis. Appl. **20**, 35–46 (2009). Springer
19. Grother, P., Ngan, M.: Face recognition vendor test (frvt) performance of face identification algorithms. Technical report (2014). http://biometrics.nist.gov/cs_links/face/frvt/frvt2013/NIST_8009.pdf
20. Beveridge, R., Bolme, D., Teixeira, M., Draper, B.: The csu face identification evaluation system user's guide: version 5.0. Technical report (2013). http://www.cs.colostate.edu/evalfacerec/data/normalization.html
21. Viola, P., Jones, M.: Robust real-time face detection. IJCV **57**, 137–154 (2004)
22. Jiang, X.D., Mandal, B., Kot, A.: Eigenfeature regularization and extraction in face recognition. IEEE PAMI **30**, 383–394 (2008)
23. Mandal, B., Jiang, X.D., Kot, A.: Dimensionality reduction in subspace face recognition. In: IEEE ICICS, pp. 1–5 (2007)
24. Liu, W., Wang, Y., Li, S.Z., Tan, T.N.: Null space approach of fisher discriminant analysis for face recognition. In: ECCV, pp. 32–44 (2004)
25. Liu, W., Wang, Y.W., Li, S.Z., Tan, T.N.: Null space-based kernel fisher discriminant analysis for face recognition. In: Proceedings of IEEE International Conference on Automatic Face and Gesture Recognition, pp. 369–374 (2004)
26. Mandal, B., Jiang, X.D., Kot, A.: Multi-scale feature extraction for face recognition. In: IEEE International Conference on Industrial Electronics and Applications (ICIEA), pp. 1–6 (2006)
27. Mandal, B., Jiang, X.D., Kot, A.: Verification of human faces using predicted eigenvalues. In: 19th International Conference on Pattern Recognition (ICPR), pp. 1–4. Tempa, Florida, USA (2008)
28. Turk, M., Pentland, A.: Eigenfaces for recognition. J. Cogn. Neurosci. **3**, 71–86 (1991)
29. Swets, D.L., Weng, J.: Using discriminant eigenfeatures for image retrieval. IEEE PAMI **18**, 831–836 (1996)
30. Moghaddam, B., Jebara, T., Pentland, A.: Bayesian face recognition. Pattern Recognit. **33**, 1771–1782 (2000)
31. Moghaddam, B., Pentland, A.: Probabilistic visual learning for object representation. IEEE PAMI **19**, 696–710 (1997)
32. Wang, X., Tang, X.: A unified framework for subspace face recognition. IEEE Trans. Pattern Anal. Mach. Intell. **26**, 1222–1228 (2004)
33. Chen, L.F., Liao, H.Y.M., Ko, M.T., Lin, J.C., Yu, G.J.: A new lda-based face recognition system which can solve the small sample size problem. Pattern Recognit. **33**, 1713–1726 (2000)
34. Mandal, B., Eng, H.L.: Regularized discriminant analysis for holistic human activity recognition. IEEE Intell. Syst. **27**, 21–31 (2012)
35. Mandal, B., Eng., H.L.: 3-parameter based eigenfeature regularization for human activity recognition. In: IEEE International Conference on Acoustics Speech and Signal Processing (ICASSP), pp. 954–957 (2010)

36. Mandal, B., Zhikai, W., Li, L., Kassim, A.: Evaluation of descriptors and distance measures on benchmarks and first-person-view videos for face identification. In: International Workshop on Robust Local Descriptors for Computer Vision, ACCV (2014)
37. Wolf, L., Hassner, T., Maoz, I.: Face recognition in unconstrained video with matched background similarity. In: IEEE CVPR, pp. 529–534 (2011)
38. Hastie, T., Tibshirani, R.: Discriminant analysis by gaussian mixtures. J. Roy. Stat. Soc. Ser. B **58**, 155–176 (1996)
39. Gkalelis, N., Mezaris, V., Kompatsiaris, I.: Mixture subclass discriminant analysis. IEEE Signal Process. Lett. **18**, 319–322 (2011)
40. Mandal, B., Ching, S., Li, L.: Werable device database (2014). https://drive.google.com/folderview?id=0B2veiY3G5XZxVkd3eC1SdEROaGM&usp=sharing
41. Google: Google glass (2014). http://www.google.com/glass/start/

Lifelog Scene Change Detection Using Cascades of Audio and Video Detectors

Katariina Mahkonen, Joni-Kristian Kämäräinen[⊠], and Tuomas Virtanen

Department of Signal Processing, Tampere University of Technology,
Tampere, Finland
{katariina.mahkonen,joni.kamarainen,tuomas.virtanen}@tut.fi

Abstract. The advent of affordable wearable devices with a video camera has established the new form of social data, lifelogs, where lives of people are captured to video. Enormous amount of lifelog data and need for on-site processing demand new fast video processing methods. In this work, we experimentally investigate seven hours of lifelogs and point out novel findings: (1) audio cues are exceptionally strong for lifelog processing; (2) cascades of audio and video detectors improve accuracy and enable fast (super frame rate) processing speed. We first construct strong detectors using state-of-the-art audio and visual features: Mel-frequency cepstral coefficients (MFCC), colour (RGB) histograms, and local patch descriptors (SIFT). In the second stage, we construct a cascade of the trained detectors and optimise cascade parameters. Separating the detector and cascade optimisation stages simplify training and results to a fast and accurate processing pipeline.

1 Introduction

Wearable devices with a video camera, such as Google Glasses, are becoming commodity hardware and it seems that consumers are willing to push personal blogs even further: video and audio logging of their lives from the first-person view (Fig. 1), *lifelogs*. The lifelog applications have recently become under active investigation, but it is still unclear how to adopt and adapt the existing video processing techniques. In addition, a huge number of video streams and on-device processing require computationally economic but fast methods.

One important application for lifeloggers is to automatically annotate important moments which can be quickly stored, indexed and shared. This can be achieved by condensing important moments into a short "skim" and thus research on video skimming (summarisation/abstraction) has recently gained momentum [1,2]. The best skimming methods are too heavy for on-device processing, but their core components, such as *scene detection* which produces the smallest data pieces for skimming, scenes, may be doable. On-device scene change detection can help fast (on-line) generation of summaries. State-of-the-art scene detection methods rely on visual information, but another important cue, audio, provides an alternative modality with orders of magnitude faster processing.

© Springer International Publishing Switzerland 2015
C.V. Jawahar and S. Shan (Eds.): ACCV 2014 Workshops, Part III, LNCS 9010, pp. 434–444, 2015.
DOI: 10.1007/978-3-319-16634-6_32

Fig. 1. Video frames from our CASA2 lifelog dataset.

Visual and audio cues are often complementary and therefore many hybrids have been proposed [3,4]. However, these works mainly concentrate on maximising accuracy and omit potential for faster computation.

What is the best approach to combine detectors using features of varying importance and even from different modalities? In machine learning literature, a particularly suitable technique for cost (computation time) sensitive learning are detector cascades introduced by Viola and Jones [5]. State-of-the-art cascade construction methods do not operate on stages [5], but simultaneously optimise the whole cascade and its parameters using all training data at once [6–8]. That sets restrictions on used detectors while in our case they can be very different and therefore joint optimisation of the methods, cascade structure and its parameters is too complicated and slow, even impossible. In this work, we take a novel approach: we adopt the cascade structure but cast the problem as a classifier combination [9]: the detectors are trained separately as "strong detectors", then cascaded based on their complexity (audio detectors first) and finally the cascade parameters are optimised similar to expert weights in [9]. Our relaxed design results to simpler cascade construction and training, allows using pre-trained detectors by others, and still our "soft cascades" achieve fast (super frame rate) processing and superior accuracy.

2   Related Work

Detector Cascades - Viola and Jones [5] is the seminal work introducing cascades as a machine learning approach to tackle the real-time requirement in face detection. Their method operated on stages each aiming at high recall and false positives passed to the next more complicated classification stage. The approach is effective but sub-optimal and recent holistic approaches optimising detectors, cascade structure and cascade parameters simultaneously with all training data can provide better cascades [6–8]. That, however, sets requirements for the detectors which we wish to avoid in our work to be able to to exploit the best available detectors. We use all training data to train a set of binary detectors, we combine them using a free-form logical rule (a fixed cascade structure) and then optimise cascade parameters by exhaustive or beam search. In that sense our model is close to combining classifiers theory [9] adapted to cascades.

Combining Audio and Video Cues - Many novel video applications based on visual features have been proposed [10,11], but combinations of audio and visual features seem always superior [3,4,12]. In contrast to the previous works, we do not explicitly engineer a combined audio-video-detector, but take a set of detectors, train them separately for the specific task, and then construct and optimise a cascade structure. That is also justified by the fact that lifelog data is different from the previously used full length movies [13], TV news [14], filmstrips [3] or the mixture TrecVid data [15]. The lifelog data is raw, abruptly moving, unedited, first-person-shot video.

Contributions – Our contributions in this work are two-fold: (1) we investigate how existing visual and audio processing methods work on lifelog data and (2) introduce "soft cascades" of strong detectors for efficient scene change detection in super frame rate. We have collected seven hours of real lifelog data and in our cascades we utilise the best performing cues from various studies: colour (RGB) histograms [16], local image patch descriptor (SIFT) based visual bag-of-words [17] and Mel-frequency cepstral coefficients (MFCCs) [18]. Our focus is on the essential low-level task in video summarising and indexing: *scene detection* [19]. We also report interesting results for shot detection [20]. We point out the following interesting findings:

- In lifelog analysis, audio cues seem to be much more important than in the previous works on movie or broadcast videos (e.g., the TRECVid campaign [20]) or camcorder recorded home videos. In our experiments, visual cues often fail in scene detection.
- Video cues, however, provide partially complementary information to audio, and that can be used to boost the detection accuracy without too much computational increase using the soft decision cascade paradigm proposed in this work.
- The cascades can be constructed easily by separating the detector parameter and cascade parameter optimisation into two separate stages.

3   Decision Cascades

The general goal of constructing cascades is to find a set of detectors (nodes) that minimise a target function consisting of penalties for accuracy loss, (computational) cost of evaluating nodes, and a regularisation term to avoid overfitting [6]. Optimisation of such target function requires inter-operability of the detectors, for example, access to the internal decision tree nodes in [6], i.e. cascade methods operate on "weak classifiers". In our case, we may have N very different type of detectors pre-trained for the same task and we wish to explicitly cascade them into the computationally fastest order. A same detector may appear multiple times but its execution is needed only once. In that sense, our approach is not consistent with the assumptions with the cascading works [6–8], but more resembles the combining classifiers ideology [9] where "strong classifiers" are trained and their combination weights optimised. Our combination, however, is a cascade structure and weights are detection thresholds.

A strong detector cascade is constructed by combining N detectors $D_i, i = 1, \ldots, N$ that map an input feature space X to a decision $t \in T$ in the decision space T. For simplicity, we may assume that t is binary, i.e. $D : X \rightarrow \{0, 1\}$. Our scene change detection is also a binary task. A cascade can be represented as a logical function, such as

$$D = (D_1 \cap D_2 \cap \ldots) \tag{1}$$

or

$$D = (D_1 \cup D_2 \cup \ldots) \tag{2}$$

or any disjunction of conjunctions. It is clear, that significant computational improvement can be achieved if the detector D_i returns 0 in (1) or 1 in (2), since then execution of D_j for $j > i$ is then unnecessary. In particular, if the detectors are indexed such that the computational complexities increase, $\Omega(D_i) \ll \Omega(D_j)$ for $i < j$, then the cascade can provide remarkable computational speedup.

The problem is that there are a large number of ways to combine outputs of the detectors, especially when the number of detectors increase, and only a few are optimal for a certain task. Moreover, the optimal configuration does not only mean an optimal form of the logical function D, but also optimal values for each detector's internal parameters Θ_i and cascade parameters Φ (detection thresholds). The only approach guaranteeing the global optimum is the exhaustive search which easily becomes unfeasible. We propose a doable but still effective optimisation procedure by the following assumptions:

- The cascade structure is built such that the complexity of detectors increase gradually: the computationally lightest detector first and heaviest last.
- The detectors are pre-trained: the parameters Θ_i are optimised independently for the given task.
- The task is to optimise the cascade parameters Φ for the fixed structure and pre-trained detectors.

The first assumption is justified by the fact that it can provide the lowest computational complexity for similar performance. If the detectors mutually correctly detect (in case of D as in Eq. (2)) and leave undetected (in case of D as in Eq. (1)) the same part of the input space, the detection performance can be even improved in addition to saving in the computational load. The second and third assumptions are justified by the fact that since the exhaustive search is not feasible, a separate optimisation of each detector still provides the best average performance and their mutual relationship is compensated on the cascade level parameter optimisation. A greedy algorithm for the optimisation is given in Algorithm 1. The algorithm is in the sense greedy that it moves thresholds one by one always selecting the threshold that provides the smallest amount of negative examples while including one more positive example. This iteration is repeated until all positive examples are covered.

Data: Target class classification scores y_n^i, for $i = 1 \ldots M$ data points and N
classifiers; logical cascade expression in the disjunctive normal form;
Result: Precision(P) – recall(R) curve and cascade parameters Θ for every
point on it.
Init: $\Theta = [\theta_1, \theta_2, \ldots, \theta_N] = [\infty, \infty, \ldots, \infty]$; // P=R=0
while $R < 1$ **do**
     for *each conjunctive* ($\wedge$) *part of the cascade* **do**
         Find the new thresholds $\hat{\theta}_j$ for participating sub-classifiers $\mathcal{D}_j$ that
         select one new positive example and count the number of negative
         examples introduced.
     end
     Set $\theta_j \leftarrow \hat{\theta}_j$ based on the component providing the smallest amount of
     negative examples;
     Store Θ;
     Compute and store P and R ;
end

4   Audio and Visual Cues

For our cascade construction in Sect. 3 we only need that a selected classifier
outputs classification scores for tested example (y_n^i). For scene detection we
selected the audio and visual cues most successful in earlier works. These cues
are shortly reviewed next.

4.1   MFCC Detector

As audio features we use Mel-frequency cepstral coefficients (MFCC) [18] which
have proved to be useful in many audio information retrieval tasks like speech
recognition [21], audio event detection [22,23] and music information retrieval [24].

The audio track is analysed in successive, non-overlapping frames (not to be
conflicted with video frames). From an audio frame at time t, one MFCC-vector
$\mathbf{x}(t)$ of length $D_\mathbf{x}$ is extracted. The context change with MFCC cut detector is
measured according to changes in distributions of vectors $\mathbf{x}$. A mean $\mu_d(t)$ and
variance $\sigma_d(t)$ of each MFCC, indexed by d, is calculated within a sliding audio
frame sequence of length T_s preceding time t. A distance between consecutive
audio frames, $L_{\mathrm{MFCC}}(t)$, for scene and shot change detection at time t is then
given by

$$L_{\mathrm{MFCC}}(t) = \sum_{d=1}^{D_\mathbf{x}} \left| \frac{\mu_d(t) - \mu_d(t + T_s)}{\sigma_d(t) + \sigma_d(t + T_s)} \right|^2 \tag{3}$$

that is slightly different to Fisher's linear discriminant, but found better in our
experiments.

4.2   Colour (RGB) Detector

Despite of its simplicity, variants of colour (RGB) histogram distance have been
used in the most state-of-the-art shot detection methods [20] and since it is also

one of the computationally cheapest visual features it was selected for our experiments. An RGB histogram is computed from each video frame. The histogram vector $\mathbf{h}(t)$ of the frame t is of length 192, containing the incidence frequencies of pixel values 1–64 on red, green and blue channel.

A distance $L_{\mathrm{RGB}}(t)$ of two consecutive RGB frames is calculated as

$$L_{\mathrm{RGB}}(t) = \left| \Delta_{\mathbf{h}}(t) - \overline{\Delta_{\mathbf{h}}(t-1)} \right| \\ + \left| \Delta_{\mathbf{h}}(t) - \overline{\Delta_{\mathbf{h}}(t+T_{\mathbf{h}})} \right|. \tag{4}$$

The idea is, that gradual change is a natural way of RGB histogram evolving. Thus we compare the L_1 change $\Delta_{\mathbf{h}}(t) = \|\mathbf{h}(t) - \mathbf{h}(t-1)\|_1$ between RGB-histograms of consecutive frames to running average change $\overline{\Delta_{\mathbf{h}}(t-1)}$ over $T_{\mathbf{h}}$ preceding frames to see whether the view has changed entirely instead of natural evolution. The second term in (4) accounts for comparing the current change to the forthcoming video frames respectively (not available for on-line processing).

4.3   SIFT Bag-of-Words Detector

This approach is computationally much slower than MFCC and RGB based detectors, but it has been the mainstream approach in detection of visual object classes [25,26]. At the core of this method are histograms of codes of local patch descriptors (SIFT) extracted from each video frame. For patch encoding, a visual codebook must be constructed from extracted descriptors. It has been reported that specific codebooks constructed from the input video perform much better than general codebooks and therefore this approach was adopted by us. The codebook is constructed from k-means clustering with a fixed k (codebook size). For each video frame, SIFT descriptors are extracted on a dense grid, assigned to the best matching codes, and the histogram of codes computed and used as a feature. To compute a shot or scene change score at time t from SIFT-histograms $\mathbf{b}$, a plain L_1-distance $L_{\mathrm{BOW}}(t) = \|\mathbf{b}(t) - \mathbf{b}(t-1)\|_1$ is used. Overall, the L_1 distance instead of the Euclidean distance for evaluating the difference between consecutive histograms, both RGB and BoW detectors, worked clearly best. The settings were selected based on the best found in unsupervised image classification using SIFT bag-of-features [27].

5   Experiments

Data, experiments, performance measures, and results for the selflog video scene detection are reported in this section. Since the same method also applies for shot detection (camera switched) we also report our shot detection results.

5.1   Captured Selflog Data Set

We have collected over 7 hours of video data for our evaluations (Fig. 1). The videos were shot with a small spy camera with the frame rate 15 frames/second and frame size of 176×144 pixels. The frames are YUV420p encoded with h263 compression and stored in an mp4 container. The stereo sound tracks are recorded by a pair of in-ear microphones with 44.1 kHz sampling rate and stored without compression.

The database contains video from 23 different types of environments (scenes), 6–16 shootings from each: amusement park, basketball game, beach, bus, cafeteria, inside car, family yard, football game, hallway, home, inside train, nature, office, outdoor festival, outdoor market, party, pub/club, railway station, restaurant, shop, sports event, at street and track'n'field.

The video was annotated for shot and scene detection. Shot detection corresponds to the situation that the scene remains the same, but the camera was turned off and turned back on in a different location in the same scene. The scene change corresponds to the situation that the user moves to another environment.

In our evaluation, the automatically found change points were compared to the known true scene and shot change times (groundtruth). If the found change point was within 0.25 seconds from a true change point, the detection was assigned correct.

5.2   Performance Measures

To compare the performances of the used shot and scene detection methods, we use precision, $P = tp/(tp + fp)$, recall $R = tp/N$ where tp stands for the number of correct shot or scene changes depending on the task, fp stands for the number of incorrectly identified change points and N is the total number of true change points in the video. A combination of R and P, an F-measure $F = 2 \cdot R \cdot P/(R + P)$, is also used as it simplifies comparison by describing the detection performance with a single value. We are also taking the computation time needed by different systems into account. The computation time CT is given as a number relative to the length of a video, i.e. for $CT = 1$ the system works tightly in real time.

5.3   Detector Parameters

To avoid overfitting to our test data, we trained the detector parameters with separate material of home videos collected before the selflog data. The data is similar to lifelog data, but does not contain the same scenes and was recorded with a standard-quality hand-held camcorder.

In the colour histogram based RGB detector the only method parameter is the length of the time interval to calculate the average change of consecutive RGB-histograms T_{RGB}. The value $T_{\mathrm{RGB}} = 10$ video frames was found best.

In the BoW detector the main method parameter is the SIFT codebook size. We also experimented different detectors and descriptors, but the dense SIFT in the VLFeat toolbox (http://vlfeat.org) was found the best. The codebook is computed from the input data and the optimal codebook size was $D_{\mathrm{SIFT}} = 100$.

Based on experiments with the homevideo data, the following MFCC parameters were selected:

- audio window length = 80 ms
- number of Mel-frequency bands = 80
- number of MFCCs, $D_{\mathbf{x}} = 20$
- audio frame sequence length, $T_s = 10$ s

The number of Mel-frequency bands and the number of used MFCCs did not make a big difference in performance. The audio frame length and the sequence length for distribution estimation were more sensitive. Another finding was that the longer the audio window and the longer the sequence length, the better is the performance. However, to be able to detect also short scenes, these parameters were restricted.

5.4   Results

Single detectors - The results of the single detectors in scene and shot detection are shown in Fig. 2. It is noteworthy that all detectors have very different behaviour with respect to precision and recall. The striking result, however, is that for selflog data the audio cue outperforms the both visual cues with clear margins and being more prominent in scene detection where it is almost twice better. The result is quite opposite to state-of-the-art results with pre-edited material such as movies and TV programs [3,13–15].

Detector Cascades - The results for various cascades are shown in Table 1 including the single detectors. The single audio MFCC detector performs surprisingly well (F-score: 0.84), but as indicated by the different behaviour of the single precision-recall curves in Fig. 2 the other detectors also provide strong complementary information about scene changes. This is evident as the optimal relationship is AND ($\cap$) and for the two cascades MFCC and RGB and MFCC and SIFT the results are 0.90 and 0.95: when two detectors make a wrong decision the third corrects it. Note that for the both cases the computation time is $2\times$ faster than real-time (super frame rate). The best scene detection accuracy is F-score 0.96 which is achieved

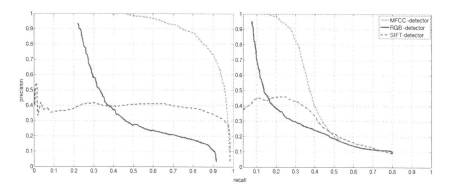

Fig. 2. Precision-recall curves for the single MFCC, RGB and SIFT detectors in scene detection (left) and shot detection (right).

Table 1. Selflog scene and shot detection cascade performances. Performances are reported as the best F-scores in the precision-recall curve with the corresponding cascade computing time (CT) (processing time in seconds per second of video).

Cascade	Scene detection		Shot detection	
	F-score	CT (s/s)	F-score	CT (s/s)
MFCC only	0.84	0.01	0.46	0.01
RGB only	0.40	0.43	0.31	0.43
SIFT only	0.52	184.00	0.37	184.00
MFCC ∪ RGB	0.85	0.44	0.46	0.44
MFCC ∩ RGB	0.90	0.02	0.49	0.03
MFCC ∪ SIFT	0.84	184.00	0.45	184.00
MFCC ∩ SIFT	0.95	0.30	0.51	0.81
RGB ∩ SIFT	0.68	1.30	0.42	1.30
MFCC ∩ RGB ∩ SIFT	0.91	0.30	0.48	0.38
(MFCC∩RGB) ∪ (MFCC∩SIFT)	0.96	0.30	0.52	0.31
(MFCC∩RGB) ∪ (MFCC∩SIFT) ∪ (RGB∩SIFT)	0.96	0.30	0.53	0.32

with classifiers trained with completely separate data and only optimising the cascade parameters. The resulting classifier is a disjunction of the two available strong conjunctions and achieves the performance with computation time 0.30 s needed to process 1.0 s of video input (>3× frame rate). It is noteworthy that the SIFT detector is essential for the performance while it is active only in very few cases as apparent by comparing its single detector and cascade detector computing times.

The same findings hold also for shot detection (best single 0.46, best cascade 0.53) which is much more difficult task in the case of lifelog data.

It should be noted that the selection of cascade parameters is not critical for good performance, since they mutually compensate each other providing smooth and intuitive performance change.

6   Conclusions

The ultimate goal of our work is fast streaming, storing, indexing, retrieval and sharing of selflog video produced by millions of users using their wearable video capturing devices. Past research on video analysis has provided effective but often too slow methods for the above tasks. In this work, we sought to improve the existing techniques with the help of two hypotheses: *multiple video modalities* provide complementary information and *cascade type processing* improves efficiency. The both assumptions were found valid in our experiments where scene and shot detection from real lifelog recordings of more than seven hours were investigated. The strikingly important role of audio, complementary of audio and video, and finally the optimised cascade structure provided us superior detection accuracy in super frame rate. These results indicate that cascades are the tools of future, fusing even more modalities (GPS, accelerometer, gyroscope, compass, barometer, proximity etc.) can be beneficial, and computationally light methods can be constructed from the existing methods. In our future work, we will follow these findings and investigate a light-weight cascade for on-line video skimming and scene indexing.

References

1. Gygli, M., Grabner, H., Riemenschneider, H., Gool, L.V.: Creating summaries from user videos (2014)
2. Zhao, B., Xing, E.: Quasi real-time summarization for consumer videos. In: Proceedings of the CVPR (2014)
3. Kyperountas, M., Kotropoulos, C., Pitas, I.: Enhanced eigen-audioframes for audiovisual scene change detection. IEEE Trans. Multimedia $9(4)$, 785–797 (2007)
4. Song, Y., Zhao, M., Yagnik, J., Wu, X.: Taxonomic classification for web-based videos. In: Proceedings of the CVPR (2010)
5. Viola, P., Jones, M.: Robust real-time face detection. Int. J. Comput. Vis. 57, 137–154 (2001)
6. Chen, M., Xu, Z., Weinberger, K., Chapelle, O., Kedem, D.: Classifier cascade for minimizing feature evaluation cost. In: AISTATS (2012)
7. Wu, T., Zhu, S.C.: Learning near-optimal cost-sensitive decision policy for object detection. In: ICCV (2013)
8. Shen, C., Wang, P., Paisitkriangkrai, S., van den Hengel, A.: Training effective node classifiers for cascade classification. Int. J. Comput. Vis. 103, 326–347 (2013)
9. Kittler, J., Hatef, M., Duin, R.P.W., Matas, J.: On combining classfiers. IEEE PAMI 20, 226–239 (1998)
10. Wang, M.: Movie2comics: towards a lively video content presentation. IEEE Trans. Multimedia 14, 858–870 (2012)
11. Yip, S.: The automatic video editor. In: ACM Multimedia, pp. 596–597 (2003)
12. Chen, S.C., Shyu, M.L., Liao, W., Zhang, C.: Scene change detection by audio and video clues. In: ICME, vol. 2, pp. 365–368 (2002)
13. Pfeiffer, S., Lienhart, R., Effelsberg, W.: Scene determination based on video and audio features. In: Multimedia Tools and Applications, pp. 685–690 (1999)
14. Jiang, H., Lin, T., Zhang, H.: Video segmentation with the assistance of audio content analysis. In: IEEE International Conference on Multimedia and Expo (III), pp. 1507–1510 (2000)
15. Smeaton, A.F., Over, P., Kraaij, W.: Trecvid: evaluating the effectiveness of information retrieval tasks on digital video. In: Proceedings of ACM Multimedia, New York, USA (2004)
16. Gargi, U., Kasturi, R., Strayer, S.H.: Performance characterization of video-shot-change detection methods. IEEE Trans. Circuits Syst. Video Technol. $10(1)$, 1–13 (2000)
17. Lowe, D.G.: Distinctive features from scale-invariant keypoints. Int. J. Comp. Vis. 60, 91–110 (2004)
18. Steven, B., Davis, P.M.: Comparison of parametric representations for monosyllabic word recognition in continuously spoken sentences. In: IEEE Transactions on Acoustics, Speech, and Signal Processing ASSP-28, pp. 357–366 (1980)
19. Fabro, M., Boszormenyi, L.: State-of-the-art and future challenges in video scene detection: a survey. Multimedia Syst. 19, 427–454 (2013)

20. Smeaton, A., Over, P., Doherty, A.: Video shot boundary detection: seven years of TRECVid activity. Comput. Vis. Image Underst. **114**, 411–418 (2010)
21. Rabiner, L., Juang, B.H.: Fundamentals of Speech Recognition. Prentice Hall, Upper Saddle River (1993)
22. Heittola, T., Measaros, A., Virtanen, T., Eronen, A.: Sound event detection in multisource environments using source separation. In: Workshop on Machine Listening in Multisource Environments, Florence, Italy, pp. 36–40 (2011)
23. Aucouturier, J.-J., Defreville, B., Pachet, F.: The bag-of-frames approach to audio pattern recognition: a sufficient model for urban soundscape but not for polyphonic music. J. Acoust. Soc. Am. **122**, 881–891 (2007)
24. Downie, J.: Music information retrieval. Ann. Rev. Inf. Sci. Technol. **37**, 295–340 (2003)
25. Sivic, J., Zisserman, A.: Video Google: a text retrieval approach to object matching in videos. In: Proceedings of the ICCV (2003)
26. Csurka, G., Dance, C., Willamowski, J., Fan, L., Bray, C.: Visual categorization with bags of keypoints. In: ECCV Workshop on Statistical Learning in Computer Vision (2004)
27. Tuytelaars, T., Lampert, C., Blaschko, M., Buntine, W.: Unsupervised object discovery: a comparison. Int. J. Comput. Vis. **88**, 284–302 (2010)

Activity Recognition in Egocentric Life-Logging Videos

Sibo Song[1]([✉]), Vijay Chandrasekhar[2], Ngai-Man Cheung[1], Sanath Narayan[3], Liyuan Li[2], and Joo-Hwee Lim[2]

[1] Singapore University of Technology and Design, Serangoon, Singapore
sibo_song@mymail.sutd.edu.sg
[2] Institute for Infocomm Research, Connexis, Singapore
[3] Indian Institute of Science, Bangalore, India

Abstract. With the increasing availability of wearable cameras, research on first-person view videos (egocentric videos) has received much attention recently. While some effort has been devoted to collecting various egocentric video datasets, there has not been a focused effort in assembling one that could capture the diversity and complexity of activities related to *life-logging*, which is expected to be an important application for egocentric videos. In this work, we first conduct a comprehensive survey of existing egocentric video datasets. We observe that existing datasets do not emphasize activities relevant to the life-logging scenario. We build an egocentric video dataset dubbed LENA (Life-logging EgoceNtric Activities) (http://people.sutd.edu.sg/~1000892/dataset) which includes egocentric videos of 13 fine-grained activity categories, recorded under diverse situations and environments using the Google Glass. Activities in LENA can also be grouped into 5 top-level categories to meet various needs and multiple demands for activities analysis research. We evaluate state-of-the-art activity recognition using LENA in detail and also analyze the performance of popular descriptors in egocentric activity recognition.

1 Introduction

With the increasing availability of wearable devices such as Google Glass, Microsoft SenseCam, Samsung's Galaxy Gear, Autographer, MeCam and LifeLogger, there is a recent upsurge of interest in lifelogging. Lifelogging is an activity of recording and documenting some portions of one's life. Typically, the recording is automatic using wearable devices. Lifelogging can potentially lead to many interesting applications, ranging from lifestyle analysis, behavior analysis, health monitoring, to stimulation for memory rehabilitation for dementia patients.

Much advancement has been made in the hardware design for life-logging devices. Figure 1 shows some of the life-logging wearable devices. For example, Microsoft SenseCam [1] (commercially available as Vicon Revue) and Autographer [2] are wearable cameras that incorporate numerous advanced sensors (accelerometer, ambient light sensor, passive infrared) to determine the appropriate time to take a photo. Google Glass is an augmented glass that can be

© Springer International Publishing Switzerland 2015
C.V. Jawahar and S. Shan (Eds.): ACCV 2014 Workshops, Part III, LNCS 9010, pp. 445–458, 2015.
DOI: 10.1007/978-3-319-16634-6_33

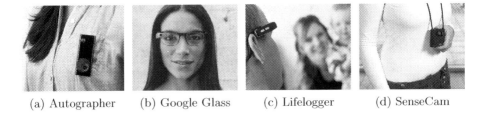

(a) Autographer   (b) Google Glass   (c) Lifelogger   (d) SenseCam

Fig. 1. A variety of life-logging wearable devices.

worn throughout the day to record first-person view video (egocentric video) at 720p HD resolution. However, algorithms for analyzing life-logging data, especially videos, need further improvement. For example, Vicon Revue [3], a wearable camera used by many serious life-loggers, supplies only primitive software that allows simple photo navigation and manual captioning/labeling of individual photos. Note that lifelogging usually generates a large amount of data. For instance, it is common for a lifelogging camera like Autographer to capture over 1000 photos a day. Likewise, several hours of lifelogging videos may be recorded by Google Glass daily. Therefore, manual processing of life-logging data could be extremely laborious. Automatic analysis of lifelog is crucial for many applications.

In the context of automatic analysis of visual lifelog, we describe in this paper an effort to advance the field with the design of an egocentric video database containing 13 categories of activity relevant to life-logging applications. These videos are recorded using Google Glass and capture the diversity and complexity of different human daily activities in first-person view as shown in Fig. 2. The dataset, dubbed LENA (Life-logging EgoceNtric Activities), can be used by the vision research community to develop or evaluate new algorithms for life-logging applications.

Compared with previously-proposed egocentric video databases, one particular feature of LENA is its hierarchical grouping of activities: top-level categorization represents broad classification of daily human activity, while second-level categorization represents finer activity distinction. We use the proposed LENA database to evaluate the performance of state-of-the-art activity recognition algorithms. We also compare the performance of algorithms with activities at top-level and second-level. This reveals the performance difference of state-of-the-art to recognize coarse and fine level human activities in the context of first-person-view video.

The rest of this paper is organized as follows. In Sect. 2 we survey egocentric video datasets. Our life-logging videos dataset is presented in Sect. 3. Approach for evaluation is explained in Sect. 4 and experiment results are in Sect. 5.

2   Survey of Datasets

We describe a survey of 12 existing egocentric video datasets in this section (see Table 1). We found that these datasets focus on different applications and have a large diversity of camera view-point, video quality, camera location, etc.

Fig. 2. Sample frames from our egocentric video database. The activities in frames are: (a),(b): watch videos, (c),(d): read, (e),(f): walk straight, (g),(h): walk up and down, (i),(j): drink, (k),(l): housework.

For instance, dataset used in [4] focuses on object recognition, datasets of [5] and [6] consist of actions in kitchen, and datasets in [7–9] are mostly for social interaction. However, none of these focuses on comprehensive recording of activities related to life-logging.

2.1   Intel Egocentric Object Recognition Dataset

The Intel Egocentric Object Recognition Dataset (IEOR) [4] mainly focuses on recognition of handled objects using a wearable camera. It has ten video sequences from two human subjects manipulating 42 everyday object instances. However, the purpose of this dataset is to study object recognition in everyday life settings from an egocentric view instead of life-logging activities classification.

2.2   CMU-MMAC Dataset

The CMU Multi-Modal Activity Database (CMU-MMAC) database [10] contains multi-modal measures of the human activity of subjects performing the tasks involved in cooking and food preparation, for example, making brownies, pizza, sandwich, etc. The CMU-MMAC database was collected in Carnegie Mellon's Motion Capture Lab. Several modalities are recorded like video, audio, motion capture, etc. However, cooking alone is obviously not adequate to represent the diversity of life activities.

Table 1. A list of existing egocentric video datasets.

	Dataset	No	Activities	Comments
1	IEOR	10	Manipulate Objects	42 objects
2	CMU-MMAC	185	Cooking	5 recipes
3	GTEA	28	Food preparation	7 types of food
4(a)	UEC QUAD	1	Ego action like run, jump, etc	11 simple action
4(b)	UEC PARK	1	Ego action like jog, twist, etc	29 simple action
5	W31	31	Walking	From metro to work
6(a)	GTEA Gaze	17	Food preparation	30 kinds of food
6(b)	GTEA Gaze+	30	Food preparation	Around 100 actions
7	ADL	20	Food, hygiene and entertainment	18 indoor activities
8	UTokyo	5	Office activities	5 office tasks
9	FPSI	113	Social interaction	6 types of activities
10	UT Ego	4	Life-logging activities	11 events
11	JPL	57	Social interaction	7 types of activities
12	EGO-GROUP	10	Social interaction	4 different scenarios

(No.: Number of videos in each dataset.)

2.3   Georgia Tech Egocentric Activities Datasets

The Georgia Tech Egocentric Activities Dataset (GTEA) [5] consists of 7 types of daily activities, Hotdog, Sandwich, Instant Coffee, Peanut Butter Sandwich, Jam and Peanut Butter Sandwich, Sweet Tea, Coffee and Honey, Cheese Sandwich. The camera is mounted on a cap worn by the subject. The GTEA dataset focuses more on food preparation, so it is useful for recognizing objects and not so much for daily life activities classification.

2.4   UEC Datasets

The UEC Datasets [11] are actually two choreographed videos. The first video (QUAD) contains 11 different simple ego-actions, for instance, jump, run, stand, walk, stand, etc. The second video (PARK) is a 25 min workout video which contains 29 different ego-action categories such as pull-ups, jog, twist, etc. The actions are very fine-grained and more related to sports instead of life-logging activities.

2.5   W31 Datasets

The W31 Dataset [12] consists of 31 videos capturing the visual experience of a subject walking from a metro station to work. It consists of 7236 images in total. This dataset is collected to detect unplanned interactions with people or objects and does not contain other activities.

2.6   Georgia Tech Egocentric Activities Gaze(+) Datasets

The Georgia Tech Egocentric Activities Gaze(+) Datasets [6] consist of two datasets which contain gaze location information associated with egocentric videos.

The GTEA Gaze dataset is recorded by Tobii eye-tracking glasses. The Tobii system has an outward-facing camera that records at 30 fps rate and 480×640 pixel resolution. While, one problem of this dataset is that it only collects the meal preparation activity. There are 30 different kinds of food and objects in the videos. And the datasets includes 17 sequences of meal preparation activities performed by 14 different subjects. Each sequence takes about 4 min on average.

The GTEA Gaze+ dataset is collected to overcome some of the GTEA Gaze dataset's shortcomings. The resolution is 1280×960 and tasks are more organized. The number of tasks and objects used in this dataset are significantly bigger. It is collected from 10 subjects and each performs a set of 7 meal preparation activities. Gaze location at each frame is recorded. Each sequence takes around 10–15 min and contains around 100 different actions like pouring, cutting, mixing, turning on/off etc.

2.7   Activities of Daily Living Dataset

The Activities of Daily Living (ADL) Dataset [13] is a set of 1 million frames of dozens of people performing unscripted, everyday activities. The data is annotated with activities, object tracks, hand positions, and interaction events. The dataset is a 10 h of video, amassed from 20 people in 20 different homes and recorded by chested-mounted cameras. The dataset is good for indoor activities classification. However, it does not involve any outdoor activities. From high level, it only has three categories: hygiene, food and entertainment.

2.8   UTokyo First-Person Activity Recognition Dataset

The UTokyo First-Person Activity Recognition Dataset [14] includes five tasks (reading a book, watching a video, copying text from screen to screen, writing sentences on paper and browsing the internet). Office activities of five subjects are recorded and each action is about two minutes. This dataset only records the office activities and five tasks.

2.9   Georgia Tech First-Person Social Interactions Dataset

The First-Person Social Interactions Dataset [9] contains day-long videos of eight subjects spending their day at Disney World Resort. The cameras are mounted on a cap worn by subjects. It is only a set of social interaction activities and recorded at Disney World Resort.

2.10 UT Egocentric Dataset

The University of Texas at Austin Egocentric (UT Ego) Dataset [15] contains four videos captured from head-mounted Looxcie cameras. Each video is about 3–5 h long, captured in a natural, uncontrolled setting. The videos are recorded at 15 fps and 320 × 480 resolution. The videos capture a variety of activities such as eating, shopping, attending a lecture, driving, and cooking. While, the dataset is not easy and perfect for activities classification task because it needs to be cut into several clips and they may have different time duration.

2.11 JPL First-Person Interaction Dataset

The Jet Propulsion Laboratory (JPL) first-person Dataset [8] contains videos of interactions between humans and the observer. A GoPro2 camera is mounted on the head of our humanoid model and participants are required to interact with the humanoid by performing activities like shaking, hugging, petting, etc. Videos were recorded continuously during human activities and they are in 320 × 240 resolution with 30 fps. The limitation of this dataset is that camera is placed on the model instead of a real person. So some head motion and noise of egocentric view are removed. Another problem is that this dataset focuses on social interaction activities.

2.12 EGO-GROUP

The EGO-GROUP dataset [7] contains 10 videos, more than 2900 frames annotated with group compositions and 19 different subjects. There are four different scenarios in the dataset: laboratory, coffee break, party and outdoor. Similar to Georgia Tech First-Person Social Interactions Dataset and JPL First-Person Interaction Dataset, the video in EGO-GROUP dataset are collected in a more social way.

We have summarized the limitations for the popular egocentric video datasets when they are used for life-logging activities classification. In what follows, we describe our proposed life-logging dataset: LENA (Life-logging EgoceNtric Activities) to overcome these limitations.

3 Google Glass Life-Logging Videos Dataset

Google Glass is a type of wearable technology with a camera and an optical display. It is relatively easier to collect egocentric videos using Google Glass than other wearable cameras. And for the existing dataset, the camera viewpoint varies considerably due to different kinds of cameras and also camera's positions (see Fig. 1). The integrated camera in Google Glass makes it very similar to first-person view and also convenient to collect life-logging activities videos.

3.1   Dataset Collection

The Google Glass Life-logging Dataset contains 13 distinct activities performed by 10 different subjects. And each subject record 2 clips for one activity. So each activity category has 20 clips. Each clip has a duration of exactly 30 seconds. The activity categories are: *watching videos, reading, using Internet, walking straight, walking back and forth, running, eating, walking up and down, talking on the phone, talking to people, writing, drinking* and *housework*. Sample frames in some of these categories are shown in Fig. 2. Subjects have been instructed to perform the activities in a natural and unscripted way.

3.2   Video Normalization

The original quality of Google Glass video is 1280×720 and the frame-rate is 30 fps. We also provide a version with dimension down-scaled to 430×240, to reduce the running time when needed. All the clips are processed with *ffmpeg* video library.

3.3   Characteristics

Firstly, LENA contains large variability in scenes and illumination. Videos are recorded both indoor and outdoor, with change in the illumination conditions

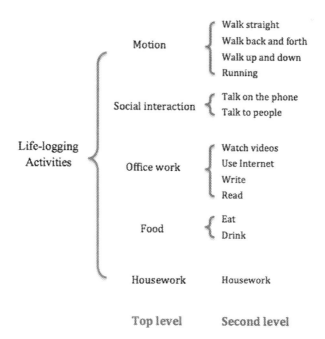

Fig. 3. Hierarchical grouping of life-logging activities

(e.g., from morning to afternoon). Videos are collected from 10 different subjects and the activity videos are captured in an uncontrolled setting. There is also considerable variability for some activities like housework and reading. For housework recording, there are several different activities like washing cups and dishes, mopping, sweeping etc.

Secondly, we build a taxonomy based on the categories as shown in Fig. 3. All 13 categories can be grouped into 5 top level types: *motion*, *social interaction*, *office work*, *food* and *housework*. This allows evaluation of new visual analysis algorithms against different levels of life-logging activity granularity. We believe that these characteristics can make LENA a very useful one for egocentric video research.

4   Activity Recognition Evaluation

We evaluate state-of-the-art trajectory-based activity recognition using our dataset. The dense trajectory approach we used has already been applied to third-person view action recognition in [16]. And object recognition is not performed for activities classification. We evaluate dense trajectories approach using LENA and trajectories are obtained by tracking densely sampled points using optical flow fields. Motion in frames and head movements are used in the recognition with this approach.

4.1   Trajectory Features

Several kinds of descriptors (*HOG*, *HOF* and *MBH*) are computed for each trajectory. *Trajectory* is a concatenation of normalized displacement vectors. The other descriptors are computed in the space-time volume aligned with the trajectory. *HOG* (histograms of oriented gradients) focuses on static appearance information. And both *HOF* (histograms of optical flow) and *MBH* (motion boundary histograms) measure motion information. *HOF* directly quantizes the orientation of optical flow vectors. *MBH* splits the optical flow into horizontal and vertical components, and quantizes the derivatives of each component.

4.2   Fisher Vector Encoding

We use Fisher vector to encode the trajectory features in the experiment. Fisher vector encodes both first and second order statistics between the video descriptors and a Gaussian Mixture Model (GMM). In our experiment, the number of Gaussians is set to $K = 256$ and randomly sample a subset of 256,000 features to estimate the GMM for building the codebook. The dimensions of the features are reduced by half using PCA. In particular, each video is represented by a $2DK$ dimensional Fisher vector, where D is the descriptor dimension after performing PCA. Finally, we apply power and $L2$ normalization to Fisher vector.

The cost parameter $C = 100$ is used for linear SVM and one-against-rest approach is utilized for multi-class classification. We use *libsvm* library [17] to implement the SVM algorithm.

5 Activity Recognition Experiment Results

In this section, we evaluate the performance of trajectory-based activity recognition with our Life-logging Egocentric Videos datasets and make a comparison using different descriptors. We apply the dense trajectory approach both on the 13 second-level categories and 5-top level categories.

5.1 Evaluation for Fine-Grained Categories

The classification results on LENA using dense trajectory approach and Fisher vector encoding is reported in Table 2. The combined descriptors result in about 80 % accuracy. Therefore, further improvement is desirable and this is the subject of future research. *HOF* and *MBH* descriptors result in better performance than *HOG*, as *HOG* only captures static information. Figure 4 shows the confusion matrix of combined descriptors on LENA.

Table 2. Comparison of all descriptors' accuracy for second-level (fine) categories.

Descriptor	*HOF*	*HOG*	*MBH*	*Trajectory*	Combined
Accuracy	76.38 %	68.15 %	78.04 %	74.46 %	81.12 %

From Fig. 4 we can see that performance of *walk up and down, read, use Internet* and *run* categories, are superior. While, for *talk on the phone, write, drink* and *eat* the performance is around 50 %. Note that for *talk on the phone, drink* and *eat*, there are hardly any objects like phone, cup and snacks in the scene, as the videos are recorded in first-person view. Thus it is more difficult for recognition, especially using the *HOG* descriptor.

We also make a comparison among different descriptors. Figure 5 shows the performance of the 4 different descriptors on second level categories. Overall, *HOF, MBH* and *Trajectory* have similar results. While, performance on *HOG* descriptor is about 10 % less than the other three. The *HOG* descriptor shows the worst performance which is 68.15 %, especially for *write* and *drink* categories.

For combined descriptor, the accuracy of *drink* is 34 % which is the lowest. Videos of *drink* action are often mis-classified into *write* and *housework*. The *MBH* descriptor alone obtains the best performance with LENA.

5.2 Evaluation for Top-Level (coarse) Categories

One feature of LENA is that we have a hierarchical grouping of life-logging activities. Then we evaluate top-level categories using dense trajectory algorithm. The dataset we used is the same as the second level categories. We only changed the ground-truth of the dataset and trained one-against-rest SVM classifiers on the training set.

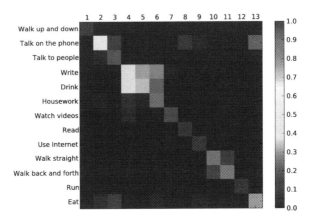

Fig. 4. Confusion matrix of combined descriptors for second-level (fine) categories.

In Table 3, the accuracy of every descriptor is much higher than results in Table 2. The performance has improved by about 10 % for *HOF*, *HOG* and *Trajectory*. However, *HOG* still performs worst among all the descriptors. Figure 6 shows the comparison between top level and second level categories. The accuracy of *HOG* descriptor has been improved most which is 8.67 %. One reason could be that the difference among the top-level categories are much more salient than second level categories. We also observe that *HOF*, *Trajectory* and combined descriptor have almost the same results. *Trajectory* even has a slight higher accuracy than combined descriptor.

We also construct confusion matrices for top level categories classification in Fig. 7. Interestingly, *motion* has nearly 100 % accuracy. It is because the difference between *motion* and the other four activities are much more obvious. The most easily mistaken pair is *food* and *office work*. While, even the *food* activity which performs the worst has an accuracy of more than 50 %. Overall, the trajectory algorithm does well on the top-level categories.

Confusion matrices of individual descriptor are also presented in Fig. 8. From the figure we can see that *motion* has a high accuracy on every kind of descriptor. One the contrary, category *food* which contains *eat* and *drink* actions has the poorest performance. The low mis-classification rate of *motion* category is quite understandable, as in *motion* recording, there are always similar head motion and body movement. While, in *office work* and *food* recording, subjects usually do not move head and body sharply as they do in *run* and *walk straight* actions recording.

Table 3. Comparison of all descriptors' accuracy for top-level categories.

Descriptor	*HOF*	*HOG*	*MBH*	*Trajectory*	Combined
Accuracy	84.00 %	77.42 %	82.46 %	84.50 %	84.23 %

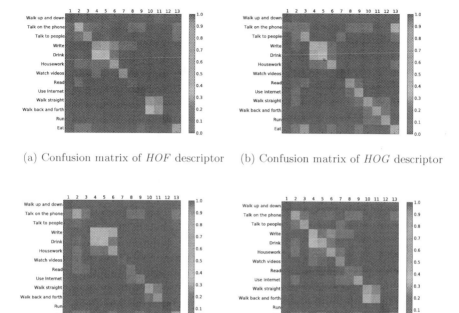

(a) Confusion matrix of *HOF* descriptor   (b) Confusion matrix of *HOG* descriptor

(c) Confusion matrix of *MBH* descriptor (d) Confusion matrix of *Trajectory* descriptor

Fig. 5. Confusion matrix of individual descriptor.

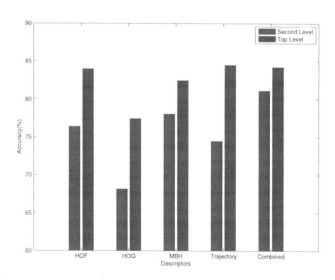

Fig. 6. Comparison between classification accuracy of top-level and second-level categories.

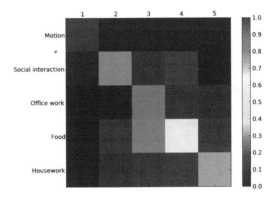

Fig. 7. Confusion matrix of combined descriptors for top-level categories.

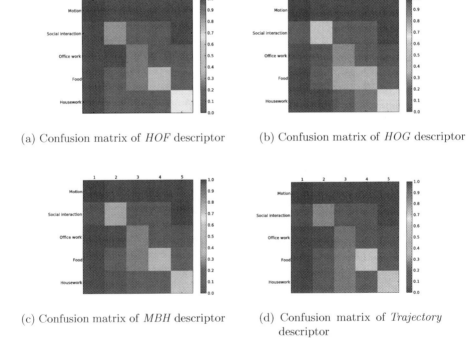

(a) Confusion matrix of *HOF* descriptor

(b) Confusion matrix of *HOG* descriptor

(c) Confusion matrix of *MBH* descriptor

(d) Confusion matrix of *Trajectory* descriptor

Fig. 8. Confusion matrix of individual descriptor.

6   Conclusions

We summarize our contributions as follows:

- We surveyed and discussed popular egocentric video datasets. Analysis of 12 existing datasets suggests that although they can address a variety of applications, their utility for daily life activities classification and analysis is inadequate in many ways.
- We presented LENA, an egocentric video dataset of life-logging activities. Recorded by Google Glass, the egocentric videos contain a variety of scenes and personal styles, capturing the diversity and complexity of life activities. The activities are organized into two levels of categorization, enabling research on coarse and fine-grained life activity analysis using a single dataset.
- We performed detailed evaluation of state-of-the-art activity recognition using LENA. We found that with dense trajectory approach the accuracy of activity recognition is around 80 %. Thus, further research is needed on life-logging activity recognition. Furthermore, thanks to the two-level categorization structure, we were able to reveal the performance gap of state-of-the-art in recognizing activity at two different granularities. We also analyzed the performance of state-of-the-art descriptors (based on gradient, optical flow, motion boundary) in egocentric activity recognition.

We believe that LENA could be valuable for the research on egocentric activities recognition and classification, especially for life activities. Future work involves understanding the importance of global motion in egocentric activity recognition and using graph-theoretic approach for life activity recognition. In addition, while our focus in this paper is activity recognition, many other visual data analysis tasks such as automatic discovery of activity topics could be investigated using our proposed dataset.

References

1. Microsoft research sensecam. http://research.microsoft.com/sensecam (2013)
2. Autographer. http://www.autographer.com/ (2013)
3. Vicon revue wearable cameras. http://viconrevue.com/ (2013)
4. Ren, X., Gu, C.: Figure-ground segmentation improves handled object recognition in egocentric video. In: 2010 IEEE Conference on Computer Vision and Pattern Recognition (CVPR), pp. 3137–3144. IEEE (2010)
5. Fathi, A., Ren, X., Rehg, J.M.: Learning to recognize objects in egocentric activities. In: 2011 IEEE Conference On Computer Vision and Pattern Recognition (CVPR), pp. 3281–3288. IEEE (2011)
6. Fathi, A., Li, Y., Rehg, J.M.: Learning to recognize daily actions using gaze. In: Fitzgibbon, A., Lazebnik, S., Perona, P., Sato, Y., Schmid, C. (eds.) ECCV 2012, Part I. LNCS, vol. 7572, pp. 314–327. Springer, Heidelberg (2012)
7. Alletto, S., Serra, G., Calderara, S., Solera, F., Cucchiara, R.: From ego to nos-vision: detecting social relationships in first-person views. In: Proceedings of the IEEE Conference on Computer Vision and Pattern Recognition Workshops, pp. 580–585 (2014)

8. Ryoo, M.S., Matthies, L.: First-person activity recognition: what are they doing to me? In: 2013 IEEE Conference on Computer Vision and Pattern Recognition (CVPR), pp. 2730–2737. IEEE (2013)
9. Fathi, A., Hodgins, J.K., Rehg, J.M.: Social interactions: a first-person perspective. In: 2012 IEEE Conference on Computer Vision and Pattern Recognition (CVPR), pp. 1226–1233. IEEE (2012)
10. CMU: Multi-modal activity database. http://kitchen.cs.cmu.edu/index.php (2010)
11. Kitani, K.M., Okabe, T., Sato, Y., Sugimoto, A.: Fast unsupervised ego-action learning for first-person sports videos. In: 2011 IEEE Conference on Computer Vision and Pattern Recognition (CVPR), pp. 3241–3248. IEEE (2011)
12. Aghazadeh, O., Sullivan, J., Carlsson, S.: Novelty detection from an ego-centric perspective. In: 2011 IEEE Conference on Computer Vision and Pattern Recognition (CVPR), pp. 3297–3304. IEEE (2011)
13. Pirsiavash, H., Ramanan, D.: Detecting activities of daily living in first-person camera views. In: 2012 IEEE Conference on Computer Vision and Pattern Recognition (CVPR), pp. 2847–2854. IEEE (2012)
14. Ogaki, K., Kitani, K.M., Sugano, Y., Sato, Y.: Coupling eye-motion and ego-motion features for first-person activity recognition. In: 2012 IEEE Computer Society Conference on Computer Vision and Pattern Recognition Workshops (CVPRW), pp. 1–7. IEEE (2012)
15. Lee, Y.J., Ghosh, J., Grauman, K.: Discovering important people and objects for egocentric video summarization. CVPR 1, 2–3 (2012)
16. Wang, H., Kläser, A., Schmid, C., Liu, C.L.: Dense trajectories and motion boundary descriptors for action recognition. Int. J. Comput. Vis. 103, 60–79 (2013)
17. Chang, C.C., Lin, C.J.: LIBSVM: a library for support vector machines. ACM Trans. Intell. Syst. Technol. 2(27), 1–27:27 (2011). http://www.csie.ntu.edu.tw/cjlin/libsvm

3D Line Segment Based Model Generation by RGB-D Camera for Camera Pose Estimation

Yusuke Nakayama[1]([✉]), Hideo Saito[1], Masayoshi Shimizu[2],
and Nobuyasu Yamaguchi[2]

[1] Graduate School of Science and Technology, Keio University,
3-14-1 Hiyoshi, Kohoku-ku, Yokohama, Japan
{nakayama,saito}@hvrl.ics.keio.ac.jp
[2] Fujitsu Laboratories Ltd.,
4-1-1 Kamikodanaka, Nakahara-ku, Kawasaki, Japan
{shimizu.masa,nobuyasu}@jp.fujitsu.com

Abstract. In this paper, we propose a novel method for generating 3D line segment based model from an image sequence taken with a RGB-D camera. Constructing 3D geometrical representation by 3D model is essential for model based camera pose estimation that can be performed by corresponding 2D features in images with 3D features of the captured scene. While point features are mostly used as such features for conventional camera pose estimation, we aim to use line segment features for improving the performance of the camera pose estimation. In this method, using RGB images and depth images of two continuous frames, 2D line segments from the current frame and 3D line segments from the previous frame are corresponded. The 2D-3D line segment correspondences provide camera pose of the current frame. All of 2D line segments are finally back-projected to the world coordinate based on the estimated camera pose for generating 3D line segment based model of the target scene. In experiments, we confirmed that the proposed method can successfully generate line segment based models, while 3D models based on the point features often fail to successfully represent the target scene.

1 Introduction

Generating 3D geometrical models of the environment is an essential technology for a lot of vision-based applications. For example, it is almost impossible to estimate camera pose for AR applications from the captured image sequence without such geometrical 3D model of the object environment. For constructing 3D models, Structure-from-Motion (SfM) approaches are often used. Traditional SfM approaches like [1] take correspondences from images, then recover 3D geometrical model of the scene and pose of the camera at each frame. To obtain these correspondences, feature points are generally detected, and then the detected features are matched between different frames. For the matching, feature point descriptors such as the scale-invariant feature transform (SIFT) [2] and speeded-up robust features (SURF) [3] are often used. However, in the situation where only a few feature points are detected, feature point matching may

© Springer International Publishing Switzerland 2015
C.V. Jawahar and S. Shan (Eds.): ACCV 2014 Workshops, Part III, LNCS 9010, pp. 459–472, 2015.
DOI: 10.1007/978-3-319-16634-6_34

fail, therefore reconstructed objects are inaccurate. In man-made environment, there are usually a lot of untextured objects which do not provide sufficient number of feature points.

Line segment features can be considered as an alternative feature to solve this problem. A lot of line segments are detected in a man-made situation even where only a few feature points are detected. Over the years, several methods about line based 3D reconstruction have been reported. Comparing with feature point based reconstruction, the number of researches about line based is very few. This is mainly because line segment matching is still challenging task. Line segments have less distinctive appearance so that a descriptor of line segment feature is difficult to be defined. Therefore, some existing methods [4–6] do not rely on appearance-based line segment matching. However, recently, researches about line segment feature descriptor such as the mean standard-deviation line descriptor (MSLD) [7] or the Line-based Eight-directional Histogram Feature(LEHF) [8] have been reported. With these line segment feature descriptors, line segment based 3D model can be generated in man-made environment.

As one more advantage of line segments, line segment matching can be robust to large viewpoint changes. This is because line segments provide more information from their length, geometry characteristics and relative positions. One of the applications for generated 3D models is the estimation of camera pose using 3D models. With 3D models which consists feature points, the camera pose estimation cannot deal with large viewpoint changes because feature point matching is not robust to changes in perspective. On the other hand, thanks to the line segment's advantage, 3D line segment based model and estimation of the camera pose from it can deal with the perspective changes. Therefore, generating 3D line segment based model is also important for the large viewpoint change situation.

In this paper, we propose a novel method for generating 3D line segment based model from an image sequence taken with a RGB-D camera. For obtaining line segments matching between images, we use Directed LEHF [9] which is an improved version of LEHF [8]. In the proposed method, 2D line segments are first detected from RGB images. Then, 2D line segments between consecutive two frames are matched by Directed LEHF. From this matching, the 2D line segments of the previous frame can be transferred to 3D line segments in the world coordinate. Then, the current frame's 2D line segments and the previous frame's 3D line segments can be corresponded. These 2D-3D correspondences give camera pose of the current frame by solving the Perspective-n-Lines (PnL) problem. With this estimated camera pose, the 2D line segments of current frame are translated to 3D line segments. This procedure is repeated and objects represented by 3D line segments are obtained as a 3D model. We have experimentally demonstrated the performance of the proposed method by comparison with other 3D reconstruction methods. The experimental result shows that our proposed method using line segment matching can generate an accurate 3D model in the situation which has few feature points. Moreover, we demonstrated that our 3D line segment based model can use for estimating camera poses in the large viewpoint changes situation.

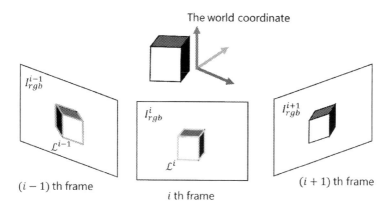

Fig. 1. Detection of 2D line segments. The blue line segments from I_{rgb}^{i-1} and the yellow line segments from I_{rgb}^i are $\mathcal{L}^{i-1}$ and $\mathcal{L}^i$, respectively (Color figure online).

2 Proposed Method

In this section, we describe the proposed method for generating a 3D line segment based model of a target scene from an image sequence captured by moving a RGB-D camera, such as Kinect. N multiple images of a target scene are captured by RGB-D camera. Suppose we have N RGB images $\{I_{rgb}^N\}$ and N depth images $\{I_d^N\}$ captured with the Kinect. The basic idea is that each frame's camera pose can be estimated from its previous frame's camera pose, and then 2D line segments on RGB Images are back-projected into the 3D world coordinate by the estimated camera pose. The camera pose at ith frame is represented a transform matrix $RT_{cw}^i = [R_i|t_i]$ containing a 3×3 rotation matrix (R_i) and 3D translation vector (t_i). In the following subsections, we describe 3D geometry of the consecutive two frames, $(i-1)$th and ith frame. Then we will describe the way of computing RT_{cw}^i using known $(i-1)$th frame's camera pose RT_{cw}^{i-1}. We set the first frame's camera pose RT_{cw}^0 as 4×4 Identity matrix, because we assume that the world coordinate is defined by the camera coordinate of the first frame.

2.1 Detection of 2D Line Segments

First, 2D line segments are detected from RGB images, by employing the fast line segment detector (LSD) [10]. Using LSD, we can obtain end points of each 2D line segment. As shown in Fig. 1, two sets of line segments detected from I_{rgb}^{i-1} and I_{rgb}^i by LSD are indicated by $\mathcal{L}^{i-1} = \{l_{i-1}^0, l_{i-1}^1, \cdots, l_{i-1}^{M_{i-1}}\}$ and $\mathcal{L}^i = \{l_i^0, l_i^1, \cdots, l_i^{M_i}\}$, respectively.

2.2 Creation of 3D Line Segments

We also have depth images, I_d^{i-1} and I_d^i. Therefore, the positions of the 3D line segments in the camera coordinate of $(i-1)$th frame and ith frame are

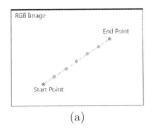 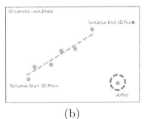 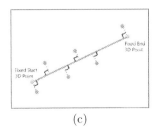

(a)   (b)   (c)

Fig. 2. The way to create 3D line segments. (a) 2D points on the line segment, (b) 3D points in the camera coordinate, (c) the 3D line segment and fixed end points.

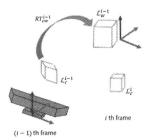 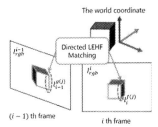

Fig. 3. Creation of the 3D line segments.   **Fig. 4.** Directed LEHF matching.

obtained, so that we can create 3D line segments in the camera coordinate. Suppose we choose one 2D line segment in $\mathcal{L}^{i-1}$ and $\mathcal{L}^i$. First, as shown in Fig. 2(a), we get equally-spaced points on the line segment between the start point and the end point. These acquired points, the start point and the end point are translated from the image coordinate to the camera coordinate using depth value. These translated 3D points are supposed to be on the same 3D line. However, because of the Kinectfs depth value error, the 3D points also have error, therefore, they are not on the same 3D line. To decrease the error, RANSAC [11] is used to find inliers of the 3D points. Thus, we can eliminate points which stay from the others. In this process, as shown in Fig. 2(b), the two points which are located at the either end of the inliers are assumed as a tentative start 3D point and end 3D point. Next, we compute a 3D line which minimizes the lengths of perpendiculars from each inlier point to the 3D line. Finally, we chose the extremities of the perpendiculars from the tentative start point and end point as a fixed start 3D point and a fixed end 3D point. Therefore, connecting these fixed start 3D point and end 3D point, we obtain a 3D line segment. Figure 2(c) shows the fixed two points and the obtained 3D line segment. By applying this procedure to every line segment in $\mathcal{L}^{i-1}$ and $\mathcal{L}^i$, we obtain $\mathcal{L}_c^{i-1} = \{L_{c,i-1}^0, L_{c,i-1}^1, \cdots, L_{c,i-1}^{M_{i-1}}\}$ and $\mathcal{L}_c^i = \{L_{c,i}^0, L_{c,i}^1, \cdots, L_{c,i}^{M_i}\}$, where $L_{c,i}$ is translated from l_i into the camera coordinate. These 3D line segments in $\mathcal{L}_c^{i-1}$ is translated from the camera coordinate to the world coordinate by known RT_{cw}^{i-1}. Therefore, we get $\mathcal{L}_w^{i-1} = \{L_{w,i-1}^0, L_{w,i-1}^1, \cdots, L_{w,i-1}^{M_{i-1}}\}$, where $L_{w,i}$ is translated from l_i into the world coordinate, as shown in Fig. 3.

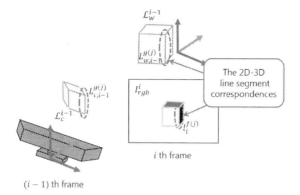

Fig. 5. The 2D-3D line segment correspondences. $l_{i-1}^{g(j)}$ from $\mathcal{L}^{i-1}$ and $l_i^{f(j)}$ from $\mathcal{L}^i$ are matched. $l_{i-1}^{g(j)}$ is transferred to $L_{c,i-1}^{g(j)}$ in $\mathcal{L}_c^{i-1}$. Then $L_{c,i-1}^{g(j)}$ is translated to $L_{w,i-1}^{g(j)}$ in $\mathcal{L}_w^{i-1}$. Therefore $L_{w,i-1}^{g(j)}$ and $l_i^{f(j)}$ are corresponded.

2.3 Matching 2D Line Segments by Directed LEHF

Next, we obtain 2D line segment matching between two images. To evaluate similarity of 2D line segments, we use Line-based Eight-directional Histogram Feature (LEHF) [8]. LEHF is a descriptor of line segments based on the gradient histogram for eight different directions along with the line segment. LEHF descriptor cannot deal with rotation change more than 180°. To get matching of the opposite direction line segments, LEHF needs to compute descriptor distances from two directions, forward direction and inverse direction. However, if the descriptor distances from the two directions are similar, mismatching is occurred. To improve the matching performance, we proposed Directed LEHF [9]. We adopt Directed LEHF as a line segment feature descriptor.

Using a method explained in [12], the Directed LEHF defines the direction of each 2D line segment by the average intensity of the gradient of the perpendicular direction with the line segment at all the points on the line segment. The sign of the average determines the direction of the line segment. The direction of the line segment determines the start point and end point of each 2D line segment in $\mathcal{L}^{i-1}$ and $\mathcal{L}^i$.

We search for 2D line segment matchings with $\mathcal{L}^{i-1}$ and $\mathcal{L}^i$ by the Directed LEHF matching. The resulting set of matching 2D line segment is represented as

$$\mathcal{LM}^i = \{(l_{i-1}^{g(j)}, l_i^{f(j)}), j = 0, 1, \cdots, K^i\}, \tag{1}$$

in which $(l_{i-1}^{g(j)}, l_i^{f(j)})$ represents a pair of matching 2D line segments $g(j) \in [0, M_{i-1}]$ and, $f(j) \in [0, M_i]$, as shown in Fig. 4.

2.4 2D-3D Line Segment Correspondences

In the 2D line segment matching $\mathcal{LM}^i$, $l_{i-1}^{g(j)}$ from $\mathcal{L}^{i-1}$ and $l_i^{f(j)}$ from $\mathcal{L}^i$ are matched. The 3D line segment back-projected from $l_{i-1}^{g(j)}$ into the world coordinate

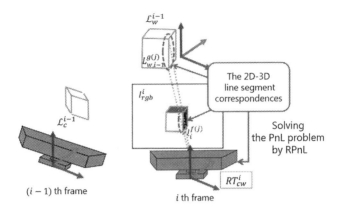

Fig. 6. Camera pose estimation by solving the PnL problem.

is $L_{w,i-1}^{g(j)}$ in $\mathcal{L}_w^{i-1}$. Then, the 3D line segment in $\mathcal{L}_w^{i-1}$ and the 2D line segment in $\mathcal{L}^i$ are brought to be correspondence with the matching of $\mathcal{LM}^i$. The set of 2D-3D line segment correspondences is represented as

$$\mathcal{LC}^i = \{(L_{w,i-1}^{g(j)}, l_i^{f(j)}), j = 0, 1, \cdots, K^i\}, \qquad (2)$$

in which $(L_{w,i-1}^{g(j)}, l_i^{f(j)})$ represents a pair of 2D-3D line correspondences $g(j) \in [0, M_{i-1}]$ and $f(j) \in [0, M_i]$. The position of the 2D line segment and the 3D line segment is shown in Fig. 5.

2.5   Solution for the PnL Problem

Given a set of 2D-3D line correspondences, we solve the PnL problem, then estimate the camera pose. (The PnL problem is a counterpart of the PnP problem for point correspondences.) However, there is a fear that $\mathcal{LC}^i$ contains some mismatches. We use a method which solves the PnL problem with an algorithm like RANSAC explained in [9], and estimate the camera pose RT_{cw}^i. This method mainly use RPnL [13] for solving the PnL problem. Suppose we have $\mathcal{LC}^i$ which is K^i sets of 2D-3D line segment correspondences, we randomly select four 2D-3D line segment correspondences from $\mathcal{LC}^i$. This is because the program of RPnL needs at least four correspondences. Let the four set of 2D-3D line segment correspondences be represented as

$$\mathcal{LC}_{four}^i = \{(L_{w,i-1}^{a(k)}, l_i^{b(k)}), k = 0, 1, 2, 3\}, \qquad (3)$$

in which $(L_{w,i-1}^{a(k)}, l_i^{b(k)})$ represents four pairs of 2D-3D line segment correspondences $a(k) \in [0, M_{i-1}]$ and $b(k) \in [0, M_i]$. Then, the rest of $(K^i - 4)$ 2D-3D line segment correspondences are represented as

$$\mathcal{LC}_{rest}^i = \{(L_{w,i-1}^{g(j)}, l_i^{f(j)})|0 \le j \le K^i, g(j) \ne a(k), f(j) \ne b(k), k = 0, 1, 2, 3\}. \qquad (4)$$

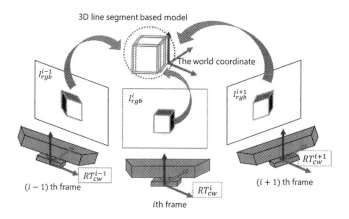

Fig. 7. The 3D line segment based model.

With $\mathcal{LC}^i_{four}$, we solve the PnL problem using RPnL and estimate the camera pose RT'_{cw}. The 3D line segments in $\mathcal{L}^i_c$ which are back-projected from $l^{f(j)}_i$ in $\mathcal{LC}^i_{rest}$ are translated from the camera coordinate into the world coordinate by RT'_{cw}. Let the 3D line segment in the world coordinate translated by RT'_{cw} be $L'^{f(j)}_{w,i}$. We calculate the error $e(j)$ between $L^{g(j)}_{w,i-1}$ from $\mathcal{LC}^i_{rest}$ and $L'^{f(j)}_{w,i}$. We define $e(j)$ as

$$e(j) = S(j)/(length_{w,i-1} + length_i), \tag{5}$$

where $S(j)$ is an area of rectangle obtained by connecting four end points of $L^{g(j)}_{w,i-1}$ and $L'^{f(j)}_{w,i}$, $lenght_{w,i-1}$ is length of $L^{g(j)}_{w,i-1}$, and $length_i$ is length of $L'^{f(j)}_{w,i}$. The total of $e(j)$ is defined as error given by RT'_{cw}.

We also randomly select another set of $\mathcal{LC}^i_{four}$ and repeat the steps explained above N_{RANSAC} times to estimate RT'_{cw}. We choose RT'_{cw} which gives the smallest total of $e(j)$ as a tentative camera pose $tentativeRT_{cw}$. Next, using $tentativeRT_{cw}$, all of the 3D line segments in $\mathcal{L}^i_c$ which are back-projected from $l^{f(j)}_i$ are translated to the 3D line segments in the world coordinate $L'^{f(j)}_{w,i}$. We calculate $e(j)$ and if $e(j)$ is less than threshold (TH_e), we save the 2D-3D line segment correspondences as inlier.

Finally, we compute the camera pose of ith frame using another algorithm for the PnL problem proposed by Kumar and Hanson [14]. This algorithm estimates the camera pose iteratively. It needs a set of 2D-3D line segment correspondences and initial camera pose as inputs. We take the inliers and $tentativeRT_{cw}$ as inputs, and obtain the camera pose of ith frame RT^i_{cw} as output of the algorithm. Figure 6 shows this procedure.

Once RT^i_{cw} is obtained, all of the 3D line segments in the i th camera coordinate $\mathcal{L}^i_c$ are translated to the 3D line segments in the world coordinate $\mathcal{L}^i_w$. These procedures discussed above are repeated in every consecutive two frame. Then, each frame's camera pose and 3D line segments in the world coordinate are obtained. The 3D line segments of every frame are 3D line segment based

Fig. 8. Parts of the input RGB images.

Fig. 9. The 3D line segment based model of proposed method.

model as an output of our proposed method. Figure 7 shows the concept of the 3D line segment based model generation.

3   Experiment

We conducted two experiments for generating 3D models. One is to show that our proposed method can generate a 3D model of the scene where other methods cannot generate accurate models. Another is for demonstrating that camera pose estimation by our proposed model can perform well in large viewpoint changes. In both experiments, we compared our proposed method with a method which uses SIFT feature point matching instead of Directed LEHF matching. For model generation by our method, we set N_{RANSAC} to 5000 and TH_e to 0.003.

3.1   Experiment 1: Generating 3D Model

First of all, we generated 3D model of a scene by our proposed method. We used 81 frames as input image sequence. Figure 8 shows some of the input RGB images. Figure 9 shows the 3D line segment based model of the scene as an output of our proposed method. Each frame's camera pose was estimated and 2D line segments detected in RGB images are back-projected into the 3D coordinate. With the estimated camera poses and each frame's depth image, we also back-projected every points from RGB images into the world coordinate. Then we obtain reconstructed object shape of the scene represented by colored point clouds. Figure 10 shows virtual viewpoint images of the point clouds. In Fig. 9, line segments on the same physical edges of the door structure detected in the different frames are almost overlapping in the 3D space. This demonstrates that the camera poses of the image sequence are correctly estimated by our method.

Figure 10 also demonstrates the same fact, because the colors rendered by back projection from different frames are almost matching so that the virtual view images are synthesized without any blur.

Fig. 10. The reconstructed object shape represented by point clouds.

Fig. 11. The 3D line segment based model using SIFT feature point matching.

Figures 9 and 10 demonstrate that our proposed method could estimate accurate camera pose of each frame and generate 3D models.

For comparison purpose, we conducted the same experiment with SIFT feature point matching instead of Directed LEHF matching for estimating camera pose of every frame. After camera pose estimation, the same procedure of generating the 3D line segment based model is also performed for detected line segments by LSD, which is also the same as our proposed method. Figure 11 shows the 3D line segment based model using the camera pose estimated by SIFT feature point matching. Compared with Figs. 9 and 10, camera poses estimated by the method with feature point matching are less accurate than proposed method. This is because the scene does not provide sufficient number of SIFT feature points.

Next, we tried to get a 3D model of the same scene by other methods. Figure 12 shows the 3D model reconstructed by Autodesk 123D Catch [15]. Autodesk 123D Catch is a free web service which reconstructs 3D object from images. We used 70 images which are included in the 81 images used in previous model generation. As shown in Fig. 12, the reconstructed model is distorted and inaccurate. Autodesk 123D Catch needs images which are captured from various angles. However, especially in terms of this kind of planar object, various viewpoint images cannot be obtained. Compared Fig. 12 with Fig. 10, our proposed method can generate better 3D model.

Figure 13 shows the reconstructed model of KinectFusion [16]. At this time, the KinectFusion which we used was the open source *Kinfu* code in the Point Cloud Library (PCL) from Willow Garage [17]. Using *Kinfu*, Kinect's RGB image, depth image and camera pose of each frame are obtained. Therefore, with

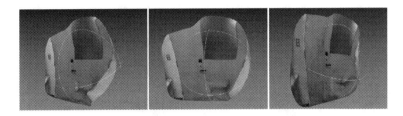

Fig. 12. The 3D model generated by 123D Catch.

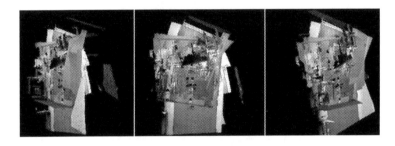

Fig. 13. The 3D model generated by KinectFusion.

the camera pose which KinectFusion estimated, we also obtain reconstructed object shape represented by colored point clouds. Figure 13 is virtual viewpoint images of the reconstructed object shape. KinectFusion estimates the camera pose using the alignment of point clouds such as ICP algorithm [18]. However, with this kind of planar object, point clouds alignment is failed, and the estimated camera pose has some error. Therefore, the reconstructed model shown in Fig. 13 is not accurate.

As shown in Figs. 9, 11, 12 and 13, in this situation, our proposed method can reconstruct an accurate 3D model.

3.2 Experiment 2: On-line Camera Pose Estimation with 3D Model

One of the possible applications of the 3D line segment based model generated by using the proposed method is on-line camera pose estimation for mobile AR. We suppose that line segment based model of the target scene for mobile AR is generated by the proposed method. In this model generation phase, the 3D line segments' position in the world coordinate and their Directed LEHF value from their projected 2D line segments on RGB images are stored as 3D line segment database. Then a mobile camera pose of each frame can be estimated on-line process as described below. First, 2D line segments and Directed LEHF values are extracted from the input image. Second, a 3D line segment which has a similar value to the 2D line segment's Directed LEHF value is searched into the 3D line segment database. Then, the 2D line segment from the input image and the 3D line segment in the database are brought to correspondences.

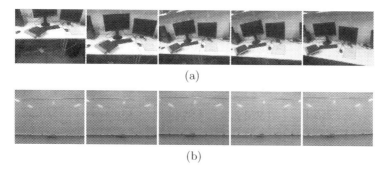

(a)

(b)

Fig. 14. Parts of the input RGB images for experiment 2, (a) the desk scene, (b) the white-board scene.

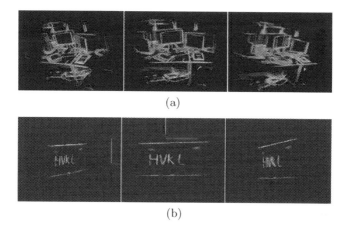

(a)

(b)

Fig. 15. The 3D line segment based models for experiment 2.

These 2D-3D line segment correspondences may have some mismatching. This mismatching is eliminated by RANSAC. Finally, the camera pose of the input image is computed from the 2D-3D matching by RPnL.

In this experiment, we captured two scenes using Kinect and generated the 3D line segment based model of each scene. One scene is about a desk on which two displays and books are put and another scene is about a white-board on which the word "HVRL" is written. We used 101 frames for the desk scene and 11 frames for the white-board scene. Some of RGB images for the two scenes are shown in Fig. 14. Note that we captured the scenes from the front view of the objects as shown in Fig. 14. Therefore, the 3D line segment based models do not contain line segments from side view of the scenes. Figure 15 shows generated 3D line segment based models of each scene. With these models, we constructed 3D line segment databases for both scenes. As well as Experiment1, we conducted the same experiment with SIFT feature point matching instead of Directed LEHF matching. Then, we also constructed SIFT feature point databases which contained 3D points' position in the world coordinate and SIFT feature value.

(a)   (b)

Fig. 16. The four points of each scene used for measuring re-projection errors.

We estimated 27 camera poses of input images for the desk scene and 86 camera poses of input images for the white-board scene with each 3D line segment database and SIFT feature point database. For evaluating the accuracy of the estimated camera poses, we re-projected four points in each scene which are shown in Fig. 16 to the input images, and their re-projection errors are measured. Figure 17 shows the images which have the re-projected four points. The images on the upper low of Fig. 17(a) and (b) show example images of the result images by our proposed method, and the images on the lower row show example images based on SIFT feature points, respectively. Moreover, the line segments and feature points which used for estimating camera pose are shown on the images in green. These green line segments and feature points include no outlier of RANSAC.

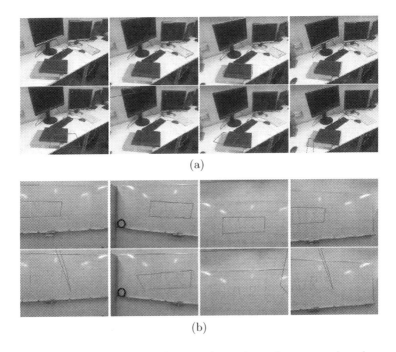

(a)

(b)

Fig. 17. The result images for evaluating the estimated camera pose. (upper row: proposed method, lower row: feature point matching)

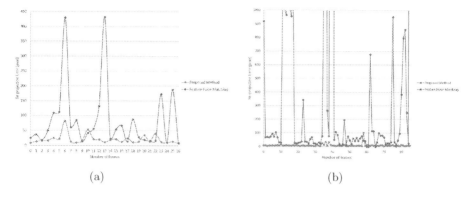

(a)   (b)

Fig. 18. Re-projection errors.

As shown in Fig. 17, in the case of feature point matching, estimated camera poses are not accurate and few feature points are used. On the other hand, our proposed method can estimate the camera poses more accurate and many line segments are used for estimation.

Figure 18 shows the re-projection errors. The camera pose estimation by SIFT feature failed in these situations due to the lack of the correspondences, however, the camera poses which are estimated by proposed method show smaller re-projection errors.

In the case of the desk scene, as shown in Fig. 17(a), the input images are taken from left side view of the desk. However, the 3D line segment based model of the scene contains line segments only from front view of the desk. Therefore, a large viewpoint change is occurred between model and input images. With this, Fig. 17 also shows that feature point matching cannot deal with a change in perspective but line segment matching with our proposed model is robust for this large viewpoint change. This is the advantage of using line segments instead of feature points.

4   Conclusion

We propose a method for generating 3D line segment based model from a RGB-D image sequence captured by RGB-D camera. In this method, 2D line segments are detected by LSD from the RGB image sequence. Each 2D line segments are then associated with depth image for defining a 3D line segment. The line segments are matched between consecutive frames using Directed LEHF for computing camera pose of each frame of the input image sequence. The camera poses of all frames finally generate a 3D model represented by the 3D line segments.

In the experiments, we demonstrate that the proposed method can generate 3D line segment based models even in the case that a few feature points can be detected. We also demonstrate the on-line camera pose estimation for mobile AR application can effectively performed by the use of the 3D line segment based model generated by the proposed method.

References

1. Sturm, P., Triggs, B.: A factorization based algorithm for multi-image projective structure and motion. In: Buxton, B.F., Cipolla, R. (eds.) ECCV 1996. LNCS, vol. 1065. Springer, Heidelberg (1996)
2. Lowe, D.G.: Object recognition from local scale-invariant features. In: Proceedings of the Seventh IEEE International Conference on Computer Vision, vol. 2, pp. 1150–1157. IEEE (1999)
3. Bay, H., Tuytelaars, T., Van Gool, L.: SURF: speeded up robust features. In: Leonardis, A., Bischof, H., Pinz, A. (eds.) ECCV 2006, Part I. LNCS, vol. 3951, pp. 404–417. Springer, Heidelberg (2006)
4. Jain, A., Kurz, C., Thormahlen, T., Seidel, H.P.: Exploiting global connectivity constraints for reconstruction of 3d line segments from images. In: IEEE Conference on Computer Vision and Pattern Recognition, IEEE, pp. 1586–1593 (2010)
5. Hofer, M., Wendel, A., Bischof, H.: Line-based 3d reconstruction of wiry objects. In: Proceedings of the 18th Computer Vision Winter Workshop (2013)
6. Hofer, M., Wendel, A., Bischof, H.: Incremental line-based 3d reconstruction using geometric constraints (2013)
7. Wang, Z., Wu, F., Hu, Z.: MSLD: a robust descriptor for line matching. Pattern Recogn. **42**, 941–953 (2009)
8. Hirose, K., Saito, H.: Fast line description for line-based slam. In: Proceedings of the British Machine Vision Conference, pp. 83.1–83.11 (2012)
9. Nakayama, Y., Honda, T., Saito, H., Shimizu, M., Yamaguchi, N.: Accurate camera pose estimation for kinectfusion based on line segment matching by LEHF. In: Proceedings of the International Conference on Pattern Recognition, pp. 2149–2154 (2014)
10. Von Gioi, R.G., Jakubowicz, J., Morel, J.M., Randall, G.: LSD: a fast line segment detector with a false detection control. IEEE Trans. Pattern Anal. Mach. Intell. **32**, 722–732 (2010)
11. Fischler, M.A., Bolles, R.C.: Random sample consensus: a paradigm for model fitting with applications to image analysis and automated cartography. Commun. ACM **24**, 381–395 (1981)
12. Fan, B., Wu, F., Hu, Z.: Line matching leveraged by point correspondences. In: IEEE Conference on Computer Vision and Pattern Recognition, IEEE, pp. 390–397 (2010)
13. Zhang, L., Xu, C., Lee, K.-M., Koch, R.: Robust and efficient pose estimation from line correspondences. In: Lee, K.M., Matsushita, Y., Rehg, J.M., Hu, Z. (eds.) ACCV 2012, Part III. LNCS, vol. 7726, pp. 217–230. Springer, Heidelberg (2013)
14. Kumar, R., Hanson, A.R.: Robust methods for estimating pose and a sensitivity analysis. CVGIP Image Underst. **60**, 313–342 (1994)
15. Autodesk 123D Catch. http://www.123dapp.com/catch
16. Newcombe, R.A., Davison, A.J., Izadi, S., Kohli, P., Hilliges, O., Shotton, J., Molyneaux, D., Hodges, S., Kim, D., Fitzgibbon, A.: Kinectfusion: real-time dense surface mapping and tracking. In: Proceedings of the 10th IEEE International Symposium on Mixed and Augmented Reality, IEEE, pp. 127–136 (2011)
17. Rusu, R.B., Cousins, S.: 3d is here: point cloud library (PCL). In: Proceedings of IEEE International Conference on Robotics and Automation, IEEE, pp. 1–4 (2011)
18. Besl, P.J., McKay, N.D.: Method for registration of 3-d shapes. In: Proceedings of SPIE 1611, Sensor Fusion IV: Control Paradigms and Data Structures, International Society for Optics and Photonics, pp. 586–606 (1992)

Integrated Vehicle and Lane Detection with Distance Estimation

Yu-Chun Chen, Te-Feng Su$^{(\boxtimes)}$, and Shang-Hong Lai

Department of Computer Science, National Tsing Hua University,
Hsinchu 30013, Taiwan, R.O.C
tfsu@cs.nthu.edu.tw

Abstract. In this paper, we propose an integrated system that combines vehicle detection, lane detection, and vehicle distance estimation in a collaborative manner. Adaptive search windows for vehicles provide constraints on the width between lanes. By exploiting the constraints, the search space for lane detection can be efficiently reduced. We employ local patch constraints for lane detection to improve the reliability of lane detection. Moreover, it is challenging to estimate the vehicle distance from images/videos captured form monocular camera in real time. In our approach, we utilize lane marker with the associated 3D constraint to estimate the camera pose and the distances to frontal vehicles. Experimental results on real videos show that the proposed system is robust and accurate in terms of vehicle and lane detection and vehicle distance estimation.

1 Introduction

The goal of Advanced Driver Assistance Systems (ADAS) is to improve traffic safety and reduce the number of road accidents. A considerable amount of research efforts on improving automotive safety with automotive vision technologies have been reported. In recent years, vision-based driving assistance systems have received more and more attentions for their low cost and capability of providing information about driving environments. In these systems, robust and reliable vehicle and lane detection is a critical step, and the detected vehicles can be used for various applications, including autonomous cruise control system, lane departure warning system, and forward collision warning systems. These applications usually require applying different techniques in computer vision, such as object detection, line detection and distance estimation. However, most of these basic tasks for ADAS were developed individually. However, the information from different tasks in ADAS can be integrated in a collaborative manner. For instance, the width of two lanes is always larger than the width of vehicle. Otherwise, lane markers should follow a specific pattern according to government rules and this pattern is very useful for vehicle distance estimation. For the reasons, we propose an integrated system that combines vehicle detection, lane detection, and distance estimation altogether in a collaborative manner in this paper.

© Springer International Publishing Switzerland 2015
C.V. Jawahar and S. Shan (Eds.): ACCV 2014 Workshops, Part III, LNCS 9010, pp. 473–485, 2015.
DOI: 10.1007/978-3-319-16634-6_35

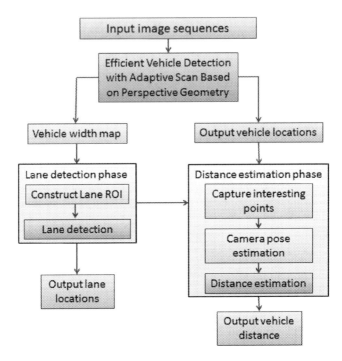

Fig. 1. The flowchart of the proposed system.

In recent years, some research groups utilized additional hardware, such as infrared or radar sensors, to capture additional depth information for the ADAS systems. Without doubt, using the additional depth information can improve the performance of distance estimation and vehicle detection. However, it requires additional cost. Vehicle video recorders have been popularly used to record the video of the frontal scene from the driver's viewpoint. Currently, it is mainly used to provide evidence for car accident. However, it potentially can be used as an ADAS that provides warning to the driver. Therefore, we focus on developing the vision-based ADAS with input videos of traffic scene acquired from a monocular camera mounted on vehicle. The system is able to detect lanes and vehicles, and estimate distances between the vehicles and the camera. In the field of vehicle detection for ADAS, reducing search space based on perspective geometry of the road is efficient to reduce the computation cost. Vehicle width model, a search space reduction manner, was proposed in [1], and our vehicle detection is also based on the same strategy. More details can be found in [1].

Lane detection from images does not seem to be a complicated problem. Tradition methods usually used Canny edge detection and Hough transform to find straight lines. Unfortunately, noisy or misleading edges frequently appear in traffic scenes. For example, shadows, buildings, or traffic signs could degrade the accuracy. On the other hands, Region of Interest (ROI) for lane detection and tracking is also effective to alleviate the problem. Our proposed lane detection method utilizes vehicle width model to reasonably set the ROI of lane. Otherwise, in order to enhance lane detection, we use patch identification to compute the associated confidence. Experiment results show that the proposed method improves the quality by reasonable ROI and patch identification. Distance estimation from road images acquired from a camera mounted in the vehicle is very challenging. Previous works on vehicle distance estimation normally require perform a camera calibration process in advance. Kosecka and Zhang [2] showed that vanishing points can be found automatically and they can be used to estimate the camera intrinsic parameters. Moreno-Noguer et al. [3] proposed to estimate the camera pose from some interesting points detected from an image with their real 3D world coordinates when the intrinsic parameters are given. In this paper, we combine the above two related works, namely, the estimation of intrinsic parameters from vanishing points and 3D camera pose estimation, in an automatic way for vehicle distance estimation. The proposed distance estimation algorithm does not require any human intervention, camera calibration or pose measurement before the distance estimation process. Figure 1 illustrates the flowchart of the proposed system. The main contribution of this paper is that we construct an integrated ADAS which contains vehicle detection, lane detection and distance estimation in a collaborative manner. The proposed ADAS system can detect vehicles and lanes, and it can estimate vehicle distance from a single road image. Our experimental results also show the proposed vehicle detection with adaptive search strategy based on perspective road geometry is superior to the standard sliding-window search in terms of both speed and accuracy. Our Lane detection is also quite robust under different situations, including complex background. The proposed distance estimation is accomplished from a single road image via 3D pose estimation with the 2D-3D point correspondences extracted from the anchor points in the lane pattern.

The main contribution of this paper is that we construct an integrated ADAS which contains vehicle detection, lane detection and distance estimation in a collaborative manner. The proposed ADAS system can detect vehicles and lanes, and it can estimate vehicle distance from a single road image. Our experimental results also show the proposed vehicle detection with adaptive search strategy based on perspective road geometry is superior to the standard sliding-window search in terms of both speed and accuracy. Our Lane detection is also quite robust under different situations, including complex background. The proposed distance estimation is accomplished from a single road image via 3D pose estimation with the 2D-3D point correspondences extracted from the anchor points in the lane pattern.

The rest of this paper is organized as follows. In Sect. 2, the related recent works on vehicle detection, lane detection, and distance estimation in traffic scenes are reviewed. In Sect. 3, we describe the proposed algorithms for lane and vehicle detection. The vehicle distance estimation algorithm is described in Sect. 4. In Sect. 5, we show some experimental results and quantitative evaluation to demonstrate the superior performance of the proposed method. Section 6 concludes this paper.

Fig. 2. An example illustrates the relation between vehicle width map and ROI for lane detection. In real traffic scene, the distance between adjacent lanes is always between 2 to 3 times of vehicle width.

2   Related Work

In the vision-based ADAS, the camera is mounted on the vehicle to capture road images. It is very challenging to detect vehicle in images, due to the wide variations of vehicles in colors, sizes, views and shapes appeared in images. Various feature extraction methods have been used for vehicle detection, such as Gabor filters [4], and HOG [5]. In these methods, SVM or Adaboost are used as the classifiers for the vehicle detectors. In order to reduce computational cost in vehicle detection, several approaches have been proposed [1,5–7].

Lane detection plays a very important role in ADAS. Detecting lanes from images usually is based on the analysis of edges extracted from images. Unfortunately, there are quite a lot of disturbing edges appeared in traffic scenes, such as vehicles, buildings, traffic signs, shadow, or skid marks. Therefore, lane detection by using traditional straight line detection techniques is prone to errors for real traffic scenes. Some approaches [8,9] first enhance images with some enhancement filters first and then apply inverse perspective mapping, which transforms images into bird's-eye view images, followed by detecting straight lines on the mapped images. Nevertheless, when lanes are occluded by vehicles, the performance is usually degraded. The determination of the inverse perspective mapping is also a problem, because road images may be acquired by different cameras on different vehicles with different poses. There are some lane detection methods based on using motion information to find the vanishing points and then select the edges that intersect at the vanishing point as the target lanes. Zhou et al. [10] utilized the lane detection result from the previous frame and updated the lane detection within a given range. However, the width of lane varies considerably under different image acquisition situations and the tracking range should be different. In [11], they set the ROI empirically. It is not applicable for general cases.

Distance estimation from a single image is very challenging in general. In [12], some samples with known distance and vehicle width in image are used to train

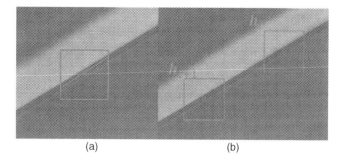

(a)   (b)

Fig. 3. (a) Patch bisection characteristics and (b) patch similarity characteristics for lane detection

a distance estimation model. However, this learning-based approach is very restricted for practical use since the testing and learning data should be acquired under the same setting. If the pose of camera changes or the testing data is captured by a different camera, it would need another setting or new training data to train a new model to overcome this problem. Muller et al. [13] utilized stereo cameras to estimate the distance. But, it requires additional cost in the hardware as well as the computation for stereo matching. Dagan et al. [14] proposed to estimate the vehicle distance by using the relation between the velocity of the camera, which is mounted on a moving vehicle, and the change of width of frontal vehicles in a short period of time. The method works under the assumptions that the velocity of the vehicle is given and the camera is calibrated. In [15], the authors proposed to measure the height and the pose of camera first, and then estimate the vehicle distance assuming the focal length is given. However, it is not easy to measure the pose of the camera mounted in the vehicle in practice. Therefore, automatic distance estimation from a single image is particularly important. Subsequently, we will describe the proposed algorithms for lane detection and distance estimation in the next section.

3   Vehicle and Lane Detection

3.1   Vehicle Detection

In our previous work [1], we proposed a vehicle width model for search space reduction in vehicle detection based on using the perspective geometry in a road image. It is under the assumption that vehicles are on the same ground plane and vehicles have the same width. Therefore, the width of vehicles on different y-coordinate in image has the following linear relationship:

$$W = a(Y - d) \tag{1}$$

where W is the vehicle width in an image, a is the parameter that controls the increment of vehicle width, and d is horizon line illustrated in Fig. 2. It is an

---

Algorithm 1. Patch Identification

Input:Each candidate H, patch size SxS pixels, divided into N patch
Output:The confidence of candidate H
Initial:$D \leftarrow 0$
Average select N patches h_1, h_2, K, h_N with size SxS, as in Fig. 3(a), on H.
Foreach $i \leftarrow 1, 2, ..., N$
    $D \leftarrow D+$ Difference(left part, right part), see Fig. 3(b)
End foreach Foreach $j \leftarrow 1, 2, ..., N - 1$
    $D \leftarrow D+$ Similarity(h_i, $h_{(i + 1)}$), see Fig. 3(a)
End foreach
Confidence = D

HOG-based vehicle detection, and the SVM with RBF kernel is chosen to be the classifier.

The system first detects vehicles by a sliding-window search for a few frames, and then it selects a number of pairs of positive detection results with its y-coordinate location and width in the image to estimate the parameters a and d in Eq. 1. Finally, the system constructs an adaptive search window strategy based on this linear prediction model. In comparison with the sliding-window search, the adaptive window strategy is more efficient and provides higher accuracy. More details of the efficient vehicle detector can be found in [1].

3.2 Lane Detection

For most cases, the distance between adjacent lanes is usually fixed and given. As we can see in Fig. 2, W is the vehicle width in Eq. 1; the red region which contains the left and right lanes is 3 times the vehicle width. For the reason, we construct an ROI for lane detection based on vehicle width. Thus, Canny edge detection and Hough transform are employed to find edges in the ROI. All line segments in the ROI are extended to the horizon and the bottom of the image and then considered as lane candidates. If two lane candidates are very close to each other in the image, then they are merged into one. Since there are several noisy and misleading line segments detected from the road image, such as brake marks or traffic sign, we propose a patch-based identification technique for the lane detection. The patch-based identification is developed based on the characteristic of lanes in images for selecting the correct lanes from the candidates. The detailed patch-based identification is described in the following.

Lane candidates are divided into two sets based on the slope sign computed from the image, because the slopes of two frontal lanes in the images are usually opposite for road scenes under normal camera settings. For each set $L = l_i$, where l_i are lane candidates, we apply patch identification to determine the confidence for each candidate as follows:

$$c_i = pi(l_i), l_i \in L \tag{2}$$

where Pi is the patch identification function described in Algorithm 1. If the lane candidate is correct, it should have the following two characteristics: First,

Fig. 4. (a) Detected line segments for three directions. (b) Estimated vanishing points for three directions.

the left part and right part of each patch (e.g. Fig. 3(a)) are quite different, and the two patches along the line located with a fixed distance (e.g. Fig. 3(b)) are similar to each other. Based on the above two characteristics, we can select the lanes with the highest patch identification score, i.e. the confidence value c_i, from all candidates. The above selection process is performed individually for the two sets of candidate lanes with positive and negative slopes, respectively.

In most situations, the lanes for the positive and negative slopes can be found. Nevertheless, when the lanes are blurred or the driver makes a turn to change lane, the lane detection may fail sometimes. Thus, we take the average of the lane width from the lane detection results in the previous frames. The average lane width can be used in the following cases. If both the positive-slope and negative-slope lanes cannot be identified, the system would preserve the lane detection result from the previous frame. If only one of the positive-slope and negative-slope lanes can be found, the system would use the average lane width to estimate the other lane.

4   Vehicle Distance Estimation from a Single Image

In this section, we describe our algorithm for vehicle distance estimation from only a single road image. Our algorithm does not need to calibrate camera or measure the camera pose in advance. We first detect the three vanishing points estimation from an image and estimate the focal length from these vanishing points by using the algorithm by Kosecka and Zhang [2]. We apply this algorithm on the first few frames to estimate the focal length. Figure 4 depicts an image of the vanishing point detection results.

In this work, we assume that the principle point is located at the image center and the image distortion is not considered. With the estimated focal length, we have the camera calibration matrix K. Then, we can estimate the camera pose from 6 2D-3D corresponding points, i.e. image coordinates and associated 3D world coordinates, by the 3D pose estimation method [3]. In the proposed method, we detect the anchor points along the dash lanes (e.g. Fig. 5, yellow points) by using the template matching method. The other three points are the corresponding points on the other lane (e.g. Fig. 5, red points), with known lane width W_1 (approximately 3.75 m) in 3D world distance. The 3D word coordinates

Fig. 5. The anchor points are used to estimate the camera pose (Color figure online).

Fig. 6. An example of distance map computed for this image. Each region corresponds to a distance range.

of these six points can be determined according to the fixed lane pattern from the local traffic regulation. Therefore, we can estimate the camera pose, i.e. rotation matrix R and translation vector T, from the six pairs of point correspondences. We combine them into the projection matrix M.

When the system detects a vehicle with the middle bottom of the window located at (u, v) in image coordinate, we can use M to estimate the position of the car in real world coordinate (X, Y, Z). Because we assume the vehicle be located on the road plane, Y is 0 at all times. Thus, we have

$$s \begin{bmatrix} u \\ v \\ 1 \end{bmatrix} = M \begin{bmatrix} X \\ 0 \\ Z \\ 1 \end{bmatrix} \tag{3}$$

The three unknowns X, Z, and s can be easily computed by solving Eq. 4. We determine the distance between the real world coordinate of the vehicle and the translation of camera T in the z-coordinate to be the frontal vehicle distance.

$$
\begin{bmatrix} m_{11} & m_{13} & -u \\ m_{21} & m_{23} & -v \\ m_{31} & m_{33} & -1 \end{bmatrix} \begin{bmatrix} X \\ Z \\ s \end{bmatrix} = - \begin{bmatrix} m_{14} \\ m_{24} \\ m_{34} \end{bmatrix} \tag{4}
$$

Sometimes, shadows or lane occlusion by frontal vehicles may cause false results in detecting the lane anchor points. In order to reduce the false distance estimation, we construct a distance map for each frame by computing the distance at each pixel below the horizon line (see as Fig. 6), and take the average of the distance maps from previous frames and use it when false distance estimation is detected.

5   Experimental Results

In our experiment, we evaluate the performance of the proposed method on several image sequences captured by us by using a monocular vehicle video recorder. The sequences were acquired under different situations, as described in Table 1. TP-Day1 video does not contain changing lane situations, and there is no shadow effect. TP-Day2 video is more challenging, because it contains changing lane and shadows of builds on the road. Shadows make the intensity of lane in the images more difficultly to detect. TP-Day3 video contains some hills, and it potentially can cause problems for lane detection, because hills will change the locations of the vanishing points. Therefore, the methods based on vanishing points may cause unpredictable problems. The ground-truth lanes are labelled manually, and the lane detection results are correct when the distance between the ground-truth lane and the detected lane computed at horizon line and the bottom of image is smaller than a threshold.

We compare our lane detection method with the vanishing- point-based method [10]. Furthermore, we apply the vanishing point-based method within the same ROI to improve their lane detection performance. Table 1 shows that the lane detection within ROI significantly improves the detection accuracy, because it reduces some unreliable edges. Our lane detection algorithm dramatically outperforms the vanishing-point-based method [10] with and without using ROI on these three road videos of different situations. This proves the proposed patch-based identification approach is quite robust for lane detection under different road conditions. For vehicle distance estimation, we fix the camera in the car for capturing our testing data, and the ground truth of distance estimation can be obtained by using some markers placed along the side lane, finding them in the image, and computing the distances at different locations in the image. Figure 7 illustrates how the ground truth is obtained. Figure 8 shows the accuracy of distance estimation at different distances by using the proposed 3D pose estimation method. In this work, we focus on the distance lower than 20 m in our experiments. The distance estimation results are compared with the ground truth distances. As expected, the farther the distance of the frontal vehicle is, the larger the distance estimation error is. Since we usually pay more attention to near vehicles during driving, the performance characteristics of the proposed

Fig. 7. It illustrates how the ground truth of distance estimation is obtained. The images are acquired from a vehicle video recorder. There are markers placed at 4, 6, 8, K, 18 m away along the side lane.

Table 1. THE accuracy of lane detection under three different conditions

Data	frames	Situation	Acc. of [10]	Acc. of [10] with ROI	Acc. of ours
TP-DAY1	500	No shadow, no changing lane	0.758	0.832	0.952
TP-DAY2	2160	Shadow, changing lane	0.690	0.784	0.835
TP-DAY3	605	Rugged, no changing lane	0.603	0.711	0.921

distance estimation algorithm is quite effective to assist drivers to obtain more information of frontal vehicles.

For the integrated ADAS system, we combine the vehicle detection, lane detection, and distance estimation in a collaborative manner. Vehicle width model is used to constrain the region of lane detection with an appropriate ROI. After we detect lanes, the anchor points along the lanes can be easily found and used for vehicle distance estimation. This combination not only reduces the cost of the pre-processing in each part of the system, but also improves the accuracy and Fig. 9 depicts some sample results of our ADAS for each sequence in each row. The results of lane detection are quite good, and the errors of distance estimation are lower than 1 m in average.

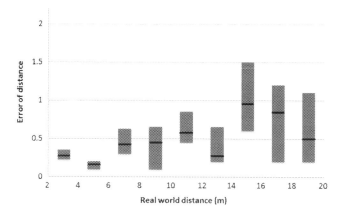

Fig. 8. Distance estimation result. The error is smaller than 2 m in all testing data, and the error is lower when the frontal vehicle is closer to the camera.

Fig. 9. Sample frames of the ADAS, each row for each sequence.

6   Conclusion

In this paper, we proposed an efficient advanced driver assistance system that combines vehicle detection, lane detection, and distance estimation. The information of each part can be used to improve the performance for another part. The proposed ADAS can be applied for different data captured from different cameras with different camera poses. The proposed system can detect lanes as well as vehicles and then estimate the distances of the detected vehicles from a single image. Therefore, the proposed approach can be widely used in general conditions. For the future work, it is practical to implement the ADAS on a multi-core platform, such as GPU, to achieve real-time performance. In addition, converting the real-time ADAS system into a program running on a mobile computing platform, such as a smart phone, is very convenient for drivers to improve driving safety.

References

1. Chen, Y.C., Su, T.F., Lai, S.H.: Efficient vehicle detection with adaptive scan based on perspective geometry. In: ICIP 2013, pp. 3321–3325 (2013)
2. Kosecká, J., Zhang, W.: Video compass. In: Heyden, A., Sparr, G., Nielsen, M., Johansen, P. (eds.) ECCV 2002, Part IV. LNCS, vol. 2353, pp. 476–490. Springer, Heidelberg (2002)
3. Lepetit, V., Moreno-Noguer, F., Fua, P.: EPnP: an accurate o(n) solution to the PnP problem. Int. J. Comput. Vis. **81**, 155–166 (2009)
4. Sun, Z., Bebis, G., Miller, R.: On-road vehicle detection using evolutionary gabor filter optimization. IEEE Trans. Intell. Transp. Syst. **6**, 125–137 (2005)
5. Cheon, M., Lee, W., Yoon, C., Park, M.: Vision-based vehicle detection system with consideration of the detecting location. IEEE Trans. Intell. Transp. Syst. **13**, 1243–1252 (2012)
6. Srinivasa, N.: Vision-based vehicle detection and tracking method for forward collision warning in automobiles. IEEE Intell. Veh. Symp. **2**, 626–631 (2002)
7. Hoiem, D., Efros, A., Hebert, M.: Putting objects in perspective. IEEE Comput. Soc. Conf. Comput. Vis. Pattern Recogn. **2**, 2137–2144 (2006)
8. Borkar, A., Hayes, M., m.T. Smith: Robust lane detection and tracking with ransac and Kalman filter. In: IEEE International Conference on Image Processing, pp. 3261–3264 (2009)
9. Borkar, A., Hayes, M., Smith, M.T.: A novel lane detection system with efficient ground truth generation. IEEE Trans. Intell. Transp. Syst. **13**, 365–374 (2012)
10. Zhou, S., Jiang, Y., Xi, J., Gong, J., Xiong, G., Chen, H.: A novel lane detection based on geometrical model and Gabor filter. In: IEEE Intelligent Vehicles Symposium, pp. 59–64 (2010)
11. Lin, Q., Han, Y., Hahn, H.: Real-time lane departure detection based on extended edge-linking algorithm. In: International Conference on Computer Research and Development, pp. 725–730 (2010)
12. Wu, C.F., Lin, C.J., Lee, C.Y.: Applying a functional neurofuzzy network to real-time lane detection and front-vehicle distance measurement. IEEE Trans. Syst. Man Cybern. Part C Appl. Rev. **42**, 577–589 (2012)

13. Muller, T., Rannacher, J., Rabe, C., Franke, U.: Feature- and depth-supported modified total variation optical flow for 3d motion field estimation in real scenes. In: IEEE Conference on Computer Vision and Pattern Recognition, pp. 1193–1200 (2011)
14. Dagan, E., Mano, O., Stein, G., Shashua, A.: Forward collision warning with a single camera. In: IEEE Intelligent Vehicles Symposium, pp. 37–42 (2004)
15. Zhang, Q., Xie, M.: Study on the method of measuring the preceding vehicle distance based on trilinear method. Int. Conf. Comput. Model. Simul. **2**, 475–479 (2010)

Collaborative Mobile 3D Reconstruction
of Urban Scenes

Attila Tanács[1]([✉]), András Majdik[1], Levente Hajder[1,2], József Molnár[1],
Zsolt Sánta[1], and Zoltan Kato[1]

[1] University of Szeged, Árpád tér 2, Szeged 6720, Hungary
{tanacs,majdik,molnarj,santazs,kato}@inf.u-szeged.hu
[2] Institute for Computer Science and Control (MTA SZTAKI),
Kende u. 13–17, Budapest 1111, Hungary
hajder.levente@sztaki.mta.hu

Abstract. Reconstruction of the surrounding 3D world is of particular interest either for mapping, civil applications or for entertainment. The wide availability of smartphones with cameras and wireless networking capabilities makes collecting 2D images of a particular scene easy. In contrast to the client-server architecture adopted by most mobile services, we propose an architecture where data, computations and results can be shared in a collaborative manner among the participating devices without centralization. Camera calibration and pose estimation parameters are determined using classical image-based methods. The reconstruction is based on interactively selected arbitrary planar regions which is especially suitable for objects having large (near) planar surfaces often found in urban scenes (*e.g.* building facades, windows, etc). The perspective distortion of a planar region in two views makes it possible to compute the normal and distance of the region w.r.t the world coordinate system. Thus a fairly precise 3D model can be built by reconstructing a set of planar regions with different orientation. We also show how visualization, data sharing and communication can be solved. The applicability of the method is demonstrated on reconstructing real urban scenes.

1   Introduction

By the explosive growth in number of digital cameras and extensive internet access, we can experience a huge increase in images taken from various scenes. Photos of the same scene are usually taken from widely different viewpoints thus yielding wide-baseline multiview images of the scene. A fundamental application is 3D reconstruction of a large scene from a collection of such images. One approach is to use thousands of images taken from the same scene available on photo sharing services such as Flickr [1, 2]. The large amount of data is processed

This research was supported by the European Union and the European Social Fund through project FuturICT.hu (grant no.: TÁMOP-4.2.2.C-11/1/KONV-2012-0013). The authors thank the work of Zoltán Molnár, Péter Rácz and Sándor Laczik in the Android implementation.

C.V. Jawahar and S. Shan (Eds.): ACCV 2014 Workshops, Part III, LNCS 9010, pp. 486–501, 2015.
DOI: 10.1007/978-3-319-16634-6_36

either on a cluster of computers [1], or on a single PC exploiting the parallel GPU architecture [2]. Current state of the art methods provide a reconstruction result in several hours on desktop PCs. The process can be considerably speeded up by taking into account various constraints, *e.g.* considering building facades with known (mainly vertical) orientations in urban scenes [3].

Similar smartphone applications are also emerging such as virtual view generation from stereo images [4] or virtual mobile tours [5]. An interesting new mobile-related approach was introduced by Google by incorporating a depth sensor into mobile devices and combine data with location and orientations sensor information used for spatial reconstruction [6]. However, this approach is focusing mainly to map the interior of rooms since depth sensors work best for objects in close range to the device and they are not reliable under direct sunlight.

3D reconstruction of buildings in urban scenes is a widely studied field in literature [7–9]. Many urban scenes contain buildings having large (near) planar facade regions. In our approach, if a planar image region ("patch") is segmented in one of the images, the task is to find its occurrence in the other image. Knowing the intrinsic calibration parameters of the cameras and the homography between corresponding planar image region pairs, the position and orientation of the 3D planar surface can be computed [10]. By having a group of such region pairs, a fast 3D reconstruction of the scene can be achieved by sequentially applying the method on the individual patches.

In an interactive mobile application, the reconstructed planar regions can be selected in one of the images and then automatically find their corresponding position in the image of another participating smartphone. This is a classical problem in computer vision usually solved by detecting and matching keypoints [11–13], which works also efficiently on mobile devices [14,15]. However, in urban scenes low rank repetitive structures are common, which makes point correspondence estimation unreliable [16]. In [17] it has been shown that due to the overlapping views the general 8 degree of freedom (DOF) of the homography mapping can be geometrically constrained to 3 DOF and the resulting segmentation/registration problem can be efficiently solved by finding the region's occurrence in the second image using pyramid representation and normalized mutual information as the intensity similarity measure.

In our approach we assume that a group of people is taking pictures of the same scene approximately at the same time. Thus, utilizing wireless networking capabilities, an ad-hoc mobile camera network can be created from the participating devices via the construction of a vision-graph [18]. The available sensory data (location, orientation) can greatly help to determine camera pairs with overlapping views. We focus on exploiting these features to solve mobile computer vision tasks in a collaborative manner. In this paper, we propose a complete processing pipeline to reconstruct planar 3D surfaces from region-based correspondences. This is the key step towards a fully distributed multiview reconstruction of a scene.

The theoretical background is summarized in Sect. 2 including the solution of the patch correspondence problem, the direct formulas for planar surface reconstruction and characterizing the reconstruction uncertainty in order to detect possibly wrong reconstructions. We also discuss possible collaborative scenarios taking into account privacy issues in Sect. 3. The proposed reconstruction pipeline is evaluated on images of real urban scenes in Sect. 4.

2   Pairwise Reconstruction Pipeline

In this section we present the key steps (see Fig. 1) of the reconstruction pipeline that is based on a pair of stereo images and the detected planar patch correspondences between them.

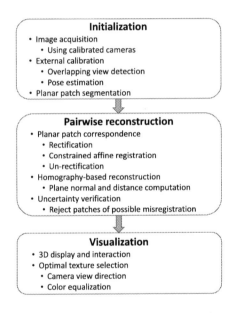

Fig. 1. Main steps of the pairwise reconstruction method.

2.1   Image Acquisition

Typical mobile camera sensors are usually cheap and small sized. Although their resolution is quite high (even 13–21 MP) they usually have problems in low light situations and introduce large amount of noise, blurriness, JPEG artifacts and color distortions, which challenges correspondence across different devices. In our test we used the following devices: HTC EVO 3D, Samsung Galaxy S, Samsung Galaxy S3, Samsung Galaxy S4, Samsung Note 3 and HTC One (M7) smartphones and Samsung Galaxy Note 10.1, LG G Pad 8.3 and Acer Iconia Tab 10 tablets. To compensate the various resolutions and computing capacities of these devices, we selected image sizes closest to 2 MP resolution in 4:3 ratio (not all devices provide exactly 2 MP resolution). Since our application scenario (collaborative data acquisition and processing) assumes that images are taken almost the same time, this provides very similar conditions and hence minimal lighting variation between participating devices. Sensory data such as position, orientation, and gravity are also stored with the images, which is subsequently used for visual graph construction and pose estimation. To detect stereo camera pairs with overlapping view in an ad-hoc mobile camera network, we can follow *e.g.* a vision-graph based approach using the sensor data of the devices stored together with images [18]. Hereafter, we will thus concentrate on a camera pair and show how to achieve 3D reconstruction of planar surface patches.

2.2 Pose Estimation

Since mobile cameras are typically equipped with a fixed focus lens, the internal camera parameters can be pre-calibrated in advance. In our tests, we used the Matlab Camera Calibration Toolbox [19] for that purpose. In order to obtain the full camera matrix, we have to compute the relative pose of the cameras.

Our processing pipeline assumes that the world coordinate system is fixed to the first camera. We thus compute the relative pose of the other camera by estimating the essential matrix acting between the cameras. For that purpose, we will establish point correspondences using ASIFT [20] and compute the fundamental matrix $\mathbf{F}$ using the *Normalized 8-point algorithm* [21]. Since the intrinsic camera parameters $\mathbf{K}_1$ and $\mathbf{K}_2$ are known, the essential matrix $\mathbf{E}$ is obtained as $\mathbf{E} = \mathbf{K}_2^T \mathbf{F} \mathbf{K}_1$ and pose parameters $\mathbf{R} \in \mathbb{SO}(3)$ and $\mathbf{t} \in \mathbb{R}^3$ are obtained by SVD decomposition of $\mathbf{E}$ and testing for four-fold ambiguity [21]. Thus the camera matrices have the following form:

$$\mathbf{P}_1 = \mathbf{K}_1[\mathbf{I} \mid \mathbf{0}] \qquad \mathbf{P}_2 = \mathbf{K}_2[\mathbf{R} \mid \mathbf{t}], \tag{1}$$

Finally, camera matrices are refined by minimizing the overall reprojection error of the triangulated point pairs:

$$\min_{\mathbf{X}_j, \mathbf{P}_2} \sum_j \|\mathbf{P}_1 \mathbf{X}_j - \mathbf{x}_j\|^2 + \|\mathbf{P}_2 \mathbf{X}_j - \mathbf{x}_j'\|^2, \tag{2}$$

where $\mathbf{P}_1 \mathbf{X}_j$ and $\mathbf{P}_2 \mathbf{X}_j$ are the backprojections of $\mathbf{X}_j$ in the cameras while $\mathbf{x}_j$ and $\mathbf{x}_j'$ denotes the true pixel coordinates on the first and the second cameras, respectively. Note that, for this process $\mathbf{P}_1$ is fixed and the 3D points are obtained by triangulating the point correspondences. To solve the above problem, we used the *Generic Sparse Bundle Adjustment* library from Lourakis *et al.* [22], written in C++. The library is using the fact that the Jacobian of the problem given by (2) is sparse, thus it achieves high computational efficiency. The demo implementation of this library is capable of dealing with several *bundle adjustment* like problems, in our tests we used the `sba_motstr_levmar` driver, with fixed intrinsic parameters.

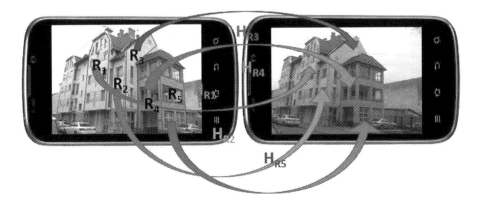

Fig. 2. Segmented planar patches and correspondences for a stereo image pair. 5 polygonal regions $R_1, \ldots, R_5$ are segmented as planar patches in the left image. The border of the regions are shown in red. We are seeking planar homographies $H_{R_1}, \ldots, H_{R_5}$ to establish correspondences. (Figure is best viewed in color.) (Color figure online)

2.3 Planar Patch Correspondence

The theoretical foundation of planar surface reconstruction [10] relies on homographies computed between segmented corresponding planar regions in the image pairs. In our interactive system, this is achieved by (interactively) segmenting an arbitrary planar region on the user's mobile and then the system finds its occurrending region in the second image on the other mobile device (see Fig. 2) via a simultaneous segmentation/registration process. The adopted algorithm is robust enough to sparse occlusions produced by *e.g.* electric wires, trees, lamp post, etc.

The segmentation of a planar patch in the first image can be accomplished in many ways and should not be a precise segmentation of a particular region. The user can *e.g.* select a polygon in the first image delineating the borders of the patch (see Fig. 2). This should not necessarily lie along object edges, can be of arbitrary shape and can even be composed of non-connected parts. This approach can be easily deployed to mobile devices by swiping near the borders or inside regions in the displayed image. Region based segmentation could also be used including region growing, graph-cut methods [23], MSER detection [24] or mean-shift/camshift filtering [25]. Another possibility is to use automatic plane detection methods [7,8], or RANSAC-based homography constraints [9]. Since the user has to decide anyway which regions he/she would like to reconstruct, we do not use these approaches due to their complexity.

It is important to note that the shape of the patch can be arbitrary. We define a pixelwise mask indicating which pixels are part of it. The segmented patch can even be composed of non-connected parts. In an interactive system, patches defined by polygonal borders are easy to specify and they also yield visually more pleasing results.

Once a candidate region is marked by the user, the matching method presented in [17] is used to find a corresponding region on the other device. The algorithm works as follows: Given a pair of cameras with overlapping views and a region $\mathcal{R}_1$ in one image, corresponding to a planar 3D patch, we are looking for the corresponding region $\mathcal{R}_2$ in the second image and the homography $\mathbf{H}$ aligning the regions such that $\mathcal{R}_2 = \mathbf{H}\mathcal{R}_1$. Since the two cameras have an overlapping view, their images are related by epipolar geometry [21]. Furthermore, we have a set of inlier ASIFT keypoint pairs and the fundamental matrix determined in the pose estimation step (see Subsect. 2.2). Hence we can transform the images into a common image plane using a rectification algorithm provided by OpenCV [26]. Rectification gives us two planar homographies denoted by $\mathbf{H}_1$ and $\mathbf{H}_2$, which are applied to the images, respectively. This yields a pair of images as if they were acquired by a standard stereo camera pair (parallel optical axes and imaging planes of the cameras coincide) [21]. As a consequence, $\mathbf{H}_1\mathcal{R}_1$ and $\mathbf{H}_2\mathcal{R}_2$ are related by an affine transformation $\mathbf{A}$ such that $\mathbf{H}_2\mathcal{R}_2 = \mathbf{A}\mathbf{H}_1\mathcal{R}_1$. Therefore, once $\mathbf{A}$ is determined, the homography $\mathbf{H}$ acting between the original image regions is obtained as $\mathbf{H} = \mathbf{H}_2^{-1}\mathbf{A}\mathbf{H}_1$. It is shown in [17], that $\mathbf{A}$ is a special affine matrix having only 3 DOF:

$$\mathbf{A} = \begin{pmatrix} a_1^1 & a_2^1 & t \\ 0 & 1 & 0 \\ 0 & 0 & 1 \end{pmatrix}.$$

Following [17], $\mathbf{A}$ is estimated by searching the best alignment of the rectified patch $\mathbf{H}_1\mathcal{R}_1$ over the second rectified image using normalized mutual information of the image intensities as the similarity metric [27]. The objective function is optimized using a variant of Powell's direction set method implemented in C++ [28]. To speed up the search, we use a three-level Gaussian-pyramid representation of the rectified images [29]. At the coarsest level first we do a full search for the best translation parameter t taking into account every possible pixel translations. Then, a 3 DOF search is started by initializing the scaling parameter $a_1^1 = 1$ and the shear parameter $a_2^1 = 0$. The optimal parameters are propagated to the finer levels. If computing time is of concern, the optimization at the finest pyramid level can be omitted. As experimental tests proved [17], it speeds up the process considerably (cca. 2.5×) causing only small degradation of precision on average. Computing time is usually around 12 s on a Samsung Galaxy Note 3 device using a single threaded implementation. To fully make use of the multi-core architecture of modern smartphones, either more patch matchings can be run in parallel or the algorithm implementation could be threaded.

2.4   Homography-Based Reconstruction

Once the homography acting between the projections of a planar object is known, the affine parameters can be accurately determined by calculating the partial derivatives of the homography. It was shown in [10] that if the camera parameters are also known then the normal of the corresponding 3D surface patch can be computed in real-time using a closed-form formula. Let us write the matrix components of the Jacobian by taking the derivatives of an estimated homography H_{ij} acting between a pair of cameras i and j as

$$J_{ij} = \begin{pmatrix} a_{11} \, a_{12} \\ a_{21} \, a_{22} \end{pmatrix} = \begin{pmatrix} \frac{h_{11}-h_{31}x_j}{r} & \frac{h_{12}-h_{32}x_j}{r} \\ \frac{h_{21}-h_{31}y_j}{r} & \frac{h_{22}-h_{32}y_j}{r} \end{pmatrix}$$

with scale factor $r = h_{31}x_i + h_{32}y_i + h_{33}$. We can form two vectors, both perpendicular to the surface normal [10]:

$$\boldsymbol{p} = [a_{22}(\nabla y_i \times \nabla x_j) - a_{11}(\nabla y_j \times \nabla x_i)] \ ,$$
$$\boldsymbol{q} = [a_{21}(\nabla x_j \times \nabla x_i) - a_{12}(\nabla y_i \times \nabla y_j)] \ .$$

The surface unit normal vector can then be obtained as:

$$\boldsymbol{n} = \frac{\boldsymbol{p} \times \boldsymbol{q}}{|\boldsymbol{p} \times \boldsymbol{q}|} \ .$$

Knowing the normal vector $\boldsymbol{n}$ of the plane and the homography allows us to determine the distance d from an observed planar patch by minimizing the geometric error of the transferred points over the image regions:

$$\arg \min_d = \sum_p ||\mathbf{H}_{ij}\boldsymbol{p} - \mathbf{A}\boldsymbol{p}||^2 \ .$$

Note that the above minimization problem has also a closed form solution obtained by looking for the vanishing point of the first derivative of the above cost function w.r.t. d [10]. These parameters fully define the spatial position and orientation of the planar patch in the 3D world coordinate system, hence providing the 3D reconstruction of a matching pair of planar image regions.

2.5   Uncertainty Verification

Of course, as any 3D reconstruction method, our pipeline may also fail in many situations (either due to degenerate camera-plane geometry or error in homography estimation). It is thus crucial for a real-life application, that these errors be detected and filtered out from the final reconstruction. A major issue in reconstruction pipelines is false matching which makes the final reconstruction wrong. Therefore, we try to make this part more robust by automatically detecting false matches. The basic idea is that region-wise homographies estimated during our region-matching step also contain the relative camera pose. Therefore, if a homography is correctly estimated then the relative pose factorized from it should match the relative pose computed from point-correspondences in the first step of our processing pipeline (see Subsect. 2.2).

Given an estimated region-wise homography by $\mathbf{H}$, it can be factorized [30] into camera and plane parameters as

$$\mathbf{K}_2^{-1}\mathbf{H}\mathbf{K}_1 = \lambda \left(\mathbf{R} + \frac{\boldsymbol{t}\boldsymbol{n}^T}{d} \right)$$

where d is the spatial distance between the first camera center and the plane, λ is the scale of the homography. The latter can be easily calculated: it is the second singular value of $\mathbf{K_2^{-1}HK_1}$ as proved in [31].

If the camera parameters $\mathbf{R}$ and t are known, the product of the patch normal n and the inverse of the distance d can optimally be estimated in the least squares sense as

$$n/d = \frac{t^T}{t^T t} \left(\mathbf{R} - \frac{1}{\lambda}\mathbf{K_2^{-1}HK_1} \right)$$

The scale of the 3D reconstruction is arbitrary due to the perspective scale ambiguity, therefore, the normal estimation can be simplified by constraining the second camera location as $t^T t = 1$. Then the normal is estimated as $n/d = t^T \left(\mathbf{R} - \mathbf{K_2^{-1}HK_1} \right)$. If the homography $\mathbf{H}$ is an outlier (*i.e.* it is mismatched), then the normal estimation is uncertain. The uncertainty can be measured as the difference of the original normalized homography $\mathbf{K_2^{-1}HK_1}$ and the one computed using the estimated normal. Thus, we define the error matrix $\epsilon_{\mathbf{ERR}}$ as

$$\epsilon_{\mathbf{ERR}} = \left(tt^T - \mathbf{I} \right) \left(\mathbf{K_2^{-1}HK_1} - \lambda\mathbf{R} \right)$$

where $\mathbf{I}$ is the 3×3 identity matrix. The final uncertainty value is the norm of $\epsilon_{\mathbf{ERR}}$ multiplied by the estimated depth d since the depth itself increases the uncertainty of the 3D reconstruction. In our pipeline, the L_2 (Frobenius) norm was used. Other norms have been tested also, but we experienced similar results.

Six test cases containing outliers are listed in Table 1. The outliers are labeled by bold characters, they are selected visually based on patch matches. The highest uncertainty value corresponds to the outlier for every test cases, except test sequence #122 where the whole reconstruction is failed. However, the differences between the values corresponding to outlier and inliers are not very high for several test sequences. It is possible that not all the outliers can be detected by the uncertainty calculation since a wrong camera pose can yield relatively low uncertainty value with relatively low probability. Nevertheless, the whole reconstruction pipeline has become more robust by the application of the proposed uncertainty measurement.

Table 1. Test results for outlier detection. Values are the Frobenius norm of the error matrix $\epsilon_{\mathbf{ERR}}$

Sequence No	Plane #1	Plane #2	Plane #3	Plane #4
108	**0.0221**	0.0115	0.014	—
121	0.0059	0.0059	**0.0193**	0.0071
122	**0.0576**	**0.0543**	**0.0830**	**0.0503**
203	0.0059	0.0050	**0.0461**	0.0051
231	0.0193	0.0145	**0.0334**	0.0154
232	0.0182	0.0112	**0.0889**	0.0112

Fig. 3. Top: Initial patches of sequence No. 232. transformed by estimated homographies. The third one is an outlier due to failed patch registration. Bottom: Two views of the reconstructed 3D scene. The placement of the third planar patch is wrong but this can be detected by validating the homographies.

Based on the above findings, if an uncertainty value is significantly higher than the median, then the reconstructed plane is labeled as an outlier (we empirically set the threshold to $1.7median$). To demonstrate the efficiency of outlier detection, 14 outlier-less test sequences were added to the ones listed in Table 1 including 69 patches. The outliers were correctly detected except the sequence #122 in which there is no inlier. However, four quasi-correct homographies in outlier-free test sequences were labeled as outliers. Therefore, visual verification is proposed after the uncertainty measurement. Note that the proposed outlier filtering method can only be applied if at least three homographies are given due to the median calculation – but this is not a strict restriction in practical applications.

Figure 3 shows an example of outlier removal, where the transformed homographies of sequence #232 are drawn on the second image of the stereo pair. The third homography is an outlier since it should be between the second and fourth walls. It does also not fit to the neighboring walls in 3D. The error for the patch is significantly larger as it is seen in the last row of Table 1.

2.6   Visualization

Once the normal vector and the distance of a planar patch is determined, for each pixel of the planar patch its 3D coordinate can be computed w.r.t. the world coordinate system defined by the first camera. Since we have polygonal planar patches, it is sufficient to transform the defining points and image data from the photos can be texture mapped over them. 3D visualization is implemented using OpenGL ES which requires triangle primitives. Also due to the perspective distortion, the area of triangles should be small relative to the display size (we set the threshold to 0.1% of the original image size). We produce the initial triangulation using Delaunay method [32], then the triangles are recursively subdivided into two smaller triangles until the size criterion is met.

Based on the established correspondences, we have two images of each planar patch from the two cameras. We can select the texture information from the one having the smaller perspective distortion based on the angle between the plane normal and the camera directions – the smaller the better. We have to take into account that the images might be acquired by different camera sensors thus color representation can be different. If texture data is selected from both images, *e.g.* color transfer between the images [33] can be applied to reduce this effect.

3   Collaborative Mobile Implementation

Of course, the proposed reconstruction method can be run on a single device. Two images from different viewpoints should be taken, planar patches to be segmented in one of the images and then the reconstruction result can be visualized. Since smartphones are becoming more powerful nowadays and network access is available, a much more interesting application scenario is to exploit this connectivity in a collaborative manner: Connected smartphones form an ad-hoc camera network where visual computations and results can be shared among the participating devices. Note that the details of the low level network communication is out of scope of this paper – we assume that the devices can send and receive data over network.

One task to solve is the pair formation: which device should talk to which other device? Although we do not deal with this problem, a possible solution based on constructing a vision graph has been proposed in [18]: From a collection of images, cameras with overlapping views can be determined based on the location and orientation sensor information and also taking into account the image content. We also remark, that the vision graph also provides the relative pose estimation between cameras, hence this step of our pipeline can be merged into the vision-graph construction step. Other approaches are also possible [34].

The pairwise collaborative reconstruction can be implemented in several ways, here we consider two possible scenarios. We assume that at least one of the devices initiates the reconstruction process, others may share data and/or computing resources. Considering privacy issues, a user can decide whether he/she wants to share image data, or keep that private. In the latter case, though, computation resources must be offered, *i.e.* image data from other devices should be accepted and the reconstructed 3D geometry of the scene (without image data) sent back. To make the situation simpler we can specify that the initiator should conform to the policy of the contacted device (send image data if the other one is not willing to). Figure 4 shows the outline of this scenario. The initiator sends the request to the contacted device which performs the computations and downloads the necessary data from the initiator. First, the keypoint+descriptor data is necessary for point pair matching. The data for one ASIFT keypoint takes 144 bytes. In our tests, the ASIFT detector produced around 7000–10000 points on average yielding 1–1.5 MB of data. The calibration matrices (9 floating point values) are necessary for the pose estimation. For matching, image patches are also needed, which can be transferred either by sending the whole image and

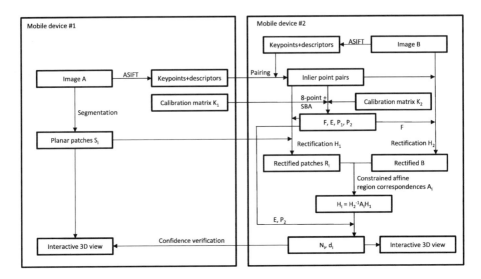

Fig. 4. A possible collaborative solution. Mobile device #1 initializes the process, Mobile device #2 keeps its image data private but provides the reconstruction computations and sends back the results. Both devices can visualize the result.

the polygonal border data (cca. 1 MB) or – if occupying less space – the patch regions as compressed images. The contacted device then performs the computations and sends the reconstruction parameters without image data. This takes another 4 floating point numbers for each patch. Notice that the contacted device can also use the results for 3D visualization. The drawback is that the initiator can only use its own image data for texturing.

If the contacted device is willing to share image data an analogous approach is possible. Here both devices can utilize the image contents of both photos for texturing. Since the relative poses can be concatenated, available pairwise reconstruction results can be further propagated in the network[1] see Figs. 5, 6). Thus planar regions from many cameras can be shared with the peers and visualized in any single mobile device. This also opens a way for a bundle adjustment involving more cameras and their views.

4   Experimental Results and Discussion

We tested the performance of the proposed reconstruction pipeline on a dataset of 52 real image pairs. Images were taken using different smartphones as described in Subsect. 2.1, each image pair consisted of photos from different devices and 2–6 planar regions were interactively selected as a polygonal boundary. The results were evaluated visually by classifying them into three groups corresponding to

---

[1] The free scale parameter of the separate reconstructions can be computed as the ratio of the distance reconstruction parameter of the same patch.

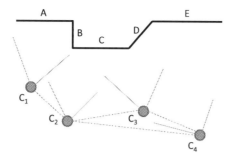

Fig. 5. Ad-hoc camera network. Red stripped lines connect cameras with overlapping view suitable for reconstruction. Note that planar region B is visible only in one camera thus cannot be reconstructed. Results of pairwise reconstruction can be further propagated in the network (Color figure online).

Fig. 6. Reconstruction result from three cameras (right) merging two pairwise reconstructions (left and middle). The planar patch that is visible in both reconstructions is marked in red border. It is used to fix the free scale parameter of the individual reconstructions.

good (at most smaller orientation errors), acceptable (visible, but acceptable errors) and failed reconstruction. About half of the cases produced good results (24 cases), 10 of them was acceptable and 18 cases failed. If we allow the patch correspondence method to do full optimization, results improve to 27 good, 12 acceptable and 13 failed cases – at the price of an increased computational time. Failure often occurred because the camera baseline was not wide enough or the ASIFT feature-based rectification was not correct. Some results are shown in Fig. 7.

We run an offline test on a Samsung Galaxy Note 3 smartphone. (Note that an Android demo implementation will be made available). ASIFT pairing took around 45 s (7 s for each keypoint detection and half a minute for pairing) and an average 12 s was necessary for solving one correspondence problem. Other parts of the method (pose estimation, rectification, reconstruction) take only fractions of a second regardless of the platform.

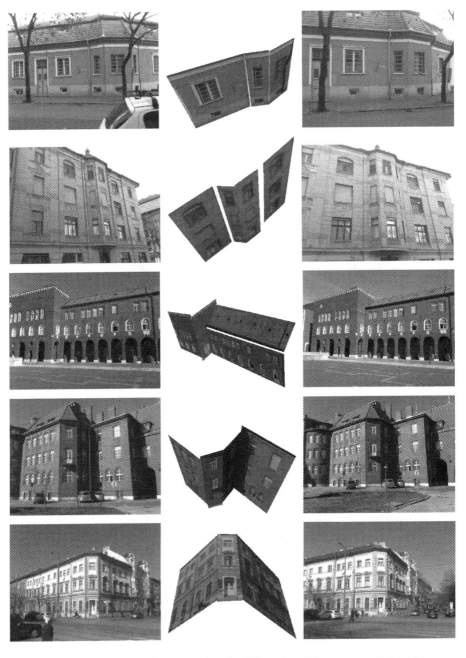

Fig. 7. 3D reconstruction of images taken by different mobile cameras. Original images (left and right columns). The outlines of the segmented patches can be seen in the original image (left column). Visualized reconstructed planar patches (center column). Further examples are available at http://www.inf.u-szeged.hu/rgvc/projects. php?pid=patchrecon.

The computing time requirement might seem excessive but the causes are platform specific implementation issues, not algorithmic ones, and there is still room for several improvements:

- The patch correspondence search can be run in parallel forking multiple threads for each patch. We experienced that a quad-core smartphone is able to execute four such processes without significantly slow down (overhead needed for the OS and background tasks was below 10 %). Note that octa-core processors are appearing right now allowing even more parallel processes.
- The region matching part could be implemented using the SIMD (Single Instruction Multiple Data) paradigm of the GPU with OpenGL ES 2.0 shaders.
- The real bottleneck of our pipeline is keypoint detection and pairing. Although the ASIFT method produces sufficient number of inliers, it is time consuming especially since the current Android implementation does not make full use of all the processor cores contrary to the PC version. Thus, there is potential for further improvement.
- Also, the keypoint detection part can be executed in the background right after taking the picture while the user is busy with planar patch selection and initiating the collaborative reconstruction. Therefore we can assume that keypoint and descriptor data are available with the image at the beginning of the reconstruction process. Other point detectors and pairing methods could also be used but our experience is that many times those produce insufficient number of pairs for accurate pose estimation.
- We can experience a constant, significant growth in the processing power of smartphones. For example, based on our reconstruction tests, the Samsung Galaxy Note 3 proved to be 1.25×, 1.58× and 4.67× faster than the Galaxy S4, S3 and the original (4 years old) Galaxy S devices, respectively. Note that we use top-of-the-line smartphones of year 2013.

5   Conclusions

In this paper we proposed an interactive reconstruction pipeline that can be used in a mobile collaborative framework. The reconstruction is based on segmented planar regions imaged from two different viewpoints. Homography correspondence is established from which spatial orientation and distance can be readily computed. An overview of the major steps was given and reconstruction performance was presented on urban scenes. Future work will concentrate on extending the reconstruction pipeline to multi-view scenarios, which would certainly improve reconstruction quality and – based on our outlier filtering – can correct false reconstructions if a better view is available in the network.

References

1. Agarwal, S., Snavely, N., Simon, I., Seitz, S.M., Szeliski, R.: Building Rome in a day. In: Proceedings of IEEE International Conference on Computer Vision, pp. 72–79 (2009)

2. Frahm, J.-M., Fite-Georgel, P., Gallup, D., Johnson, T., Raguram, R., Wu, C., Jen, Y.-H., Dunn, E., Clipp, B., Lazebnik, S., Pollefeys, M.: Building rome on a cloudless day. In: Daniilidis, K., Maragos, P., Paragios, N. (eds.) ECCV 2010, Part IV. LNCS, vol. 6314, pp. 368–381. Springer, Heidelberg (2010)
3. Shum, H.Y., Han, M., Szeliski, R.: Interactive construction of 3D models from panoramic mosaics. In: Proceedings of IEEE Computer Vision and Pattern Recognition, pp. 427–433 (1998)
4. Wei, C.-H., Chiang, C.-K., Sun, Y.-W., Lin, M.-H., Lai, S.-H.: Novel multi-view synthesis from a stereo image pair for 3d display on mobile phone. In: Park, J.-I., Kim, J. (eds.) ACCV Workshops 2012, Part II. LNCS, vol. 7729, pp. 568–579. Springer, Heidelberg (2013)
5. Chen, S.C., Hsu, C.W., Huang, D.Y., Lin, S.Y., Hung, Y.P.: TelePort: Virtual touring of Dun-Huang with a mobile device. In: Proceedings of IEEE International Conference on Multimedia and Expo Workshops, pp. 1–6 (2013)
6. Google: Project Tango (2014). https://www.google.com/atap/projecttango/#project. Accessed 10 June 2014
7. Werner, T., Zisserman, A.: New techniques for automated architecture reconstruction from photographs. In: Proceedings of European Conference on Computer Vision, vol. 2, pp. 541–555 (2002)
8. Furukawa, Y., Curless, B., Seitz, S., Szeliski, R.: Manhattan-world stereo. In: Proceedings of IEEE Computer Society Conference on Computer Vision and Pattern Recognition, pp. 1422–1429 (2009)
9. He, Q., Chu, C.H.: Planar surface detection in image pairs using homographic constraints. In: Bebis, G., Boyle, R., Parvin, B., Koracin, D., Remagnino, P., Nefian, A., Meenakshisundaram, G., Pascucci, V., Zara, J., Molineros, J., Theisel, H., Malzbender, T. (eds.) ISVC 2006. LNCS, vol. 4291, pp. 19–27. Springer, Heidelberg (2006)
10. Molnár, J., Huang, R., Kato, Z.: 3D reconstruction of planar surface patches: a direct solution. In: Jawahar, C.V., Shan, S. (eds.) ACCV 2014 Workshops, Part I. LNCS, vol. 9008, pp. 286–300. Springer, Heidelberg (2014)
11. Lowe, D.: Distinctive image features from scale-invariant keypoints. Int. J. Comput. Vis. **60**, 91–110 (2004)
12. Bay, H., Ess, A., Tuytelaars, T., Gool, L.: SURF: Speeded up robust features. Comput. Vis. Image Underst. **110**, 346–359 (2008)
13. Mikolajczyk, K., Schmid, C.: A performance evaluation of local descriptors. IEEE Trans. Pattern Anal. Mach. Intell. **27**, 1615–1630 (2004)
14. Miksik, O., Mikolajczyk, K.: Evaluation of local detectors and descriptors for fast feature matching. In: Proceedings of the International Conference on Pattern Recognition, pp. 2681–2684 (2012)
15. Juhász, E., Tanács, A., Kato, Z.: Evaluation of point matching methods for wide-baseline stereo correspondence on mobile platforms. In: Proceedings of the International Symposium on Image and Signal Processing and Analysis, pp. 806–811. IEEE, Trieste (2013)
16. Torii, A., Sivic, J., Pajdla, T., Okutomi, M.: Visual place recognition with repetitive structures. In: Proceedings of IEEE Conference on Computer Vision and Pattern Recognition, pp. 883–890. IEEE (2013)
17. Tanács, A., Majdik, A., Molnár, J., Rai, A., Kato, Z.: Establishing correspondences between planar image patches. In: Proceedings of the International Conference on Digital Image Computing: Techniques and Applications, Wollongong, Australia (2014)

18. Kovács, L.: Processing geotagged image sets for collaborative compositing and view reconstruction. In: Proceedings of ICCV Workshop on Computer Vision for Converging Perspectives, Sydney, Australia, pp. 460–467 (2013)

19. Bouguet, J.: Camera Calibration Toolbox for Matlab (2013). http://www.vision.caltech.edu/bouguetj/calib_doc/. Accessed 10 June 2014

20. Morel, J., Yu, G.: ASIFT: A new framework for fully affine invariant image comparison. SIAM J. Imaging Sci. **2**, 438–469 (2009)

21. Hartley, R., Zisserman, A.: Multiple View Geometry in Computer Vision. University Press, Cambridge (2003)

22. Lourakis, M.A., Argyros, A.: SBA: A software package for generic sparse bundle adjustment. ACM Trans. Math. Softw. **36**, 1–30 (2009)

23. Shi, J., Malik, J.: Normalized cuts and image segmentation. IEEE Trans. Pattern Anal. Mach. Intell. **22**, 888–905 (2000)

24. Matas, J., Chum, O., Urban, M., Pajdla, T.: Robust wide baseline stereo from maximally stable extremal regions. In: Proceedings of British Machine Vision Conference, pp. 384–396 (2002)

25. Comaniciu, D., Meer, P., Member, S.: Mean shift: A robust approach toward feature space analysis. IEEE Trans. Pattern Anal. Mach. Intell. **24**, 603–619 (2002)

26. Hartley, R.I.: Theory and practice of projective rectification. Int. J. Comput. Vis. **35**, 115–127 (1999)

27. Cahill, N.D.: Normalized measures of mutual information with general definitions of entropy for multimodal image registration. In: Fischer, B., Dawant, B.M., Lorenz, C. (eds.) WBIR 2010. LNCS, vol. 6204, pp. 258–268. Springer, Heidelberg (2010)

28. Press, W., Teukolsky, S., Vetterling, W., Flannery, B.: Numerical Recipes in C: The Art of Scientific Computing, 2nd edn. Cambridge University Press, New York (1992)

29. Burt, P., Adelson, E.: The Laplacian pyramid as a compact image code. IEEE Trans. Commun. **31**, 532–540 (1983)

30. Faugeras, O., Lustman, F.: Motion and structure from motion in a piecewise planar environment. Technical Report RR-0856, INRIA (1988)

31. Zhang, Z., Hanson, A.R.: 3D reconstruction based on homography mapping. In: Proceedings of ARPA Image Understanding Workshop, pp. 0249–6399 (1996)

32. de Berg, M., van Kreveld, M., Overmars, M., Schwarzkopf, O.: Computational Geometry: Algorithms and Applications, 2nd edn. Springer, Heidelberg (2000)

33. Reinhard, E., Ashikhmin, M., Gooch, B., Shirley, P.: Color transfer between images. IEEE Comput. Graph. Appl. **21**, 34–41 (2001)

34. Kang, H., Efros, A.A., Hebert, M., Kanade, T.: Image matching in large scale indoor environment. In: Proceedings of IEEE Computer Society Conference on Computer Vision and Pattern Recognition Workshops, pp. 33–40 (2009)

Workshop on Human Identification for Surveillance (HIS)

Gaussian Descriptor Based on Local Features for Person Re-identification

Bingpeng Ma[1]([✉]), Qian Li[2], and Hong Chang[3]

[1] School of Computer and Control Engineering, University of Chinese Academy of Sciences, Beijing 100049, China
bpma@ucas.ac.cn

[2] School of Computer Science and Technology, Huazhong University of Science and Technology, Wuhan 430074, China
qian.li@vipl.icat.ac.cn

[3] Key Lab of Intelligent Information Processing of Chinese Academy of Sciences (CAS), Institute of Computing Technology, CAS, Beijing 100190, China
changhong@ict.ac.cn

Abstract. This paper proposes a novel image representation for person re-identification. Since one person is assumed to wear the same clothes in different images, the color information of person images is very important to distinguish one person from the others. Motivated by this, in this paper, we propose a simple but effective representation named Gaussian descriptor based on Local Features (GaLF). Compared with traditional color features, such as histogram, GaLF can not only represent the color information of person images, but also take the texture and spatial structure as the supplement. Specifically, there are three stages in extracting GaLF. First, pedestrian parsing and lightness constancy methods are applied to eliminate the influence of illumination and background. Then, a very simple 7-d feature is extracted on each pixel in the person image. Finally, the local features in each body part region are represented by the mean vector and covariance matrix of a Gaussian model. After getting the representation of GaLF, the similarity between two person images are measured by the distance of two set of Gaussian models based on the product of Lie group. To show the effectiveness of the proposed representation, this paper conducts experiments on two person re-identification tasks (VIPeR and i-LIDS), on which it improves the current state-of-the-art performance.

1 Introduction

The task of person re-identification can be defined as finding the correspondences between a *probe set* of images representing a single person and those from a *gallery set*. In recent years, person re-identification has attracted a lot of attentions because of its importance in many real-world applications, such as video surveillance. Depending on the number of available images per individual (the size of the probe set), the scenarios in person re-identification can be categorized as: (a) single-shot [1,2], if only one frame per individual is available in both probe and gallery sets; and (b) multiple-shot [1,2], if multiple frames per

© Springer International Publishing Switzerland 2015
C.V. Jawahar and S. Shan (Eds.): ACCV 2014 Workshops, Part III, LNCS 9010, pp. 505–518, 2015.
DOI: 10.1007/978-3-319-16634-6_37

individual are available in both probe and gallery sets. In this paper, we just care about the single-shot scenario.

One key issue of person re-identification is how to represent the human images in the bounding boxes. Coming from different cameras in a distributed network or from the same camera at different time, the human images may have great variations in illumination, pose, viewpoint, background, partial occlusions and resolution. These variations increase the difficulty of person re-identification. To this end, researchers have proposed a lot of representation methods which are somewhat robust to the above variations. Generally speaking, these representations are based on (i) color [1], usually encoded within histograms of RGB or HSV values [1], (ii) shape, e.g. using HOG based signature [3,4], (iii) texture, often represented by Gabor filters [5–7], differential filters [7], Harr-like representations [8] and Co-occurrence Matrices [4], (iv) interest points, e.g. SURF [9] and SIFT [10,11] and (v) image regions [1,3].

Since these elementary features (color, shape, texture, etc.) capture different aspects of the information contained in the images, they are often combined to give a richer signature. For example, [5] combines 8 color features with 21 texture filters (Gabor and differential filters). [1,2] combine the descriptors of Maximally Stable Colour Regions (MSCR) with weighted Color Histograms (wHSV), achieving the state-of-the-art results on several widely-used person re-identification datasets. Similarly, [12–14] combine their features with wHSV and MSCR, respectively.

Among these representations, the color features of the human images, as simple but efficient visual signatures, are the most commonly used representation. In the task of person re-identification, it is often assumed that person wears the same clothes in different images. Under this assumption, compared with other representations, the color feature of the person images, such as the histograms of the different color channels, is the primary key to distinguish one person from the others. However, on one side, the color features change easily with illumination variations and camera parameters. On the other side, the color histograms lose the information of the texture and spatial structures of the human images. These drawbacks limit the applications of the color features in the real person re-identification systems.

To overcome the above drawbacks of the color features, this paper presents a novel image representation named Gaussian descriptor based on Local Features, GaLF for short. Specially, GaLF includes three stages. In the first stage, pedestrian parsing and lightness constancy method are applied to the human images to eliminate the influence of the background and illumination, respectively. In the second stage, a very simple 7-dimensional local feature is extracted on each pixel in the human images. The 7-d features include the information of color, texture and spatial structure at the same time. In the third stage, the distribution of local features in each body part region are modeled by a Gaussian model. It is easy to know that the mean vector can keep the color information and spatial structure while the covariance matrix keep the texture as the supplement. Therefore, by integration of the mean vector and the covariance matrix, GaLF can keep the information of color, texture and spatial structure at the same time.

After getting the representations of GaLF, the similarity measurement of two human images can be obtained by computing the similarity between two sets of part-wise Gaussian models. In this paper, based on product of Lie group, both the difference of the mean vectors and the LOG-Euclidean distance of the covariance matrices are computed in the Riemannian space. Finally, the similarity of two human images are decided by the sum of the difference of the mean vectors and the LOG-Euclidean distance of the covariance matrices of two Gaussian sets. Similar to the combination of the different features to give a richer signature, GaLF can be also combined with some other representations. GaLF and its combination are experimentally validated on two challenging public datasets for person re-identification: VIPeR and i-LIDS. On both databases, the results of the proposed representations outperform the current state-of-the-art.

There are three main contributions in this paper. First, we propose a novel pixel-wise feature representation which can express color, texture and spatial structure at the same time. Second, we propose local Gaussian descriptors, GaLF, based on pedestrian parsing and lightness constancy results, and the similarity between two sets of local Gaussian descriptors is measured properly based on the product of Lie group. Third, the proposed representation method is successfully applied to the task of person re-identification, achieving even higher performance than state-of-the-art methods.

The remaining of this paper is organized as follows: Sect. 2 describes the proposed representation in details. Experimental validations are given in Sect. 3. Finally, Sect. 4 concludes the paper and some future works are also discussed.

2   Gaussian Descriptor Based on Local Features

This section introduces the proposed novel image representation: GaLF. GaLF is a three stage representation. In the following parts, we introduce each stage of GaLF in details, followed by how to improve the performance of GaLF by combining it with other representations.

2.1   Data Processing

Considering the traditional color features are easily varied with the variations of illuminations and camera parameters, in the first stage of GaLF, we use the pedestrian parsing method to discard the background and the lightness constancy method to eliminate the influence caused by the illumination, respectively.

Specially, for each image, we use a method named Deep Decompositional Network (DDN) [15] for parsing pedestrian images into six semantic regions, such as hair, head, body, arms, legs and background. DDN is able to effectively characterize the boundaries of body parts and accurately estimate complex pose variations with good robustness to occlusions and background clutters. The further details about DDN can be found in [15].

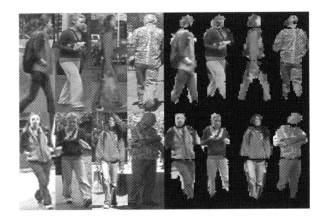

Fig. 1. The images after data processing. The images in the same column belong to the same person. The images in the left are the origin image in the database and the images in the right are the images after data processing.

After getting the human body regions, we use Gray World (GW), a very simple lightness constancy adaptation approach, to eliminate the influence of the illumination. GW makes us perceive the objects as medium gray which reflect the average luminance of a scene. In terms of histogram properties, it seeks to equalize the mean of the different channels and this corresponds to a level distribution which has its center mass around the middle value. Generally speaking, GW can eliminate some global chromatic dominant.

We show the images after person parsing and color constancy processing in Fig. 1. The image pair in the same column belongs to the same person. The images in the left sub-figure are the original images in the database. We can find that there are the great variations of background and illumination between the original images of the same person. It is easy to understand that these variations increase the difficulty of the task of person re-identification. The images in the right sub-figure are the images after data processing. For the images after data processing, background is eliminated roughly and the similarity of the same person's images is improved greatly. On the whole, the images in Fig. 1 affirm that data processing can improve the similarity of the images belong to the same person.

2.2   Local Features

In the second stage of GaLF, we extract the local features on each pixel in the human images. There are some traditional local features in computer vision, such as SIFT. They have gained the great success in many tasks. But in GaLF, considering the computation complex and the importance of the cloth color information in person re-identification, we design a very simple 7-dimensional local features to keep more color information.

Specially, in this paper, we extract a 7-dimensional local feature $f(x, y)$ for the pixel at position (x, y). The 7-d local descriptor can be computed as:

$$f(x, y) = [y, L(x, y), A(x, y), B(x, y),$$
$$d_{L_y}(x, y), d_{A_y}(x, y), d_{B_y}(x, y)] \tag{1}$$

In $f(x, y)$, y is the coordinate in the vertical direction, which can be used to keep the spatial structure of the body. $L(x, y)$, $A(x, y)$ and $B(x, y)$ are the intensity of L, A and B color channels at position (x, y), respectively. They can be taken as the color information. $d_{L_y}(x, y)$, $d_{A_y}(x, y)$ and $d_{B_y}(x, y)$ are the gradient in the vertical direction for the different channels and can be taken as the texture information. So, $f(x, y)$ includes the simple information about position, color and textures.

In Eq. 1, we only use the information in the vertical direction, not the information in the horizontal direction. The reason can be attributed to the misalignment. Since the size of the human image is the same in the database and the body is full of the image in the vertical direction, we can argue that human body has been aligned coarsely in the vertical direction. But, in the horizontal direction, the misalignment happens because the human bodies have the pose variations or the images are captured at the different views. We believe that the position information in the horizontal direction can not be used directly in Eq. 1. To validate our idea, we design the experiments in Sect. 3. The results confirm our argues.

2.3 Gaussian Models

In the third stage of GaLF, we use a Gaussian model to model the 7-d local features in the same semantic body region, which is gained by DDN. Then, the representation of GaLF is the concatenation of the mean vector and the covariance matrix of the Gaussian model.

Specially, for the local features in the n-th body region of image i, their mean vector and covariance matrix can be computed and denoted by μ_{in} and Σ_{in}, respectively. In GaLF, we use the parameter $\mathbf{g}_{in} = (\mu_{in}, \Sigma_{in})$ of the Gaussian model to represent the n-th region of image i. Finally, for human image i, its GaLF representation $\mathbf{D}_i$ is represented by a set of Gaussian models: $\{\mathbf{g}_{in} = (\mu_{in}, \Sigma_{in}), n = 1, \cdots, N\}$, where N is the total regions of the human images.

After getting the representation of GaLF, the distance between region I_{in} and I_{jn} can be obtained by computing the distance between their representations $\mathbf{g}_{in}$ and $\mathbf{g}_{jn}$. Recently, by regarding the space of Gaussians into as a product of Lie groups, [16] measure the intrinsic distance between Gaussians in the underlying Riemannian manifold. They have gained the great success in the image retrieval task. This paper follows their ways and the similarity of Gaussian model $\mathbf{g}_{in}$ and $\mathbf{g}_{jn}$ can be computed by:

$$d_\theta(\mathbf{g}_{in}, \mathbf{g}_{jn}) = (1 - \theta)a(\mathbf{g}_{in}, \mathbf{g}_{jn}) + \theta b(\mathbf{g}_{in}, \mathbf{g}_{jn})$$
$$a(\mathbf{g}_{in}, \mathbf{g}_{jn}) = \left((\mu_{in} - \mu_{jn})^T (\Sigma_{in}^{-1} + \Sigma_{jn}^{-1})(\mu_{in} - \mu_{jn})\right)^{1/2} \tag{2}$$
$$b(\mathbf{g}_{in}, \mathbf{g}_{jn}) = \| \log(\Sigma_{in}) - \log(\Sigma_{jn}) \|_F$$

In Eq. 2, θ is a constant in the range $[0,1]$ to balance the weight of $a(\mathbf{g}_{in}, \mathbf{g}_{jn})$ and $b(\mathbf{g}_{in}, \mathbf{g}_{jn})$. In this paper, θ is set to the constant 0.4. $a(\mathbf{g}_{in}, \mathbf{g}_{jn})$ measures the difference of mean vectors μ_{in} and μ_{jn}. According to the method of the Lie group, the distance between μ_{in} and μ_{jn} should be appropriately weighted by the associative covariance matrices Σ_{in} and Σ_{jn}. $b(\mathbf{g}_{in}, \mathbf{g}_{jn})$ measures the Log-Euclidean distance between covariance matrix Σ_{in} and Σ_{jn}, which is the geodesic distance in the Riemannian space. $\| \cdot \|_F$ denotes the matrix Frobenius norm. Satisfying all the metric axioms, $d_\theta(\mathbf{g}_{in}, \mathbf{g}_{jn})$ is the metric between $\mathbf{g}_{in}$ and $\mathbf{g}_{jn}$.

Finally, the distance between image I_i and I_j is obtained by sum the similarity between their regions:

$$d(\mathbf{D}_i, \mathbf{D}_j) = \sum_{n=1}^{n} d_\theta(\mathbf{g}_{in}, \mathbf{g}_{jn}) \tag{3}$$

Compared with one Gaussian Model, Gaussian Mixed Model (GMM) can represent the region more well and has been widely used. However, in GaLF, for each region, we only use one Gaussian model. Three reasons make us believe that it is possible to just use one Gaussian model to represent the body region. First, in this paper, we just put our attentions on the single-shot scenario. Under this setting, there is just one image in the galley set for each person. Second, since the body region is gained by the pedestrian method, the pixels in the same region are very similar to each other. Finally, compared with GMM, the computational complex of one Gaussian model is decreased greatly.

2.4   Enriched GaLF

As mentioned in Sect. 1, better person re-identification performance was usually obtained by combining different type of image descriptors. In this paper, we follow the same methodology and combine our GaLF representation with other two representations: (a) MSCR, as defined in [1] and (b) Local Descriptors encoded by Fisher Vectors (LDFV) [13]. For symbolic simplicity, we name this combination as eGaLF (enriched eGaLF). In eGaLF, the difference between two image signatures $\mathbf{eD}_1 = (GaLF_1, MSCR_1, LDFV_1)$ and $\mathbf{eD}_2 = (GaLF_2, MSCR_2, LDFV_2)$ is computed as:

$$\begin{aligned} d_{eGaLF}(\mathbf{eD}_1, \mathbf{eD}_2) &= \frac{1}{3} d(GaLF_1, GaLF_2) \\ &+ \frac{1}{3} d_{MSCR}(MSCR_1, MSCR_2) \\ &+ \frac{1}{3} d_{LDFV}(LDFV_1, LDFV_2) \end{aligned} \tag{4}$$

Obviously, improvements could be obtained by optimizing the weighs based on additional information like class labels, but as we are looking for an unsupervised method, we let them fixed once for all. Regarding the definition of d_{MSCR} and d_{LDFV}, we use the one given in [1] and [13], respectively.

Fig. 2. Some images in the VIPeR database. The images in the same column are belonging to the same person.

3   Experiment

In this section, the proposed representations are experimentally validated on the VIPeR [17] database and the i-LIDS database [11].

3.1   Pedestrian Re-identification on VIPeR Database

The VIPeR database has been widely used and is considered to be one of the benchmarks for person re-identification. It contains $1,264$ images of 632 pedestrians. There are exactly two views per pedestrian, taken from two non overlapping viewpoints. All the images have been normalized to 128×48 pixels. The VIPeR database contains a high degree of viewpoint and illumination variations: most of the examples contain a viewpoint change of 90 degrees, as it can be seen in Fig. 2.

We use Cumulative Matching Characteristic (CMC) curve [18], the standard performance measurements for person re-identification, to show the performance of the proposed representation. CMC measures the expectation of the correct match being at rank r. For the fair comparison, we follow the same experimental protocol [1] and report the average performance over 10 different random sets of 316 pedestrians.

Firstly, we compare GaLF with the following single representations: wHSV [1], MSCR [1], PartsSC [19], gBiCov [20], bLDFV [13] and SDC [14]. wHSV is the histogram of HSV channels and can be seen as the baseline in this task. MSCR are extracted from the body region and described by their area, centroid, second moment matrix and average color of the region. bLDFV is the global representation integrated by the local features based on the Fisher Vectors method.

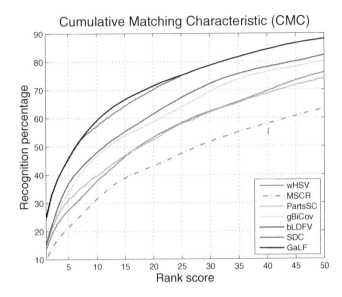

Fig. 3. VIPeR database: CMC curves of the single representations.

SDC is a method based on unsupervised salience learning [14], which can be seen as the state-of-the-art on VIPeR database for the single representation.

Figure 3 shows the CMC plots of GaLF as well as other single representations. We just show the CMC curve of SDC from rank 1 to rank 25 as it shown in [14]. From the figure, we can see that first, the accuracies of GaLF are better than those of wHSV about 15 % at each rank. These obvious advantage shows that by using the textures and spatial structure as the supplement, compared with wHSV, GaLF can gain the better results compared with wHSV. Then, for the different single representations, the performance of GaLF are the best of all. The rank 1 matching rate for GaLF is 25.08 % while that of bLDFV is 15.40 %. The rank 10 matching rate for GaLF is 59.18 % while that of bLDFV is 47.93 %. Though the rank 1 matching rates of SDC is very close to that of GaLF, the rank 10 matching rate for SDC is around 56 %, which is less about 3 % compared with GaLF. These scenes show the good performance of the proposed representation.

Secondly, we compare eGaLF with the representations of SDALF [1], Comb [19], eBiCov [20], eLDFV [13], and eSDC [14]. As shown in Table 1, those representations are all the combinations of the different features. From the table, we can know that for other representations, they often select wHSV(Hist) as their color component. But for eGaLF, we use the proposed GaLF as the color component, not the color histogram.

Figure 4 shows the CMC plots of eGaLF as well as other representations. From Fig. 4, we can know that the results of eGaLF are much better than those of other methods. The rank 1 matching rate for eGaLF is 28.34 % while that of eLDFV and eSDC is 22.34 % and 26.74 %, respectively. The rank 10 matching

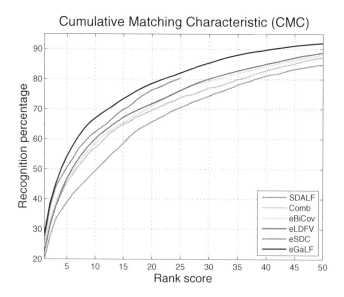

Fig. 4. VIPeR database: CMC curves of the combination of the different features.

Table 1. The combination of the different features.

Representations	Features
SDALF [1]	wHSV, MSCR, RHSP
Comb [19]	Hist, PartsSC, Cov
eBiCov [12]	wHSV, MSCR, gBiCov
eLDFV [13]	wHSV, MSCR, bLDFV
eSDC [14]	wHSV, MSCR, SDC
eGaLF	GaLF, MSCR, bLDFV

rate for eGaLF is 66.94 % while that of eLDFV and eSDC is 60.04 % and 62.37 %, respectively. These results show the good performance of the proposed representation. Specially, compared with eLDFV, eGaLF just use GaLF to replace the histogram feature wHSV in eLDFV. So, the advantage of eGaLF shows that GaLF is much better than the histogram features again.

As shown in Eq. 1, in GaLF, we only use the information in the vertical direction, not the horizontal direction. We argue that the position information of the body in the horizontal direction can not be used directly because the misalignment often happens in the task of person re-identification. To validate this idea, similar to Eq. 1, we also design a 11-d local feature which includes the information in the horizontal direction:

$$f'(x, y) = [f(x, y), x, d_{L_x}(x, y), d_{A_x}(x, y), d_{B_x}(x, y)] \tag{5}$$

In $f'(x, y)$, x is the coordinate in the horizontal direction. $d_{L_x}(x, y), d_{A_x}(x, y)$ and $d_{B_x}(x, y)$ are the gradient in the horizontal direction for the different channels.

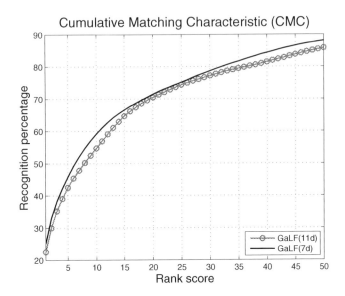

Fig. 5. VIPeR database: CMC curve using the different local features.

We repeat the experiment by using $f'(x, y)$ to replace $f(x, y)$ in GaLF. In Fig. 5, we show the CMC curve of the 7-d and 11-d local features while at the same time. From the figure, we can know that the performance of GaLF is decreased when using the information in the horizontal direction. This scene is accorded with our judgement. So, based on this result, we discard the information in the horizontal direction in GaLF.

In Eq. 2, parameter θ decides the weight of the similarity of mean vectors and covariance matrixes in the distance of two Gaussian models. To show the influence of θ to the performance, we also repeat the experiments by using the different θ. The results are shown in Fig. 6. In the figure, $\theta = 0$ means only using the similarity of $a(\cdot)$ while $\theta = 1$ using $b(\cdot)$. From the figure, we can know that from 0 to 0.4, the performance is improved gradually. When θ is set to 0.4, GaLF can get the best performance. Then, the performance is decreased gradually when θ varies from 0.4 to 1. Based on the results of GaLF using the different θ, in this paper, we set θ to 0.4 for simplicity. We also can know that the performance of $a(\cdot)$ is better than that of $b(\cdot)$ though the advantage is not so obviously. Considering the color information is kept in $a(\cdot)$, the good performance of $a(\cdot)$ also show the color information is the key issue in GaLF.

3.2   Person Re-identification on i-LIDS Database

Besides VIPeR database, we also test the proposed GaLF and eGaLF on the i-LIDS database. The i-LIDS database has been captured by multiple non-overlapping cameras at a busy airport arrival hall. There are 119 pedestrians with total 476 images. All the images are normalized to the size of 128×64 pixels.

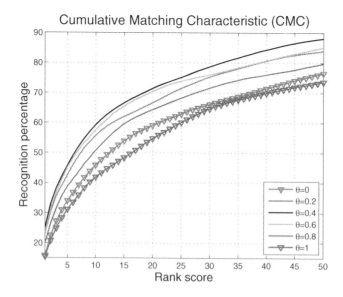

Fig. 6. VIPeR database: CMC curve of GaLF using the different θ.

Fig. 7. Some images in the i-LIDS database. The images in the same column are belonging to the same person.

Many of these images undergo quite large illumination changes and occlusions (see Fig. 7).

We follow the same experimental settings of [1,2] and test the proposed descriptors in the single-shot scenario. Considering there are 4 images on average for each pedestrian, we randomly select one image for each pedestrian to build the gallery set, while the rest (357 images) form the probe set. We repeat this procedure 10 times and compute the average CMC. Figure 8 shows the CMC curves given by GaLF, eGaLF, SDALF [1], Custom Pictorial Structures (PS) [2], gBiCov [20] and SCR [21]. On the i-LIDS database, the best single-shot published performance is obtained by a covariance-based technique (SCR).

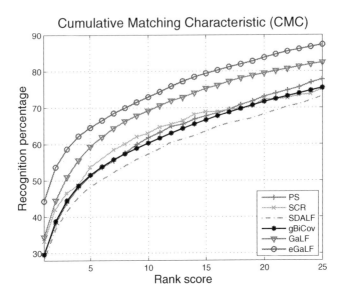

Fig. 8. i-LIDS database: CMC curves of the different methods in the single shot scenario.

From Fig. 8, we can know that GaLF and eGaLF outperform other representations on this database. The rank 1 matching rate for GaLF and eGaLF is 34.50 % and 44.34 %, respectively, while that of SCR is about 30 %. The rank 10 matching rate for GaLF and eGaLF is 69.18 % and 72.94 %, respectively, while that of SCR is around 63 %. Compared with the results on the VIPeR database, the advantages of the proposed GaLF and eGaLF are more obvious. These results show the good performance of the proposed representations again.

4   Conclusion

This paper propose a novel image representation named GaLF for the task of person re-identification. In contrast to the pervious color features, such as color histogram, using the Gaussian model to represent the local features of the body region, GaLF can keep the color features of the input human image while make the texture and spatial structure as the supplement. One important advantage of the proposed representation is its simplicity. Experiments on two pedestrian databases (VIPeR and i-LIDS) show that the proposed GaLF can achieve the state-of-the-art performances in unsupervised setting.

There are several aspects to be further studied in the future. First, in this paper, we just care about the single-shot scenario and use one Gaussian model to represent the human region. For the multi-shot scenario, since there are many samples for one person, we can use GMMs to model the regions of the same person. But this idea need to be validated by experiments. Then, the proposed representations is an unsupervised method. How to use the metric learning method to improve the performance of the proposed representations should be researched further.

Acknowledgment. This paper is partially supported by National Natural Science Foundation of China under Contract nos. 61173065, 61272319 and 61332016 and the President Fund of UCAS.

References

1. Farenzena, M., Bazzani, L., Perina, A., Murino, V., Cristani, M.: Person re-identification by symmetry-driven accumulation of local features. In: Proceedings of IEEE Conference on Computer Vision and Pattern Recognition, pp. 2360–2367 (2010)
2. Cheng, D., Cristani, M., Stoppa, M., Bazzani, L., Murino, V.: Custom pictorial structures for re-identification. In: Proceedings of British Machine Vision Conference (2011)
3. Oreifej, O., Mehran, R., Shah, M.: Human identity recognition in aerial images. In: Proceedings of IEEE Conference on Computer Vision and Pattern Recognition, pp. 709–716 (2010)
4. Schwartz, W., Davis, L.: Learning discriminative appearance based models using partial least squares. In: Brazilian Symposium on Computer Graphics and Image Processing (2009)
5. Prosser, B., Zheng, W., Gong, S., Xiang, T.: Person re-identification by support vector ranking. In: Proceedings of British Machine Vision Conference (2010)
6. Zhang, Y., Li, S.: Gabor-LBP based region covariance descriptor for person re-identification. In: International Conference on Image and Graphics, pp. 368–371 (2011)
7. Gray, D., Tao, H.: Viewpoint invariant pedestrian recognition with an ensemble of localized features. In: Forsyth, D., Torr, P., Zisserman, A. (eds.) ECCV 2008, Part I. LNCS, vol. 5302, pp. 262–275. Springer, Heidelberg (2008)
8. Bak, S., Corvee, E., Bremond, F., Thonnat, M.: Person re-identification using haar-based and DCD-based signature. In: Proceedings of International Workshop on Activity Monitoring by Multi-camera Surveillance Systems (2010)
9. Gheissari, N., Sebastian, T., Tu, P., Rittscher, J., Hartley, R.: Person reidentification using spatiotemporal appearance. In: Proceedings of IEEE Conference on Computer Vision and Pattern Recognition 2, pp. 1528–1535 (2006)
10. Kai, J., Bodensteiner, C., Arens, M.: Person re-identification in multi-camera networks. In: Proceedings of IEEE Conference on Computer Vision and Pattern Recognitio Workshops, pp. 55–61 (2011)
11. Zheng, W., Gong, S., Xiang, T.: Associating groups of people. In: Proceedings British Machine Vision Conference (2009)
12. Ma, B., Su, Y., Jurie, F.: Bicov: a novel image representation for person re-identification and face verification. In: Proceedings of British Machine Vision Conference, pp. 57.1–57.11 (2012)
13. Ma, B., Su, Y., Jurie, F.: Local descriptors encoded by fisher vectors for person re-identification. In: Fusiello, A., Murino, V., Cucchiara, R. (eds.) ECCV 2012 Ws/Demos, Part I. LNCS, vol. 7583, pp. 413–422. Springer, Heidelberg (2012)
14. Zhao, R., Ouyang, W., Wang, X.: Unsupervised salience learning for person re-identification, pp. 3586–3593 (2013)
15. Luo, P., Wang, X., Tang, X.: Pedestrian parsing via deep decompositional neural network. In: Proceedings of IEEE International Conference on Computer Vision (2013)

16. Li, P., Wang, Q., Zhang, L.: A novel earth mover's distance methodology for image matching with Gaussian mixture models. In: Proceedings of IEEE International Conference on Computer Vision (2013)
17. Gray, D., Brennan, S., Tao, H.: Evaluating appearance models for recognition, reacquisition, and tracking. In: IEEE International Workshop on Performance Evaluation of Tracking and Surveillance (2007)
18. Moon, H., Phillips, P.: Computational and performance aspects of PCA-based face-recognition algorithms. Perception **30**, 303–321 (2001)
19. Kviatkovsky, I., Adam, A., Rivlin, E.: Color invariants for person reidentification. IEEE Trans. Pattern Anal. Mach. Intell. **35**, 1622–1634 (2013)
20. Ma, B., Su, Y., Jurie, F.: Covariance descriptor based on bio-inspired features for person re-identification and face verification. Image Vis. Comput. **32**, 379–390 (2014)
21. Bak, S., Corvee, E., Bremond, F., Thonnat, M.: Person re-identification using spatial covariance regions of human body parts. In: Proceedings of International Conference on Advanced Video and Signal-Based Surveillance (2010)

Privacy Preserving Multi-target Tracking

Anton Milan[1(✉)], Stefan Roth[2], Konrad Schindler[3], and Mineichi Kudo[4]

[1] School of Computer Science, University of Adelaide, Adelaide, Australia
anton.milan@adelaide.edu.au
[2] Department of Computer Science, TU Darmstadt, Darmstadt, Germany
[3] Photogrammetry and Remote Sensing, ETH Zürich, Zürich, Switzerland
[4] Division of Computer Science, Hokkaido University, Sapporo, Japan

Abstract. Automated people tracking is important for a wide range of applications. However, typical surveillance cameras are controversial in their use, mainly due to the harsh intrusion of the tracked individuals' privacy. In this paper, we explore a privacy-preserving alternative for multi-target tracking. A network of infrared sensors attached to the ceiling acts as a low-resolution, monochromatic camera in an indoor environment. Using only this low-level information about the presence of a target, we are able to reconstruct entire trajectories of several people. Inspired by the recent success of offline approaches to multi-target tracking, we apply an energy minimization technique to the novel setting of infrared motion sensors. To cope with the very weak data term from the infrared sensor network we track in a continuous state space with soft, implicit data association. Our experimental evaluation on both synthetic and real-world data shows that our principled method clearly outperforms previous techniques.

1 Introduction

Tracking multiple people in indoor environments has many important applications, including customer behavior analysis in retail, crowd flow estimation for building design and planning of evacuation routes, or assistance in daily living for elderly people. While standard surveillance cameras can be employed to address this task, they also have several disadvantages. First, depending on the exact setup, a camera network may be too costly to install and to maintain. Second, standard RGB cameras are highly sensitive to lighting changes and do not work in dark environments. Finally, and most importantly, surveillance cameras are often seen as an intrusion into a person's privacy because they enable a clear identification of the observed person and provide rich visual information about the appearance, pose and exact action of each subject [1,2].

In this paper we present an affordable, privacy-preserving alternative to address multi-target tracking. To that end, we employ a network of infrared

Electronic supplementary material The online version of this chapter (doi:10.1007/978-3-319-16634-6_38) contains supplementary material, which is available to authorized users. Videos can also be accessed at http://www.springerimages.com/videos/978-3-319-16633-9.

© Springer International Publishing Switzerland 2015
C.V. Jawahar and S. Shan (Eds.): ACCV 2014 Workshops, Part III, LNCS 9010, pp. 519–530, 2015.
DOI: 10.1007/978-3-319-16634-6_38

motion sensors that are attached to the ceiling in an indoor scenario. Each sensor is activated whenever a person passes underneath it, yielding a set of sparse measurements that is then used to infer the exact location of each target. To reconstruct the individual trajectories we rely on recent advances in multiple object tracking, *e.g.*, [3–5]. Although the sensory system mounted overhead does not suffer from occlusion, a number of other challenges must be addressed. First, the number of sensors is considerably lower than the number of pixels in a video, leading to a very sparse signal that provides a rather crude approximation of true target locations. Second, a binary sensor response is the only available source of evidence about the presence of a target. Therefore, high-level cues such as a person's appearance or a continuous-valued likelihood of an object detector cannot be exploited. Third, each sensor can be simultaneously activated by several targets, while one single target can activate multiple neighboring sensors when passing between them. Hence we also have to allow many-to-one and one-to-many assignments. Note that the majority of multi-target tracking approaches cannot be directly applied to the present setting because of their common implicit assumptions that each measurement may originate from at most one target and that each target can cause at most one measurement. The strategy we present here is able to handle both cases. Our main contribution is twofold:

- We introduce a novel infrared tracking dataset including the measurements and manually annotated ground truth. The dataset consists of three synthetic and three real world sequences, covering various levels of difficulty.
- We demonstrate how a recent multiple target tracking approach developed for regular cameras [4] can be adopted to address the challenges in this novel setting.

In contrast to previous work in the realm of infrared-sensor tracking [6,7], we present a simple and more robust tracking method and evaluate its performance using standard tracking metrics. To the best of our knowledge, this is the first time that a global tracking approach is applied to infrared sensor responses. Experimental results show the superiority of our method on several real-world sequences. We make all our data as well as the source code publicly available.[1]

2   Related Work

Research on the automated tracking of multiple targets originated several decades ago in the realm of aerial and naval navigation with radar and sonar sensors. Some of the most notable early works include the multiple hypothesis tracker (MHT) [8] and the joint probabilistic data association (JPDA) [9], which are only rarely used nowadays, as their computation times scale exponentially with the number of targets; these methods hence quickly reach their limits in crowded environments. Such strategies usually apply filtering techniques, for example the Kalman filter [10], in order to estimate the true target locations from noisy

---

[1] http://research.milanton.net/irtracking.

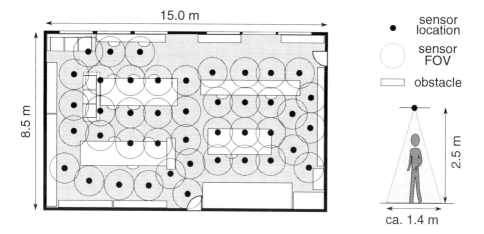

Fig. 1. Overview of our infrared ceiling sensor network.

observations. More recent approaches follow an offline strategy, where a batch of frames is analyzed at once [3–5,11–13]. The main motivation behind this is that potential errors may be corrected once more observation steps are available, making these methods more robust against localization noise, false measurements, and target drift.

While RGB cameras or radar/sonar equipment have been the typical sensor choice in the past, tracking results in the literature based on infrared sensors are rather limited. The scheme proposed by Luo *et al.* [7] utilizes a Kalman filter to estimate the location of a single target. However, a rather complex hardware array, where each node consists of five individual sensors equipped with specialized Fresnel lenses, is employed in their setup. Unfortunately, only synthetically generated simulation results are presented. The tracking algorithm described by Hosokawa *et al.* [6] relies on a more complex target localization scheme and includes several ad-hoc procedures to resolve ambiguities. Tao *et al.* [14] also follow a similar setup, but concentrate on activity recognition, in particular fall detection, rather than on tracking individuals in a multi-person scenario.

In contrast, we present an affordable and flexible framework with minimal calibration effort that allows us to robustly keep track of several individuals in an indoor scenario while preserving the individuals' privacy. Quantitative and qualitative evaluation on both synthetic and real-world data demonstrates encouraging results.

3   Infrared Ceiling Sensor Network

We build on the sensor network originally proposed by Hosokawa *et al.* [6]. The entire network consists of 43 nodes attached to the ceiling in a large room of approximately 15.0×8.5 m (*cf.* Fig. 2). Each node is a *pyroelectric infrared sensor*, often simply referred to as an *infrared motion sensor*. The sensor is activated

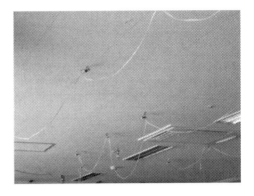

Fig. 2. A sensor node *(left)* and the entire setup *(right)*.

whenever it detects an abrupt temperature change within its range. We exploit this behavior to detect a person moving underneath it. To obtain measurements that are more precisely localized, the detection cone is narrowed to about a 70 cm radius on the ground. The sensors are distributed across the entire room such that they cover most of the area of interest and their fields-of-view do not overlap substantially (*cf.* Fig. 1). The nodes do not have to be perfectly aligned or arranged in a specific way during installation. An approximate location of each node is sufficient for an accurate calibration, which makes the deployment of such a system rather easy. Note that there is no other information available, such that disambiguating the identity based on visual features is infeasible. The sampling rate of the sensors can be adjusted for a specific application and is set to 2 Hz in our setting. The cost of each sensor is as low as few US Dollars.

4   Multi-target Tracking

Most modern multi-target tracking approaches follow the so-called tracking-by-detection strategy [4,5,12,15,16], which we also adopt here. In this two-stage strategy a set of measurements is first obtained for each frame independently, forming the target hypotheses. These observations, which are prone to noise and potentially contain false measurements, then serve as input to a tracking algorithm. Moreover, our method also belongs to the class of off-line (non-recursive) state estimation techniques, where, instead of processing one frame at a time as done, *e.g.*, in particle filtering [17], a larger time interval is analyzed in one step. This significantly improves robustness and does not pose a serious limitation, since a slight delay of only a few seconds (in our case 20 frames) is acceptable in practice for the potential applications of our system.

We formulate tracking as minimization of a continuous energy function, which we argue is particularly appropriate for the novel setting that we address here. In particular, we follow the recent work of Milan *et al.* [4] and demonstrate how it can be adapted to this rather different kind of imagery. The state vector $\mathbf{X}$

consists of all (X, Y) coordinates of all targets on the ground plane. We will denote a location of person i in frame t with $\mathbf{X}_i^t$, and the location of the sensor node g with $\mathbf{S}_g$. The set of active nodes at time t is denoted $G(t)$. Finally, N is the total number of targets, and s_i and e_i mark the temporal start and end points of each target, respectively. The energy

$$E = E_{\text{det}} + a E_{\text{dyn}} + b E_{\text{exc}} + c E_{\text{per}} + d E_{\text{reg}}, \tag{1}$$

consisting of a data term, three physically-based (soft) constraints, and a regularizer is then minimized in order to find a locally optimal solution. The approach offers two important advantages. First, trajectories are reconstructed in continuous space such that the low spatial resolution of the sensor network is mitigated. Second, data association is only solved implicitly and not restricted to one-to-one correspondence between observations and target locations. In other words, it is possible that the same measurement may in fact originate from two separate targets, and that one single target can activate two sensors simultaneously. Both situations frequently occur in the observed data and are correctly captured by our model. The individual components are defined as follows:

Observation. The observation term E_{det} keeps the resulting trajectories close to the obtained measurements. To reflect the localization uncertainty of the infrared sensors, we use an inverse Cauchy-like function

$$E_{\text{det}}(\mathbf{X}) = \sum_{i=1}^{N} \sum_{t=s_i}^{e_i} \left[\lambda - \sum_{g \in G(t)} \frac{s^2}{\|\mathbf{X}_i^t - \mathbf{S}_g\|^2 + s^2} \right], \tag{2}$$

where s controls the size of the lobe. Given that the sensors' field-of-view covers a circular area of approximately 1.4 m in diameter, we employ this value in all our experiments. A uniform penalty λ is applied to all targets to prevent false trajectories without measurements nearby.

Dynamics. The data acquired by infrared sensors is rather limited and exceedingly noisy. A dynamic model is therefore important to bridge missing observations and to restrict data association to plausible solutions. Here, we rely on a constant velocity assumption, and penalize acceleration using

$$E_{\text{dyn}}(\mathbf{X}) = \sum_{i=1}^{N} \sum_{t=s_i+1}^{e_i-1} \|\mathbf{X}_i^{t+1} - 2\mathbf{X}_i^t + \mathbf{X}_i^{t-1}\|^2. \tag{3}$$

Exclusion. Accurately modeling target exclusion is important for several reasons. On one hand, it is desirable to obtain a physically plausible solution without inter-target collisions. On the other hand, we must take into account that a sensor response may be caused by more than one target and that one single target can activate more than one sensor. A continuous exclusion term

$$E_{\text{exc}}(\mathbf{X}) = \sum_{i \neq j} \sum_{t=\max\{s_i, s_j\}}^{\min\{e_i, e_j\}} \frac{1}{\|\mathbf{X}_i^t - \mathbf{X}_j^t\|^2} \tag{4}$$

that directly penalizes situations when two targets come too close to one another serves this purpose. It pushes two trajectories away from one another just enough to avoid a collision, but not too far, since a single measurement should be allowed to explain two targets.

Persistence. Assuming that targets cannot appear or disappear in the middle of the tracking area, the term

$$E_{\text{per}}(\mathbf{X}) = \sum_{\substack{i=1,\dots,N \\ t\in\{s_i,e_i\}}} \frac{1}{1 + \exp\big(-q \cdot b(\mathbf{X}_i^t) + 1\big)} \tag{5}$$

enforces persistent trajectories and reduces the number of fragmentations. The parameter $q = 1/35\,\text{cm}$ controls the entrance margin and $b(\cdot)$ computes the distance of a trajectory to the border of the tracking area.

Regularization. Finally, we add a regularizer to keep the number of total targets low and enforce longer trajectories:

$$E_{\text{reg}}(\mathbf{X}) = N + \mu \sum_{i=1}^{N} |F(i)|^{-1}\,, \tag{6}$$

where $F(i) := e_i - s_i + 1$ is the total life span of the ith trajectory.

4.1   Optimization

Each energy component is differentiable in closed form, making the entire formulation well suited for gradient-based minimization. We choose to apply a standard conjugate gradient descent to minimize Eq. (1) locally. However, given the highly non-convex nature of the energy, a purely gradient-based optimization would be very susceptible to initialization. Therefore, we add a set of jump moves, as in [4]. These non-local jumps in the energy landscape change trajectory lengths and potentially the number of targets, thus allowing a more flexible probing of the solution space to escape weak minima. Upon convergence of the gradient descent, one of six jump moves described below is executed in a greedy fashion; then the gradient descent restarts.

Growing and shrinking. Each trajectory can be extended by linear extrapolation for an arbitrary number of time steps both forward and backward in time. Similarly, a track is shortened by discarding a fragment of a certain length from either end. Growing is useful for finding new targets, while shrinking weeds out false positives that may have been introduced by noise or during intermediate optimization steps.

Merging and splitting. Two existing trajectories are merged into one if the merge lowers the energy. Note that the individual energy components, in particular the dynamics and the exclusion terms, assert that this step will not cause physically implausible situations with intersecting trajectories or unlikely motion

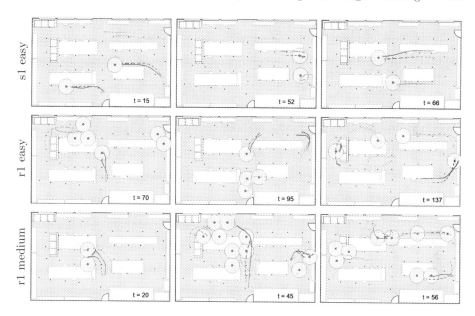

Fig. 3. Qualitative results on synthetic (top row) and real data (middle and bottom rows). 10 frames of both the recovered trajectories (solid) and the ground truth (dashed lines) are shown at three example time steps for each sequence. Note that despite the extreme amount of noise present in the observations (large circles), our method is able to successfully recover most of the targets' trajectories.

patterns. A single track may also be split into two at a specific point in time. Both these moves provide a method to bridge over regions with missing sensor responses and to reduce fragmentation of tracks and identity swaps.

Adding and removing. These two moves operate on entire trajectories. Removing a false positive target from the current solution may decrease the overall energy because it results in a more plausible explanation of the data. On the other hand, it is important to allow for inserting new tracks around active sensor locations that do not have a target nearby. This is done conservatively by adding a short tracklet of only three frames. Note that it can grow and merge with other existing trajectories at a later optimization step.

5   Experiments

We present an experimental evaluation of our method on both synthetic and real-world data. Quantitatively assessing the performance of multi-target tracking is an inherently challenging task [18]. Here, we follow the most common strategy and present the evaluation using a set of standard metrics. Next to recall and precision, we compute the *CLEAR MOT* [19] metrics consisting of *MOTA* (Multiple-Object Tracking Accuracy) and *MOTP* (Multiple-Object Tracking Precision). The MOTA includes all possible error types – spurious trajectories or false positives *(FP)*, missed targets or false negatives *(FN)*, and mismatches or identity switches *(ID)* – and is normalized such that 100 % corresponds to no errors. The MOTP directly measures the performance of location estimation by computing the average distance between the true target and the inferred location, again normalized to 100 %. We use a 1.5 m hit/miss threshold on the ground plane. The weights a through d for the individual energy terms in Eq. (1) were determined empirically and kept fixed for all experiments at $\{.0006, .8, .08, .02\}$. The additional parameters λ and μ were set to .004 and 1, respectively.

5.1   Datasets

Synthetic data. Acquiring large amounts of accurate ground truth for multi-target tracking is tiresome and costly. Therefore, we first test our presented method on a synthetic dataset. The data is created by simulating plausible trajectories and the generated sensor responses. Trajectories are spline interpolations between sparse key points corresponding to typical motion patterns. An average target speed of 1 m/s with Gaussian noise is assumed for our purpose. A sensor is set to 'active' if at least one trajectory passes within a distance close than its range of operation, which amounts to 70 cm. Three sequences (*s1 easy/medium/hard*) with two, four and six targets, respectively, present a reasonable variability in person count and density.

Real-world data. While simulated data may be comparatively easy to acquire, it typically does not fully capture the complexity of real observations. Thus, it is essential to also test a system on real sensor data. To that end, we recorded three sequences of approximately two minutes each. Six persons were moving freely around the entire walkable space inside the lab. The ground truth was annotated manually, relying on videos from two cameras that served as reference (*cf.* Fig. 4). Note that a precise localization of each person is ambiguous – and sometimes even impossible – with the available setup, due to low resolution and/or occlusions. Nonetheless, this new dataset is a sensible basis to quantitatively evaluate the tracking performance. The entire dataset consists of 974 frames, which amounts to over eight minutes of data. The average person count across all sequences is 3.2.

Qualitative results of our proposed method are shown in Figure 3. Each row depicts three frames from one particular sequence, while in each frame the currently active sensors are indicated with large circles, and the estimated and true trajectories are plotted with solid, respectively dashed lines for the past ten time steps. Note the extremely noisy observations, including spurious activations and missing signal.

Our method is able to correctly recover the number of targets and their trajectories in cases where the targets remain well separated (see top row). Note that a precise localization is not always possible due to the scarcity of the available data, as can be seen, *e.g.*, for the red target. In realistic environments with more targets and sensor noise, there is a certain drop in performance, as expected. The main cause of failure are from ambiguous measurements, which typically occur

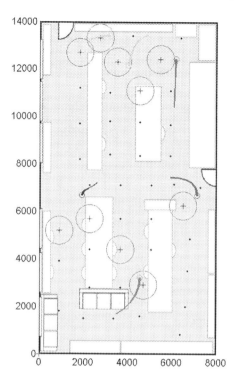

Fig. 4. A screenshot of the data acquisition setup. The video captured by two cameras on the right serves as reference for annotating each individual's location on the ground plane. The large red circles indicate active sensors in the current frame (Color figure online).

when people walk close to each other for a prolonged period of time (*cf.* bottom left frame). In such cases, the sensors are unable to provide enough information to robustly resolve the ambiguity.

5.2   Quantitative Evaluation

Table 1 shows quantitative results of our proposed method on all six sequences (three synthetic (s*) and three real (r*) ones). Note that we achieve near perfect precision and only few identity switches with simulated data, i.e. in the absence of sensor noise. The performance decreases for real data, but still stays above 50 % MOTA on average. Table 2 lists average results of our method compared to those of three other strategies. One is a linear location estimation technique specifically designed for inrared motion sensors, where the number of targets is determined by connected components of sensor responses [14].[2] While it can recover most targets without producing too many false positives, the number

---

[2] Implementation provided by the authors.

Table 1. Quantitative results on synthetic *(top)* and real *(bottom)* data.

Sequence	Recall	Precision	MOTA	MOTP	FP	FN	ID
s1 easy	93.3 %	98.9 %	88.2 %	76.6 %	2	13	8
s1 medium	82.3 %	97.1 %	76.2 %	73.1 %	9	64	13
s1 hard	71.0 %	94.3 %	63.5 %	71.0 %	25	170	19
Mean (s*)	**82.2 %**	**96.8 %**	**76.0 %**	**73.6 %**	**12**	**82**	**13**
r1 easy	84.2 %	81.0 %	57.5 %	53.8 %	143	114	50
r1 medium	74.5 %	83.2 %	52.9 %	56.9 %	88	149	38
r1 hard	74.4 %	85.3 %	55.4 %	53.1 %	86	172	42
Mean (r*)	**77.7 %**	**83.2 %**	**55.3 %**	**54.6 %**	**106**	**145**	**43**

Table 2. Comparison to other methods averaged over synthetic *(top)* and real *(bottom)* sequences. The best average performance for each measure is highlighted in bold face.

	Method	Recall	Precision	MOTA	MOTP	FP	FN	ID
synth.	Linear [14]	81.0 %	**99.8 %**	66.6 %	64.6 %	**1**	**81**	58
	DP [20]	78.5 %	92.0 %	55.9 %	65.3 %	27	81	57
	KSP [12]	78.9 %	97.5 %	75.5 %	67.5 %	6	83	**6**
	Ours	**82.2 %**	96.8 %	**76.0 %**	**73.6 %**	12	82	13
real	Linear [14]	79.3 %	71.5 %	9.3 %	50.1 %	212	137	252
	DP [20]	71.6 %	62.7 %	9.6 %	47.3 %	281	188	128
	KSP [12]	**89.4 %**	63.7 %	31.1 %	48.3 %	337	**70**	48
	Ours	77.7 %	**83.2 %**	**55.3 %**	**54.6 %**	**106**	145	**43**

of ID switches is quite high. The second baseline is the globally optimal approach by Pirsiavash *et al.* [20] based on dynamic programming (DP). It is able to better keep correct identities over time, but struggles to handle the extremely noisy measurements leading to a high number of false alarms and missed targets. Finally, the third method is the k-shortest paths (KSP) approach [12], where targets are tracked on a discrete grid. Its power to robustly keep target identities over long time periods is unfolded in the present setting with overhead sensors and in absence of occlusion, particularly with noiseless synthetic data. However, the coarse discretization of the grid is not able to handle real-world noisy measurements, leading to many false tracks. The continuous state representation in combination with the soft assignment strategy of our method clearly outperforms previous techniques with respect to the most relevant tracking metrics (MOTA, MOTP).

Finally, Table 3 shows the mean average people count error of the four methods, again split into two groups of synthetic and real data. Although all four methods show similar performance in the absence of noise, our proposed approach provides the most accurate estimate on the number of people present in the scenes with realistic data.

Table 3. Person count estimation. Per-frame mean absolute error (MAE) and standard deviation is shown for synthetic and real data.

Method	Linear [14]	DP [20]	KSP [12]	**Ours**
MAE (synth.)	0.57 ± 0.78	0.62 ± 0.75	0.57 ± 0.75	$\mathbf{0.54 \pm 0.81}$
MAE (real)	1.00 ± 0.95	1.25 ± 1.16	1.52 ± 1.06	$\mathbf{0.76 \pm 0.76}$

6   Discussion and Conclusion

We have presented a method for tracking multiple people while fully preserving their privacy. A ceiling infrared sensor network serves as a low resolution camera providing only a sparse binary signal about the presence of moving targets. Building on recent advances in multiple target tracking, we follow an energy minimization strategy to localize walking people and reconstruct their trajectories in spite of the impoverished observation data. Our framework outperforms other methods measured with respect to widely used tracking metrics, both on synthetic and real-world data.

The low spatial resolution and high noise level of the signal provided by the sensors clearly limits the ability to resolve all identity ambiguities. A precise localization as well as targets that remain still for longer time periods causing a sensor to become inactive still remain challenges that should be addressed in future work.

References

1. Babaguchi, N., Koshimizu, T., Umata, I., Toriyama, T.: Psychological study for designing privacy protected video surveillance system: PriSurv. In: Senior, A. (ed.) Protecting Privacy in Video Surveillance, pp. 147–164. Springer, London (2009)
2. Norris, C., Armstrong, G.: CCTV and the social structuring of surveillance. In: Painter, K., Tilley, N. (eds.) Surveillance of Public Space. Crime Prevention Studies, vol. 10, pp. 157–178. Criminal Justice Press, Monsey (1999)
3. Jiang, H., Fels, S., Little, J.J.: A linear programming approach for multiple object tracking. In: IEEE Conference on Computer Vision and Pattern Recognition (CVPR) (2007)
4. Milan, A., Roth, S., Schindler, K.: Continuous energy minimization for multitarget tracking. IEEE Trans. Pattern Anal. Mach. Intell. **36**, 58–72 (2014)
5. Zhang, L., Li, Y., Nevatia, R.: Global data association for multi-object tracking using network flows. In: IEEE Conference on Computer Vision and Pattern Recognition (CVPR) (2008)
6. Hosokawa, T., Kudo, M., Nonaka, H., Toyama, J.: Soft authentication using an infrared ceiling sensor network. Pattern Anal. Appl. **12**, 237–249 (2009)
7. Luo, X., Shen, B., Guo, X., Luo, G., Wang, G.: Human tracking using ceiling pyroelectric infrared sensors. In: 2009 IEEE International Conference on Control and Automation, ICCA 2009, pp. 1716–1721 (2009)
8. Reid, D.B.: An algorithm for tracking multiple targets. IEEE Trans. Autom. Control **24**, 843–854 (1979)

9. Fortmann, T.E., Bar-Shalom, Y., Scheffe, M.: Multi-target tracking using joint probabilistic data association. In: 19th IEEE Conference on Decision and Control including the Symposium on Adaptive Processes, vol. 19, pp. 807–812 (1980)

10. Kalman, R.E.: A new approach to linear filtering and prediction problems. Trans. ASME-J. Basic Eng. **82**, 35–45 (1960)

11. Andriyenko, A., Schindler, K., Roth, S.: Discrete-continuous optimization for multi-target tracking. In: IEEE Conference on Computer Vision and Pattern Recognition (CVPR) (2012)

12. Berclaz, J., Fleuret, F., Türetken, E., Fua, P.: Multiple object tracking using k-shortest paths optimization. IEEE Trans. Pattern Anal. Mach. Intell. **33**, 1806–1819 (2011)

13. Butt, A.A., Collins, R.T.: Multi-target tracking by Lagrangian relaxation to min-cost network flow. In: IEEE Conference on Computer Vision and Pattern Recognition (CVPR) (2013)

14. Tao, S., Kudo, M., Nonaka, H.: Privacy-preserved behavior analysis and fall detection by an infrared ceiling sensor network. Sensors **12**, 16920–16936 (2012)

15. Andriluka, M., Roth, S., Schiele, B.: People-tracking-by-detection and people-detection-by-tracking. In: IEEE Conference on Computer Vision and Pattern Recognition (CVPR) (2008)

16. Roshan Zamir, A., Dehghan, A., Shah, M.: GMCP-tracker: global multi-object tracking using generalized minimum clique graphs. In: Fitzgibbon, A., Lazebnik, S., Perona, P., Sato, Y., Schmid, C. (eds.) ECCV 2012, Part II. LNCS, vol. 7573, pp. 343–356. Springer, Heidelberg (2012)

17. Breitenstein, M.D., Reichlin, F., Leibe, B., Koller-Meier, E., Van Gool, L.: Robust tracking-by-detection using a detector confidence particle filter. In: IEEE International Conference on Computer Vision (ICCV) (2009)

18. Milan, A., Schindler, K., Roth, S.: Challenges of ground truth evaluation of multi-target tracking. In: Proceedings of the CVPR 2013 Workshop on Ground Truth - What is a Good Dataset? (2013)

19. Bernardin, K., Stiefelhagen, R.: Evaluating multiple object tracking performance: the CLEAR MOT metrics. Image Video Process. **2008**, 1–10 (2008)

20. Pirsiavash, H., Ramanan, D., Fowlkes, C.C.: Globally-optimal greedy algorithms for tracking a variable number of objects. In: IEEE Conference on Computer Vision and Pattern Recognition (CVPR) (2011)

Full-Body Human Pose Estimation
from Monocular Video Sequence
via Multi-dimensional Boosting Regression

Yonghui Du, Yan Huang[✉], and Jingliang Peng

School of Computer Science and Technology, Shandong University, Jinan, China
yonghuid@gmail.com, {yan.h,jpeng}@sdu.edu.cn

Abstract. In this work, we propose a scheme to estimate two-dimensio-nal full-body human poses in a monocular video sequence. For each frame in the video, we detect the human region using a support vector machine, and estimate the full-body human pose in the detected region using multi-dimensional boosting regression. For the human pose estimation, we design a joints relationship tree, corresponding to the full hierarchical structure of joints in a human body. Further, we make a complete set of spatial and temporal feature descriptors for each frame. Utilizing the well-designed joints relationship tree and feature descriptors, we learn a hierarchy of regressors in the training stage and employ the learned regressors to determine all the joint's positions in the testing stage. As experimentally demonstrated, the proposed scheme achieves outstanding estimation performance.

1 Introduction

Human pose estimation is an important research topic with many potential applications, *e.g.*, image- and video-based event detection, interactive video gaming and human-computer interaction. Nevertheless, accurate and efficient human pose estimation has been a challenging problem. Challenges mainly come from the fact that the human body is an articulated object with many degrees of freedom and there are too many complicating factors like clothing, lighting and occlusion.

During the recent years, intensive research has been conducted in human pose estimation. However, most of the published schemes work on a single still image, and many of them conduct pose estimation for only the upper human body. There have been very few works on estimating human poses in video sequences. In particular, two-dimensional (2D) full-body human poses estimation in monocular video sequences is largely underrepresented in the research, to the best of our knowledge.

In this work, we propose a scheme to estimate two-dimensional (2D) full-body poses of a human in a monocular video sequence, which extends an existent scheme [1] for estimating upper-body human poses in still images. Compared with [1], major extensions of our work include: (1) we employ the support vector

© Springer International Publishing Switzerland 2015
C.V. Jawahar and S. Shan (Eds.): ACCV 2014 Workshops, Part III, LNCS 9010, pp. 531–544, 2015.
DOI: 10.1007/978-3-319-16634-6_39

machine (SVM) method and the histogram of oriented gradients (HOG) descriptor to detect the human region in each frame, (2) we design a full-body JRT as the basic structure for pose representation and estimation, (3) we propose a motion feature descriptor and use it together with the spatial one to describe local image features, and (4) we construct a database of videos each with annotated full-body human regions and poses, which we use for both training and testing result verification.

The rest of this paper is organized as follows. Related work is introduced in Sect. 2, the proposed scheme is described in Sect. 3 and experimental results are given and analyzed in Sect. 4. Finally, this work is concluded in Sect. 5.

2   Related Work

Human pose estimation has been intensively researched in the past decade or so. We briefly review recent work in this section, while comprehensive surveys on earlier algorithms can be found in references [2,3].

A large class of algorithms is based on the pictorial structures (PS) model [4–9], which represents the human body as a series of rigid parts and a set of relations between certain pairs of parts. Approaches extending the PS model have also been proposed, such as the deformable structures model [10,11] and the cascade of pictorial structure models [12]. These structure-model-based algorithms require a large number of constraints and correspondingly intensive computation.

Methods have been proposed which use machine learning techniques for human pose estimation. Okada and Soatto [13] propose a piecewise linear regression method for human pose estimation. In their approach, they train several local linear regressors (which are based on pose clusters generated by K-means) and a support vector machine for estimating the human pose from the histogram of oriented gradients (HOG) feature vector. However, the accuracy of linear regression method is limited, due to the diversity of human poses. Dantone *et al.* [8] employ two-layered random forests as joint regressors, but this method needs a large search space, negatively affecting its computational efficiency. Hara and Chellappa [1] propose to use multidimensional output regression tree with dependency graph for upper-body human pose estimation. Their algorithm breaks a complex problem of human pose estimation down into a sequence of local pose estimation problems which are less complex. As a result, it achieves a good tradeoff between accuracy and efficiency.

Research has also been conducted recently on estimating the human pose from a single depth image. Some methods base on the random forest have been proposed, such as [14–16]. These methods work efficiently, assuming that the positions of human parts or labels of pixels are independent. Sun *et al.* [17] present a conditional regression forest model which takes the dependency relationships among the human parts into account. However, this method requires prior knowledge about torso orientation, human height, *etc.*

All the above-reviewed algorithms work on still images. By contrast, there has been far less work published on video-based human pose estimation. In particular, the work which can estimate two-dimensional (2D) full-body human poses in monocular video sequences is largely underrepresented in the research, to the best of our knowledge. Bissacco *et al.* [18] propose multi-dimensional boosting regression with appearance and motion to address the problem of three-dimensional (3D) human pose estimation in video sequences. They utilize both 2D videos and 3D motion data to train the regressors and the trained regressors map the Haar features of a detected human region directly to an entire set of 3D joint angles representing the full-body pose. Zuffi *et al.* [11] combine the dense optical flow with the deformable structures model [10] to address the upper-body human pose estimation in monocular video sequences. However, significant computation and memory use is introduced by the dense optical flow.

3   Proposed Scheme

We assume as input to our scheme a monocular video sequence of a human. Our aim is to, for each video frame, estimate the full-body human pose that is represented as 2D positions of pre-defined human joints.

To the best of our knowledge, most of the researches based on regression for full-body human pose estimation are to learn a mapping function from the features computed from a local image region that contains an entire human body to a human pose. Those methods have a defect that the local image region should be large enough to contain the human body and thus may contain a large background region as well, increasing the complexity of human pose estimation. Therefore, we choose to extend the work by Hara and Chellappa [1] since it has achieved a good balance between accuracy and efficiency for estimating upper-body human poses in still images.

At each video frame, we first detect a rectangular region containing the human's image, utilizing the SVM method and the HOG descriptor. Thereafter, we estimate the full-body human pose in this detected region. Specifically, we design a joints relationship tree (JRT) corresponding to the hierarchy of joints in a full human body (see Fig. 1). The root joint is always fixed at 1/2 width and 1/3 height of the human's bounding rectangle. For each of the other joints, its position relative to its parent is determined via regression on the features of a temporally and spatially local region around its parent. The hierarchy of regressors at all the non-root nodes are learned in a training stage (see Fig. 2). Details of the proposed scheme are given in the following subsections.

3.1   Human Region Detection

We utilize a linear SVM classifier on HOG features [19,20] to detect the human region in each video frame. In the training stage, we randomly sample rectangular regions of $N \times M$ (we use 64×128 in experiments) in the video frames. If a sample contains a human or part of a human, we label it as positive; otherwise, we label it as negative. Next, we extract all the sample regions' HOG features, and train a linear SVM classifier on those HOG features. In the testing stage, We slide a rectangular window of $N \times M$ over each video frame, from top to bottom and

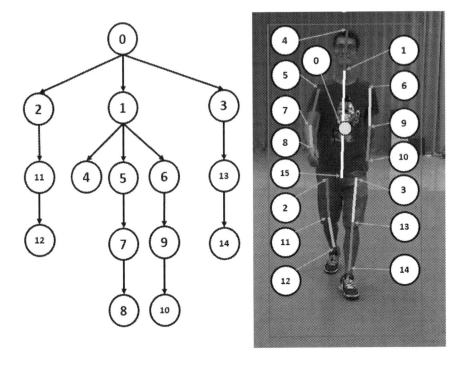

Fig. 1. Left: Joints Relationship Tree (JRT); Right: the joints marked in an image corresponding to the nodes of the JRT. The red rectangle represents the detected human region; a human body pose is represented as 10 sticks each connecting two joints in the image, which are head, torso, upper and lower arms and upper and lower legs; the yellow circle is at $1/2$ width and $1/3$ height of the human's bounding rectangle. Node 15 is at the center of Node 2 and Node 3.

from left to right. Each window is classified as positive or negative, utilizing the trained SVM classifier. The tight bounding rectangle of all the positive samples then give the detected human region in each frame. As examples, the results of human region detection for several video frames are illustrated in Fig. 3.

It is worth mentioning that Hara and Chellappa [1] do not conduct human region detection in their algorithm but directly use the annotated human region information in their test image datasets.

3.2   Joints Relationship Tree

We design a joints relationship tree for the full human body, which extends the dependency graph in reference [1]. The JRT we construct is illustrated in Fig. 1, where the left part shows the JRT structure and the right part marks the joints in an image corresponding to the nodes in the JRT. As shown in Fig. 1, there are totally 15 nodes in the JRT, which are numbered from 0 to 14 and correspond to a root location and 14 joints in a human body.

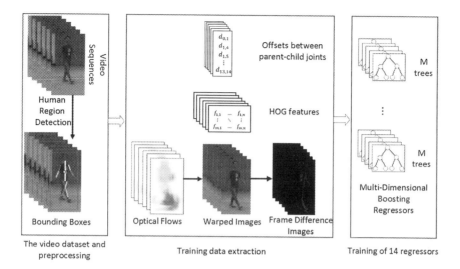

Fig. 2. The process of training. **Left**: The video dataset and preprocessing. In the preprocessing stage, we detect a rectangular region containing the human's image and annotate the full-body human poses. **Center**: Training data extraction. Here, $d_{i,j}$ represents the normalized offset vector between each parent-child joint pair based on the JRT in each video frame, and $f_{i,1} \cdots f_{i,n}$ represent the HOG feature vector for the i-th joint, computed from the subimage centered on its parent joint, in each video frame. The bottom part represents the optical flows computed, the images warped according to the optical flows, and the absolute frame difference images computed. More HOG features are further computed on the frame difference images. **Right**: Training of 14 regressors. For each edge in the JRT (See Fig. 1), we train a regressor that maps the local features around the parent joint to an offset between the parent and the child. The training of regressors is based on the extracted training data, *i.e.*, HOG features and offsets, as illustrated in **Center**.

Fig. 3. Human regions detected on several video frames using the SVM and HOG approach.

Denoting the image of the t-th frame as I_t, and the estimated position of the i-th joint in I_t as $J_{t,i}$. With each parent-child pair, (i,j), in the JRT, we associate a mapping function, $G_{i,j}(X_{t,i})$, which gives the normalized offset vector from $J_{t,i}$ to $J_{t,j}$ based on the spatially and temporally local feature vector, $X_{t,i}$, around $J_{t,i}$ in I_t. That is,

$$J_{t,j} = G_{i,j}(X_{t,i}) \cdot S_t + J_{t,i} \tag{1}$$

In Eq. 1, S_t is the normalizing factor used when training $G_{i,j}(\cdot)$. We define S_t as proportional to the width of the detected human region in I_t and $S_t = W_t/K$ ($K = 64$ in our method) where W_t is the width of the detected human region in I_t. The mapping function, $G_{i,j}(\cdot)$, is the regression function that we learn from manual annotations of the joint positions in the training video frames. How the feature vector, $X_{t,i}$, is computed and how the regression function, $G_{i,j}(\cdot)$, is learned are described in the following subsections.

3.3   Local Features

As described in Sect. 3.2, the regression between each parent-child joint pair is based on spatially and temporally local features around the parent. Therefore, for the i-th joint in the t-th frame, we need to compute a local feature vector, $X_{t,i}$, around $J_{t,i}$, as detailed below.

In order to characterize the spatially local features around $J_{t,i}$, we take a $K \times K$ ($K = 64$ in our method) appearance patch, $I_{t,i}$, centered on $J_{t,i}$ from I_t, and then compute the HOG feature vector [20], $H_{t,i}$, of $I_{t,i}$ as

$$H_{t,i} = [f_t(i,1), f_t(i,2), \cdots, f_t(i,n)] \tag{2}$$

where n is the dimensionality of the vector and $f_t(i,j)$ $(j = 1, \cdots, n)$ is an item of the HOG feature descriptors.

As we known from previous researches, motion feature can enhance the performance of still image based pose estimation methods by utilizing the temporal correlation between temporally close frames. Motion features are particularly helpful in two cases that are often hard for still image based methods: (1) occluded body part, and (2) coloring/illumination similarity between a body part and its background. For these two cases, motions from previous frames will provide further hints on the affected joint's position in the problematic frame.

In order to characterize the temporally local features around $J_{t,i}$, we first compute Lucas-Kanade [21] optical flows between two frames. Denote the optical flow field from frame I_t to frame I_{t-n} as $U_{t,t-n}$ ($n < t$). We obtain the warped image I_{t-n}^t which is the frame I_{t-n} warped to the frame I_t by using bilinear interpolation with the flow field $U_{t,t-n}$. Then we compute an array, $M_{t,i}$, of motion patches as:

$$M_{t,i} = \begin{bmatrix} |I_{t,i} - I_{t-1,i}^{t,i}| \\ |I_{t,i} - I_{t-2,i}^{t,i}| \end{bmatrix} \tag{3}$$

where $I_{t-n,i}^{t,i}$ $(n = 1, 2)$ is the warped image patch from $I_{t-n,i}$ to $I_{t,i}$ by using the optical flow field $U_{t,t-n}$ from I_t to I_{t-n}. As an example, we illustrate in Fig. 4 the process of computing a motion patch from two adjacent video frames.

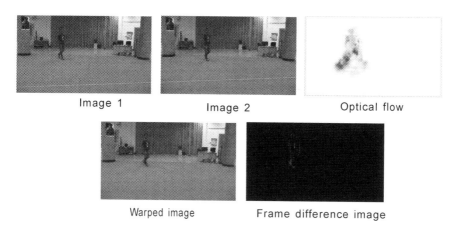

Image 1   Image 2   Optical flow

Warped image   Frame difference image

Fig. 4. The process of motion patch computation. The first row shows two adjacent video frames and their optical flow image obtained by the Lucas-Kanade optical flows algorithm; the second row shows the warped image from image 1 using the optical flow and the absolute frame difference image between image 2 and the warped image.

After the motion patch array, $M_{t,i}$, is obtained, we compute the HOG feature vectors, $H'_{t,i}$ and $H''_{t,i}$, from $M_{t,i}(1)$ and $M_{t,i}(2)$, respectively. Thereafter, we normalize $H_{t,i}$, $H'_{t,i}$ and $H''_{t,i}$. Still denoting the normalized HOG feature vectors as $H_{t,i}$, $H'_{t,i}$ and $H''_{t,i}$, we finally set $X_{t,i} = (H_{t,i}, H'_{t,i}, H''_{t,i})$. It is worth noting that $H_{t,i}$ is the spatial feature descriptor that has also been used by Hara and Chellappa [1], while $H'_{t,i}$ and $H''_{t,i}$ form the motion feature descriptor that we propose in this work.

In general, a frame closer to the t-th in time has a higher impact on the pose estimation accuracy for the t-th frame. Therefore, we make higher-resolution HOG descriptions for frames closer to the t-th, which is achieved by controlling cell and block sizes in the HOG computation. Specifically, we set the cell sizes to 8×8, 16×16 and 32×32 for the computation of $H_{t,i}$, $H'_{t,i}$ and $H''_{t,i}$, respectively, while setting the block size to 2×2 for all the three HOGs. In each cell, the number of orientation bins with signed gradients is set to 9. As a result, we get a 1,764-dimensional $H_{t,i}$, a 324-dimensional $H'_{t,i}$, a 36-dimensional $H''_{t,i}$, and a 2,124-dimensional $X_{t,i}$.

3.4   Multi-dimensional Boosting Regression

We use a training set of manually annotated video sequences to learn the regression function between any parent-child pair in the JRT. All the human joints' 2D positions are marked on each frame of each training video sequence. We assume that all the training video sequences contain N frames in total, forming a training frame set, T.

Let us focus on learning the regression between one parent-child pair, (m, n), while the same process applies to all the other parent-child pairs as well. From each image, $I_i \in T$ $(1 \le i \le N)$, we compute the local feature vector,

X_i, around the m-th joint using the method described in Sect. 3.3, and compute the normalized offset between the two annotated joints positions, $J_{i,m}$ and $J_{i,n}$, as $Y_i = (J_{i,n} - J_{i,m})/S_i$. Now that we have a set of training samples $\{Y_i, X_i\}_{i=1}^N$, the learning is conducted via a standard multi-dimensional boosting regression process, as described in the following.

In general, given a set of training samples $\{Y_i, X_i\}_{i=1}^N$, where $Y \in R^v$ is the output vector and $X \in R^u$ is the input vector. The regression function can be theoretically sought by:

$$F^*(X) = \arg\min_{F(X)} \sum_{i=1}^N \omega_i \Psi(Y_i, F(X_i)) \tag{4}$$

where ω_i is the weight of the i-th training sample and $\Psi(\cdot)$ is the loss function.

In order to achieve the goal of Eq. 4, we may construct the strong regressor $F(X)$ as an ensemble of weak regressors $h(X; A_m, R_m)$:

$$F(X) = \sum_{m=0}^M h(X; A_m, R_m) \tag{5}$$

where $h(X; A_m, R_m) = \sum_{l=1}^L (A_{ml} \cdot 1_{R_{ml}}(X \in R_{ml}))$ is a regression tree with indicator function $1_{R_{ml}}(X \in R_{ml})$, vectors $A_m = \{A_{m1}, A_{m2}, \cdots, A_{mL}\}$ and input space partitioning $R_m = \{R_{m1}, R_{m2}, \cdots, R_{mL}\}$, where each A_{ml} is the average of the output vectors of the training samples that fall into space partition R_{ml}. In the training stage, the space partitioning is conducted iteratively. At each step, denoting the current space partitioning as $R_m = \{R_{m1}, R_{m2}, \cdots, R_{ml'}\}$ and the corresponding average output vectors as $A_m = \{A_{m1}, A_{m2}, \cdots, A_{ml'}\}$, we select one from the l' leaves with the largest sum of squared error E_{ml} for further partitioning, following the method in [1,18]. E_{ml} is defined as:

$$E_{ml} = \sum_{X_i \in R_{ml}} \omega_i \|Y_i - A_{ml}\|_2^2 \tag{6}$$

Further, we apply the Gradient TreeBoost algorithm [18] as follows:

$$F_m(X) = F_{m-1}(X) + vh(X; A_m, R_m) \quad (m \geq 1) \tag{7}$$

where v $(0 < v < 1)$ is a shrinkage parameter and can control the learning rate, and $F_0(X) = mean\{Y_i\}_{i=1,2,\cdots,N}$. The parameters, (A_m, R_m), of a weak regressor is determined by:

$$(A_m, R_m) = \arg\min_{A,R} \sum_{i=1}^N \omega_i \Psi(Y_i, F_{m-1}(X_i)$$
$$+ vh(X_i; A, R)) \tag{8}$$

To summarize, the eventual regressor we construct is

$$F(X) = F_0(X) + v \sum_{m=1}^M h(X; A_m, R_m) \tag{9}$$

Algorithm 1. Multi-dimensional Gradient Boosting Regression

Input: A set of training samples $\{Y_i, X_i\}_{i=1}^N$
Output: The strong regressor $F_M(X)$
1: $F_0(X) = mean\{Y_i\}_{i=1,2,\cdots,N}$
2: **for** $m = 1$ to M **do**
3:    $\tilde{Y}_i = Y_i - F_{m-1}(X_i), i = 1, \ldots, N$
4:    $(A_m, R_m) = \arg\min_{A,R} \sum_{i=1}^N \omega_i ||\tilde{Y}_i - h(X_i; A, R)||_2^2$
5:    $F_m(X) = F_{m-1}(X) + \upsilon h(X; A_m, R_m)$
6: **end for**
7: **return** $F_M(X)$

---

and the overall multi-dimensional boosting regression process is put in Algorithm 1.

4   Experimental Results

4.1   Dataset and Metric

Due to the unavailability of annotated monocular video database for full-body human pose estimation, we take video sequences with a DV camcorder to construct our own dataset. Each video sequence is taken of one person. The dataset contains 1,200 image frames in total, each sized at 960×540 pixels and annotated. In those videos the actors perform many different full-body actions such as walk, parade step, run, jump, one-hand wave, two-hand wave and so on. Some samples of the image frames in our dataset are shown in Fig. 5. To annotate a video frame, we run human region detection algorithm in it and manually mark the positions of the 14 joints (in accordance with the JRT structure) inside the detected human region. As illustrated in Fig. 1, a human body pose is then represented as 10 sticks each connecting two joints in the image, which are head, torso, upper and lower arms and upper and lower legs. Of all the 1,200 image frames, we apply the 5-fold cross validation and use 800 for training and 400 for testing.

As the performance metric, we adopt the percentage of correctly estimated body parts(PCP) tool [1,5,22,23]. With a PCP_t metric, it is considered correct if the estimated stick's endpoints lie within $100t\,\%$ the length of the ground-truth stick from their ground-truth (annotated) locations.

4.2   Settings and Results

We need to train 14 regressors according to the JRT. In our experiments, for each boosting regression model, we set the number of trees as $M = 1000$, the number of leaves in each tree as $L = 5$ and the shrinkage parameter as $\nu = 0.1$.

Fig. 5. Some samples of original image frames from videos in our dataset, corresponding to the motions of walk, parade step, run, jump, one-hand wave, and two-hand wave.

In our experiments, we test four types of local feature vectors for the regression. In addition to the 2,124-dimensional HOG and optical-flow-temporal-difference(HOG-OFTD) features as introduced in Sect. 3.3, we also test other three feature vectors: one-scale-spatial (OSS), multi-scale-spatial (MSS) [1] and HOG-temporal-difference(HOG-TD) feature vectors. OSS computes the HOG of the local appearance patch with a cell size of 8×8 and a block size of 2×2, resulting in a 1,764-dimensional feature vector. MSS computes the HOG of the local appearance patch with a block size of 2×2 and cell sizes of 8×8, 16×16, 32×32, and concatenate the feature vectors for all these cell sizes to form a 2,124-dimensional feature vector. In order to test it for full body pose estimation, we run the original method in [1] on our proposed JRT structure but still use the MSS HOG features as used in [1]. HOG-TD computes the HOG of the local appearance patch with a cell size of 8×8 and a block size of 2×2, and the HOG of the motion patch, $M_{t,i} = |I_{t,i} - I_{t+1,i}|$, with a block size of 2×2 and cell size of 16×16, resulting in a 2,048-dimensional feature vector.

Using the four types of local feature vectors and $PCP_{0.5}$ as the performance metric, we obtain the statistics as given in Table 1 for head, upper arms and forearms, torso and upper and lower legs. From Table 1, we observe that HOG-OFTD yields the best performance on left upper arm, right and left forearms, left upper leg, right and left lower legs. Statistics in Table 1 demonstrates that, for most of the body parts, introducing the optical flow and frame difference as motion features leads to improved results over pure spatial features.

Further, we give in Table 2 the statistics about the average $PCP_{0.5}$ on our dataset. From Table 2, we see that HOG-OFTD leads to the best average estimation accuracy.

Table 1. $PCP_{0.5}$ statistics for four types of features (OSS, MSS, HOG-TD and HOG-OFTD) on our dataset. Results are given for 10 human body parts: head, right and left upper arms and forearms, torso, right and left upper legs and lower legs. R and L stands for right and left, respectively; u.a and l.a standards for upper and lower arm, respectively; u.l and l.l standards for upper and lower leg, respectively.

	Head	R.u.a	R.l.a	L.u.a	L.l.a	Torso	R.u.l	R.l.l	L.u.l	L.l.l
OSS	98.18	78.73	46	79.64	34.54	100	96.18	76	92.36	76.36
MSS [1]	98.73	83.82	19.82	82.91	43.27	100	96.36	77.27	93.45	77.45
HOG-TD	**99.78**	**88**	21.78	83.78	36	100	**97.56**	77.11	98.22	88.22
HOG-OFTD	98.45	82.72	**58.32**	**86.11**	**53.44**	100	96.44	**80.22**	**98.75**	**90.74**

Table 2. Average $PCP_{0.5}$ for four types of features (OSS, MSS, HOG-TD and HOG-OFTD) on our dataset.

Features	Average $PCP_{0.5}$
OSS	77.80
MSS [1]	77.31
HOG-TD	79.04
HOG-OFTD	**84.52**

Table 3. Time per image frame excluding human detection for two types of features (MSS, HOG-OFTD) on our dataset.

Features	Time/frame
MSS [1]	0.9 s
HOG-OFTD	1.2 s

In the testing phase, the running time of regression on each tree of height h is $O(h)$, since we follow a simple path down the regression tree. So the running time for each boosting regression model with M trees is $O(Mh)$. We give in Table 3 the statistics about the average timing per frame excluding human detection on our dataset. We implemented our scheme in matlab language. Running our implementation on a desktop computer with an Intel Core(TM)i5 3.10 GHz CPU and 4 GB memory.

Visual results of the full-body pose estimation using HOG-OFTD on a selected set of video frames are shown in Fig. 6.

From this figure, we see that the overall human poses are estimated with a good accuracy, though some failure cases exist locally. Those failure cases mainly happen in regions with self-occlusion and/or fast motion, for which insufficient information can be obtained and/or more randomness exists, adding to the difficulty of accurate estimation.

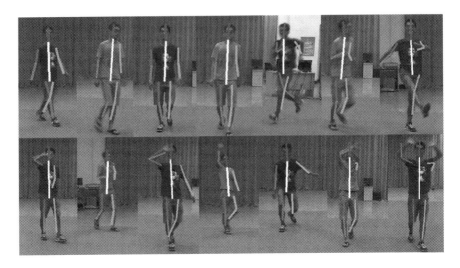

Fig. 6. Visual estimation results of our scheme with HOG-OFTD on selected frames.

5   Conclusion and Future Work

In this paper, we propose a scheme to estimate 2D full-body human poses from monocular video sequences. At each frame, it detects the human region using an SVM and HOG human detection algorithm and then estimates the human pose in the detected region through multi-dimensional boosting regression. Specifically, we design a joint relationship tree reflecting the hierarchical structure of joints in a human body. In the training stage, we learn a regressor for each parent-child pair, which estimates the child joint's offset vector based on spatially and temporally local features around the parent; in the testing stage, we first fix the location of the root node relative to the detected human region, and then traverse the JRT in a depth-first order to estimate all the joints' positions utilizing the learned regressors. As experimentally demonstrated, the proposed scheme achieves outstanding estimation performance.

In the future, while further improving the estimation accuracy, we will increase the diversity of our datasets and seek to accelerate the computation for potential use in real-time applications.

Acknowledgement. This work is partially supported by Shandong Provincial Natural Science Foundation, China (Grant No. ZR2011FZ004), the National Natural Science Foundation of China (Grants No. 61472223, U1035004 and 61303083), the Scientific Research Foundation for the Excellent Middle-Aged and Youth Scientists of Shandong Province of China (Grant No. BS2011DX017) and the Program for New Century Excellent Talents in University (NCET) in China.

References

1. Hara, K., Chellappa, R.: Computationally efficient regression on a dependency graph for human pose estimation. In: Computer Vision and Pattern Recognition, pp. 3390–3397 (2013)
2. Moeslund, T.B., Hilton, A., Krüger, V.: A survey of advances in vision-based human motion capture and analysis. Comput. Vis. Image Underst. **104**, 90–126 (2006)
3. Poppe, R.: Vision-based human motion analysis: an overview. Comput. Vis. Image Underst. **108**, 4–18 (2007)
4. Felzenszwalb, P.F., Huttenlocher, D.P.: Pictorial structures for object recognition. Int. J. Comput. Vis. **61**, 55–79 (2005)
5. Eichner, M., Marin-Jimenez, M., Zisserman, A., Ferrari, V.: 2d articulated human pose estimation and retrieval in (almost) unconstrained still images. Int. J. Comput. Vis. **99**, 190–214 (2012)
6. Andriluka, M., Roth, S., Schiele, B.: Pictorial structures revisited: people detection and articulated pose estimation. In: Computer Vision and Pattern Recognition, pp. 1014–1021 (2009)
7. Sapp, B., Jordan, C., Taskar, B.: Adaptive pose priors for pictorial structures. In: Computer Vision and Pattern Recognition, pp.422–429 (2010)
8. Dantone, M., Gall, J., Leistner, C., Van Gool, L.: Human pose estimation using body parts dependent joint regressors. In: Computer Vision and Pattern Recognition, pp.3041–3048 (2013)
9. Pishchulin, L., Andriluka, M., Gehler, P., Schiele, B.: Strong appearance and expressive spatial models for human pose estimation. In: The IEEE International Conference on Computer Vision, pp. 3487–3494 (2013)
10. Zuffi, S., Freifeld, O.,Black, M.J.: From pictorial structures to deformable structures. In: Computer Vision and Pattern Recognition, pp. 3546–3553 (2012)
11. Zuffi, S., Romero, J., Schmid, C., Black, M.J.: Estimating human pose with flowing puppets. In: The IEEE International Conference on Computer Vision, pp. 3312–3319 (2013)
12. Sapp, B., Toshev, A., Taskar, B.: Cascaded models for articulated pose estimation. In: Daniilidis, K., Maragos, P., Paragios, N. (eds.) ECCV 2010, Part II. LNCS, vol. 6312, pp. 406–420. Springer, Heidelberg (2010)
13. Okada, R., Soatto, S.: Relevant feature selection for human pose estimation and localization in cluttered images. In: Forsyth, D., Torr, P., Zisserman, A. (eds.) ECCV 2008, Part II. LNCS, vol. 5303, pp. 434–445. Springer, Heidelberg (2008)
14. Girshick, R., Shotton, J., Kohli, P., Criminisi, A., Fitzgibbon, A.: Efficient regression of general-activity human poses from depth images. In: The IEEE International Conference on Computer Vision, pp. 415–422 (2011)
15. Shotton, J., Fitzgibbon, A., Cook, M., Sharp, T., Finocchio, M., Moore, R., Kipman, A., Blake, A.: Real-time human pose recognition in parts from single depth images. In: Computer Vision and Pattern Recognition, pp. 1297–1304 (2011)
16. Shotton, J., Sharp, T., Kipman, A., Fitzgibbon, A., Finocchio, M., Blake, A., Cook, M., Moore, R.: Real-time human pose recognition in parts from single depth images. Commun. ACM **56**, 116–124 (2013)
17. Sun, M., Kohli, P., Shotton, J.: Conditional regression forests for human pose estimation. In: Computer Vision and Pattern Recognition, pp. 3394–3401 (2012)
18. Bissacco, A., Yang, M.H., Soatto, S.: Fast human pose estimation using appearance and motion via multi-dimensional boosting regression. In: Computer Vision and Pattern Recognition, pp. 1–8 (2007)

19. Pang, Y., Yuan, Y., Li, X., Pan, J.: Efiicient HOG human detection. Sign. Process. **91**, 773–781 (2011)
20. Dalal, N., Triggs, B.: Histograms of oriented gradients for human detection. In: Computer Vision and Pattern Recognition, pp. 886–893 (2005)
21. Lucas, B.D., Kanade, T., et al.: An iterative image registration technique with an application to stereo vision. IJCAI **81**, 674–679 (1981)
22. Ferrari, V., Marin-Jimenez, M., Zisserman, A.: Progressive search space reduction for human pose estimation. In: Computer Vision and Pattern Recognition, pp. 1–8 (2008)
23. Yang, Y., Ramanan, D.: Articulated pose estimation with flexible mixtures-of-parts. In: Computer Vision and Pattern Recognition, pp. 1385–1392 (2011)

Improve Pedestrian Attribute Classification by Weighted Interactions from Other Attributes

Jianqing Zhu[✉], Shengcai Liao, Zhen Lei, and Stan Z. Li

Center for Biometrics and Security Research and National Laboratory
of Pattern Recognition, Institute of Automation, Chinese Academy of Sciences,
Beijing, China
jianqingzhu@foxmail.com,
{scliao,zlei,szli}@cbsr.ia.ac.cn

Abstract. Recent works have shown that visual attributes are useful in a number of applications, such as object classification, recognition, and retrieval. However, predicting attributes in images with large variations still remains a challenging problem. Several approaches have been proposed for visual attribute classification; however, most of them assume independence among attributes. In fact, to predict one attribute, it is often useful to consider other related attributes. For example, a pedestrian with *long hair* and *skirt* usually imply the *female* attribute. Motivated by this, we propose a novel pedestrian attribute classification method which exploits interactions among different attributes. Firstly, each attribute classifier is trained independently. Secondly, for each attribute, we also use the decision scores of other attribute classifiers to learn the attribute interaction regressor. Finally, prediction of one attribute is achieved by a weighted combination of the independent decision score and the interaction score from other attributes. The proposed method is able to keep the balance of the independent decision score and interaction of other attributes to yield more robust classification results. Experimental results on the Attributed Pedestrian in Surveillance (APiS 1.0) [1] database validate the effectiveness of the proposed approach for pedestrian attribute classification.

1   Introduction

The smart video surveillance technologies [2–4] including object detection [5,6], object tracking [7] and object classification [8], have attracted more and more attentions in public security field. Pedestrian related technique is one of the hottest topic in this field. Pedestrian detection [6,9], pedestrian tracking [10], behavior analysis [11] and clothing recognition [12] are all extensively studied in these years.

In this paper, we focus on pedestrian attribute classification [1] to present a more comprehensive description of pedestrian. As shown in Fig. 1, pedestrian attribute classification is to predict the presence or absence of several attributes. Pedestrian attributes used in surveillance application include *gender*, *hair*, *clothing appearance* and *carrying thing*, etc. The pedestrian attribute classification

© Springer International Publishing Switzerland 2015
C.V. Jawahar and S. Shan (Eds.): ACCV 2014 Workshops, Part III, LNCS 9010, pp. 545–557, 2015.
DOI: 10.1007/978-3-319-16634-6_40

can be used to provide useful information for applications such as pedestrian tracking, re-identification [13] and retrieval, etc. As shown in Fig. 2, attributes such as *male, hand carrying* and *back bag* can be effectively applied to assist locating the desired target in the pedestrian retrieval application.

Fig. 1. Pedestrian attribute classification describes pedestrians with a list of visual attributes.

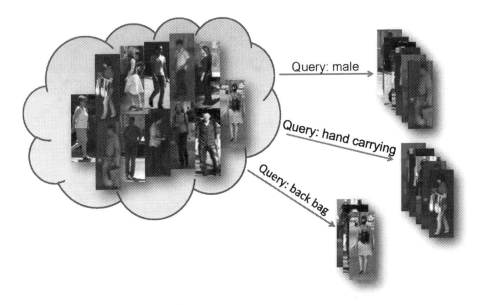

Fig. 2. Pedestrian retrieval based on attributes. Attributes such as *male, hand carrying* and *back bag* can be effectively applied to assist locating the desired target.

1.1   Related Work

Attributes are powerful to infer high-level semantic knowledge. There are many computer vision applications based on attribute, such as face verification [14], object recognition [15], clothing description [16], image retrieval [17] and scene classification [18], etc. The successes of these applications rely heavily on the accuracy of predicted attribute values (i.e. the decision scores of separated attribute classifiers). Kumar et al. [14] used semantic attributes as mid-level features to aid face verification. In this application, the prediction model of each attribute on an input image is first learned, and the supervised object models on top of those attribute predictions are then built.

The most popular method for attribute classification is to extract low-level features from an image, and then train classifier for each attribute separately. Daniel et al. [19] proposed an attribute-based people searching system in surveillance environments. In this application, people are identified by a series of attribute detectors. Each attribute detector is independently trained from large amounts of training data. Layne et al. [13] utilized scores from 15 attribute classifiers as mid-level representations to aid person re-identification, where each attribute classifier is independently trained by the SVM algorithm. The main drawback of separately training each attribute classifier is that it ignores the interactions between different attributes which are helpful for improving classification performance. In fact, there is the interaction issue among pedestrian attributes. For instance, if the *long hair* and *skirt* of a pedestrian have shown, the attribute *male* is unlikely to appear.

In order to build interaction models among different attributes, Chen et al. [16] explored the mutual dependencies between attributes by applying a Conditional Random Field (CRF) with the SVM margins from the separately trained attribute classifiers. Each attribute function is used as a node and the edge connecting every two attribute nodes reflects the joint probability of these two attributes. Their CRF is fully connected, which means that all attribute nodes are pairwise connected. In the CRF, it is assumed that the observed feature f_i is independent of all other features once the attribute a_i is known. However, this assumption is not always held, because two attributes may appear in the same region.

Bourdev et al. [20] used SVM algorithm to explore interactions between different attributes. Firstly, each attribute classifier is separately trained by SVM algorithm on a set of Poselets [21]. Then, they used the SVM algorithm learning on the scores of all separately trained attribute classifiers to capture interactions between different attributes. In other words, the final decision score of an attribute is constructed by linearly combining all decision scores that come from separately trained attribute classifiers and the linear coefficients are learned by SVM. However, since an attribute is most relevant to itself, the final decision score of an attribute in this interaction model will heavily rely on the decision score of its own attribute classifier, resulting in the role of other attributes is ignorable.

For that, a more effective pedestrian attribute classification is proposed in this paper by exploiting interactions among different attributes. The proposed method linearly combines the independent decision score and the interaction score of an attribute to yield the final decision score of attribute classification. The independent decision score is produced by an independently trained classifier. The interaction score of an attribute is learned by using Lasso regression algorithm on all independent decision scores excluding its own independent decision scores. For each attribute, the proposed approach introduces a weight parameter to control the contribution of interaction score, achieving more robust classification results. Experiment results on the APiS 1.0 database show that the proposed approach can exploit the interactions among different attributes more effectively than the interaction model proposed in [20].

The remaining of this paper is organized as follows. Section 2 introduces the details of the proposed pedestrian attribute classification method. Section 3 shows the experiment results on APiS 1.0 database. Finally, Sect. 4 concludes the paper.

2   Modeling Attribute Interaction for Pedestrian Attribute Classification

2.1   Feature Extraction

We apply a sliding window strategy for feature extraction. In each sub-window, a joint color histogram, a MB-LBP histogram and a Histogram of Oriented Gradient (HOG) are extracted. The color histogram has 8, 3 and 3 bins in the H, S, and V color channels, respectively. The MB-LBP [22] histogram includes 30 bins, 10 from 3×3 scale descriptor, 10 from 9×9 scale one, and 10 from 21×21 scale one. For the HOG feature extraction, each sub-window is equally divided into 2×2 sub-regions, and in each sub-region a histogram of oriented gradient is extracted with 9 orientation bins. The HOG feature associated with each sub-window is obtained by concatenating the above four histograms into a 36-dimensional vector. The diagram of feature extraction is shown in Fig. 3. The details of feature extraction are described in [1].

2.2   Independent Attribute Classifier

In this work, the Gentle AdaBoost [23] algorithm is chosen to independently train each attribute classifier. More specifically, we first concatenate the color, MB-LBP and HOG features, and then use Gentle AdaBoost to construct classifiers. We select the stump classifier with the minimum square error as the weak classifier in the Gentle AdaBoost algorithm.

2.3   Interaction Model

In order to exploit the interactions among different attributes, an interaction model is required. The interaction model firstly proposed in [20] is shown in Fig. 4. From this figure, we can find that each combined attribute classifier directly connected with all independent attribute classifiers. It means that each combined attribute classifier is constructed by linearly combining all independent attribute classifiers.

Assume that there are m attributes to predict; x is a testing sample; h_i is the i-th ($i \in \{1, 2, ..., m\}$) attribute classifier which is independently trained by

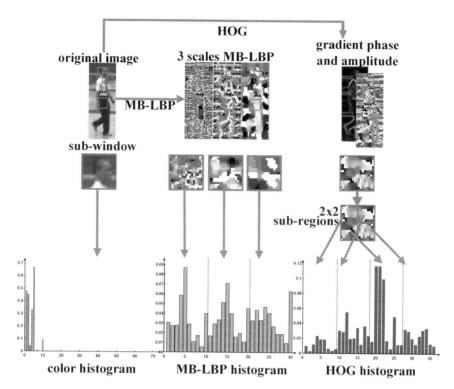

Fig. 3. The diagram of feature extraction.

using the Gentle AdaBoost algorithm; H_i represents the combined classifier for i-th attribute. H_i is calculated as follows:

$$H_i(\boldsymbol{x}) = \sum_{j=1}^{m} w_{ij} h_j(\boldsymbol{x}), \tag{1}$$

where the linear coefficients $\boldsymbol{w}_i = \{w_{ij}, j = 1, 2, ..., m\}$ can be learned by using the SVM algorithm. However, the problem is that an attribute is most relevant to itself, which may bring about such a fact that the combined decision score of an attribute in this method heavily relies on the independent decision score produced by its own independent attribute classifier, thus ignoring the role of other attributes. This drawback will be validated in our following experiments.

To address this problem, we propose our interaction model, as shown in Fig. 5. Different from that in Fig. 4, an combined attribute classifier consists of an independent classifier and an interaction regressor trained on other independent attribute decision scores. We further introduce parameters to control the weight of the interaction regressors.

In the proposed interaction model, H_i is learned as follows:

$$H_i(\boldsymbol{x}) = h_i(\boldsymbol{x}) + \lambda_i G_i(\boldsymbol{x}), \tag{2}$$

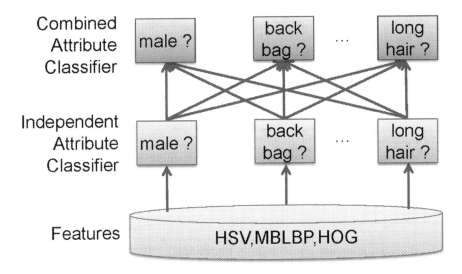

Fig. 4. The interaction model proposed in [20]. A combined attribute classifier is learned on the independent decision scores produced by all separated attribute classifiers.

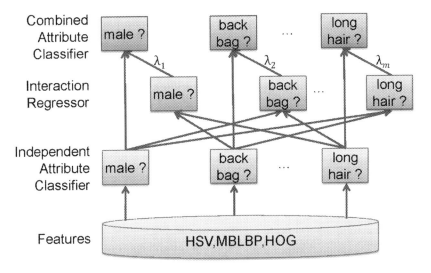

Fig. 5. The proposed interaction model. $\{\lambda_1, \lambda_2, ..., \lambda_m\}$ are used to control the weight of interactions.

where

$$G_i(\boldsymbol{x}) = \sum_{j=1, j \neq i}^{m} w_{ij} h_j(\boldsymbol{x}). \tag{3}$$

From Eq. (2), we can find that the interaction regressor of the i-th attribute G_i only involves $m-1$ attribute classifiers, excluding the i-th attribute classifier h_i. This strategy can directly avoid the problem of the combined decision score relying on h_i too heavily and capture the interactions among the rest attributes. Meanwhile, the parameter λ_i is used to keep the balance between h_i and G_i. If $\lambda_i = 0$, H_i will degrade into h_i.

Generally speaking, one attribute may only be related to a part of the rest attributes. Therefore, the linear coefficients $\boldsymbol{w}_i = \{w_{ij}, j = 1, 2, ..., m \text{ and } j \neq i\}$ in Eq. (3) should be sparse. With consideration of this potential sparse characteristic, the following objective formulation is designed:

$$\boldsymbol{w}_i = \arg\min_{w}\{\frac{1}{2}\|A_i\boldsymbol{w} - \boldsymbol{y}_i\|_2^2 + \gamma_i\|\boldsymbol{w}\|_1\}, \tag{4}$$

where A_i represents the independent decision scores of $\{h_j, j = 1, ..., m \text{ and } j \neq i\}$ on training set; $\boldsymbol{y}_i$ represents the i-th attribute labels of the training set; γ_i is a parameter used to control the sparseness of $\boldsymbol{w}_i$. The larger γ_i is, the more sparsely $\boldsymbol{w}_i$ will be. Assume that the training set includes n samples, then A_i is organized as a matrix with $n \times (m-1)$ dimensions and $\boldsymbol{y}_i$ is a column vector with $n \times 1$ dimensions.

Equation (4) is the Lasso [24] problem, which formulates the least-square estimation problem with l_1-norm penalty to approximate the sparse representation solution. If $\gamma_i = 0$, Eq. (4) will degrade into the least-square estimation problem. In our implementation, the Dual Augmented Lagrangian (DAL) algorithm [25] is used to solve Eq. (4). In our experiments, we determinate λ_i and γ_i with cross validation, because cross validation is a simple and general way used to find appropriate parameters.

3   Experiments

To evaluate the performance of the proposed interaction model, it is compared with the baseline [1] and the interaction model proposed in [20] on Attributed Pedestrian in Surveillance (APiS 1.0) [1] database under the same evaluation protocol. The APiS 1.0 database has 3661 images, and each image is labeled with 11 binary attribute annotations. The linear coefficients $\boldsymbol{w}$ in Eqs. (1) and (3) are learned by the Lasso algorithm. Since APiS 1.0 database does not provide validation sub-set, we randomly divide the APiS 1.0 database into 5 equal sized sub-sets (the partition is different with that in [1]) for the selection of λ and γ. Specially, the best pair of parameters is the one whose corresponding result has the largest Area Under ROC curve (AUC). Based on the color, MB-LBP and HOG features extraction described in Sect. 2.1, each attribute classifier independently trained by the Gentle AdaBoost algorithm includes 3,000 weak classifiers as suggested in [1]. Note that both the interaction model proposed in [20] and our proposed interaction model are built on attribute scores predicted from the same feature representation.

3.1   Average Recall Rate Comparison

Table 1 lists the comparison of average recall rates when the average false positive rates are 0.1. We can find that there are only 3 of 11 attributes achieve higher average recall rates than the baseline method [1] when using the interaction model proposed in [20], and the biggest improvement increased by only 2.78% recall rate for *long pants* attribute. However, the proposed model offers higher average recall rates for 9 of 11 attributes and 6 of them have obvious improvements (1.90% for *M-S pants*, 8.45% for *long pants*, 6.25% for *skirt*, 3.04% for *male*, 3.18% for *long hair* and 4.47% for *S-S bag*).

From Table 1, we can find that the proposed method fails to improve the average recall rates of *long jeans* and *hand carrying* attributes; however, the proposed method has equal performance with the baseline method [1] for *hand carrying* attribute and very minor degradations for *long jeans* attribute. These results validate that the strategy of introducing parameters to control the weight of interaction score is robust. Though it may not improve performance for some attributes, it will not cause significant performance degradation. In addition, compared with the baseline method, the average improvements in recall rates obtained by the interaction model proposed in [20] and the proposed model are 0.30% and 2.58%, respectively. We also tried a least square solution of our model, corresponding to $\gamma_i = 0$ in Eq. (4). As a result, the sparse solution has a marginal mean accuracy improvement (0.47%) over the least square solution. Nevertheless, the Lasso model has a better interpretation of finding the most

Table 1. The comparison of average recall rates when the average false positive rates are 0.1. Where *M-S pants* is the abbreviation of Medium and Short pants, and *S-S bag* is the abbreviation of Single Shoulder bag.

Attribute	Recall rate(%)		
	Baseline [1]	Interaction model in [20]	The proposed
long jeans	**89.85**	89.74	89.18
M-S pants	78.65	78.65	**80.55**
long pants	76.68	79.46	**85.13**
skirt	68.23	67.71	**74.48**
male	58.30	57.53	**61.34**
back bag	56.16	56.16	**56.51**
T-shirt	55.22	56.36	**56.47**
long hair	55.15	55.15	**58.33**
shirt	54.62	54.22	**54.82**
hand carrying	**52.14**	**52.14**	**52.14**
S-S bag	38.45	39.64	**42.92**
average improvements(%)	-	0.30	**2.58**

relevant interaction between attributes. To sum up, our method can achieve better recall rate performances compared with the interaction model proposed in [20].

3.2 Average ROC Comparison

Figure 6 compares the average ROC curves of all attribute defined in APiS 1.0 database. The AUC sum of 11 attributes obtained by the baseline method [1], the interaction model proposed in [20] and the proposed model are 9.5371, 9.5372 and 9.5935, respectively. This shows that the interaction model proposed in [20] almost does not obtain performance improvements with respect to the baseline method [1], while the proposed method achieves performance improvements. It can be seen that the proposed model offers larger AUC for 9 of 11 attributes and 5 attributes (*long pants, M-S pants, skirt, long hair* and *male*) obtain obvious improvements. For *back bag* and *shirt* attributes, the proposed method fails to improve their performances. However, the propose method has equal performance with the baseline method [1] for *back bag* attribute and very minor degradations for *shirt* attribute. These results also validate that the strategy of introducing parameters to control the weight of interaction part is robust.

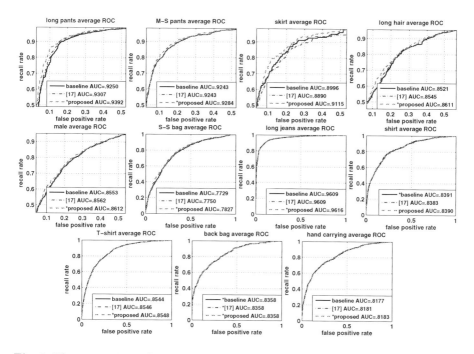

Fig. 6. The comparison of average ROC curves. Where *M-S pants* is the abbreviation of Medium and Short pants, and *S-S bag* is the abbreviation of Single Shoulder bag; '*' indicates the corresponding average ROC curve has maximum AUC value.

3.3   Interaction Analysis

We provide visualizations of the coefficients learned by the interaction model proposed in [20] and our method, as shown in Figs. 7 and 8. From Fig. 7, we can find that each attribute is most relevant to itself, because the corresponding absolute coefficient has the maximum value. This indicates that the combined decision score of a given attribute heavily relies on the independent decision score produced by its corresponding attribute classifier, thus ignoring the role of other attributes.

As shown in Fig. 8, in our model, the interaction part of an given attribute excludes the corresponding attribute classifier and only involves 10 other attribute classifiers. This strategy effectively avoids the situation that the combined decision score of a given attribute heavily relies on its independent decision score, and then capture the interactions from the rest attributes. Therefore, from Fig. 8 we can find that *long jeans* attribute is highly positive correlated with *long pants* attribute; *long hair* attribute is highly negative correlated with *male* attribute and *skirt* attribute is highly positive correlated with *longhair* attribute.

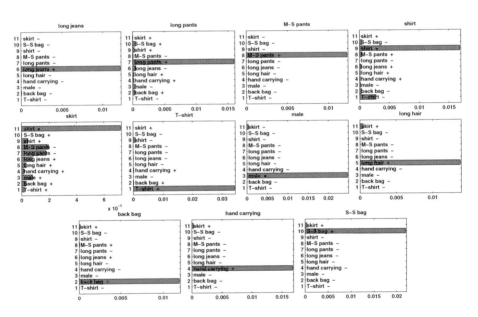

Fig. 7. The absolute coefficients learned by the interdependency model proposed in [20]. Where *M-S pants* is the abbreviation of Medium and Short pants, and *S-S bag* is the abbreviation of Single Shoulder bag; '+' and '−' represent the two attributes hold positive correlation and negtive correlation relationship, respectively.

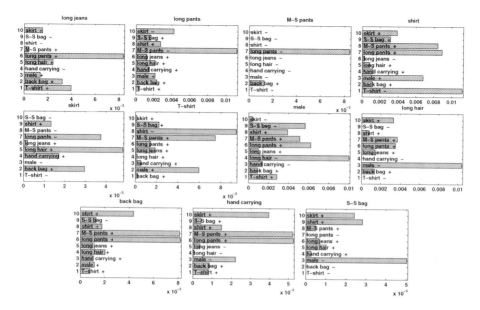

Fig. 8. The absolute coefficients learned by the proposed model. Where *M-S pants* is the abbreviation of Medium and Short pants, and *S-S bag* is the abbreviation of Single Shoulder bag; '+' and '−' represent the two attributes hold positive correlation and negtive correlation relationship, respectively.

4   Conclusion

This paper has proposed a novel method for pedestrian attribute classification which exploits interactions among different attributes. In our method, prediction of one attribute is achieved by a weighted combination of the independent decision score and the interaction score. The independent decision score of an attribute is obtained from a classifier independently trained by using the Gentle AdaBoost algorithm. The interaction score of an attribute is obtained from a regressor trained by using Lasso algorithm on all the rest independent decision scores. The proposed method further introduces weight parameters to keep the balance of the independent decision score and the interaction score. Experimental results on APiS 1.0 database have shown that the interactions among different attributes have effectively improved the attribute classification performance.

Acknowledgement. This work was supported by the Chinese National Natural Science Foundation Projects #61105023, #61103156, #61105037, #61203267, #61375037, #61473291, National Science and Technology Support Program Project #2013BAK02B01, Chinese Academy of Sciences Project No. KGZD-EW-102-2, and AuthenMetric R&D Funds.

References

1. Zhu, J., Liao, S., Lei, Z., Yi, D., Li, S.Z.: Pedestrian attribute classification in surveillance: database and evaluation. In: ICCV Workshop on Large-Scale Video Search and Mining (2013)
2. Shu, C.F., Hampapur, A., Lu, M., Brown, L., Connell, J., Senior, A., Tian, Y.: IBM smart surveillance system (s3): a open and extensible framework for event based surveillance. In: IEEE Conference on Advanced Video and Signal Based Surveillance, pp. 318–323 (2005)
3. Hampapur, A., Brown, L., Connell, J., Ekin, A., Haas, N., Lu, M., Merkl, H., Pankanti, S.: Smart video surveillance: exploring the concept of multiscale spatiotemporal tracking. IEEE Signal Process. Mag. **22**, 38–51 (2005)
4. Sedky, M.H., Moniri, M., Chibelushi, C.C.: Classification of smart video surveillance systems for commercial applications. In: IEEE Conference on Advanced Video and Signal Based Surveillance, pp. 638–643 (2005)
5. Gu, C., Arbeláez, P., Lin, Y., Yu, K., Malik, J.: Multi-component models for object detection. In: Fitzgibbon, A., Lazebnik, S., Perona, P., Sato, Y., Schmid, C. (eds.) ECCV 2012, Part IV. LNCS, vol. 7575, pp. 445–458. Springer, Heidelberg (2012)
6. Yan, J., Lei, Z., Yi, D., Li, S.Z.: Multi-pedestrian detection in crowded scenes: a global view. In: IEEE Conference on Computer Vision and Pattern Recognition, pp. 3124–3129 (2012)
7. Yang, B., Nevatia, R.: Online learned discriminative part-based appearance models for multi-human tracking. In: Fitzgibbon, A., Lazebnik, S., Perona, P., Sato, Y., Schmid, C. (eds.) ECCV 2012, Part I. LNCS, vol. 7572, pp. 484–498. Springer, Heidelberg (2012)
8. Hariharan, B., Malik, J., Ramanan, D.: Discriminative decorrelation for clustering and classification. In: Fitzgibbon, A., Lazebnik, S., Perona, P., Sato, Y., Schmid, C. (eds.) ECCV 2012, Part IV. LNCS, vol. 7575, pp. 459–472. Springer, Heidelberg (2012)
9. Dalal, N., Triggs, B.: Histograms of oriented gradients for human detection. In: IEEE Conference on Computer Vision and Pattern Recognition, pp. 886–893 (2005)
10. Guan, Y., Chen, X., Wu, Y., Yang, D.: An improved particle filter approach for real-time pedestrian tracking in surveillance video. In: International Conference on Information Science and Technology Applications, Atlantis Press (2013)
11. Jackson, S., Miranda-Moreno, L.F., St-Aubin, P., Saunier, N.: A flexible, mobile video camera system and open source video analysis software for road safety and behavioural analysis. In: Transportation Research Board 92nd Annual Meeting (2013)
12. Yang, M., Yu, K.: Real-time clothing recognition in surveillance videos. In: IEEE International Conference on Image Processing, pp. 2937–2940 (2011)
13. Layne, R., Hospedales, T., Gong, S., Mary, Q.: Person re-identification by attributes. In: British Machine Vision Conference (2012)
14. Kumar, N., Berg, A.C., Belhumeur, P.N., Nayar, S.K.: Attribute and simile classifiers for face verification. In: IEEE Conference on International Conference on Computer Vision, pp. 365–372 (2009)
15. Farhadi, A., Endres, I., Hoiem, D., Forsyth, D.: Describing objects by their attributes. In: IEEE Conference on Computer Vision and Pattern Recognition, pp. 1778–1785 (2009)
16. Chen, H., Gallagher, A., Girod, B.: Describing clothing by semantic attributes. In: Fitzgibbon, A., Lazebnik, S., Perona, P., Sato, Y., Schmid, C. (eds.) ECCV 2012, Part III. LNCS, vol. 7574, pp. 609–623. Springer, Heidelberg (2012)

17. Yu, F.X., Ji, R., Tsai, M.H., Ye, G., Chang, S.F.: Weak attributes for large-scale image retrieval. In: IEEE Conference on Computer Vision and Pattern Recognition, pp. 2949–2956 (2012)

18. Li, L.-J., Su, H., Lim, Y., Fei-Fei, L.: Objects as attributes for scene classification. In: Kutulakos, K.N. (ed.) ECCV 2010 Workshops, Part I. LNCS, vol. 6553, pp. 57–69. Springer, Heidelberg (2012)

19. Vaquero, D., Feris, R., Tran, D., Brown, L., Hampapur, A., Turk, M.: Attribute-based people search in surveillance environments. In: IEEE Workshop on Applications of Computer Vision (2009)

20. Bourdev, L., Maji, S., Malik, J.: Describing people: A poselet-based approach to attribute classification. In: IEEE International Conference on Computer Vision, pp. 1543–1550 (2011)

21. Bourdev, L., Malik, J.: Poselets: Body part detectors trained using 3d human pose annotations. In: IEEE International Conference on Computer Vision, pp. 1365–1372 (2009)

22. Liao, S.C., Zhu, X.X., Lei, Z., Zhang, L., Li, S.Z.: Learning multi-scale block local binary patterns for face recognition. In: Lee, S.-W., Li, S.Z. (eds.) ICB 2007. LNCS, vol. 4642, pp. 828–837. Springer, Heidelberg (2007)

23. Friedman, J., Hastie, T., Tibshirani, R.: Additive logistic regression: a statistical view of boosting (with discussion and a rejoinder by the authors). Ann. Stat. **28**, 337–407 (2000)

24. Tibshirani, R.: Regression shrinkage and selection via the lasso. J. Roy. Stat. Soc. Ser. B (Methodological) **58**, 267–288 (1996)

25. Tomioka, R., Suzuki, T., Sugiyama, M.: Super-linear convergence of dual augmented-lagrangian algorithm for sparsity regularized estimation (2009). arXiv preprint arXiv:0911.4046

Face Recognition with Image Misalignment via Structure Constraint Coding

Ying Tai, Jianjun Qian, Jian Yang$^{(\boxtimes)}$, and Zhong Jin

School of Computer Science and Engineering, Nanjing University of Science
and Technology, Nanjing, China
tyshiwo@gmail.com, qjjtx@126.com, {csjyang,zhongjin}@njust.edu.cn

Abstract. Face recognition (FR) via sparse representation has been widely studied in the past several years. Recently many sparse representation based face recognition methods with simultaneous misalignment were proposed and showed interesting results. In this paper, we present a novel method called structure constraint coding (SCC) for face recognition with image misalignment. Unlike those sparse representation based methods, our method does image alignment and image representation via structure constraint based regression simultaneously. Here, we use the nuclear norm as a structure constraint criterion to characterize the error image. Compared with the sparse representation based methods, SCC is more robust for dealing with illumination variations and structural noise (especially block occlusion). Experimental results on public face databases verify the effectiveness of our method.

1 Introduction

Face recognition is a classical problem in computer vision. Given a face image, we know that its appearance may be affected by many variances, such as illumination, pose, facial expression, noise (i.e. occlusion, corruption and disguise) and so on. More recently, a new face recognition framework called sparse representation based classification (SRC) was proposed [1], which casts the recognition problem as seeking a sparse linear representation of the query image over the training images. Generally speaking, SRC shows good robustness to many of the above problems, and its success inspires many extensive works [2–4]. However, SRC needs the images in both training set and testing set to be well-aligned. It means that the performance of SRC will deteriorate a lot when dealing with the images with misalignment. Additionally, many other FR methods also suffer from misalignment. Therefore, face alignment plays an important role in a practical face recognition system.

A lot of work has been done toward the face alignment problem, where face images are aligned to a fixed canonical template. The work in [5] is proposed to seek an optimal set of image domain transformations such that the matrix of transformed image can be decomposed as the sum of a sparse matrix of errors and a low-rank matrix of recovered aligned images. The work in [6] is derived from [7]. Since [5] needs to readjust all the transformations of previous images to minimize

© Springer International Publishing Switzerland 2015
C.V. Jawahar and S. Shan (Eds.): ACCV 2014 Workshops, Part III, LNCS 9010, pp. 558–573, 2015.
DOI: 10.1007/978-3-319-16634-6_41

the rank when a new image is coming, which is very time-consuming when the image set is large. In [6], an optimal alignment for the newly arriving image was sought so that after alignment the new image could be linearly reconstructed by previously well-aligned image basis. However, these two methods are just for alignment. In [7], Wagner et al. proposed a novel method for face alignment and recognition. They sought the transformation of test image via subject-by-subject exhaustive search and got some impressive results, while it is proved to be time-consuming. Yang et al. [8] presented a novel face recognition method, named misalignment robust representation (MRR). MRR sought the optimal alignment through an efficient two-step optimization with a coarse-to-fine search strategy. It uses l_1-norm constraint on the representation residual, which is regarded as a reason for its effectiveness [9]. As a work derived from [7], MRR achieves similar results but much faster. More recently, Zhuang et al. [10] sought additional illumination examples of face images from other subjects to form an illumination dictionary for single-sample face alignment and recognition.

However, the models mentioned above are all vector-based models, which need to convert images into vectors before dealing with 2D images in the form of matrices. In the process of converting, some structural information (e.g. the rank of matrix) might be lost. As mentioned in [11], Yang et al. proposed a model named nuclear norm based matrix regression (NMR) and employed nuclear norm constraint as a criterion to make full use of the low-rank structural information caused by some occlusion and illumination changes. They presented some interesting results in [11], which reveals nuclear norm constraint is a better choice than l_1-norm or l_2-norm constraint when dealing with structural noise. However, NMR also concentrates on the recognition problem of the aligned face images. In this paper, we also perform the nuclear norm constraint on the error image and propose a method called structure constraint coding (SCC) for face recognition with image misalignment. Compared with NMR, the main novelties of our method are: (1) we extend NMR to deal with the misaligned images; (2) we further analyze the advantages of nuclear norm from the viewpoint of distributions of the error images and its singular values. An observation that the distribution of the singular values of some structural noise (e.g., the block occlusion, sunglass or scarf) approximates Laplacian distribution is presented in this paper. As we know, the distribution of sparse noise approximates Laplacian distribution [12], which explains the good performance brought by l_1-norm constraint on the error image, because from the viewpoint of maximum likelihood estimation (MLE), the l_1-norm constraint on the error image assumes it follows Laplacian distribution. We know that the nuclear norm constraint on the error image is equal to calculate the sum of its singular values. Since singular values are non-negative, the nuclear norm constraint on the error image can be seen as the l_1-norm constraint on the singular values of the error image. This observation explains the strength of our method when dealing with illumination variations and structural noise.

Similar as in [7,8,10], we perform face alignment and representation simultaneously. The main difference between SCC and those sparse representation based

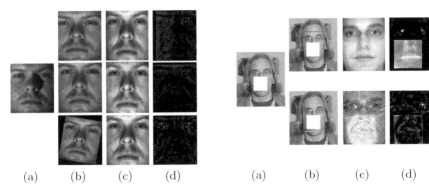

(a)   (b)   (c)   (d)   (a)   (b)   (c)   (d)

Fig. 1. Some misalignment instances. **Right top:** an artificial translation of 15 pixels in x direction is introduced to the test image. **Right middle:** an artificial translation of 15 pixels in y direction is introduced to the test image. **Right bottom:** an artificial rotation of 10 degrees is introduced to the test image. (a) training image, (b) test images with artificial deformation, (c) the reconstructed images, (d) the representation residuals.

Fig. 2. An intuitive comparison between MRR and SCC with an occluded example image from the CMU Multi-PIE database. The green boxes are the initial face locations and the red boxes show the alignment results. **Right top:** results of SCC, **Right bottom:** results of MRR, (a) test image with block occlusion, (b) the alignment process, (c) the reconstructed images, (d) the representation residuals (Color figure online).

methods is that SCC keeps structural information of images, while they ignore it. We show that the nuclear norm constraint is a good choice for misalignment problem (seen from Fig. 1), and a better choice for alleviating the effect of illumination and removing the structural noise (seen from Fig. 2). Figure 1 gives some misalignment instances from the Extended Yale B face database [13] and Fig. 2 presents an intuitive comparison between MRR and SCC with an occluded example image from the CMU Multi-PIE database [14]. From Fig. 1, we can see that our method could handle 2D deformation well. From Fig. 2, it is obvious that the reconstructed image generated by our method looks more clearly than the image generated by MRR. Apparently the results caused by MRR cannot lead to a correct identity, while our method can. Extensive experiments on the benchmark face databases will further demonstrate the robustness of SCC to those problems mentioned above.

The rest of this paper is organized as follows. Section 2 briefly reviews the related method MRR. Section 3 presents our model and algorithm. Section 4 conducts experiments and Sect. 5 offers our conclusions.

2   Review of Misalignment-Robust Representation (MRR)

In this section, we briefly review the MRR [8]. Suppose that y is the query face image which is warped due to misalignment and a training set $A = [a_1, \ldots, a_n]$,

MRR lets both the image y and set A be aligned to a template y_t and assumes that the deformation-recovered image $y_0 = y \circ \tau$ has a sparse representation over the well-aligned training set $y_0 = \hat{A}\alpha + e$, here $\hat{A} = A \circ T$, $T = [t_1, \ldots, t_n]$ is a transformation set, which is estimated offline first through an alignment method named RASL [5], τ represents some kind of spatial transformation, α is the sparse coding coefficient and e is the representation residual vector. This correspondence-based representation could make the face space spanned by the training face images as close to the true face space as possible, which can help to prevent the simultaneous alignment and representation from falling into a bad local minimum. The model of MRR is

$$\min_{e,\alpha,\tau,T} \|e\|_1, \ s.t. \quad y \circ \tau = (A \circ T)\alpha + e \tag{1}$$

To accelerate the algorithm, MRR rewrites the dictionary via singular value decomposition: $A \circ T = U \Sigma V^T$, where U, V are orthogonal matrixes and Σ is a diagonal matrix with descending-order diagonal values. Let $\beta = \Sigma V^T \alpha$, since only the first several elements of β will have big absolute values, the model of MRR is approximated as

$$\hat{\tau} = \min_{e,\beta_\eta,\tau} \|e\|_1, \ s.t. \quad y \circ \tau = U_\eta \beta_\eta + e \tag{2}$$

where U_η is formed by the first η column vectors of U. After optimizing Eq. (2), the representation coefficient α could be solved by

$$\hat{\alpha} = \min_\alpha \|e\|_{l_p} + \lambda \|\alpha\|_2^2, \ s.t. \quad y \circ \hat{\tau} = (A \circ T)\alpha + e \tag{3}$$

where $p = 1$ for face image with occlusion and $p = 2$ for face image without occlusion.

Compared with [7], MRR has shown impressive results with lower computation cost. However, its speed is established only when the query images are clean. If the query images are with occlusion, p in Eq. (3) is set as 1, which is also proved to be time-consuming. In addition, the l_1-norm minimization of representation residual is not robust enough to deal with images with occlusion and illumination variations. Experiments in Sect. 4 also demonstrate our view.

3   Structure Constraint Coding

3.1   Justification for Nuclear Norm Based Constraint

Given an image, the nuclear norm constraint performed on the error image $E \in R^{p \times q}$ calculates the sum of its singular values, which can be shown as $\|E\|_* = \Sigma_{i=1}^{min\{p,q\}} \sigma_i(E)$. As we mentioned before, the motivation of performing nuclear norm constraint on the error image lies in two aspects. The first is that as a 2D image matrix based model, our model adopts image matrix directly to keep the structural information, which can be described by nuclear norm. The second is based on the observation that the distribution of the singular values of some structural noise (e.g., the block occlusion, sunglass or scarf) approximates Laplacian distribution. Here, we further present explanation through some figures and give a discussion on the difference between nuclear norm constraint and l_1-norm constraint on the error image. As we can see from Fig. 3, an example image from AR database is introduced. To exhibit the difference between nuclear norm constraint and l_1-norm or l_2-norm constraint on the 2D image matrix intuitively, we rearrange the example image by exchanging the locations of some pixels. The structural information of the image is changed after this operation. However, we can find that the l_1-norm or l_2-norm values of the two images in Fig. 3(a) and (b) keep the same, while the nuclear norm values make a change. Actually, the rearrangement over the example image can be of any form and the main motivation is to show that the nuclear norm is sensitive to the changes of images' structural information. Figure 4 gives the distribution of the error term of an occluded image from the AR database. Figure 4(c) is the error image of the occluded image Fig. 4(a) and the reconstruction image Fig. 4(b). Figure 4(d) illustrates the error image fitted by two different distributions, Gaussian and Laplacian distribution, which are both far away from the empirical distribution. However, Fig. 4(e), which illustrates the distribution of the singular values of the error image, shows that the empirical distribution approximates the Laplacian distribution. It verifies our analysis that the nuclear norm based constraint is a suitable way to characterize the structural noise and the poor performance of adding l_1-norm constraint on the error image in this situation could also be explained since the distribution of the structural noise itself follows no rules. In general, the l_1-norm constraint is a better choice to handle sparse noise while the nuclear norm constraint is a better choice for structural noise, which may not be sparse.

3.2 Problem Formulation

Given a set of n images $A_1, \ldots, A_n \in R^{p \times q}$ including all subjects and a query warped image $Y \in R^{p \times q}$, here n is the number of all training images and every image is stacked as a matrix. If the query image and the training set were well-aligned to each other, then Y could be represented by $A_1, \ldots, A_n$ linearly

$$Y = \alpha_1 A_1 + \alpha_2 A_2 +, \ldots, +\alpha_n A_n + E \qquad (4)$$

where $\alpha_1, \ldots, \alpha_n$ is the set of representation coefficients and E is the representation residual matrix. However, the query image Y is warped here. Just like [8], we align both the query image and training images to a well cropped and centered face template y_t first. After that the structure of Y is corresponded well to the training set. As a bridge, the template y_t does not need to be obtained explicitly [8]. The proposed correspondence-based model is

$$Y \circ \tau = \alpha_1(A_1 \circ t_1) + \alpha_2(A_2 \circ t_2) +, \ldots, +\alpha_n(A_n \circ t_n) + E \qquad (5)$$

where the operations $Y \circ \tau$ and $A_i \circ t_i, i = 1, 2, \ldots, n$ align the query image Y and each training image A_i to y_t via the transformation τ and $t_i, i = 1, 2, \ldots, n$, respectively. It should be noted that the transformation τ or t_i is the same as in [5–8].

Suppose $A \circ T$ is a well-aligned training set, where $A = [vec(A_1), \ldots, vec(A_n)]$, $vec(A_i)$ is an operator converting the matrix A_i into a vector and $T = [t_1, t_2, \ldots, t_n]$ is a set of transformation parameters. Let's define a linear mapping from R^n to $R^{p \times q}$:

$$(A \circ T)(\alpha) = \Sigma_{i=1}^n \alpha_i (A_i \circ t_i) = \alpha_1 (A_1 \circ t_1) + \alpha_2 (A_2 \circ t_2) +, \ldots, + \alpha_n (A_n \circ t_n) \quad (6)$$

then Eq. (5) can be rewritten as $Y \circ \tau = (A \circ T)(\alpha) + E$. The nuclear norm constraint is performed on the representation residual so as to increase the robustness of SCC to illumination variations and structural noise. And our proposed model can be formulated as

$$\min_{E,\alpha,\tau,T} \|E\|_* + \frac{\lambda}{2} \|\alpha\|_2^2, \; s.t. \quad Y \circ \tau = (A \circ T)(\alpha) + E. \quad (7)$$

Besides, just like Ridge regression, we add a l_2-norm regularization term to avoid overfitting.

Among the four parameters in our model, the transformation set T can be estimated offline. There are several alignment methods [5,6,15–17] having been proposed and here we choose RASL [5], which is fast and effective. We should note that in [5] the training images come from the same subject while in our method the training images come from all subjects. We estimate T offline and then get the aligned training samples via $\hat{A} = A \circ \hat{T}$. To better understand our model, we rewrite Eq. (7) as

$$\min_{E,\alpha,\tau} \|E\|_* + \frac{\lambda}{2} \|\alpha\|_2^2, \; s.t. \quad Y \circ \tau = \hat{A}(\alpha) + E. \quad (8)$$

However, we still need to deal with three parameters. In general, the optimization of those parameters is very time-consuming since the size of training set A is too large. To reduce the computational costs of SCC, we adopt a simple and traditional way as a filtering step before conducting our model, which is to choose S nearest subjects relative to the test image and then build a new smaller dictionary. We will introduce the filtering step in the next subsection.

3.3   The Filtering Step of SCC

The motivation of this filtering step is to reduce the large-scale dataset into a small subset. Here, we require an efficient method and ensure that after this step the correct subject is still in the reduced dictionary in most cases. Fortunately, the work in [8] gives us some enlightenment. As mentioned in Sect. 2, Yang et al. [8] adopts a coarse search model to find S candidates with the smallest residuals relative to the test image. They rewrote the dictionary via singular value decomposition(SVD): $\hat{A} = U \Sigma V^T$ and the coarse search model is

$$\min_{e,\beta,\tau} \|e\|_1, \; s.t. \quad y \circ \tau = U\beta + e \quad (9)$$

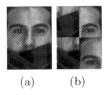

(a)   (b)

Fig. 3. (a) An example image from the AR database, (b) the rearranged image.

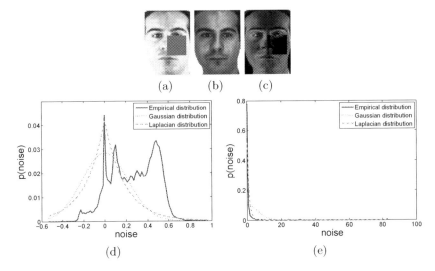

(a)   (b)   (c)

(d)   (e)

Fig. 4. (a) An occluded example image from the AR database, (b) the reconstruction image, (c) the error image, (d) the empirical distribution and the fitted distributions of the error image, (e) The empirical distribution and the fitted distributions of the singular values of error image.

where $\beta = \Sigma V^T \alpha$. Since only the first several elements of β have big absolute values, we could use $U_\eta \beta_\eta$ to approximate $U\beta$, where U_η is formed by the first η column vectors of U, and this operation will significantly speed up the method. Motivated by this idea, in our method the filtering model is conducted as

$$\min_{E,\beta,\tau} \|E\|_*, \ s.t. \quad Y \circ \tau = U(\beta) + E \tag{10}$$

here, we perform nuclear norm constraint on the representation residual. And Eq. (10) could also be approximated as

$$\hat{\tau}_1 = \min_{E_1,\beta_\eta,\tau_1} \|E_1\|_*, \ s.t. \quad Y \circ \tau_1 = U_\eta(\beta_\eta) + E_1. \tag{11}$$

After we get the estimated $\hat{\tau}_1$, the representation coefficient α (regularized by l_2-norm as [18]) could be solved by

$$\hat{\alpha}_1 = \min_{\alpha_1} \|E\|_* + \frac{\lambda}{2} \|\alpha_1\|_2^2, \ s.t. \quad Y \circ \hat{\tau}_1 = \hat{A}(\alpha_1) + E \tag{12}$$

Equation (12) can be optimized using the alternating direction method of multipliers to solve. Then we define the corresponding class reconstruction error as

$$r_i(Y) = \|\hat{Y} - \hat{Y}_i\|_* = \|\hat{A}(\hat{\alpha}_1) - \hat{A}(\delta_i(\hat{\alpha}_1))\|_* \tag{13}$$

here, $\hat{Y} = \hat{A}(\hat{\alpha}_1)$ is the reconstructed image of Y and $\hat{Y}_i = \hat{A}(\delta_i(\hat{\alpha}_1))$ is the reconstructed image of Y in Class i, where $\delta_i : R^n \to R^n$ is the characteristic function that selects the coefficients associated with the i-th class and $\delta_i(\hat{\alpha}_1)$ is a vector whose only nonzero entries are the entries in $\hat{\alpha}_1$ that associated with class i. We choose the top S candidates $k_1, \ldots, k_S$ with the smallest residuals to build a new training dictionary $D_f = [\hat{A}_{k_1}, \ldots, \hat{A}_{k_n}]$.

3.4   Classification

After the filtering step, we can get a smaller dictionary, and then we use Eq. (8) to get the transformation $\hat{\tau}_2$ and coefficient $\hat{\alpha}_2$ together. The corresponding model here is

$$< \hat{\tau}_2, \hat{\alpha}_2 >= \min_{\tau_2, \alpha_2, E_2} \|E_2\|_* + \frac{\lambda}{2}\|\alpha_2\|_2^2, \ s.t. \ \ Y \circ \tau_2 = D_f(\alpha_2) + E_2. \tag{14}$$

It should be noted that the initial value of τ_2 is $\hat{\tau}_1$, which is estimated from the filtering step. What's more, the final representation is performed on D_f. Then just like the setting in Sect. 3.3, the identity of the test image is classified as

$$identity(Y) = arg\min_i \|\hat{Y} - \hat{Y}_i\|_* = \|D_f(\hat{\alpha}_2) - D_f(\delta_i(\hat{\alpha}_2))\|_*. \tag{15}$$

We will discuss how to solve Eq. (8) in the following subsection.

3.5   Algorithm

The alternating direction method of multipliers (ADMM) or the augmented Lagrange multipliers (ALM) method has been widely applied to solve the nuclear norm optimization problems [19,20]. While ADMM is suitable in the case with alternating between two terms and its convergence has been well established for various cases [21,22]. Recently, Peng et al. [5] uses ALM to solve the certain three-term alternation (see also [7,8]) efficiently, without giving the convergence analysis. Here, we provide the process of using ALM to solve Eq. (8).

There is still a problem in Eq. (8). The objective function is non-convex due to the constraint $Y \circ \tau = \hat{A}(\alpha) + E$. Inspired by the work in [5–7], we solve the problem via an iterative convex optimization framework, which iteratively linearizes the current estimate of τ and seek for representations like:

$$\min_{E, \alpha, \Delta\tau} \|E\|_* + \frac{\lambda}{2}\|\alpha\|_2^2, \ s.t. \ \ Y \circ \tau + J\Delta\tau = \hat{A}(\alpha) + E \tag{16}$$

where $J = \frac{\partial}{\partial \tau} Y \circ \tau$ is the Jacobian of $Y \circ \tau$ with respect to the transformation parameters τ, and $\Delta\tau$ is the step in τ. This linearized formulation is now a convex programming and thus can be solved efficiently.

The augmented Lagrangian function is defined by

$$
\begin{aligned}
L_\mu(E, \alpha, \Delta\tau, Z) =& \|E\|_* + \frac{\lambda}{2}\|\alpha\|_2^2 + Tr(Z^T(Y \circ \tau + J\Delta\tau - \hat{A}(\alpha) - E)) \\
& + \frac{\mu}{2}\|Y \circ \tau + J\Delta\tau - \hat{A}(\alpha) - E\|_F^2
\end{aligned}
\tag{17}
$$

where $\mu > 0$ is a penalty parameter, Z is the Lagrange multiplier matrix, $Tr(\cdot)$ is the trace operator and $\|\cdot\|_F$ denotes the Frobenius norm. The ALM algorithm iteratively estimates both the Lagrange multiplier and the optimal solution by iteratively minimizing the augmented Lagrange function

$$
\begin{aligned}
(E^{k+1}, \alpha^{k+1}, \Delta\tau^{k+1}) &= arg\min L_{\mu^k}(E, \alpha, \Delta\tau, Z^k) \\
Z^{k+1} &= Z^k + \mu^k(Y \circ \tau + J\Delta\tau^{k+1} - \hat{A}(\alpha^{k+1}) - E^{k+1}).
\end{aligned}
\tag{18}
$$

However, the first step in the above iteration (18) is difficult to solve directly. So typically, people adopt an alternating strategy, which minimizes the function against the three unknowns $E, \alpha, \Delta\tau$ one at a time, to minimize the Lagrangian function approximately:

$$
\alpha^{k+1} = arg\min_\alpha L_{\mu^k}(E^k, \alpha, \Delta\tau^k, Z^k)
\tag{19}
$$

$$
E^{k+1} = arg\min_E L_{\mu^k}(E, \alpha^{k+1}, \Delta\tau^k, Z^k)
\tag{20}
$$

$$
\Delta\tau^{k+1} = arg\min_{\Delta\tau} L_{\mu^k}(E^{k+1}, \alpha^{k+1}, \Delta\tau, Z^k)
\tag{21}
$$

$$
Z^{k+1} = Z^k + \mu^k(Y \circ \tau + J\Delta\tau^{k+1} - \hat{A}(\alpha^{k+1}) - E^{k+1})
\tag{22}
$$

each step of the above iteration involves solving a convex program, which has a simple closed-form solution. Hence, each step can be solved efficiently. For convenience, let's rewrite the augmented Lagrangian function in a different form. We will give a simple expression for the last two items in Eq. (17) as

$$
\begin{aligned}
& Tr(Z^T(Y \circ \tau + J\Delta\tau - \hat{A}(\alpha) - E)) + \frac{\mu}{2}\|Y \circ \tau + J\Delta\tau - \hat{A}(\alpha) - E\|_F^2 \\
& = \frac{\mu}{2}\|Y \circ \tau + J\Delta\tau - \hat{A}(\alpha) - E + \frac{1}{\mu}Z\|_F^2 - \frac{1}{2\mu}\|Z\|_F^2
\end{aligned}
\tag{23}
$$

then we can have

$$
\begin{aligned}
L_\mu(E, \alpha, \Delta\tau, Z) =& \|E\|_* + \frac{\lambda}{2}\|\alpha\|_2^2 \\
& + \frac{\mu}{2}\|Y \circ \tau + J\Delta\tau - \hat{A}(\alpha) - E + \frac{1}{\mu}Z\|_F^2 - \frac{1}{2\mu}\|Z\|_F^2
\end{aligned}
\tag{24}
$$

based on Eq. (24), it is easy to solve the problems in Eqs. (19–22). Specifically, Eq. (19) can be expressed as

$$
\alpha^{k+1} = arg\min_\alpha (\frac{\mu}{2}\|Y \circ \tau + J\Delta\tau^k - \hat{A}(\alpha) - E^k + \frac{1}{\mu^k}Z^k\|_F^2 + \frac{\lambda}{2}\|\alpha\|_2^2)
\tag{25}
$$

Equation (20) can be expressed as

$$E^{k+1} = arg\min_{E}(\frac{\mu}{2}\|Y \circ \tau + J\Delta\tau^k - \hat{A}(\alpha^{k+1}) - E + \frac{1}{\mu^k}Z^k\|_F^2 + \|E\|_*) \quad (26)$$

Equation (21) can be expressed as

$$\Delta\tau^{k+1} = arg\min_{\Delta\tau}(\|Y \circ \tau + J\Delta\tau - \hat{A}(\alpha^{k+1}) - E^{k+1} + \frac{1}{\mu^k}Z^k\|_F^2). \quad (27)$$

Next we will spell out how to solve these problems above. Now, we consider Eq. (25). Letting $H = [Vec(\hat{A}_1), \dots, Vec(\hat{A}_n)]$, we can rewrite $\hat{A}(\alpha) = \Sigma_{j=1}^n \alpha_j \hat{A}_j$ into the matrix form $H\alpha$. Denote $g = Vec(Y \circ \tau + J\Delta\tau^k - E^k + \frac{1}{\mu^k}Z^k)$. Therefore, Eq. (25) is equivalent to

$$\alpha^{k+1} = arg\min_{\alpha}(\frac{\mu}{2}\|H\alpha - g\|_F^2 + \frac{\lambda}{2}\|\alpha\|_2^2). \quad (28)$$

It is obviously that Eq. (28) is a standard regression model, so we can get its closed-form solution

$$\alpha^{k+1} = (H^T H + \frac{\lambda}{\mu}I)^{-1}H^T g. \quad (29)$$

Next, we consider Eq. (26), which is equivalent to

$$E^{k+1} = arg\min_{E}(\frac{1}{2}\|E - (Y \circ \tau + J\Delta\tau^k - \hat{A}(\alpha^{k+1}) + \frac{1}{\mu^k}Z^k)\|_F^2 + \frac{1}{\mu}\|E\|_*) \quad (30)$$

the optimal solution can be computed via the singular value thresholding algorithm [23]. To spell out the solution, let us define the soft-thresholding or shrinkage operator for scalars as follows:

$$S_\xi[x] = sign(x) \cdot \max(|x| - \xi, 0) \quad (31)$$

where $\xi > 0$. When applied to vectors and matrices, the shrinkage operator acts element-wise. With the shrinkage operator, we can write the solution of Eq. (30) as

$$(U, \Sigma, V) = svd(Y \circ \tau + J\Delta\tau^k - \hat{A}(\alpha^{k+1}) + \frac{1}{\mu^k}Z^k))$$
$$E^{k+1} = US_{\frac{1}{\mu^k}}[\Sigma]V^T. \quad (32)$$

Then we consider Eq. (27). Like Eqs. (25), (27) is also a regression model. Denote $f = Vec(\hat{A}(\alpha^{k+1}) - Y \circ \tau + E^{k+1} - \frac{1}{\mu^k}Z^k)$. Therefore, Eq. (27) is equivalent to

$$\Delta\tau^{k+1} = arg\min_{\Delta\tau}(\|f - J\Delta\tau\|_F^2) \quad (33)$$

and its closed-form solution is

$$\Delta\tau^{k+1} = (J^T J)^{-1}J^T f. \quad (34)$$

In summary, the core of ALM algorithm for our method involves three subproblems: the ridge regression, the singular value thresholding and the least square estimation. The entire algorithm is summarized in Algorithm 1.

Algorithm 1. SCC Algorithm via ALM

Input: A set of aligned training matrices $\hat{A}_1, \ldots, \hat{A}_n$ and a query matrix $Y \in R^{p \times q}$, the model parameters λ, μ and initial transformation τ_0 of Y.

1. Let $H = [Vec(\hat{A}_1), \ldots, Vec(\hat{A}_n)]$ and define $S_\xi[x] = sign(x) \cdot \max(|x| - \xi, 0)$;

2. **While** not converged ($k = 0, 1, \ldots$) **do**

3. Updating α: Let $g = Vec(Y \circ \tau + J\Delta\tau^k - E^k + \frac{1}{\mu^k} Z^k)$, $\alpha^{k+1} = (H^T H + \frac{\lambda}{\mu} I)^{-1} H^T g$;

4. Updating E: $(U, \Sigma, V) = svd(Y \circ \tau + J\Delta\tau^k - \hat{A}(\alpha^{k+1}) + \frac{1}{\mu^k} Z^k)$, $E^{k+1} = U S_{\frac{1}{\mu^k}}[\Sigma] V^T$;

5. Updating $\Delta\tau$: Let $f = Vec(\hat{A}(\alpha^{k+1}) - Y \circ \tau + E^{k+1} - \frac{1}{\mu^k} Z^k)$, $\Delta\tau^{k+1} = (J^T J)^{-1} J^T f$;

6. Updating Z: $Z^{k+1} = Z^k + \mu^k (Y \circ \tau + J\Delta\tau^{k+1} - \hat{A}(\alpha^{k+1}) - E^{k+1})$;

7. **End while**

Output: Solution $\alpha^*, E^*, \Delta\tau^*$ to Eq. (16)

---

4   Experiments

As a face recognition method, since the transformation τ used in our model is the same as in [8], the alignment effect between our method and MRR differs little. Therefore, we mainly focus on the recognition performance in our experiments. Three databases are used here, including the CMU Multi-PIE database [14], the Extended Yale B database [13] and the LFW (Labeled Faces in the Wild) database [24]. The outer eye corners of all the face images are manually marked as the ground truth for registration. We first compare our method with some related work. Then we test the robustness of SCC on different databases. It should be noted that in our experiments there are 4 parameters. Apart from γ in estimating (we use the default value of γ in [5]), we still need to set λ, η and S beforehand. Among them, η and S are relatively easy to set. Here we set $\eta = 35$ and $S = 8$ for all of the experiments. Last but not least, λ is a very important parameter in our experiments, which affects the value of alignment error. Figure 5 intuitively illustrates its effect. We can see that the bigger the λ is, the more information the residual can get. However, when λ is too big, the reconstructed image may lose lots of discriminative information. Generally speaking, for the experiment of robustness to structural noise, λ is set as a big value (i.e., 0.5), while for clean images, λ is fixed as 0.05. Besides, we should note that due to the lack of the code, the results of RASR in some comparative experiments are cited from [7,8].

4.1   Comparison with Related Work

We first compare SCC with some related work using the CMU Multi-PIE database [14] and the Extended Yale B database [13]. The Multi-PIE database contains images of 337 subjects captured in four sessions with simultaneous variations in pose, expression and illumination, while the Extended Yale B database contains 38 persons under 9 poses and 64 illumination conditions. Here, we compare SCC with four state-of-the-art methods, SRC [1], Huang's method (H's) [25], RASR [7] and MRR [8]. In Multi-PIE, As in [7,8], all the subjects in Session 1, each of which

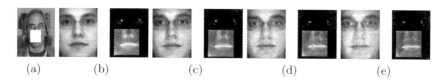

(a)   (b)   (c)   (d)   (e)

Fig. 5. The reconstructed image and representation residual are presented respectively in different value of λ. (a) The query image, (b) $\lambda = 5$, (c) $\lambda = 0.5$, (d) $\lambda = 0.05$, (e) $\lambda = 0.005$.

has 7 frontal images with extreme illuminations $\{0, 1, 7, 13, 14, 16, 18\}$ and neutral expression, are used for training and the subjects from Session 2 with illumination $\{10\}$ are used for testing. The images are cropped to 80×64 and an artificial translation of 5 pixels (in both x and y directions) is introduced to the test image. For the settings in Extended Yale B, as in [7, 8, 25], 20 subjects are selected and for each subject 32 frontal images (selected randomly) are used for training, with the remaining 32 images for testing. An artificial translation of 10 pixels (in both x and y directions) is introduced to the test image and the images are cropped to 88×80. The results are shown in Table 1. We note that SRC is very sensitive to misalignment. What's more, compared with other methods, SCC achieves the best results, which shows that our method does well in dealing with 2D deformation.

Table 1. Recognition rates (%) with translations on the Multi-PIE and Extended Yale B database

Methods	SRC [1]	RASR [7]	H's [25]	MRR [8]	SCC
Multi-PIE	24.1	92.2	67.5	92.8	**94.0**
Extended Yale B	51.1	93.7	89.1	93.6	**95.4**

4.2   Robustness to Illumination Variations

We evaluate the robustness of SCC to deal with illumination variations on the CMU Multi-PIE database [14]. The first 249 subjects in Session 1 and Sessions 2–4 are used as the training and testing sets, respectively. Here, we fix pose and expression, only choose images from different illumination conditions. More specifically, for each subject, 7 frontal images with the same illuminations as those in Sect. 4.1 are used for training. We conduct three tests here and 10 frontal images selected from Sessions 2–4 are used for testing, respectively. The images are resized to 80×64. The classification results of MRR and SCC are shown in Table 2. It can be seen that SCC is better than MRR in all cases (about 2.6 %, 1.8 % and 1.9 % improvement in the cases of Session 2, 3, 4, respectively), which shows the robustness of our method to deal with illumination variations.

Table 2. Recognition rates (%) of MRR and SCC vs. illumination variations on the Multi-PIE database

Sessions	Session 2	Session 3	Session 4
MRR	91.5	91.4	92.4
SCC	**94.1**	**93.2**	**94.3**

4.3   Robustness to the Number of Training Samples

From the previous section, we know that SCC is robust on illumination variations. Here we evaluate SCC's robustness to the number of training samples in comparison with MRR [8] and RASR [7] on the CMU Multi-PIE database [14]. As in [7,8], the first 100 subjects in Session 1 and Session 3 are used as the training and testing sets, respectively. For each subject, 7 frontal images with the same illuminations as those in Sect. 4.1 are used for training, while 4 frontal images with illuminations {3, 6, 11, 19} are used for testing. The images are resized to 80×64 and three tests with the first 3, 5 and 7 training samples per subject are performed. The recognition results versus the number of training samples are shown in Table 3. SCC performs best in all cases. We can see that the performances of MRR [8] and RASR [7] degrade fast when the number of training samples is changed from 5 to 3, while SCC seems more stable.

Table 3. Recognition rates (%) vs. the number of training samples on the Multi-PIE database

Sample number	3	5	7
RASR	78.2	95.8	96.8
MRR	82.0	97.5	97.5
SCC	**90.8**	**97.9**	**98.2**

4.4   Robustness to Structural Noise

In this section, we evaluate the robustness of SCC to deal with various levels of structural noise, specifically block occlusion here, on the CMU Multi-PIE database [14]. In this experiment, a randomly located block of the face image is replaced by the image Baboon. The training set remains the same as in Sect. 4.1, while the frontal images with illuminations {3, 6, 11, 19} from Session 1 are used for testing here. The images are cropped to 80×64. Table 4 presents the recognition rates of MRR and SCC with the variations of different occlusion rates. As we discussed in Sect. 1, SCC is much better than MRR in this case, especially when the occlusion rate is high. The performance of MRR drops rapidly when the occlusion level is up to 20 %, while our method still performs well until the occlusion level is up to 30 %. Actually, it doesn't matter what the occlusion is.

It can not only be the block occlusion addressed here, but also the real-disguise, such as sunglass or scarf. And our method still performs better than those vector-based methods.

Table 4. The recognition rates (%) of MRR and SCC vs. different occlusion rates on the Multi-PIE database

Percent occluded	10 %	20 %	30 %	40 %	50 %
MRR	95.5	83.0	67.0	43.8	21.8
SCC	**98.8**	**93.0**	**84.8**	**74.8**	**61.8**

4.5   Experiment on the LFW Database

Both the CMU Multi-PIE database and the Extended Yale B database are collected under controlled environment. In this section, we want to evaluate the effectiveness of our method under uncontrolled environment, such as the LFW (Labeled Faces in the Wild) database [24]. Unlike the controlled images, these images are collected from the Internet and exhibit significant variations in pose and facial expression, in addition to changes in illumination and occlusion. In LFW, since there are only 24 subjects contain more than 35 samples, to make full use of the comparative methods, we choose 20 of them and construct an adequate training dictionary in this experiment. For each subject, 20 samples (selected randomly) are used for training, while the remaining for testing. Here, except for MRR, we test some popular classifiers, such as LRC [26], SRC [1] and CRC [18]. For MRR and SCC, the images are resized to 80×64, while for the other three methods, the face images are automatically detected by using Viola and Jone's face detector [27]. The recognition results are listed in Table 5. SCC achieves the best result among all the methods. In addition, because of the variations in pose, facial expression, illumination and so on, the alignment FR methods significantly outperform those misalignment methods with at least 17.7 % improvement.

Table 5. Recognition rates (%) of LRC, SRC, CRC, MRR and SCC on the LFW database

LRC	SRC	CRC	MRR	SCC
44.1	55.7	60.3	78.0	**82.0**

5   Conclusions

This paper proposes a novel method called structure constraint coding (SCC) for face recognition with image misalignment. It does image alignment and image representation via structure constraint based regression simultaneously. Unlike the vector-based models, SCC keeps structural information of images through performing the nuclear norm constraint on the error matrix. We conduct experiments on three popular face databases. Experimental results clearly demonstrate that SCC performs better than those vector-based methods when handling face misalignment problem coupled with illumination variations and structural noise.

References

1. Wright, J., Yang, A., Ganesh, A., Sastry, S., Ma, Y.: Robust face recognition via sparse re-presentation. IEEE PAMI **31**, 210–227 (2009)
2. Elhamifar, E., Vidal, R.: Robust classification using structured sparse representation. In: CVPR (2011)
3. Yang, M., Zhang, L., Yang, J., Zhang, D.: Robust sparse coding for face recognition. In: CVPR (2011)
4. Zhou, Z., Wagner, A., Mobahi, H., Wright, J., Ma, Y.: Face recognition with contiguous occlusion using markov random fields. In: ICCV (2009)
5. Peng, Y., Ganesh, A., Wright, J., Xu, W., Ma, Y.: Rasl: Robust alignment by sparse and low-rank decomposition for linearly correlated images. In: CVPR (2010)
6. Wu, Y., Shen, B., Ling, H.: Online robust image alignment via iterative convex optimization. In: CVPR (2012)
7. Wagner, A., Wright, J., Ganesh, A., Zhou, Z., Mobahi, H., Ma, Y.: Towards a practical face recognition system: robust alignment and illumination by sparse representation. IEEE PAMI **34**, 372–386 (2012)
8. Yang, M., Zhang, L., Zhang, D.: Efficient misalignment-robust representation for real-time face recognition. In: Fitzgibbon, A., Lazebnik, S., Perona, P., Sato, Y., Schmid, C. (eds.) ECCV 2012, Part I. LNCS, vol. 7572, pp. 850–863. Springer, Heidelberg (2012)
9. Yang, J., Zhang, L., Xu, Y., Yang, J.: Beyond sparsity: the role of l1-optimizer in pattern classification. Pattern Recogn. **45**, 1104–1118 (2012)
10. Zhuang, L., Yang, A., Zhou, Z., Sastry, S., Ma, Y.: Single-sample face recognition with image corruption and misalignment via sparse illumination transfer. In: CVPR (2013)
11. Yang, J., Qian, J., Luo, L., Zhang, F., Gao, Y.: Nuclear norm based matrix regression with applications to face recognition with occlusion and illumination changes (2014). arXiv:1207.0023
12. Yang, M., Zhang, L., Yang, J., Zhang, D.: Regularized robust coding for face recognition. IEEE TIP **22**, 1753–1766 (2013)
13. Georghiades, A., Belhumeur, P., Kriegman, D.: From few to many: Illumination cone models for face recognition under variable lighting and pose. IEEE PAMI **23**, 643–660 (2001)
14. Gross, R., Matthews, I., Cohn, J., Kanade, T., Baker, S.: Multi-pie. Image Vis. Comput. **28**, 807–813 (2010)
15. Cootes, T., Edwards, G., Taylor, C.: Active appearance models. IEEE PAMI **23**, 681–685 (2001)
16. Cootes, T., Taylor, C.: Active shape models - smart snakes. In: BMVC (1992)
17. Huang, G., Jain, V., Learned-Miller, E.: Unsupervised joint alignment of complex images. In: ICCV (2007)

18. Zhang, L., Yang, M., Feng, X.: Sparse representation or collaborative representation which helps face recognition? In: ICCV (2011)
19. Lin, Z., et al.: The augmented lagrange multiplier method for exact recovery of corrupted low-rank matrices (2010). arXiv:1009.5055v2
20. Hansson, A., et al.: Subspace system identification via weighted nuclear norm optimization (2012). arXiv:1207.0023
21. Glowinski, R., Marroco, A.: Sur lapproximation, par elements finis dordre un, et la resolution, par penalisationdualite, dune classe de problemes de dirichlet non lineares. Revuew Francaise dAutomatique, Informatique et Recherche Operationelle **9**, 41–76 (1975)
22. Gabay, D., Mercier, B.: A dual algorithm for the solution of nonlinear variational problems via finite element approximations. Comput. Math. Appl. **2**, 17–40 (1976)
23. Cai, J., Cands, E., Shen, Z.: A singular value thresholding algorithm for matrix completion. SIAM J. Optim. **20**(4), 1956–1982 (2010)
24. Huang, G., Ramesh, M., Berg, T., Learned-Miller, E.: Labeled faces in the wild: A database for studying face recognition in unconstrained environments. Technical Report 07–49, University of Massachusetts (2007)
25. Huang, J., Huang, X., Metaxas, D.: Simultaneous image transformation and sparse representation recovery. In: CVPR (2008)
26. Naseem, I., Togneri, R., Bennamoun, M.: Linear regression for face recognition. IEEE PAMI **32**, 2106–2112 (2010)
27. Viola, P., Jones, M.: Robust real-time face detection. Int. J. Comput. Vis. **57**, 137–154 (2004)

People Re-identification Based on Bags
of Semantic Features

Zhi Zhou[1]($\boxtimes$), Yue Wang[2], and Eam Khwang Teoh[1]

[1] School of Electrical and Electronic Engineering, Nanyang Technological University,
Singapore 639798, Singapore
zzhou5@e.ntu.edu.sg
[2] Visual Computing Department, Institute for Infocomm Research (I2R),
Singapore 138632, Singapore

Abstract. People re-identification has attracted a lot of attention recently. As an important part in disjoint cameras based surveillance system, it faces many problems. Various factors like illumination condition, viewpoint of cameras and occlusion make people re-identification a difficult task. In this paper, we exploit the performance of bags of semantic features on people re-identification. Semantic features are mid-level features that can be directly described by words, such as hair length, skin tone, race, clothes colors and so on. Although semantic features are not as discriminative as local features used in existing methods, they are more invariant. Therefore, good performance on people re-identification can be expected by combining a set of semantic features. Experiments are carried out on VIPeR dataset. Comparison with some state-of-the-art works is provided and the proposed method shows better performance.

1 Introduction

Multi-camera based surveillance system is more and more popular in our daily life. How to monitor the environment and collaboration with such a surveillance system becomes a challenge. Considering there are usually 20–40 cameras in a surveillance system, it is costing and inefficient if only rely on human visual inspectors. Thus, a system which is able to automatically detect and identify the people appears in cameras and understand his/her behavior is highly desiered. One of the tasks is to identify the person when he/she disappears from one camera and appears in another one, known as people re-identification. Efficient people re-identification is an important part for people tracking in a surveillance system with disjoint located cameras and it has become a popular research topic.

Some problems exist in people re-identification, such as illumination change and view-point change. Inappropriate selection of features could lead to poor performance. In this paper, we exploit the performance of semantic features on people re-identification. A set of semantic features are selected from all over the body. Experiment is carried out on VIPeR dataset [1]. Some image samples in VIPeR dataset are shown in Fig. 1. Comparison with some state-of-the-art methods is provided.

© Springer International Publishing Switzerland 2015
C.V. Jawahar and S. Shan (Eds.): ACCV 2014 Workshops, Part III, LNCS 9010, pp. 574–586, 2015.
DOI: 10.1007/978-3-319-16634-6_42

Fig. 1. Some examples of image pairs from VIPeR dataset [1].

The remaining structure of this paper is arranged as follows. Section 2 gives a brief introduction on related works. The proposed method is detailed in Sect. 3. Experimental results and discussions are presented in Sect. 4. Finally, the conclusion of this paper is given out in Sect. 5.

2   Related Works

Lots of works have been done on people re-identification in the past several years. Some works tried to represent the target by extracting color information as feature. One of the earliest works is from Javed [2], one color histogram is used to represent a person. Then information combining color, position and time is feed into a Bayesian model to track a person in disjoint cameras. Bird [3] divided the human body into ten horizontal regions. The descriptor is formed by cascading values of median color in HSV color spaces from each region. Kao [4] employed color information from several images to form a hierarchical color structure. Then the similarity of two images is measured based on Bayesian decision. In Madden's work [5], an on-line clustering method is used to classify pixels belonging to a person. Then color spectrum histogram is used to represent the person.

Besides color, some works exploited texture information. Farenzena [6] partitioned the person into head, torso and leg. Then for each part, local features describing texture like Maximally Stable Color Regions and Recurrent High-Structured Patches are used together with color to represent this person. Berdugo [7] combined background subtraction method and saliency map to segment the people first. Three texture features–oriented gradients, color ratio and color saliency were employed to obtain a discriminative descriptor.

Other local features such as key-points are used as well. Gheissari [8] combined interest point matching and model fitting in corresponding body parts for people re-identification. Hamdoun [9] extracted interest points from body and use descriptors of interest points to model this person in a K-D tree. People re-identification is done by matching interest points and searching in the K-D tree. Wang [10] introduced local shape information and appearance context for the representation of people. Histogram of Gradient (HOG) in the log-RGB color space and spatial distribution of the appearance are used as local descriptors. In Azizs [11] method, SIFT, SURF and Spin images are used as descriptors to represent people. Bak [12] extracted Mean Riemannian Covariance (MRC) patches during the track to model the people detected by HOG. The combination of MRC patches was then used for re-identification. Zheng [13] made use of information from close neighborhood to identify a person in a group of people. Salient features are used with unsupervised method in Zhao's method [14].

Some researchers treat people re-identification as a recognition problem, they improved the performance by enhancing the similarity calculation or by exploiting machine learning methods. Zheng [15] proposed a Probabilistic Relative Distance Comparison (PRDC) model to optimize the calculation of similarity between a pair of images. Gray [16] selected a set of local features including eight color channels and texture features like Schmid and Gabor. AdaBoost is used to train samples to get the optimal weight for each feature. Similarly, Prosser [17] used SVM to train samples and same features are used. In Baks [18] method, two kinds of local features– haar-like features and dominant color descriptor (DCD) are extracted. AdaBoost is used to choose the most distinctive features.

Recently, some semantic features like soft biometrics have been used to solve face identification problems. Dantcheva and Dugelay [19] proposed a face re-identification method based on three soft biometric traitshair, skin and clothes. Vaquero [20] introduced a new people identification method by using semantic features in the head area such as facial hair type, type of eyewear and hair type. Classifier is used to train selected attributes. Layne [21,22] defined soft-biometrics as attributes and used machine learning methods like matric learning or SVM. Selected attributes are trained and binary results are given for a test image.

3   The Proposed Method

In this paper, we conduct people re-identification with bags of semantic features. Features used cover the whole body area, including hair length, colors of clothes and pants, length of sleeves and pants, clothes patterns, intensity contrast between clothes and pants. Though each feature is not as discriminative as local features, a discriminative descriptor can be obtained by combining these features. In this section, details of extracting these semantic features will be introduced, followed by the similarity calculation when we conduct the re-identification experiment on the VIPeR dataset [1].

A flow chart of the proposed method is shown is Fig. 2. Images in both gallery set and probe set are first processed to segment the foreground, which means the

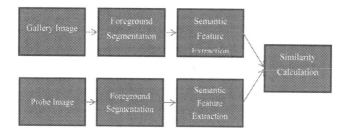

Fig. 2. Flow chart of the proposed method.

Fig. 3. Examples of foreground extraction.

human body. Then semantic features are extracted from three parts–head, torso and leg. Then similarity between two images are calculated based on extracted semantic features.

3.1 Pre-processing

Since images in VIPeR [1] contain some background information around the people. The involvement of background information always downgrade the extraction of human descriptors. Therefore, Stel Component Analysis [23] is used to segment the human body and remove background for better performance. Some segmentation results are shown in Fig. 3. As we can see, human body can be well segmented from clustering background, though minor background still exist.

3.2 Semantic Feature Extraction

To obtain a more discriminative descriptor, semantic features that cover the whole body of people are selected. In the head area, soft biometrics [24] can be used. The rest of body is divided into torso and leg areas. Semantic features related to the apparel of people can be extracted in both two areas.

Hair Length. A lot of soft biometric traits are listed in [24], such as glasses, hair length, hats, facial hair and so on. However, due to the low resolution in images, especially in images from surveillance cameras, some of those traits are hard to be extracted and do not perform well in people re-identification. In this

Fig. 4. Examples of hair segmentation.

paper, only hair length is selected because it is observable, even in images with low resolution and various view angles. To extract hair length, Wang's method [25] is adopted, it's a morphological method and demands the segmentation of hair first. Mean shift segmentation [26] is used to separate hair from the other parts of head area. The hair length is then calculated and categorized into short hair and long hair. Some examples of the segmentation result are shown in Fig. 4.

Color from Torso and Leg. Color is considered in both torso and leg areas. Color is the most common used feature in computer vision tasks. However, it usually suffers from varying factors like illumination. To constrain the effect of illumination change and different color specifications in cameras, colors are represented in nine predefined categories in the proposed method, instead of using color histograms. Nine common color categories are defined as black, gray, white, red, green, blue, cyan, magenta and yellow. Color histogram with nine bins is calculated in HSV color space, and the color bin with maximum vote is picked.

Color Intensity Contrast. Even represented in pre-defined categories, the same object may still appear in different colors in different cameras. Therefore, as a complementary feature to color, the color intensity contrast between torso area and leg area is selected as one of the semantic features. Color intensity contrast is not discriminative, but the link between two color intensities is invariant to those changing factors. The color intensity histograms from torso area and leg area are compared and the contrast is classified into four categories – High contrast with brighter torso, High contrast with darker torso, Low contrast with brighter torso and Low contrast with darker torso.

Clothes Pattern. Besides color, clothes pattern is also used in torso area. Under circumstances like illumination change, the color may change, but the pattern of clothes remains the same. Similar with color, five predefined categories are used– plain color, horizontal stripe, vertical stripe, checker and complex patterns. The first four are common clothes patterns in our daily life while the rest is classified into the fifth category. As shown in Zhou's work [27], clothes pattern can be obtained by using histograms of magnitude and orientation of gradient in the torso area. Plain color is decided only by the histogram of magnitude of gradient. Since plain color clothes usually form few edges in the image, large proportion of low magnitude pixels exist in the histogram of magnitude of gradient, while in histograms of the other categories are not. Furthermore, by using the histogram of orientation, we can detect horizontal stripes and vertical stripes because they

Fig. 5. Examples of clothes pattern. From left to right are plain colors, horizontal stripes and checkers. Vertical stripe sample does not exist in VIPeR.

Fig. 6. Examples of skin detection.

cause intensive response in horizontal and vertical directions respectively. The rest categories are classified by both of the two histograms. Some examples of defined categories are shown in Fig. 5.

Length of Sleeves and Pants. One of the common ways for human to describe the apparel of a people is the length of sleeves or pants. Therefore, the length of sleeves and pants are used in the proposed method as semantic features. A simple way to detect the length of sleeves or pants is not to detect directly. Since color of skin is usually gathering in a small range of color space [28] and easy to be detected, we can extract the length of sleeves or pants by detect nude arms and legs instead. HSV color space is used in the proposed method and a compact range in H and S plane [29] is used to detect skins. Some examples of skin detection is shown in Fig. 6. Then the length of sleeves or pants can be decided by extracting the length of nude arms or legs. Two categories are used in both the detection of sleeve length and pant length.

3.3 Similarity Calculation

After all semantic features extracted, similarity between two images are calculated as weighted combination of features' similarities

$$S\left(I_p, I_q\right) = \sum_s h_s\left(s_p, s_q\right) \tag{1}$$

where $h_s(s_p, s_q)$ represents the weighted similarity of one feature between two images s_p and s_q, and it is calculated as

$$h_s(s_p, s_q) = \begin{cases} w_c, if & s_p = s_q = c \\ 0, & else \end{cases} \qquad (2)$$

Here, w_c is the weight for a specific category of a semantic feature. Weights are calculated beforehand with some samples, which will not be included into the images used for people re-identification. For each sample, images in both gallery set and probe set of the dataset are used. For each feature, every category is assigned a weight, based on the probability of its appearance in sample images. It is calculated as

$$w_c = \frac{n_{ab}}{n_a + n_b - n_{ab}} \times \frac{n_s}{\sqrt{n_a n_b}} \qquad (3)$$

where n_{ab} is the number of samples with both images in gallery set and probe set belong to this category, n_a and n_b are the number of samples at least one of the images belongs to this category, n_s is the number of samples used. In this way, those common categories will receive lower weights while categoreis seldom appear receive high weights. Finally, weights within one semantic feature are normalized.

4   Experiment and Discussion

To test the performance of the proposed method on people re-identification, we use the public dataset VIPeR [1]. VIPeR contains image pairs of 632 people. For each people, two images are captured from cameras with different viewpoint and put into two subsets respectively. This is a challenging dataset since these images suffer from severe illumination changes and changes of viewpoint. Even worse, some occlusion is included. In the experiment, images are grouped into gallery set and probe set. For each person, one image is in gallery set and the other is in probe set. Then, for one image in the probe set, the similarities with images in gallery set are calculated, and an ID is assigned to it

$$ID_j = arg \max_{i \in 1, \dots N} S(g_i, p_j) \qquad (4)$$

To evaluate the people re-identification performance, we use Cumulative Matching Characteristic (CMC) curves and Synthetic Recognition Rate (SRR) curves. For an image in probe set, a rank score is obtained by matching with images in gallery set. As the rank score increase, CMC curve records the probability of finding right matches within the rank score. It is a cumulative curve that increases as the rank score increases. For SRR curve, a specific number of people are first selected, then matching is done on these people. SRR curve records the probability that finds the right matches in the first rank. This curve decreases as the number of people selected increases. Besides these two curves, Area Under

Table 1. Detection rates for using single features (in percentage).

Features	Hair length	Clothes color	Clothes pattern	Sleeve length	Pant color	Pant length	Color contrast
Detection rate	53.8	51.42	59.34	76.58	56.01	79.59	51.58

Curve (AUC) for CMC curve and some statistics at lower rank scores are also provided.

First, we compare the performance of the combination of semantic features with performances of single features. Detection rates of using only single features are shown in Table 1. The detection rate of sleeve length and pant length achieve 70 % above accuracy, while other features just have detection rate slightly higher than 50 % due to the dramatic changes in illumination and viewpoint. Table 2 listed the number of categories of each feature, and weights of semantic features contributed to the combined descriptor matching.

Table 2. Number of categories of each feature and weights (in percentage) contributed to the descriptor matching.

Features	Hair length	Clothes color	Clothes pattern	Sleeve length	Pant color	Pant length	Color contrast
Categories of feature	2	9	4	2	9	2	4
Weights contributed	10.34	28.43	6.07	9.1	25.47	4.85	15.74

In Fig. 7, the CMC curves show that the recognition rates of using single features are low. However, when these features are combined, the recognition rate is significantly improved. AUCs are listed in Table 3. Figure 8 shows the SRR curve. The features combined method performs better than single feature method.

In addition, we compare the proposed method with some state-of-the-art works. SDALF [6] and PRDC [15] are selected. All codes of these methods are provided by authors. Compared with these methods, the proposed method does not need training beforehand for a specific database, because all semantic features selected can be extracted with simple methods. Since training is needed for RDC, half of 632 peopl are used as training samples. The rest are used for people re-identification.

Figure 9 shows the CMC curves. 316 pairs of images are used in people re-identification. The proposed method shows superior performance than the other two. Area Under Curve (AUC) for CMC curve is shown in Table 4. The proposed method achieves the best AUC. Some statistics at lower rank scores of CMC curve are shown in Table 5. At lower rank scores, the proposed method achieves the best recognition rate. SRR curves are shown in Fig. 10.

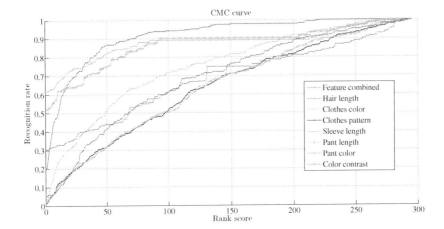

Fig. 7. Comparison on CMC curves between features combined and single features.

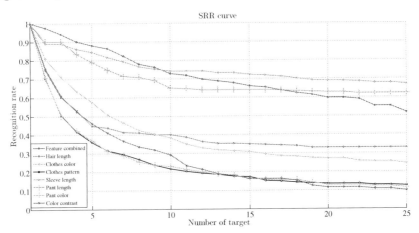

Fig. 8. Comparison on SRR curves between features combined and single features.

Table 3. Comparison of AUCs (in percentage), the best is show in bold.

Features	AUCs	Features	AUCs
Combined	**87.85**	Hair length	68.73
Clothes color	75.24	Clothes pattern	63.89
Sleeve length	86.04	Pant color	64.11
Pant length	84.92	Color contrast	67.78

Table 4. Comparison of AUCs (in percentage), the best is shown in bold.

	SDALF [6]	PRDC [15]	Proposed method
AUCs	87.85	91.71	**91.83**

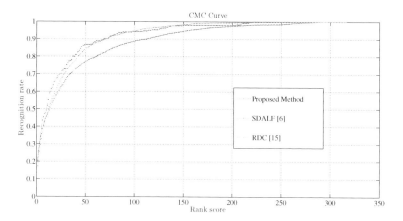

Fig. 9. Comparison on CMC curves between the proposed method and SDALF [6], PRDC [15].

Table 5. Comparison of recognition rate (in percentage) in lower rank scores, the best is shown in bold.

	SDALF [6]	PRDC [15]	Proposed method
Rank 1	13.54	17.32	**20.68**
Rank 5	32.82	35.69	**38.64**
Rank 10	45.95	45.39	**49.15**
Rank 25	66.87	62.68	**70.17**
Rank 50	84.05	76.78	**86.78**

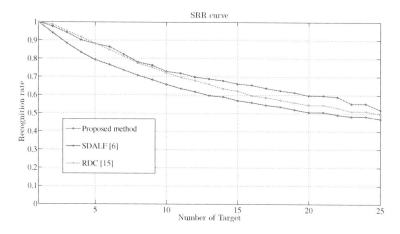

Fig. 10. Comparison on SRR curves between the proposed method and SDALF [6], PRDC [15].

Some of the reasons undermine the performance of the proposed method are discussed as below.

(1) The detection of hair length is mostly affected by different viewpoints of cameras. For example, the hair length could be different when observing from the front view and from the back view. Because in the front view, long hair could be occluded by head and body and be mistaken as short hair. Since segmentation is performed to divide face and hair in the proposed method, the performance of segmentation may also undermine the detection rate of hair length.

(2) The detections of clothes color 'and pant color are severely affected by the dramatic illumination change in VIPeR dataset. In some images, the illumination is too dark or too bright to differentiate colors (Fig. 11). The severe lighting condition also affects the extraction of length of sleeves and pants, because skin can not be detected under some extreme lighting conditions and lead to the failure on detecting short sleeves and pants.

(3) Different viewpoints of cameras raise the problem that, some people carry backpack and have different colors or patterns with the clothes. In this case, features from front view and back view are mismatched.

Fig. 11. Examples of image pairs suffer from severe illumination change.

5   Conclusion

In this paper, we propose a people re-identification method with bags of semantic features. 7 features are selected, which cover the whole body. Without using learning method, each feature can be extracted with simple method. The experiment shows that the combination of these semantic features outperforms single feature. Experimental results also show that the proposed method achieves good performance compared with state-of-the-art methods.

Semantic features used in the proposed method are far from enough for robust and discriminative description. In our future work, we will try to include more semantic features. Also, we will exploit the combination of robustness from semantic features and discriminative from local features. Another way to improve the description is adopting an advanced method to conduct more detailed and accurate segmentation on the human body.

References

1. Gray, D., Brennan, S., Tao, H.: Evaluating appearance models for recognition, reacquisition, and tracking. In: IEEE International Workshop on Performance Evaluation of Tracking and Surveillance (2007)
2. Javed, O., Rasheed, Z., Shafique, K., Shah, M.: Tracking across multiple cameras with disjoint views. In: Proceedings of Ninth IEEE International Conference on Computer Vision (ICCV), pp. 952–957 (2003)
3. Bird, N.D., Masoud, O., Papanikolopoulos, N.P., Isaacs, A.: Detection of loitering individuals in public transportation areas. IEEE Trans. Intell. Transp. Syst. **6**, 167–177 (2005)
4. Kao, J.-H., Lin, C.-Y., Wang, W.-H., Wu, Y.-T.: A unified hierarchical appearance model for people re-identification using multi-view vision sensors. In: Huang, Y.-M.R., Xu, C., Cheng, K.-S., Yang, J.-F.K., Swamy, M.N.S., Li, S., Ding, J.-W. (eds.) PCM 2008. LNCS, vol. 5353, pp. 553–562. Springer, Heidelberg (2008)
5. Madden, C., Cheng, E.D., Piccardi, M.: Tracking people across disjoint camera views by an illumination-tolerant appearance representation. Mach. Vis. Appl. **18**, 233–247 (2007)
6. Farenzena, M., Bazzani, L., Perina, A., Murino, V., Cristani, M.: Person re-identification by symmetry-driven accumulation of local features. In: IEEE Conference on Computer Vision and Pattern Recognition (CVPR), pp. 2360–2367 (2010)
7. Berdugo, G., Soceanu, O., Moshe, Y., Rudoy, D., Dvir, I.: Object reidentification in real world scenarios across multiple non-overlapping cameras. In: Proceedings of the European Signal Processing Conference, pp. 1806–1810 (2010)
8. Gheissari, N., Sebastian, T.B., Hartley, R.: Person reidentification using spatiotemporal appearance. In: IEEE Conference on Computer Vision and Pattern Recognition (CVPR), vol. 2, pp. 1528–1535 (2006)
9. Hamdoun, O., Moutarde, F., Stanciulescu, B., Steux, B.: Person re-identification in multi-camera system by signature based on interest point descriptors collected on short video sequences. In: Second ACM/IEEE International Conference on Distributed Smart Cameras (ICDSC), pp. 1–6 (2008)
10. Wang, X., Doretto, G., Sebastian, T., Rittscher, J., Tu, P.: Shape and appearance context modeling. In: Proceedings of IEEE International Conference on Computer Vision (ICCV) (2007)
11. Aziz, K.-E., Merad, D., Fertil, B.: Person re-identification using appearance classification. In: Kamel, M., Campilho, A. (eds.) ICIAR 2011, Part II. LNCS, vol. 6754, pp. 170–179. Springer, Heidelberg (2011)
12. Bak, S., Corvee, E., Bremond, F., Thonnat, M.: Boosted human re-identification using riemannian manifolds. Image Vis. Comput. **30**, 443–452 (2012)
13. Zheng, W.S., Gong, S., Xiang, T.: Associating groups of people. In: Proceedings of the British Machine Vision Conference (BMVC) (2009)
14. Zhao, R., Ouyang, W., Wang, X.: Unsupervised salience learning for person re-identification. In: IEEE Conference on Computer Vision and Pattern Recognition (CVPR), pp. 3586–3593 (2013)
15. Zheng, W.S., Gong, S., Xiang, T.: Reidentification by relative distance comparison. IEEE Trans. Pattern Anal. Mach. Intell. **35**, 653–668 (2013)
16. Gray, D., Tao, H.: Viewpoint invariant pedestrian recognition with an ensemble of localized features. In: Forsyth, D., Torr, P., Zisserman, A. (eds.) ECCV 2008, Part I. LNCS, vol. 5302, pp. 262–275. Springer, Heidelberg (2008)

17. Prosser, B., Zheng, W.S., Gong, S., Xiang, T., Mary, Q.: Person re-identification by support vector ranking. In: Proceedings of the British Machine Vision Conference (BMVC), vol. 1, p. 5 (2010)
18. Bak, S., Corvee, E., Brémond, F., Thonnat, M.: Person re-identification using haar-based and dcd-based signature. In: IEEE International Conference on Advanced Video and Signal Based Surveillance (AVSS), pp. 1–8 (2010)
19. Dantcheva, A., Dugelay, J.L.: Frontal-to-side face re-identification based on hair, skin and clothes patches. In: IEEE International Conference on Advanced Video and Signal-Based Surveillance (AVSS), pp. 309–313 (2011)
20. Vaquero, D.A., Feris, R.S., Tran, D., Brown, L., Hampapur, A., Turk, M.: Attribute-based people search in surveillance environments. In: Workshop on Applications of Computer Vision (WACV), pp. 1–8 (2009)
21. Layne, R., Hospedales, T.M., Gong, S., et al.: Person re-identification by attributes. In: BMVC, vol. 2, p. 3 (2012)
22. Layne, R., Hospedales, T.M., Gong, S.: Towards person identification and re-identification with attributes. In: Fusiello, A., Murino, V., Cucchiara, R. (eds.) ECCV 2012 Ws/Demos, Part I. LNCS, vol. 7583, pp. 402–412. Springer, Heidelberg (2012)
23. Jojic, N., Perina, A., Cristani, M., Murino, V., Frey, B.: Stel component analysis: modeling spatial correlations in image class structure. In: IEEE Conference on Computer Vision and Pattern Recognition (CVPR), pp. 2044–2051 (2009)
24. Dantcheva, A., Velardo, C., Dangelo, A., Dugelay, J.L.: Bag of soft biometrics for person identification. Multimedia Tools Appl. 51, 739–777 (2011)
25. Wang, Y., Zhou, Z., Teoh, E.K.: Human hair segmentation and length detection for human appearance model. In: International Conference on Pattern Recognition (ICPR) (2014)
26. Comaniciu, D., Meer, P.: Mean shift: a robust approach toward feature space analysis. IEEE Trans. Pattern Anal. Mach. Intell. 24, 603–619 (2002)
27. Zhou, Z., Wang, Y., Teoh, E.K.: People apparel model for human appearance description. In: International Conference on Information, Communications and Signal Processing (ICICS), pp. 1–5 (2013)
28. Vezhnevets, V., Sazonov, V., Andreeva, A.: A survey on pixel-based skin color detection techniques. Proc. Graphicon. 3, 85–92 (2003)
29. Sobottka, K., Pitas, I.: A novel method for automatic face segmentation, facial feature extraction and tracking. Sig. Process. Image Commun. 12, 263–281 (1998)

Tracking Pedestrians Across Multiple Cameras via Partial Relaxation of Spatio-Temporal Constraint and Utilization of Route Cue

Toru Kokura[1]($\boxtimes$), Yasutomo Kawanishi[2],
Masayuki Mukunoki[3], and Michihiko Minoh[3]

[1] Graduate School of Informatics, Kyoto University, Kyoto, Japan
kokura@mm.media.kyoto-u.ac.jp
[2] Institute of Innovation for Future Society, Nagoya University, Nagoya, Japan
kawanishiy@murae.m.is.nagoya-u.ac.jp
[3] Academic Center for Computing and Media Studies,
Kyoto University, Kyoto, Japan
{mukunoki,minoh}@mm.media.kyoto-u.ac.jp

Abstract. We tackle multiple people tracking across multiple non-overlapping surveillance cameras installed in a wide area. Existing methods attempt to track people across cameras by utilizing appearance features and spatio-temporal cues to re-identify people across adjacent cameras. @ However, in relatively wide public areas like a shopping mall, since many people may walk and stay arbitrarily, the spatio-temporal constraint is too strict to reject correct matchings, which results in matching errors. Additionally, appearance features can be severely influenced by illumination conditions and camera viewpoints against people, making it difficult to match tracklets by appearance features. These two issues cause fragmentation of tracking trajectories across cameras. We deal with the former issue by selectively relaxing the spatio-temporal constraint and the latter one by introducing a route cue. We show results on data captured by cameras in a shopping mall, and demonstrate that the accuracy of across-camera tracking can be significantly increased under considered settings.

1 Introduction

We address the problem of tracking pedestrians across multiple non-overlapping surveillance camera views in a wide area. It has a lot of commercial applications in practice and great importance for business growth. For example, in shopping malls, global tracking results can be valuable for finding similar spots and planning shop reallocation for sales growth. Since acquiring such global tracking results demands enormous time, labor and cost, a method for doing it automatically is required.

To this end, existing methods attempt to track pedestrians across camera views by utilizing appearance features and certain spatio-temporal cue to re-identify and associate pedestrians across adjacent camera views [1–5]. Appearance features are usually based on color histograms and texture descriptors, and

© Springer International Publishing Switzerland 2015
C.V. Jawahar and S. Shan (Eds.): ACCV 2014 Workshops, Part III, LNCS 9010, pp. 587–601, 2015.
DOI: 10.1007/978-3-319-16634-6_43

the spatio-temporal cue is commonly based on the travel time between adjacent camera views, respectively.

In a shopping mall, since many pedestrians walk and stay arbitrarily no matter whether they are under the view of some camera(s) or outside of the views of all the cameras, their travel time between adjacent camera views varies from time to time. Additionally, appearance features can be influenced by camera viewpoint changes and illumination condition variations. These issues result in fragmented trajectories making difficult the global tracking across camera views, so that the existing methods fail to work when pedestrians walk and stay arbitrarily in wide public areas like a shopping mall. We deal with the former issue by relaxing the spatio-temporal cue selectively, and the latter one by introducing a route cue. Our goal is to make possible the global tracking in such important real scenarios.

2   Reated Work

In this paper, we assume that we have already acquired cropped pedestrian image sequences by intra-camera pedestrian tracking for each camera. Each sequence of pedestrian images is called a *tracklet*. Tracklets are gained by conducting tracking within a camera view [6–8].

Tracking pedestrians across multiple camera views can be achieved if for each pair of tracklets acquired in adjacent camera views, we are able to correctly judge whether they correspond to the same pedestrian or not, in another word, to successfully perform person re-identification. To this end, Farenzena *et al.* [2] proposed an effective feature which accumulates various kinds of information including clothings' color and texture.

Javed *et al.* [3] presented a method for tracking pedestrians across two adjacent camera views by utilizing a spatio-temporal cue. The spatio-temporal cue consists of a spatio-temporal likelihood and a spatio-temporal constraint. The former is the likelihood of travel time between two adjacent camera views, and is described as a probability distribution. The spatio-temporal likelihood is used to compare similarity between tracklets. The latter is the constraint that the travel time in which a pedestrian moves from a camera view to an adjacent camera view must be within a time span, namely, between a given *minimum travel time* and a given *maximum travel time* for the camera pair. The spatio-temporal constraint is used to reduce matching candidates for computational efficiency.

Liana *et al.* [4] increased the accuracy of tracking across two adjacent camera views by optimizing matching of pedestrians. They track multiple pedestrians simultaneously and acquire the optimal matching. If we can match tracklets across each pair of adjacent camera views, tracking across multiple camera views will be possible, however, once a matching error occurs, a tracking trajectory may get fragmented or wrongly connected, which leads to the failure in tracking over a wide area.

Song *et al.* [5] proposed a method for tracking pedestrians across multiple camera views. This method achieved higher matching accuracy between each camera pair by utilizing additional information; pedestrian appearance and observing time of tracklets acquired by other cameras than the camera pair, however, it requires vast computational cost. Chen *et al.* [1] presented a method for tracking pedestrians across multiple camera views with less computational complexity by restricting matching candidates with a spatio-temporal constraint.

Alahi *et al.* [9] introduced a concept of data association [10] for the across-camera tracking, and proposed a method to optimize all trajectories. In the method, optimal trajectories are acquired by calculating the posterior probability of each considerable trajectory.

3   Tracking Pedestrians Using a Spatio-Temporal Cue

For each tracklet $r_i (i \in \mathbb{N})$, those methods [1,5] focus on only a set of tracklets which were acquired before r_i was acquired. They choose tracklets for the set using a spatio-temporal constraint. By using the spatial constraint, they exclude tracklets acquired by a camera view not adjacent to the camera view where tracklet r_i was acquired. They also exclude tracklets whose travel time is not in a given time span. The time span is characteristic to each camera pair.

The left tracklets after the exclusion process are the candidate set for the tracklet r_i denoted by H_i. They compute similarities between tracklet r_i and every tracklet in H_i. For the computation of the similarity, they use two kinds of information; one is the spatio-temporal likelihood, more precisely, the likelihood of transition time; and the other is the appearance likelihood, i.e., the similarity of appearance features. If the similarity between tracklet r_i and its most similar tracklet in the candidate set H_i is greater than a given threshold, the two tracklets are considered to be matched. Otherwise, no matching is found and the tracklet r_i is considered to be the first tracklet of a trajectory, that is, the starting tracklet of a pedestrian in the camera network.

Such kind of methods have the following two drawbacks.

Firstly, when an observed pedestrian is significantly delayed between a pair of adjacent camera views, the travel time may go out of the given time span. A delay often happens when there are places visitable between the adjacent camera views, such as stores, restrooms, signboards, smoking areas, and exhibitions.

Secondly, except in a very simple environment such as a straight road, it is often difficult to install cameras to observe pedestrians from similar viewpoints. Thus, each pedestrian's appearance varies across camera views. In general, additionally, lighting condition varies with observation time because of weather, light intensity and existence of the sunlight. Since these factors may cause significant appearance change, they may lead to the wrong judgment that the corresponding tracklet does not exist in the candidate set, even though it actually does.

We give examples of these two problems. To make it easy for understanding, we focus on only three cameras from a possible complex camera network as shown in Fig. 1. The views of Cameras 1 and 2, and those of Cameras 2 and 3 are adjacent, and there is no direct path linking the views of Cameras 1 and 3 without passing other views.

The two problems we've concerned are summarized below:

- Problem(i): Because of the delay, the tracklet which should be matched is excluded from the candidate set.
- Problem(ii): Because of appearance variation, the tracklet which should be matched is considered not to exist in the candidate set, even though it does.

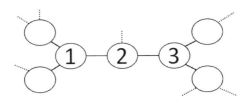

Fig. 1. Adjacency of camera views in a camera network

An example of the problem(i) is shown in Fig. 2. In the figure, each tracklet is shown as a set of stacked images of a pedestrian. The figure indicates that the tracklets r_1 and r_2 are acquired when the pedestrian P_A passed through the views of Camera 2 and Camera 1 respectively. The pedestrian corresponding to the other two tracklets are different from the pedestrian P_A. For tracklet r_2, the candidate set H_2 consists of tracklets existing in the rectangle in the figure. H_2 is the set to an extent satisfying the spatio-temporal constraint of the tracklets. The pedestrian got delayed, and therefore tracklet r_1 is not included in the rectangle. This results in a matching error.

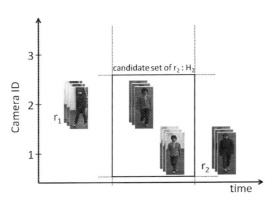

Fig. 2. Problem(i): delay

An example of the problem(ii) is illustrated in Fig. 3. The figure indicates that the tracklets r_3, r_4 and r_5 are acquired when the pedestrian P_B passed through the views of Cameras 1, 2 and 3 respectively. The pedestrian corresponding to the rest one tracklet is different from the pedestrian P_B. In this case, the candidate set H_5 consists of tracklets existing in the rectangle as marked in the figure. Although the tracklet which should match to r_5 is actually included in the candidate set, the pedestrian looks different because she is observed from different viewpoints by Camera 2 and Camera 3. Thus, appearances of the tracklets r_4 and r_5 are not similar enough, making the matching of them very difficult.

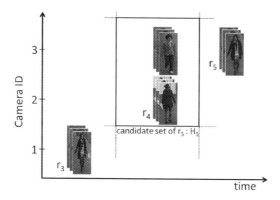

Fig. 3. Problem(ii): appearance variation

4   Tracking Pedestrians via Partial Relaxation of Spatio-Temporal Cue and Utilization of Route Cue

4.1   How to Solve the Problems

The proposed method copes with the two problems via the following two ideas respectively;

- Idea(i): By partially relaxing the spatio-temporal constraint only against pairs of tracklets whose appearance similarities are significantly high, we include the corresponding tracklets in the candidate set. The tracklet pairs are chosen from every camera in the camera network, not only from adjacent cameras.
- Idea(ii): By utilizing a route cue, we predict that the corresponding tracklet actually exists in the candidate set.

Details of the ideas are given below. Let r_i be a tracklet denoted by $r_i = (\mathbf{f}_i, s_i, e_i, c_i)$, where $\mathbf{f}_i$ is the sequence of the pedestrian's appearance features extracted from the cropped pedestrian image sequence within the view of camera c_i, s_i is the time when the first frame of this sequence is captured, e_i is the time when the last frame of the sequence is captured. Using this notation, the candidate set H_i of a tracklet r_i mentioned in the previous section can be defined as

$$H_i = \{r_m \mid (c_i, c_m) \in E,$$
$$t_{\min}(c_i, c_m) < s_i - e_m < t_{\max}(c_i, c_m)\}, \tag{1}$$

where $t_{\min}(c_i, c_m)$ denotes the minimum travel time between the views of camera c_i and camera c_m, and $t_{\max}(c_i, c_m)$ denotes the maximum travel time between the two cameras, and E denotes a set of camera pairs which have a direct path between them.

Here, we assume that when a pedestrian moved from one camera view to another camera view, a tracklet r_j is acquired from the former camera view and another tracklet $r_i(i \neq j)$ is from the latter. We focus on the situations when the problem(i) or problem(ii) occurs between the two tracklets.

Firstly, by selectively relaxing the spatio-temporal constraint in matching tracklets, we tackle the problem(i), where the corresponding tracklet deviates from the candidate set. When a tracklet pair that does not fulfill the spatio-temporal constraint because of delay, the tracklet r_j is excluded from the candidate set H_i. If the similarity of the two tracklets r_i and r_j is significantly high, we relax the spatio-temporal constraint for the tracklet r_j, namely setting $t_{\max}$ as ∞, to include the tracklet into the candidate set H_i. In this case, the spatio-temporal likelihood is not used to calculate similarity between tracklets r_i and r_j, and only appearance likelihood is used. This enables us to match two tracklets regardless of the duration of the delay.

The example of this process is shown in Fig. 4. The situation of this figure is the same as the situation of Fig. 2. In this figure, by relaxing the spatio-temporal constraint, the tracklet r_1 will be found in the relaxed candidate set H_2. Since appearance similarity of tracklets r_1 and r_2 is significantly high, they can be matched correctly.

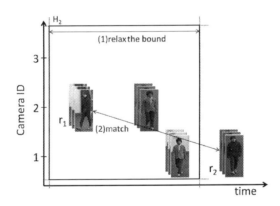

Fig. 4. Idea(i): partial relaxation of spatio-temporal cue

Secondly, we tackle the appearance variation by introducing a route cue. The route cue is a novel constraint that when a pedestrian is observed by two cameras whose views are not adjacent, the pedestrian should be observed by the other cameras whose views exist on a must-pass route between the two camera views. This constraint is an enhanced version of the spatio-temporal constraint, and ensures that the tracklet r_j actually exists in the candidate set. Here, let r_k, r_j

and r_i be the tracklets acquired when a pedestrian P_C passed through the views of three cameras respectively, and $s_k < s_j < s_i$. This process narrows the number of elements of H_i, and reduces the existence probability of a tracklet $r_l \in H_i$ where similarity(r_i, r_l) and similarity(r_j, r_l) are higher than similarity(r_i, r_k) and similarity(r_j, r_k). Here, similarity$()$ means the similarity between two tracklets. Purging such a tracklet r_l from the candidate set makes it possible to match r_j correctly. Additionally, when the existence of the tracklet r_j is ensured, it can be matched even if r_j's similarity against r_i and r_k is lower than the matching threshold. Even if appearance change happens, r_j belongs to the same person as r_k and r_j, thus r_j's similarity against r_i and r_k is still high to some extent.

Here, we will describe the process in detail. We consider the situation where there is a tracklet r_k acquired in the view of a camera $c_k (\neq c_i)$ before a tracklet r_i is acquired, and the views of cameras c_i and c_k are not adjacent. Let r_m be a candidate tracklet of the matching partner of r_i, the tracklet r_m must fulfill following two constraints;

- s_m and e_m of the tracklet r_m must fulfill following inequities:

$$\begin{cases} s_m - e_k > t_{\min}(c_k, c_m) \\ s_i - e_m > t_{\min}(c_m, c_i). \end{cases} \tag{2}$$

These inequities mean that the tracklet r_m is acquired at a time between the time when the tracklets r_k and r_i are acquired with consideration of the minimum travel time between pairs of the cameras.
- The view of camera c_m must exist on a route from the view of camera c_k to the view of camera c_i.

We consider only the tracklets fulfilling these two constraints as the candidate set. The tracklet which has the highest similarity to r_i in the candidate set H_i is matched with r_i.

The tracklet r_k very similar to r_i can be found by executing the matching with partial relaxation of the spatio-temporal constraint, if the similarity between r_i and r_k is higher than a given threshold. If the camera views which the tracklets c_k and c_i are acquired are not adjacent, we conduct the matching mentioned above.

By following the two constraints, we can reduce the number of elements in the candidate set H_i. Additionally, we can ensure that the candidate set H_i includes the tracklet r_j, thus even if similarity(r_i, r_j) and similarity(r_k, r_j) are lower than the given threshold, we can match the tracklets.

An example of this process is shown in Fig. 5. The situation of this figure is as same as the situation of Fig. 3. Because the appearances of tracklet r_3 and r_5 are similar enough, they are matched by relaxed spatio-temporal constraint. Since the views of camera $c_3 (= 1)$ and $c_5 (= 3)$ are not adjacent, by using the route cue, the corresponding tracklet to the tracklet r_5 must exist in the area enclosed by the rectangle in the figure. By optimally selecting the tracklet which has the highest similarity to r_3 and r_5 from the candidate set H_5, the tracklet r_4 can be correctly matched even if the similarity of the tracklet r_5 and r_4 is lower than the given threshold.

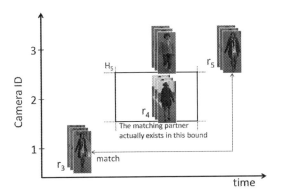

Fig. 5. Idea(ii): utilization of route cue

4.2   Proposed Tracking Procedure

Including these two ideas, we propose a novel tracking method. One of the novelties of the method is its deliberate procedure. We describe the procedure below.

We apply matching procedures using idea (i)(ii) to a tracklet set before applying the existing method [1]. Here, in order to apply matching with utilization of the route cue, the tracklet r_k must be found beforehand. Therefore, we should first perform matching with partial relaxation of the spatio-temporal cue. Before the two above-mentioned procedures, pairs of tracklets which have neither problem(i) nor problem(ii) should be matched with each other. This is easily achieved by searching for tracklet pairs whose spatio-temporal likelihood and appearance similarity are high. Here, the spatio-temporal likelihood between two tracklets means how similar the two tracklets are when considering a transition time distribution between two cameras which acquired the two tracklets.

Accordingly, the steps of matching should be the followings;

- Step 1: For each pair of tracklets acquired by adjacent cameras, if the wighted average of their spatio-temporal likelihood and appearance likelihood are higher than a given threshold, match them.
- Step 2: By relaxing the spatio-temporal constraint for all tracklets, if there are pairs of tracklets whose similarity of appearance is significantly high, match them.
- Step 3: Execute matchings utilizing the route cue.
- Step 4: By applying the existing method [1], try to match tracklets while fixing the matched pairs in above steps.

We empirically tuned the thresholds used in the step 1 and step 2 so strictly that false matchings don't happen. Matching tracklets with low similarity invokes false matchings, thus strict thresholding can prevent them.

Each process conducts locally optimal matching, and after all 4 steps finish, an optimal solution is obtained.

5   Benchmark Datasets

To evaluate the robustness of the proposed method against the delay and the appearance change, we should evaluate on many datasets collected from various environments which have different rates of the delay and appearance change.

However, these datasets are difficult to prepare, especially when controlling the rates of the delay and the appearance change are considered. Instead of trying hard to build such "real" datasets, we generate multiple "virtual" datasets and use them for evaluation. If we prepare some "real" datasets, very similar datasets must be included in the virtual datasets.

Here, we describe how to generate the virtual datasets from the Shinpuhkan 2014 dataset collected from surveillance cameras mounted in a real shopping mall [11]. We denote the original Shinpuhkan 2014 dataset by $\tilde{D}$. This dataset $\tilde{D}$ consists of tracklets $\tilde{R}$ of 24 pedestrians, which are captured by 16 non-overlapping camera views. These cameras cover both shade areas and sunny areas within the mall.

We consider the rate of the delay and the appearance change happen as parameters of a virtual dataset and generate various virtual datasets which contain tracklets generated virtually while changing the parameters. Here, we first describe the parameters for the virtual dataset, and then we describe how to generate a virtual dataset with these parameters.

5.1   Parameters for the Virtual Dataset

We want to generate virtual datasets with different rates of the delay and appearance change. We define a virtual dataset by $D(\beta, N_{cam}, N_{sun}; \tilde{D})$, where β denotes the parameter of the delay happening, N_{cam} denotes the number of camera views we use in the dataset and N_{sun} denotes the parameter controlling the appearance change.

To simulate the delay of pedestrians, we parametrize the probability of the delay happening at $\beta \in [0, 1]$. In the virtual dataset, the pedestrians delay at a probability of β. Once a pedestrian delays, the pedestrian spent more t_{delay} seconds between two camera views. t_{delay} is sampled randomly from a given uniform distribution.

To simulate the appearance change, we parameterize the probability of the appearance change happening by the number of camera views N_{sun} which are under the sunlight. The greater the value of N_{sun}/N_{cam} is, the more often appearance changes happen. It is because the appearance of a pedestrian is heavily affected by the illumination condition, especially whether it is under the sunlight or not. Though also the direction variation of people against cameras causes appearance variations, we didn't control it. It is because the dataset $\tilde{D}$ originally contains the direction variation, which makes the direction variation automatically contained in generated datasets, and it is difficult to control how much is contained.

If a pedestrian is observed in two camera views which observe shaded areas, appearances of the pedestrian in the two images captured by the two cameras are similar. However, if the pedestrian is observed under the sunlight in one camera view, appearance of the pedestrian will be different from the one observed under the other camera. We show those examples in Fig. 6. Three images in Fig. 6 represent the same person. Figure 6(a) is captured in the shaded area and Fig. 6(b) and (c) are captured under the sunlight. We can see the big difference between Fig. 6(a) and the others. Appearance variation also exists between Fig. 6(b) and (c) caused by the direction of the sunlight against the person.

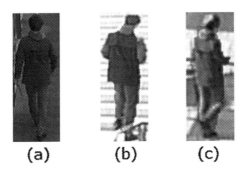

(a)   **(b)**   **(c)**

Fig. 6. Illumination change and appearance variation

How to Generate a Virtual Dataset. We generate a virtual dataset by simulating pedestrians' walk over a virtual camera network. Therefore, the dataset generation consists of two parts: the virtual camera network generation and the simulation of pedestrians' walk, namely, the virtual tracklets generation.

Virtual Camera Network Generation. First, we divide the vertices $\tilde{V}$ in the camera network $\tilde{G}$ of the original dataset $\tilde{D}$ into two groups $\tilde{V}_{sun}$ and $\tilde{V}_{shade}$ by lighting condition of the camera. We select N_{sun} vertices from $\tilde{V}_{sun}$ and $N_{cam} - N_{sun}$ vertices from $\tilde{V}_{shade}$ and we denote the set of the selected vertices by V.

Then, we generate N_{path} pedestrian path candidates over V. For each subset $V_1, \ldots, V_{N_{path}}$ of V, we generate a Hamiltonian paths $\Pi_1, \ldots, \Pi_{N_{path}}$. Letting all the edges contained in the generated Hamiltonian paths be an edge set E, we get a camera network $G = (V, E)$.

Virtual Tracklets Generation. The original dataset $\tilde{D}$ consists of the tracklets $\tilde{R}$ of 24 pedestrians. We generate a set of virtual tracklets R by generating virtual tracklets for each pedestrian in $\tilde{D}$.

We randomly ordered the pedestrians and denote them as $\{p_1, \ldots, p_{24}\}$. For each pedestrian $p_k \in \{p_1, \ldots, p_{24}\}$, we randomly selected a path $\pi_k \in \{\Pi_1, \ldots, \Pi_{N_{path}}\}$. For each camera view corresponding to a vertex in the path π_k, we select a tracklet of the person p_k, and we select a sequence of real tracklets of the pedestrian p_k along the path π_k from $\tilde{R}$.

To generate virtual tracklets, we need to change the observation time for each tracklets. The observation time of tracklets of a pedestrian p_k is determined by the time point $t_0^{p_k}$ when the pedestrian is observed in the camera network at the first time, the time span $\Delta t_i^{p_k}$ in which the pedestrian passes through the camera view corresponding to the vertex $v_i^{p_k}$, and the time span $\Delta t_{i,i+1}^{p_k}$ in which the pedestrian travels from the camera view corresponding to the vertex $v_i^{p_k}$ to the next camera view corresponding to the vertex $v_{i+1}^{p_k}$ of the camera network along the path $\pi_k = (v_1^{p_k}, \ldots, v_{N_{\pi_k}}^{p_k})$.

We set a given time point to $t_0^{p_0}$. Then we determine the time points $t_0^{p_k}$ iteratively by sampling the time span $(t_0^{p_k} - t_0^{p_{k+1}})$ from a given gamma distribution. Gamma distribution is often used to model a time span in which something occurs.

For the time span $\Delta t_i^{p_k}$, we simply used the original time span which the pedestrian p_k actually spent to pass through the camera view.

To determine the time span $\Delta t_{i,i+1}^{p_k}$, we need to model the travel time of pedestrians between two camera views. Here, we model it by sum of the time span $t_{ordinary}(v_i^{p_k}, v_{i+1}^{p_k})$, which pedestrians ordinary spent to travel from the camera view denoted by the vertex $v_i^{p_k}$ to the next camera view denoted by the vertex $v_{i+1}^{p_k}$ and the time span t_{delay} which denotes the delay time as;

$$\Delta t_{i,i+1}^{p_k} = t_{ordinary}(v_i^{p_k}, v_{i+1}^{p_k}) + \delta t_{delay}, \tag{3}$$

where $t_{ordinary}(v_i^{p_k}, v_{i+1}^{p_k})$ is sampled from a given gamma distribution for each pair $(v_i^{p_k}, v_{i+1}^{p_k})$ and t_{delay} is sampled from a given uniform distribution between 5 min to 60 min. δ is a controlling parameter sampled from a Bernoulli distribution with the parameter β.

Then we can calculate the observation time for each virtual tracklet from $t_0^{p_k}$, $\Delta t_i^{p_k}$, and $\Delta t_{i,i+1}^{p_k}$, that is, we can calculate the entrance time s_i and the exit time e_i of a virtual tracklet r_i. Finally, we get a set of virtual tracklets R.

6   Evaluation

6.1   Experiment Configurations

We generated various virtual datasets by the procedure introduced in the previous section to show the effectiveness of our proposed method.

Virtual Dataset Parameters. We changed the delay parameter β from 0 % to 100 % by 10 %. Camera networks were randomly generated for each virtual dataset with the parameters $N_{cam} = 5$ and $N_{path} = 4$. We had two paths randomly generated and the rest two set to be inverse of them. We set the start vertices of the former two paths to be the same vertex, and also set the end vertices of the paths to be the same vertex. We also changed the appearance parameter N_{cam} from 0 to 5. Parameters of every distribution were determined empirically. We generated 50 different datasets for each parameter set of β and N_{sun}.

Calculation of Similarity Between Tracklets. We calculated the similarity between tracklets as a weighted average of the appearance likelihood and the temporal likelihood.

We calculated appearance likelihood $l_{app}(r_i, r_j)$ between tracklets r_i and r_j as follows;

$$l_{app}(r_i, r_j) = \min_{a,b} d(\mathbf{f}_i^a, \mathbf{f}_j^b), \tag{4}$$

where $\mathbf{f}_i^a$ is the appearance feature extracted the a_{th} frame of tracklet r_i, and d is the Bhattacharyya coefficient, which is used to calculate distance between histograms or probability distributions. We described the appearance feature of pedestrian images by using Weighted Color Histogram [2].

Methods for Comparison. We compared our method with the traditional method [1] which was proposed by Chen et al.. The method is equivalent to the step 4 of our method. The spatio-temporal constraint works well under low rate of delay happening, but it makes the result worse under high rate of delay happening. To evaluate the effect of the spatio-temporal constraint, we compared the two methods with 2 different settings: with/without the spatio-temporal constraint in the step 4. For the "without the spatio-temporal constraint" setting, we set the maximum travel time t_{max} to ∞.

- traditional F The traditional method [1] without the spatio-temporal constraint.
- proposed F The proposed method without the spatio-temporal constraint.
- traditionalST F The traditional method [1] with the spatio-temporal constraint.
- proposedST F The proposed method with the spatio-temporal constraint.

6.2   Evaluation Criterion

We evaluated the tracking results by measuring the correctness for each sequence of tracklets. As the evaluation criterion, we utilized F-measure, which is usually used for the accuracy evaluation in the field of information retrieval. To define F-measure in this evaluation, we introduce the label series l over a tracking result, where matched tracklets should have the same label and not matched tracklets should have different labels.

F-measure is defined as the harmonic average of the precision and the recall. Let the series of these labels $l = (l^1, \ldots, l^N)$, where l^i represents the label assigned to a tracklet r_i. We defined F-measure as follows;

$$F\text{-}measure(l) = \frac{2 \cdot precision(l) \cdot recall(l)}{precision(l) + recall(l)}. \tag{5}$$

We defined the precision as the percentage of correct matchings, and defined the recall as the percentage of conducted matchings in ground truth as follows;

$$precision(l) = \frac{|TP(l)|}{|L(l)|}, \tag{6}$$

$$recall(\boldsymbol{l}) = \frac{|TP(\boldsymbol{l})|}{|L(\hat{\boldsymbol{l}})|}. \tag{7}$$

where $TP(\boldsymbol{l})$ and $FP(\boldsymbol{l})$ denote the set of true positives and the set of false positives respectively, and $L(\boldsymbol{l})$ is the linkage set. $\hat{\boldsymbol{l}}$ is the ground truth of the label series.

$TP(\boldsymbol{l})$ and $FP(\boldsymbol{l})$ are defined using the linkage set $L(\boldsymbol{l})$ as follows respectively;

$$TP(\boldsymbol{l}) = \{x | x \in L(\boldsymbol{l}) \wedge x \in L(\hat{\boldsymbol{l}})\}, \tag{8}$$

$$FP(\boldsymbol{l}) = \{x | x \in L(\boldsymbol{l}) \wedge x \notin L(\hat{\boldsymbol{l}})\}. \tag{9}$$

We defined the linkage set $L(\boldsymbol{l})$ as follows;

$$L(\boldsymbol{l}) = \{x = (g_q^j, g_q^{j+1}) | \, q < \max_i l^i \wedge (g_q^j, g_q^{j+1}) \in G_q\}, \tag{10}$$

where

$$G_q = \{i | l^i = q\}. \tag{11}$$

Based on the definition, we calculated F-measure against all of the 4 methods listed in the previous subsection.

6.3   Results

We show the results of the case $N_{sun} = 0$ and $N_{sun} = 2$ in Figs. 7 and 8 respectively. Both figures are plotted with F-measure as the vertical axis and the delay probability β as the horizontal axis. Each line represents the average of the results of 50 trials.

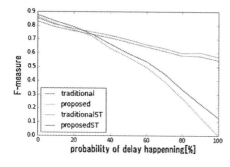
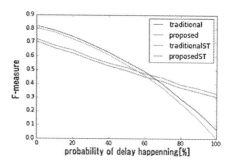

Fig. 7. The result where $N_{sun} = 0$ **Fig. 8.** The result where $N_{sun} = 2$

Whatever value the delay probability β is, F-measure of the proposed method exceeds that of the existing method. The difference of F-measure between the proposed method and the existing method grows greater as the delay probability β gets larger. This suggests that our method can deal with the problem (i): the problem of delay happening.

Furthermore, whether there are cameras observing the sunshine in the datasets or not, F-measure of the proposed method is greater than that of the existing method. From this, we confirmed that our method can deal with the problem (ii): the problem of appearance variation.

Based on the above results, we can conclude that our tracking method can deal with both problems of delay happening and appearance changes.

7 Conclusion

This paper has presented a method for tracking pedestrians across multiple non-overlapping camera views by selectively relaxing the spatio-temporal cue and introducing the route cue. We have shown that the proposed method can consistently improve the tracking accuracy under different parameter settings of delay and appearance change.

The possible future work includes modeling the travel time statistically in the situation where a person has a heavy delay and setting the criteria of relaxation ratio for the spatio-temporal cue.

Acknowledgement. This work was supported by "R&D Program for Implementation of Anti-Crime and Anti-Terrorism Technologies for a Safe and Secure Society," Funds for integrated promotion of social system reform and research and development of the Ministry of Education, Culture, Sports, Science and Technology, the Japanese Government.

References

1. Chen, K.Y., Huang, C.L., Hsu, S.C., Chang, I.C.: Multiple objects tracking across multiple non-overlapped views. In: Proceedings of Pacific-Rim Symposium on Image and Video Technology, pp. 128–140 (2011)
2. Farenzena, M., Bazzani, L., Perina, A., Murino, V., Cristani, M.: Person re-identification by symmetry-driven accumulation of local features. In: IEEE Conference on Computer Vision and Pattern Recognition, pp. 2360–2367 (2010)
3. Javed, O., Rasheed, Z., Shafique, K., Shah, M.: Tracking across multiple cameras with disjoint views. In: Proceedings of International Conference on Computer Vision, pp. 952–957 (2003)
4. Lian, G., Lai, J.H., Zheng, W.S.: Spatial-temporal consistent labeling of tracked pedestrians across non-overlapping camera views. Pattern Recogn. **44**, 1121–1136 (2011)
5. Song, B., Chowdhury, A.K.R.: Robust tracking in a camera network: a multi-objective optimization framework. IEEE J. Sel. Top. Sig. Process. **2**(4), 582–596 (2008)
6. Hofmann, M., Haag, M., Rigoll, G.: Unified hierarchical multi-object tracking using global data association. In: IEEE International Workshop on Performance Evaluation of Tracking and Surveillance, pp. 22–28 (2013)
7. Zhang, L., Li, Y., Nevatia, R.: Global data association for multi-object tracking using network flows. In: IEEE Conference on Computer Vision and Pattern Recognition, pp. 1–8 (2008)

8. Pirsiavash, H., Ramanan, D., Fowlkes, C.C.: Globally-optimal greedy algorithms for tracking a variable number of objects. In: IEEE Conference on Computer Vision and Pattern Recognition, pp. 1201–1208 (2011)
9. Alahi, A., Ramanathan, V., Fei-Fei, L.: Socially-aware large-scale crowd forecasting. In: Proceedings of the IEEE Conference on Computer Vision and Pattern Recognition, pp. 2203–2210 (2013)
10. Bar-Shalom, Y., Tse, E.: Tracking in a cluttered environment with probabilistic data association. Automatica **11**, 451–460 (1975)
11. Kawanishi, Y., Wu, Y., Mukunoki, M., Minoh, M.: Shinpuhkan 2014: a multi-camera pedestrian dataset for tracking people across multiple cameras. In: 20th Korea-Japan Joint Workshop on Frontiers of Computer Vision, pp. 232–236 (2014)

Discovering Person Identity via Large-Scale Observations

Yongkang Wong[1]([✉]), Lekha Chaisorn[1], and Mohan S. Kankanhalli[1,2]

[1] Interactive and Digital Media Institute, National University of Singapore,
Singapore, Singapore
`yongkang.wong@nus.edu.sg`
[2] School of Computing, National University of Singapore, Singapore, Singapore

Abstract. Person identification is a well studied problem in the last two decades. In a typical automated person identification scenario, the system always contains the prior knowledge, either person-based model or reference mugshot, of the person-of-interest. However, the challenge of automated person identification would increase by multiple folds if the prior information is not available. In today's world, rich and large quantity of information are easily attainable through the Internet or closed-loop surveillance network. This provides us an opportunity to employ an automated approach to perform person identification with minimum prior knowledge, presume that there are sufficient amount of observations. In this paper, we propose a dominant set based person identification framework to learn the identity of a person through large-scale observations, where each observation contains instances from various modality. Through experiments on two challenging face datasets we show the potential of the proposed approach. We also explore the conditions required to obtain satisfy performance and discuss the potential future research directions.

1 Introduction

Today we are living in a world of big data. With the recent advance in hardware technology, telecommunication protocol, and the growing popularity of social media, we are blessed with the rich and large quantity of data that are easily attainable through the Internet or closed-loop surveillance network. However, while it is relatively easy, albeit expensive, to install servers to handle the increasing demand on storage and computational performance, it is quite another issue to adequately monitor and analyze the data, big data. On the other hand, big data has provided the research community new research opportunities and directions. Therefore, the question is how can we utilize the big data for better innovations.

In the last two decades, person identification has received a lot of attention from the computer vision and machine learning community [1–6]. For example, iris recognition [1], fingerprint recognition [2], face recognition [3–5], and gait recognition [6]. In the literature, any of the person identification problem can

© Springer International Publishing Switzerland 2015
C.V. Jawahar and S. Shan (Eds.): ACCV 2014 Workshops, Part III, LNCS 9010, pp. 602–616, 2015.
DOI: 10.1007/978-3-319-16634-6_44

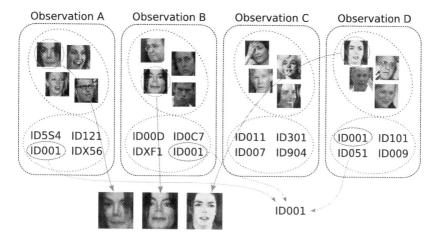

Fig. 1. Conceptual example of instance association based person identification with multiple observations. Given four unique observations, where each observation contains a number of instances of the "personal ID" and "facial image". We can analyze the IDs to select a subset of salient observations (i.e., A, B and D) for ID0001, followed by identify the candidate images with high internal coherency. Note that the relationship between the identity labels and images are unknown.

be generalized into three distinct configurations: closed-set identification, open-set identification, and verification [7]. The task of closed-set identification is to classify a given face as belonging to one of the K previously seen persons in a gallery, where as the open-set identification take into account the possibilities of impostor attack with an additional "unknown person" class. The task of verification is to determine if two given samples belong to the same individual, where one or both identities may not have been observed beforehand [3].

In this work, we aim to address a different category of identification problem, namely *instance association* based person identification, where the identity of each previously unseen individual is learned through single observation [8,9] or a large number of observations. The identification problem on hand is a weakly label learning problem, where each observation contains multiple instances and labels without given relationships. For example, assume that we can access to a large number of articles (*observations*) and each article contains multiple facial images and name-entities (*instances*). If there exist a genuine subset of salient observations, such that these observations all contain name-entities from a unique individual (denoted as salient ID) and no other name name-entities co-occur in the exact set of observations, it is intuitive to assume that the same facial image that co-occurs in these observations represent the salient ID. Under this scenario, the identification task can only be performed if enough samples are observed at several observations. Note that if the association between a name-entity and the correspond facial image is provided as prior knowledge, i.e., a facial image and its genuine identity are given, it will be classified as open/closed-set identification problem. A conceptual example of instance association based identification problem with multiple observations is shown in Fig. 1.

Here, we illustrate two real-world surveillance applications. Assume that we are provided with video feeds from Closed-Circuit Television (CCTV), as well as the datalog from the wireless access points and access control system, we can form a large collection of spatial-temporal observations over time, where each observation contains video footage and electronic signals appear at a specific location and duration. In an event of criminal act, a security officer can shortlist a number of candidates (e.g., electronic signal detected at point A), then the system can analyze the achieved observations to short-list the visual images of these candidates. Another practical example is to associate the visual image of the holder of a stolen access card. It is important to note that CCTV generally does not cover all the area. Hence, the identification on hand would need to consider the problem of incomplete observations from visual data. We will address this realistic scenario in the future work.

In this work, we proposed a dominant set based person identification framework to learn the identity of a person through multiple observations, where each observation contains instances from two modalities (i.e., name-entity and facial image). We consider the learning problem as a graph labeling problem, where the instances in each observation are considered as a vertex set of graph. By using an intuitive salient observations detection stage, the problem of graph labeling can be cast as a dominant set clustering problem. Given the dominant set clustering output and consider the structure of observations we proposed three instance selection approaches to perform person identification. The proposed framework is quantitatively and qualitatively evaluated on two challenging face datasets, where one of the dataset contains large number of face images obtained from the Internet. To understand the limitation of this problem, we simulated a number of variation of co-occurrence rate in the salient observations.

Contributions. In this work, we describe a person identification problem, namely *instance association based person identification*, which assume the identity of a person is not directly given but available in a weakly label manners. The described problem is realistic and applicable to several surveillance applications. We propose to use dominant set based approach to addressed this identification task and mathematically outline the problem as a graph labeling problem. Experimental results on two challenging face datasets under various constraints shows the potential of the proposed approach.

We continue the paper as follows. Section 2 describes related work. We formulate the problem of instance association based person identification in Sect. 3, where the proposed framework is presented and discussed in Sect. 4. Section 5 is devoted to experiments and Sect. 6 provides the main findings and future directions.

2   Related Work

Person identification, or identity inference, has received significant attentions over the last two decades. Generally, the person identification task always assumes the mugshots of person-of-interest are always known, where the task of identification is to either classify a given face as belonging to one of the previously seen persons in a gallery (i.e., closed-set identification), or belonging to an unknown individual (i.e., open-set identification) [7]. In the literature, there exist very little work to perform person identification without the gallery of persons. One related work is to associate people appearing in news video with their names [8–10]. The task of this work is to find the video segments where a person appears and associate the person with a name. Satoh *et al.* [8] explored the co-occurrence of facial images, video caption, and transcript in a news video. The underlying idea is that the (similar) faces that frequently co-occur with a certain name are likely to match the name and vice versa. Houghton [10] designed an automated system to create a named faces database, which utilize a similar approach as [8] to analyze the content on websites and news videos. Yang and Hauptmann explored the *features* and *constraints*, which reveal the relationships among the names of different people, to perform name association in news video [9]. We note that the aforementioned problem is focus on a single observation scenario. In addition, the validity of co-occurrence might not whole for all observation. For example, a news video can contains many faces without mentioned all the names, or without any facial images of the identity of interest. In contrast, the assumption of co-occurrence is more likely to hold with large number of observations, ideally on a scale of big data. However, the difficulty of the learning problem is also increased dramatically.

Different from the names and face association application, Cho *et al.* [11,12] proposed a system to dynamically associate the personal identification, obtained via RFID system, fingerprint, iris recognition, etc., to the persons observed by the visual sensors. The system assumed there exists a gate region to collect the personal identity and each individual is continuously detected and tracked in the premises. In this system, the association is achieved with heterogeneous sensors of visual sensors and identification sensors, which required carefully calibrated information to associate the identity to the individual in a particular location. Our proposed problem does not have this constraint.

The instance association based person identification problem is related to Multi-Instance Multi-Label (MIML) learning problem [13,14]. Under the formulation of MIML [13], each object is described by multiple instances and associated with multiple class labels. The goal is to learn a classifier to classify the genuine instance for each label. This formulation is very useful for complex object with multiple semantic meaning, such as scene recognition, text categories classification, etc. In the domain of this person identification problem, the instances (a.k.a. personal attribute) can be extracted from various modality. Specifically, the personal attributes can be categorized into biometrics attributes or non-biometrics attribute. The biometric attribute include gait, iris, facial image, fingerprint, hand-geometry, etc., where the non-biometric attribute are name, age, contact information (such as email, phone number, etc.), and personal affiliation. For future research direction, this person identification learning problem can be modeled as a multi-modality instances learning problem. We will address this in future work.

3   Problem Formulation

3.1   Instance Association Based Person Identification via Graph Labeling

In the context of big data, we assume that each individual can be observed with a variety of sensors. Through the observations from these sensors, the data can be modeled as a super-set of collection $\mathbb{O}$, which comprises of a finite number of local observations $\boldsymbol{O}^m$ for $m = 1, 2, \ldots, M$. Each of the m-th observation consist of a collection of instances from a specific spatial-temporal subspace. In addition, the instances collected in each observation belong to a dedicated modality and the relationship between the instances are unknown. For example, $\boldsymbol{O}^m$ can consists of all the faces and audio recording segments collected over a short period of time at hotel lobby, or the faces and names from a newspaper.

We first consider the graph labeling problem using instances from a single modality. For example, gait, iris, facial image, etc. Hence, each observation $\boldsymbol{O}^m$ consists of one bag of instance $\mathbb{X}^{m,s} = \{\boldsymbol{x}_1^{m,s}, \boldsymbol{x}_2^{m,s}, \ldots, \boldsymbol{x}_N^{m,s}\}$, where $\boldsymbol{x} \in \mathbb{R}^d$ is a d-dimensional vector from modality s. Each instance is considered as a vertex and connected to its neighbors through undirected edges that having positive weights. Given a local observation set $\mathbb{O} = \{\boldsymbol{O}^1, \boldsymbol{O}^2, \ldots, \boldsymbol{O}^M\}$, the task of a graph label $\varphi : \mathbb{O} \mapsto \widehat{\mathcal{X}}$ is to produce $\widehat{\mathcal{X}} = \{\widehat{\boldsymbol{x}}_1, \widehat{\boldsymbol{x}}_2, \ldots, \widehat{\boldsymbol{x}}_c\}$, where ideally instances in $\widehat{\mathcal{X}}$ belong to the same target $\mathcal{T}$.

Now, lets consider the scenario of multi-modality graph labeling problem. Here, the m-th observation is re-written as $\boldsymbol{O}^m = \{\mathbb{X}^{m,1}, \mathbb{X}^{m,2}, \ldots, \mathbb{X}^{m,S}\}$ where $\mathbb{X}^{m,s}$ is a bag of instances from modality s. Follow the scenario of the single modality approach, the task is to employ a graph label function $\varphi : \mathbb{O} \mapsto \widehat{\mathcal{X}}$ to produce $\widehat{\mathcal{X}}$. Differing from the single modality scenario, we are now facing the problem of matching instances from different modality. Note that the latent relationship of instances from different modality may be unknown in our problem, e.g., a voice pattern may not be associated to any face images in an observation. In this work, we formulate the multi-modality person identification problem as iterative multi-stage clustering problem, where each iteration is dedicated to identify a randomly selected individual. Specifically, we perform the following task in each iteration. Given the mega set of observations $\mathbb{O}$, we employ a salient observation classifier $\mathcal{F} : \mathbb{O} \mapsto \mathbb{O}^l$ to produce $\mathbb{O}^l$, where each observation in $\mathbb{O}^l$ contains at least one instance that belong to target $\mathcal{T}^l$. Note that $\mathbb{O}^l$ can be extracted with non-biometrics instances, such as electronics signal or name-entity, which give high confidence decision when compared to biometrics instances (e.g., facial images or gait). Given the newly extracted $\mathbb{O}^l$, the graph labeling problem is now become a dominant-set detection problem, where the dominant set can be extract with $\varphi_{\mathrm{DS}} : \mathbb{O}^l \mapsto \widehat{\mathcal{X}}$.

4   Proposed Method

In this section, we elaborate the proposed framework for the instance association based person identification problem. In this work, we evaluate the propose

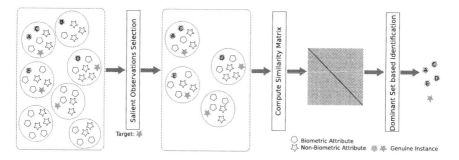

Fig. 2. Conceptual diagram of the proposed instance association based person identification framework. In this example, the system is assigned to perform person identification for a particular target.

framework with two types of personal attribute, i.e., facial image (biometric attribute) and person ID (non-biometric attribute), where the generalization to other types of modality is discussed in Sect. 6. To reduce the complexity of this work, we assume that the face image and ID of the same individual will always co-occur in the same observation, where the scenario of missing data will be addressed in future work. The proposed framework is an iterative person identification process and each iteration comprised of three components: **(1)** salient observations detection, **(2)** compute similarity matrix with biometrics instances, and **(3)** person identification with dominant set analysis. Given a finite set of observations, $\mathbb{O}$, the task of the proposed framework is to iteratively analyze each of the available person IDs, and perform person identification task if the detection criteria (see Sect. 4.1) is satisfied. A conceptual overview of the proposed method is shown in Fig. 2.

4.1 Salient Observations Detection

Given a finite set of observations, $\mathbb{O}$, and an ID-of-interest (IDoI), l, the task is to find a subset of salient observations, $\mathbb{O}^l$, which satisfy two criteria: **(1)** all observations should contain at least one instance of IDoI, and **(2)** the IDs that not belong to IDoI should not be observed in the same set of observations. This is to ensure that each salient observations has at least one genuine face image from each of the salient observations. Given the detected $\mathbb{O}^l$, we refer to the face images that belong to IDoI as genuine and the remaining images as imposter.

To measure the quality of the detected $\mathbb{O}^l$ with respect to the person identification problem, we define two properties, namely *observability* and *observation noise*. The *observability* is a binary attribute which combines the two aforementioned criteria. If the observability of a l is false, this indicate that $\mathbb{O}^l$ can not be used to find the associated genuine image of l. This generally happen if IDoI only appear in one observation or there exist other ID(s) that co-occurs in the same set of observations. The *observation noise*, denoted as ON, is a scalar attribute that correlate the rate of an imposter ID that co-occur in $\mathbb{O}^l$. ON $= 1.0$ indicates there exist an imposter that co-occur in all observations in $\mathbb{O}^l$. Lower value of ON lead to better chance of successful person identification.

4.2   Similarity Matrix with Face Verification

Given the IDoI l and the extracted salient observations $\mathbb{O}^l = \{\boldsymbol{O}_1^l, \boldsymbol{O}_2^l, \dots, \boldsymbol{O}_M^l\}$ where the m-th observation has $|\boldsymbol{O}_m^l|$ face images, we can represent the face images as an undirected edge-weighted (similarity) graph with no self-loops $G = (\boldsymbol{V}, \boldsymbol{E}, \boldsymbol{w})$. Here, $\boldsymbol{V} = \{1, \dots, n\}$ is the vertex set, $\boldsymbol{E} \subseteq \boldsymbol{V} \times \boldsymbol{V}$ is the edge set, and $\boldsymbol{w} : E \to \mathbb{R}_+^*$ is the (positive) weight function. We define a $N_\mathbb{O} \times N_\mathbb{O}$ symmetric matrix $\boldsymbol{W}$ of pairwise similarities between candidate face images, where $N_\mathbb{O} = \sum_{m=1}^M |\boldsymbol{O}_m^l|$ is the total number of face images in the salient observations. Using the symmetric similarity matrix, the goal is to perform graph-based clustering method (see Sect. 4.3) to retrieve a dominant set of instances, where the dominant instances form the most coherent subset.

Image Representation: In this work, we employ a local feature-based face representation, namely Locally Sparse Encoded Descriptor (LSED), which has shown good robustness against various alignment errors and pose mismatches with various face dataset, as well as its simplicity in implementation [3]. Briefly, a given face image is first split into R fixed size regions, followed by a secondary split into small overlapping blocks with a size of 8×8 pixels. Each block is represented by a low-dimensional texture descriptor, $\boldsymbol{y}$, followed by a sparse coding based encoding method to obtain a block level sparse descriptor. The r-th region descriptor, $\boldsymbol{h}_r$, is computed by pooling the block level sparse descriptor from region r with average pooling operation. Due to the relaxed spatial constraints within each region, it allows some movement and/or deformations of the face components and leads to a degree of inherent robustness to expression and pose changes [3,15]. In this work, we select the implicit sparse encoding via probabilistic approach in [3] due to its robust performance under various image conditions and light weight computational cost. The i-th block level sparse descriptor in region r can now be computed via:

$$\boldsymbol{h}_{r,i} = \left[\frac{w_1 p_1\left(\boldsymbol{y}_{r,i}\right)}{\sum_{g=1}^G w_g p_g\left(\boldsymbol{y}_{r,i}\right)}, \cdots, \frac{w_1 p_G\left(\boldsymbol{y}_{r,i}\right)}{\sum_{g=1}^G w_g p_g\left(\boldsymbol{y}_{r,i}\right)} \right]^T \tag{1}$$

where the g-th element in $\boldsymbol{h}_{r,i}$ is the posterior probability of $\boldsymbol{y}_{r,i}$ according to the g-th components of a Gaussian Mixture Models, and w is the associated weight. We direct the user to [3] for details descriptions. A conceptual example of LSED face descriptor can be found on Fig. 3.

Similarity Matrix: Considering a pair of face images A and B, the similarity is then defined as the inverse of a cohort normalization [16] based distance, written as

$$w_{A,B} = \frac{1}{d_{\mathrm{norm}}(A, B)} \tag{2}$$

where

$$d_{\mathrm{norm}}(A, B) = \frac{d_{\mathrm{raw}}(A, B)}{\sum_{i=1}^{N_C} s_{\mathrm{raw}}(A, C_i) + \sum_{i=1}^{N_C} s_{\mathrm{raw}}(B, C_i)} \qquad (3)$$

The cohort faces C_i are assumed to be reference faces that are different form the images of persons A or B. The raw distance, $d_{\mathrm{raw}}(A, B)$, is obtained by compare the corresponding face regions of the two images using Euclidean distance.

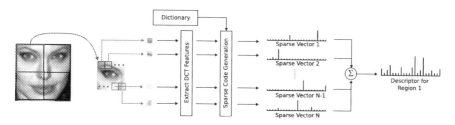

Fig. 3. Conceptual example of Locally Sparse Encoded Descriptor for face images. The local image patches are encoded locally via sparse representation based encoding, where the region level pooling enforce the regional structure information of face images.

4.3 Dominant Set Based Person Identification

Based on the discussion from Sect. 4.1, it is intuitive that each observation in $\mathbb{O}^l$ contains at least one instance of face image that belong to the IDoI l. Therefore, the person identification problem can be re-casted as a dominant set clustering problem. Following [17], the cluster of vertices can be associated to a $N_{\mathbb{O}}$-dimensional vector, $\boldsymbol{x}_{\mathrm{DS}}$, where its components express the *participation* of nodes in the cluster. Intuitively, the dominant components in $\mathbb{O}^l$ should be belong to the IDoI l, where the associated components will have large value in $\boldsymbol{x}_{\mathrm{DS}}$. One way to define the *cohesiveness* of a cluster is given by the following quadratic form [17]:

$$f(\boldsymbol{x}_{\mathrm{DS}}) = \boldsymbol{x}_{\mathrm{DS}}^T \boldsymbol{A} \boldsymbol{x}_{\mathrm{DS}} \qquad (4)$$

where the element i, j of $\boldsymbol{A}$, $a_{i,j}$, is equal to $w(i, j)$ if $(i, j) \in \boldsymbol{E}$. Note that $\boldsymbol{A}$ is an undirected graph with no self-loops, therefore all the element in the main diagonal of $\boldsymbol{A}$ is zero. Now, the dominant set clustering problem can be solve by finding an optimum vector $\boldsymbol{x}_{\mathrm{DS}}$ to maximize f.

One way to find the optimal solution is with the *replicator equation* [18]. First, each element of $\boldsymbol{x}_{\mathrm{DS}}$ is initialized to $1/|N_{\mathbb{O}}|$, followed by iteratively solve the following model:

$$\boldsymbol{x}_{\mathrm{DS},i}(t+1) = \boldsymbol{x}_{\mathrm{DS},i}(t) \frac{(\boldsymbol{A}\boldsymbol{x}_{\mathrm{DS}})_i}{\boldsymbol{x}_{\mathrm{DS}}(t)^T \boldsymbol{A}\boldsymbol{x}_{\mathrm{DS}}(t)} \qquad (5)$$

The algorithm terminates when $f(\boldsymbol{x}_{\mathrm{DS}}(t+1)) - f(\boldsymbol{x}_{\mathrm{DS}}(t)) < \epsilon$, where the parameter ϵ is the stopping criterion. In addition to the aforementioned termination criteria, the algorithm will be terminated after T iterations.

Given the optimal x_{DS} that maximize Eq. 4, we are able to perform human identification with three candidate selection approaches.

- *local observation analysis:* From Sect. 4.1, we know that each observation contains at least one genuine candidate. Therefore, the baseline approach is to select the instances with the highest participation in their corresponding observation. The dilemma of this approach is that if the observations contain more than one genuine instance the recall will reduce. We term this approach *top selection approach.*
- *maximize internal coherence:* The selection of dominant candidates is conducted by selecting the instances with high participation to the dominant set cluster, which can be obtained if $x_{DS,i} > \tau$ and the threshold τ determine the strength of the participation. However, the genuine instances will be ignored if the corresponding value in x_{DS} is lower than τ. This is particular obvious when the face image is exposed to multiple types of environment variations and pose changes. We note this could be a potential problem for this approach but can be solved with the improvement in face matching algorithm. We term this approach *threshold-based approach.*
- *fusion approach:* This approach combine the above selection approaches. Given the optimum x_{DS}, we first extract all candidates that satisfy $x_{DS,i} > \tau$, followed by select the instance with the highest participation in their corresponding observation if no instance is classified as genuine in threshold stage. We term this approach *fusion-based approach.*

5 Experiments

In this section, we examine the performance of the proposed framework for instance association based human identification problem. We first provide an overview of the image datasets and protocol used in the experiments, followed by the evaluation metrics. Then, we quantitatively analyze the performance under various configurations and provide qualitative comparisons.

5.1 Image Dataset and Protocol

Experiments were conducted on three datasets: Yale Face Dataset B [19], extended Yale Face Dataset B [20], and the Labeled Faces in the Wild (LFW) dataset [21]. The first two datasets are comprised of facial images captured in a laboratory configuration with various illumination variations, where the LFW dataset contains real-world facial images obtained by automatic crawling the Internet.

The Yale Face dataset B was explicitly created to study the face recognition performance under the influence of illumination variations. We combined the frontal view images of both Yale face dataset B, denoted as YaleB, and produce a total of 2,455 images (with 64 illuminations conditions) of 38 individuals. The cropped grayscale facial images were extracted with the manual labeled eye coordinates, and have the size of 64×64 pixels. Each of the images has zero degree in-plane rotation and the inter-ocular distance was 32 pixels (located at $(15, 19)$ and $(47, 19)$). This dataset was used to evaluate the proposed framework under small variation of illumination conditions and various level of ON. The face images were divided into two sets: training set and evaluation set. We randomly selected 8 images from each individual to form the evaluation set, where the remaining images were assigned to the training set. The training set is dedicated for LSED's dictionary training and cohort selection. For the evaluation set, we randomly created 10 set of observations for each individual with each observation contains 8 face images. In total, we generated 4 evaluation sets with different level of ON (i.e., 0.2, 0.4, 0.6, and 0.8)[1].

The LFW dataset was designed to provide a platform to study face recognition performance under uncontrolled environments. It contains 13,233 face images of 5,749 individuals. Among all the subjects, we selected 158 individual with 10 facial images of more as our genuine set and the remaining individual are used as impostors. In this work, we cropped the originally detected face images (i.e., without using additional algorithm to correct the alignment errors) by a fixed bounding box with coordinates $(62, 71)$ to $(187, 196)$ and rescale to 96×96 pixels. We used the training set from LFW view 1 for dictionary training and cohort selection. We created an evaluation set of 1,200 observations where each observation contains 8 instances of facial images and names. The evaluation set contains 158 subset of salient observations that fulfill the observability criterion and ON was limited to 0.2.

We qualitatively report the performance of the proposed algorithm with F-measure metric, which is

$$F_1 = 2 \cdot \frac{\text{precision} \cdot \text{recall}}{\text{precision} + \text{recall}} \tag{6}$$

where the precision and recall are defined as $TP/(TP + FP)$ and $TP/(TP + FN)$, respectively. The notation TP, FP, and FN are the total number of true positive, false positive, and false negative identification (computed with each set of salient observations), respectively. In addition, we also report the Receiver Operating Characteristic (ROC) curve, where the True Positive Rate (TPR), False Positive Rate (FPR) are defined as $TP/(TP+FN)$ and $FP/(FP+TN)$, respectively. The notation FN is the total number of false negative identification.

5.2   Performance Evaluation

We evaluate the proposed method with three candidate selection approaches (see Sect. 4.3 for details), namely top selection approach, threshold-based approach, and fusion-based approach. The effect of selection threshold τ is discussed in this section. Based on preliminary experiments, the proposed approach used the following parameters: the face images are divided into 5×5 regions for YaleB dataset and 3×3 regions for LFW dataset. Dimension of each DCT-based texture

---

[1] The protocol will be publicly available.

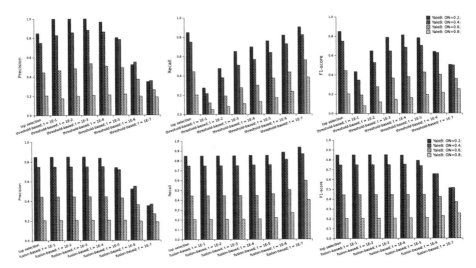

Fig. 4. Performance comparison of various threshold τ on the YaleB dataset. Top row is the performance of the threshold-based approach whereas the bottom is with the fusion-based approach. The first column in each plot are the performance with top selection approach.

descriptor is 15, and the number of visual words in the dictionary is 1024. The stopping criterion ϵ is set to 0.001.

In experiment 1, we first quantitatively evaluate the performance with F-measure over four configuration of ON. The evaluation is conducted on YaleB dataset as the facial images are well aligned and captured under strict controlled. The average precision, recall, and F_1-score are shown in Fig. 4. As shown in the figure, the threshold-based approach achieved precision of 1.0 (ON = 0.2) but lower value in recall when τ is larger than 0.001. The best F_1-score is obtained when $\tau = 0.0001$. The similar performance pattern is observed across all variance of ON, where the performance with ON = 0.8 is considered as noise. We visually evaluate the output of the identification and found that the classified images are generally belong to the same impostor. This indicates that the influence of ON is signification and is the core challenge in our problem. For potential application in real-world deployment of such system, the automated system should avoid identification if the noise level is too high. For the fusion-based approach, the performance when τ is higher than 0.0001 is identical to the top selection approach. This is expected as the top participant is each observation is selected when no candidates satisfy the selection threshold. The inclusion of this clearly improve the performance. The performance is consistently better than the threshold-based approach in recall and F_1-score.

In addition to the F-measure metric, we compare the performance of threshold-based approach and fusion-based approach with ROC curve on both datasets. As shown in Fig. 5, the performance reduced when we increased the observation noise ON. As expected, the performance with LFW dataset (ON was limited to 0.2)

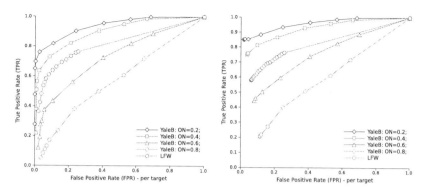

Fig. 5. ROC curve of person identification with YaleB and LFW dataset using: (*left*) threshold-based approach and (*right*) fusion-based approach.

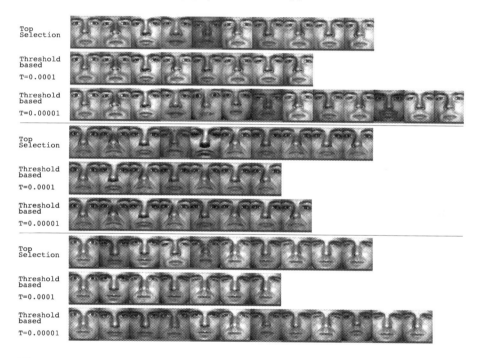

Fig. 6. Qualitative comparison of the proposed method with YaleB dataset. Blue and indicate correct and incorrect identification, respectively (Color figure online).

is lower than YaleB dataset. This is acceptable due to the variations in image quality and capture conditions. The qualitative comparison will be discussed in next section. The analysis also shows that the area under the ROC curve with fusion-based approach is generally higher than threshold-based approach. The only exception is when ON is equal to 0.8.

In experiment 2, we qualitatively compare the performance of the proposed method on YaleB and LFW dataset, where the observation noise ON was limited to 0.2. We compare the identification output with the top selection approach

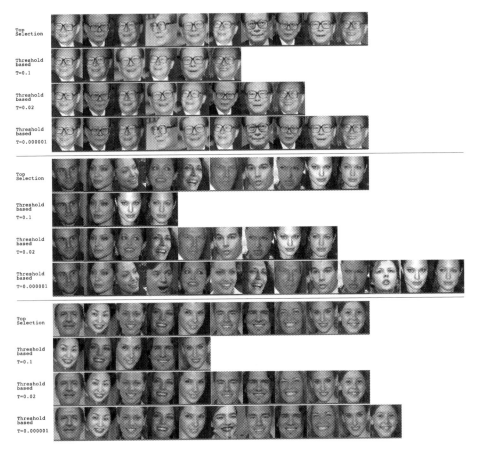

Fig. 7. Qualitative comparison of the proposed method with LFW dataset. Blue and indicate correct and incorrect identification, respectively (Color figure online).

and threshold-based approach (with a selected number of τ). Results are shown in Figs. 6 and 7. Through the analysis on the YaleB dataset, we found that the optimum value of τ vary across different individual. Unlike the closed/open-set identification problem, it is impractical for us to tune this parameter as the image conditions and facial expression vary across different real-world application. We note that the best approach is to use a better face descriptor (or face matching algorithm) to stabilize the impact from these factors. A good similarity score normalization algorithm can be considered. Another observation we made is that the eye region of the individual change when the illumination conditions is different. For example, participant might close his eye when the flash level is high (shown in the last row of Fig. 6). For the LFW dataset, we learned that the head gear (e.g., spectacle, hat, etc.) plays an important role in our evaluation (see the first results on Fig. 7). We also select an example where all the predictions are wrong (see the bottom row in Fig. 7). In this particular example, the prediction is heavily affected by the facial expression, which also affect to the most confidence selection scenario for the threshold-based approach (i.e., $\tau = 0.1$). Another possible future work is to employ fusion algorithm to utilize the strength of multiple algorithms. A detail study will be shown in future publications.

6   Main Findings and Future Directions

In this paper, we propose a novel problem for real-world person identification application, namely *instance association based person identification*, which is motivated with the increasing number of real-world data from the cyber-physical space and the improve accessibility to these data. Despite the large number of literature in person identification problem, most of them assume the mugshots and the identity associated information is given, where the goal is to identify if a current probe image belong to the known person or not. In this paper, we assume that this information is unknown and the goal is to learn the real identity of an image via large-number of observations. Formally, we divide the data into multiple spatial-temporal constrained observations, where each observation contains a finite number of instances from different modalities. We formulate the necessary components for the instance association based person identification problem. Through the observation of non-biometric attribute in the observations, we extracted the salient observations which satisfy the conditions to learn the genuine identity of an attribute-of-interest. We shown that the problem can be formulated as a dominant set clustering problem. Performance on two challenging face datasets, i.e., Yale dataset B and the Label Faces in the Wild (LFW) dataset, shows promising performance for the person identification problem on hand.

For future research directions, we would like to cast the person identification problem on hand as the multi-instance multi-label learning problem [13,14]. In particularly, we would like to extend the existing work to simultaneously associate instances from various biometrics (i.e., multi-modality). Another research direction is to address the missing data problem (i.e., without the co-occurrence assumption) in the observations, this problem is deliberately ignored in this paper. Last but not least, we would like to emphasize that the proposed problem is practical and envisage the potential to apply this problem to the other real-world applications.

Acknowledgment. This research was carried out at the NUS-ZJU Sensor-Enhanced Social Media (SeSaMe) Centre. It is supported by the Singapore National Research Foundation under its International Research Centre @ Singapore Funding Initiative and administered by the Interactive Digital Media Programme Office.

References

1. Bowyer, K.W., Hollingsworth, K., Flynn, P.J.: A survey of iris biometrics research: 2008–2010. In: Burge, M.J., Bowyer, K.W. (eds.) Handbook of Iris Recognition, pp. 15–54. Springer, London (2013)

2. Maltoni, D., Maio, D., Jain, A.K., Prabhakar, S.: Handbook of Fingerprint Recognition, 2nd edn. Springer, London (2009)
3. Wong, Y., Harandi, M.T., Sanderson, C.: On robust face recognition via sparse coding: the good, the bad and the ugly. IET Biometrics **3**, 176–189 (2014)
4. Zhang, X., Gao, Y.: Face recognition across pose: a review. Pattern Recogn. **42**, 2876–2896 (2009)
5. Zhao, W.Y., Chellappa, R., Phillips, P.J., Rosenfeld, A.: Face recognition: a literature survey. ACM Comput. Surv. **35**, 399–458 (2003)
6. Liu, L.-F., Jia, W., Zhu, Y.-H.: Survey of gait recognition. In: Huang, D.-S., Jo, K.-H., Lee, H.-H., Kang, H.-J., Bevilacqua, V. (eds.) ICIC 2009. LNCS, vol. 5755, pp. 652–659. Springer, Heidelberg (2009)
7. Cardinaux, F., Sanderson, C., Bengio, S.: User authentication via adapted statistical models of face images. IEEE Trans. Sig. Process. **54**, 361–373 (2006)
8. Satoh, S., Nakamura, Y., Kanade, T.: Name-it: naming and detecting faces in news videos. IEEE MultiMedia **6**, 22–35 (1999)
9. Yang, J., Hauptmann, A.G.: Naming every individual in news video monologues. In: Proceedings of ACM International Conference on Multimedia, pp. 580–587 (2004)
10. Houghton, R.: Named faces: putting names to faces. IEEE Intell. Syst. **14**, 45–50 (1999)
11. Cho, S.H., Hong, S., Nam, Y.: Association and identification in heterogeneous sensors environment with coverage uncertainty. In: IEEE International Conference on Advanced Video and Signal Based Surveillance, pp. 553–558 (2009)
12. Cho, S.H., Hong, S., Moon, N., Park, P., Oh, S.J.: Object association and identification in heterogeneous sensors environment. EURASIP J. Adv. Sig. Process. **2010**, 18 p. (2010). Article ID 591582
13. Zhou, Z., Zhang, M., Huang, S., Li, Y.: Multi-instance multi-label learning. Artif. Intell. **176**, 2291–2320 (2012)
14. Yang, S., Jiang, Y., Zhou, Z.: Multi-instance multi-label learning with weak label. In: International Joint Conference on Artificial Intelligence (2013)
15. Heisele, B., Ho, P., Wu, J., Poggio, T.: Face recognition: component-based versus global approaches. Comput. Vis. Image Underst. **91**, 6–21 (2003)
16. Doddington, G.R., Przybocki, M.A., Martin, A.F., Reynolds, D.A.: The NIST speaker recognition evaluation - overview, methodology, systems, results, perspective. Speech Commun. **31**, 225–254 (2000)
17. Pavan, M., Pelillo, M.: Dominant sets and pairwise clustering. IEEE Trans. Pattern Anal. Mach. Intell. **29**, 167–172 (2007)
18. Weibull, J.W.: Evolutionary Game Theory, 1st edn. The MIT Press, Cambridge (1997)
19. Georghiades, A.S., Belhumeur, P.N., Kriegman, D.J.: From few to many: illumination cone models for face recognition under variable lighting and pose. IEEE Trans. Pattern Anal. Mach. Intell. **23**, 643–660 (2001)
20. Lee, K.C., Ho, J., Kriegman, D.J.: Acquiring linear subspaces for face recognition under variable lighting. IEEE Trans. Pattern Anal. Mach. Intell. **27**, 684–698 (2005)
21. Huang, G.B., Ramesh, M., Berg, T., Learned-Miller, E.: Labeled Faces in the Wild: A database for studying face recognition in unconstrained environments. Technical report 07–49, University of Massachusetts, Amherst (2007)

Robust Ear Recognition Using Gradient Ordinal Relationship Pattern

Aditya Nigam[1]([⊠]) and Phalguni Gupta[2]

[1] School of Computing and Electrical Engineering,
Indian Institute of Technology Mandi, Mandi 175005, India
`aditya@iitmandi.ac.in`
[2] Department of Computer Science and Engineering,
Indian Institute of Technology Kanpur, Kanpur 208016, India
`pg@cse.iitk.ac.in`

Abstract. A reliable personal recognition based on ear biometrics is highly in demand due to its vast application in automated surveillance, law enforcement *etc.* In this paper a robust ear recognition system is proposed using gradient ordinal relationship pattern. A reference point based normalization is proposed along with a novel ear transformation over normalized ear, to obtain robust ear representations. Ear samples are enhanced using a local enhancement technique. Later a dissimilarity measure is proposed that can be used for matching ear samples. Two publicly available ear databases IITD and UND-E are used for the performance analysis. The proposed system has shown very promising results and significant improvement over the existing state of the art ear systems. The proposed system has shown robustness against small amount of illumination variations and affine transformations due to the virtue of ear transformation and tracking based matching respectively.

1 Introduction

Personal authentication plays an important role in the society. Every application requires at least some level of security to assure personal identity. Hence an automated and accurate human access control mechanism plays an important role in several social applications such as law enforcement, secure banking, immigration control *etc.* Security can be realized at one of the three levels.

1. **Level-1 [Possession]:** The user possesses something which is required to be produced for successful authentication. For example, key of a car or room.
2. **Level-2 [Knowledge]:** The user knows something which is required to be entered correctly for successful authentication. For example, PIN (personal identification number), credit card CVV (card verification value) *etc.*
3. **Level-3 [Biometrics]:** The user owns certain physiological and behavioral characteristics which are required to be acquired and matched for successful authentication. For example, face, iris, fingerprint, signature, gait *etc.*

Authors would like to acknowledge the funding and support provided by IIT, Mandi.

C.V. Jawahar and S. Shan (Eds.): ACCV 2014 Workshops, Part III, LNCS 9010, pp. 617–632, 2015.
DOI: 10.1007/978-3-319-16634-6_45

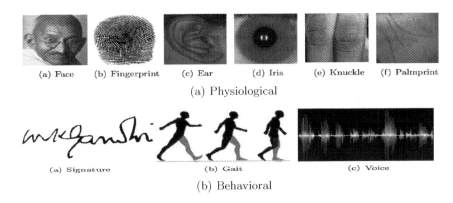

(a) Face   (b) Fingerprint   (c) Ear   (d) Iris   (e) Knuckle   (f) Palmprint

(a) Physiological

(a) Signature   (b) Gait   (c) Voice

(b) Behavioral

Fig. 1. Biometric traits

Biometrics can be used for personal authentication using physiological (such as face [1,2], fingerprint [3,4], iris [5–7], palmprint [8], ear [9,10], knuckleprint [11–13] *etc.*) and behavioral (such as gait, speech *etc.*) characteristics as shown in Fig. 1, which are assumed to be unique for each individual. However, each trait has its own challenges and trait specific issues hence none of the biometric trait can be considered as the best one. The biometric trait selection completely depends upon the application where ultimately it is going to be deployed. Ear can be considered as a significant biometric because of its passive nature, as it can be acquired without much user cooperation unlike iris, fingerprint. In [14], manually measured twelve features extracted from 10,000 ear samples suggested that ear may contain unique characteristics. Also it is inherently different from the invasive biometrics such as face which changes with expression, age, illumination, pose and artifacts such as beard, sunglasses *etc.* Moreover the structure and size of a ear is assumed to be invariant for wide age ranges. Ear size is bigger than fingerprint and iris and smaller than face hence can be easily acquired and processed. Still, scale and affine transformation remains to be the major challenges for any ear based recognition system.

In this paper an automated reference point based ear normalization is proposed to handle scale and affine variations. Also an image transformation along with a feature extraction and matching algorithm is proposed that can compliment each other for better performance. The system is tested over two publicly available ear databases *viz.* IITD [15] and UND-E [16]. It has shown significantly better performance than the state of the art ear authentication systems, presented in all these recent ear identification journal works [10,15,17]. The paper is organized as follows. In Sect. 2 the literature available on ear recognition is reviewed. The proposed ear recognition system is discussed in Sect. 3. The experimental results are shown in Sect. 4. Conclusions are discussed in the last Section.

Table 1. Summarized literature review

Approach	Number of Images	Performance (Rank 1)
Force Field Transformation and NKFDA [18]	711	75.3 %
Sparse Representation [19]	192	100 %
Locally Linear Embedding (LLE) [20]	1501	80 %
SIFT Landmark Points over Ear Model Face [21]	1060	95.32 %
SIFT features from various color models [22]	800	96.93 %
Partitioned Iterated Function System(PIFS) [23]	228	61 %
Gabor Wavelet and GDA [24]	308	99.1 %
Local Binary Pattern and Wavelet [25]	308	100 %

2   Literature Review

Several ear recognition and authentication systems have been proposed in last decade. Force field transformation [18] is used along with null space based kernel fisher discriminant analysis $NKFDA$ to perform multi-pose ear recognition. Sparse representations [19] are used to develop a dictionary during training stage. A linear system of equations is obtained from probe images and is solved using sparsity of probe image vector and l_1 minimization. Ear recognition is done using local linear embedding LLE and its improved version in [20]. Feature level fusion of multiple ear samples is done for better results in [21]. $SIFT$ features are fused to obtain a single fused template from all training images. In [22] regions with consistent color are used for matching using $SIFT$ features with GMM based skin modeling. Human ear recognition against occlusion ($HERO$) is proposed in [23] which is based on fractals. It can deal with synthetic and natural occlusion. They indexed images using partitioned iterated function system ($PIFS$). Gabor filter [24] based feature extraction and general discriminant analysis GDA are used together for ear recognition.

Some of the current state of the art systems are discussed as follows. In [9], connected component analysis of the graph constructed using edge-map is used to segment ear and SURF features are matched using nearest neighborhood ratio matching. Automated ear segmentation [10] is done using Fourier descriptors and ear identification is performed using log-gabor filters. Later in [17], they have utilized the phase information using 2D quadrature filtering to improve their results. Finally in [15], they have proposed sparse feature coding scheme using localized radon transformation in order to produce their best result. Table 1 summarizes the most significant work done in the field of ear recognition.

3   Proposed Method

The proposed ear recognition system consist of following steps, $viz.$ ROI extraction, Reference point detection (Sect. 3.1), Ear Normalization (Sect. 3.2),

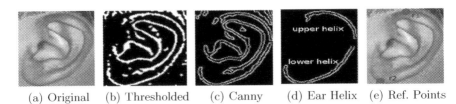

(a) Original   (b) Thresholded   (c) Canny   (d) Ear Helix   (e) Ref. Points

Fig. 2. Reference points

Ear Enhancement (Sect. 3.3), Ear Transformation (Sect. 3.4), Feature Extraction (Sect. 3.5) and Matching (Sect. 3.6). The ear ROI from any profile face image is cropped using the connected component graph based algorithm proposed in [26]. The cropped ear is automatically normalized using the proposed method that first detect two reference points and later use them for ear normalization. The top and bottom-most ear points are chosen as reference points as shown in Fig. 2(e). The normalized ear is further enhanced and transformed to obtain robust ear representations. Finally the corner features are extracted and matched using the tracking based dissimilarity measure named as Incorrectly Tracked Corners (*ITC*).

3.1   Reference Point Detection

The major problem in ear recognition is of scale variations. Hence to handle it ear samples are normalized using two reference points r_1 and r_2. Hence these two reference points that are considered as top and bottom ear points (as shown in Fig. 2(e)) are detected automatically. The cropped ear ROI image is preprocessed using an adaptive thresholding [27] and the noise is removed by applying median filtering. The edges are detected using Canny edge detection algorithm [28], and the output is divided into two halves. From both halves the largest contour is extracted, which is considered as the outer helix part of the ear as shown in Fig. 2(d). The two reference points r_1 (over upper helix) and r_2 (over lower helix) are the points which are at maximum distance over these helical structures as shown in Fig. 2(e) and Eq. 1. In Eq. 1, x, y represents all points belonging to upper and lower helix respectively, r_1 and r_2 are the required reference points and $||x, y||$ represents the euclidean distance between x and y point.

$$\{r_1, r_2\} = argMax_{\forall (x \in UP.Helix)\ and\ (y \in LW.Helix)} ||x, y|| \tag{1}$$

3.2   Ear Normalization

The issues of affine transformations such as rotation, scaling and alignment, are overpowered by the using the proposed automated ear normalization. All database profile face images are scaled to a predefined size and are registered using two reference points *viz.* (r_1) and (r_2), which are detected as discussed

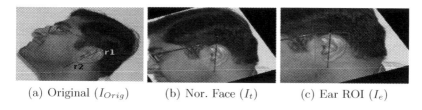

(a) Original (I_{Orig})   (b) Nor. Face (I_t)   (c) Ear ROI (I_e)

Fig. 3. Ear normalization

above and shown in Figs. 2(e), 3(a). The reference points are detected in cropped ear ROI but actual profile image is normalized to ensure that:

1. The points r_1 and r_2 should lie on same coordinates in all the images (*i.e.* (x_1^t, y_1^t) and (x_2^t, y_2^t) respectively).
2. The line joining r_1 and r_2 should be vertically aligned as shown in Fig. 3(c).

Let the co-ordinates of r_1 and r_2 are (x_1^o, y_1^o) and (x_2^o, y_2^o) respectively, in original image I_{orig} as shown in Fig. 3(a). The image I_{orig} is scaled to I_s by a scaling factor $S = \frac{d}{distance(r_1, r_2)}$ with respect to r_1, so as to set a predefined fixed distance (d) between the points r_1 and r_2 using the scaling matrix $[T_s]$, as given in Eq. (2). The scaled image I_s is then rotated to I_r by an angle ϕ with respect to r_1, where ϕ is the angle between the vertical direction (*i.e.* y-axis) and the line-segment r_1 and r_2 using the rotation matrix $[T_r]$, as given in Eq. (2). Now I_r is translated to I_t (*i.e* Nor. Face, Fig. 3(b)), by $t_x = x_1^t - x_1^o$ and $t_y = y_1^t - y_1^o$ units, in x and y directions using the translation matrix $[T_t]$. The combined image transformation matrix $[T]$ is given below.

$$T = \underbrace{\begin{bmatrix} S & 0 & x_1^o \cdot (1-S) \\ 0 & S & y_1^o \cdot (1-S) \\ 0 & 0 & 1 \end{bmatrix}}_{\text{Scaling Matrix } [T_s] \ w.r.t. \text{ point } r_1} \underbrace{\begin{bmatrix} cos\phi & -sin\phi & x_1^o \cdot (1-cos\phi) + y_1^o \cdot sin\phi \\ sin\phi & cos\phi & y_1^o \cdot (1-cos\phi) - x_1^o \cdot sin\phi \\ 0 & 0 & 1 \end{bmatrix}}_{\text{Rotation Matrix } [T_r] \ w.r.t. \text{ point } r_1} \underbrace{\begin{bmatrix} 1 & 0 & t_x \\ 0 & 1 & t_y \\ 0 & 0 & 1 \end{bmatrix}}_{\text{Translation } [T_t]}$$

$$(2)$$

In order to extract a consistent ear ROI, height-width ratio ($\delta = 1.4$) is used that is computed heuristically from a set of randomly selected 100 images (50 images each from IITD and UND-E dataset). As the height of a ear is normalized to (d in I_t), by using the above derived transformation matrix T (as given in Eq. (2)) over I_{orig}, the width of the ear box to be cropped can be computed as $w = \frac{\delta}{d} + offset$. Here $offset$ is the left margin along each box side with the line joining the reference points exactly at the midpoint of the box, as shown in Fig. 3(c).

3.3   Ear Enhancement

The extracted region of interest (ROI) of an ear contains texture information but generally is of poor contrast. Suitable image enhancement technique is required

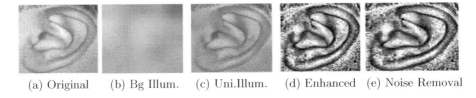

(a) Original   (b) Bg Illum.   (c) Uni.Illum.   (d) Enhanced   (e) Noise Removal

Fig. 4. Ear image enhancement

to apply on the ROI. The ear texture is enhanced in such a way that it increases its richness as well as its discriminative power. The ear ROI is divided into blocks and the mean of each block is considered as the coarse illumination of that block. This mean is expanded to the original block size as shown in Fig. 4(b). Selection of block size plays an important role. It should be such that the mean of the block truly represents the illumination effect of the block. So, larger block may produce improper estimate. Since the ear sample images are of size 160×200, it is observed that a block size of 40×40 is the best choice for our experiment. The estimated illumination of each block is subtracted from the corresponding block of the original image to obtain the uniformly illuminated ROI as shown in Fig. 4(c). The contrast of the resultant ROI is enhanced using Contrast Limited Adaptive Histogram Equalization ($CLAHE$) [29]. It removes the artificially induced blocking effect using bilinear interpolation and enhances the contrast of image without introducing much external noise. Finally, Wiener filter [30] is applied to reduce constant power additive noise to obtain the enhanced ear texture as shown in Fig. 4(e).

3.4   Ear Transformation

The normalized and enhanced ear samples are transformed using two proposed encoding schemes Gradient Ordinal Relation Pattern ($GORP$) and STAR GORP ($SGORP$). The gradient of any edge pixel is positive if it lies on an edge created due to light to dark shade (*i.e. high to low gray value*) transition else it will be having negative gradient or zero value. Hence all the edge pixels can be divided into three classes of $+ve$, $-ve$ or zero gradient values as shown in Fig. 5(b,c). The *sobel* x-direction kernel of size 3×3 and 9×9 are applied to

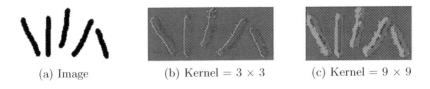

(a) Image   (b) Kernel $= 3 \times 3$ (c) Kernel $= 9 \times 9$

Fig. 5. Red: -ve grad; Green: +ve grad.; Blue: zero grad (Kernel size represents the size of sobel kernel used to compute gradient.) (Color figure online)

obtain Fig. 5(b,c) respectively. Bigger size kernel produces coarse level features. The *sobel* kernel lacks rotational symmetry hence more consistent *scharr* kernel [31] which is obtained by minimizing angular error is applied. This gradient augmented information of an edge pixel can be more discriminative and robust. The proposed encoding schemes $GORP$ and $SGORP$ precisely uses this information to calculate a 8-bit code for each pixel by using x and y-direction derivatives of its 8 neighboring pixels to obtain four codes *viz.* $vcode^{GORP}$, $hcode^{GORP}$, $vcode^{SGORP}$ and $hcode^{SGORP}$ as shown in Fig. 7(b,c,d,e) respectively. The vertical/horizontal code (*i.e.* *vcode* and *hcode*) signifies that whether x-direction or y-direction derivatives are used while encoding.

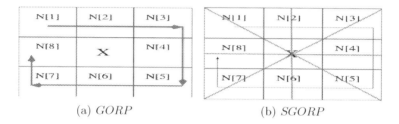

(a) $GORP$ (b) $SGORP$

Fig. 6. Neighborhood structure used to compute the gradient and the robust encodings (Color figure online).

Let $P_{i,j}$ be the $(i,j)^{th}$ pixel of an ear sample P and $N[l]$, $l = 1, 2, ...8$ (as shown in Fig. 6(a)) are the gradients of 8 neighboring pixels centered at pixel $P_{i,j}$ that are obtained by applying *scharr* kernel.

[a] GORP based Encoding: An eight bit encoding *viz.* *gorp_code* for every pixel is computed whose k^{th} bit can be defined as:

$$gorp\_code[k] = \begin{cases} 1 & \text{if } N[k] > 0 \\ 0 & \text{otherwise} \end{cases} \qquad (3)$$

In $vcode^{GORP}$ or $hcode^{GORP}$ as shown in Fig. 7 (b,c) every pixel is represented by its *gorp_code* instead of its gray value which is obtained using the gradient of its 8 neighboring pixels computed using x and y direction *scharr* kernel respectively.

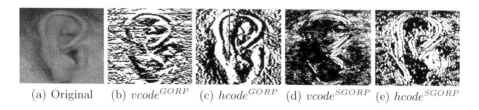

(a) Original   (b) $vcode^{GORP}$ (c) $hcode^{GORP}$ (d) $vcode^{SGORP}$ (e) $hcode^{SGORP}$

Fig. 7. Ear transformation using local gradient

[b] SGORP based Encoding: The *sgorp_code* encodes the ordinal relationships between diagonally opposite neighbors along with top and bottom one (as shown in Fig. 6(b) in red color). Due to this star structure it is termed as STAR GORP (*i.e SGORP*). In $vcode^{SGORP}$ or $hcode^{SGORP}$ as shown in Fig. 7 (d,e) every pixel is represented by its *sgorp_code* instead of its gray value. It is a 4 bit code represented in 8 bits as follows:

$$
\begin{aligned}
sgorp\_code[1] = sgorp\_code[2] = 1 \ \textbf{if} \ abs(N[1] - N[5]) > T \ \textbf{else} \ 0 \\
sgorp\_code[3] = sgorp\_code[4] = 1 \ \textbf{if} \ abs(N[2] - N[6]) > T \ \textbf{else} \ 0 \\
sgorp\_code[5] = sgorp\_code[6] = 1 \ \textbf{if} \ abs(N[3] - N[7]) > T \ \textbf{else} \ 0 \\
sgorp\_code[7] = sgorp\_code[8] = 1 \ \textbf{if} \ abs(N[4] - N[8]) > T \ \textbf{else} \ 0
\end{aligned}
\tag{4}
$$

where T, is empirically determined threshold that depends upon the size of *scharr* kernel used to compute gradient. The pattern of edges within a neighborhood can be assumed to be robust; hence each pixel's encodings (*i.e. gorp_code* or *sgorp_code*) are considered, that only used sign of the derivative within its specified neighborhood which is their ordinal relationships. Hence they ensures the robustness of the proposed transformation under illumination variation.

Justifications: The subsequent samples of same ear will be having varying illumination and affine transformations. The proposed $GORP$ and $SGORP$ transformation can handle illumination variations. In Fig. 8(a) the first ear sample is shown in six significantly different illumination conditions that are artificially created to observe the robustness of the proposed transformation. The Fig. 8(b) and (c) shows their corresponding $hcode^{GORP}$ and $vcode^{GORP}$. One can clearly observe that they are not varying much even under such drastic illumination variations.

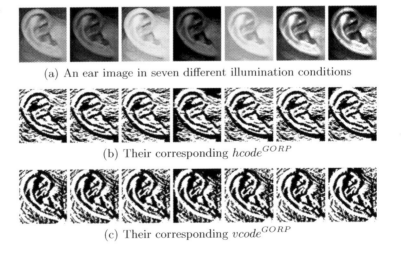

(a) An ear image in seven different illumination conditions

(b) Their corresponding $hcode^{GORP}$

(c) Their corresponding $vcode^{GORP}$

Fig. 8. Robustness of $GORP$ against varying illumination

3.5 Feature Extraction

We have considered corner points as features because of their repeatability and discrimination. The autocorrelation matrix M [32], can be used to calculate corner points that are having strong orthogonal derivatives. The matrix M is a 2×2 matrix and hence it can have two eigen values λ_1 and λ_2 such that $\lambda_1 \geq \lambda_2$ with e_1 and e_2 as the corresponding eigenvectors. All pixels having $\lambda_2 \geq T_q$ (smaller eigen value greater than a threshold) are considered as corner points.

3.6 Feature Matching

Tracking algorithms can be used to achieve robustness against affine transformations as it assumes that any feature can move within a small neighborhood. Tracking can also handle partial ears, as they are occluded mostly by hairs or ear-rings. They can easily be detected and masked using skin-tone detection and feature over them are ignored while tracking. A constrained version of KL tracking [33], has been used to perform matching between two ear samples E_a and E_b. The KL tracking make use of three assumption namely brightness consistency, temporal persistence and spatial coherency as defined below, hence its performance depends completely on how well these three assumptions are satisfied.

Let there be a feature at location (x, y) at a time instant t with intensity $I(x, y, t)$ and this feature has moved to the location $(x + \delta x, y + \delta y)$ at the time instant $t + \delta t$. Three basic assumptions that are used by KL Tracking to perform tracking successfully are:

– **Brightness Consistency**: Features in a frame do not change much for small change in the value of δt, $i.e$

$$I(x, y, t) \approx I(x + \delta x, y + \delta y, t + \delta t) \tag{5}$$

– **Temporal Persistence**: Features in a frame moves only within a small neighborhood. Using the Taylor series and neglecting the high order terms, one can estimate $I(x + \delta x, y + \delta y, t + \delta t)$ with the help of Eq. (5) as:

$$I_x V_x + I_y V_y = -I_t \tag{6}$$

where V_x, V_y are the components of the optical flow velocity for pixel $I(x, y, t)$ and I_x, I_y and I_t are the derivatives in the corresponding directions.
– **Spatial Coherency**: Estimating unique flow vector from Eq. 6 for every feature point is an ill-posed problem. Hence KL tracking estimates the motion of any feature by assuming local constant flow (i.e. a patch of pixels moves coherently). An over-determined system of linear equations have been finally obtained and solved using least square method to estimate the flow vector for each pixel. Finally this vector is used to estimate the new position of that feature.

Justification for Tracking based Matching: In this work we expect that these assumptions are more likely to be satisfied while tracking is performed between features of same subject (genuine matching) and degrades substantially for others (imposter matching). The above fact hold because brightness consistency is assured by $GORP$ and $SGORP$ transformation while the temporal persistence and spatial coherency are satisfied only for genuine matching. Therefore one can infer that the tracking performance of KL tracking algorithm will be good (or at-least better) for genuine matching as compared to imposter.

However, all the tracked corner features may not be the true matches because of the noise, local non-rigid distortions and less difference in inter class matching as compared to intra class matching. Hence a notion of consistent optical flow is proposed to handle this.

Consistent Optical Flow: The true corner matches have their optical flow that can be aligned with the actual affine transformation between the images being matched. Hence the estimated optical flow direction is quantized into eight directions (*i.e.* at an angular difference of $\frac{\pi}{8}$) and the most consistent direction (MCD) is selected as the one which has the maximum number of correctly tracked corner features. Any corner matching pair (*i.e* corner and its corresponding corner) having optical flow direction other than the most consistent direction (MCD) is considered as false matching pair and have to be discarded.

A dissimilarity measure ITC (Incorrectly Tracked Corners) has been proposed to estimate the KL-tracking performance by evaluating simple geometrical and statistical quantities defined as:

Locality: Euclidean distance between any corner and its estimated tracked location should be less than or equal to an empirically selected threshold TH_d. The parameter TH_d depends upon the amount of translation and rotation in the sample images. High TH_d signifies more translation.

Dissimilarity: The tracking error is defined as pixel-wise sum of absolute difference between a local patch centered at current corner feature and that of its estimated tracked location patch, that should have to be less than or equal to TH_e. This error should have to be less than or equal to an empirically selected threshold TH_e. The parameter TH_e ensures that the matching corners must have similar neighborhood patch around it.

Correlation: Phase only correlation [34] between a local patch centered at any feature and that of its estimated tracked location patch should be at-least equal to an empirically selected threshold TH_{cb}.

Any corner is considered as tracked successfully if it satisfies the above defined three constraints. These three parameters (*viz.* TH_d, TH_e and TH_{cb}) are experimentally determined over a small validation set (including only 200 images, 100 each from both dataset) and are optimized *w.r.t* performance. The values for which the optimized performance is achieved are $TH_d = 13$, $TH_e = 750$ for both $GORP$ and $SGORP$ databases while $TH_{cb} = 0.1$ for $GORP$ and 0.4 for $SGORP$. The $SGORP$ transformation requires more correlation because it is more sparse as compared to $GORP$ as shown in Fig. 7 here correlation and path wise error are computed using a block size of 5×5. Let E_a and E_b are two ear sample images that have to be matched and $(I_A^{v^L}, I_B^{v^L})$, $(I_A^{h^L}, I_B^{h^L})$, $(I_A^{v^{SL}}, I_B^{v^{SL}})$ and

---

Algorithm 1. $ITC(E_a, E_b)$

Require:

(a) Let E_a and E_b be two ear images and $(I_A^{v^L}, I_B^{v^L})$, $(I_A^{h^L}, I_B^{h^L})$, $(I_A^{v^{SL}}, I_B^{v^{SL}})$ and $(I_A^{h^{SL}}, I_B^{h^{SL}})$ are their four corresponding $vcode^{GORP}$, $hcode^{GORP}$, $vcode^{SGORP}$ and $hcode^{SGORP}$ pairs respectively.

(b) $N_A^{v^L}, N_A^{h^L}, N_A^{v^{SL}}$, and $N_A^{h^{SL}}$ are the number of corners in $I_A^{v^L}, I_A^{h^L}, I_A^{v^{SL}}$, and $I_A^{h^{SL}}$ respectively.

Ensure: Return $ITC(E_a, E_b)$.

1: Track individually corners within the 4 ear encoding pairs $(I_A^{v^L}, I_B^{v^L})$, $(I_A^{h^L}, I_B^{h^L})$, $(I_A^{v^{SL}}, I_B^{v^{SL}})$ and $(I_A^{h^{SL}}, I_B^{h^{SL}})$.

2: Obtain four set of corners that are successfully tracked in,

   (a) $(I_A^{v^L}, I_B^{v^L})$ pair tracking (*i.e.* stc^{v^L}),

   (b) $(I_A^{h^L}, I_B^{h^L})$ pair tracking (*i.e.* stc^{h^L}),

   (c) $(I_A^{v^{SL}}, I_B^{v^{SL}})$ pair tracking (*i.e.* $stc^{v^{SL}}$),

   (d) $(I_A^{h^{SL}}, I_B^{h^{SL}})$ pair tracking (*i.e.* $stc^{h^{SL}}$),

   that have their tracked position within TH_d and their local patch dissimilarity under TH_e also the patch-wise correlation is at-least equal to TH_{cb}.

3: Quantize optical flow direction for each successfully tracked corner set into only eight directions (*i.e.* at an interval of $\frac{\pi}{8}$) to obtain 4 histograms H^{v^L}, H^{h^L}, $H^{v^{SL}}$ and $H^{h^{SL}}$ using these four corner set $stc^{v^L}, stc^{h^L}, stc^{v^{SL}}$ and $stc^{h^{SL}}$ respectively as computed above.

4: For each histogram, out of 8 bins the bin (*i.e.* direction) having the maximum corners is considered as the consistent optical flow direction. The maximum value obtained from each histogram is termed as corners tracked correctly (*i.e.* ctc value) represented as $ctc^{v^L}, ctc^{h^L}, ctc^{v^{SL}}$ and $ctc^{h^{SL}}$.

5: **return** $ITC(E_a, E_b) = 1 - \dfrac{\frac{ctc^{v^L}}{N_A^{v^L}} + \frac{ctc^{h^L}}{N_A^{h^L}} + \frac{ctc^{v^{SL}}}{N_A^{v^{SL}}} + \frac{ctc^{h^{SL}}}{N_A^{h^{SL}}}}{4};$

$(I_A^{h^{SL}}, I_B^{h^{SL}})$ are their four corresponding $vcode^{GORP}$, $hcode^{GORP}$, $vcode^{SGORP}$ and $hcode^{SGORP}$ pairs respectively. The Algorithm 1 can be used to compute the $ITC(E_a, E_b)$ based dissimilarity score between two ear sample images E_a and E_b.

4 Experimental Analysis

4.1 Database

The proposed system is tested over two widely used publicly available ear databases IITD [10,15,17] and UND-E [16]. The IITD ear database have 493 images collected from 125 subjects (age range 14–58 years) over a period of 9 months in indoor environment with out extra illumination. Each subject has given at-least 3 left ear images. The UND-E database consists of 443 profile face images collected from 114 subjects with 3 to 9 images per subject. Images are collected on different days with different pose and illumination. It can be noted that there exist a huge intra-class variation in these images due to pose variation and different imaging conditions.

4.2 Testing Strategy

The testing strategy used in this work is exactly the same as adopted in [10, 15, 17] so as to compare our result against these state of the art systems. The test protocol A (*i.e.* single image in training and testing) computes the average of 3 tests in which only initial 3 images per subject are used with one as the *test image* while remaining images as *training images* (one by one). The test protocol B (*i.e.* all to all) computes the average performance when all images from all subjects are used as *training images* (except itself). For IITD dataset using protocol A, 125×3 and $124 \times 125 \times 3$ genuine and imposter matchings are considered respectively while using protocol B, 493 genuine and 124×493 imposter matchings are performed. Similarly for UND-E dataset using protocol A, 110×3 and $109 \times 110 \times 3$ genuine and imposter matchings are considered respectively while using protocol B, 443 genuine and 109×443 imposter matchings are performed.

4.3 Performance Analysis

The performance of the system is measured using correct recognition rate (CRR) in case of identification and equal error rate (EER) for verification. The CRR (*i.e.* the **Rank 1** accuracy) of any system is defined as the ratio of the number of correct (Non-False) top best match of ear ROI and the total number of ear ROI in the query set. At any given threshold, the probability of accepting the impostor is known as false acceptance rate (FAR) and the probability of rejecting the genuine user known as false rejection rate (FRR). Equal error rate (EER) is the value of FAR for which FAR and FRR are equal.

$$EER = \{FAR | FAR = FRR\} \tag{7}$$

For all these experimentations average results in terms of EER, CRR and receiver operating characteristics (ROC) [35] curves are reported in Table 2 and Fig. 10. In Table 2 the proposed system is compared with six existing most recent state of the art algorithms proposed in [10, 15, 17] and is found to be performing much better. The proposed system outperforms non negative formulation (NNG) [15] in both CRR and EER parameters over both databases and using both testing protocol A and B. The prime reasons behind the proposed system's superior performance includes:

- In [10, 15, 17] they have only considered left half of the ear and discarded most of the right half as shown in Fig. 9. But in this work we have utilized the information from both halves of ear ROI.
- In this work $GORP$ and $SGORP$ based transformations have been proposed that can handle some amount of illumination variations in the subsequent ear samples as shown in Fig. 8.

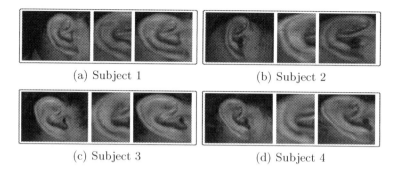

(a) Subject 1   (b) Subject 2

(c) Subject 3   (d) Subject 4

Fig. 9. Ear images. Each subject's first image is the raw ear image, second is used by [15] while third image which is full ear ROI is used in this work.

Table 2. Comparative performance analysis

Algorithm	Protocol A		Protocol B	
	IITD	UND	IIITD	UND
	CRR %, EER %	CRR %, EER %	CRR %, EER %	CRR %, EER %
Hessian [15]	95.47, 5.28	86.06, 9.84	95.94, 4.64	85.91, 9.97
LG [10]	95.73, 4.32	87.88, 8.18	95.74, 4.06	88.91, 7.40
Q.Log.G [17]	96.53, 3.73	90.30, 7.27	96.55, 3.78	90.99, 7.16
LRT [15]	94.40, 4.27	84.84, 7.07	94.93, 3.87	84.76, 7.01
s-LRT [15]	97.07, 2.38	91.52, 5.45	97.57, 2.03	92.61, 5.08
NNG [15]	96.80, 1.87	91.52, 5.15	97.16, 1.62	92.38, 5.31
Proposed	**98.93, 1.05**	**98.18, 2.42**	**99.20, 1.43**	**99.20, 4.78**

- The ear samples are enhanced so as to obtain robust texture information as shown in Fig. 4.
- Matching is done using corner feature tracking, hence it can handle some amount of affine transformations.
- To handle partial and occluded ear by hairs or ear ring, simple Gaussian model based skin tone detection [36] is done and features over occluded regions are discarded while tracking.

From Table 2, one can observe that the CRR as well as EER of the proposed system has been found to be much better than the non negative formulation (NNG) [15] in the case of UND-E database because the proposed systems can handle affine variations better than NNG. Also database enlargement has not been done as performed in [10,15,17] which makes the system very efficient.

In Fig. 10 the ROC curves that plots FAR Vs FRR for all thresholds, are plotted for the proposed ear recognition system. Since FAR and FRR both are error hence the curve which is lower and towards both axis is considered as better. The curve shown in red color is the EER curve and the point where it intersect

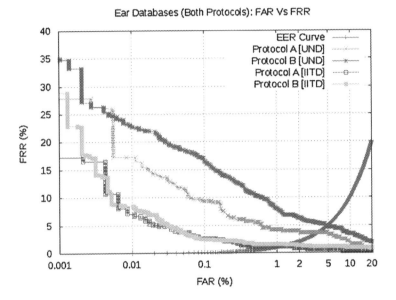

Fig. 10. The Average ROC for IITD and UND Databases. (x-axis is in log scale) (Color figure online)

any curve is the equal error rate of that system. One can clearly observe that the performance of the propose system is better over IIT ear database under either testing protocol. This is because UND-E ear database is much more challenging and contains ear samples with in-plane rotations.

5   Conclusion

In this paper an end to end automated ear based recognition system is proposed. All steps in the proposed ear recognition system $viz.$ ROI extraction, Reference point detection, Ear Normalization, Ear Enhancement, Ear Transformation, Feature Extraction and Matching are proposed and discussed. Two publicly available ear databases IITD [15] and UND-E [16] are considered for performance analysis. The proposed system has shown very promising results and significant improvement over the six recently proposed state of the art ear authentication systems [10,15,17]. The superior performance is achieved by utilizing full (both halves) of ear ROI's and by using the robust encoding scheme ($GORP$ and $SGORP$) along with the tracking based measure that can handle affine transformations effectively. The partial ear and occlusion due to hair and ear ring are handled by using Gaussian model based skin tone detection.

References

1. Nigam, A., Gupta, P.: A new distance measure for face recognition system. In: International Conference on Image and Graphics, ICIG, pp. 696–701 (2009)

2. Nigam, A., Gupta, P.: Comparing human faces using edge weighted dissimilarity measure. In: International Conference on Control, Automation, Robotics and Vision, ICARCV, pp. 1831–1836 (2010)
3. Cappelli, R., Ferrara, M., Maltoni, D.: Minutia cylinder-code: a new representation and matching technique for fingerprint recognition. IEEE Trans. Pattern Anal. Mach. Intell. **32**, 2128–2141 (2010)
4. Singh, N., Nigam, A., Gupta, P., Gupta, P.: Four slap fingerprint segmentation. In: Huang, D.-S., Ma, J., Jo, K.-H., Gromiha, M.M. (eds.) ICIC 2012. LNCS, vol. 7390, pp. 664–671. Springer, Heidelberg (2012)
5. Nigam, A., Gupta, P.: Iris recognition using consistent corner optical flow. In: Lee, K.M., Matsushita, Y., Rehg, J.M., Hu, Z. (eds.) ACCV 2012, Part I. LNCS, vol. 7724, pp. 358–369. Springer, Heidelberg (2013)
6. Nigam, A., Anvesh, T., Gupta, P.: Iris classification based on its quality. In: Huang, D.-S., Bevilacqua, V., Figueroa, J.C., Premaratne, P. (eds.) ICIC 2013. LNCS, vol. 7995, pp. 443–452. Springer, Heidelberg (2013)
7. Bendale, A., Nigam, A., Prakash, S., Gupta, P.: Iris segmentation using improved hough transform. In: Huang, D.-S., Gupta, P., Zhang, X., Premaratne, P. (eds.) ICIC 2012. CCIS, vol. 304, pp. 408–415. Springer, Heidelberg (2012)
8. Nigam, A., Gupta, P.: Palmprint recognition using geometrical and statistical constraints. In: Babu, B.V., Nagar, A., Deep, K., Pant, M., Bansal, J.C., Ray, K., Gupta, U. (eds.) SocProS 2012. AISC, vol. 236, pp. 1303–1315. Springer, Heidelberg (2012)
9. Prakash, S., Gupta, P.: An efficient ear recognition technique invariant to illumination and pose. Telecommun. Syst. **52**, 1435–1448 (2013)
10. Kumar, A., Wu, C.: Automated human identification using ear imaging. Pattern Recogn. **45**, 956–968 (2012)
11. Nigam, A., Gupta, P.: Finger knuckleprint based recognition system using feature tracking. In: Sun, Z., Lai, J., Chen, X., Tan, T. (eds.) CCBR 2011. LNCS, vol. 7098, pp. 125–132. Springer, Heidelberg (2011)
12. Badrinath, G.S., Nigam, A., Gupta, P.: An efficient finger-knuckle-print based recognition system fusing sift and surf matching scores. In: Qing, S., Susilo, W., Wang, G., Liu, D. (eds.) ICICS 2011. LNCS, vol. 7043, pp. 374–387. Springer, Heidelberg (2011)
13. Nigam, A., Gupta, P.: Quality assessment of knuckleprint biometric images. In: International Conference on Image Processing, ICIP, pp. 4205–4209 (2013)
14. Iannarelli, A.: Ear Identification. Paramount Publishing Company, Fremont (1989)
15. Kumar, A., Chan, T.S.: Robust ear identification using sparse representation of local texture descriptors. Pattern Recogn. **46**, 73–85 (2013)
16. Chang, K., Bowyer, K.W., Sarkar, S., Victor, B.: Comparison and combination of ear and face images in appearance-based biometrics. IEEE Trans. Pattern Anal. Mach. Intell. **25**, 1160–1165 (2003)
17. Chan, T.S., Kumar, A.: Reliable ear identification using 2-D quadrature filters. Pattern Recogn. Lett. **33**, 1870–1881 (2012)
18. Dong, J., Mu, Z.: Multi-pose ear recognition based on force field transformation. In: Second International Symposium on Intelligent Information Technology Application, IITA, vol. 3, pp. 771–775 (2008)
19. Naseem, I., Togneri, R., Bennamoun, M.: Sparse representation for ear biometrics. In: Bebis, G., Boyle, R., Parvin, B., Koracin, D., Remagnino, P., Porikli, F., Peters, J., Klosowski, J., Arns, L., Chun, Y.K., Rhyne, T.-M., Monroe, L. (eds.) ISVC 2008, Part II. LNCS, vol. 5359, pp. 336–345. Springer, Heidelberg (2008)

20. Xie, Z., Mu, Z.: Ear recognition using lle and idlle algorithm. In: 19th International Conference on Pattern Recognition, ICPR, pp. 1–4 (2008)
21. Badrinath, G., Gupta, P.: Feature level fused ear biometric system. In: 7th International Conference on Advances in Pattern Recognition, ICAPR, pp. 197–200 (2009)
22. Kisku, D.R., Mehrotra, H., Gupta, P., Sing, J.K.: Sift-based ear recognition by fusion of detected keypoints from color similarity slice regions. In: International Conference on Advances in Computational Tools for Engineering Applications, ACTEA, pp. 380–385. IEEE (2009)
23. De Marsico, M., Michele, N., Riccio, D.: Hero: human ear recognition against occlusions. In: IEEE Computer Society Conference on Computer Vision and Pattern Recognition Workshops, CVPRW, pp. 178–183 (2010)
24. Wang, X., Yuan, W.: Gabor wavelets and general discriminant analysis for ear recognition. In: 8th World Congress on Intelligent Control and Automation, WCICA, pp. 6305–6308 (2010)
25. Zhi-qin, W., Xiao-dong, Y.: Multi-scale feature extraction algorithm of ear image. In: International Conference on Electric Information and Control Engineering, ICEICE, pp. 528–531 (2011)
26. Prakash, S., Jayaraman, U., Gupta, P.: Connected component based technique for automatic ear detection. In: 16th IEEE International Conference on Image Processing, ICIP, pp. 2741–2744 (2009)
27. Shafait, F., Keysers, D., Breuel, T.M.: Efficient implementation of local adaptive thresholding techniques using integral images. In: Electronic Imaging, International Society for Optics and Photonics, p. 681510 (2008)
28. Canny, J.: A computational approach to edge detection. IEEE Trans. Pattern Anal. Mach. Intell. **8**, 679–698 (1986)
29. Pizer, S.M., Amburn, E.P., Austin, J.D., Cromartie, R., Geselowitz, A., Greer, T., ter Haar Romeny, B., Zimmerman, J.B., Zuiderveld, K.: Adaptive histogram equalization and its variations. Comput. Vis. Graph. Image Process. **39**, 355–368 (1987)
30. Wiener, N.: Extrapolation, Interpolation, and Smoothing of Stationary Time Series. The MIT Press, Cambridge (1964)
31. Scharr, H.: Optimal operators in digital image processing. PhD thesis (2000)
32. Shi, J., Tomasi, C.: Good features to track. In: Computer Vision and Pattern Recognition, CVPR, pp. 593–600 (1994)
33. Lucas, B.D., Kanade, T.: An iterative image registration technique with an application to stereo vision. In: 7th International Joint Conference on Artificial Intelligence, IJCAI, pp. 674–679 (1981)
34. Meraoumia, A., Chitroub, S., Bouridane, A.: Fusion of finger-knuckle-print and palmprint for an efficient multi-biometric system of person recognition. In: IEEE International Conference on Communications, ICC (2011)
35. Badrinath, G., Gupta, P.: Palmprint based recognition system using phase-difference information. Future Gener. Comput. Syst. **28**, 287–305 (2012)
36. Cai, J., Goshtasby, A.: Detecting human faces in color images. Image Vis. Comput. **18**(1), 63–75 (1999)

Gait-Assisted Person Re-identification in Wide Area Surveillance

Apurva Bedagkar-Gala and Shishir K. Shah$^{(\boxtimes)}$

Quantitative Imaging Laboratory, Department of Computer Science,
University of Houston, Houston, TX 77204-3010, USA
{avbedagk,sshah}@central.uh.edu

Abstract. Gait has been shown to be an effective feature for person recognition and could be well suited for the problem of multi-frame person re-identification (Re-ID). However, person Re-ID poses very unique set of challenges, with changes in view angles and environments across cameras. Thus, the feature needs to be highly discriminative as well as robust to drastic variations to be effective for Re-ID. In this paper, we study the applicability of gait to person Re-ID when combined with color features. The combined features based Re-ID is tested for short period Re-ID on dataset we collected using 9 cameras and 40 IDs. Additionally, we also investigate the potential of gait features alone for Re-ID under real world surveillance conditions. This allows us to understand the potential of gait for long period Re-ID as well as under scenarios where color features cannot be leveraged. Both combined and gait-only features based Re-ID is tested on the publicly available, SAIVT Soft-Bio dataset. We select two popular gait features, namely Gait Energy Images (GEI) and Frame Difference Energy Images (FDEI) for Re-ID and propose a sparsified representation based gait recognition method.

1 Introduction

Person re-identification (Re-ID) is a fundamental task in multi-camera tracking and other surveillance applications. Re-ID is defined as a problem of person recognition across videos or images of the person taken from different cameras. Beyond surveillance, it has applications in robotics, multimedia, and more popular utilities like automated photo/video tagging or browsing [1]. Re-ID has been a topic of intense research in the past six years [2–8]. Most of the current approaches rely on appearance-based similarity between images to establish correspondences. The typical features used to quantify appearance are low-level color and texture extracted from clothing. A review of appearance descriptors for Re-ID is presented in [9]. However, such appearance features are only stable over short time intervals as people dress differently on different days. In case of large temporal separation or long-period Re-ID, more stable person descriptions based on unique features like biometrics are needed. Biometrics like face and gait have been shown to have tremendous utility in person recognition/ identification applications [10,11]. However, leveraging biometrics for Re-ID has its own challenges.

© Springer International Publishing Switzerland 2015
C.V. Jawahar and S. Shan (Eds.): ACCV 2014 Workshops, Part III, LNCS 9010, pp. 633–649, 2015.
DOI: 10.1007/978-3-319-16634-6_46

Re-ID data comes from uncontrolled environments with non-cooperative sub-jects where face of a person will not always be visible. Further, the data is often low quality due to low sensor resolutions and low frame rates. If the face is visi-ble, it varies greatly in pose, facial expressions, and illumination conditions. All these factors make capturing reliable facial data and subsequent face recognition very difficult. Even though the state-of-the-art face recognition techniques yield high recognition rates, it is important to note that these results are obtained on high resolution data captured under controlled lighting and pose settings. Automated facial recognition on low resolution images under variations in pose, age and illumination conditions is still an open problem [10,12].

(a)   (b)

Fig. 1. Images from SAIVT SoftBio dataset and extracted silhouettes of the same person from different cameras, (a) images with visible face regions, (b) images without visible face regions.

Gait is a behavioral biometric that has been effective for human identifi-cation [11]. Gait is especially suited for Re-ID as it can be extracted by non-obtrusive methods and does not require co-operative subjects. Figure 1 shows images and corresponding silhouettes of the same person taken from different cameras. In Fig. 1(a) the person's face is visible but not in Fig. 1(b). This situ-ation occurs frequently in surveillance videos as the camera angles are uncon-trolled. Further, we can see that in sequence (a) the face resolution is really low and hence extracting usable facial features is challenging. However, the detected silhouettes can be used to extract gait features for Re-ID, even with low resolu-tion and low frame rate data. The availability of video data makes gait feature extraction feasible as gait is extracted over multiple frames. On the other, gait is sensitive to view angles and walking poses. Surveillance cameras usually have a wide field-of-view (FOV) and people often tend to change their walking pose during the duration of observation. This greatly increases the probability of common walking views across different camera views, which can be leveraged for gait recognition. Nonetheless, silhouette extraction errors due to occlusions, illumination variations as shown in Fig. 2, can affect gait feature potency.

In this paper, we study the impact of incorporating gait features extracted from real world surveillance videos along with color (clothing-based appearance) features on Re-ID performance. If the number of frames available for a subject are not enough or silhouette extraction is faulty, then the gait features are not used

Fig. 2. Example images from SAIVT SoftBio dataset and erroneous extracted silhouettes.

and the Re-ID is only based on color features. The following are the contributions of our work:

- Investigate if gait extracted from real world video sequences can be successfully leveraged as an additional feature along with appearance features for person Re-ID in the context of surveillance.
- Identify robust gait features that can be extracted from noisy and incomplete silhouettes and yet retain discriminative capability for Re-ID.
- Propose a sparsified representation-based gait recognition method, where probe gait features are represented as a linear combination of gallery gait features and reconstruction residuals are used for recognition.

The proposed combined feature method is tested for Re-ID and potential of purely gait features is also tested to analyze its usability for long period Re-ID. We restrict the definition of long period Re-ID as Re-ID using videos captured on different days or where appearance features like clothing color break down. The paper is organized as follows: the next section gives a short overview of Re-ID techniques and gait features. Section 3 introduces the color and gait features and gait recognition methods adopted. Section 4 presents each of the datasets, experimental results and discussion. Section 5 concludes the paper.

2   Related Work

Person Re-ID has received much attention in the past few years [2–7]. In general, recent approaches have focused on two aspects of the problem: (1) design of discriminative, descriptive and robust visual descriptors to characterize a person's appearance [6,13,14]; and (2) learning suitable distance metrics that maximize the chance of a correct correspondence [5,7,15]. All of these methods are based on clothing-based appearance features, hence applicable to short period Re-ID. Gala *et al.* [16] have investigated the incorporation of low resolution facial features as potential descriptor but they are combined with clothing color, making them suitable only for short period Re-ID. There has been some work [17] that leveraged gait information for Re-ID. However, it combines gait features with motion phase and camera topology, which requires a training phase. Moreover, the datasets on which the method was demonstrated were fairly simplified videos. Kawai *et al.* [18] propose a view dependent gait and color similarity fusion technique however it required explicit view information regarding gallery and probe poses, which is unrealistic in Re-ID. Color depth cameras are used to extract height temporal information which is combined with color features in [19], however this approach is not applicable with standard surveillance cameras.

2.1   Gait Features

Gait features are divided into two categories, *model-based* and *model-free* [20]. Model-based features like stride and cadence require an explicit model construction, hence, are more sensitive to the accuracy of silhouette extraction techniques and model fitting requires large computational cost. On the other hand, they are invariant to view angle changes and scale. Model-free features capture changes in silhouette shapes or body motion over time. This makes them partially robust to errors in silhouette extraction process but are more sensitive to variations due to pose and scale changes. Since silhouette extraction on surveillance video can be very challenging due to illumination variations, complex backgrounds, occlusions and unconstrained environments [21], model-free gait features are well-suited for Re-ID.

Han *et al.* [11] extract a gait feature called Gait Energy Image (GEI) that is robust to silhouette errors and computationally easy to extract. GEI captures the spatio-temporal description of person's walking pattern into a single image template by averaging silhouette over time, however it fails to retain the dynamic changes in the pattern. GEI represented using Gabor features [22] and wavelet features [23] extracted from different scales, orientations and x-y co-ordinates have been shown to be effective gait features. In order to incorporate the temporal information for gait recognition, Gait History Image (GHI) was proposed in [24]. It used difference between consecutive frames to retain frequency of motion within a GEI. All of these features are robust to silhouettes shapes but not as much to incomplete silhouette detections. The effect of incomplete silhouettes is partially alleviated by the Frame Difference Energy Image (FDEI) [25] which retains only positive portions from consecutive frame differences and combines them with GEIs to retain both static and dynamic portions of the person's walking pattern. In this paper, GEI and FDEI features are selected for gait representation.

2.2   Cross-View Gait Recognition

Gait recognition across different view angles is also an active area of research [26–28]. A view transformation model (VTM) is learnt in [28,29] that learns a mapping between different view angles to transform data from probe and gallery sequences in the same view angle. A model that projects both the gallery and probe features into a subspace where they are highly correlated is learned to improve gait recognition in [27]. The model is learnt using canonical correlation analysis and combined with a Gaussian process classifier trained for view angle recognition for effective cross-view gait recognition. Gait dynamics from different views are synchronized into fixed length stances using non-linear similarity based HMMs in [26] and multi-linear projection model is learned to help gait recognition. All of these methods require some way to quantify the view angles from either gallery sequence, probe sequence or both, and involves a projection stage that introduces gait feature errors. Due to low quality data available for gait feature extraction such methods prove challenging to leverage for Re-ID. In order, to deal with view angle changes some methods [22,23,30] use locality preserving constraints are leveraged during gait recognition. We propose using sparsified representation based gait recognition technique that requires no explicit common view angle synthesis or projection subspace learning and we discuss it in detail in the next section.

3   Approach

Gait is described as the walking characteristic of a person and in order to understand the applicability of gait features to Re-ID they are combined with color features to create a person model for recognition. A person's walking pattern is periodic in nature, thus motion patterns during a walking sequence repeat themselves at regular time intervals. Thus, the features are extracted over a sequence of frames over a gait period. Gait period is defined as the time between two successive strikes of the same foot. Before gait period extraction is performed a silhouette sequence selection step is used to ensure that gait can be reliably extracted. To detect if silhouettes extracted are faulty or do not have enough information, a simple constraint is applied to number of non-zero pixels in the silhouette i.e. they should be more than 40 % of the image size. For a sequence, if gait period cannot be detected, then gait features cannot be extracted. This serves as a natural silhouette sequence selection step. We adopt the method proposed in [31] for gait period estimation of arbitrary walking sequences. The aspect ratio of silhouette over a sequence of frames is used to estimate the gait period. Hence the assumption is that silhouettes are already extracted and preprocessed. Pre-processing involves size normalization and horizontally aligning the silhouettes. Figure 3 depicts the step by step results of the gait period estimation method, which is as follows:

- The aspect ratio of the silhouette over a sequence of frames is represented as a 1D temporal signal. The signal is z-normalized (subtracting the signal mean and dividing by signal's standard deviation) and smoothed using a moving average filter.
- Peaks in the aspect ratio signal are magnified by computing its auto-correlation sequence and the first derivative of the auto-correlation signal is used to detect zero-crossings.
- Zero crossing positions of positive and negative peaks are used to compute distance between prominent peaks, average of distances between consecutive peaks result in the gait period in number of frames.

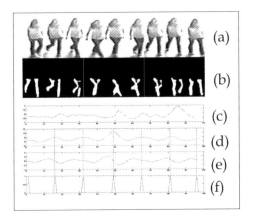

Fig. 3. Gait Period Estimation: (a) Sample image sequence. (b) Sample silhouette sequence. (c) Bounding box aspect ratio after average filtering. (d) Auto-correlation signal of aspect ratio signal. (e) First order derivative of auto-correlation. (f) Detected peak positions.

Most of the dynamic information about walking motion can be deduced from the lower half of the silhouette region [11], hence gait period estimation is performed using aspect ratio of lower body silhouettes.

3.1   Gait Energy Image (GEI)

The first of the two gait features used is called the GEI first proposed in [11]. GEI is a spatio-temporal representation of a person's walking characteristic condensed into a single image and it captures the changes in the shape of the silhouette over a sequence of images. GEI is computed by averaging the silhouettes in the spatial domain over a gait cycle. Given, a sequence of silhouettes, the GEI is given by,

$$GEI(i,j) = \frac{1}{N} \sum_{t=1}^{N} S_t(i,j) \tag{1}$$

where, S_t is the silhouette at frame t, (i, j) are the spatial image co-ordinates and N is the estimated gait cycle period. A given sequence of images can contain multiple gait cycles and hence can be represented by multiple GEIs. It has been shown that GEIs are robust to errors in silhouettes in individual frames [11] and are computationally efficient. Figure 4 shows GEI extracted from a sample sequence from both the SAIVT SoftBio dataset and the dataset we acquired, which we refer to as the multi-camera Re-ID (MCID) dataset. For the MCID dataset, we only used lower body silhouettes to extract gait features due to challenging conditions for silhouette extraction.

3.2   Frame Difference Energy Image (FDEI)

The second gait feature used is called Frame Difference Energy Image (FDEI) and was proposed in [25]. This feature is designed to deal with incomplete

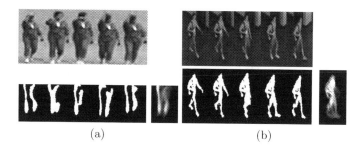

(a)   (b)

Fig. 4. GEI features extracted from sample sequences from the 2 datasets, (a) Lower body GEI from MCID dataset and (b) GEI from the SAIVT Soft-Bio dataset.

silhouettes at the same time retaining the shape and motion changes. This is done by constructing multiple thresholded GEIs from a single gait cycle and then summing these with differences between consecutive silhouettes along the sequence. The difference is taken in such a manner that only the positive values are retained and negative values are discarded so that only parts of the silhouette that are missing or contain motion information is retained. Having estimated the gait cycle, it is further divided into sub-cycles with equal number of frames. The silhouettes within each sub-cycle are averaged using Eq. 2 as follows:

$$GEI_c(i,j) = \frac{1}{N_c} \sum_{t=1}^{N_c} S_t(i,j),$$ (2)

where N_c denotes the period of the sub-cycle. This creates GEIs per sub-cycle of every gait cycle. These GEIs are thresholded, only retaining the pixels with value greater than or equal to 80 % of the maximum value, in order to remove noise due to silhouette errors. These are referred to as dominant GEIs (DGEIs) for each sub-cycle and are computed using Eq. 3 as follows:

$$DGEI_c(i,j) = \begin{cases} GEI_c(i,j), & \text{if } GEI_c(i,j) \geq (0.8xmax(GEI_c)) \\ 0, & \text{otherwise} \end{cases}$$ (3)

The frame difference is computed by subtracting consecutive silhouettes and only the positive portion of the difference is included in the feature as follows:

$$DS_t(i,j) = \begin{cases} 0, & \text{if } S_t(i,j) \geq S_{t-1}(i,j) \\ S_{t-1}(i,j) - S_t(i,j), & \text{otherwise} \end{cases}$$ (4)

The FDEI is generated by summation of the positive frame difference and corresponding sub-cycles dominant GEI as follows,

$$FDEI(i,j) = DS_t(i,j) + DGEI_c(i,j)$$ (5)

If the silhouette in frame t is incomplete, and one at frame $t-1$ is complete, then DS_t contains the incomplete portion of the silhouette and summation of difference image with the dominant GEI compensates for missing portion. Figure 5 shows FDEI extracted from a sample sequence from both the datasets.

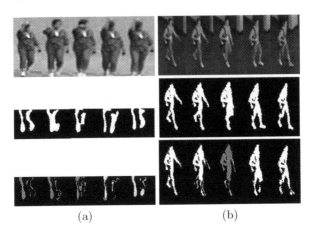

(a) (b)

Fig. 5. FDEI features extracted from sample sequences from the 2 datasets, (a) Lower body FDEI from MCID dataset and (b) FDEI from the SAIVT Soft-Bio dataset.

3.3 Feature Matching for Re-ID

Color Features. The distribution of colors of the person's clothing is characterized using weighted HSV histogram proposed in [6]. The silhouette is divided into three body parts corresponding to head, torso and legs regions by detection of one vertical axis of symmetry and two horizontal axes of asymmetry. The histogram for each body part is weighted by the distance from the axis of symmetry. The histograms from each body part are concatenated channel-wise to generate a single color feature descriptor.

Combined Similarity Measure for Re-ID. The similarity measure between a gallery subject G and a probe subject P is a weighted sum of color feature-based similarity and gait features based similarity and is defined as:

$$dist(G, P) = w_{color}.d_{color}(G, P) + w_{gait}.d_{gait}(G, P) \qquad (6)$$

where, d_{color} is the color feature-based similarity and d_{gait} is the gait features based similarity. The color similarity is obtained using the Bhattacharyya distance [32]. Since, a sequence of frames are available for every subject, color similarity for a given probe-gallery pair is simply the maximum similarity among all the probe-gallery frame pairs. Gait similarity is computed as the re-construction error between probe gait features and the gallery gait model, and is explained in detail in the next section.

Further, w_{color} and w_{gait} are the weights given to color and gait similarity, respectively, such that: $w_{gait} = 1 - w_{color}$. If good quality silhouette sequences are available for both gallery and probe subject and gait features can reliably be extracted, gait similarity is incorporated in the overall similarity measure by setting w_{gait} to a non-zero value. On the other hand, in the absence of usable gait features, Re-ID is established using only color features, i.e., $w_{gait} = 0$ and $w_{color} = 1$.

In this paper, we define the acceptable quality of the silhouettes simply as to whether or not the number of positive pixels is greater than 40 % of the total pixels in the image. Gait period estimation process provides another gait feature selection mechanism. If the silhouettes are too noisy or the sequences are too short, then gait period cannot be reliably estimated.

3.4 Gait Recognition by Sparsified Representation

In order to compute the gait similarity/recognition we utilize a sparsified representation [33]. Sparsified representation is used here as it has been shown to be robust to incomplete or missing data or occlusions in data. The underlying implication is that given a dictionary matrix built using labeled training features of several subjects, a test feature of i-th subject is the linear combination of the training images of only the same subject from the dictionary. Following this implication, we construct a dictionary which is simply a matrix $V = [v_1, ..., v_n]$, where each column is obtained by vectorizing the gait features belonging to all the gallery subjects. Since, each gallery subject can have multiple gait features, either GEIs or FDEIs, multiple columns in the matrix V may belong to the same gallery subject. We build a separate dictionary for each gallery subject, hence the dictionary of a given gallery subjects consists of different gait cycles from the same subject. This ensures that, given a dictionary matrix V, if the probe subject is as close as possible the gallery subject in both identity and view angle, the probe image will lie approximately in a linear span defined by only a subset of the gait features that constitute the matrix V. This implies that given a probe gait feature, for example, GEI_P, it can be expressed as $GEI_P = V.\alpha$ and the intent is to find the sparest α that generated GEI_P in V. Thus, among all possible solutions of α, we want the sparsest. This amounts to solving the following ℓ_1-minimization:

$$\hat{\alpha} = \arg \min \|\alpha\|_1 \quad s.t. \quad GEI_P = V.\alpha \tag{7}$$

This optimization is solved using linear programming that leverages augmented Lagrange multipliers [34]. Thus, d_{gait} is given by Eq. 8 and is an estimate of how well $\hat{\alpha}$ reproduces GEI_P.

$$d_{gait}(G, P) = \|GEI_P - V\hat{\alpha}\|_2 \tag{8}$$

Further, for a given probe subject, there can be multiple GEIs/FDEIs and the gait similarity is simply the minimum among all the probe GEIs/FDEIs.

4 Results and Discussion

The combined gait and color features based Re-ID is tested on two datasets, our multi-camera Re-ID (MCID) dataset and SAIVT SoftBio dataset [35]. The SAIVT SoftBio dataset was selected as it provides real word multi-camera surveillance videos with multiple frames per person from different camera views.

4.1   SAIVT SoftBio Dataset

Combined Color and Gait Features Model for Re-ID. The dataset contains 150 subjects in up to 8 camera views. Only 122 subjects are seen in at least 2 camera views. In our experiments, the gallery is formed by different cameras and so are the probe subjects, so it is as close to the real world Re-ID scenario as possible. The only constraint is that the same subjects gallery and probe camera views are different. The dataset provides background images for each subject per camera view, hence simple background subtraction followed by low level image processing are sufficient to extract silhouette sequences for gait feature extraction. For each subject, in a given camera view, multiple frames are available. Only 5 randomly selected frames from a sequence are used to extract the color features, while all frames are used to extract gait features. Table 1 shows the Re-ID performance of only color, color & GEI (color+GEI) and color & FDEI (color+FDEI) in terms of rank 1, 5, and 20 Re-ID accuracy and normalized AUC extracted from cumulative matching characteristic (CMC) curves using all the 122 subjects. When gait features cannot be extracted, Re-ID relies only on color features. If gait is available, then a weighted sum of color and gait similarity is used to establish a match. The table shows the performance variations with different color and gait weight combinations. We can see that overall using combination of color and gait performs better than only color. As we increase the weight assigned to gait features, we notice a significant boost in performance. The improvement in rank 1 matching accuracy is most notable with changes in weights suggesting that the gait features can be more effective than color features, when available.

Table 1. Rank 1, 5 and 20 matching accuracy and nAUC measures for color, color+GEI and color+FDEI with varying weights.

	Rank 1(%)	Rank 5(%)	Rank 20(%)	nAUC(%)
Color model	0.82	3.28	18.85	56.55
Color(0.8)+GEI(Sparse,0.2)	0.82	3.28	21.31	59.09
Color(0.8)+GEI(NN,0.2)	0.82	3.28	20.49	56.45
Color(0.5)+GEI(Sparse,0.5)	0.82	4.92	27.05	62.34
Color(0.5)+GEI(NN,0.5)	0.82	3.27	22.13	56.61
Color(0.1)+GEI(Sparse,0.9)	**4.10**	**9.84**	**36.06**	**64.75**
Color(0.1)+GEI(NN,0.9)	0.82	4.09	23.77	57.06
Color(0.8)+FDEI(Sparse,0.2)	0.82	3.28	21.31	58.98
Color(0.8)+FDEI(NN,0.2)	0.82	3.28	18.85	55.04
Color(0.5)+FDEI(Sparse,0.5)	0.82	5.74	27.05	62.24
Color(0.5)+FDEI(NN,0.5)	0.82	3.28	19.67	53.86
Color(0.1)+FDEI(Sparse,0.9)	**4.10**	**11.48**	**36.06**	**64.82**
Color(0.1)+FDEI(NN,0.9)	0.82	4.09	19.67	53.63

For comparison, the gait similarity is also computed using nearest neighbor technique [36] which uses the euclidean distance as the underlying distance metric. The nearest neighbor gait similarity is combined with color similarity as described in the approach section. From the above table we can see that with any weight combination of color and sparsified representation gait similarity based Re-ID outperforms the color and gait nearest neighbor similarity. Figure 6 shows the CMC curves obtained by using only color, combined color and GEI, and combined color and FDEI. The gallery size = 122 subjects. Figure 6(a), (b) and (c) show the CMC curves obtained by varying weights given to color and GEI/FDEI. We see that the difference in performance using GEI and FDEI is negligible. Thus, gait features even when extracted from imperfect silhouettes and varying viewpoints provide discriminative value. They add value to color features even in short period Re-ID where clothing color is a reasonable feature.

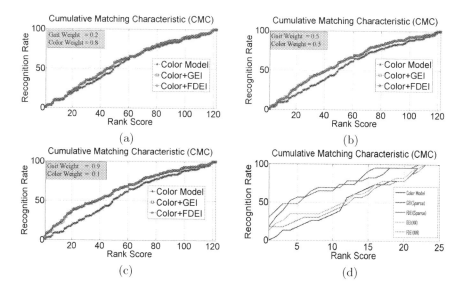

Fig. 6. CMC Curves on the SAIVT SoftBio dataset, (a), (b) and (c) show performance combining color and gait features with varying weights for Re-ID and (d) shows performance using only color, only GEI and only FDEI features for Re-ID (Gallery size = 23).

Only Gait Features Model for Re-ID. In order to better understand the potential of gait features, we performed another experiment with the same dataset. Re-ID was performed using only gait features without the help of color and then compared with pure color based performance. Since only gait features are used the number of subjects for which usable gait features available across views were limited to 23 subjects. Figure 6(d) shows the CMC curves obtained using color, GEI and FDEI features. From the figure, the power of gait features is even more prominent. Either of the gait features outperform the color features significantly. FDEI features yeild a much better Rank 1 matching accuracy when compared to GEI features, yet overall their performance is still comparable. Once

Table 2. Rank 1, 5 and 20 matching accuracy and nAUC measures for color, GEI and FDEI features.

	Rank 1(%)	Rank 5(%)	Rank 20(%)	nAUC(%)
Color model	0	17.40	78.26	49.72
GEI (Sparse)	17.40	**56.52**	95.65	72.78
FDEI (Sparse)	**30.43**	47.83	**95.65**	**73.16**
GEI (NN)	13.04	34.78	82.6	55.95
FDEI (NN)	17.4	26.08	78.26	51.42

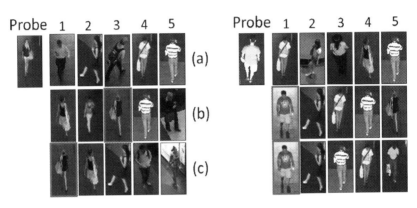

Fig. 7. Examples of Re-ID gallery ranking, showing the first 5 ranked on SAIVT Soft-Bio dataset using: (a) only color, (b) color+GEI and (c) color+FDEI, for two different IDs. The images highlighted by the red border denote the true match (Color figure online).

again, the effectiveness of sparsified representation for gait recognition is evident. Re-ID performance using sparsified representation is clearly better than nearest neighbor gait recognition. Table 2 summarizes the rank 1, 5, and 20 matching accuracy and nAUC measures for the 3 features using both gait recognition techniques. From the table we can see that in terms of rank 1 and nAUC, FDEI features perform better than GEI. These results can be viewed as performance of gait features for long period Re-ID as they do not utilize any clothing or appearance features. This speaks to the ability of gait features for long period Re-ID and also demonstrates the robustness of the proposed gait recognition method for cross view angle gait matching. Figure 7 shows the first 5 ranked images from the gallery set given two different probes using the 3 models, we can see that in both cases the color only Re-ID cannot find an ID match to the probe in the top 5 ranked IDs. In the probe ID shown in the left column, the color+GEI Re-ID located the true ID in the top 5 ranked IDs. On the other hand, the color+FDEI Re-ID rank 1 ID is the correct ID as the probe.

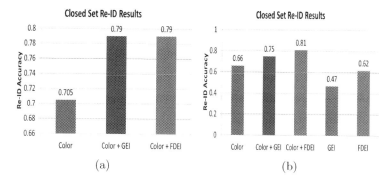

Fig. 8. Closed set Re-ID performance: (a)Bar graph shows the results obtained by our color, color+GEI and color+FDEI on the entire MCID dataset and (b) Bar graph shows the results obtained by our color, color+GEI, color+FDEI, only GEI and only FDEI on subset of MCID dataset.

4.2   MCID Dataset

The MCID dataset is acquired using a camera network consisting of 9 cameras in and around a building. The re-identification data consists of 40 subjects, 19 of them are seen in multiple cameras. As the re-identification is in the context of multi-camera tracking, Re-ID is established for each camera pair. Consistent with the previous dataset experiments, only 5 randomly selected frames from a sequence are used to extract the color features, while all frames are used to extract gait features. Due to complex and uncontrolled conditions of acquisition, traditional background modeling techniques did not perform well for silhouette extraction. Gray level thresholding combined with morphological operations were used to extract lower body silhouettes. Thus, both gait features were extracted only from lower body dynamics. For this dataset, we empirically set $w_{gait} = 0.2$ and $w_{color} = 0.8$ as this gives the best possible Re-ID accuracy.

Combined Color and Gait Features Model for Re-ID. Since Re-ID is performed between camera pairs, all the subjects in the probe set might not be included in the gallery set. Thus, for each camera pair, only a subset of the probe set or closed probe set, i.e., an intersection between the probe and gallery set IDs is used to establish re-identification. This is a closed set experiment, in the sense that it is only possible to have correct matches or mismatches and results are evaluated using matching accuracy, i.e., number of probe subjects matched correctly. Again, closed set experiment is consistent with our experiments on the SAIVT SoftBio dataset. Figure 8(a) shows the Re-ID performance using all 3 models: one based on only color, color+GEI and color+FDEI. The value of adding gait features to color is very evident, as we observe a significant improvement in the re-identification accuracy. One important observation here is that for the gait features of the MCID dataset are extracted from only lower body silhouettes or truncated GEIs. Even so the truncated gait features combined with color provide a significant boost.

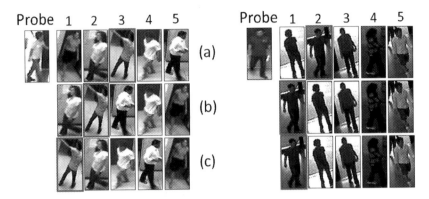

Fig. 9. Examples of Re-ID gallery ranking, showing the first 5 ranked on MCID dataset using: (a) only color, (b) color+GEI and (c) color+FDEI, for two different IDs. The images highlighted by the red border denote the true match (Color figure online).

Only Gait Features Model for Re-ID. To better understand the role of gait features in Re-ID performance, we perform another experiment, using only gait features without the help of color and then compared with pure color based performance. Since only gait features are used, only subjects from each camera pair for which usable gait features are available for Re-ID. The results of this experiment is shown in Fig. 8(b). This figure provides balanced view of performance of all the features, alone and combined. Both purely GEI and FDEI perform reasonably well as compared to the color even though they do not outperform color. It should be noted that only lower body silhouettes are used to generate gait features. The Re-ID accuracy using FDEI comes within 4% of the color accuracy. Combining either GEI or FDEI boosts the Re-ID accuracy significantly. Incorporating FDEI with color helps enhance Re-ID performance to 81% from 66%, a considerable boost. Purely GEI or FDEI features performs better on SAIVT SoftBio dataset compared to MCID, as the MCID features are truncated gait features, which slightly affects the performance using pure gait features. However, we argue that combining gait features with color is an effective strategy to improve Re-ID performance. Figure 9 shows the first 5 ranked images from the gallery set given two different probes using the 3 models, we can see that in both cases the color only Re-ID finds the true ID match to the probe in the top 5 ranked IDs at ranks 3 and 2 for the 2 IDs, respectively. In the probe ID shown in the left column, the color+GEI Re-ID located the true ID at rank 2. On the other hand, the color+FDEI Re-ID rank 1 ID is the correct ID as the probe in both probe IDs.

5   Conclusions

In this paper, we studied the applicability of gait features for Re-ID. The potential of gait features was explored in combination with color features as well as gait as the only feature for Re-ID. A sparsified representation based cross view gait recognition method is proposed that does not require view angle estimation or gait feature reconstruction for gait matching. We demonstrated its effectiveness for person Re-ID for short and long period Re-ID. We would like to point out that this was the first study of its kind offering a strategy for incorporating gait features for Re-ID and extensive analysis. The most important observation is that in spite of silhouette imperfections and view angle variations, gait can be effectively leveraged for Re-ID and is very well equipped for long period Re-ID. One of the future directions will be to explore different strategies for gait and color similarity measure fusion. Another area that needs more study is to estimate definitively the conditions under which gait features become unusable for Re-ID.

References

1. Sivic, J., Zitnick, C.L., Szeliski, R.: Finding people in repeated shots of the same scene. In: Proceedings of the British Machine Vision Conference, pp. 909–918 (2006)
2. Gheissari, N., Sebastian, T., Hartley, R.: Person reidentification using spatiotemporal appearance. In: IEEE Computer Society Conference on Computer Vision and Pattern Recognition (2006)
3. Wang, X., Doretto, G., Sebastian, T., Rittscher, J., Tu, P.: Shape and appearance context modeling. In: International Conference on Computer Vision, pp. 1–8 (2007)
4. Bedagkar-Gala, A., Shah, S.: Multiple person re-identification using part based spatio-temporal color appearance model. In: The Eleventh International Workshop on Visual Surveillance (in conjunction with ICCV 2011) (2011)
5. Prosser, B., Zheng, W.S., Gong, S., Xiang, T.: Person re-identification by support vector ranking. In: Proceedings of the British Machine Vision Conference, pp. 21.1–21.11 (2010)
6. Bazzani, L., Cristani, M., Murino, V.: Symmetry-driven accumulation of local features for human characterization and re-identification. Comput. Vis. Image Underst. **117**, 130–144 (2013)
7. Zheng, W.S., Gong, S., Xiang, T.: Reidentification by relative distance comparison. IEEE Trans. Pattern Anal. Mach. Intell. **35**, 653–668 (2013)
8. Bedagkar-Gala, A., Shah, S.K.: A survey of approaches and trends in person re-identification. Image Vis. Comput. **32**, 270–286 (2014)
9. Satta, R.: Appearance descriptors for person re-identification: a comprehensive review. CoRR (2013)
10. Zhao, W., Chellappa, R., Phillips, J., Rosenfeld, A.: Face recognition: a literature survey. ACM Comput. Surv. **35**, 399–458 (2003)
11. Han, J., Bhanu, B.: Individual recognition using gait energy image. IEEE Trans. Pattern Anal. Mach. Intell. **28**, 316–322 (2006)
12. Ao, M., Yi, D., Lei, Z., Li, S.: Face recognition at a distance: system issues. In: Tistarelli, M., Li, S., Chellappa, R. (eds.) Handbook of Remote Biometrics, pp. 155–167. Springer, London (2009)
13. Cheng, D., Cristani, M., Stoppa, M., Bazzani, L., Murinog, V.: Custom pictorial structures for re-identification. In: Proceedings of the British Machine Vision Conference, pp. 68.1–68.11 (2011)

14. Bak, S., Corvee, E., Bremond, F., Thonnat, M.: Person re-identification using spatial covariance regions of human body parts. In: IEEE International Conference on Advanced Video and Signal Based Surveillance, pp. 435–440 (2010)

15. Dikmen, M., Akbas, E., Huang, T.S., Ahuja, N.: Pedestrian recognition with a learned metric. In: Kimmel, R., Klette, R., Sugimoto, A. (eds.) ACCV 2010, Part IV. LNCS, vol. 6495, pp. 501–512. Springer, Heidelberg (2011)

16. Bedagkar-Gala, A., Shah, S.K.: Part-based spatio-temporal model for Multi-Person re-identification. Pattern Recogn. Lett. **33**(14), 1908–1915 (2011)

17. Roy, A., Sural, S., Mukherjee, J.: A hierarchical method combining gait and phase of motion with spatiotemporal model for person re-identification. Pattern Recogn. Lett. **33**, 1891–1901 (2012)

18. Kawai, R., Makihara, Y., Chunsheng, H., Iwama, H., Yagi, Y.: Person re-identification using view-dependent score-level fusion of gait and color features. In: International Conference on Pattern Recognition (ICPR), pp. 2694–2697 (2012)

19. John, V., Englebienne, G., Krose, B.: Person re-identification using height-based gait in colour depth camera. In: IEEE International Conference on Image Processing (ICIP), pp. 3345–3349 (2013)

20. Wang, J., She, M., Nahavandi, S., Kouzani, A.: A review of vision-based gait recognition methods for human identification. In: International Conference on Digital Image Computing: Techniques and Applications (DICTA), pp. 320–327 (2010)

21. Bouwmans, T.: Recent advanced statistical background modeling for foreground detection: a systematic survey. Recent Pat. Comput. Sci. **4**, 147–176 (2011)

22. Xu, D., Huang, Y., Zeng, Z., Xu, X.: Human gait recognition using patch distribution feature and locality-constrained group sparse representation. IEEE Trans. Image Process. **21**, 316–326 (2012)

23. Hu, H.: Multiview gait recognition based on patch distribution features and uncorrelated multilinear sparse local discriminant canonical correlation analysis. IEEE Trans. Circ. Syst. Video Technol. **24**, 617–630 (2014)

24. Liu, J., Zheng, N.: Gait history image: a novel temporal template for gait recognition. In: IEEE International Conference on Multimedia and Expo, pp. 663–666 (2007)

25. Chen, C., Liang, J., Zhao, H., Hu, H., Tian, J.: Frame difference energy image for gait recognition with incomplete silhouettes. Pattern Recogn. Lett. **30**, 977–984 (2009)

26. Hu, M., Wang, Y., Zhang, Z., Zhang, Z.: Multi-view multi-stance gait identification. In: IEEE International Conference on Image Processing (ICIP), pp. 541–544 (2011)

27. Bashir, K., Xiang, T., Gong, S.: Cross view gait recognition using correlation strength. In: Proceedings of the British Machine Vision Conference (2010)

28. Kusakunniran, W., Wu, Q., Li, H., Zhang, J.: Multiple views gait recognition using view transformation model based on optimized gait energy image. In: IEEE 12th International Conference on Computer Vision Workshops (ICCV Workshops) (2009)

29. Makihara, Y., Sagawa, R., Mukaigawa, Y., Echigo, T., Yagi, Y.: Gait recognition using a view transformation model in the frequency domain. In: Leonardis, A., Bischof, H., Pinz, A. (eds.) ECCV 2006, Part III. LNCS, vol. 3953, pp. 151–163. Springer, Heidelberg (2006)

30. Tee, C., Goh, M., Teoh, A.: Gait recognition using sparse grassmannian locality preserving discriminant analysis. In: IEEE International Conference on Acoustics, Speech and Signal Processing (ICASSP), pp. 2989–2993 (2013)

31. Wang, L., Tan, T., Ning, H., Hu, W.: Silhouette analysis-based gait recognition for human identification. IEEE Trans. Pattern Anal. Mach. Intell. **25**, 1505–1518 (2003)
32. Bhattacharyya, A.: On a measure of divergence between two multinomial populations. Sankhy: Indian J. Stat. (1933–1960) **7**, 401–406 (1946)
33. Wright, J., Yang, A., Ganesh, A., Sastry, S., Ma, Y.: Robust face recognition via sparse representation. IEEE Trans. Pattern Anal. Mach. Intell. **31**(2), 210–227 (2009)
34. Yang, J., Zhang, Y.: Alternating direction algorithms for l1-problems in compressive sensing. Technical report (2009)
35. Bialkowski, A., Denman, S., Sridharan, S., Fookes, C., Lucey, P.: A database for person re-identification in multi-camera surveillance networks. In: Digital Image Computing: Techniques and Applications, pp. 1–8 (2012)
36. Bishop, C.M.: Pattern Recognition and Machine Learning. Springer-Verlag New York Inc., Secaucus (2006)

Cross Dataset Person Re-identification

Yang Hu, Dong Yi, Shengcai Liao, Zhen Lei, and Stan Z. Li[⊠]

National Laboratory of Pattern Recognition, Center for Biometrics
and Security Research, Institute of Automation, Chinese Academy
of Sciences (CASIA), 95 Zhongguancun East Road, Beijing 100190, China
{yhu,dong.yi,scliao,zlei,szli}@nlpr.ia.ac.cn

Abstract. Until now, most existing researches on person re-identification
aim at improving the recognition rate on single dataset setting. The training data and testing data of these methods are form the same source.
Although they have obtained high recognition rate in experiments, they
usually perform poorly in practical applications. In this paper, we focus on
the cross dataset person re-identification which make more sense in the real
world. We present a deep learning framework based on convolutional neural
networks to learn the person representation instead of existing hand-crafted
features, and cosine metric is used to calculate the similarity. Three different datasets Shinpuhkan2014dataset, CUHK and CASPR are chosen as
the training sets, we evaluate the performances of the learned person representations on VIPeR. For the training set Shinpuhkan2014dataset, we also
evaluate the performances on PRID and iLIDS. Experiments show that
our method outperforms the existing cross dataset methods significantly
and even approaches the performances of some methods in single dataset
setting.

1 Introduction

Person re-identification has attracted more and more attention in recent years.
It aims to recognize individuals through person images taken from two or more
non-overlapping camera views. As an important and basic component in surveillance system, person re-identification is closely related to many other applications, such as cross-camera tracking, behavior analysis, object retrieval and so
on. However, the person re-identification problem is a very challenging task. The
person images of existing datasets are captured from surveillance cameras which
usually set to work in the wide-angle mode to cover a wider area, therefore the
resolution of person images is very low even using the high-def cameras. Moreover, changes in illumination, viewpoint, background, pose, camera parameter
and occlusion under different camera views make the person re-identification a
difficult problem: (1) lack of samples to generate the true distributions of various
classes; (2) the distributions of the intra classes and inter classes are unstable
due to the diversities and ambiguities of the samples; (3) above all, the samples
of person re-identification datasets are inseparable.

In the past few years, researchers have done a lot meaningful work to advance
the development of person re-identification. Many difficulties of the task have

© Springer International Publishing Switzerland 2015
C.V. Jawahar and S. Shan (Eds.): ACCV 2014 Workshops, Part III, LNCS 9010, pp. 650–664, 2015.
DOI: 10.1007/978-3-319-16634-6_47

been solved to some extent by discriminative hand-crafted features and metric learning methods. However, the cross dataset setting is different, which has been ignored by most existing methods. The testing sets and training sets are not the same, we don't know the testing data as well as the acquisition conditions, many important factors such as illumination condition, viewpoint and occlusion are totally unconstrained therefore make the task more challenging.

Most existing person re-identification methods conduct their experiments on several public datasets. The most common way is spitting the dataset into two parts, one part for training, the other part for testing. In this setting, the training data and testing data are from the same source. However in real applications, it is very difficult to get a training set which has a similar scenario of the testing set. Therefore unchanged view information and similar data construction make most learning based methods have bad generalization and easy to over-fitting. For feature design based methods, it is also difficult to ensure that the designed features to be effective for new coming data. In a work, most existing methods do not perform well.

Since the performances of single dataset person re-identification methods are improving rapidly, researchers should pay more attention to the cross dataset person re-identification task which make more sense to the real applications. In this paper, we try to solve the cross dataset person re-identification problem by the deep learning (DL) framework which has got huge successes in speech recognition and vision. DL is especially suitable for dealing with large training sets, most existing public person re-identification datasets are small, both in the number of subjects and the number of images per subject. However, with the development of person re-identification, more and more datasets have emerged. The scales of these datasets are getting larger although are still not comparable to other fields [1,2]. Three datasets Shinpuhkan2014dataset [3], CUHK02 [4,5] and CASPR are selected as the training sets, experiments will show the performances of the learned person representations as well as the impact of different training set structures. Compared to the standard feedforward neural networks, convolutional neural networks (CNNs) have much fewer connections and parameters and easier to train. Therefore, we choose CNNs to learn a generic representation of person images.

In this paper, we make the following contributions: (1) We present a CNN framework to learn effective features for person re-identification which has not been paid much attention before; (2) As a different and important aspect of person re-identification, the cross dataset person re-identification advocated by this paper is very meaningful for real applications; (3) The person representations obtained by the proposed method have good performances and generalize well to many datasets.

2   Related Work

The recognition rates of person re-identification have increased a lot over the lase several years. Among these methods, metric learning (ML) approaches have played very important roles [6–13]. Weinberger et al. [6] proposed the LMNN method to learn a Mahanalobis distance metric for k-nearest neighbor (kNN) classification by semidefinite programming. Subsequently, in [7], a method similar to LMNN called Large Margin Nearest Neighbor with Rejection (LMNN-R) was proposed and achieved significant improvement. Davis et al. [8] presented an approach called LTML to formulate the learning of Mahalanobis distance function the problem as that of minimizing the differential relative entropy between two multivariate Gaussians under constraints on the distance function. Zheng et al. [9] formulated person re-identification as a distance learning problem, which aimed to learn the optimal distance that can maximize the probability that a pair of true match having a smaller distance than a wrong match pair. Koestinger et al. [10] proposed the KISSME method to learn a distance metric from equivalence constraints based on a statistical inference perspective. Li et al. [11] proposed the Locally-Adaptive Decision Functions (LADF) method to learn a decision function for person verification that can be viewed as a joint model of a distance metric and a locally adaptive thresholding rule. However, most of these methods have shown to be sensitive to parameters selecting and very easily over-fitting, the performances are not satisfactory in real applications.

Another type of person re-identification methods try to tackle the problem by seeking feature representations which are both distinctive and stable for describing the appearance [14–17]. Farenzena et al. [18] proposed the Symmetry-Driven Accumulation of Local Features (SDALF) method, multiple features were combined considering the symmetry and asymmetry property in pedestrian images to handle view variation. Malocal et al. [19] turned the local descriptors into Fisher Vector to produce a global representation of the image. Cheng et al. [20] utilized Pictorial Structures for person re-identification. Color information and color displacement within the whole body were extracted per-part, and the extracted descriptors were then used in a matching step. Salience were gradually applied in person re-identification as well [21–23]. However, most of hand-crafted features are not distinctive and stable enough and may lose efficacy due to various factors, such as illumination, viewpoint and occlusion, especially when the data sources have changed.

There are also other person re-identification methods. Gray and Tao [24] proposed to use AdaBoost to select good features out of a set of color and texture features. Prosser et al. [25] formulated the person re-identification problem as a ranking problem and applied the Ensemble RankSVM to learn a subspace where the potential true match get the highest rank. In [26], visual features of an image pair from different views are first locally aligned by being projected to a common feature space and then matched with softly assigned metrics which are locally optimized. Liu et al. [27] allowed users to quickly refine their search, which achieved significant improvement. In conclusion, most existing person re-identification methods pay their attentions on single dataset setting, they all contribute a lot to promote the development of this field, but the performances in real applications are still not good enough.

Fortunately, some researchers have noticed the cross dataset setting and have done some meaningful work. Ma et al. [28] proposed a Domain Transfer Ranked

Support Vector Machines (DTRSVM) method for re-identification under target domain cameras which utilized the image pairs of the source domain as well as the unmatched (negative) image pairs of the target domain. Although this method cannot be totally considered as cross dataset person re-identification (used the information of the target domain), it inspired us a lot and got impressive results.

Yi et al. [29] proposed a method called Deep Metric Learning (DML) which learn the metric by a "siamese" deep neural network. The network had a symmetry structure with two sub-networks which were connected by a cosine layer. In this paper, the cross dataset person re-identification experiment were conducted, the author utilized the CUHK Campus dataset as the training set and the ViPeR dataset for testing. Big improvement has been made by DML compared with the DTRSVM on VIPeR. Moreover, DML is the first to conduct experiments on cross dataset person re-identification, although further research is needed, it has promoted the development of cross dataset person re-identification a lot.

Due to the obvious superiorities on large amount of training data and the development of computation resources such as GPU, more and more researchers are carrying on researches on CNN. Therefore, CNN has achieved great success in many fields of computer vision, for instance, DeepFace [1,2] have exhibited impressive results on face recognition and image classification. Inspired by DeepFace, we will address the cross dataset person re-identification problem by learning the person representation. In the following sections, the implementation details will be described, the comparison experiments and the discussions will be reported as well.

3   Person Representation

In recent years, as more data has become available, learning-based methods have started to outperform hand-crafted features, because they can discover and optimize features for the specific task. For cross dataset person re-identification, the influence of resolution, illumination and pose changes are much greater. Moreover, the change of data sources may make the well designed features perform well on one dataset but worse on another. Here, we address the problem by learning the person representation through deep neutral network.

3.1   The Architecture of CNN

The flowchart of our method is shown in Fig. 1. Given a person image, it is first normalized to 48 by 128 pixels, and then divided into three overlapped parts of size 48 by 48 pixels respectively. For each person part, we train a CNN and extract features of this part. Finally, we concatenate the three learned features to get the final person representation.

As the way used for handwritten digit recognition, we train our CNN in a multi-class classification manner. The architecture of our neural network is summarized in Fig. 2. The input of our network is the raw 3-channels (RGB) person

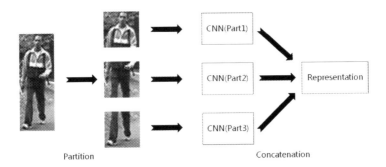

Fig. 1. The flowchart of the proposed method. Learn the representation for each part independently and concatenate them to get the final representation.

image part of size 48 by 48 pixels and send to a convolutional layer (C1). The number of filters in C1 is 32 and the size of the filters is 5×5. Then the 32 feature maps are fed to a max-pooling layer (S2) which takes the max value over 3×3 spatial neighborhoods with a stride of 2 for each channel. Followed by the S2 is another convolutional layer (C3) with 32 filters and the size of the filters is 5×5. The first three layers can extract low-level features such as simple edges and texture. Max-pooling layers play an important role in dealing with the local rotations and transformations, we apply a max-pooling layer after each convolutional layer to make the network more robust. In some other CNN framework, the max-pooling layer is only applied in the first convolutional layer to avoid losing information about the precise position of detailed structure and micro-textures. However, in our task, persons have a variety of changes in pose and the range of variation is very wide. Moreover, no alignment is applied to the unconstrained persons. Therefore, we apply a max-pooling layer after each convolutional layer to make sure the learned representations more general and robust.

The structure of the subsequent layers are just like the former layers. There are two convolutional layers C3 and C5 with 32 filters and 64 filters respectively, the size of the filters is 5×5. Here, we will introduce why we choose this structure. In many other work, local layers are used to ensure that every location in the feature map learns a different set of filters. There are two reasons why local layers are applied: (1) In an aligned image, different regions may have different local statistics and discrimination abilities, the spatial stationarity assumption of convolution cannot hold; (2) Large labeled datasets are available which can afford large locally connected layers such as in the face recognition field. Therefore, the local connected layers are reasonable. However, in person re-identification field, person alignment is a very difficult problem and there is no existing effective algorithm to use. On the other hand, the scales of datasets are much smaller which are not enough to learn the huge number of parameters of local connected layers. Therefore, we choose convolutional layers instead of local layers which sharing the weight for the cross dataset person re-identification task. It may sacrifice some discriminations but is a safe solution to make the learned representation more general and robust.

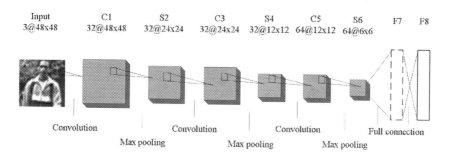

Fig. 2. The structure of the CNN used in our method.

The top two layers (F7 and F8) are fully connected. The first fully connected layer (F7) is optional, the output of the last fully connected layer (F8) is fed to a K-way softmax which produces a distribution over K class labels. If the training set has a small number of subjects, F7 is removed and the output of F8 (after softmax) is used as the feature vector which has a dimension of the number of subjects. If the training set has a relatively large number of subjects, F7 is needed for dimension reduction to avoid over-fitting, and the output of F7 in the network will be used as feature vector. More details will be discussed in the experiments.

3.2 Training and Verification Metric

The training aims to maximize the probability of the correct person identity by maximizing the multinomial logistic regression objective. We achieve this by minimizing the cross-entropy loss for each training sample. If k is the index of the true label for a given input, the loss is: $L = -log p_k$. The loss is minimized by stochastic gradient descent with the size of batch 128. The gradients are computed by standard backpropagation of the error [30]. Fast learning has a great influence on the performance of models trained on large datasets. Therefore, we choose the ReLU [31] to be our activation function: $max(0, x)$ for each layer.

After we get the person representation, we can apply a simple metric, such as the cosine distance, Euclidean distance and Battacharrya distance, to calculate the similarity of two input instances. In this paper, we pay attention to learn general and robust person representation for cross dataset person re-identification, the classifier design is not the focus. Therefore, for the proposed method, we use the cosine distance. Experiments can show that even with simple metric, our learned person representation can perform well.

4   Experiments

In this section, we first introduce the datasets used in our experiments and then the training details, after that we present the comparison with the state-of-the-art, some findings and discussions of the experiments are presented as well.

4.1   Datasets

Three datasets Shinpuhkan2014dataset, CUHK02 and CASPR are chosen as the training sets. We get three representations from the CNN models trained by these three datasets respectively. Then, we evaluate the performances of the representations on VIPeR [32] compared with two other existing cross dataset person re-identification methods DTRSVM [28] and DML [29]. After that, we fix the training set to Shinpuhkan2014dataset, and evaluate performance of the learned representation on PRID [33] and i-LIDS [9].

Shinpuhkan2014dataset consists of more than 22,000 images of 24 people which are captured by 16 cameras installed in a shopping mall "Shinpuh-kan". All images are manually cropped and resized to 48×128 pixels, grouped into tracklets and added annotation. The number of tracklets of each person is 86. This dataset contains multiple tracklets in different directions for each person within a camera. The greatest advantage of this dataset is that all the persons have appeared in 16 cameras. Some image samples can be seen in Fig. 3.

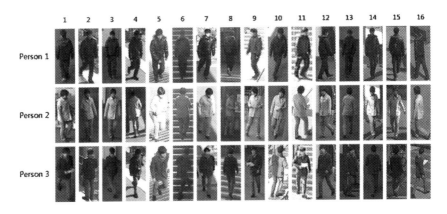

Fig. 3. Some image samples of three persons selected from the Shinpuhkan2014dataset, images from the same column indicate that they are from the same camera view. We can see that for each person, the appearances are different in 16 cameras and there are many kinds of changes.

CASPR is a person re-identification dataset collected by ourselves. We capture six videos in a research institution and segment 7414 images of 200 persons from these videos. Note that each of the 200 person has at least 2 associated camera views, some of them have appeared in 5 camera views. All segmented images are scaled to 48×128 pixels. Each subject has at least 7 images and at most 93 images. Figure 4 shows some example pairs of images from the collected database. It can be observed that there is a large variation in the observed color, and there are also lighting changes and viewpoint changes that challenge the matching of persons across cameras. We are expanding this dataset and it will be released in the near future.

Fig. 4. Example pairs of images from the collected database. Images in the same column come from the same person.

CUHK02 contains 1,816 persons captured from 5 pairs of cameras (P1-P5, ten camera views). They have 971, 306, 107, 193 and 239 persons respectively. Each identity has two samples per camera view. It has the largest number of subjects so far. Samples from this dataset can be found in Fig. 5.

Fig. 5. Example pairs of images from the CUHK02 database. CUHK02 has five pairs of camera views denoted with P1-P5. Here, at lest two exemplar persons are shown for each pair of views. Images in the same column represent the same person.

VIPeR. The Viewpoint Invariant Pedestrian Recognition database is one of the earliest single-shot datasets, and it is the most widely used benchmark so far in person re-identification field. It contains 632 pairs of pedestrians and images in VIPeR suffer greatly from illumination and viewpoint changes, making it a very challenging dataset.

4.2   Training on the Datasets

Firstly, we train the deep neural network on the Shinpuhkan2014dataset by the Tesla GPU. As the number of subjects is very small (24 persons), we take out the first fully connected layer (F7) and consider the output of the last fully connected layer (F8, after a 24-way softmax) to be our feature vector. We initialize the weights in each layer from a normal distribution with mean zero and standard deviation 0.0001, and the biases are set to 0.5. The weight decay of the F8 is set to 3. The stride is set to 1 and the padding is set to 2. During the training, we flip the images to double the number of training samples, and use the training set to test every 5 epoches. On our experience, as the number of classes (subjects) in this network is small which makes the cost of the test set drop easily, we set the epoch to 100 for this dataset with a small learning rate (0.001 for weight and 0.002 for bias).

For the CUHK02 dataset, the first fully connected layer (F7) is needed and the output of F7 is used as our feature vector. We set the dimension of F7 to 24 and we can get a 24-dimension feature vector for each part of a input image (see the division of a person in Fig. 1). Most of the parameters are same with the parameters when we train the model on Shinpuhkan2014dataset except the number of epoches. As the numbers of subjects are bigger and the numbers of images per subject are smaller, the error rate is much more difficult to drop. Therefore, we set the epoch to 250 for this dataset.

For the CASPR dataset, the number of subjects and the number of images per subject is the middle of that of the Shinpuhkan2014dataset and the CUHK02. The error rate is also hard to drop. We apply almost the same structure and parameters of CUHK02 except the epoches. A more efficient way is to apply a two-step strategy to train the model which is also feasible for the CUHK02 dataset. The first step, we set a relatively large learning rate (0.01 for weight and 0.02 for bias) to make the error rate drop, when the it start to drop, we change the learning rate to a smaller value (0.001 for weight and 0.002 for bias) to make the error rate drop slowly and smoothly. The number of epoch is 60 and 100 for step one and step two respectively. This strategy can save the training time as well as preventing the oscillation of error rate which make it hard to drop to a more optimal point.

4.3   Results on VIPeR

For each of the three training sets, we show the performances of the learned person representations on VIPeR. We split VIPeR into testing set with 316 subjects randomly, and repeat the process 11 times. The first split is used for parameter tuning, the other 10 splits are used for reporting the results.

Figures 6, 7 and 8 show the rank curves of the three parts and their fusion. The recognition rates are summarized in Table 1 as well as the comparison with DTRSVM and DML. The most difficult point is the difference in data distribution of different datasets, the model learned on one dataset probably lose

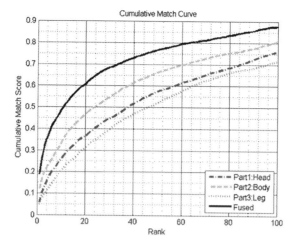

Fig. 6. The CMC curves on the VIPeR dataset, the training set is Shinpuhkan2014 dataset. Performances of each body parts and the fusion are shown.

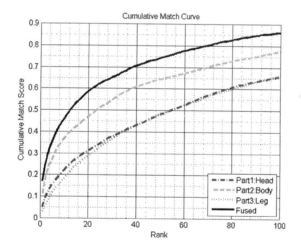

Fig. 7. The CMC curves on the VIPeR dataset, the training set is CUHK02.

efficacy on new data. From the results we can see that the cross dataset evaluation accuracies of person re-identification are currently very low. For our method, although the training sets are different, it develops a general feature representation that can be directly applied in different scenarios, which is very important for practical applications. Our method gets very impressive results and even approaches the performance of some methods in single database setting, such as ELF [24] and PRDC [9]. Shinpuhkan2014dataset has large variations such as viewpoint, background, illumination, pose and deformation due to the 16 different camera views, the number of images is large as well. Although the number of

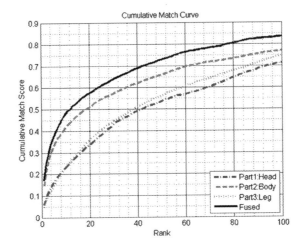

Fig. 8. The CMC curves on the VIPeR dataset, the training set is CASPR.

subjects is not large enough, this dataset is very suitable for learning the person representation.

Table 1. Comparison with the DTRSVM and DML on VIPeR

Methods	Training sets	Rank1 (%)	Rank10 (%)	Rank20 (%)	Rank30 (%)
DTRSVM [28]	i-LIDS	8.26	31.39	44.83	53.88
DTRSVM [28]	PRID	10.90	28.20	37.69	44.87
DML [29]	CUHK Campus	16.17	45.82	57.56	64.24
Ours	CUHK02	16.90	45.89	58.58	65.32
Ours	CASPR	17.22	48.01	57.56	64.15
Ours	Shinpuhkan2014dataset	**18.64**	**48.54**	**60.63**	**68.48**

We also compared with unsupervised feature design methods SDALF [18] and eBicov [34] to show the efficacy of our method. Shinpuhkan2014dataset has been chosen as our training set, we conduct experiments on VIPeR with the same data partition provided by SDALF, and also conduct experiments on i-LIDS with the same protocol of SDALF. The performances are summarized in Table 2. The results show that the person representation learned by our method has good generalization performance. Moreover, the proposed method directly extract feature from the original image without silhouette mask.

4.4   Results on PRID and i-LIDS

Since using Shinpuhkan2014dataset as training set can get the best result, we conduct two more cross dataset experiments on PRID and i-LIDS by the same

Table 2. Comparison with the unsupervised feature design methods

Methods	Test sets	Rank1 (%)	Rank5 (%)	Rank10 (%)	Rank20 (%)
SDALF [18]	VIPeR	19.87	38.89	49.37	65.73
eBiCov [34]	VIPeR	20.66	42.00	**56.18**	68.00
Ours	VIPeR	**21.27**	**43.99**	56.11	**69.72**
SDALF [18]	i-LIDS	28.49	48.21	57.28	68.26
Ours	i-LIDS	**29.83**	**50.76**	**61.04**	**73.64**

models when testing the VIPeR dataset. PRID dataset consists of person images from two static surveillance cameras. Total 385 persons are captured by camera A, while 749 persons captured by camera B. The first 200 persons appeared in both cameras, and the remainders only appear in one camera. 100 out of the 200 image pairs are randomly taken out while and the others for testing. We repeat this for 10 times. The recognition rates are summarized in Table 3 as well as the comparison with DTRSVM.

Table 3. Comparison with the DTRSVM on PRID 2011

Methods	Training sets	Rank1 (%)	Rank10 (%)	Rank20 (%)	Rank30 (%)
DTRSVM [28]	VIPeR	4.6	17.25	22.9	28.1
DTRSVM [28]	i-LIDS	3.95	18.85	26.6	33.2
Ours	Shinpuhkan2014dataset	16.80	43.30	52.40	56.80

From the results we can see that the PRID is a very difficult dataset, the recognition rates of DTRSVM are very low. The chromatic aberration of the images in camera B folder might be the reason. However, our method still can get impressive results which outperform DTRSVM significantly.

The i-LIDS contains 476 person images from 119 persons, 80 persons are randomly chosen for testing. We choose one image from each person randomly to consist the gallery set, the remaining images are used as the probe set. This procedure is repeated 10 times. The CMC curves are shown in Fig. 9, the recognition rates are summarized in Table 4 as well as the comparison with other state-of-the-art methods.

In Fig. 9, we can see that the performance of the head part almost equals that of the fusion and the performance of the leg part is poor. The reason is that the i-LIDS dataset has many occlusions, the leg parts are often obscured by suitcases. The segmentation is not precise as other datasets, only the head parts are relatively stable. The results presented in Table 4 are exciting, our method outperforms most learning based methods without any training procedure of the testing set. Given all the cross dataset experiments we have conducted, we can see that the person representation learned by our method can perform well on different datasets without prior information of the testing sets.

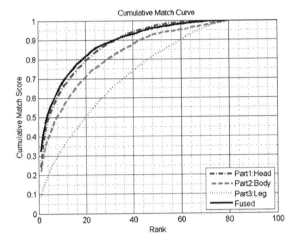

Fig. 9. The CMC curves on i-LIDS, the training set is Shinpuhkan2014dataset.

Table 4. Comparison with the state-of-the-art on i-LIDS with p = 80

Methods	Rank1(%)	Rank5 (%)	Rank10 (%)	Rank20 (%)	Rank30 (%)
MCC [35]	12.00	33.66	47.96	67.00	NA
ITM [8]	21.67	41.80	55.12	71.31	NA
Adaboost [24]	22.79	44.41	57.16	70.55	NA
LMNN [6]	23.70	45.42	57.32	70.92	NA
Xing's [36]	23.18	45.24	56.90	70.46	NA
L1-norm	26.73	49.04	60.32	72.07	NA
Bhat	24.76	45.35	56.12	69.31	NA
PRDC [9]	**32.60**	54.55	65.89	78.30	NA
Ours	32.41	**55.19**	**67.70**	**81.54**	**88.81**

5   Conclusions

The cross dataset person re-identification problem has not been paid much attention before, but it is very important for practical applications. This paper proposed a feature learning method by using CNN for the cross dataset person re-identification. The structure and training process were described in detail. Three public datasets, VIPeR, PRID and i-LIDS, were tested to evaluate the performance of the learned person representation. Extensive results illustrated that the learned representation had good generalization.

Acknowledgment. This work was supported by the Chinese National Natural Science Foundation Projects #61105023, #61103156, #61105037, #61203267, #61375037,

National Science and Technology Support Program Project #2013BAK02B01, Chinese Academy of Sciences Project No. KGZD-EW-102-2, and AuthenMetric R&D Funds.

References

1. Taigman, Y., Yang, M., Ranzato, M., Wolf, L.: DeepFace: closing the gap to human-level performance in face verification. In: Conference on Computer Vision and Pattern Recognition (CVPR)
2. Krizhevsky, A., Sutskever, I., Hinton, G.E.: Imagenet classification with deep convolutional neural networks. In: NIPS, pp. 1106–1114 (2012)
3. Kawanishi, Y., Wu, Y., Mukunoki, M., Minoh, M.: Shinpuhkan 2014: A multi-camera pedestrian dataset for tracking people across multiple cameras. In: Proceedings of the 20th Korea-Japan Joint Workshop on Frontiers of Computer Vision (FCV) (2014)
4. Li, W., Zhao, R., Wang, X.: Human reidentification with transferred metric learning. In: Lee, K.M., Matsushita, Y., Rehg, J.M., Hu, Z. (eds.) ACCV 2012, Part I. LNCS, vol. 7724, pp. 31–44. Springer, Heidelberg (2013)
5. Li, W., Wang, X.: Locally aligned feature transforms across views. In: CVPR, pp. 3594–3601 (2013)
6. Weinberger, K., Blitzer, J., Saul, L.: Distance metric learning for large margin nearest neighbor classification. Adv. Neural Inf. Process. Syst. **18**, 1473 (2006)
7. Dikmen, M., Akbas, E., Huang, T.S., Ahuja, N.: Pedestrian recognition with a learned metric. In: Kimmel, R., Klette, R., Sugimoto, A. (eds.) ACCV 2010, Part IV. LNCS, vol. 6495, pp. 501–512. Springer, Heidelberg (2011)
8. Davis, J.V., Kulis, B., Jain, P., Sra, S., Dhillon, I.S.: Information-theoretic metric learning. In: Proceedings of the 24th International Conference on Machine Learning, pp. 209–216, ACM (2007)
9. Zheng, W.S., Gong, S., Xiang, T.: Person re-identification by probabilistic relative distance comparison. In: CVPR, pp. 649–656 (2011)
10. Kostinger, M., Hirzer, M., Wohlhart, P., Roth, P.M., Bischof, H.: Large scale metric learning from equivalence constraints. In: 2012 IEEE Conference on Computer Vision and Pattern Recognition (CVPR), pp. 2288–2295. IEEE (2012)
11. Li, Z., Chang, S., Liang, F., Huang, T.S., Cao, L., Smith, J.R.: Learning locally-adaptive decision functions for person verification. In: 2013 IEEE Conference on Computer Vision and Pattern Recognition (CVPR), pp. 3610–3617. IEEE (2013)
12. Guillaumin, M., Verbeek, J., Schmid, C.: Is that you? metric learning approaches for face identification. In: 2009 IEEE 12th International Conference on Computer Vision, pp. 498–505. IEEE (2009)
13. Gong, S., Cristani, M., Yan, S., Loy, C.C.: Person Re-identification. Springer, London (2014)
14. Hu, Y., Liao, S., Lei, Z., Yi, D., Li, S.Z.: Exploring structural information and fusing multiple features for person re-identification. In: 2013 IEEE Conference on Computer Vision and Pattern Recognition Workshops (CVPRW). pp. 794–799, IEEE (2013)
15. Gheissari, N., Sebastian, T.B., Hartley, R.: Person reidentification using spatiotemporal appearance. In: CVPR, vol. 2, pp. 1528–1535 (2006)
16. Hamdoun, O., Moutarde, F., Stanciulescu, B., Steux, B.: Person re-identification in multi-camera system by signature based on interest point descriptors collected on short video sequences. In: ICDSC, pp. 1–6 (2008)

17. Wang, X., Doretto, G., Sebastian, T., Rittscher, J., Tu, P.H.: Shape and appearance context modeling. In: ICCV, pp. 1–8 (2007)
18. Farenzena, M., Bazzani, L., Perina, A., Murino, V., Cristani, M.: Person re-identification by symmetry-driven accumulation of local features. In: CVPR, pp. 2360–2367 (2010)
19. Ma, B., Su, Y., Jurie, F.: Local descriptors encoded by fisher vectors for person re-identification. In: Fusiello, A., Murino, V., Cucchiara, R. (eds.) ECCV 2012 Ws/Demos, Part I. LNCS, vol. 7583, pp. 413–422. Springer, Heidelberg (2012)
20. Cheng, D.S., Cristani, M., Stoppa, M., Bazzani, L., Murino, V.: Custom pictorial structures for re-identification. In: BMVC, vol. 2, p. 6 (2011)
21. Zhao, R., Ouyang, W., Wang, X.: Unsupervised salience learning for person re-identification. In: 2013 IEEE Conference on Computer Vision and Pattern Recognition (CVPR), pp. 3586–3593. IEEE (2013)
22. Zhao, R., Ouyang, W., Wang, X.: Person re-identification by salience matching. In: ICCV (2013)
23. Liu, Y., Shao, Y., Sun, F.: Person re-identification based on visual saliency. In: 2012 12th International Conference on Intelligent Systems Design and Applications (ISDA), pp. 884–889. IEEE (2012)
24. Gray, D., Tao, H.: Viewpoint invariant pedestrian recognition with an ensemble of localized features. In: Forsyth, D., Torr, P., Zisserman, A. (eds.) ECCV 2008, Part I. LNCS, vol. 5302, pp. 262–275. Springer, Heidelberg (2008)
25. Prosser, B., Zheng, W.S., Gong, S., Xiang, T., Mary, Q.: Person re-identification by support vector ranking. In: BMVC, vol. 1, p. 5 (2010)
26. Li, W., Wang, X.: Locally aligned feature transforms across views. In: 2013 IEEE Conference on Computer Vision and Pattern Recognition (CVPR), pp. 3594–3601. IEEE (2013)
27. Liu, C., Loy, C.C., Gong, S., Wang, G.: Pop: person re-identification post-rank optimisation. In: International Conference on Computer Vision (2013)
28. Ma, A., Yuen, P., Li, J.: Domain transfer support vector ranking for person re-identification without target camera label information. In: 2013 IEEE International Conference on Computer Vision (ICCV), pp. 3567–3574 (2013)
29. Yi, D., Lei, Z., Liao, S., Li, S.Z.: Deep metric learning for person re-identification. In: International Conference on Pattern Recognition (2014)
30. Lecun, Y., Bottou, L., Bengio, Y., Haffner, P.: Gradient-based learning applied to document recognition. In: Proceedings of the IEEE, pp. 2278–2324 (1998)
31. Nair, V., Hinton, G.E.: Rectified linear units improve restricted boltzmann machines. In: ICML, pp. 807–814 (2010)
32. Gray, D., Brennan, S., Tao, H.: Evaluating appearance models for recognition, reacquisition, and tracking. In: IEEE International Workshop on Performance Evaluation of Tracking and Surveillance, Citeseer (2007)
33. Hirzer, M., Beleznai, C., Roth, P.M., Bischof, H.: Person re-identification by descriptive and discriminative classification. In: Heyden, A., Kahl, F. (eds.) SCIA 2011. LNCS, vol. 6688, pp. 91–102. Springer, Heidelberg (2011)
34. Ma, B., Su, Y., Jurie, F., et al.: Bicov: a novel image representation for person re-identification and face verification. In: British Machive Vision Conference (2012)
35. Globerson, A., Roweis, S.T.: Metric learning by collapsing classes. In: NIPS (2005)
36. Xing, E.P., Ng, A.Y., Jordan, M.I., Russell, S.J.: Distance metric learning with application to clustering with side-information. In: NIPS, pp. 505–512 (2002)

Spatio-Temporal Consistency for Head Detection in High-Density Scenes

Emanuel Aldea[1](✉), Davide Marastoni[2], and Khurom H. Kiyani[3]

[1] Autonomous Systems Group, Université Paris Sud, Gif-sur-Yvette, France
emanuel.aldea@u-psud.fr
[2] Università di Pavia, Pavia, Italy
[3] Communications and Signal Processing Group, Imperial College London,
London, UK

Abstract. In this paper we address the problem of detecting reliably a subset of pedestrian targets (heads) in a high-density crowd exhibiting extreme clutter and homogeneity, with the purpose of obtaining tracking initializations. We investigate the solution provided by discriminative learning where we require that the detections in the image space be localized over most of the target area and temporally stable. The results of our tests show that discriminative learning strategies provide valuable cues about the target localization which may be combined with other complementary strategies in order to bootstrap tracking algorithms in these challenging environments.

1 Introduction

One of the strongest recent developments in computer vision has been related to the analysis of crowded scenes. The interest that this specific field has raised may be explained from two different perspectives. In terms of applicability, continuous surveillance of public and sensitive areas has benefited from the advancements in hardware and infrastructure, and the bottleneck moved towards the processing level, where human supervision is a laborious task which often requires experienced operators. Other circumstances involving the analysis of dense crowds are represented by large scale events (sport events, religious or social gatherings) which are characterized by very high densities (at least locally) and an increased risk of congestions. From a scientific perspective, the detection of pedestrians in different circumstances, and furthermore the interpretation of their actions involve a wide range of branches of computer vision and machine learning.

A rough but quite consistent indicator of the difficulty of analyzing a crowded scene is represented by the number of pixels associated to individual targets (pedestrians). For large objects clearly exhibiting body parts at least sporadically, the detection and tracking algorithms have advanced significantly in the last decade [1]. The aim of the present work is to investigate contexts in which the scale of the scene or other logistical or practical constraints impose a small target size; this is typically the case of large scale, high-density crowds. In these circumstances, research efforts have focused primarily on holistic approaches

© Springer International Publishing Switzerland 2015
C.V. Jawahar and S. Shan (Eds.): ACCV 2014 Workshops, Part III, LNCS 9010, pp. 665–679, 2015.
DOI: 10.1007/978-3-319-16634-6_48

for analysis, which involve primarily the extraction of coarse-level information, such as flow patterns or texture. Although these parameters may be sufficient for characterizing the crowd up to a certain scale, they are unable to grasp finer variations in local dynamics which are not consistent with the global flow, or in local density. However, these fine scale phenomena are essential, not only for security considerations, but also for understanding better the interactions among targets at high density levels and their influence on the dynamics of the crowd.

Single camera analysis represents the typical setup for a broad range of applications related to detection and risk prevention in public and private environments. Although some camera networks may contain thousands of units, it is quite common to perform processing tasks separately in each view. However, single view analysis is limited by the field of view of individual cameras and furthermore by the spatial layout of the scene; also, frequent occlusions in crowded scenes hamper the performance of standard detection algorithms and complexify the tracking task.

Multiple camera analysis has the potential to overcome problems related to occluded scenes, long trajectory tracking or coverage of wider areas. Among the main scientific challenges, these systems require mapping different views to the same coordinate system; also, solutions for the novel problems they address (detection in dense crowds, object and track association, re-identification etc.) may not be obtained simply by employing and extending previous strategies used in single camera analysis.

In order to perform large scale crowd analysis supported by dense tracking, a multiple camera approach is imperative in order to cope with strong, frequent occlusions. Nevertheless, in order to initialize the tracks, a hypothesis about the location of the targets has to be formulated in single camera views; then this hypothesis may be refined in multiple projections. In our work, we study the problem of providing a preliminary initialization of a target density map in high-density strongly occluded environments. The aim of the method we propose is thus not to provide a perfect, exhaustive detection of targets, but rather to bootstrap the tracking process with a detection process which may be somewhat tolerant to false positives, since temporal and multiple camera cues enforced by a full tracking framework would have the ability to perform further filtering.

In order to formulate detection hypotheses in the image space, we rely on a discriminative learning process. This solution has been used extensively and is de facto the algorithm employed for pedestrian detection in non crowded environments, and recently in applications where visibility is often reduced to upper body parts. Our work shows that, among other strategies that are necessary for tackling the problem of person detection in high-density scenes, discriminative learning performs reliably and may be employed in order to initialize a large scale tracking process. The value of such a study rests on the need for better solutions for studying crowded human urban environments in order to improve the security of the flows involved, and the supporting infrastructure as to increase and not diminish the comfort of participants.

Our paper is structured as follows. The next section presents the related work which is relevant for the problem we address, and underlines the relevance of our investigation in the context of identifying low-resolution targets with frequent occlusions. Section 3 highlights the main steps performed in the discriminative learning based classification of image content, at pixel level. Section 4 presents a preliminary filtering strategy that allows for taking into account the spatial and temporal coherence that is expected from true positives in a video sequence. Section 5 illustrates an application of the proposed algorithm to the analysis of a highly crowded scene, and Sect. 6 presents the conclusions of our study and future directions of work.

2   Related Work

The growth of the cities and their evolution towards megalopolises have transformed the foundations of our society. The worlds population is projected to grow from the current 7 billion to around 9 billion by 2050, and hence to increase the burden on resources and on the associated demands on public transport. In the light of these concerns, and also on the grounds of safety improvement during mass events [2,3], there is an urgent imperative to study in detail the phenomena occurring in high-density crowds.

The interest surrounding the study of crowd phenomena spanned during the last decade across multiple fields, including physics, sociology, simulation, visualization and computer vision; among them, computer vision has an essential role of linking the theoretical field with the actual phenomenon (i.e. *calibration* and *validation*) through video analysis (denoted also in other fields as empirical data collection). Indeed, models used in simulations have not been either proposed or validated for high-density crowd scenarios. In the case of real data i.e. recordings of dense crowd movement, the extraction of pedestrian trajectories has been performed either by human operators, a process which is time consuming and cumbersome, or in an unsupervised manner but only in specific conditions i.e. vertical cameras and using primitive methods. In both cases, a major hindrance is the strong occlusion among pedestrians which makes extracting accurate trajectories or accurate local density information nearly impossible.

As the density of a crowded environment increases, conventional approaches used in video analysis stop working, since supporting hypotheses (visibility of body parts, occlusion level, presence of background, presence of ground plane etc.) are not valid anymore. Most importantly, the behaviour of people involved in the crowd changes in order to adapt to the space constraints and the available degrees of freedom. A high-density environment is considered a scene where density is higher than approximately 4 people/m^2. The immediate consequences of this density are:

- heads are the only visible body parts (except occasionally shoulders)
- there is no static background
- occlusions are frequent and persistent
- the image content is rather homogeneous.

Noncrowded scenes have represented for a long time the main area of interest for the computer vision community, and pedestrian detection algorithms evolved

significantly in the last decade, addressing complex applications such as identification of people, grouping analysis, estimation of body parts, gesture based
and trajectory based action analysis etc. However, as it has been already highlighted many times [4], all these methods are not appropriate when high-density
crowd analysis is performed, and new methods must be designed in order to cope
with extreme clutter. Actually, clutter is indeed the main difficulty, but practical
considerations also raise difficult questions. Technical difficulties widen the gap
between proof-of-concept experiments aimed at high-density crowded scenes and
functional solutions. The size of the interest objects, the accessibility to areas of
interest, the size of the problem raise as well novel fundamental research challenges that require significant innovations with regards to established methods.

Single camera analysis. When coping with pedestrian tracking, the established
approach is based on the HOG detector [5], as this representation is adapted for
the detection of upright subjects which are at least partially visible. The major
applications of computer vision research that are responsible for the advancements in the field are the intelligent transportation and the surveillance industry. Secondarily, advancements in machine learning supported studies focusing
on multi-target tracking and models of social behaviour, which are aimed very
often at scenes with few subjects and consistent interactions. Among these three
applicative domains, surveillance has naturally shifted the most towards the
analysis of denser scenes.

Some initial attempts [6,7], managed to initialize tracking of occluded subjects and proposed an effective approach based on mean-shift [8], or relied on 3D
human models integrated into a Bayesian framework, but these methods cannot
handle properly persistent occlusions or multiple close-by subjects. In [9], local
and global features are used in a probabilistic framework in order to estimate
the reliability of a detection; again, this method is sensitive to occlusions and
does not scale properly to dense crowds.

It has been shown already that the temporal information may be used in
order to analyse the coherence of the movement through clustering and assist the
detection process [10]; again, these methods attain their limits for dense scenes
because of occlusions, lack of background, homogeneity and similar movement.

Very recently, the detection of the particular head-shoulder shape ("Ω-shape")
has been addressed specifically [11–14]. The common characteristics of these
studies are the use of the HOG descriptor, of discriminative classifiers and the
exploitation of additional image features related to local higher order statistics.
Focusing on the detection of the Ω-shape has strong benefits: heads remain visible in crowded scenes, and it generalizes quite well the human appearance from
different perspectives. However, the main concern regarding this solution is that
additional work has to be done in order to increase the robustness against occlusions ([14] being a promising approach). Secondly, it is not yet clear what would be
the minimal size of objects required in order to maintain a good level of detection;
as we will see in the following section, in large scale analysis the size of objects is
much smaller than the one reported in these works.

It is worth mentioning at this point a fundamentally different approach to crowd analysis which has been popularized by [15,16], which uses the spatial organization of the flow field in order to integrate local and global dynamics, and prior behaviour knowledge. The benefit of this approach is that no training data or notions of appearance models are employed at all. The main disadvantage is that different streams are modelled as entities with quasi-constant size and density. Fine-level analysis as the one we intend to perform requires the individual movement of all pedestrians and is able to resolve the dynamics on an individual scale.

Other recent works perform opportunistic tracking in high-density crowds, which relies on the salient appearance of some pedestrians, and manage to track individuals on impressive distances in very difficult environments [17,18]. The essential aspect of this type of approach is that not only salient targets are tracked reliably, but also their tracking process may propagate to neighboring targets. Even though in our work we do not make any assumptions about the saliency of parts of the scene, we consider that exploiting highly salient objects if they are present is relevant for tracking in high-density crowds in order to add constraints to the detection space. The drawback of this approach is that color information has to be present, and for small targets the penalty of using color sensors which degrade the sharpness of image gradients is significant. Also, it seems that in terms of dense analysis, the community is getting close to a performance limit which is mainly set by the occlusion level, and a fundamental shift is necessary in order to improve the results significantly.

Multiple camera strategies. The problem of occlusion cannot be solved robustly by employing single camera recordings. As the interest of the computer vision community extended gradually from single pedestrian tracking in uncluttered scenes to crowd analysis, it has become clear that multiple camera networks are required. The use of multiple cameras for video analysis (mainly surveillance) is an extensive topic, which we cannot cover in detail, but fundamental insights may be found in [19,20]. We underline though a small scale experiment proposed in [21] which proves the potential of multiple camera tracking in occluded scenes. This study also proposes an effective solution for exploiting jointly hypotheses related to the presence of a head in multiple cameras, but the consistency is evaluated using the pixel intensity information, and extending this type of solution to large scale scenarios raises several difficult scientific and technical questions.

Bootstrapping high-density crowd analysis. The characteristics of a crowd that challenges usual analysis strategies are the absence of a background, the occlusions and the homogeneity in terms of appearance and dynamics of the moving mass. These are the main reasons that make the crowd analysis problem still be considered challenging currently [22].

The strategy we propose aims to bootstrap (initialize) a complex analysis process by mapping and refining a probability density to the single view image space. Then, either data fusion or tracking algorithms may be employed in order to benefit from multiple data inputs in the form of multiple camera views.

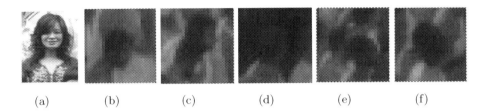

$$\text{(a)} \qquad \text{(b)} \qquad \text{(c)} \qquad \text{(d)} \qquad \text{(e)} \qquad \text{(f)}$$

Fig. 1. In 1(a) we present for comparison an image used in [12] for discriminative learning of the head-shoulder shape. In the related literature, descriptors are computed on patches of sizes varying between 32×32 to 48×64. 1(b) shows a typical well contrasted head in our dataset. Beside significantly lower resolutions per target, the data we use exhibit often low contrast between close targets 1c or between targets and the dynamic background 1(d), and strong occlusions 1(e), 1(f).

A common prerogative of these algorithms is that they require by themselves an initialization procedure. In the following sections, we show that even in the difficult conditions characterizing high-density crowds, we may obtain head detection maps in single camera views that may be used subsequently either for initializing tracking algorithms, or for extending the detection process within a multiple-camera network.

3   Learning for Head Identification

In the following section, we will start by detailing the classification process that we use in order to obtain a probability estimation for the presence of a target (a head) in the image space, and then we will also motivate the interest of the benchmark we use, and which is correlated to the purpose of the detection process i.e. initializing a tracker and/or extending the analysis to multiple views.

The descriptor and the learning process. In order to perform discriminative learning, we rely on the HOG descriptor initially proposed in [5]. Compared to other studies taking interest in the detection of heads in crowded settings, or at least in environments where the rest of the body is barely visible [11–13], we formulate two fundamentally different assumptions. Firstly, we assume that the size of the targets is significantly smaller - approximately a disk of a three-four pixel radius in the image space - which makes the analysis slightly uncomfortable even for a human (see Fig. 1 for an illustration of typical objects sizes used in [12] compared to the targets considered in our work). Secondly, we assume that occlusions are frequent and strong.

Under these circumstances, the parameter which has a significant impact on the classifier performance is the extraction window size. This has to be large enough in order to allow for a reliable characterization of the target using its immediate context, but at the same time small enough in order not to bias the learning process towards non-local learning and towards the detection of joint groups of targets.

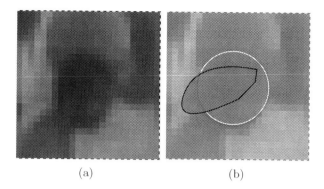

(a)   (b)

Fig. 2. In 2(a) we present a typical area to be labeled by an user in order to obtain ground truth data. In 2(b) we show the green area around the clicked spot which is considered true positive, as well as a red area which could be the result of a classifier detection (Color figure online).

For the learning task we rely on an SVM classifier, and we consider two different kernel functions. For HOG descriptors h_1 and h_2, we consider a linear classifier $K_L(h_1, h_2) = \langle h_1, h_2 \rangle$, and also the Histogram Intersection Kernel (HIK) function which has shown consistently good performance particularly in the context of pedestrian detection using HOG descriptors [23]:

$$K_I(h_1, h_2) = \sum_{i=1}^{dim} \min[h_1(i), h_2(i)] \qquad (1)$$

We perform a pixel-wise classification and then, for each pixel $I(i, j)$ we transfer the binary classifier decision into a probability estimation $p_{i,j}$ [24].

Benchmark design. We are interested in obtaining a dense probability map P over the image domain I, which should highlight the maxima associated to the presence of heads ideally through a local plateau, and not only through an unstable peak. This behavior highly desirable taking into account the fact that the objective of the method we present is to act as an *initializer* module for a tracking algorithm. Thus we can tolerate a non-localized response and a certain amount of false positives which may be filtered out by the tracker.

The scoring we propose for evaluating our initializer reflects this aim, and works in the following way. The ground truth data is represented by image content (distinct from training data) where a human user clicks exhaustively and as accurately as possible in the center of the targets. We then expect that *all* pixels located in discretized disks of radius r around ground truth points be classified as positives (the green area in Fig. 2(b) corresponding to a click performed in Fig. 2(a)). For a certain threshold τ, we consider that probability estimates $p_{i,j} \geq \tau$ are the detected positives (the red area in Fig. 2(b)). Then we define the following:

- the true positives (**TP**): detected positives located inside the disks (intersection between red and green areas)
- the false positives (**FP**): detected positives located outside the disks (red area outside the green disk)
- the false negatives (**FN**): detected negatives located inside the disks (green pixels outside the red area)
- the true negatives (**TN**): the rest of the ground truth domain.

Theoretically, when we vary the r parameter from $r = 0$ to the typical radius of targets, the performance should increase monotonically, and then decrease for higher values of r. In practice, given the imprecisions of the classifier but most importantly the fact that the targets are far from having a perfectly circular shape close in radius to r, the performance tends to decrease with r. Although this methodology is overly pessimistic, it is helpful for visualizing the performance of the classifier within a local neighborhood of the ground truth points. Consequently, this signature characterizes the ability of the method we propose to make a compromise between precise localization and robust detection, and thus to provide a spatio-temporal persistence of the detection.

4   Exploiting Temporal and Spatial Cues

One of the challenges raised by pixel-wise classification is that the local gradient varies on a high-dimensional manifold, and during the movement of a head through the crowd the classifier response for its constituent pixels is noisy, both on a temporal scale (a moving pixel representing the same head area might exhibit occasionally low detection probability) or on a spatial scale (certain pixels inside a compact head region might exhibit occasionally low detection probability).

In the following paragraphs, we propose a solution for introducing temporal consistency in the probability map based on its temporal evolution. The main assumption that supports our approach is that short-term variations in the probability values should be small for pixels belonging genuinely to targets. Secondarily, we assume that positive responses should be locally high since a target consists in multiple connected pixels, so we would like to encourage clustered responses in the probability distribution.

We underline the fact that with respect to a veritable tracking algorithm, this process is fundamentally different since we do not infer at object level, and thus we limit the consistency check to a limited time interval, and at the immediate pixel neighborhood. However, this is completely in line with our objective of providing a reliable *pixel-wise* label for head detection.

Explicit temporal consistency check. In order to associate pixel measurements related to the same entity, we use dense optical flow recursively to project the current pixel in the previous and next N images of the video sequence.

For a detection threshold τ and for the pixel $I_{i,j}^t$ present in the video at coordinates (i, j) at time t, let us consider a corresponding projection $I_{i,j}^{t+k}$, where

$-N \leq k \leq N$. If we consider the probability $p_{i,j}^{t+k}$ as well as the probabilities of all its neighbors (in 8-adjacency), we perform maximal voting in order to obtain the label $l_{i,j}^{t+k}$ of $I_{i,j}^{t+k}$. The objective of this process is to sample temporal information regarding the analyzed pixel by regularizing spatially at the same time in the immediate neighborhood of the projections.

Finally, we perform a maximal vote on the set

$$L_{i,j}^t = \{l_{i,j}^{t+k}\}_{-N \leq k \leq N}$$

consisting in the $2N+1$ projection labels we collected, and we assign the resulting label to the current pixel $I_{i,j}^t$.

Explicit spatial regularization. In order to perform spatial regularization explicitly in the current probability map, we refine a posteriori the pixel classification, by assuming a Markov random field (MRF) over the pixel states. However, this time we consider a basic symmetric neighborhood structure based on 4-adjacency, i.e.

$$N_{i,j}^t = \{I_{i-1,j}^t, I_{i+1,j}^t, I_{i,j-1}^t, I_{i,j+1}^t\}$$

and we consider as observation set the current probability map associating to the pixel $I_{i,j}^t$ the values $p_{i,j}^t \in [0,1]$ provided by the classifier.

5   Experimental Results

We tested our head identification method on high-density images acquired at Makkah during very congested times of the Hajj period, in October 2012. For training, we used data from multiple images, amounting for 1032 positive and negative examples. The window size for the HOG descriptor was set to 24×24, according to the considerations we underlined in Sect. 3.

General observations. We trained a linear and a HIK based SVM classifier, and the two algorithms selected 241 and 343 examples as support vectors respectively. A first observation related to the high-density crowd analysis is that the cluttered context gives rise to a significant degradation in the classifier performance. Figure 3 shows a straightforward detection obtained by applying the linear classifier for each pixel in two different regions of the same image - one which is very cluttered and one where the head density is moderate. The final step consists in obtaining a detection probability map, thresholding it and performing non maximal suppression locally in order to recover only the strongest responses. The moderate density detection illustrated in Fig. 3(a) shows that a direct approach is able to provide an acceptable detection result, which could be fed directly to a tracking algorithm for initialization. The cluttered scene however presents an entirely different kind of panorama, with a fair number of peaks that are associated to a head, but also with a high number of misses and a significant number of false detections.

Under these circumstances, the solution we propose is to postpone in the decision process the techniques that lead to loss of information such as thresholding

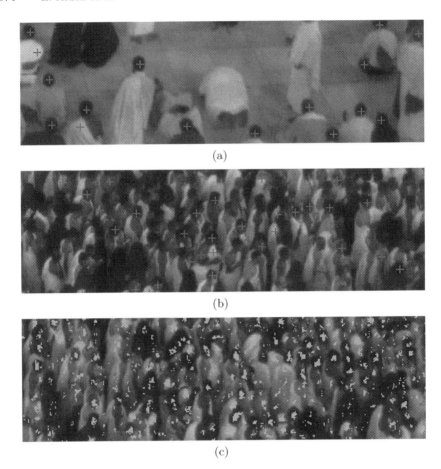

(a)

(b)

(c)

Fig. 3. In Fig. 3(a) we present the results of a straightforward detection involving thresholding and non-maximal suppression on a non-crowded area of the scene free of occlusions, with good results. The same algorithm fails to exhibit the same good performance in a cluttered environment - see Fig. 3(b). We argue that in these cases is to refine the probability map over the image space (illustrated in Fig. 3(c) with high probability areas highlighted in green) rather than to perform operations such as thresholding or non-maximal suppression which involve loss of information (Color figure online).

or non-maximal suppression, and focus the computational effort on improving the consistency of the detection probability map. We present therefore in 3(c) the probability density map for a cluttered area; pixels highlighted in green for visualization purposes exhibit probability values higher than a relatively low threshold. This time we note that despite the fact that there is a certain amount of false positives, the detection map manages to cover most of the targets, while filtering out at the same time a good amount of non-relevant areas.

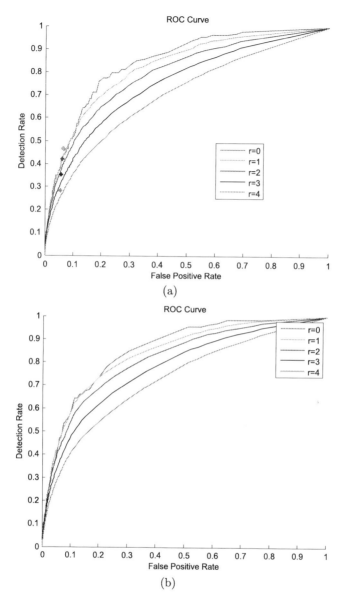

Fig. 4. In Fig. 4(a) we present the performance of the linear classifier for the different values of r; the locations indicated by a cross in the corresponding color show the performance in terms of FPR-TDR of the classifier output which has been regularized using a MRF approach. Figure 4(b) presents the performance for the same base classifier where we introduced the temporal consistency check.

ROC analysis of the detector. In the following part of the section, we will try to analyze quantitatively the interest of exploiting the density map and improving its consistency. In order to have access to numerical estimates of the detection

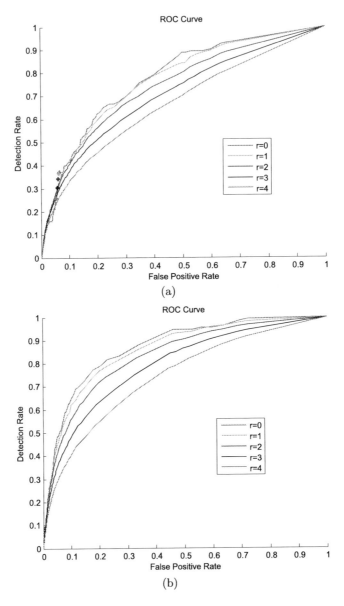

Fig. 5. In Fig. 5(a) we present the performance of the HIK classifier for the different values of r; the locations indicated by a cross in the corresponding color show the performance in terms of FPR-TDR of the classifier output which has been regularized using a MRF approach. Figure 5(b) presents the performance for the same base classifier where we introduced the temporal consistency check (Color figure online).

performance, we define a ground truth set consisting in image content where we identify exhaustively and as accurately as possible the persons which are present; the ground truth set amounts for a number of 132 targets. Then we apply the benchmarking strategy detailed in Sect. 3.

Table 1. Area under the ROC curve for the different classification strategies proposed.

Kernel type	Consistency	$r = 0$	$r = 1$	$r = 2$	$r = 3$	$r = 4$
Linear	-	0.8446	0.8237	0.7948	0.7259	0.7035
Linear	Temporal	0.8548	0.8409	0.8164	0.7771	0.7296
HIK	-	0.7742	0.7673	0.7413	0.7059	0.6666
HIK	Temporal	0.8998	0.8565	0.8316	0.7897	0.7420

For each of the testing scenarios which are illustrated in Fig. 5, we consider a ground truth radius r around the pixels clicked by the human user ranging from $r = 0$ (we consider only the clicked points) to $r = 4$, which is close to the upper limit for a head radius in our image set. Then for each of these five values for r, we vary the detection threshold in the interval $\tau \in [0, 1]$ and we compute the False Positive (FPR) and True Detection (TDR) rates, which are used for plotting the corresponding ROC curves.

Figure 4(a) illustrates the performance of the linear classifier for the different values of r; the locations indicated by a cross in the corresponding color show the performance in terms of FPR-TDR of the classifier output which has been regularized using a MRF approach as depicted in Sect. 4. We note that the performance evaluated using this metric is located above the corresponding ROC curves. Figure 4(b) presents the performance for the same base classifier where we introduced the temporal consistency check presented in Sect. 4. In this case, we did not perform the explicit spatial regularization since the output of temporal consistency check is a binary labeling, and the extra information gain a posteriori is insignificant.

Finally, Figs. 5(a) and (b) illustrate the same metrics in the case of the HIK classifier. Table 1 allows for a precise quantitative comparison among the proposed strategies in terms of the area under the different ROC curves.

Discussion. The performance of the classifiers underlines the fact that discriminative learning may be employed, even in extremely cluttered environments, to provide target cues to tracking algorithms. Depending on the trade-off that we prefer between the risk of target miss and the presence of false positives, the ROC curves should assist in finding the appropriate decision boundaries.

One important observation that underlines the applicability of this approach compared to other analysis strategies in high-density environments is that we do not make any assumption about the presence of salient objects, either in terms of color or shape. The strategy we presented exploits only the implicit saliency of the head shape compared to its immediate environment. Although the discriminative characteristic of the classifier is moderate, the overall performance indicates that this characteristic is sufficient for target identification.

Secondly, we note the good performance of the HIK classifier when the temporal consistency of the detection is taken into account; this classifier seems to be more sensitive to fine variations in the descriptor content and therefore it benefits more from a regularization framework and also from a consistent training database.

Finally, we note that for increasing values of r the performance decreases relatively slowly at the beginning, which shows that this family of classifiers has a stable response above the target area. The classifier performance as well as the testing accuracy could benefit from a pixel-level annotation of the targets but beside the fact that the effort required is significant, the approximation of the target area as a small disk region around the clicked pixels is consistent with the results. We also highlight the fundamental difficulty of building objective pixel-wise ground truth data for this range of detection tasks, which may further hint at the current lack of standardized high-density crowd data available for research.

6   Conclusion and Future Work

In this paper we investigate the applicability of discriminative learning strategies for detecting and initializing tracking targets in high-density crowds exhibiting extreme clutter and homogeneity. By avoiding approaches based primarily on thresholding and non-maximal suppression of the detections, we show that we can build consistent detection probability maps which present a plateau response in target locations. These maps are suitable for spatio-temporal regularization, and also for offering an application adapted compromise between the desired rate of false positive detections and the target miss rate.

As future work, we would like to carry out spatio-temporal regularization jointly in a MRF framework, as well as to estimate the limitation of the classifiers determined by changes in the appearance related to the topology. This would allow us ultimately to apply these methods jointly in multiple cameras and validate a probability density map using independent data sources.

Acknowledgement. K. Kiyani would like to acknowledge the Qatar QNRF under the grant NPRP 09-768-1-114.

References

1. Ferryman, J., Ellis, A.L.: Performance evaluation of crowd image analysis using the PETS2009 dataset. Pattern Recogn. Lett. **44**, 3–15 (2014). Pattern Recognition and Crowd Analysis
2. Helbing, D., Johansson, A., Al-Abideen, H.Z.: Dynamics of crowd disasters: an empirical study. Phys. Rev. E **75**, 046109 (2007)
3. Krausz, B., Bauckhage, C.: Loveparade 2010: automatic video analysis of a crowd disaster. Comput. Vis. Image Underst. **116**, 307–319 (2012)
4. Zhan, B., Monekosso, D., Remagnino, P., Velastin, S., Xu, L.Q.: Crowd analysis: a survey. Mach. Vis. Appl. **19**, 345–357 (2008)
5. Dalal, N., Triggs, B.: Histograms of oriented gradients for human detection. In: Proceedings of the 2005 IEEE Computer Society Conference on Computer Vision and Pattern Recognition (CVPR 2005) - Volume 1, CVPR 2005, vol. 1, pp. 886–893. IEEE Computer Society, Washington, DC (2005)

6. Zhao, T., Nevatia, R.: Bayesian human segmentation in crowded situations. In: 2003 IEEE Computer Society Conference on Computer Vision and Pattern Recognition, 2003, Proceedings, vol. 2, pp. II-459–466 (2003)

7. Zhao, T., Nevatia, R.: Tracking multiple humans in crowded environment. In: Proceedings of the 2004 IEEE Computer Society Conference on Computer Vision and Pattern Recognition, 2004, CVPR 2004, vol. 2, pp. 406–413. IEEE (2004)

8. Comaniciu, D., Meer, P., Member, S.: Mean shift: a robust approach toward feature space analysis. IEEE Trans. Pattern Anal. Mach. Intell. **24**, 603–619 (2002)

9. Leibe, B., Seemann, E., Schiele, B.: Pedestrian detection in crowded scenes. In: CVPR, pp. 878–885 (2005)

10. Rabaud, V., Belongie, S.: Counting crowded moving objects. In: 2006 IEEE Computer Society Conference on Computer Vision and Pattern Recognition, vol. 1, pp. 705–711 (2006)

11. Li, M., Zhang, Z., Huang, K., Tan, T.: Estimating the number of people in crowded scenes by mid based foreground segmentation and head-shoulder detection. In: 19th International Conference on Pattern Recognition, 2008, ICPR 2008, pp. 1–4 (2008)

12. Li, M., Bao, S., Dong, W., Wang, Y., Su, Z.: Head-shoulder based gender recognition. In: 2013 20th IEEE International Conference on Image Processing (ICIP), pp. 2753–2756 (2013)

13. Ye, Q., Gu, R., Ji, Y.: Human detection based on motion object extraction and headshoulder feature. Optik - Int. J. Light Electron Opt. **124**, 3880–3885 (2013)

14. Wang, S., Zhang, J., Miao, Z.: A new edge feature for head-shoulder detection. In: 2013 20th IEEE International Conference on Image Processing (ICIP), pp. 2822–2826 (2013)

15. Ali, S., Shah, M.: A lagrangian particle dynamics approach for crowd flow segmentation and stability analysis. In: IEEE Conference on Computer Vision and Pattern Recognition, 2007, CVPR 2007, pp. 1–6 (2007)

16. Ali, S., Shah, M.: Floor fields for tracking in high density crowd scenes. In: Forsyth, D., Torr, P., Zisserman, A. (eds.) ECCV 2008, Part II. LNCS, vol. 5303, pp. 1–14. Springer, Heidelberg (2008)

17. Moore, B.E., Ali, S., Mehran, R., Shah, M.: Visual crowd surveillance through a hydrodynamics lens. Commun. ACM **54**, 64–73 (2011)

18. Idrees, H., Warner, N., Shah, M.: Tracking in dense crowds using prominence and neighborhood motion concurrence. Image Vis. Comput. **32**, 14–26 (2014)

19. Aghajan, H., Cavallaro, A.: Multi-camera Networks: Principles and Applications. Academic Press, London (2009)

20. Javed, O., Shah, M.: Automated Multi-camera Surveillance: Algorithms and Practice. The International Series in Video Computing, vol. 10. Springer, New York (2008)

21. Eshel, R., Moses, Y.: Tracking in a dense crowd using multiple cameras. Int. J. Comput. Vis. **88**, 129–143 (2010)

22. Wang, X.: Intelligent multi-camera video surveillance: a review. Pattern Recogn. Lett. **34**, 3–19 (2013)

23. Maji, S., Berg, A., Malik, J.: Classification using intersection kernel support vector machines is efficient. In: IEEE Conference on Computer Vision and Pattern Recognition, 2008, CVPR 2008, pp. 1–8(2008)

24. Lin, H.T., Lin, C.J., Weng, R.: A note on platts probabilistic outputs for support vector machines. Mach. Learn. **68**, 267–276 (2007)

Multi-target Tracking with Sparse Group Features and Position Using Discrete-Continuous Optimization

Billy Peralta[1]($\boxtimes$) and Alvaro Soto[2]

[1] Universidad Católica de Temuco, Temuco, Chile
bperalta@uct.cl
[2] Pontificia Universidad Católica de Chile, Santiago, Chile
asoto@ing.puc.cl

Abstract. Multi-target tracking of pedestrians is a challenging task due to uncertainty about targets, caused mainly by similarity between pedestrians, occlusion over a relatively long time and a cluttered background. A usual scheme for tackling multi-target tracking is to divide it into two sub-problems: data association and trajectory estimation. A reasonable approach is based on joint optimization of a discrete model for data association and a continuous model for trajectory estimation in a Markov Random Field framework. Nonetheless, usual solutions of the data association problem are based only on location information, while the visual information in the images is ignored. Visual features can be useful for associating detections with true targets more reliably, because the targets usually have discriminative features. In this work, we propose a combination of position and visual feature information in a discrete data association model. Moreover, we propose the use of group Lasso regularization in order to improve the identification of particular pedestrians, given that the discriminative regions are associated with particular visual blocks in the image. We find promising results for our approach in terms of precision and robustness when compared with a state-of-the-art method in standard datasets for multi-target pedestrian tracking.

1 Introduction

Automatic multi-target tracking is the computational task of detecting the trajectories of objects in a sequence of images. It has many and diverse applications in the real world, e.g. surveillance [1,2], sports [3] and sensor networks [4]. In this work, we focus on multi-target tracking of pedestrians, where recent research has shown significant progress. Nonetheless, current techniques only offer good performance with easy conditions i.e. a static background and separated pedestrians. As the area of inspection becomes more crowded, the performance of these algorithms tends to decrease and they are clearly outperformed by humans.

Successful tracking methods are based on the premise of the presence of previous detections i.e. target people are detected by a generic pedestrian detector.

© Springer International Publishing Switzerland 2015
C.V. Jawahar and S. Shan (Eds.): ACCV 2014 Workshops, Part III, LNCS 9010, pp. 680–694, 2015.
DOI: 10.1007/978-3-319-16634-6_49

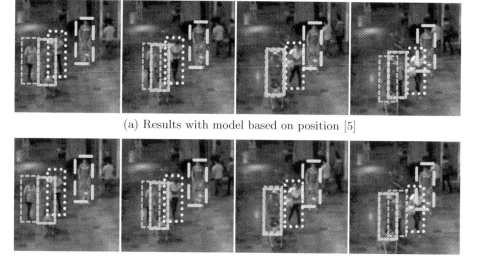

(a) Results with model based on position [5]

(b) Results with our model based on sparse grouped features

Fig. 1. A comparison of our method with a state-of-the-art model [5]. In (a), the model based on location information tends to confuse as it only considers position, as shown by the yellow and light-blue tracklets. In contrast, our model (b) uses visual feature information with sparse selection of groups of features to select the more important visual components of the tracklets. In our case, all the tracklets are correctly identified (Color figure online).

In fact, this scheme can be combined with an online model to deal with appearance variation or scene lighting [6]. Some advantages of using pedestrian detections as input as compared to a generic tracker are that this does not require initialization of the trackers and avoids model drifting. In single target tracking the task is usually accomplished by fitting a trajectory prediction function according to evidence given by detections; in multi-target tracking by contrast, we have multiple possible identities that complicate trajectory prediction by trackers. We have two specific problems: (i) the assignation of a unique identity to each detection, which is also called the data association problem, and (ii) trajectory estimation of each target (Fig. 1).

Great emphasis is currently placed on the data association problem, since many discrete optimization techniques has been developed over several decades, even though this problem is NP-hard [7]. In the case of trajectory estimation, it is usually computed assuming known correct labelings. In order to solve the multi-target tracking problem, a key observation of Andriyenko et al. [5] is given by the complementarity between data association, which is a discrete optimization problem as the assignation of identities is nominal; and trajectory estimation, which is a continuous optimization problem as the position variable is numeric. In this case, the two problems are solved jointly using a discrete-continuous alternating optimization under a Markov Random Field (MRF) framework. Although

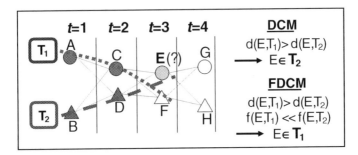

Fig. 2. The difference between our model (FDCM) and a state-of-the-art model (DCM) [5]. We have two objects that are detected in four moments ($t = 1$ to 4) originating 8 nodes in MRF (node A to node H). We assume two known trackers T_1 and T_2 and that the geometric and feature distances (d and f) are defined. With DCM, the node E has label T_2 because it is geometrically near to trajectory of tracker T_2: $d(E, T_1) > d(E, T_2)$. With FDCM, the node E has label T_1 as it considers feature distance between nodes and trackers: $d(E, T_1) + f(E, T_1) < d(E, T_2) + f(E, T_2)$. The feature distance between node E and T_1 is small because T_1 is formed by circles and node E is a circle; on contrast, T_2 is composed by triangles. In similar way, node F belongs to trackers T_2.

this method is natural for this problem, it does not consider visual feature information and may potentially lose valuable information about the targets.

In multi-target tracking of pedestrians, we observe that if the targets are well separated, position information is usually enough to assign a correct label to each detection. In a more challenging setting, as in the case of crowded scenes, visual feature information can be very valuable for distinguishing between multiple individuals by using particular features such as the type or color of their clothes. In the present work, we embed a feature-dependent function inside the association potential component of a Markov Random Field. We use a multi-class logistic regression model to represent the relation between features of detections and the identity of trackers. Furthermore, we expect that in a frame sequence, a person will usually have a particular discriminative region of image; therefore, we also propose the use of group Lasso regularization. Figure 2 shows the intuition of our idea, if we use geometric distance node E has label T_2, but if we consider feature and geometric distance, node E has label T_1. Finally, the estimation of parameters of this function assumes a fixed labeling and adopts a Markov Random Field solution based on a pseudo-likelihood approximation.

2   Background

2.1   Related Work

Multi-target tracking research has made significant progress in recent years. Kuo et al. [8] presented a discriminative appearance model for multiple targets using an online learning scheme based on the AdaBoost algorithm. Training samples are selected considering a sliding time window and spatial-temporal constraints.

Although this model appears effective, it does not consider position information which can be useful in case of occlusion or poor lighting conditions. Berclaz et al. [9] proposed a mathematical model based on constrained linear programming to address data association. This model is based on discretization given by an occupancy map where each cell can be filled with a target. A disadvantage of this method is that it can involve a very large number of variables i.e. if we assume a square occupancy map, the number of nodes is related quadratically to the length of the side of the square. Benfold and Reid [1] described a multi-target tracking system considering accurate estimations of head locations. They solve data association using a MCMC scheme combined with KLT tracking and HOG pedestrian detections in a multithreaded application. This system appears efficient because of its real-time performance. Nonetheless, a potential problem with this model is its complexity due to the combination of multiple algorithms.

Yang and Nevatia [10] proposed an online model based on Conditional Random Fields (CRF) to find and discriminate multiple targets. Although this model is efficient, the non-submodularity of energy function hinders the CRF inference provoking the use of heuristic solutions, and therefore, there is no guarantee that they will find an optimal solution. Butt and Collins [11] address the problem of multiple tracking by using detections in triplets, where a triplet is defined as three consecutive detections of a pedestrian. The use of triplets gives natural access to motion information i.e. with this information they estimate the speed of a particular candidate target in the triplet. Nevertheless, this model is supported by the reliability of the triplets, which can be uncertain as they may overlap other triplets. They use an heuristic process in order to avoid conflicts between nearby triplets, using the nearest detections. Hoffman et al. [12] proposed a hierarchical model for multiple tracking. They present a probabilistic model that finds the paths for each level of hierarchy. This path-finding problem is usually solved by the Hungarian algorithm. Nevertheless, due to the complexity of the hierarchical model, they require some heuristics for post-processing the results, where the parameters can be difficult to obtain.

An interesting work is presented by Andriyenko et al. [5], in which data association and trajectory estimation problems are addressed respectively by a joint discrete and continuous optimization, on a Markov Random Field where the nodes set is given by the pedestrian detections and the edge set is given by the pairs of nearby detections. The discrete optimization assumes that the trajectory and label costs are known; then they estimates the unary and pairwise terms of MRF; they subsequently apply a graph cut algorithm based on α-expansion as the energy function remains sub-modular. In this MRF, the unary term is proportional to the distance between detections and the spline associated with a specific label. In the case of the pair-wise term of two nearby nodes, it is greater than zero if the labels of the two nodes are different; otherwise it is zero. The continuous optimization of trajectories is hard given the presence of label cost. They assumes the labels as known, disregard the label cost, and fit a cubic spline model over the targets positions considering one label at a time in order to estimate the trajectories, where a change is accepted only if the global energy function is decreased. After the trajectory fitting, they calculate the label cost. This discrete-continuous optimization process is done alternately until convergence of the MRF energy is reached.

The work of Andriyenko et al. [5] has the disadvantage that does not consider the visual feature information available, which may be helpful in differentiating between different trackings in complex environments. It is unlikely that two people will have the same appearance and clothing, and in this case an algorithm could use their local characteristics to facilitate the tracking process. Our proposal is based on a non-trivial augmentation of the association potential function by incorporating a term that indicates the consistency between the features of candidate detections and the average features of a candidate target. We will now describe the model of [5].

2.2 Discrete-Continuous Model

Following the notation of [5], we assume an input set of M detections $D = (d_1, ..., d_M)$ and a set of labels represented by variable f with nominal values $f = (1, ..., N)$; this labeling variable assigns each detection to one of N trajectories $T = (T_1, ..., T_N)$ and identifies a false alarm using the outlier label $\oslash$. Each index detection d is associated solely with one ordered pair (i, d), where i is the detection index in a frame in relative time index t. Using this convention, each detection d is associated with a position p_i^t in the image. Using a MRF framework, the model graph is identified by Q where the nodes D are given by the set of detections inside a MRF framework and the pair of nearby nodes represents the edge set V. Specifically the edge set is defined by a temporal restriction between pairs of detections: $V(d_t^j, d_{t+1}^k) = 1$, if only $\left\| p_t^j - p_{t+1}^k \right\|^2 < \tau$. The energy of this MRF model is defined by the following equation:

$$E_d^T(f) = \sum_{d \in D} U_d(f_d, T) + \sum_{d, d' \in V} S_{d, d'}(f) + \sum_{i=1}^{N} \hat{E}_v^{te}(T_i) + \kappa h_f(T) + lnZ \quad (1)$$

In this case, the first term U_d represents the association potential function, the second term $S_{d, d'}$ represents the interaction potential function; and together they make up the discrete data association. The third term $\hat{E}_v^{te}(T_i)$ represents the continuous trajectory model. And finally, the fourth term $h_f(T)$ represents the label cost term, where this term penalizes complex configurations of MRF solutions. The term Z represents the partition function of the energy; in [5], the partition function is not stated as they do not need to estimate any parameter, however, in our model we need operate with it as we require to learn the weight of the features.

The association potential function component U is defined as:

$$U_{d_j^t}(l, T) = c_j^t \left\| p_j^t - T_l(t) \right\|^2 \quad (2)$$

If the detection is labeled as an outlier, it is penalized with a constant outlier cost $\oslash$, which is modulated by c_j^t. The interaction potential function component S is given by the following Potts model:

$$S_{d_t^j, d_{t+1}^k}(f) = S_{d_t^j, d_{t+1}^k}(f_{d_t^j}, f_{d_{t+1}^k}) = \eta \delta(f_{d_t^j} - f_{d_{t+1}^k}) \qquad (3)$$

The continuous trajectory model evaluates the smoothing degree of trajectories. In this case, this model uses a cubic B-spline to fit the trajectories of targets. In our model, we do not alter this component as it is unrelated with visual feature information.

The label cost function penalizes excessive complexity of trajectories. This function is the sum of five terms: The dynamic cost h_i^{dyn}, which penalizes complex splines by adding the cubic coefficients of splines, $C_i(r, 3)$, i.e. it prefers simpler curves. The persistence cost h_i^{per}, which penalizes unreliable trajectories by adding the distance to border of image ($\bar{b}$) and the inverse size of the same trajectory. The high-order data fidelity cost h_i^{fid}, which punishes trajectories that are far away from detections over a long period by adding the square of the cardinal of the subsequences of such trajectories G_k; we use a quadratic penalty instead of cubic as in [5] as we obtained better experimental results. The mutual exclusion cost h_i^{col}, which penalizes collisions between trajectories by considering physical constraints i.e. two objects cannot be in the same position. Finally, the regularization cost, which penalizes the cardinality of the set of trajectories. For more details, please see [5].

3   Our Approach

In our work, we augment the association potential term by including visual feature information with an embedded feature selection based on image regions. We consider an input set of M detections $D = (d_1, ..., d_M)$ with a corresponding set of features $X = (x_1, ..., x_M)$ where each component $x_i \in \Re^K$ is given by $x_i = (x_i^1, ..., x_i^K)$. As in the case of the detection variable, each feature vector x_d is associated with a unique ordered pair (t, j). By considering N possible trackers, we add an auxiliary class variable depending on a particular label variable $l \in (1, ..., N)$, $Y = (y_1^l, ..., y_M^l)$ where each component $y_i^l \in \{-1, +1\}$ depends on the assumed detection label f_i and is given by the indicatrix formula $y_i^l = 2 * I(l = f_i) - 1$. Moreover, we use a logistic regression model for connecting class and feature variables, Y and X, given by $Y = \sigma(W^T X)$, with $W \in \Re^{K x N}$ and σ as sigmoid function. Our task is to find the best set of weights W that explains the relation between the set of features X and set of class variables Y in order to maximize the likelihood of the MRF.

3.1   MRF Model

The association potential function is enhanced with the addition of a term related to the classification of detections according to tracker label information. As the association potential function U depends on label l, the parameter of the logistic regressor is a vector $W_l \in \Re^K$. We regularize this vector considering sparsity and natural groups with group Lasso penalization [13, 14]. The number of non-overlapping groups of visual features is G. By using the bijective correspondence between an index of detections d and an ordered pair (t, j), the association potential function is reformulated by considering the regularization terms in $R(w_l)$ as:

$$U_{d_t^j}(l, T, x_t^j) = c_j^t(\left\|p_t^j - T_l(t)\right\|^2 + \alpha \, log(1 + exp(-y_l^d(w_l^T x_t^j)))) + R(w_l) \quad (4)$$

$$R(w_l) = \lambda_S \left\|w_l\right\|^1 + \lambda_G \sum_{g=1}^{G} \left\|w_l^g\right\|^2 \quad (5)$$

MRF training is hard to solve as it needs to calculate the value of the partition function, which demands an exponential number of configurations. As we require to find an optimal set of weights W, we cannot avoid this problem as in [5]. To do this, we use the pseudolikelihood approximation [15], where the energy function is approximated using a local partition function instead of the global partition function:

$$E_{PL:d}^T(f) = \sum_{d \in D} U_d(f_d, T) + \sum_{d,d' \in V} S_{d,d'}(f) + \sum_{i=1}^{N} \hat{E}_v^{te}(T_i) + \kappa h_f(T) + ln Z_{loc} \quad (6)$$

The majority of terms are defined in Eq. 1 (Subsect. 2.2). The local partition function (Z_{loc}) is given for each node and represents the sum of the energies under all the possible labels [15] which has the advantage of being much more maneuverable than the global partition function. We also apply the same strategy as [5] and solve the model by alternating between optimization of the trajectory set, T, and the labelings set, f. By fixing T, we can ignore this term and have the pseudolikelihood energy expressed as a function of w given by:

$$E_{PL:d}^T(w_l) = \sum_{d \in D} \left\{ \left[c_j^t(\left\|p_t^j - T_l(t)\right\|^2 + \alpha \, log(1 + exp(-y_l^d(w_l^T x_t^j))) + \lambda_S \left\|w_l\right\|^1 \right. \right.$$
$$\left. \left. + \lambda_G \sum_{g=1}^{G} \left\|w_l^g\right\|^2) + \sum_{d' \in V_d} S_{d,d'}(f_d, f_{d'}) \right] - ln(Z_{loc}^d) \right\} \quad (7)$$

where the local function partition $Z_{loc}^d(w_l)$, considering $Q_t^j = \left\|p_t^j - T_m(t)\right\|^2$, is given by:

$$Z_{loc}^d(w_l) = \sum_m exp \left\{ c_j^t(Q_t^j + \alpha \, log(1 + exp(-y_m^d(w_l^T x_t^j))) + \sum_{d' \in V_d} S_{d,d'}(f, f_{d'}) \right\} \quad (8)$$

By fixing the terms independent of w_l with an auxiliary variable K_d and the regularization terms as $R(w_l)$, the optimization equation is re-expressed as a function of w_l^* by:

$$w_l^* = \arg\min_{w_l} \sum_{d \in D} \left\{ K_d + c_j^t \alpha \, log(1 + exp(-y_l^d(w_l^T x_t^j))) - ln(Z_{loc}^d) \right\} + R(w_l) \quad (9)$$

3.2   Estimation of MRF Parameters

Equation 9 is hard to solve due to the presence of the local partition function and the regularization term. To solve Eq. 9, we heuristically follow a strategy similar to Lee et al. [16], who first solve an unregularized version of the cost function and then add the sparsity constraints. In our case, we use the additivity property of gradient and follow a three-step strategy: (i). First, we solve an unregularized version of the energy function given by Eq. 9 where the component referring to local partition function can be separated. (ii). As we have a logistic regression term that approximates to the unregularized version of the energy function, we add the Lasso and Group-Lasso regularization terms in order to have a standard logistic regression model with Group-Lasso regularization [13,14]. In this step, we estimate the gradient of this model by using the direction of the solution given by the code based on the SLEP package [17]. (iii). Finally, we approximate the total gradient of Eq. 9 by summing the results of the gradients given in steps (i) and (ii), and subtracting the gradient of the version of Eq. 9 without the local partition function and the regularization term, which is simple to solve.

In the first step, we solve an unregularized versions of energy $(E_{PL:d}^{T(u)})$ using an optimization based on Newton method. The gradient is given by:

$$\nabla_{w_l} E_{PL:d}^{T(u)}(w_l) = -\alpha \sum_{d \in D} c_j^t \left[p_d(\bar{y}_l/d; w_l) y_l^d x_t^j - \left\langle p_d(\bar{y}_m/d; w_l) y_m^d x_t^j \right\rangle_{p_m(y_m/d; w_l)} \right] \quad (10)$$

In the previous equation, we use the notation $p_d(\bar{y}_m/d; w_l) = 1 - p_d(y_m/d; w_l)$. We accelerate the calculations by using the Hessian of unregularized energy function that is given by:

$$H_{w_l}(E_{PL:d}^{T(u)}(w_l)) = -\alpha \sum_{d \in D} \left\{ c_j^t x_t^j x_t^{j\,T} \left[p_d(y_l/d; w_l)(1 - p_d(y_l/d; w_l)) \right. \right.$$

$$\left. \left. - \left\langle (1 - p_d(y_l/d; w_l)) p_d(y_m/d; w_l) \right\rangle_{p_m(y_m/d; w_l)} \right] \right\} \quad (11)$$

The equation requires a known labeling. We use an iterative scheme based on hard partitioning according to labels [15]: first we fix the labeling and calculate the weight. Once the weights are fixed, we calculate the labeling using the same scheme as Andriyenko using the max-flow solution. In this case, after labeling we also optimize the continuous trajectory model and the label cost. We continue this process until we reach a convergence criterion. We also use the random change heuristic in the configuration of splines and accept them if only they decrease the energy function [5].

4   Experiments

4.1   Datasets

Our method is evaluated on four public datasets of pedestrian video sequences and two own datasets. The first three datasets come from TUD dataset, they are:

Campus(TUD-CAMP), Crossing (TUD-CROS) and Stadtmitte (TUD-STAD) subsets [18]. These videos have 91, 201 and 179 frames respectively, and show pedestrians in street scenes. The low viewpoint means that the targets often occlude one another. The fourth dataset is PEDS-2009-S2L1 which has 795 frames and considers a high viewpoint. The detections are performed using a SVM classifier based on Histogram of Gradients [19]. The fifth and sixth datasets (U-HALL-1 and U-HALL-2) were obtained from an indoor environment in a university campus; the first is a easy case of multi-tracking where the people are usually separated. In order to have a variety of experimental settings, we simulate a perfect pedestrian classifier using human labeling. The feature-based methods of bibliographic revision have not available their code, for such reason, we only compare with [5]. In this work, our main goal is to improve the pure MRF framework for multi-tracking tasks considering feature information. Table 1 shows the details of each dataset.

4.2   Implementation Details

We compare our feature-based discrete-continuous method (FDCM) with the discrete-continuous model (DCM) of [5], which is a state-of-art method of multi-target tracking. For FCDM, we use two combined features based on texture and color information: (i) dense HOG over a 6×3 grid with 576 resulting features and (ii) 3-level pyramidal RGB features producing 315 features. As our model uses a group sparse regularization [13], we consider the same arbitrary groups of features for all datasets. They correspond to a 2×2 non-overlapping spatial grid considering texture and color features as separated, with a total of eight groups. In relation to free cost parameters of a particular dataset, we use the best results in another similar own dataset and apply over such particular dataset. The sparsity regularization free parameters (λ_S, λ_G) are obtained by the best result in an independent dataset where these parameters result to be $\lambda_S = 0.1$ and $\lambda_G = 0.2$ and are applied to all datasets. In relation to performance metrics, we need to reduce the effects of random initial solutions. Therefore, we run the optimization procedure with 20 randoms seeds, and pick the three results with lowest energy and average the metrics. On the other hand, we do not initialize the algorithms with online individual trackers because this input can mask the real performance of each algorithm.

4.3   Metrics

For quantitative analysis, we use CLEAR MOT metrics [20]. We use Multi-Object Tracking Accuracy (MOTA), which combines all errors (number of missing targets, false positives and identity switches) into a normalized score between zero and one; and Multi-Object Tracking Precision (MOTP), which measures the bounding box overlap between tracked targets and ground truth detections with a normalized score between zero and one. Additionally, we use a variant of MOTA that penalizes the logarithm of the number of identity switches (MOTAL). In order to consider the performance on rough detections, we also measure the

Table 1. Datasets details.

Dataset name	# frames	# persons	Scene
TUD-CAMP	91	10	Street
TUD-CROS	201	13	Street
TUD-STAD	179	8	Street
PETS	795	19	Outdoor
U-HALL-1	200	3	Hall
U-HALL-2	200	9	Hall

classical precision and recall. Considering [5], we also show the metric identity switches (ID-SW), which accumulates the identity changes of people; and the rate of false detections per frame (FAR).

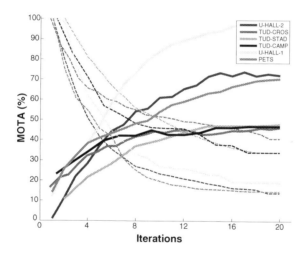

Fig. 3. Convergence of the optimization. The energy keeps decreasing for all iterations (dashed lines, rescaled for visualization). The decreasing of energies is generally reflected in the improving of tracking accuracies (solid lines).

4.4   Results

First, we analyze the convergence behaviour of FDCM by examining the relation between energy and multi-tracking accuracy, MOTA. Figure 3 shows that the most significant performance increase generally appears within the first few iterations, however, the optimization procedure is still able to find better configurations in posterior steps. The energy behaviour is variable and depends on each dataset, for example, while U-HALL and TUD-CAMP tend to converge in terms of accuracy, the energy function appears first stabilized in TUD-CAMP dataset. Now, we are going to analyze and compare multi-tracking performance for the tested algorithms.

Table 2 shows the results for each dataset, comparing the metrics. As measured by MOTAL and MOTA, our technique is able to clearly outperform the discrete-continuous method by an average of 4 and 3 %, respectively. As measured by MOTP, the two techniques show similar results with a slight advantage of 0.4 %. FDCM outperforms DCM in terms of Precision and Recall by an average of 3 % and 0.7 %, respectively. By analyzing more qualitative metrics, FDCM slightly beats DCM in the rate of false detections by an average of 0.26. In terms of change of identities, FDCM is superior by 6 less switches of identity on average. We observe that the two techniques achieve a similar results in terms of MOTP and Recall, it is natural as both metrics are related; this contrasts strongly with the MOTA and Precision results where the advantage of FDCM is notorious. These results are explained because FDCM is dependent on visual feature information, which ensures greater precision in tracking due to the use of more information; however, this process is fed by a list of detections given mainly by the MRF model, where spatial distance is decisive: two detections with similar visual features have not opportunity to be labelled with the same tracker if they are geometrically far. Moreover, the results in the number of identity switches and the rate of false detections confirm the advantages of our proposal in the multi-target tracking problem.

Table 2. The results of normalized scores with a larger margin over 1 % are shown in bold font; those with a smaller margin are shown in italic font in order to differentiate the level of the margins. The best results of unnormalized scores (FAR and ID-SW) are shown in bold font. In MOTAL, MOTA, Precision, FAR and ID-SW, FDCM shows better results than DCM. In MOTP and Recall, FDCM is slightly better than DCM.

Dataset	Metric						
	MOTAL	MOTA	MOTP	PREC	REC	FAR	ID-SW
TUD-CAMP (DCM)	47.9	43.7	68.9	77.4	66.7	1.01	16
TUD-CAMP (*FDCM*)	**49.6**	**45.9**	*69.2*	**80.1**	**68.2**	**0.85**	**15**
TUD-CROS (DCM)	47.2	42.8	74.1	73.5	73.6	1.47	63
TUD-CROS (*FDCM*)	**53.0**	**47.4**	*74.4*	**78.3**	*74.1*	**1.12**	**50**
TUD-STAD (DCM)	48.2	46.3	75.3	78.1	71.7	1.51	43
TUD-STAD (*FDCM*)	**54.4**	**51.1**	*75.4*	**81.2**	*71.9*	**1.05**	**41**
PETS (DCM)	84.4	83.7	73.5	98.5	85.7	0.07	34
PETS (FDCM)	**87.2**	**86.9**	*73.8*	*98.8*	85.6	**0.05**	**29**
U-HALL-1 (DCM)	94.6	94.5	77.2	95.2	99.8	0.11	1
U-HALL-1 (*FDCM*)	**99.1**	**99.1**	77.2	**99.3**	99.8	**0.01**	**0**
U-HALL-2 (DCM)	80.4	75.3	76.2	87.5	94.0	1.12	87
U-HALL-2 (*FDCM*)	**82.1**	*76.1*	**77.2**	*88.4*	*94.7*	**1.03**	**81**

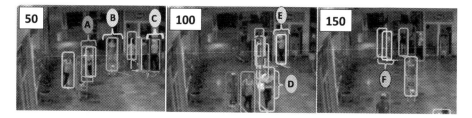

(a) Tracking results according to DCM

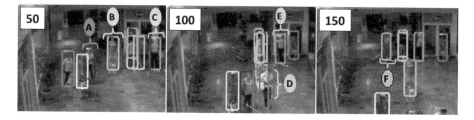

(b) Tracking results according to FDCM

Fig. 4. Multi-target tracking results for U-HALL dataset in frames 50, 100 and 150. In frame 50, the persons A, B and C have one correct detection with FDCM; whereas they have two detections with DCM. In frame 100, the person E has a correct detection with FDCM; whereas it has two detections with DCM. In frame 150, person F has a correct detection with FDCM; whereas it is detected twice with DCM. Nonetheless, there are some mistakes in FDCM, for example, person D is detected twice with FDCM; whereas it is correctly detected with DCM.

Finally, Fig. 4 shows some examples of the results of our technique, FCDM, in comparison to continuous-discrete model, DCM, for the U-HALL dataset. It can be seen that our technique achieves greater precision in various detections. Visual feature information helps FCDM to identify the target correctly. Nonetheless, our technique has some failures as in frame 100, where the person D has two detections with FDCM and one correct detection with DCM. A possible explanation is uncertainty associated with the visual feature information in each detection. Nonetheless, by jointly considering quantitative and qualitative analysis, our method stresses the usefulness of visual feature information.

5   Conclusions

In this work, we present a method of introducing visual feature information inside a Markov Random Field model. Our model optimizes data association, appearance discrimination and trajectory estimation through an alternating optimization procedure in the discrete and continuous components of MRF energy. In particular, in the discrete component we obtain a sparse set of weights for weighting the features; this considers the natural group of features in order to optimize the global energy function, using an approximation based on the pseudo-likelihood function. The data association is solved using a graph cut algorithm based on an α-expansion. On the other hand, trajectory estimation is solved using analytic fitting of splines to assigned detections. We show that the proposed technique outperforms the base technique based on discrete-continuous optimization. In future work, we plan to use max-margin optimization in order to obtain better multi-target tracking results. Another possible avenue is the use of more detailed features which can be related to semantic parts of pedestrians.

Acknowledgement. This work was funded by FONDEF grant D10I1054.

A   Derivation of Gradient and Hessian of Unregularized Energy Function

In our method, we must minimize the unregularized version of energy function given by Eq. A.3, this problem is given by:

$$w_l^* = \arg\min_{w_l} E_{PL:d}^{T(u)}(w_l) = \arg\min_{w_l} \sum_{d \in D} \left\{ K_d + c_j^t \alpha \, log(1 + exp(-y_l^d w_l^T x_t^j)) \right\} \quad (A.1)$$

By calling the gradient $\nabla_{w_l} E_{PL:d}^{T(u)}(w_l)$ as $G(w_l)$, this term is calculated as following:

$$G(w_l) = \frac{\partial}{\partial w_l} \sum_{d \in D} \left\{ K_d + c_j^t(\alpha \, log(1 + exp(-y_l^d(w_l^T x_t^j)))) - ln(Z_{loc}^d) \right\}$$

$$G(w_l) = \sum_{d \in D} \left\{ c_j^t \alpha \frac{\partial}{\partial w_l} log(1 + exp(-y_l^d(w_l^T x_t^j))) - \frac{\partial}{\partial w_l} ln(Z_{loc}^d) \right\}$$

$$G(w_l) = \sum_{d \in D} \left\{ c_j^t \alpha \frac{\frac{\partial}{\partial w_l} \left[1 + exp(-y_l^d(w_l^T x_t^j)) \right]}{1 + exp(-y_l^d(w_l^T x_t^j))} - \frac{\frac{\partial}{\partial w_l} \left[Z_{loc}^d \right]}{Z_{loc}^d} \right\}$$

$$G(w_l) = \sum_{d \in D} \left\{ c_j^t \alpha \frac{exp(-y_l^d(w_l^T x_t^j)) - y_l^d(x_t^j)}{1 + exp(-y_l^d(w_l^T x_t^j))} \right.$$
$$\left. - \frac{\sum_m exp\left(E_{PL:d}^{T(u)}(w_l, y_m) \right) \frac{\partial}{\partial w_l} \left[c_j^t \alpha log(1 + exp(-y_m^d(w_l^T x_t^j))) \right]}{Z_{loc}^d} \right\}$$

$$G(w_l) = \sum_{d \in D} \left\{ c_j^t \alpha \frac{exp(-y_l^d(w_l^T x_t^j)) - y_l^d(x_t^j)}{1 + exp(-y_l^d(w_l^T x_t^j))} \right.$$
$$\left. - \sum_m \frac{exp\left(E_{PL:d}^{T(u)}(w_l, y_m) \right)}{Z_{loc}^d} \left[c_j^t \alpha \frac{exp(-y_m^d(w_l^T x_t^j))}{1 + exp(-y_m^d(w_l^T x_t^j))} - y_m^d(x_t^j) \right] \right\}$$

$$G(w_l) = -\alpha \sum_{d \in D} \left\{ c_j^t \left[p_d(\bar{y}_l/d; w_l) y_l^d(x_t^j) \right] - \sum_m c_j^t \, p_m(y_m/d; w_l) \left[p_d(\bar{y}_m/d; w_l) y_m^d(x_t^j) \right] \right\}$$

$$G(w_l) = -\alpha \sum_{d \in D} \left\{ c_j^t \left[p_d(\bar{y}_l/d; w_l) y_l^d(x_t^j) - \left\langle p_d(\bar{y}_m/d; w_l) y_m^d(x_t^j) \right\rangle_{p_m(y_m/d; w_l)} \right] \right\} \quad (A.2)$$

We call the Hessian $H_{w_l}(E_{PL:d}^{T(u)}(w_l))$ as $H(w_l)$ and calculate it from gradient expression as following:

$$H(w_l) = -\alpha \frac{\partial}{\partial w_l^T} \sum_{d \in D} \left\{ c_j^t \left[p_d(y_l/d; w_l) y_t^d(x_t^j) - \left\langle p_d(y_m/d; w_l) y_m^d(x_t^j) \right\rangle_{p_m(y_m/d; w_l)} \right] \right\}$$

$$H(w_l) = -\alpha \sum_{d \in D} \left\{ c_j^t x_t^j {x_t^j}^T \left[p_d(y_l/d; w_l)(1 - p_d(y_l/d; w_l)) \right. \right.$$
$$\left. \left. - \langle (1 - p_d(y_l/d; w_l)) p_d(y_m/d; w_l) \rangle_{p_m(y_m/d; w_l)} \right] \right\} \tag{A.3}$$

With both terms, we can optimize the Eq. A.1 with Newton's method in order to estimate the optimal value of weight vector w.

References

1. Benfold, B., Reid, I.: Stable multi-target tracking in real-time surveillance video. In: Conference on Computer Vision and Pattern Recognition (CVPR), pp. 3457–3464 (2011)
2. Bar-Shalom, Y., Fortmann, T.: Tracking and Data Association. Academic Press, Boston (1988)
3. Liu, J., Carr, P., Collins, R., Liu, Y.: Tracking sports players with context-conditioned motion models. In: Conference on Computer Vision and Pattern Recognition (CVPR), pp. 1–8 (2013)
4. Liu, J., Chu, M., Liu, J., Reich, J., F.Zhao: Distributed state representation for tracking problems in sensor networks. In: Third International Symposium on Information Processing in Sensor Networks, pp. 234–242 (2004)
5. Andriyenko, A., Schindler, K., Roth, S.: Discrete-continuous optimization for multi-target tracking. In: Conference on Computer Vision and Pattern Recognition, pp. 1926–1933 (2012)
6. Grabner, H., Leistner, C., Bischof, H.: Semi-supervised on-line boosting for robust tracking. In: Forsyth, D., Torr, P., Zisserman, A. (eds.) ECCV 2008, Part I. LNCS, vol. 5302, pp. 234–247. Springer, Heidelberg (2008)
7. Collins, J., Uhlmann, J.: Efficient gating in data association with multivariate gaussian distributed states. IEEE Trans. Aerosp. Electron. Syst. **28**, 909–916 (1992)
8. Kuo, C.H., Huang, C., Nevatia, R.: Multi-target tracking by on-line learned discriminative appearance models. In: Conference on Computer Vision and Pattern Recognition (CVPR), pp. 685–692 (2010)
9. Berclaz, J., Fleuret, F., Turetken, E., Fua, P.: Multiple object tracking using k-shortest paths optimization. IEEE Trans. Pattern Anal. Mach. Intell. (TPAMI) **33**, 1806–1819 (2011)
10. Yang, B., Nevatia, R.: An online learned crf model for multi-target tracking. In: Conference on Computer Vision and Pattern Recognition (CVPR), pp. 2034–2041 (2012)
11. Butt, A.A., Collins, R.T.: Multiple target tracking using frame triplets. In: Lee, K.M., Matsushita, Y., Rehg, J.M., Hu, Z. (eds.) ACCV 2012, Part III. LNCS, vol. 7726, pp. 163–176. Springer, Heidelberg (2013)

12. Hofmann, M., Haag, M., Rigoll, G.: Unified hierarchical multi-object tracking using global data association. In: International Workshop on Performance Evaluation of Tracking and Surveillance (PETS), pp. 16–18 (2013)

13. Meier, L., Van De Geer, S., Bühlmann, P.: The group lasso for logistic regression. J. Roy. Stat. Soc.: Ser. B (Statistical Methodology) **70**, 53–71 (2008)

14. Simon, N., Friedman, J., Hastie, T., Tibshirani, R.: A sparse-group lasso. J. Comput. Graph. Stat. **22**, 231–245 (2013)

15. Li, S.: Markov Random Field Modeling in Image Analysis. Springer-Verlag New York Inc., Secaucus (2001)

16. Lee, S.I., Lee, H., Abbeel, P., Ng, A.: Efficient l1 regularized logistic regression. In: National Conference on Artificial Intelligence, pp. 1–9 (2006)

17. Liu, J., Ji, S., Ye, J.: SLEP: Sparse learning with efficient projections, vol. 1813. Arizona State University (2009)

18. Andriluka, M., Roth, S., Schiele, B.: Discriminative appearance models for pictorial structures. Int. J. Comput. Vis. **99**, 259–280 (2012)

19. Dalal, N., Triggs, B.: Histograms of oriented gradients for human detection. In: Conference on Computer Vision and Pattern Recognition, pp. 886–893 (2005)

20. Stiefelhagen, R., Bernardin, K., Bowers, R., Garofolo, J.S., Mostefa, D., Soundararajan, P.: The CLEAR 2006 evaluation. In: Stiefelhagen, R., Garofolo, J.S. (eds.) CLEAR 2006. LNCS, vol. 4122, pp. 1–44. Springer, Heidelberg (2007)

Hybrid Focal Stereo Networks for Pattern Analysis in Homogeneous Scenes

Emanuel Aldea[1,3]([⊠]) and Khurom H. Kiyani[2,3]

[1] Autonomous Systems Group, Université Paris Sud,
Gif-sur-Yvette, France
emanuel.aldea@u-psud.fr
[2] Communications and Signal Processing Group,
Imperial College London, London, UK
[3] AquaMed Research and Education, Doha, Qatar

Abstract. In this paper we address the problem of multiple camera calibration in the presence of a homogeneous scene, and without the possibility of employing calibration object based methods. The proposed solution exploits salient features present in a larger field of view, but instead of employing active vision we replace the cameras with stereo rigs featuring a long focal analysis camera, as well as a short focal registration camera. Thus, we are able to propose an accurate solution which does not require intrinsic variation models as in the case of zooming cameras. Moreover, the availability of the two views simultaneously in each rig allows for pose re-estimation between rigs as often as necessary. The algorithm has been successfully validated in an indoor setting, as well as on a difficult scene featuring a highly dense pilgrim crowd in Makkah.

1 Introduction

The problem of multiple camera calibration has been a central topic for the pattern recognition and robotics communities since their inception. Moreover, the use of camera networks has become pervasive in our society; beside their use in surveillance and security enforcement, cameras are heavily relied upon in application domains related to entertainment and sports, geriatrics and elderly care, the study of natural and social phenomena, etc. Motivated by all these developments, a large body of work has been devoted to the problem of estimating accurately the camera network topology, i.e. camera positions and orientations in a common reference system. Inferring the topology in camera networks with non-overlapping fields of view (FOV) is a topic specific to wide-area tracking relying more on high-level image processing and statistical inference and will not be addressed in the current work; the focus of the current article is on estimating the geometric topology for cameras with overlapping FOV. Although such a network may be composed of a large number of cameras firmly attached to a mobile object such as a robot, car, or UAV, most commonly camera networks are static and point towards a specific scene of interest. In these cases, multiple camera calibration is performed by using a specific calibration pattern

© Springer International Publishing Switzerland 2015
C.V. Jawahar and S. Shan (Eds.): ACCV 2014 Workshops, Part III, LNCS 9010, pp. 695–710, 2015.
DOI: 10.1007/978-3-319-16634-6_50

or object [1–3], which is deployed and moved in the scene during a dedicated calibration phase. If the use of a calibration object is not possible, scene based calibration may be performed by exploiting visible interest points in methods based on pose refinement [4], or if applicable by using dynamic silhouettes, such as in [5].

A homogeneous scene is defined as an environment lacking completely salient features which would allow their association in different camera views, and which would thus allow for scene based calibration. Typical examples are liquid flow, vapour flow, plant canopy (crops, jungle) or high-density crowds, the latter being the context we have chosen for illustrating our work. The analysis of a homogeneous scene which is not directly accessible for setting up markers, or where the use of calibration objects is not feasible, raises a problem which is not solved by the common methods employed for multiple camera calibration. We approach this problem by replacing cameras with hybrid static stereo rigs, where a long focal camera is used for analysis and a large FOV camera is used for registration with other rigs. By proposing this solution, we avoid using active cameras which require complex models for the dynamic evolution of their intrinsic parameters. Other benefits of possessing simultaneous large and small FOV images of the scene are the fact that the registration does not assume anything about the analysed scene, the fact that the salient features do not have to be static as long as the cameras are accurately synchronized, but then if they are static they can be used to re-estimate continuously the pose and correct phenomena such as camera shaking.

The outline of the paper is as follows. In Sect. 2 we illustrate the fundamental problem that we address, and discuss related work and alternative solutions. Then, Sect. 3 recalls the fundamental notions which are required for scene based calibration and for the understanding of the proposed algorithm, which is presented in Sect. 4. Section 5 illustrates a small scale experiment as well as an application of the proposed algorithm to the analysis of a highly crowded scene, and Sect. 6 presents the conclusions.

2   Motivation and Related Work

Based on the simple pinhole projection model (also recalled in Sect. 3), let us illustrate an issue related to the representation of a homogeneous region of interest in a camera sensor (for all the following tests and examples we will employ Sony ICX274 sensors with a 8.923 mm diagonal and an effective pixel resolution of 1624×1234). We have acquired from the same position and with the same camera three shots using lenses with 4, 8 and 12 mm focals respectively. In the left column of Fig. 1 we present from top to bottom three 50×50 pixel patches from the shots taken with increasing focal lengths. The adjacent images from left to right show areas from these patches (of initial size 20×30 and zoomed for visualization purposes with no interpolation applied). In this case, the long focal lenses are required for retrieving with enough detail entities such as body parts, bags etc. which are essential for a wide range of tasks related to action understanding, monitoring, tracking and surveillance.

Fig. 1. Left column, from top to bottom: 50×50 pixel patches from shots taken with increasing focal lengths ($f = 4\,\mathrm{mm}$, $f = 8\,\mathrm{mm}$ and $f = 12\,\mathrm{mm}$) from the same position. The following images, from left to right: interest areas from the three previous patches, of initial size 20×30 and zoomed for visualization purposes with no interpolation applied. Detailed features essential for scene analysis are not retrieved below a certain focal length.

From the above illustrations, we note that a wide FOV is beneficial for accurate registration in a camera network, whilst a narrow FOV is beneficial for retrieving the details from the area of interest. By lacking salient features the narrow FOV is not able to estimate robustly or at all the relative pose between multiple cameras.

A calibration pattern visible from all views set on the area of interest can solve the relative pose problem. However, there are multiple applications where this solution is not practical. The area of interest may be far and thus quite large, or it may be inaccessible. During the analysis, the camera poses might change accidentally due to shocks, periodically due to vibrations, or by design (mobile observers); all these scenarios require frequent relative pose estimation updates. In the following paragraphs, we recall briefly some works that are relevant for the problem of multiple view detailed analysis *and* relative pose estimation, highlighting their respective benefits and shortcomings for this scenario.

Zooming cameras. One possible solution is to deploy a network of cameras which use motorized zoom lenses. One major consequence of the zooming process is the variation of intrinsic and distortion parameters of the cameras, which have to be re-estimated. Various solutions for zooming recalibration based on the scene have been proposed [6–12]; these solutions are often denoted as self-calibration methods. Beside the fact that all these methods make simplifying assumptions about a subset of the varying parameters, the fundamental limitation is that they still require the continuous presence of salient features, usually interest points or straight lines, in order to operate.

Pan-Tilt-Zoom (PTZ) cameras. PTZ cameras have a built-in zooming function and are specifically designed for live monitoring. However, tasks such as surveillance or auto tracking do not require necessarily accurate self-calibration. In the area of PTZ camera network calibration, scene based solutions have also been proposed [13,14] but they have the same fundamental limitation i.e. requiring the presence of distinctive landmarks in the field of view.

The strategies recalled up to this point propose interesting solutions for self-calibration and camera registration in the presence of a sufficient number of salient features, but they are not applicable for a camera view if the lens is zoomed on a *homogeneous* and/or *dynamic* scene. Although these scenarios are less common, examples of possible applications abound in the study of crowds and of different types of flows encountered in natural phenomena. The underlying idea for the solution we propose is about transferring pose information in a scene-independent manner to the zoomed camera from a secondary camera able to infer its pose. This leads to a straightforward minimal solution based on a rigid stereo rig featuring two cameras, one with a small FOV used for analysis, and one with a large FOV used for registration within a network of such rigs.

Surprisingly, this solution has not been applied to the analysis of homogeneous scenes. Even considering a broader range of applications, the use of hybrid stereo systems featuring large and narrow FOV is limited. We recall here the setup deployed by the STEREO solar observation mission [15,16], where the hybrid imagers are nonetheless registered accurately using a star catalogue i.e. am inertial frame of reference. In robotics [17] employ a fisheye and a perspective camera on a UAV. More recently, in [18] the hybrid stereo strategy is employed in order to estimate accurately the pose of a moving binocular in order to insert virtual objects realistically. The fundamental difference is that the design of the system proposed in [18] is tailored for minor pose variations (see the use of the IMU and the small error assumptions), which determine small perspective changes in the appearance of the distant scene, thus greatly simplifying the visual odometry. In contrast, our work addresses a problem where large changes in perspective do not allow for such convenient associations in the large FOV cameras (and for any association at all in the narrow FOV cameras).

With respect to the previous works, the solution based on a hybrid stereo system has clear benefits for the analysis of dynamic homogeneous scenes. We do not have to adopt any simplifying assumptions about the variations of intrinsic parameters, and the calibration precision will be maintained at the optimal level provided by state of the art calibration algorithms. Secondly, the extrinsic parameters of each stereo rig can be estimated independently of the scene. In contrast to the scenario of a zooming camera, the availability at each instant of an accurately registered pair of images allows for accurate pose re-estimation between rigs as often as necessary, overcoming the effect of movement and vibrations.

3   Background on Scene Based Pose Estimation

The projection model. In the following, we will briefly recall the pinhole camera and optical distortion models that we employ. A point in 3D space $\mathbf{X} = [X\ Y\ Z]^\mathrm{T}$ projects within the image space into a pixel $\mathbf{x} = [x\ y]^\mathrm{T}$ according to:

$$\begin{pmatrix} \mathbf{x} \\ 1 \end{pmatrix} = \lambda \mathbf{K} \begin{bmatrix} \mathbf{R} \mid -\mathbf{RC} \end{bmatrix} \begin{pmatrix} \mathbf{X} \\ 1 \end{pmatrix} \tag{1}$$

with λ being an undetermined scale factor, $\mathbf{R}$ the orientation of the camera and $\mathbf{C}$ the location of its optical center in world coordinates (we also note $\mathbf{t} = -\mathbf{RC}$), and $\mathbf{K}$ the intrinsic parameters:

$$\mathbf{K} = \begin{bmatrix} f_x & s & c_x \\ 0 & f_y & c_y \\ 0 & 0 & 1 \end{bmatrix} \tag{2}$$

Above, f_x and f_y are the focal lengths, $[c_x \ c_y]^{\mathrm{T}}$ represents the principal point, and the skew parameter s is considered 0.

In order to switch to different coordinate frames, we rely on elements of SE(3), the group of rigid body transformations in $\mathbb{R}^3$. A transformation matrix $\mathbf{E}$ takes the form:

$$\mathbf{E} = \begin{bmatrix} \mathbf{R} & \mathbf{t} \\ 0 & 1 \end{bmatrix} \tag{3}$$

Element multiplication amounts to transitive chaining coordinate frame transformations: $\mathbf{E}^{CA} = \mathbf{E}^{CB} \mathbf{E}^{BA}$ would transfer a 3D point in homogeneous coordinates from reference system A to reference system C.

In order to account for radial distortion, the extension of the pinhole model assumes that if the 3D point $\mathbf{X}$ is projected to $[\tilde{x} \ \tilde{y} \ 1]^{\mathrm{T}}$ under the initial assumptions, then $\mathbf{X}$ would be actually imaged to the distorted location $[x_d \ y_d]^{\mathrm{T}}$:

$$\begin{pmatrix} x_d \\ y_d \end{pmatrix} = \left(1 + \sum_{i=1}^{3} \kappa_i \tilde{r}^{2i}\right) \begin{pmatrix} \tilde{x} \\ \tilde{y} \end{pmatrix} \tag{4}$$

where $\tilde{r} = (\tilde{x}^2 + \tilde{y}^2)^{1/2}$. Thus, $(f_x, f_y, c_x, c_y, \kappa_1, \kappa_2, \kappa_3)$ is in most scenarios the suitable parameter set for a full intrinsic calibration.

Epipolar geometry. One tool that we will employ in the following sections is the epipolar constraint, which is a direct implication of the projective geometry between two views. It is worth noting that this constraint is independent of the scene structure, depending exclusively on the intrinsic parameters and the relative pose - as long as the the salient features of the scene are static, or as long as the cameras are accurately synchronized.

Considering two projections $\mathbf{x}_1$ and $\mathbf{x}_2$ of the same point $\mathbf{X}$ in cameras C_1 and C_2, the epipolar constraint defines the relationship between the projections as $\mathbf{x}_2^{\mathrm{T}} \mathbf{F} \mathbf{x}_1 = 0$. $\mathbf{F}$ is known as the fundamental matrix [19], which depends explicitly on the calibration parameters in the following way: $\mathbf{F} = \mathbf{K}_2^{-\mathrm{T}} \mathbf{t}_\times \mathbf{R} \mathbf{K}_2^{-1}$ where $\mathbf{t}_\times$ is the skew-symmetric matrix associated to $\mathbf{t}$.

The main interest of the epipolar constraint is that it does not make any assumptions about the 3D structure of the scene. Thus, compared to other optimisation algorithms that are commonly employed to estimate the relative pose, the determination of $(\mathbf{R}, \mathbf{t})$ using the epipolar geometry provides a practical minimal parametrization and does not require an initialization. However, the result may be used for the initialization of more complex optimisations, such as the bundle adjustment procedure, briefly recalled in the following paragraph.

Bundle adjustment (BA). Assuming a zero-mean Gaussian distribution of the corner detection errors, bundle adjustment [4,20] is the Maximum Likelihood Estimator for the joint estimation problem of relative camera poses and of observed 3D point locations. The BA procedure will minimize the following reprojection error:

$$\min_{\hat{P}^i, \hat{\mathbf{X}}_j} \sum_{i,j} d\left(\hat{P}^i(\hat{\mathbf{X}}_j), \mathbf{x}_j^i \right)^2 \tag{5}$$

In the error function above, $\hat{\mathbf{X}}_j$ is the location hypothesis for a point observed by the ith camera. The projection function $\hat{P}^i$ related to the pinhole model (accounting for radial distortion too) depends on the ith camera pose; we consider that the intrinsic parameters are known and are not part of the optimization problem. BA will thus minimize jointly for all the possible camera-point pairs (i, j) the distance between the reprojection $\hat{P}^i(\hat{\mathbf{X}}_j)$ and the actual measurement $\mathbf{x}_j^i$. Solving this optimization problem is studied in depth in the literature, and it generally boils down to exploiting the sparsity of its Hessian matrix and to employing an adapted LS algorithm such as Levenberg-Marquardt [21].

Although BA seems like an ideal solution for multiple view pose estimation, it does have some well-known shortcomings that we will briefly discuss in connection with our specific aim. One common criticism is related to the computational requirements, but this issue is more prevalent in large scale robotics applications, especially if there are real-time constraints to take into account. For a relatively small camera network, the size of the problem is reasonable even for frequent updates. Another important aspect is related to the initialization, which has to be relatively accurate in order to allow the problem to converge to the correct solution. In order to cope with this, we will rely on an initialization based on the epipolar constraint discussed above, but other options are possible too (see for example [22], or [23] if 3D information about some scene features is available). Finally, some practical aspects are equally relevant. Given the high number of parameters which are usually involved, constraining the relative pose variables is more effective if the adjacent camera views for the large FOV cameras are close enough in order to allow for a significant FOV overlap. Stability is also improved if the corner correspondences are spread onto the common field of view.

4   The Proposed Algorithm

Outline. Let us consider a network of N hybrid stereo rigs, the ith rig featuring a small FOV camera C_i^s used for analysis, and a large FOV camera C_i^l employed for pose estimation in a global frame. The aim of the following procedure is to align accurately the cameras $\{C_1^s, C_2^s, \ldots, C_N^s\}$. We assume that for each rig, the cameras C_i^s and C_i^l have been calibrated. In the following, we will denote by $\mathbf{E}_i^{sl}$ the transform that transfers a point from the large FOV camera to the analysis camera on the ith stereo rig. Also, $\mathbf{E}_{ji}^l$ and $\mathbf{E}_{ji}^s$ are transforms that transfer points from the ith rig to the jth between the large FOV cameras, and respectively between the analysis cameras.

The fact that the stereo rigs are passive allows for a precise intrinsic and extrinsic calibration which can be performed independently of the scene in a controlled environment. Thus the intrinsic parameters $\mathbf{K}_i^s$, $\mathbf{K}_i^l$ as well as the rigid transform $\mathbf{E}_i^{sl}$ that projects a 3D point from the pose estimation camera of the rig to the analysis camera are considered as known.

For the next step, let us consider a pair of spatially adjacent rigs (i, j); in most scenarios, cameras are spread as much as possible, and thus it is necessary to consider adjacent pairs in order to obtain enough reliable interest point matches. Due to initialization requirements, we cannot apply BA directly in order to estimate $\mathbf{E}_{ji}^l$ between the two large FOV cameras on the rigs. We perform SIFT detection and matching [24], and use the normalized 8-point algorithm [25] with RANSAC [26] for robustness to outliers. For the matching step, we employ two filtering strategies based on the uniqueness assumption (the ratio τ of the similarity scores for the top two candidates [24]) and on married matching (both features are the top candidate for each other [27]). Then, we decompose the fundamental matrix [20, chap. 9] and choose the correct solution based on the chirality constraint [28]. Let us denote $\tilde{\mathbf{E}}_{ji}^l$ the rigid transformation estimated after this step. Using $\tilde{\mathbf{E}}_{ji}^l$, and based on the inlier set of matches that were validated during the RANSAC procedure, we build a set of 3D points $\tilde{\mathbf{X}}_{ji}$ by linear triangulation [29].

At this point, we can employ BA using $\tilde{\mathbf{E}}_{ji}^l$ and $\tilde{\mathbf{X}}_{ji}$ as initial estimates, and we obtain a refined relative pose estimation $\hat{\mathbf{E}}_{ji}^l$ for the large FOV cameras in the pair of rigs (i, j). Ideally, BA involving more than a pair of rigs should be performed afterwards whenever possible; however, in a typical setting, cameras are spaced as much as possible around a scene, the limit being imposed by common FOV considerations and the performance of the interest point matching procedure. Thus, we may assume that in most situations non-adjacent rigs will have difficulties for the matching procedure, and will have matches corresponding to disjoint sets of 3D points, which effectively yields the BA problems independent. A particular setting is that of a scene surrounded in a full circle by rigs, and in this case a full BA may be beneficial.

Having the BA estimations, it is trivial to express the C_i^l poses in a common reference system; in the following, we set this reference system as depicted by the position and orientation of C_1^l. Let $\hat{\mathbf{E}}_i^l$ be the rigid transform that links C_1^l to C_i^l. For any two rigs (i, j), we can now use the extrinsic calibrations $\mathbf{E}_i^{sl}$, $\mathbf{E}_j^{sl}$ and the global allignment of the large FOV cameras in order to infer the global allignment of the analysis cameras in the same reference system, as well as their relative pose:

$$\mathbf{E}_i^s = \mathbf{E}_i^{sl}\hat{\mathbf{E}}_i^l; \, \mathbf{E}_j^s = \mathbf{E}_j^{sl}\hat{\mathbf{E}}_j^l; \qquad (6)$$

$$\mathbf{E}_{ji}^s = \mathbf{E}_j^{sl}\hat{\mathbf{E}}_j^l \left(\mathbf{E}_i^{sl}\hat{\mathbf{E}}_i^l\right)^{-1} \qquad (7)$$

Enforcing a common scale. BA can estimate accurately the relative pose up to an unknown scale factor. This limitation applies to the $\hat{\mathbf{E}}_{ji}^l$ estimates only; the values $\mathbf{E}_i^{sl}$ that specify the baseline for cameras on the same rig are not concerned as long as a known size calibration pattern is used for stereo extrinsic calibration. Since the different BA procedures depicted in the following paragraphs are typically independent, we have to enforce a common scale factor among all optimizations using additional information. Depending on the application, it is easier to adopt one of the following strategies. For a small sized scene, we may add a known size object in the common FOV of C_i^l and C_j^l; we thus use $\tilde{\mathbf{X}}_{ji}$ in order to impose a metric scale to the reconstruction. For a large scene, we may either use a similar approach as for the previous setting, or if it is not applicable we may measure the distance between C_i^l and C_j^l (using for example a laser rangefinder), thus using $\tilde{\mathbf{t}}_{ji}$ in order to impose a metric scale to the reconstruction.

5   Experimental Results

A small scale scenario. We have created a simple example in an indoor environment, using LEGO figurines placed closely in the middle of a homogeneous surface. We have used two hybrid stereo rigs and taken a snapshot of the figurines and surrounding environment. The resulting images are presented in Fig. 2: the upper and lower rows show the views from the large FOV (C_1^l, C_2^l) and small FOV (C_1^s, C_2^s) cameras respectively. We have also highlighted the results of the matching procedures; the first matching set ($\tau = 0.4$ for uniqueness) is required for the matching step of the algorithm, while the second set (a more permissive value $\tau = 0.75$ has been used in order to have enough matches) is not used in the algorithm - as the scene is supposed to be poor in salient features - but it is used *exclusively as ground truth for validating the result of the algorithm.* We apply the steps highlighted in the previous Section in order to compute $\mathbf{E}_{21}^s$: estimation of $\mathbf{E}_{21}^l$ using SIFT matching followed by decomposition of $\mathbf{F}_{21}^l$ and BA, then exploitation of $\mathbf{E}_1^{sl}$ and $\mathbf{E}_2^{sl}$ provided by stereo calibration, and also the setup of the right scale by using an object of known size (the long brick of length 79.8 mm).

In order to estimate numerically the quality of the rigid transform $\mathbf{E}_{21}^s$ obtained, we have exploited the matches that we were able to determine directly between the small FOV cameras in this example. In homogeneous scenes, interest points may be completely absent, or the scarcity of matches may have a detrimental effect on the stability of the estimation of $\mathbf{F}$. Therefore, in our example we set up a base BA problem between C_1^s and C_2^s where we initialize the system by the decomposition of $\mathbf{F}_{21}^s$. Alternatively, we use the rigid transform $\mathbf{E}_{21}^s$ as initialization for the triangulation of matches and for the BA procedure. The resulting

(a)   (b)

Fig. 2. A set of images used for pose estimation in a simple indoor environment; the images in (a) correspond to C_1^l and C_2^l, and the images in (b) show the images captured by C_1^s and C_2^s. Both pairs of images have been matched using SIFT; the first set of matches are necessary for the algorithm, whilst the second set is used *exclusively as ground truth for validating the result of the algorithm.*

solutions and mean reprojection errors for these two scenarios are presented in Table 1.

As we notice, the two optimization problems converge towards the same solution, but $\mathbf{E}_{21}^s$ brings the optimization much closer to the objective in terms of mean reprojection error. This result is interesting for a number of reasons. Firstly, even though we do not have a case of optimization stuck in a local minimum due to the worse initialization, this is a good example of coarse to fine resolution of the relative pose estimation. This approach is helpful for robotics applications in case of unstable optimizations (few matches in the small FOV cameras), and also interesting for the computation gain due to a faster convergence of BA (25 iterations with initial subpixel mean reprojection error compared to 37 iterations). Secondly, and most importantly, this example shows that in cases where we can not compute $\mathbf{E}_{21}^s$ directly due to the complete absence of salient features, we are able using this algorithm to infer the unknown rigid transform from the adjacent large FOV cameras with a high level of accuracy.

Table 1. Relative poses between C_1^s and C_2^s. The Euler angles are expressed in degrees, and the mean reprojection errors in pixels. Tilde values represent estimations prior to the BA procedure, and hat values denote estimations refined by BA. The difference between the two rows consists in the initialization of BA; in the first case we use the SIFT matches depicted in Fig. 2(b), whilst in the second case we use the result of our algorithm.

$(\tilde{\psi}; \tilde{\theta}; \tilde{\phi})$	$\tilde{C}$	$\tilde{\epsilon}$	$(\hat{\psi}; \hat{\theta}; \hat{\phi})$	$\hat{C}$	$\hat{\epsilon}$	Iter.	Observations
$\begin{pmatrix} 24.13 \\ 21.04 \\ 10.67 \end{pmatrix}$	$\begin{pmatrix} -0.89 \\ -0.30 \\ 0.33 \end{pmatrix}$	37.16	$\begin{pmatrix} 23.95 \\ 21.13 \\ 3.74 \end{pmatrix}$	$\begin{pmatrix} -0.79 \\ -0.25 \\ 0.55 \end{pmatrix}$	0.199	37	Base solution
$\begin{pmatrix} 23.85 \\ 16.42 \\ 3.53 \end{pmatrix}$	$\begin{pmatrix} -0.77 \\ -0.23 \\ 0.59 \end{pmatrix}$	0.489	$\begin{pmatrix} 23.95 \\ 21.13 \\ 3.74 \end{pmatrix}$	$\begin{pmatrix} -0.79 \\ -0.25 \\ 0.55 \end{pmatrix}$	0.199	25	Init. by $\mathbf{E}_{21}^s$

Pose estimation for high-density crowds. We have deployed two hybrid stereo rigs at the grand mosque in Makkah during very congested times of the Hajj period, in October 2012. The access constraints to the site impose a large perspective change between the two points of observation. As a result, neither SIFT nor even ASIFT [30] algorithms were capable to provide any correct matches which are required as inputs for the algorithm we propose. Consequently, we had to rely on manual matching of salient structures in order to bootstrap the algorithm.

In Fig. 3 we present the data our algorithm processed and registered; images in (a) and (b) correspond to C_1^l and C_1^s, and (c) and (d) correspond to C_2^l and C_2^s respectively. The large FOV cameras contain enough common salient features, although the perspective variation does not allow for automated matching, and human intervention is necessary. Figure 3(e) presents such a user specified correspondence; in total we have used 34 user specified correspondences, of which 26 have been considered inliers for the fundamental matrix evaluation. For visualization purposes, Figs. 3(f) and 3(g) present the central structure with the manually matched features, and the 3D structure of the scene with the camera axis aligned and an approximate representation of the ground plane.

The numerical results of the algorithm for this setting are presented in Table 2; the relative rotations are expressed in degrees using Euler angles, the relative center position is expressed as a unit $\mathbb{R}^3$ vector, and mean reprojection errors are expressed in pixels. Also, as specified in the algorithm outline (Sect. 4), tilde values represent estimations prior to the BA procedure, and hat values denote estimations refined by BA. The first row corresponds to Step (iii) of the algorithm, the pose estimation between C_1^l and C_2^l. The output values are consistent with the actual location of the cameras; the large angle displacements emphasize the difficulty of the task, and explain as well the limitation of the automated matching procedure in this case.

We have also refined the relative positions of the cameras within the individual rigs. These values are provided by the stereo calibration procedure, and we validated them by performing SIFT matching between the large and small FOV cameras, and by using the stereo calibration pose as an initializer for BA (rows 2 and 3 in Table 2). The threshold for uniqueness filtering has been set as $\tau = 0.3$. However, the stereo calibration performed on site could not be done in optimal conditions. As an alternative solution, we used as pose initializations values that we obtained in the same way as for the first row of Table 2, by estimating and decomposing the fundamental matrix. The solutions obtained are presented on rows 4 and 5 in Table 2. These solutions were more accurate, and finally they have the advantage of requiring only the intrinsic camera parameters. It is worth noting that the stereo baseline is approximately 6 cm, while the distance to the scene is three orders of magnitude higher, and in these circumstances the relative angles and not the camera center relative positions will be the most relevant for scene based estimation.

Having thus obtained all the necessary relative poses (rows 1, 4 and 5 in Table 2), we are able to estimate the relative pose between the long focal cameras. Ground truth estimations are not possible, but in order to estimate the accuracy

Fig. 3. A set of images used for pose estimation; the images in (a) and (b) correspond to C_1^l and C_1^s, and (c) and (d) correspond to C_2^l and C_2^s respectively. An example of user specified correspondences is illustrated in (e). In (f) we present the interest points used in the central region of one of the images, and in (g) the inferred camera orientation (RGB axis for XYZ), with the approximate ground plane highlighted in green, for easier visualization.

of the result we have located in the crowd a number of salient elements (either distinctive heads, or distinctive configurations of people) and we illustrate the

Table 2. Relative poses between analysis cameras placed on different rigs (first row), and between cameras placed on the same rig (rows 2–5). The Euler angles are expressed in degrees, and the mean reprojection errors in pixels. Tilde values represent estimations prior to the BA procedure, and hat values denote estimations refined by BA. The difference between the rows 2–3 and 4–5 consists in the initialization of the BA; in the first case we use the stereo calibration, whilst in the second case we use directly the images, in the same way as for the first row initialization.

Cam. pair	$(\tilde{\psi}; \tilde{\theta}; \tilde{\phi})$	$\tilde{C}$	$\tilde{\epsilon}$	$(\hat{\psi}; \hat{\theta}; \hat{\phi})$	$\hat{C}$	$\hat{\epsilon}$	Observations
$C_1^l \Rightarrow C_2^l$	$\begin{pmatrix} 53.27 \\ 71.52 \\ 32.94 \end{pmatrix}$	$\begin{pmatrix} -0.78 \\ -0.19 \\ 0.59 \end{pmatrix}$	4.00	$\begin{pmatrix} 59.68 \\ 69.11 \\ 42.75 \end{pmatrix}$	$\begin{pmatrix} -0.81 \\ -0.18 \\ 0.55 \end{pmatrix}$	0.25	Manual Init.
$C_1^l \Rightarrow C_1^s$	$\begin{pmatrix} -0.37 \\ -0.58 \\ 0.51 \end{pmatrix}$	$\begin{pmatrix} 0.79 \\ 0.00 \\ -0.62 \end{pmatrix}$	1.017	$\begin{pmatrix} -0.33 \\ -0.34 \\ 0.48 \end{pmatrix}$	$\begin{pmatrix} 0.11 \\ -0.01 \\ -0.99 \end{pmatrix}$	0.076	Stereo Calib. Init.
$C_2^l \Rightarrow C_2^s$	$\begin{pmatrix} -0.19 \\ 0.72 \\ 0.23 \end{pmatrix}$	$\begin{pmatrix} 0.94 \\ 0.05 \\ -0.34 \end{pmatrix}$	1.661	$\begin{pmatrix} -0.09 \\ 0.71 \\ 0.12 \end{pmatrix}$	$\begin{pmatrix} 0.57 \\ 0.23 \\ 0.78 \end{pmatrix}$	0.252	Stereo Calib. Init.
$C_1^l \Rightarrow C_1^s$	$\begin{pmatrix} -0.36 \\ -0.23 \\ 0.43 \end{pmatrix}$	$\begin{pmatrix} 0.01 \\ -0.06 \\ 0.99 \end{pmatrix}$	0.096	$\begin{pmatrix} -0.33 \\ -0.28 \\ 0.47 \end{pmatrix}$	$\begin{pmatrix} 0.06 \\ -0.02 \\ -0.99 \end{pmatrix}$	0.084	SIFT Matching
$C_2^l \Rightarrow C_2^s$	$\begin{pmatrix} -0.16 \\ 0.74 \\ 0.10 \end{pmatrix}$	$\begin{pmatrix} -0.02 \\ -0.01 \\ -0.99 \end{pmatrix}$	0.097	$\begin{pmatrix} -0.13 \\ 0.75 \\ 0.10 \end{pmatrix}$	$\begin{pmatrix} -0.01 \\ 0.00 \\ -0.99 \end{pmatrix}$	0.087	SIFT Matching

result by drawing for each feature the epipolar line, and judging by its proximity to the corresponding feature in the other image. In Fig. 4, the upper row corresponds to elements identified in C_1^s (Fig. 3b) and the lower row presents the same elements identified in C_2^s (Fig. 3d). The following remarks are necessary at this point. Firstly, the perspective change makes the correspondence search very tedious even for a human. Secondly, the drawing of the epipolar line has actually assisted us in pinpointing most of these correspondences, and we are confident that the method will be helpful in automating these tasks.

Discussion of the dense crowd results. Overall, the distance in the image space between the corresponding element and the corresponding epipolar line is in the range of a few pixels. The major factors responsible for these misalignments are the inaccuracies in estimating the intrinsic parameters, as well as the errors related to the relative pose estimations - but for the dimensions of the scene involved in the experiment, we argue that the results are very promising.

Moreover, some areas of the scene exhibit near perfect alignments. The first four matches presented in Fig. 4 (the white cap man in (a) positioned under the epiline, in the left part of the patch; the person in (b) looking slightly towards the

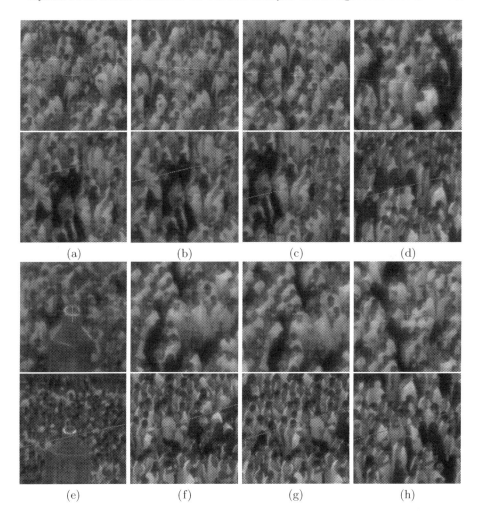

Fig. 4. A number of pixel-epipolar line correspondences between the two analysis cameras presented in Fig. 3(b) and 3(d). Ideally, the correspondent of a point highlighted by the red cross in the upper row should be situated along the blue epipolar line visible in the lower row image. These results are discussed in Sect. 5.

left; the woman in (c) wearing a white veil, and positioned in front of two other women similarly dressed; the woman in (d) wearing a white veil, and positioned with the back towards the second camera) are very accurate, in spite of the fact that in one of the images the first three persons are located near the border, a fact which potentially increases radial distortion related errors.

We could also identify the following correspondences which exhibit small but visible misalignments: the shiny circumference of the Station of Ibrahim, depicted in (e); another two men wearing white caps, presented in (f) and (g); a distinctively bearded man presented in (h).

The fact that the epipolar line does not pass precisely through the corresponding element is not detrimental for the purpose of association and tracking in the crowd. Assuming that the person is not occluded, using this extra information we would not only be able to trim down the research space to a band along the epipolar line, but also if we were able to position the ground plane within the same coordinate system we would further reduce the research space to a fraction of the band. Of course, in order to do dense matching reliably we still need a neighborhood based similarity measure that has to be resilient to major perspective change and occlusions; this is a promising direction of research that we intend to follow in order to benefit from the relative pose algorithm we propose, and ultimately in order to perform dense associations.

Finally, the present results of the proposed algorithm on this type of data are also encouraging as they illustrate the potential of multi-camera systems in extremely crowded environments. In the current research context, this application field has been associated mostly with single camera systems [31], but paradoxically it would greatly benefit from multi-view systems given the frequent occlusions and scene clutter that characterize it.

6   Conclusion

In this paper we propose a new method for aligning multiple cameras analysing a homogeneous scene. Our method addresses the settings where for practical reasons calibration pattern/object based registrations are not possible. By employing stereo rigs featuring a long focal analysis camera and a short focal registration camera, the proposed solution alleviates the requirement to get access to the studied scene. The fact that we are using a large FOV simultaneously allows us to avoid making any assumptions about the homogeneous region we analyse, such as the presence of shades, silhouettes etc. A first experiment has been conducted in an indoor environment and has shown, by using interest point correspondences in the analysis area as ground truth, that this method can guide the relative pose estimation for scenes poor in salient features in a coarse-to-fine manner supported by hardware. The second test has shown that in the absence of any salient features, the method is capable of providing a full calibration of the analysis cameras in a difficult, large scale scenario.

In the future, we would like to investigate the applicability of the proposed hybrid stereo solution in two frequently recurring settings. We intend to employ this method as a preprocessing step for a wide range of homogeneous pattern analysis applications, such as those related to the extraction of accurate models for highly dense crowd dynamics. Secondly, we would like to evaluate further the potential of this solution in specific applications such as autonomous robot navigation or image alignment and stitching, which employ pyramid based coarse-to-fine optimizations; our setup augments these systems by supplementing the image pyramid with a level provided by an independent data source.

Acknowledgements. This work was funded by QNRF under the grant NPRP 09-768-1-114. The authors acknowledge A. Gutub and his team at the Centre of Research

Excellence in Hajj and Omrah; and O. Gazzaz, F. Othman, A. Fouda and B. Zafar at the Hajj Research Institute for their organisation and logistical support in the video data collection at the grand mosque in Makkah, as well as for useful discussions.

References

1. Zhang, Z., Zhang, Z.: A flexible new technique for camera calibration. IEEE Trans. Pattern Anal. Mach. Intell. **22**, 1330–1334 (1998)
2. Baker, P., Aloimonos, Y.: Complete calibration of a multi-camera network. In: IEEE Workshop on Omnidirectional Vision, 2000, Proceedings, pp. 134–141 (2000)
3. Svoboda, T., Martinec, D., Pajdla, T.: A convenient multi-camera self-calibration for virtual environments. PRESENCE: Teleoperators Virtual Environ. **14**, 407–422 (2005)
4. Triggs, B., McLauchlan, P.F., Hartley, R.I., Fitzgibbon, A.W.: Bundle adjustment - a modern synthesis. In: Workshop on Vision Algorithms, pp. 298–372 (1999)
5. Sinha, S., Pollefeys, M.: Camera network calibration and synchronization fromsilhouettes in archived video. Int. J. Comput. Vision **87**, 266–283 (2010)
6. Sturm, P.F.: Self-calibration of a moving zoom-lens camera by pre-calibration. Image Vision Comput. **15**, 583–589 (1997)
7. Ahmed, M., Farag, A.: Nonmetric calibration of camera lens distortion: differential methods and robust estimation. Trans. Image Proc. **14**, 1215–1230 (2005)
8. Wang, A., Qiu, T., Shao, L.: A simple method of radial distortion correction with centre of distortion estimation. J. Math. Imaging Vis. **35**, 165–172 (2009)
9. Kukelova, Z., Pajdla, T.: A minimal solution to radial distortion autocalibration. IEEE Trans. Pattern Anal. Mach. Intell. **33**, 2410–2422 (2011)
10. Lourakis, M.I., Deriche, Rachid, D.: Camera Self-calibration Using the Singular Value Decomposition of the Fundamental Matrix: From Point Correspondences to 3D Measurements. Technical report RR-3748, INRIA (1999)
11. Josephson, K., Byrod, M.: Pose estimation with radial distortion and unknown focal length. In: IEEE Conference on Computer Vision and Pattern Recognition, 2009, CVPR 2009, pp. 2419–2426 (2009)
12. Dang, T., Hoffmann, C., Stiller, C.: Continuous stereo self-calibration by camera parameter tracking. Trans. Image Proc. **18**, 1536–1550 (2009)
13. Sinha, S.N., Pollefeys, M.: Towards calibrating a pan-tilt-zoom camera network. In: OMNIVIS (2004)
14. Bimbo, A.D., Dini, F., Lisanti, G., Pernici, F.: Exploiting distinctive visual landmark maps in pan-tilt-zoom camera networks. Comput. Vis. Image Underst. **114**, 611–623 (2010)
15. Eyles, C., Harrison, R., Davis, C., Waltham, N., Shaughnessy, B., Mapson-Menard, H., Bewsher, D., Crothers, S., Davies, J., Simnett, G., et al.: The heliospheric imagers onboard the stereo mission. Sol. Phys. **254**, 387–445 (2009)
16. Brown, D., Bewsher, D., Eyles, C.: Calibrating the pointing and optical parameters of stereo heliospheric imagers. Sol. Phys. **254**, 185–225 (2009)
17. Eynard, D., Vasseur, P., Demonceaux, C., Frémont, V.: Real time uav altitude, attitude and motion estimation from hybrid stereovision. Auton. Rob. **33**, 157–172 (2012)
18. Oskiper, T., Sizintsev, M., Branzoi, V., Samarasekera, S., Kumar, R.: Augmented reality binoculars. In: 2013 IEEE International Symposium on Mixed and Augmented Reality (ISMAR), pp. 219–228 (2013)

19. Faugeras, O.D.: What can be seen in three dimensions with an uncalibrated stereo rig? In: Sandini, Giulio (ed.) ECCV 1992. LNCS, vol. 588, pp. 563–578. Springer, Heidelberg (1992)

20. Hartley, R.I., Zisserman, A.: Multiple View Geometry in Computer Vision, 2nd edn. Cambridge University Press, Cambridge (2004). ISBN: 0521540518

21. Lourakis, M.A., Argyros, A.: SBA: a software package for generic sparse bundle adjustment. ACM Trans. Math. Softw. **36**, 1–30 (2009)

22. Nistér, D.: An efficient solution to the five-point relative pose problem. IEEE Trans. Pattern Anal. Mach. Intell. **26**, 756–777 (2004)

23. Kneip, L., Scaramuzza, D., Siegwart, R.: A novel parametrization of the perspective-three-point problem for a direct computation of absolute camera position and orientation. In: CVPR, pp. 2969–2976 (2011)

24. Lowe, D.G.: Distinctive image features from scale-invariant keypoints. Int. J. Comput. Vis. **60**, 91–110 (2004)

25. Hartley, R.: In defense of the eight-point algorithm. IEEE Trans. Pattern Anal. Mach. Intell. **19**, 580–593 (1997)

26. Fischler, M.A., Bolles, R.C.: Random sample consensus: a paradigm for model fitting with applications to image analysis and automated cartography. Commun. ACM **24**, 381–395 (1981)

27. Nistér, D., Naroditsky, O., Bergen, J.R.: Visual odometry. In: CVPR, vol. 1, pp. 652–659 (2004)

28. Hartley, R.I.: Chirality. Int. J. Comput. Vis. **26**, 41–61 (1998)

29. Hartley, R.I., Sturm, P.F.: Triangulation. Comput. Vis. Image Underst. **68**, 146–157 (1997)

30. Yu, G., Morel, J.M.: ASIFT: an algorithm for fully affine invariant comparison. Image Processing On Line 2011 (2011)

31. Wang, X.: Intelligent multi-camera video surveillance: a review. Pattern Recogn. Lett. **34**, 3–19 (2013)

Author Index

.

Printed in the United States
By Bookmasters